ADVANCES IN APPLIED MATHEMATICS AND GLOBAL OPTIMIZATION

IN HONOR OF GILBERT STRANG

Advances in Mechanics and Mathematics

VOLUME 17

Aims and Scope

Mechanics and mathematics have been complementary partners since Newton's time, and the history of science shows much evidence of the beneficial influence of these disciplines on each other. The discipline of mechanics, for this series, includes relevant physical and biological phenomena such as: electromagnetic, thermal, quantum effects, biomechanics, nanomechanics, multiscale modeling, dynamical systems, optimization and control, and computational methods.

Driven by increasingly elaborate modern technological applications, the symbiotic relationship between mathematics and mechanics is continually growing. The increasingly large number of specialist journals has generated a complementarity gap between the partners, and this gap continues to widen. *Advances in Mechanics and Mathematics* is a series dedicated to the publication of the latest developments in the interaction between mechanics and mathematics and intends to bridge the gap by providing interdisciplinary publications in the form of monographs, graduate texts, edited volumes, and a special annual book consisting of invited survey articles.

ADVANCES IN APPLIED MATHEMATICS AND GLOBAL OPTIMIZATION

IN HONOR OF GILBERT STRANG

Edited By

David Y. Gao
Virginia Polytechnic Institute, Blacksburg, VA

Hanif D. Sherali
Virginia Polytechnic Institute, Blacksburg, VA

Springer

Editors:
David Y. Gao
Department Mathematics
Virginia Polytechnic Institute
Blacksburg, VA 24061
gao@vt.edu

Hanif D. Sherali
Department Mathematics
Virginia Polytechnic Institute
Blacksburg, VA 24061
hanifs@vt.edu

Series Editors:
David Y. Gao
Department of Mathematics
Virginia Polytechnic Institute
Blacksburg, VA 24061
gao@vt.edu

Ray W. Ogden
Department of Mathematics
University of Glasgow
Glasgow, Scotland, UK
rwo@maths.gla.ac.uk

ISBN 978-0-387-75713-1
e-ISBN 978-0-387-75714-8
DOI 10.1007/978-0-387-75714-8

Library of Congress Control Number: 2009921139

Mathematics Subject Classification (2000): 90-00, 49-00, 74-00, 65, 74, 81, 92

Printed on acid-free paper

springer.com

This book is dedicated to
Professor Gilbert Strang
on the occasion of his 70th birthday

Gilbert Strang

Gil Strang in his MIT office

Contents

Series Preface

As any human activity needs goals, mathematical research needs problems.
—David Hilbert

Mechanics is the paradise of mathematical sciences.
—Leonardo da Vinci

Mechanics and mathematics have been complementary partners since Newton's time, and the history of science shows much evidence of the beneficial influence of these disciplines on each other. Driven by increasingly elaborate modern technological applications, the symbiotic relationship between mathematics and mechanics is continually growing. However, the increasingly large number of specialist journals has generated a duality gap between the partners, and this gap is growing wider.

Advances in Mechanics and Mathematics (AMMA) is intended to bridge the gap by providing multidisciplinary publications that fall into the two following complementary categories:

1. An annual book dedicated to the latest developments in mechanics and mathematics;
2. Monographs, advanced textbooks, handbooks, edited volumes, and selected conference proceedings.

The AMMA annual book publishes invited and contributed comprehensive research and survey articles within the broad area of modern mechanics and applied mathematics. The discipline of mechanics, for this series, includes relevant physical and biological phenomena such as: electromagnetic, thermal, and quantum effects, biomechanics, nanomechanics, multiscale modeling, dynamical systems, optimization and control, and computation methods. Especially encouraged are articles on mathematical and computational models and methods based on mechanics and their interactions with other fields. All contributions will be reviewed so as to guarantee the highest possible scientific standards. Each chapter will reflect the most recent achievements in the area. The coverage should be conceptual, concentrating on the methodological thinking that will allow the nonspecialist reader to understand it. Discussion of possible future research directions in the area is welcome. Thus, the annual volumes will provide a continuous documentation of the most recent developments in these active and important interdisciplinary fields. Chapters published in this series could form bases from which possible AMMA monographs or advanced textbooks could be developed.

Volumes published in the second category contain review/research contributions covering various aspects of the topic. Together these will provide an overview of the state-of-the-art in the respective field, extending from an introduction to the subject right up to the frontiers of contemporary research. Certain multidisciplinary topics, such as duality, complementarity, and symmetry in mechanics, mathematics, and physics are of particular interest.

The *Advances in Mechanics and Mathematics* series is directed to all scientists and mathematicians, including advanced students (at the doctoral and postdoctoral levels) at universities and in industry who are interested in mechanics and applied mathematics.

David Y. Gao
Ray W. Ogden

Preface

Complementarity and duality are closely related, multidisciplinary topics that pervade all natural phenomena, and form the basis for solving many underlying nonconvex analysis and global optimization problems that arise in science and engineering. During the last forty years, much research has been devoted to the development of mathematical modeling, theory, and computational methods in this area. The field has now matured in convex systems, especially in linear programming, engineering mechanics and design, mathematical physics, economics, optimization, and control. However, in nonconvex systems many fundamental problems still remain unsolved.

In view of the importance of complementarity–duality theory and methods in applied mathematics and mathematical programming, and in order to bridge the ever-increasing gap between global optimization and engineering science, the *First International Conference on Complementarity, Duality, and Global Optimization* (CDGO) was held at Virginia Tech, Blacksburg, August 15–17, 2005, under the sponsorship of the National Science Foundation. This conference brought together more than 100 world-class researchers from interdisciplinary fields of industrial engineering, operations research, pure and applied math, engineering mechanics, electrical engineering, psychology, management science, civil engineering, and computational science. This conference spawned some new trends in optimization and engineering science, and has stimulated young faculty and students to venture into this rich domain of research.

This AMMA Annual contains eleven chapters from selected lectures presented at the First International Conference on Complementarity, Duality, and Global Optimization (CDGO) in August 2005 and three invited chapters by experts in computational mathematics and quantum computations. These chapters deal with fundamental theory, methods, and applications of complementarity, duality, and global optimization in multidisciplinary fields of global optimization, nonconvex mechanics, and computational science, as well as the very contemporary topic of quantum computing, which is at the forefront of the scientific and technological research and development of the twenty-first century.

This special volume is dedicated to Gilbert Strang on the occasion of his 70th birthday. Professor Strang is a world-renowned mathematician not only by his scientific contributions, but also by his personal character which exemplifies what a real scientist should possess. During his exceptional academic career and social activities spanning almost a half-century, Dr. Strang has had a profound influence on the development of interdisciplinary fields in applied mathematics, mechanics, and engineering science, including the field of complementary duality in calculus of variations, optimization, numerical methods, and mathematical education. The unified beauty of duality can be viewed throughout his celebrated textbooks, lecture notes, essays, and scientific publications, which will continue to influence several generations in the broad field of mathematical sciences.

Credit for this special volume is to be shared by all the eminent contributing authors. As editors, we are deeply indebted to them. Our special thanks also go to Ann Kostant and her team, and especially to Elizabeth Loew at Springer, for all their great enthusiasm and professional help in expediting the publication of this annual volume.

May 2008 *David Y. Gao*
Blacksburg, VA *Hanif D. Sherali*

Constrained Optimism

Gilbert Strang

The editors have kindly invited me to write a few words of introduction to this volume. They even expressed the hope that I would go beyond mathematics, to say something about my own life experiences. I think every reader will recognize how hard it is (meaning impossible) to do that properly. If I choose a single word to describe an approach to the complications of life (and of mathematics too), it would be "optimism." Eventually I realized that, if you allow that word in its mathematical sense too, this whole book is for optimists.

If I may give one instance of my own optimism, it has come from writing textbooks. I enjoy the hopeless effort to express simple ideas clearly. Beyond that, I have come to expect (without knowing any reason, perhaps this defines an optimist) that the connections between all the pieces of the book will somehow appear. Suddenly a topic fits into its right place. This irrational certainty may also be the experience of a hopeful novelist who doesn't know how the characters will interact and how the plot will turn out.

Looking seriously at this approach, to applied mathematics or to life, an unconstrained optimism is hard to justify. Mathematically, an immediate constraint on all of us is that we are "not Gauss." Far wiser to accept constraints, and continue to optimize. The connection that did finally bring order to my own thinking and writing about applied mathematics and computational engineering was constrained optimization. I now call that the "Fundamental Problem of Scientific Computing."

Examples are everywhere, or those words would not be justified. So many problems involve three steps, and flows in networks are a good model. The potentials at the nodes, and the currents on the edges, are the unknowns (somehow dual). A first step goes from potentials to potential differences (by an edge-node matrix A). The second step relates potential differences to flows (by a matrix C). Ohm's law is typical, or Hooke's law, or any constitutive law: linear at first but not forever. The third step is the essential constraint of conservation or continuity or balance of forces, as in Kirchhoff's current law. This involves the transpose matrix A'.

The dual role of A and A' is at first a miracle. A reason begins to emerge through minimization and Lagrange multipliers. If we minimize a quadratic energy with a linear constraint $A'w = f$, the optimality conditions lead to a saddle point matrix ("KKT matrix"):

$$\textbf{Optimization with constraint:} \qquad \begin{bmatrix} C^{-1} & A \\ A' & 0 \end{bmatrix} \begin{bmatrix} w \\ u \end{bmatrix} = \begin{bmatrix} b \\ f \end{bmatrix}.$$

One way to solve this fundamental problem is to eliminate w. The three matrices combine into $A'CA$, symmetric and positive definite in the best case. This is the stiffness matrix of the finite element method, or the Laplacian matrix of finite differences and graph theory. It appears everywhere and we don't know the best way to solve the equation. As a differential equation it is in divergence form with $A'CA = \text{div}(c \text{ grad})$. When C is piecewise linear we have mathematical programming, where the primal–dual method has come to the front. The real problems of mechanics and biology (and life) are not linear at all. But remarkably often they still have this form with $A'C\,(Au)$.

May I thank the editors and authors and readers of the present book. I hope you will accept constraints as inevitable, and go forward.

Biographical Summary of Gilbert Strang

Education

1. 1952–1955 William Barton Rogers Scholar, M.I.T. (S.B. 1955)
2. 1955–1957 Rhodes Scholar, Oxford University (B.A., M.A. 1957)
3. 1957–1959 NSF Fellow, UCLA (Ph.D. 1959)

Positions Held

1. 1959–1961 C.L.E. Moore Instructor, M.I.T.
2. 1961–1962 NATO Postdoctoral Fellow, Oxford University
3. 1962–1964 Assistant Professor of Mathematics, M.I.T.
4. 1964–1970 Associate Professor of Mathematics, M.I.T.
5. 1970– Professor of Mathematics, M.I.T.

Awards and Duties

1. Alfred P. Sloan Fellow (1966–1967)
2. Chairman, M.I.T. Committee on Pure Mathematics (1975–1979)
3. Chauvenet Prize, Mathematical Association of America (1976)
4. Council, Society for Industrial and Applied Mathematics (1977–1982)
5. NSF Advisory Panel on Mathematics (1977–1980) (Chairman 1979–1980)
6. CUPM Subcommittee on Calculus (1979–1981)
7. Fairchild Scholar, California Institute of Technology (1980–1981)
8. Honorary Professor, Xian Jiaotong University, China (1980)
9. American Academy of Arts and Sciences (1985)
10. Taft Lecturer, University of Cincinnati (1977)
11. Gergen Lecturer, Duke University (1983)
12. Lonseth Lecturer, Oregon State University (1987)
13. Magnus Lecturer, Colorado State University (2000)
14. Blumberg Lecturer, University of Texas (2001)
15. AMS-SIAM Committee on Applied Mathematics (1990–1992)
16. Vice President for Education, SIAM (1991–1996)
17. MAA Science Policy Committee (1992–1995)

18. Committee on the Undergraduate Program in Mathematics, MAA (1993–1996)
19. President of SIAM (1999–2000)
20. Chair, Joint Policy Board for Mathematics (1999)
21. Chair, SIAM Committee on Science Policy (2001–2002)
22. Honorary Fellow, Balliol College, Oxford (1999)
23. Honorary Member, Irish Mathematical Society (2002)
24. US National Committee on Mathematics (2001–2004, Chair 2003–2004)
25. Award for Distinguished Service to the Profession, SIAM (2003)
26. Graduate School Teaching Award, MIT (2003)
27. Abel Prize Committee, Oslo (2003–2005)
28. Von Neumann Prize Medal, US Association for Computational Mechanics (2005)
29. Ford Prize for "Pascal Matrices" with Alan Edelman, Mathematical Association of America (2005)
30. Distinguished University Teacher of Mathematics, New England Section, Mathematical Association of America (2006)
31. Franklin and Deborah Tepper Haimo Prize, MAA (2006)
32. Su Buchin Prize, International Congress of Industrial and Applied Mathematics (ICIAM, Zurich, 2007).
33. Henrici Prize, (ICIAM, Zurich, 2007).

Journal Editor

1. *Numerische Mathematik* (Honorary Editor from 1996)
2. *International Journal for Numerical Methods in Engineering*
3. *Archive for Rational Mechanics and Analysis* (to 1990)
4. *Studies in Applied Mathematics*
5. *Computer Methods in Applied Mechanics and Engineering* (to 2004)
6. *SIAM Journal on Numerical Analysis* (to 1977)
7. *Numerical Functional Analysis and Optimization*
8. *Physica D: Nonlinear Phenomena* (to 1986)
9. *Communications in Numerical Methods in Engineering*
10. *SIAM Journal on Matrix Analysis and Applications* (to 1993)
11. *Acta Applicandae Mathematicae*
12. *Proceedings of the Edinburgh Mathematical Society* (to 1993)
13. *Mathematical Modelling and Numerical Analysis*
14. *Japan Journal of Applied Mathematics*
15. *Structural Optimization*
16. *Royal Society of Edinburgh, Proceedings A*
17. *Computational Optimization and Applications*
18. *SIAM Review* (to 2001)
19. *COSMOS, National University of Singapore*
20. *Numerical Algorithms* (to 2004)

List of Publications of Gilbert Strang

Books (with Russian, Japanese, and Chinese translations of the first two)

1. *An Analysis of the Finite Element Method*, with George Fix, Prentice-Hall (1973). Second Edition: Wellesley-Cambridge Press (2008).
2. *Linear Algebra and Its Applications*, Academic Press (1976). Second Edition: Harcourt Brace Jovanovich (1980). Third Edition: Brooks/Cole (1988). Fourth Edition: Brooks/Cole (2006).
3. *Introduction to Applied Mathematics*, Wellesley-Cambridge Press (1986).
4. *Nonlinear Partial Differential Equations in Applied Science*, H. Fujita, P. Lax, G. Strang, editors, Lecture Notes in Numerical and Applied Analysis 5, Kinokuniya/North Holland (1982).
5. *Topics in Nonsmooth Mechanics*, J.J. Moreau, P.D. Panagiotopoulos, G. Strang, editors, Birkhäuser (1988).
6. *Calculus*, Wellesley-Cambridge Press (1991).
7. *Introduction to Linear Algebra*, Wellesley-Cambridge Press (1993). Second Edition (1998). Third Edition (2003).
8. *Wavelets and Filter Banks*, with Truong Nguyen, Wellesley-Cambridge Press (1996).
9. *Linear Algebra, Geodesy, and GPS*, with Kai Borre, Wellesley-Cambridge Press (1997).
10. *Computational Science and Engineering*, Wellesley-Cambridge Press (2007).

Papers in Journals and Books

1. An improvement on the Holzer table based on a suggestion of Rayleigh's, with S.H. Crandall, *J. Appl. Mechanics*, Paper 56-A27 (1957).
2. On the order of convergence of the Crank-Nicolson procedure, *Journal of Mathematics and Physics* **38** (1959) 141–144.
3. Difference methods for mixed boundary-value problems, *Duke Math. J.* **27** (1960) 221–232.
4. On the Kantorovich inequality, *Proc. Amer. Math. Soc.* **11** (1960) 468.

5. A note on the joint spectral radius, with G.-C. Rota, *Proc. Netherlands Acad.* **22** (1960) 379–381.
6. Finite difference techniques for a boundary problem, with L. Ehrlich, J. Riley, and B.A. Troesch, *J. Soc. Ind. Appl. Math.* (1961).
7. Eigenvalues of Jordan products, *Amer. Math. Monthly* **69** (1962) 37–40.
8. Trigonometric polynomials and difference methods of maximum accuracy, *Journal of Mathematics and Physics* **41** (1962) 147–154.
9. Polynomial approximation of Bernstein type, *Trans. Amer. Math. Soc.* **105** (1962) 525–535.
10. Comparison theorems for supremum norms, with H. Schneider, *Numerische Math.* **4** (1962) 15–20.
11. Accurate partial difference methods I: Linear Cauchy problems, *Arch. Rat. Mech. Anal.* **12** (1963).
12. Accurate partial difference methods II: Non-linear problems, *Numerische Math.* **6** (1964) 37–46.
13. Wiener-Hopf difference equations, *J. Math. Mechanics* **13** (1964) 85–96.
14. Unbalanced polynomials and difference methods for mixed problems, *SIAM J. Numer. Anal.* **2** (1964) 46–51.
15. Necessary and insufficient conditions for well-posed Cauchy problems, *J. Diff. Eq.* **2** (1966) 107–114.
16. Matrix theorems for partial differential and difference equations, with J. Miller, *Math. Scand.* **18** (1966) 113–133.
17. Implicit difference methods for initial-boundary value problems, *J. Math. Anal. Appl.* **16** (1966) 188–198.
18. On strong hyperbolicity, *J. Math. Kyoto Univ.* **6** (1967) 397–417.
19. A variant of Caratheodory's problem, *Proc. Edinburgh Math. Soc.* **16** (1968) 43–48.
20. The nucleus of a set, *Canad. Math. Bull.* **11** (1968) 65–72.
21. On the construction and comparison of difference schemes, *SIAM J. Numer. Anal.* **5** (1968) 506–517.
22. Approximating semigroups and the consistency of difference schemes, *Proc. Amer. Math. Soc.* **20** (1969) 1–7.
23. Hyperbolic initial-boundary value problems in two unknowns, *J. Diff. Eq.* **6** (1969) 161–171.
24. On numerical ranges and holomorphic semigroups, *J. d'Analyse Math.* **22** (1969) 299–318.
25. On multiple characteristics and the Levi-Lax conditions for hyperbolicity, *Arch. Rat. Mech. Anal.* **33** (1969) 358–373.
26. Fourier analysis of the finite element method in Ritz-Galerkin theory, with G. Fix, *Stud. Appl. Math.* **48** (1969) 265–273.
27. Toeplitz operators in a quarter-plane, *Bull. Amer. Math. Soc.* **76** (1970) 1303–1307.
28. The correctness of the Cauchy problem, with H. Flaschka, *Adv. Math.* **6** (1971) 347–349.

29. The finite element method and approximation theory, *SYNSPADE Proceedings*, Academic Press (1971) 547–584.
30. The change in solution due to change in domain, with A. Berger, *AMS Symposium on Partial Differential Equations*, Berkeley (1971) 199–206.
31. Approximation in the finite element method, *Numerische Math.* **19** (1972) 81–98.
32. Approximate boundary conditions in the finite element method, with R. Scott and A. Berger, *Symposia Mathematica X, Istituto Nationale di Alta Matematica* (1972) 295–313.
33. Variational crimes in the finite element method, *The Mathematical Foundations of the Finite Element Method*, ed. by A.K. Aziz, Academic Press (1973) 689–710.
34. A Fourier analysis of the finite element variational method, with G. Fix, *Constructive Aspects of Functional Analysis*, Edizioni Cremonese, Rome (1973) 795–840.
35. Piecewise polynomials and the finite element method, *AMS Bulletin* **79** (1973) 1128–1137.
36. Optimal conditioning of matrices, with C. McCarthy, *SIAM J. Numer. Anal.* **10** (1973) 370–388.
37. The dimension of piecewise polynomial spaces and one-sided approximation, *Proc. Conference on Numerical Analysis*, Dundee, Springer Lecture Notes 363 (1974) 144–152.
38. *One-Sided Approximation and Plate Bending*, Lecture Notes in Computer Science 11, Springer-Verlag (1974) 140–155.
39. One-sided approximation and variational inequalities, with U. Mosco, *Bull. Amer. Math. Soc.* **80** (1974) 308–312.
40. The finite element method—linear and nonlinear applications, *Proc. Inter. Congress of Mathematicians*, Vancouver (1974).
41. Free boundaries and finite elements in one dimension, with W. Hager, *Math. Comp.* **29** (1975) 1020–1031.
42. A homework exercise in finite elements, *Int. J. Numer. Meth. Eng.* **11** (1977) 411–418.
43. Some recent contributions to plasticity theory, *J. Franklin Institute* **302** (1977) 429–442.
44. Discrete plasticity and the complementarity problem, *Proceedings U.S.-Germany Symposium: Formulations and Computational Algorithms in Finite Element Analysis*, M.I.T. Press (1977) 839–854.
45. Uniqueness in the theory of variational inequalities, *Adv. Math.* **22** (1976) 356–363.
46. A minimax problem in plasticity theory, *Functional Analysis Methods in Numerical Analysis*, ed. M.Z. Nashed, Springer Lecture Notes 701, Springer (1979) 319–333.
47. A family of model problems in plasticity, *Proc. Symp. Computing Methods in Applied Sciences*, ed. R. Glowinski and J.L. Lions, Springer Lecture Notes 704, Springer (1979) 292–308.

48. The saddle point of a differential program, with H. Matthies and E. Christiansen, *Energy Methods in Finite Element Analysis*, ed. by R. Glowinski, E. Rodin, and O.C. Zienkiewicz, John Wiley (1979).
49. The solution of nonlinear finite element equations, with H. Matthies, *Int. J. Numer. Meth. Eng.* **14** (1979) 1613–1626.
50. Mathematical and computational methods in plasticity, with H. Matthies and R. Temam, *Proc. IUTAM Symp. on Variational Methods in the Mechanics of Solids*, S. Nemat-Nasser, ed., Pergamon (1980) 20–28.
51. Spectral decomposition in advection-diffusion analysis by finite element methods, with R. Nickell and D. Gartling, Proc. FENOMECH Symp., Stuttgart (1978); *Comput. Meth. Appl. Mech. Eng.* **17** (1979) 561–580.
52. Existence de solutions relaxes pour les equations de la plasticite, with R. Temam, *Comptes Rendus Acad. Sc. Paris* **287** (1978) 515–519.
53. Functions of bounded deformation, with R. Temam, *Arch. Rat. Mech. Anal.* **75** (1980) 7–21.
54. Numerical computations in nonlinear mechanics, with H. Matthies, Paper 79-PVP-103, *Amer. Soc. Mech. Eng.* (1979); *Proceedings of the 4th Symposium on Computing Methods in Applied Sciences and Engineering*, ed. R. Glowinski and J.L. Lions, 517–525, North-Holland (1980).
55. Duality and relaxation in the variational problems of plasticity, with R. Temam, *J. Mecanique* **19** (1980) 1–35.
56. The quasi-Newton method in finite element calculations, Chapter 20 in *Computational Methods in Nonlinear Mechanics*, J.T. Oden, ed., North-Holland (1980).
57. The application of quasi-Newton methods in fluid mechanics, with M. Engelman and K.J. Bathe, *Int. J. Numer. Meth. Eng.* **17** (1981) 707–718.
58. A problem in capillarity and plasticity, with R. Temam, *Nondifferentiable and Variational Techniques in Optimization*, D.C. Sorenson, R.J.B. Wets, eds., *Mathematical Programming Study* **17** (1982) 91–102.
59. Optimal design for torsional rigidity, with R. Kohn, *Proc. Int. Symp. on Mixed and Hybrid Finite Element Methods*, Atlanta (1981).
60. Optimal design of cylinders in shear, with R. Kohn, *MAFELAP Conference*, Brunel (1981).
61. The width of a chair, *Amer. Math. Monthly* **89** (1982) 529–534.
62. Structural design optimization, homogenization, and relaxation of variational problems, with R. Kohn, *Proceedings of Conference on Disordered Media*, Lecture Notes in Physics 154, Springer-Verlag (1982).
63. Hencky–Prandtl nets and constrained Michell trusses, with R. Kohn, Conference on Optimum Structural Design, Tucson (1981), *Comput. Meth. Appl. Mech. Eng.* **36** (1983) 207–222.
64. The optimal accuracy of difference schemes, with A. Iserles, *Trans. Amer. Math. Soc.* **277** (1983) 770–803.
65. Duality in the classroom, *Amer. Math. Monthly* **91** (1984) 250–254.
66. Maximal flow through a domain, *Math. Program.* **26** (1983) 123–143.

67. Barriers to stability, with A. Iserles, *SIAM J. Numer. Anal.* **20** (1983) 1251–1257.
68. L^1 and L^∞ approximation of vector fields in the plane, *Nonlinear Partial Differential Equations in Applied Science*, H. Fujita, P. Lax, and G. Strang, eds., Lecture Notes in Num. Appl. Anal. 5 (1982) 273–288.
69. Notes on softening and local instability, with M. Abdel-Naby, in *Computational Aspects of Penetration Mechanics*, Springer Lecture Notes in Engineering 3, J. Chandra and J. Flaherty, eds. (1983).
70. A negative results for nonnegative matrices, *J. Xian Jiaotong Univ.* **17** (1983) 69–72.
71. Numerical and biological shape optimization, with A. Philpott, in *Unification of Finite Element Methods*, Math. Studies 94, H. Kardestuncer, ed., North-Holland (1984).
72. Explicit relaxation of a variational problem in optimal design, with R. Kohn, *Bull. Amer. Math. Soc.* **9** (1983) 211–214.
73. Optimal design and relaxation of variational problems, with R. Kohn, *Commun. Pure Appl. Math.* **39** (1986) 113–137 (Part I), 139–182 (Part II), 353–377 (Part III).
74. The constrained least gradient problem, with R. Kohn, in *Non-Classical Continuum Mechanics*, R. Knops and A. Lacey, eds., Cambridge University Press (1987).
75. The optimal design of a two-way conductor, with R. Kohn, in *Nonsmooth Mechanics*, P.D. Panagiotopoulos et al, eds., Birkhäuser (1987).
76. Fibered structures in optimal design, with R. Kohn, *Ordinary and Partial Differential Equations*, B. Sleeman and R. Jarvis, eds., Pitman Research Notes 157, Longman (1987).
77. Optimal design in elasticity and plasticity, with R. Kohn, *Int. J. Numer. Meth. Eng.* **22** (1986) 183–188.
78. A framework for equilibrium equations, *SIAM Rev.* **30** (1988) 283–297.
79. Karmarkar's algorithm in a nutshell, *SIAM News* **18** (1985) 13.
80. Karmarkar's algorithm and its place in applied mathematics, *Math. Intelligencer* **9** (1987) 4–10.
81. A proposal for Toeplitz matrix calculations, *Stud. Appl. Math.* **74** (1986) 171–176.
82. The Toeplitz-circulant eigenvalue problem, with A. Edelman, pp. 109–117 in *Oakland Conf. on PDE's and Applied Mathematics*, L. Bragg and J. Dettman, eds., Longman (1987).
83. Patterns in linear algebra, *Amer. Math. Monthly* **96** (1989) 105–117.
84. Paradox lost: Natural boundary conditions in the Ritz- Galerkin method, with J. Storch, *Int. J. Numer. Meth. Eng.* **26** (1988) 2255–2266.
85. Dual extremum principles in finite elastoplastic deformation, with Y. Gao, *Acta Appl. Math.* **17** (1989) 257–268.
86. Toeplitz equations by conjugate gradients with circulant preconditioner, with R. Chan, *SIAM J. Sci. Stat. Comp.* **10** (1989) 104–119.

87. Geometric nonlinearity: Potential energy, complementary energy, and the gap function, with Y. Gao, *Quart. Appl. Math.* **47** (1989) 487–504.
88. Teaching modern engineering mathematics, *Appl. Mech. Rev.* **39** (1986) 1319–1321; SEFI Proceedings, L. Rade, ed., Chartwell-Bratt (1988).
89. Sums and differences vs. integrals and derivatives, *College Math. J.* **21** (1990) 20–27.
90. Wavelets and dilation equations: A brief introduction, *SIAM Rev.* **31** (1989) 614–627.
91. Inverse problems and derivatives of determinants, *Arch. Rat. Mech. Anal.* **114** (1991) 255–265.
92. A thousand points of light, with D. Hardin, *Third Conference on Technology in Collegiate Mathematics* (1990).
93. A chaotic search for i, *College Math. J.* **22** (1991) 3–12.
94. The optimal coefficients in Daubechies wavelets, *Physica D* **60** (1992) 239–244.
95. Polar area is the average of strip areas, *Amer. Math. Monthly* **100** (1993) 250–254.
96. The fundamental theorem of linear algebra, *Amer. Math. Monthly* **100** (1993) 848–855.
97. Wavelet transforms versus Fourier transforms, *Bull. Amer. Math. Soc.* **28** (1993) 288–305.
98. Graphs, matrices, and subspaces, *College Math. J.* **24** (1993) 20–28.
99. The asymptotic probability of a tie for first place, with B. Eisenberg and G. Stengle, *Ann. Appl. Prob.* **3** (1993) 731–745.
100. Continuity of the joint spectral radius: Applications to wavelets, with C. Heil, *Linear Algebra for Signal Processing*, A. Bojanczyk and G. Cybenko, eds., IMA 69 (1994) Springer-Verlag.
101. Convolution, reconstruction, and wavelets, *Advances in Computational Mathematics: New Delhi*, H.P. Dikshit and C.A. Micchelli, eds. (1994), World Scientific.
102. Short wavelets and matrix dilation equations, with V. Strela, *IEEE Trans. Signal Process.* **43** (1995) 108–115.
103. Orthogonal multiwavelets with vanishing moments, with V. Strela, *Proc. SPIE Conference on Mathematics of Imaging*, J. Optical Eng. 33 (1994) 2104–2107.
104. Wavelets, *Amer. Sci.* **82** (1994) 250–255.
105. Every unit matrix is a LULU, *Linear Alg. Appl.* **265** (1997) 165–172.
106. Finite element multiwavelets, with V. Strela, *Proc. Maratea NATO Conference*, Kluwer (1995).
107. Approximation by translates of refinable functions, with C. Heil and V. Strela, *Numerische Math.* **73** (1996) 75–94.
108. The cascade algorithm for the dilation equation, *Proc. Argonne Conference on Wavelets* (1994).
109. Eigenvalues and convergence of the cascade algorithm, *IEEE Trans. Signal Process.* **44** (1996), 233–238.

110. The application of multiwavelet filter banks for data compression, with P. Heller, V. Strela, P. Topiwala, and C. Heil, *IEEE Trans. Image Process.* **8** (1999) 548–563.
111. Asymptotic analysis of Daubechies polynomials, with Jianhong Shen, *Proc. Amer. Math. Soc.* **124** (1996) 3819–3833.
112. *Biorthogonal Multiwavelets and Finite Elements*, with V. Strela, preprint (1996).
113. Condition numbers for wavelets and filter banks, *Comput. Appl. Math. (Brasil)* **15** (1996) 161–179.
114. Eigenvalues of Toeplitz matrices with 1 x 2 blocks, *Zeit. Angew. Math. Mech.* **76** (1996) 37–39.
115. Asymptotic structures of Daubechies scaling functions and wavelets, with Jianhong Shen, *Appl. Comp. Harmonic Anal.* **5** (1998) 312–331.
116. Wavelets from filter banks, *The Mathematics of Numerical Analysis AMS-SIAM Park City Symposium*, J. Renegar, M. Shub, and S. Smale, eds. (1996), 765–806.
117. Filter banks and wavelets, in *Wavelets: Theory and Applications*, G. Erlebacher, M. Y. Hussaini, L. Jameson, eds., Oxford Univ. Press (1996).
118. Creating and comparing wavelets, *Numerical Analysis: A. R. Mitchell Anniversary Volume*, D. Griffiths, ed. (1996).
119. Writing about mathematics, *SIAM News* (June 1996).
120. The mathematics of GPS, *SIAM News* (June 1997).
121. Wavelets, *Iterative Methods in Scientific Computing*, pp. 59–110, R. Chan, T. Chan, and G. Golub, eds., Springer (1997).
122. *The First Moment of Wavelet Random Variables*, with Y. Ma and B. Vidakovic, preprint (1997).
123. The search for a good basis, *Numerical Analysis 1997*, D. Griffiths, D. Higham, and A. Watson, eds., Addison Wesley Longman (1997).
124. The asymptotics of optimal (equiripple) filters, with Jianhong Shen, *IEEE Trans. Signal Process.* **47** (1999) 1087–1098.
125. Inhomogeneous refinement equations, with Ding-Xuan Zhou, *J. Fourier Anal. Appl.* **4** (1998) 733–747.
126. Autocorrelation functions in GPS data processing: Modeling aspects, with Kai Borre, *ION Conference* (1997).
127. A linear algebraic representation of the double entry accounting system, with A. Arya, J. Fellingham, J. Glover, and D. Schroeder, Manuscript (1998).
128. The discrete cosine transform, block Toeplitz matrices, and filter banks, *Advances in Computational Mathematics*, Z. Chen, Y. Li, C. Micchelli, and Y. Xu, eds., Marcel Dekker—Taylor and Francis (1998).
129. The discrete cosine transform, *SIAM Rev.* **41** (1999) 135–147.
130. The limits of refinable functions, with Ding-Xuan Zhou, *Trans. Amer. Math. Soc.* **353** (2001) 1971–1984.
131. The potential theory of several intervals and its applications, with J. Shen and A. Wathen, *Appl. Math. Opt.* **44** (2001) 67–85.

132. Row reduction of a matrix and $A = CaB$, with S. Lee, *Amer. Math. Monthly* **107** (8) (October 2000), 681–688.
133. On wavelet fundamental solutions to the heat equation: Heatlets, with J. Shen, *J. Differential Eqns.* **161** (2000) 403–421.
134. Compactly supported refinable functions with infinite masks, with V. Strela and Ding-Xuan Zhou, in *The Functional and Harmonic Analysis of Wavelets and Frames*, L. Baggett and D. Larson, eds., American Math. Soc. Contemporary Mathematics 247 (1999) 285–296.
135. Trees with Cantor eigenvalue distribution, with Li He and Xiangwei Liu, *Stud. Appl. Math.* **110** (2003) 123–136.
136. Eigenstructures of spatial design matrices, with D. Gorsich and M. Genton, *J. Multivariate Anal.* **80** (2002) 138–165.
137. On factorization of M-channel paraunitary filter banks, with X.Q. Gao and T. Nguyen, *IEEE Trans. Signal Process.* **49** (2001) 1433–1446.
138. Detection and short-term prediction of epileptic seizures from the EEG signal by wavelet analysis and gaussian mixture model, with Lingmin Meng, Mark Frei, Ivan Osorio, and Truong Nguyen, to appear.
139. Laplacian eigenvalues of growing trees, with Li He and Xiangwei Liu, *Proc. Conf. on Math. Theory of Networks and Systems*, Perpignan (2000).
140. *Teaching and learning on the Internet*, Mathematical Association of America, H. Pollatsek et al., eds. (2001).
141. The joint spectral radius, Commentary on Paper #5, *Gian-Carlo Rota on Analysis and Probability, Selected Papers*, J. Dhombres, J.P.S. Kung, and N. Starr, eds., Birkhäuser (2003).
142. Localized eigenvectors from widely spaced matrix modifications, with Xiangwei Liu and Susan Ott, *SIAM J. Discrete Math.* **16** (2003) 479–498 .
143. IMACS Matrices, *Proceedings of 16th IMACS World Congress* (2000).
144. Signal processing for everyone, *Computational Mathematics Driven by Industrial Problems*, Springer Lecture Notes in Mathematics 1739, V. Capasso, H. Engl, and J. Periaux, eds. (2000).
145. A study of two-channel complex-valued filter banks and wavelets with orthogonality and symmetry properties, with X.Q. Gao and T. Nguyen, *IEEE Trans. Signal Process.* **50** (2002) 824–833.
146. Binomial matrices, with G. Boyd, C. Micchelli, and D.X. Zhou, *Adv. Comput. Math.* **14** (2001) 379–391.
147. Block tridiagonal matrices and the Kalman filter, *Wavelet Analysis: Twenty Years' Developments*, D.X. Zhou, ed., World Scientific (2002).
148. Smoothing by Savitzky-Golay and Legendre filters, with Per-Olof Persson, in *Mathematical Systems Theory*, MTNS 2002, IMA Volume edited by J. Rosenthal and D. Gilliam, Springer (2002).
149. Too Much Calculus, *SIAM Linear Algebra Activity Group Newsletter* (2002).

150. Pascal matrices, with Alan Edelman, *Amer. Math. Monthly* **111** (2004) 189–197
151. The Laplacian eigenvalues of a polygon, with Pavel Greenfield, *Comput. Math. Appl.* **48** (2004) 1121–1133.
152. A simple mesh generator in MATLAB, with Per-Olof Persson, *SIAM Rev.* **46** (2004) 329–345.
153. The interplay of ranks of submatrices, with Tri Nguyen, *SIAM Rev.* **46** (2004) 637–646.
154. Circuit simulation and moving mesh generation, with Per-Olof Persson, *Proceedings Int. Symp. Comm. & Inf. Technology (ISCIT)*, Sapporo (2004).
155. Linear algebra: A happy chance to apply mathematics, *Proc. Int. Congress on Math. Education (ICME-10)*, Denmark (2004).
156. Book review: The SIAM 100-digit Challenge, *Science* **307** (2005) 521–522.
157. Peter Lax wins Abel Prize, *SIAM News* **38** (2005).
158. A remarkable eye for out-of-the-ordinary mathematics (interview with L. Mahadevan), *SIAM News* **38** (2005).
159. Matrices with prescribed Ritz values, with B. Parlett, *Linear Alg. Appl.* **428** (2008) 1725–1739.
160. Maximum flows and minimum cuts in the plane, *J. Global Optim.*, to appear (2008).
161. Maximum area with Minkowski measures of perimeter, *Proc. Roy. Soc. Edinburgh* **138A** (2008) 189–199.

Chapter 1
Maximum Flows and Minimum Cuts in the Plane

Gilbert Strang

Summary. A continuous maximum flow problem finds the largest t such that $\operatorname{div} v = t\,F(x,y)$ is possible with a capacity constraint $\|(v_1,v_2)\| \le c(x,y)$. The dual problem finds a minimum cut ∂S which is filled to capacity by the flow through it. This model problem has found increasing application in medical imaging, and the theory continues to develop (along with new algorithms). Remaining difficulties include explicit streamlines for the maximum flow, and constraints that are analogous to a directed graph.

Key words: Maximum flow, minimum cut, capacity constraint, Cheeger

1.1 Introduction

This chapter returns to a special class of problems (partial differential equations with inequality constraints) in continuous linear programming. They describe flow through a domain Ω, in analogy with flow along the edges of a graph. The flow is maximized subject to a capacity constraint. The key to the solution is the dual problem, which looks for a set $S \subset \Omega$ from which no more flow is possible. The boundary of S is the *minimum cut*, and it is filled to capacity by the maximum flow.

In the discrete case, Kirchhoff's current law that "*flow in* = *flow out*" must hold at every interior node of the network. The *maximum flow* is the largest flow from source to sink, subject to Kirchhoff's equation at the nodes and capacity constraints on the edges. This fits the standard framework of linear programming, and Kirchhoff's incidence matrix (of 1s, −1s, and 0s) has remarkable properties that lead to an attractive theory. Our purpose is

Gilbert Strang
Department of Mathematics, Massachusetts Institute of Technology, Cambridge, MA 02139, U.S.A., e-mail: gs@math.mit.edu

D.Y. Gao, H.D. Sherali, (eds.), *Advances in Applied Mathematics and Global Optimization*
Advances in Mechanics and Mathematics 17, DOI 10.1007/978-0-387-75714-8_1,

to point to a *maximum flow–minimum cut theorem* in the continuous case, and to introduce new questions.

The principal unknown is the vector $v(x,y)$ that gives the magnitude and direction of the flow. On a plane domain this is $v = (v_1(x,y), v_2(x,y))$. The analogue of Kirchhoff's matrix is the divergence operator:

$$\text{Conservation:} \quad \operatorname{div} v = \frac{\partial v_1}{\partial x} + \frac{\partial v_2}{\partial y} = tF(x,y) \quad \text{in } \Omega. \tag{1.1}$$

That source/sink term $tF(x,y)$ might be zero or nonzero in the interior of the flow domain Ω. There may also be a source term $tf(x,y)$ on the boundary $\partial\Omega$, in closer analogy with the discrete source and sink in classical network flow. With n as the unit normal vector to $\partial\Omega$, the (possible) boundary sources and sinks are given by a Neumann condition:

$$\text{Boundary sources:} \quad v \cdot n = tf(x,y) \quad \text{on } \partial\Omega_N. \tag{1.2}$$

Our examples involve F but not f. We only note the case $\partial\Omega_N = \partial\Omega$, when f is prescribed on the whole boundary. Then the divergence theorem $\iint \operatorname{div} v \, dx\, dy = \int v \cdot n \, ds$ imposes a compatibility condition on F and f:

$$\text{Compatibility:} \quad \iint_{\Omega} F(x,y)\, dx\, dy = \int_{\partial\Omega} f(x,y)\, ds \quad \text{if} \quad \partial\Omega_N = \partial\Omega. \tag{1.3}$$

Now comes the key inequality, a limit on the flow. The vector field $v(x,y)$ is subject to a capacity constraint, which makes the problem nonlinear. In our original paper [31] this constraint measured (v_1, v_2) always in the ℓ^2 norm at each point:

$$\text{Capacity:} \quad |v(x,y)| = \sqrt{v_1^2 + v_2^2} \le c(x,y) \quad \text{in } \Omega. \tag{1.4}$$

A more general condition would require $v(x,y)$ to lie in a convex set $K(x,y)$:

$$v(x,y) \in K(x,y) \quad \text{for all } x, y \text{ in } \Omega. \tag{1.5}$$

A typical maximal flow problem in the domain Ω is

Maximize t subject to (1.1), (1.2), and (1.4).

In returning to this maximal flow problem, our goal is to highlight four questions that were not originally considered. Fortunately there has been good progress by several authors, and partial answers are available. But the new tools are not yet all-powerful, as we illustrate with a challenge problem (uniform source $F = 1$ and capacity $c = 1$ with Ω = unit square). This continues to resist explicit solution for the velocity vector v:

$$\text{Challenge:} \quad \text{Maximize } t \text{ so that } \operatorname{div} v = t \text{ with } |v| \le 1 \text{ in } \Omega. \tag{1.6}$$

The intriguing aspect of this problem is that we can identify the minimal cut. Therefore we know the maximal flow factor $t = 2 + \sqrt{\pi}$, from the capacity across that cut ∂S. Determining ∂S is a *constrained isoperimetric problem* that is pleasant to solve (and raises new questions).

What we do not know is the flow vector v inside the square! Optimality tells us the magnitude and direction of v only along the cut, described below. We apologize for the multiplication of new challenges, when the proper goal of a chapter should be new solutions.

1.2 New Questions and Applications

The continuous maximal flow problem is attracting a small surge of interest. We mention recent papers that carry the problem forward in several directions:

1.

Grieser [16] shows how max flow–min cut duality leads to an elegant proof of Cheeger's inequality, giving the lower bound in (1.18) on the first eigenvalue of the Laplacian on Ω. The eigenfunction has $u = 0$ on $\partial\Omega$, so $\partial\Omega_N$ is empty:

$$\text{Cheeger:} \quad \lambda_1 \geq \frac{1}{4}h^2 \quad \text{where } h(\Omega) = t_{\mathsf{max}} \text{ with } F \equiv 1. \tag{1.7}$$

The Cheeger constant h is found from the constrained isoperimetric problem that arises for the minimal cut ∂S:

$$\text{Definition:} \quad h(\Omega) = \inf_{S \subset \Omega} \frac{\text{perimeter of } S}{\text{area of } S}. \tag{1.8}$$

As in the particular case of our challenge problem, $h(\Omega)$ is often computable. For the unit square we note in (1.24) that the inequality (1.7) is far from tight.

2.

Appleton and Talbot [2] have proposed an algorithm for computing the maximum flow vector v from a sequence of discrete problems. Their motivation is to study image segmentation with medical applications (see especially [3, 4]). The same techniques are successful in stereo matching [26]. Their paper is rich in ideas for efficient computations and an excellent guide to the literature.

The algorithm approaches the maximum flow field as $T \to \infty$, by introducing a "Maxwell wave equation" with capacity c and internal source $F = 0$:

$$\text{Appleton–Talbot:} \qquad \frac{\partial E}{\partial T} = -\operatorname{div} v, \qquad \frac{\partial v}{\partial T} = -\operatorname{grad} E, \qquad |v| \le c. \quad (1.9)$$

The potential is E, and the first equation $\partial E/\partial T = -\operatorname{div} v$ is a relaxation of the conservation constraint $\operatorname{div} v = 0$ (Kirchhoff's law). Appleton and Talbot prove that the energy $\iint(|E|^2 + |v|^2)$ is decreasing in every subset S of Ω. At convergence, the optimal cut is the boundary of a level set of E.

The equations (1.9) are discretized on a staggered grid. This corresponds to Yee's method (also called the FTDT method) in electromagnetics. The algorithm has a weighting function to model the effect of source terms, and the experiments with image segmentation are very promising.

Because primal–dual interior point algorithms have become dominant in optimization, we conjecture that those methods can be effective also here in the approximation of continuous by discrete maximal flows.

3.

Nozawa [23] took a major step in extending the max flow–min cut theorem from the simple isotropic condition $|v| \le 1$ in (1.4) toward the much more general capacity condition (1.5). This step can be illustrated already in our challenge problem, by changing from the ℓ^2 norm of $v(x, y)$ to the ℓ^1 or ℓ^∞ norm:

ℓ^1 challenge: Maximize t so that $\operatorname{div} v = t$ with $|v_1| + |v_2| \le 1$ in Ω. (1.10)

ℓ^∞ challenge: Maximize t so that $\operatorname{div} v = t$ with $|v_1| \le 1, |v_2| \le 1$ in Ω. (1.11)

In the isoperimetric problem (1.8), this changes the definition of perimeter. The dual norm (in this case ℓ^∞ or ℓ^1) becomes the measure of arc-length (dx, dy). Then this dual norm enters the computation of $|\partial S|$:

$$\text{Perimeter (in } \mathbf{R}^2\text{):} \quad |\partial S| = \int_{\partial S} |(dx, dy)|. \qquad (1.12)$$

The coarea formula from geometric measure theory [12], on which the proof of duality rests, continues to apply with the new definition.

As in the ℓ^2 case, the maximal t can be computed! So we have new flow fields to find, reaching bounds that duality says are achievable.

It is intriguing to connect maximal flow with the other central problem for networks and continua: the *transportation problem*. This asks for shortest paths. The original work of Monge and Kantorovich on continuous flows has been enormously extended by Evans [11], Gangbo and McCann [15], Rachev and Rüschendorf [25], and Villani [33].

Our challenge problem requires the movement of material $F(x,y)$ from Ω to $\partial\Omega$. The bottleneck is in moving from the interior of S to the minimal cut ∂S. The distribution of material is uniform in S, and its destination is uniform along ∂S, to use all the capacity allowed by $|v| \leq 1$. How is the shortest path (Monge) flow from S to ∂S related to the maximum flow?

4. Directed Graphs and Flows.

Chung [8, 9] has emphasized that Cheeger's theory (and the Laplacian itself) is not yet fully developed for directed graphs. For maximal flow on networks, Ford and Fulkerson [13] had no special difficulty when the edge capacities depend on the direction of flow. The problem is still a linear program and duality still holds.

For directed continuous flows we lack a correctly formulated duality theorem. The capacity would be a constraint $v(x,y) \in K(x,y)$ as in (1.5). Nozawa's duality theorem in [23] quite reasonably assumed that zero is an interior point of K. Then a flow field exists for sufficiently small t (the feasible set is not empty). The continuous analogue of direction-dependent capacities seems to require analysis of more general convex sets $K(x,y)$, when zero is a boundary point. In [22], Nozawa illustrated duality gaps when his hypotheses were violated.

Finally we mention that all these questions extend to domains Ω in $\mathbf{R}^n$. The constrained isoperimetric problems generalize to higher dimensions as well as different norms. The one simplification in the plane is the introduction of a stream function $s(x,y)$, with $(v_1, v_2) = (\partial s/\partial y, -\partial s/\partial x)$ as the general solution to $\operatorname{div} v = 0$. Our survey [30] formulated the corresponding primal and dual problems for $s(x,y)$ as L^1 and L^∞ approximations of planar vector fields, where Laplace's equation corresponds to L^2.

The remaining sections of this chapter discuss the topics outlined above. We compute the minimum cuts in the three versions of the challenge problem on the unit square. We also mention an isoperimetric problem (with a different definition of perimeter) to which we return in a later paper [32].

1.3 Duality, Coarea, and Cheeger Constants

The maximum flow is limited by the capacity $c(x,y)$:

Primal problem: Maximize t subject to

$$\operatorname{div} v = tF(x,y) \text{ in } \Omega, \quad v \cdot n = tf(x,y) \text{ on } \partial\Omega_N, \quad |v(x,y)|_2 \le c(x,y) \text{ in } \Omega.$$

Nozawa's duality theorem requires a proper choice of function spaces and boundary conditions, in this problem and in its dual for $u(x,y)$ in $\mathrm{BV}(\Omega)$. Where the primal involves the divergence, the dual involves the gradient. Kohn and Temam [19] extended Green's formula $\iint u \operatorname{div} v = -\iint v \cdot \operatorname{grad} u$ to allow functions $u(x,y)$ of bounded variation.

We show that the optimal $u(x,y)$ in the dual problem is the characteristic function of a set S with finite perimeter. This $u(x,y)$ is not smooth, but it lies in BV. The dual problem does not initially ask for a minimum cut.

$$\text{Dual problem:} \quad \text{Minimize } \|u\|_{\mathrm{BV},c} \text{ with } \ell(u) = 1 \quad \text{or} \quad \text{Minimize } \frac{\|u\|_{\mathrm{BV},c}}{|\ell(u)|}$$

$$\|u\|_{\mathrm{BV},c} = \iint\limits_{\Omega} c(x,y)|\operatorname{grad} u|_2 \, dx\, dy \qquad \ell(u) = \int\limits_{\partial\Omega_N} uf\, ds - \iint\limits_{\Omega} uF\, dx\, dy. \tag{1.13}$$

The key step toward the solution is to recognize the extreme points of the unit ball in this weighted BV norm $\|u\|_{\mathrm{BV},c} = \iint c|\operatorname{grad} u|_2\, dx\, dy$. Those extreme points are characteristic functions $u = \chi_S$ of open subsets S of Ω:

$$\chi_S(x,y) = \{1 \text{ for } x,y \text{ in } S, \quad 0 \text{ otherwise}\}.$$

The BV norm of χ_S is the weighted perimeter $\int c\, ds$ of S, because the gradient is a measure (a line of delta functions) supported only on that boundary ∂S.

The coarea formula gives the BV norm of u (weighted by the capacity c) as an integral over the norms of characteristic functions of level sets $S(t)$ of u:

$$\text{Coarea:} \quad \|u\|_{\mathrm{BV},c} = \int \|\chi_{S(t)}\|_{\mathrm{BV},c}\, dt \quad \text{with} \quad S(t) = \{x,y \mid u(x,y) < t\}. \tag{1.14}$$

Consider the case with $F \ge 0$ and no boundary sources f. Specializing in (1.13) to the characteristic functions $u = \chi_S$, our dual problem reduces to an isoperimetric problem for S and the minimum cut ∂S:

$$\min_{u \in \mathrm{BV}} \frac{\iint c|\operatorname{grad} u|_2\, dx\, dy}{\iint Fu\, dx\, dy} = \min_{S \subset \Omega} \frac{\text{weighted perimeter } \int_{\partial S} c\, ds}{\text{weighted area } \iint_S F\, dx\, dy}. \tag{1.15}$$

Choosing $c(x,y) = 1$ and $F(x,y) = 1$, this computes the Cheeger constant.

$$\text{Cheeger constant:} \quad h(\Omega) = \inf_{S \subset \Omega} \frac{|\partial S|}{|S|}. \tag{1.16}$$

Weak duality $h \ge t$ is the inequality $\int c\, ds \ge t \iint F\, dx\, dy$ for every feasible t in the primal problem. This is just the divergence theorem when $\operatorname{div} v =$

$tF(x,y)$ and $|v| \le 1$:

$$\text{Weak duality } h \ge t: \quad \int_{\partial S} c\, ds \ge \int_{\partial S} v\cdot n\, ds = \iint_S \operatorname{div} v\, dx\, dy = t \iint_S F\, dx\, dy. \quad (1.17)$$

Duality says that equality holds for the maximal flow v and the minimal cut ∂S.

Historically, the key inequality given by Cheeger [7] was a lower bound on the first eigenvalue λ_1 of the Laplacian on the domain Ω. Grieser [16] observed how neatly and directly this bound follows from Green's formula, when $F = 1$ and $|v| \le 1$. We expect to see the Schwarz inequality in the step from problems in L^1 and L^∞ to the eigenvalue problem in L^2:

$$\begin{aligned} t \iint u^2 &= \int (\operatorname{div} v) u^2 = -\iint v (\operatorname{grad} u^2) \\ &\le 2 \iint |u|\,|\operatorname{grad} u| \le 2 \left[\iint u^2 \iint |\operatorname{grad} u|^2 \right]^{1/2}. \end{aligned}$$

Thus any feasible t gives a lower bound $t^2/4$ to the Rayleigh quotient for any $u(x,y)$ with $u = 0$ on $\partial\Omega$:

$$\frac{t^2}{4} \le \frac{\iint |\operatorname{grad} u|^2\, dx\, dy}{\iint u^2\, dx\, dy}. \quad (1.18)$$

The minimum of the right side is $\lambda_1(\Omega)$, and the maximum of the left side is $h^2/4$. Cheeger's inequality becomes $h^2/4 \le \lambda_1(\Omega)$.

A widely studied paper [10] of Diaconis and Stroock introduces another very useful measure of the "bottleneck" that limits flow on a graph.

1.4 The Challenge Problems

When we use the ℓ^2 norm of the flow vector $v = (v_1, v_2)$ at each point, the constraint $|v(x,y)| \le c(x,y)$ is isotropic. Other norms of v give constraints that come closer to those on a discrete graph. The edges of the graph might be horizontal and vertical (from a square grid) or at 45° and −45° (from a staggered grid). We use the challenge problem with $F = c = 1$ on a unit square as an example that allows computation of the minimal cut in all three cases.

ℓ^2 constraint: $v_1^2 + v_2^2 \le 1$. (1.19)

ℓ^1 constraint: $|v_1| + |v_2| \le 1$. (1.20)

ℓ^∞ constraint: $\max(|v_1|, |v_2|) \le 1$. (1.21)

The ℓ^1 and ℓ^∞ norms give problems in linear programming.

The dual (minimum cut) problems use the dual norms. For ℓ^2 we had the usual BV norm $\iint |\operatorname{grad} u|\, dx\, dy$ and the usual measure $|\partial S| = \int ds$ of the perimeter. For the ℓ^1 and ℓ^∞ constraints, the BV norms change and the perimeters reflect those changes (coming from the coarea formula in the new norms):

$|v|_1 \le 1$ leads to

$$\|u\|_{\mathrm{BV}} = \iint |\operatorname{grad} u|_\infty\, dx\, dy \quad \text{and} \quad |\partial S|_\infty = \int_{\partial S} \max(|dx|, |dy|),$$

$|v|_\infty \le 1$ leads to

$$\|u\|_{\mathrm{BV}} = \iint |\operatorname{grad} u|_1\, dx\, dy \quad \text{and} \quad |\partial S|_1 = \int_{\partial S} |dx| + |dy|.$$

The perimeter of a square changes as the square is rotated, because the norm of (dx, dy) is changing. In each case the dual problem looks for the minimum cut as the solution to a constrained isoperimetric problem.

$$\text{Duals of } \ell^2, \ell^1, \ell^\infty: \quad \underset{S \subset [0,1]^2}{\text{Minimize}} \; \frac{|\partial S|}{|S|} \quad \text{and} \quad \frac{|\partial S|_\infty}{|S|} \quad \text{and} \quad \frac{|\partial S|_1}{|S|}. \tag{1.22}$$

In all cases the optimal S will reach the boundary $\partial\Omega$ of the square. (If S is stretched by a factor c, the areas in the denominators of (1.22) are multiplied by c^2 and the numerators by c.) The symmetry of the problem ensures that the optimal ∂S contains four flat pieces of $\partial\Omega$, centered on the sides of the square (Figure 1.1). The only parameter in the three optimization problems is the length L of those boundary pieces.

Figure 1.1a shows the solution for the ℓ^2 problem, where the "unconstrained" parts of the cut ∂S are circular arcs. This follows from the classical isoperimetric problem, and it is easy to show that the arcs must be tangent to the square. The four arcs would fit together in a circle of radius r. With $L = 1 - 2r$, the optimal cut solves the Cheeger problem:

$$t_{\mathsf{max}} = h(\Omega) = \min \frac{\text{perimeter of } S}{\text{area of } S} = \min \frac{4(1-2r) + 2\pi r}{1 - 4r^2 + \pi r^2}. \tag{1.23}$$

The derivative of that ratio is zero when

$$(1 - 4r^2 + \pi r^2)(8 - 2\pi) = (4 - 8r + 2\pi r)(8r - 2\pi r).$$

Cancel $8 - 2\pi$ to reach $1 - 4r + (4 - \pi) r^2 = 0$. Then $r = 1/(2 + \sqrt{\pi}) \approx .265$. The Cheeger constant $h(\Omega)$ is the ratio $|\partial S|/|S| = 1/r = 2 + \sqrt{\pi}$.

A prize of 10,000 yen was offered in [30] for the flow field that achieves $\operatorname{div} v = 2 + \sqrt{\pi}$ with $|v| \le 1$. Lippert [20] and Overton [24] have the strongest

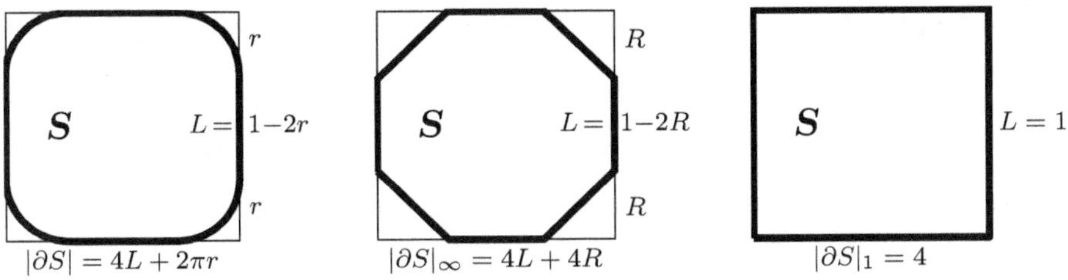

Fig. 1.1 The minimum cuts ∂S for ℓ^2, ℓ^1, and ℓ^∞ constraints on $v(x,y)$.

claim on the prize, by computing a close approximation to v. The discrete velocity clearly confirms the cut in Figure 1.1a as the set where $|v| = 1$.

The eigenfunctions of the Laplacian on the unit square are $(\sin \pi x)(\sin \pi y)$ and the lowest eigenvalue is $\lambda_1 = 2\pi^2$. Cheeger's inequality $\lambda_1 \geq h^2/4$, which other authors have tested earlier, is far from tight.

$$\text{Unit square:} \quad 2\pi^2 > (2+\sqrt{\pi})^2/4 \quad \text{or} \quad 19.74 > 3.56\,. \tag{1.24}$$

The second challenge problem has $|v_1| + |v_2| \leq 1$ leading to the measure $|\partial S|_\infty$ of the perimeter in the dual. Now the unconstrained isoperimetric problem is solved by a diamond with $|n_1| = |n_2| = 1/\sqrt{2}$ on all edges. The optimal cut ∂S in Figure 1.1b is a union of boundary pieces and diamond edges. The edge length $\sqrt{2}R$ is multiplied by $1/\sqrt{2}$ from $|n_1| = |n_2|$ to give $|\partial S|_\infty = 4L + 4R = 4 - 4R$. Then the minimum cut has $R = (2-\sqrt{2})/2 \approx .3$:

$$\min \frac{|\partial S|_\infty}{|S|} = \min_R \frac{4-4R}{1-2R^2} = \frac{\sqrt{2}}{\sqrt{2}-1} \approx 3.5\,.$$

For the flow field v in this ℓ^1 problem, the prize is reduced to 5000 yen. Lippert has reached the computational prize also in ℓ^1. This is linear programming and interior point methods soundly defeated the simplex method.

We cannot afford the prize in the ℓ^∞ problem, whose solution is simply $v = (2x-1, 2y-1)$ on the square $0 \leq x, y \leq 1$ with $\operatorname{div} v = 4 = t_{\mathsf{max}}$.

The minimum cut for the ℓ^∞ problem is the whole boundary of the square. This coincides with the unconstrained isoperimetric solution when the perimeter is measured by $\int |dx| + |dy|$. The minimizing set S would have horizontal and vertical sides wherever the constraint $S \subset \Omega$ is inactive, and here it is active everywhere on $\partial S = \partial \Omega$. The Cheeger constant in this norm is $h = 4/1$.

In [32] we prove that the unit ball in the dual norm (rotated by $\pi/2$) is isoperimetrically optimal. Here that ball is a circle or a diamond or a square. This *isoperimetrix* was discovered by Busemann [5] using the Brunn–Minkowski theorem in convex geometry (the Greeks knew much earlier about the circle). Our proof finds a simple linear equation for the support function of the optimal convex set S.

Acknowledgment This research was supported by the Singapore–MIT Alliance.

Added in proof Z. Milbers (unpublished thesis, Köln Universität, 2006) has found the flow field in our challenge example (described at the end of Section 1.1)! New applications of minimum cuts and maximum flow have also appeared in landslide modeling, the L^1 Laplacian, and especially image segmentation.

References

1. N. Alon, *Eigenvalues and expanders*, Combinatorica **6** (1986) 86–96.
2. B. Appleton and H. Talbot, *Globally minimal surfaces by continuous maximal flows*, IEEE Trans. Pattern Anal. Mach. Intell. **28** (2006) 106–118.
3. Y. Boykov and V. Kolmogorov, *An experimental comparison of min-cut/max-flow algorithms for energy minimization in vision*, IEEE Trans. Pattern Anal. Mach. Intell. **26** (2004) 1124–1137.
4. Y. Boykov, O. Veksler, and R. Zabih, *Fast approximate energy minimization via graph cuts*, IEEE Trans. Pattern Anal. Mach. Intell. **23** (2001) 1222–1239.
5. H. Busemann, *The isoperimetric problem in the Minkowski plane*, Amer. J. Math. **69** (1947) 863–871.
6. J. D. Chavez and L. H. Harper, *Duality theorems for a continuous analog of Ford-Fulkerson flows in networks*, Adv. Appl. Math. **14** (1993) 369–388.
7. J. Cheeger, *A lower bound for the smallest eigenvalue of the Laplacian*, Problems in Analysis, 1970, 195–199.
8. F. R. K. Chung, *Spectral graph theory*, CBMS Regional Conference Series in Mathematics, vol. 92, 1997.
9. F. R. K. Chung, *Laplacians and the Cheeger inequality for directed graphs*, Ann. Combinatorics **9** (2005) 1–19.
10. P. Diaconis and D. W. Stroock, *Geometric bounds for eigenvalues of Markov chains*, Ann. Appl. Probab. **1** (1991) 36–61.
11. L. C. Evans, *Survey of applications of PDE methods to Monge-Kantorovich mass transfer problems*, `www.math.berkeley.edu/~evans` (earlier version: Current Developments in Mathematics, 1997).
12. W. Fleming and R. Rishel, *An integral formula for total gradient variation*, Archiv der Mathematik **11** (1960) 218–222.
13. L. R. Ford Jr. and D. R. Fulkerson, *Flows in Networks*, Princeton University Press, 1962.
14. L. R. Ford Jr. and D. R. Fulkerson, *Maximal flow through a network*, Canad. J. Math. **8** (1956) 399–404.
15. W. Gangbo and R. McCann, *Optimal maps in Monge's mass transport problem*, C.R. Acad. Sci. Paris. Ser. I. Math. **325** (1995) 1653–1658.
16. D. Grieser, *The first eigenvalue of the Laplacian, isoperimetric constants, and the max flow min cut theorem*, Archiv der Mathematik **87** (2006) 75–85.
17. T. C. Hu, *Integer Programming and Network Flows*, Addison-Wesley, 1969.
18. M. Iri, *Theory of flows in continua as approximation to flows in networks*, Survey of Mathematical Programming **2** (1979) 263–278.
19. R. Kohn and R. Temam, *Dual spaces of stresses and strains*, Appl. Math. and Opt. **10** (1983) 1–35.
20. R. Lippert, *Discrete approximations to continuum optimal flow problems*, Stud. Appl. Math. **117** (2006) 321–333.
21. J. S. Mitchell, *On maximum flows in polyhedral domains*, Proc. Fourth Ann. Symp. Computational Geometry, 341–351, 1988.

22. R. Nozawa, *Examples of max-flow and min-cut problems with duality gaps in continuous networks*, Math. Program. **63** (1994) 213–234.
23. R. Nozawa, *Max-flow min-cut theorem in an anisotropic network*, Osaka J. Math. **27** (1990) 805–842.
24. M. L. Overton, *Numerical solution of a model problem from collapse load analysis*, Computing Methods in Applied Science and Engineering VI, R. Glowinski and J. L. Lions, eds., Elsevier, 1984.
25. S. T. Rachev and L. Rüschendorf, *Mass Transportation Problems I, II*, Springer (1998).
26. S. Roy and I. J. Cox, *A maximum-flow formulation of the n-camera stereo correspondence problem*, Proc. Int. Conf. Computer Vision (1988) 492–499.
27. F. Santosa, *An inverse problem in photolithography*, in preparation.
28. R. Sedgewick, *Algorithms in C*, Addison-Wesley, 2002.
29. G. Strang, *A minimax problem in plasticity theory*, Functional Analysis Methods in Numerical Analysis, Z. Nashed, ed., Lecture Notes in Mathematics 701, Springer, 1979.
30. G. Strang, *L^1 and L^∞ approximation of vector fields in the plane*, Lecture Notes in Num. Appl. Anal. **5** (1982) 273–288.
31. G. Strang, *Maximal flow through a domain*, Math. Program. **26** (1983) 123–143.
32. G. Strang, *Maximum area with Minkowski measures of perimeter*, Proc. Royal Society of Edinburgh **138** (2008) 189–199.
33. C. Villani, *Topics in Optimal Transportation*, Graduate Studies in Mathematics 58, American Mathematical Society, 2003.

Chapter 2
Variational Principles and Residual Bounds for Nonpotential Equations

Giles Auchmuty

Dedicated to Gil Strang for his 70th birthday

Summary. Solutions of nonsymmetric linear equations whose symmetric part is positive definite are first characterized as saddle points of a strictly convex–concave quadratic function. The associated primal problem is shown to be equivalent to a weighted quadratic minimum residual optimization problem. An a posteriori error estimate for approximate solutions is derived. Similar results are then obtained for semilinear finite-dimensional systems of equations. These include global optimization problems for the solutions and existence results based on min-max theorems. Under further assumptions, uniqueness theorems are proven using saddle point theorems.

2.1 Introduction

In this chapter, some unconstrained global optimization problems for the solutions of finite-dimensional systems of equations that are not of potential type are described and analyzed. That is, the equations need not be obtained directly as the critical points of a differentiable function.

This approach is first illustrated by considering the case of a linear equation involving a matrix that is nonsymmetric but has positive definite symmetric part. In Section 2.2 the solutions of such equations are shown to be the saddle point of certain functions, and the existence and uniqueness of these saddle points are proven directly. The dual variational principles associated with this saddle point problem are studied. In Section 2.3, the primal problem associated with this saddle problem is shown to be a weighted quadratic

---

Giles Auchmuty
Department of Mathematics, University of Houston, Houston, TX 77204-3008 U.S.A.
e-mail: auchmuty@uh.edu

This work was partially performed while the author was working for the Division of Mathematical Sciences, National Science Foundation, Arlington, VA.

D.Y. Gao, H.D. Sherali, (eds.), *Advances in Applied Mathematics and Global Optimization*
Advances in Mechanics and Mathematics 17, DOI 10.1007/978-0-387-75714-8_2,

residual minimization problem. Thus the variational principle provides elementary error estimates for the solution in terms of the value of the function being minimized. These results extend to nonsymmetric problems some of the results that are well known for the solutions of symmetric positive definite problems. See the recent texts of Ainsworth and Oden [1] and Han [7] for applications of such results to the analysis of finite element simulations.

Sections 2.4 and 2.5 describe extensions of these results to two very different classes of semilinear finite-dimensional equations. In each case, a min-max characterization of the solutions is described. These need not be saddle point problems but there is a primal functional whose global minima provide solutions of the original problem. A generalized Young's inequality and results from minimax theory are used to prove existence results and obtain characterizations of the solutions. The problem in Section 2.4 may again be interpreted as a minimum residual problem and residual bounds for approximate solutions are described. The problem in Section 2.5 uses similar methods but makes quite different assumptions on the nonlinear term to obtain existence and, under further assumptions, uniqueness results.

It is a particular pleasure to contribute this chapter to these conference proceedings in honor of Gilbert Strang as essentially all of the background material for these results has been beautifully treated in his texts on linear algebra and applied mathematics.

2.2 Saddle Point Characterizations of Nonsymmetric Linear Equations

Let A be an $n \times n$ real matrix that satisfies the *coercivity* (or *ellipticity*) condition that there is an $a_0 > 0$ such that

$$\langle Ax, x \rangle \ \geq \ a_0 \, \|x\|^2 \qquad \text{for all} \quad x \in \mathbb{R}^n. \tag{2.1}$$

Here the brackets indicate the usual Euclidean inner product and the norm is the usual 2-norm. In general a function $V : \mathbb{R}^n \to \mathbb{R}$ is said to be *coercive* provided

$$\|x\|^{-1} \, V(x) \to \infty \qquad \text{as } \|x\| \to \infty.$$

In this section, it is shown that, when A is coercive, the solutions of the linear equation

$$Ax \ = \ f \tag{2.2}$$

for given $f \in \mathbb{R}^n$, may be characterized as saddle points of a quadratic convex–concave function.

When A is real symmetric and coercive, it is well known that the solution of (2.2) is the unique minimizer of the energy functional $\mathcal{E}(x)$ on $\mathbb{R}^n$ defined by

$$\mathcal{E}(x) \;:=\; \langle Ax, x\rangle \;-\; 2\langle f, x\rangle. \tag{2.3}$$

A variety of different aspects of this problem is treated in Strang [8].

When A is coercive, but not real symmetric, then Auchmuty [3] showed that the solution of (2.2) may be characterized as the saddle point of a convex–concave function. Here a different saddle function is used that provides minimum residual variational principles and computable error bounds for approximate solutions of (2.2).

Let $A_S := (A + A^T)/2$ and $B := (A - A^T)/2$ be the symmetric and skew symmetric parts of A, D be a diagonal positive definite matrix, and $C := A_S - D$. Then C is a real symmetric matrix and equation (2.2) may be written

$$(B + C + D)x \;=\; f. \tag{2.4}$$

When C_1, C_2 are real symmetric matrices, we say that $C_2 \leq C_1$ (respectively, $C_2 < C_1$) provided $\langle (C_1 - C_2)x, x\rangle \geq 0$ for all $x \in \mathbb{R}^n$, (or $\langle (C_1 - C_2)x, x\rangle > 0$ for all nonzero $x \in \mathbb{R}^n$).

Consider the function $\mathcal{L} : \mathbb{R}^n \times \mathbb{R}^n \to \mathbb{R}$ defined by

$$\mathcal{L}(x,y) \;:=\; \langle (A_S - D/2)x, x\rangle + \langle f, y - x\rangle - \langle (B+C)x, y\rangle - \frac{1}{2}\langle Dy, y\rangle. \tag{2.5}$$

A point $(\hat{x}, \hat{y}) \in \mathbb{R}^n \times \mathbb{R}^n$ is said to be a *saddle point* of $\mathcal{L}$ provided

$$\mathcal{L}(\hat{x}, y) \;\leq\; \mathcal{L}(\hat{x}, \hat{y}) \;\leq\; \mathcal{L}(x, \hat{y}) \qquad \text{for all } (x,y) \in \mathbb{R}^n \times \mathbb{R}^n.$$

The following result shows that, provided D is small enough, the function $\mathcal{L}$ defined above has a unique saddle point, and that this saddle point characterizes the solutions of the linear equation (2.2).

Theorem 2.1. *Assume A, B, C, D as above,* (2.1) *holds, and $D < 2A_S$. Then the function $\mathcal{L}$ defined by* (2.5) *has a unique saddle point $(\hat{x}, \hat{x}) \in \mathbb{R}^n \times \mathbb{R}^n$ with $\hat{x}$ being the unique solution of* (2.2).

Proof. Given y in $\mathbb{R}^n$, $\mathcal{L}(.,y)$ is continuous. When $D < 2A_S$, this function is strictly convex and there is a $\delta > 0$ and a function c such that

$$\mathcal{L}(x,y) \;\geq\; \delta\|x\|^2 - [\|f\| + \|(C - B)y\|]\;\|x\| + c(y).$$

Hence $\mathcal{L}(.,y)$ is coercive for each $y \in \mathbb{R}^n$. Similarly $-\mathcal{L}(x,.)$ is continuous, strictly convex, and coercive for each $x \in \mathbb{R}^n$. The usual minimax theorem (see Theorem 49A in Zeidler [9] for example) implies that $\mathcal{L}$ has a saddle point in $\mathbb{R}^n \times \mathbb{R}^n$.

This function $\mathcal{L}$ is continuously differentiable, so the saddle point is a solution of the system

$$\nabla_x \mathcal{L}(x,y) \;=\; (2A_S - D)x \;+\; (B - C)y \;-\; f \;=\; 0 \tag{2.6}$$

$$\nabla_y \mathcal{L}(x,y) \;=\; f \;-\; (B + C)x \;-\; Dy \;=\; 0. \tag{2.7}$$

Here ∇ is the usual gradient operator. Add these equations to obtain

$$(C + D - B)(x - y) \;=\; 0.$$

Take inner products with $(x - y)$; then

$$\langle A_S(x-y), x-y\rangle \;\geq\; a_0\|x-y\|^2 \;=\; 0$$

as $A_S = C + D$, (2.1) holds, and $\langle Bz, z\rangle = 0$ for all z. Thus the saddle point must have the form $(\hat{x}, \hat{x})$ and (2.7) shows that the equation satisfied by $\hat{x}$ is (2.2). The uniqueness of solutions follows from (2.1). □

Note that this is a (quite different) existence-uniqueness proof for (2.2) when (2.1) holds. It is based on elementary calculus and the minimax theorem. The analysis in Auchmuty [3] used a different saddle function that did not involve a matrix of the form D. The introduction of this matrix leads to expressions that are more practical for numerical computation than those described in [3]. Strictly speaking, it is not necessary that D be diagonal; it suffices that D be symmetric and positive definite and that the inverse D^{-1} be known explicitly for use in formulae to be described in the next section.

2.3 Variational Principles for Nonsymmetric Linear Equations

A convex–concave saddle problem defines a pair of associated dual variational principles. See Auchmuty [2], Section 3 or Ekeland and Temam [5], Chapter III for general descriptions of such constructions and their properties.

Define the function $\mathcal{G} : \mathbb{R}^n \to \mathbb{R}$ by

$$\mathcal{G}(x) \;:=\; \sup_{y \in \mathbb{R}^n} \mathcal{L}(x, y). \tag{2.8}$$

The primal problem associated with $\mathcal{L}$ is to minimize $\mathcal{G}$ on $\mathbb{R}^n$. The maximization of $\mathcal{L}$ with respect to y is straightforward and yields

$$\mathcal{G}(x) \;=\; \frac{1}{2}\langle (A_S - D/2)x, x\rangle \;-\; \langle f, x\rangle \;+\; \frac{1}{2}\langle D^{-1}(f-(B+C)x), f-(B+C)x\rangle. \tag{2.9}$$

The essential results about this minimization problem may be summarized as follows.

Theorem 2.2. *Assume A, B, C, D as above and* (2.1) *holds. Then the function $\mathcal{G}$ defined by* (2.9) *is strictly convex and coercive on $\mathbb{R}^n$. It has a unique minimizer $\hat{x}$ on $\mathbb{R}^n$ with $\hat{x}$ being the unique solution of* (2.2).

Proof. First note that D^{-1} is diagonal with positive entries on the diagonal, as this holds for D. Straightforward algebra leads to the formula

$$\mathcal{G}(x) = \frac{1}{2}\langle D^{-1}(f - Ax), f - Ax\rangle \tag{2.10}$$

$$= \frac{1}{2}\langle D^{-1}Ax, Ax\rangle - \langle A^T D^{-1} f, x\rangle + \frac{1}{2}\langle D^{-1} f, f\rangle. \tag{2.11}$$

From (2.1) and Cauchy's inequality one has $\|Ax\| \geq a_0 \|x\|$. Let d_M be the largest entry in the diagonal matrix D; then

$$\mathcal{G}(x) \geq \frac{a_0}{2d_M}\|x\|^2 - c_1\|f\|\,\|x\| + c_2 \qquad \text{for all } x \in \mathbb{R}^n \tag{2.12}$$

where c_1, c_2 are constants. These formulae imply that $\mathcal{G}$ is strictly convex and coercive on $\mathbb{R}^n$ as it is quadratic. It is continuous so $\mathcal{G}$ attains its infimum and the minimizer is unique. The expression (2.11) is G-differentiable with

$$\nabla\mathcal{G}(x) = A^T D^{-1}(Ax - f).$$

This must be zero at a minimizer. Because A and D are nonsingular, this implies that the minimizers satisfy (2.2) as claimed. □

This function $\mathcal{G}$ was defined from general considerations of duality associated with the function $\mathcal{L}$. Let $r(x) := f - Ax$ be the residual of x with respect to the equation (2.2); then (2.11) implies that

$$\mathcal{G}(x) = \frac{1}{2}\langle D^{-1}r, r\rangle \geq \frac{1}{2d_M}\|r(x)\|^2 \qquad \text{for all } x \in \mathbb{R}^n. \tag{2.13}$$

This shows that the variational principle of minimizing $\mathcal{G}$ on $\mathbb{R}^n$ is a weighted (or preconditioned) minimum residual principle for the problem of solving (2.2). This enables us to obtain an a posteriori error estimate for solutions of (2.2) in terms of the values of $\mathcal{G}(x)$. When $\hat{x}$ is the solution of (2.2) and $x \in \mathbb{R}^n$, then $A(x - \hat{x}) = r$, so upon taking inner products with $x - \hat{x}$ and using (2.1) one sees that

$$a_0\|x - \hat{x}\|^2 \leq |\langle r, x - \hat{x}\rangle| \leq \|r\|\,\|x - \hat{x}\|$$

Rearrange this; then (2.13) implies

$$\|x - \hat{x}\| \leq {a_0}^{-1}\|r\| \leq {a_0}^{-1}\sqrt{2d_M\,\mathcal{G}(x)} \qquad \text{for all } x \in \mathbb{R}^n. \tag{2.14}$$

This estimate does not require knowledge of a condition number for A; just a bound on a_0. Note that D here is any positive definite diagonal matrix, but the values of $\mathcal{G}(x)$ depend on both D, D^{-1}. This suggests that in practice one may wish to investigate the dependence of these bounds on the choice of D for a particular matrix A in (2.2).

The dual problem associated with the saddle point problem for $\mathcal{L}$ is to maximize $\mathcal{H} : \mathbb{R}^n \to \mathbb{R}$ defined by

$$\mathcal{H}(y) := \inf_{x\in\mathbb{R}^n} \mathcal{L}(x, y). \tag{2.15}$$

The explicit formula for $\mathcal{H}$ analogous to (2.9) is

$$\mathcal{H}(y) = -\frac{1}{2}\langle(2C+D)^{-1}(f+(C-B)y), f+(C-B)y\rangle + \langle f,y\rangle - \frac{1}{2}\langle Dy,y\rangle. \tag{2.16}$$

The use of this function requires the evaluation of $(2C+D)^{-1}$ which usually is more effort than the determination of D^{-1}, so we just concentrate on use of the primal problem. This dual functional has very similar properties to $-\mathcal{G}$.

2.4 Variational Principles for Semilinear Equations I

The preceding analysis generalizes to large classes of semilinear finite-dimensional systems of equations. First consider an equation of the form

$$Ax = F(x), \tag{2.17}$$

where A is a real $n \times n$ matrix which is coercive but need not be symmetric, and $F : \mathbb{R}^n \to \mathbb{R}^n$ is a continuous function. Let B, C, and D be matrices defined as before, so that this equation may be written as

$$F(x) - (B+C)x = Dx = \nabla q(x) \qquad \text{with } q(x) := \frac{1}{2}\langle Dx, x\rangle. \tag{2.18}$$

Consider the function $\mathcal{G} : \mathbb{R}^n \to \mathbb{R}$ defined by

$$\mathcal{G}(x) := \frac{1}{2}\langle Dx,x\rangle + \frac{1}{2}\langle D^{-1}(F(x)-(B+C)x), F(x)-(B+C)x\rangle + \langle Cx - F(x), x\rangle \tag{2.19}$$

and the variational problem of minimizing $\mathcal{G}$ on $\mathbb{R}^n$. Let the value of this problem be

$$\alpha(\mathcal{G}) := \inf_{x\in\mathbb{R}^n} \mathcal{G}(x).$$

This is a generalization of the variational principle described in Section 2.3 with $F(x)$ replacing f. The essential properties of this optimization problem may be summarized as follows.

Theorem 2.3. *Assume A, B, C, D, and F as above. Then the function $\mathcal{G}$ defined by* (2.19) *is continuous and has value $\alpha(\mathcal{G}) \geq 0$. A point $\hat{x} \in \mathbb{R}^n$ is a solution of* (2.17) *if and only if $\hat{x}$ minimizes $\mathcal{G}$ on $\mathbb{R}^n$ and $\mathcal{G}(\hat{x}) = 0$.*

Proof. Let $q(x)$ be the quadratic form defined in (2.18); then q is strictly convex on $\mathbb{R}^n$ and its conjugate function is $q^*(z) := \frac{1}{2}\langle D^{-1}z, z\rangle$. Then from the generalized Young inequality, (see Proposition 51.2 in Zeidler [9] or Section 2.5 in Han [7]), one sees that

$$q(x) + q^*(F(x) - (B+C)x) - \langle F(x) - (B+C)x, x \rangle \geq 0 \qquad \text{for all } x \in \mathbb{R}^n. \tag{2.20}$$

This and the fact that B is skew-symmetric implies that $\mathcal{G}(x) \geq 0$ for all $x \in \mathbb{R}^n$. Hence $\alpha(\mathcal{G}) \geq 0$. Equality holds in (2.20) if and only if

$$F(x) - (B + C)x = Dx.$$

This is equation (2.17) and at such a point $\mathcal{G}(x) = 0$. □

This result provides a variational principle for the possible solutions of our equation that requires not only that the points be global minimizers but also that the value of the problem be zero. It is a straightforward computation to verify that the analogue of equation (2.13) remains valid for this nonlinear problem. Specifically let $r(x) := Ax - F(x)$ be the residual of this equation at a point $x \in \mathbb{R}^n$, then

$$\mathcal{G}(x) = \frac{1}{2} \langle D^{-1} r(x), r(x) \rangle \geq \frac{1}{2d_M} \|r(x)\|^2 \qquad \text{for all } x \in \mathbb{R}^n. \tag{2.21}$$

Thus the function $\mathcal{G}$ is again a weighted residual function for this problem and $\mathcal{G}(x)$ small implies that the residual is small. Without further conditions on the nonlinearity F there is no guarantee that the residual being small implies that a point x is close to a solution of the orginal equation. To obtain such a condition on the function F we use an associated min-max problem.

Consider the function $\mathcal{M} : \mathbb{R}^n \times \mathbb{R}^n \to \mathbb{R}$ defined by

$$\mathcal{M}(x,y) := \langle (A_S - D/2)x, x \rangle + \langle F(x), y - x \rangle - \langle (B+C)x, y \rangle - \frac{1}{2}\langle Dy, y \rangle. \tag{2.22}$$

A point $(\hat{x}, \hat{y}) \in \mathbb{R}^n \times \mathbb{R}^n$ is said to be a *min-max point* of $\mathcal{M}$ provided

$$\mathcal{M}(\hat{x}, \hat{y}) = \inf_{x \in \mathbb{R}^n} \sup_{y \in \mathbb{R}^n} \mathcal{M}(x,y). \tag{2.23}$$

Obviously a saddle point of $\mathcal{M}$ is a min-max point but the converse need not hold; see Auchmuty [4], Section 2 for descriptions of this. First note that this expression for $\mathcal{M}$ implies that

$$\mathcal{G}(x) = \sup_{y \in \mathbb{R}^n} \mathcal{M}(x,y), \tag{2.24}$$

so $(\hat{x}, \hat{y})$ is a min-max point of $\mathcal{M}$ implies that $\hat{x}$ is a minimizer of $\mathcal{G}$.

Here, and in the next section, we need a general result about the existence of such min-max points. The theorem that is used may be stated as follows.

Theorem 2.4. *Let K be a nonempty closed convex set in $\mathbb{R}^n$ and assume that $\mathcal{M} : K \times K \to \mathbb{R}$ satisfies*

(i) $\mathcal{M}(x,x) \leq 0$ *for all x in K.*

(ii) *For each x in K, $\mathcal{M}(x,.)$ is concave on K.*
(iii) *For each y in K, $\mathcal{M}(.,y)$ is l.s.c. on K.*
(iv) *For some y_0 in K, the set $E_0 := \{x \in K : \mathcal{M}(x,y_0) \le 0\}$ is bounded in K.*

Then there is an $x_0 \in K$ satisfying $\sup_{y\in K} \mathcal{M}(x_0,y) \le 0$.

This result is a specialization of Theorem 3 in Auchmuty [4] and is esssentially due to Ky Fan [6].

Theorem 2.5. *Suppose A, B, D, and F as above and the set of points that satisfy*

$$\langle (A_S - D/2)x, x\rangle \ - \ \langle F(x), x\rangle \ \le \ 0 \tag{2.25}$$

is bounded in $\mathbb{R}^n$. Then $\mathcal{M}$ has a min-max point $(\hat{x},\hat{y})$ with $\mathcal{G}(\hat{x}) = 0$ and $\hat{x}$ is a solution of (2.17).

Proof. This is proved using Fan's theorem with $K = \mathbb{R}^n$ and $\mathcal{M}$ defined by (2.22). Condition (i) holds with $\mathcal{M}(x,x) = 0$ as $A_S - D/2 = C + D/2$ and B is skew-symmetric; (ii) holds as D is positive definite; and (iii) holds as $\mathcal{M}$ is continuous. Take $y_0 = 0$ in condition (iv); then the criterion for E_0 to be bounded is that the set of points for which (2.25) holds is bounded. Thus Theorem 2.4 yields that there is an $\hat{x} \in \mathbb{R}^n$ for which $\mathcal{G}(\hat{x}) \le 0$. From Theorem 2.3, $\mathcal{G}(x) \ge 0$ for all x, so there is a minimizer of $\mathcal{G}$ with $\mathcal{G}(\hat{x}) = 0$ and it is a solution of (2.17). □

Suppose the nonlinear mapping F satisfies the following condition.

Condition C1: *There is an $R > 0$ and an a_1 such that $a_1 < a_0$ and*

$$\langle F(x), x\rangle \ \le \ a_1\|x\|^2 \qquad \textit{for all } \|x\| \ge R. \tag{2.26}$$

This condition holds both in the linear case where $F(x) = f$ is constant or when F is nonlinear and bounded on $\mathbb{R}^n$.

Corollary 2.6. *Assume A satisfies (2.1), F is continuous and C1 holds. Then there is at least one solution $\hat{x}$ of (2.17) and $\hat{x}$ minimizes $\mathcal{G}$ on $\mathbb{R}^n$.*

Proof. When (2.26) holds then the set of points for which (2.25) holds is bounded provided D is chosen to be sufficiently small. Thus Theorem 2.5 yields this result. □

2.5 Variational Principles for Semilinear Equations II

There is another class of semilinear, but not necessarily potential, equations for which there are useful variational principles whose minima provide solutions of the equations.

When a function V is continuous and coercive on $\mathbb{R}^n$, then it is bounded below on $\mathbb{R}^n$ and its *conjugate function* $V^* : \mathbb{R}^n \to \mathbb{R}$ defined by

$$V^*(z) := \sup_{x \in \mathbb{R}^n} [\langle z, x \rangle - V(x)]$$

is convex, lower semicontinuous, and finite for all $z \in \mathbb{R}^n$.

Consider the problem of solving equations of the form

$$Ax + \nabla V(x) = 0, \tag{2.27}$$

where A is a real $n \times n$ matrix and V is assumed to be C^1 and coercive on $\mathbb{R}^n$. This may be regarded as a special case of the system treated in the previous section with $F(x) = -\nabla V(x)$ but now A need not be coercive.

Define $\mathcal{J} : \mathbb{R}^n \to \mathbb{R}$ by

$$\mathcal{J}(x) := V(x) + V^*(-Ax) + \langle A_S x, x \rangle. \tag{2.28}$$

When V, A are as above then this function is finite for each $x \in \mathbb{R}^n$ and lower semicontinuous (l.s.c.). Consider the problem of minimizing $\mathcal{J}$ on $\mathbb{R}^n$ and finding

$$\alpha(\mathcal{J}) := \inf_{x \in \mathbb{R}^n} \mathcal{J}(x).$$

This problem has similar properties to that of the problem described in Section 2.4.

Theorem 2.7. *Assume V is C^1 and coercive on $\mathbb{R}^n$ and A is a real $n \times n$ matrix. Then the function $\mathcal{J}$ defined by* (2.28) *is lower semicontinuous and finite at each x and $\alpha(\mathcal{J}) \geq 0$. A point $\hat{x} \in \mathbb{R}^n$ is a solution of* (2.27) *if and only if it minimizes $\mathcal{J}$ on $\mathbb{R}^n$ and $\mathcal{J}(\hat{x}) = 0$.*

Proof. The function V^* is finite-valued, convex, and lower semicontinuous from our assumptions on V and because it is the supremum of a family of such functions. Thus $\mathcal{J}$ is finite and l.s.c. on $\mathbb{R}^n$. The generalized Young's inequality implies that $\mathcal{J}(x) \geq 0$ for all x and that $\mathcal{J}(x) = 0$ if and only if x satisfies

$$-Ax \in \partial V(x).$$

Here $\partial V(x)$ is the subdifferential of V at x. Because V is G-differentiable, $\partial V(x) = \{\nabla V(x)\}$, so the result follows. □

Here again, $\mathcal{J}$ provides a variational principle for the solutions of (2.27) for which the minimizers of $\mathcal{J}$ must satisfy an extra condition to actually be a solution of the equation. This result did not require that V be convex. When V is also convex, then some existence results for this problem can be obtained using min-max methods. Consider the function $\mathcal{W} : \mathbb{R}^n \times \mathbb{R}^n \to \mathbb{R}$ defined by

$$\mathcal{W}(x, y) := V(x) + \langle A_S x, x \rangle - \langle Ax, y \rangle - V(y). \tag{2.29}$$

This function had the property that

$$\mathcal{J}(x) = \sup_{y \in \mathbb{R}^n} \mathcal{W}(x, y). \tag{2.30}$$

The following theorem describes conditions on the function V that guarantee existence of a minimizer of $\mathcal{J}$ that obeys the criteria of Theorem 2.7.

Theorem 2.8. *Suppose V is convex, C^1, and coercive on $\mathbb{R}^n$, and the set of points satisfying*

$$V(x) + \langle A_S x, x \rangle \leq V(0) \tag{2.31}$$

is bounded in $\mathbb{R}^n$. Then $\mathcal{W}$ has a min-max point $(\hat{x}, \hat{y})$ and $\mathcal{J}(\hat{x}) = 0$. If, in addition, A satisfies (2.1), then $\mathcal{W}$ is convex–concave, $(\hat{x}, \hat{y})$ is a saddle point of $\mathcal{W}$, $\hat{x}$ is the unique minimizer of $\mathcal{J}$, and there is a unique solution of (2.27).

Proof. To prove the first part of this, the conditions of Theorem 2.4 are verified for $\mathcal{W}$. Our assumptions on V imply that (i), (ii), and (iii) all hold. Take $y_0 = 0$ in (iv); then the criteria above guarantee that the set of points which satisfy (2.31) is bounded. Thus Theorem 2.4 says that there is an $\hat{x}$ such that $\mathcal{J}(\hat{x}) \leq 0$. From Theorem 2.7, $\mathcal{J}(x) \geq 0$ for all x, so $\mathcal{J}(\hat{x}) = 0$. For a given $\hat{x}$, the supremum of $\mathcal{W}(\hat{x}, .)$ is attained as V is convex and coercive, hence there is a min-max point. When A satisfies (2.1), $\mathcal{W}(., y)$ is strictly convex on $\mathbb{R}^n$ for each y and the remaining results follow from the usual saddle point theorem. □

The condition (2.31) says that provided

$$\liminf_{\|x\| \to \infty} [V(x) + \langle A_S x, x \rangle] > V(0) \tag{2.32}$$

then the equation (2.27) has at least one solution, without requiring coercivity of A. In this case, coercivity of A yields uniqueness of the solution and dual variational principles for the problem may be described.

The results described here can be generalized to variational principles and solvability results for equations between a real reflexive Banach space X and its dual space X^* using similar constructions and proofs. The relevant saddle point theorems in [9], or the min-max theorems of [4] hold in this generality. When natural conditions are imposed on A, F so that these theorems hold, results analogous to those described here for the finite-dimensional case may be stated.

References

[1] M. Ainsworth and J.T. Oden, *A Posteriori Error Estimation in Finite Element Analysis*, John Wiley & Sons, New York, 2000.

[2] G. Auchmuty, Duality for Non-Convex Variational Principles, *J. Diff. Eqns.* **50** (1983), 80–145.
[3] G. Auchmuty, Saddle Point Methods, and Algorithms, for Non-Symmetric Linear Equations, *Numer. Funct. Anal. Optim.* **16** (1995), 1127–1142.
[4] G. Auchmuty, Min-Max Problems for Non-Potential Operator Equations, *Optimization Methods in Partial Differential Equations* (S. Cox and I. Lasiecka, eds.), Contemporary Mathematics, vol. 209, American Mathematical Society, Providence, 1997, pp. 19–28.
[5] I. Ekeland and R. Temam, *Convex Analysis and Variational Problems*, North-Holland, Amsterdam, 1976.
[6] K. Fan, A Generalization of Tychonoff's Fixed Point Theorem, *Math. Ann.* **142** (1961), 305–310.
[7] W. Han, *A Posteriori Error Analysis via Duality Theory,* Springer Science, New York, 2005.
[8] G. Strang, *An Introduction to Applied Mathematics*, Cambridge-Wellesley Press, 1986.
[9] E. Zeidler, *Nonlinear Functional Analysis and Its Applications III, Variational Methods and Optimization*, Springer Verlag, New York, 1985.

Chapter 3
Adaptive Finite Element Solution of Variational Inequalities with Application in Contact Problems

Viorel Bostan and Weimin Han

Summary. In this chapter, we perform a posteriori error analysis for the adaptive finite element solution of several variational inequalities, including elliptic variational inequalities of the second kind and corresponding quasistatic variational inequalities. A general framework for a posteriori error estimation is established by using duality theory in convex analysis. We then derive a posteriori error estimates of residual type and of recovery type, through particular choices of the dual variable present in the general framework. The error estimates are guaranteed to be reliable. Efficiency of the error estimators is theoretically investigated and numerically validated. Detailed derivation and analysis of the error estimates are given for a model elliptic variational inequality. Extensions of the results can be made straightforward in solving other elliptic variational inequalities of the second kind, and we present such an extension for a problem arising in frictional contact. Moreover, we use a quasistatic contact problem as an example to illustrate how to extend the a posteriori error analysis in solving time-dependent variational inequalities. Numerous numerical examples are included to illustrate the effectiveness of the a posteriori error estimates in adaptive solutions of the variational inequalities.

Key words: A posteriori error estimation, adaptive finite element solution, elliptic variational inequality, quasistatic variational inequality, frictional contact, duality, reliability, efficiency

3.1 Introduction

In this chapter, we present some theoretical and numerical results on a posteriori error estimation and adaptive finite element solution of elliptic

Department of Mathematics, University of Iowa, Iowa City, IA 52242, U.S.A.
e-mail: vbostan@math.uiowa.edu, whan@math.uiowa.edu

D.Y. Gao, H.D. Sherali, (eds.), *Advances in Applied Mathematics and Global Optimization*, Advances in Mechanics and Mathematics 17, DOI 10.1007/978-0-387-75714-8_3, variational inequalities of the second kind, as well as corresponding quasistatic variational inequalities, especially those arising in frictional contact problems.

The general framework for a posteriori error estimation of the chapter works for any Galerkin solutions of the variational inequalities. However, we specifically choose the finite element method for approximation, as the method today is the dominant numerical method for solving most problems in structural and fluid mechanics. It is widely applied to both linear and nonlinear problems. General mathematical theory of finite element methods can be found in [4, 26, 27, 58, 64], among others. The textbook [51] offers an easily accessible mathematical introduction of finite element methods, whereas the two recent textbooks [17, 18] provide deeper mathematical theory together with more recent and current research development such as the multigrid methods. Traditionally, convergence of finite element solutions is achieved through mesh refinement with the use of a piecewise low degree polynomial. Because h is usually used to denote the mesh size, the traditional finite element method is also termed the h-version finite element method.

On the other hand, convergence of the method can also be achieved by using piecewise increasingly higher degree polynomials over relatively coarse finite element meshes, leading to the p-version finite element method. Detailed discussion of the p-version finite element method can be found in [66]. The p-version method is more efficient in areas where the solution is smooth, so it is natural to combine the ideas of the p-version and the h-version to make the finite element method very efficient on many problems. A well-known result regarding the h-p-version finite element method is the exponential convergence rate for solving elliptic boundary value problems with corner singularities, under proper combinations of local polynomial degrees and element sizes.

Comprehensive mathematical theory of the p-version and h-p-version finite element methods with applications in solid and fluid mechanics can be found in [63]. Mixed and hybrid finite element methods are often used in solving boundary value problems with constraints and higher-order differential equations. Mathematical theory of these methods can be found in [19, 62]. Several monographs are available on the numerical solution of Navier–Stokes equations by the finite element method (see, e.g., [33]). Theory of the finite element method for solving parabolic problems can be found in [67] and more recently in [68]. Finally, we list a few representative engineering books on the finite element method, [11, 50, 76, 77]. The reader is referred to two historical notes [57, 75] on the development of the finite element method.

For practical use of a numerical method, one important issue is the assessment of the reliability and accuracy of the numerical solution. The reliability of the numerical solution hinges on our ability to estimate errors after the solution is computed; such an error analysis is called a posteriori error analysis. A posteriori error estimates provide quantitative information on the accuracy of the solution and are the basis for the development of automatic, adaptive solution procedures.

The research on a posteriori error estimation and adaptive mesh refinement for the finite element method began in the late 1970s. The pioneering work on the topic was done in [5, 6]. Since then, a posteriori error analysis and adaptive computation in the finite element method have attracted many researchers, and a variety of different a posteriori error estimates have been proposed and analyzed. In a typical a posteriori error analysis, after a finite element solution is computed, the solution is used to compute element error indicators and an error estimator. The element error indicator represents the contribution of the element to the error in the computation of some quantity by the finite element solution, and is used to indicate if the element needs to be refined in the next adaptive step. The error estimator provides an estimate of the error in the computation of the quantity of the finite element solution, and thus can be used as a stopping criterion for the adaptive procedure. Often, the error estimator is computed as an aggregation of the element error indicators, and one usually only speaks of error estimators. Most error estimators can be classified into residual type, where various residual quantities (residual of the equation, residual from derivative discontinuity, residual of material constitutive laws, etc.) are used, and recovery type, where a recovery operator is applied to the (discontinuous) gradient of the finite element solution and the difference of the two is used to assess the error. Error estimators have also been derived based on the use of hierarchic bases or equilibrated residual. Two desirable properties of an a posteriori error estimator are reliability and efficiency. Reliability requires the actual error to be bounded by a constant multiple of the error estimator, up to perhaps a higher-order term, so that the error estimator provides a reliable error bound. Efficiency requires the error estimator to be bounded by a constant multiple of the actual error, again perhaps up to a higher-order term, so that the actual error is not overestimated by the error estimator. The study and applications of a posteriori error analysis is a current active research area, and the related publications grow fast. Some comprehensive summary accounts can be found, in chronicle order, in [70, 1, 7].

Initially, a posteriori error estimates were mainly developed for estimating the finite element error in the energy norm. In the recent years, error estimators have also been developed for goal-oriented adaptivity. The goal-oriented error estimators are derived to specifically estimate errors in quantities of interest, other than the energy norm errors. Chapter 8 of [1] is devoted to such error estimators. The latest development in this direction is depicted in [10, 32].

Most of the work so far on a posteriori error analysis has been devoted to ordinary boundary value problems of partial differential equations. In applications, an important family of nonlinear boundary value and initial boundary value problems is that associated with variational inequalities, that is, problems involving either differential inequalities or inequality boundary conditions. Mechanics is a rich source of variational inequalities (see, e.g., [59]), and some examples of problems that give rise to variational inequalities are obstacle problems, contact problems, plasticity and viscoplasticity problems, Stefan problems, unilateral problems of plates and shells, and non-Newtonian flows involving Bingham fluids. An early comprehensive reference on the topic is [29], where many nonlinear boundary value problems in mechanics and physics are formulated and studied in the framework of variational inequalities. A concise introduction to the mathematical theory of some variational inequalities can be found in [54]. Numerical approximations of general variational inequalities are studied in detail in [34, 35]. Numerical methods for some variational inequalities arising in mechanics are the subject of [47, 48]. Mathematical analysis and numerical approximations of variational inequalities arising in contact mechanics are presented in [53] (for elastic materials) and [46] (for viscoelastic and viscoplastic materials). In [43, 44], elastoplasticity problems are formulated and analyzed in the form of variational inequalities.

Although several standard techniques have been developed to derive and analyze a posteriori error estimates for finite element solutions to problems in the form of variational equations, they do not work directly for a posteriori error analysis of numerical solutions to variational inequalities due to the inequality feature of the problems. Nevertheless, numerous papers can be found on a posteriori error estimation of finite element solutions of obstacle problems, for example, [2, 25, 49, 55, 56, 69] (these papers consider numerical solutions on convex subsets of finite element spaces), as well as [31, 52] (these papers use a penalty approach for discrete solutions). Obstacle problems are so-called variational inequalities of the first kind; that is, they are inequalities involving smooth functionals and are posed over convex subsets. We also note that a posteriori error estimation is discussed in [12, 13, 65], although the derivations of the estimates in these papers are arguable.

In the context of elastoplasticity with hardening, computable a posteriori error estimates are derived in [3, 20, 22] for the primal problem, which is a variational inequality of the second kind; that is, the inequality arises as a result of the presence of a nondifferentiable functional. These works deal extensively also with a priori estimates, and in the latter work a number of numerical examples are presented. Residual type error estimators were studied for an elliptic variational inequality of the second kind in [15, 16].

In this chapter, we derive and study some a posteriori error estimates for finite element solutions of elliptic variational inequalities of the second kind and corresponding quasistatic variational inequalities. The basic mathematical tool we use is the duality theory in convex analysis (cf. [30, 73]). The duality theory has been applied to derive efficient a posteriori error estimates for mathematical idealizations of physical and engineering problems (see, e.g., [37, 38]), as well as for some numerical procedures for solving nonlinear problems, such as the regularization techniques in [36, 42, 45], and the Kačanov iteration method in [40, 41]. A summary account of these can be found in

[39]. In [61, 60], the technique of the duality theory was used to derive a posteriori error estimates of the finite element method in solving boundary value problems of some nonlinear equations. In these papers, the error bounds are shown to converge to zero in the limit; however, no efficiency analysis of the estimates is given.

For convenience, we recall here a representative result on the duality theory (see [30]). Let V, Q be two normed spaces, and denote by V^*, Q^* their dual spaces. Assume there exists a linear continuous operator $\Lambda \in \mathcal{L}(V,Q)$, with transpose $\Lambda^* \in \mathcal{L}(Q^*, V^*)$. Let F be a functional mapping $V \times Q$ into the extended real line $\overline{\mathbb{R}} \equiv \mathbb{R} \cup \{\pm\infty\}$. Consider the minimization problem:

$$\inf_{v \in V} F(v, \Lambda v). \tag{3.1}$$

Define its dual problem by

$$\sup_{q^* \in Q^*} \left[-F^*(\Lambda^* q^*, -q^*) \right], \tag{3.2}$$

where F^* is the conjugate functional of F:

$$F^*(v^*, q^*) = \sup_{\substack{v \in V \\ q \in Q}} \left[\langle v, v^* \rangle + \langle q, q^* \rangle - F(v,q) \right]. \tag{3.3}$$

Then we have the following theorem.

Theorem 3.1. *Assume*

(1) V *is a reflexive Banach space and* Q *a normed space,*
(2) $F : V \times Q \to \overline{\mathbb{R}}$ *is a proper, lower semicontinuous, convex function,*
(3) $\Lambda : V \to Q$ *is a linear bounded operator with its adjoint* $\Lambda^* : Q^* \to V^*$,
(4) $\exists\, u_0 \in V$ *with* $F(u_0, \Lambda u_0) < \infty$ *and* $q \mapsto F(u_0, q)$ *continuous at* Λu_0,
(5) $F(v, \Lambda v) \to +\infty$ *as* $\|v\| \to \infty \quad \forall\, v \in V$.

Then the problem (3.1) *has a solution* $u \in V$, *its dual* (3.2) *has a solution* $p^* \in Q^*$, *and*

$$F(u, \Lambda u) = -F^*(\Lambda^* p^*, -p^*).$$

Furthermore, if F *is strictly convex, then a solution* u *of problem* (3.1) *is unique.*

The rest of the chapter is organized as follows. In Section 3.2 we introduce a model elliptic variational inequality of the second kind and its finite element approximation. We provide detailed derivation and analysis of a posteriori error estimates of the finite element solutions for the model problem. In Section 3.3 we formulate a dual problem for the model, and use the dual problem to establish a general a posteriori error estimate for any approximation of the solution of the model elliptic variational inequality. The general a posteriori error estimate features the presence of a dual variable. Different a posteriori error estimates can be obtained with different choices of the dual variable. In Section 3.4, we make a particular choice of the dual variable that leads to a residual-based error estimate of the finite element solution of the model elliptic variational inequality, and explore the efficiency of the error estimate. In Section 3.5, we make another choice of the dual variable and obtain a recovery-based error estimate of the finite element solution of the model elliptic variational inequality. We also study the efficiency of the error estimator. In Section 3.6, we present some numerical results to illustrate the effectiveness of the estimates in adaptive solution of the elliptic variational inequalities. In Section 3.7, we extend the discussion to solving a steady-state frictional contact problem.

Then we turn to an extension of the discussion in adaptively solving time-dependent variational inequalities, taking a model quasistatic variational inequality as an example. We begin with an abstract quasistatic variational inequality, introduced in Section 3.8, which contains as special cases several application problems in contact mechanics and hardening plasticity. A backward Euler discretization is used to approximate the time derivative in the quasistatic variational inequality, leading to a sequence of semidiscretized elliptic variational inequalities. An error estimate of the semidiscrete solution is derived. We then focus on a model quasistatic contact problem and derive a posteriori error estimates for finite element solutions of its semidiscretized approximations in Section 3.9, providing both residual type and recovery type error estimates. Finally, the numerical result showing the effectiveness of the error estimates in the adaptive solution of the model quasistatic contact problem is reported in Section 3.10.

We now list some notations used repeatedly in the chapter. Let Ω be a bounded domain in $\mathbb{R}^d$, $d \geq 1$, with Lipschitz boundary $\Gamma = \partial\Omega$. For any open subset ω of Ω with Lipschitz boundary $\partial\omega$, we denote by $H^m(\omega)$, $L^2(\omega)$, and $L^2(\partial\omega)$ the usual Sobolev and Lebesgue spaces with the standard norms $\|\cdot\|_{m;\omega} := \|\cdot\|_{H^m(\omega)}$, $\|\cdot\|_{0;\omega} := \|\cdot\|_{L^2(\omega)}$, and $\|\cdot\|_{0;\partial\omega} := \|\cdot\|_{L^2(\partial\omega)}$. Also, we make use of the standard seminorm $|\cdot|_{m,\omega}$ on $H^m(\omega)$. Throughout this chapter we use the same notation v to denote both $v \in H^1(\Omega)$ and its trace $\gamma v \in L^2(\Gamma)$ on the boundary. We reserve the symbol γ for the element sides.

3.2 Model Elliptic Variational Inequality and Its Finite Element Approximation

In this section, we introduce a model elliptic variational inequality of the second kind. We comment that the ideas and techniques presented for a posteriori error analysis in solving the model problem can be extended to other elliptic variational inequalities of the second kind; in particular, in Section

3.7 we provide a posteriori error analysis for the finite element solution of a steady-state frictional contact problem.

Let Ω be a domain in $\mathbb{R}^d$, $d \geq 1$, with a Lipschitz boundary Γ. Let $\Gamma_1 \subset \Gamma$ be a relatively closed subset of Γ, and denote $\Gamma_2 = \Gamma \backslash \Gamma_1$ the remaining part of the boundary. We allow the extreme situation with $\Gamma_1 = \varnothing$ (i.e., $\Gamma_2 = \Gamma$) or $\Gamma_1 = \Gamma$ (i.e., $\Gamma_2 = \varnothing$). Because the boundary Γ is Lipschitz continuous, the unit outward normal vector $\boldsymbol{\nu}$ exists a.e. on Γ. We use $\partial/\partial\nu$ to denote the outward normal differentiation operator, that exists a.e. on Γ. Assume $f \in L^2(\Omega)$ and $g > 0$ are given. Over the space

$$V = H^1_{\Gamma_1}(\Omega) = \{v \in H^1(\Omega) : v = 0 \text{ a.e. on } \Gamma_1\}, \tag{3.4}$$

we define a bilinear form and two functionals:

$$\begin{aligned} a(u,v) &= \int_\Omega (\nabla u \cdot \nabla v + u\,v)\,dx, \\ \ell(v) &= \int_\Omega f\,v\,dx, \\ j(v) &= \int_{\Gamma_2} g\,|v|\,ds. \end{aligned}$$

In the space V, we use the $H^1(\Omega)$-norm. The model problem is the following elliptic variational inequality of the second kind

$$u \in V, \qquad a(u, v-u) + j(v) - j(u) \geq \ell(v-u) \qquad \forall\, v \in V. \tag{3.5}$$

This model is a so-called simplified friction problem following [34], as it can be viewed as a simplified version of a frictional contact problem in linearized elasticity (cf. Section 3.7).

The bilinear form $a(\cdot,\cdot)$ is continuous and V-elliptic, the linear functional $\ell(\cdot)$ is continuous, and the functional $j(\cdot)$ is proper, convex, and continuous, and therefore, by a standard existence and uniqueness result for elliptic variational inequalities of the second kind (see [34, 35]), the variational inequality (3.5) has a unique solution. Moreover, due to the symmetry of the bilinear form $a(\cdot,\cdot)$, the variational inequality (3.5) is equivalent to the minimization problem: find $u \in V$ such that

$$J(u) = \inf_{v \in V} J(v), \tag{3.6}$$

where J is the energy functional:

$$J(v) = \frac{1}{2} a(v,v) + j(v) - \ell(v). \tag{3.7}$$

The minimization problem (3.6) also has a unique solution.

In the analysis of a posteriori error estimators later, we need the following characterization of the solution u of (3.5): there exists a unique $\lambda \in L^\infty(\Gamma_2)$

such that

$$a(u,v) + \int_{\Gamma_2} g\,\lambda\,v\,ds = \ell(v) \qquad \forall\, v \in V, \tag{3.8}$$

$$|\lambda| \le 1, \quad \lambda\,u = |u| \qquad \text{a.e. on } \Gamma_2. \tag{3.9}$$

The function λ can be viewed as a Lagrange multiplier. A proof of this characterization in the case $\Gamma_2 = \Gamma$ can be found in [34]. The argument there can be extended straightforwardly to the more general situation considered here; see also the proof of Theorem 3.3.

It follows from (3.8) that the solution u of (3.5) is the weak solution of the boundary value problem

$$-\Delta u + u = f \quad \text{in } \Omega,$$

$$u = 0 \quad \text{on } \Gamma_1,$$

$$\frac{\partial u}{\partial \nu} + g\lambda = 0 \quad \text{on } \Gamma_2.$$

We now turn to finite element approximations of the model problem. For simplicity, we suppose that Ω has a polyhedral boundary Γ. In order to define the finite element method for (3.5) we introduce a family of finite element spaces $V_h \subset V$, which consist of continuous piecewise polynomials of certain degree, corresponding to partitions $\mathcal{P}_h$ of $\overline{\Omega}$ into triangular or tetrahedral elements (other kinds of elements, such as quadrilateral elements, or hexahedral or pentahedral elements, can be considered as well). The partitions $\mathcal{P}_h$ are compatible with the decomposition of Γ into Γ_1 and Γ_2. In other words, if an element side lies on the boundary, then it belongs to one of the sets $\overline{\Gamma_1}$ or $\overline{\Gamma_2}$. For every element $K \in \mathcal{P}_h$, let h_K be the diameter of K and ρ_K be the diameter of the largest ball inscribed in K. For a side γ of the element K, we denote by h_γ the diameter of γ. We assume that the family of partitions $\mathcal{P}_h$, $h > 0$, satisfies the shape regularity assumption; that is, the ratio h_K/ρ_K is uniformly bounded over the whole family by a constant C. Note that the shape regularity assumption does not require that the elements be of comparable size and thus locally refined meshes are allowed. We use $\mathcal{E}_h$ for the set of the element sides; $\mathcal{E}_{h,\Gamma}$, $\mathcal{E}_{h,\Gamma_1}$, and $\mathcal{E}_{h,\Gamma_2}$ for the subsets of the element sides lying on Γ, Γ_1, and $\overline{\Gamma_2}$, respectively; and $\mathcal{E}_{h,0} = \mathcal{E}_h \backslash \mathcal{E}_{h,\Gamma}$ for the subset of the element sides that do not lie on Γ. Let $\mathcal{N}_h$ be the set of all nodes in $\mathcal{P}_h$ and $\mathcal{N}_{h,0} \subset \mathcal{N}_h$ the set of free nodes; that is, those nodes that do not lie on Γ_1. For a given element $K \in \mathcal{P}_h$, $\mathcal{N}(K)$ and $\mathcal{E}(K)$ denote the sets of the nodes of K and sides of K, respectively.

The patch $\widetilde{K}$ associated with any element K from a partition $\mathcal{P}_h$ consists of all elements sharing at least one vertex with K; that is, $\widetilde{K} = \bigcup\{K' \in \mathcal{P}_h : \overline{K'} \cap \overline{K} \neq \varnothing\}$. Similarly, for any side $\gamma \in \mathcal{E}_h$, the patch $\widetilde{\gamma}$ consists of the elements sharing γ as a common side. Note that in the case where the side γ lies on the boundary Γ, the patch $\widetilde{\gamma}$ consists of only one element. For a given element $K \in \mathcal{P}_h$, $\boldsymbol{\nu}_K$ denotes the unit outward normal vector to the sides of K. When a side γ lies on the boundary Γ, $\boldsymbol{\nu}_\gamma$ denotes the unit outward normal vector to Γ. For a side γ in the interior, $\boldsymbol{\nu}_\gamma$ is taken to be one of the two unit normal vectors. In what follows, for any piecewise continuous function φ and any interior side $\gamma \in \mathcal{E}_{h,0}$, $[\varphi]_\gamma$ denotes the jump of φ across γ in the direction $\boldsymbol{\nu}_\gamma$; that is,

$$[\varphi]_\gamma(\boldsymbol{x}) = \lim_{t\to 0+} (\varphi(\boldsymbol{x} + t\,\boldsymbol{\nu}_\gamma) - \varphi(\boldsymbol{x} - t\,\boldsymbol{\nu}_\gamma)) \qquad \forall\, \boldsymbol{x} \in \gamma.$$

In the derivation of a posteriori error estimates we use the so-called weighted Clément-type interpolation operator. There are several variants (see, e.g., [8, 21, 23, 24, 71]) of the interpolation operator introduced by Clément [28]. The main difference among these interpolants lies in the way the interpolation is performed near the boundary. In this chapter we follow the approach used in [23].

Corresponding to the partition $\mathcal{P}_h$, we denote $\mathcal{N}_v \subset \mathcal{N}_h$ to be the set of the element vertices, $\mathcal{N}_{v,\Gamma_1} \subset \mathcal{N}_v$ the subset of the element vertices lying on Γ_1, and $\mathcal{N}_{v,0} = \mathcal{N}_v \cap \mathcal{N}_{h,0}$ the subset of the interior vertices. Given $\boldsymbol{a} \in \mathcal{N}_v$, let $\varphi_{\boldsymbol{a}}$ be the linear element nodal basis function associated with $\boldsymbol{a}$. For each fixed vertex $\boldsymbol{a} \in \mathcal{N}_{v,\Gamma_1}$, choose $\boldsymbol{\xi}(\boldsymbol{a}) \in \mathcal{N}_{v,0}$ to be an interior vertex of an element containing $\boldsymbol{a}$. Let $\boldsymbol{\xi}(\boldsymbol{a}) = \boldsymbol{a}$ if $\boldsymbol{a} \in \mathcal{N}_{v,0}$. For each node $\boldsymbol{a} \in \mathcal{N}_{v,0}$ define the class $I(\boldsymbol{a}) = \{\tilde{\boldsymbol{a}} \in \mathcal{N}_v : \boldsymbol{\xi}(\tilde{\boldsymbol{a}}) = \boldsymbol{a}\}$. In this way, the set of all the vertices $\mathcal{N}_v$ is partitioned into $\mathrm{card}(\mathcal{N}_{v,0})$ classes of equivalence. For each $\boldsymbol{a} \in \mathcal{N}_{v,0}$ set

$$\psi_{\boldsymbol{a}} = \sum_{\tilde{\boldsymbol{a}} \in I(\boldsymbol{a})} \varphi_{\tilde{\boldsymbol{a}}}.$$

Notice that $\{\psi_{\boldsymbol{a}} : \boldsymbol{a} \in \mathcal{N}_{v,0}\}$ is a partition of unity. Let $\widetilde{K}_{\boldsymbol{a}} = \mathrm{supp}(\psi_{\boldsymbol{a}})$ and $h_{\boldsymbol{a}} = \mathrm{diam}(\widetilde{K}_{\boldsymbol{a}})$. The set $\widetilde{K}_{\boldsymbol{a}}$ is connected and $\psi_{\boldsymbol{a}} \neq \varphi_{\boldsymbol{a}}$ implies that $\Gamma_1 \cap \widetilde{K}_{\boldsymbol{a}}$ has a positive surface measure.

For a given $v \in L^1(\Omega)$, let

$$v_{\boldsymbol{a}} = \frac{\int_{\widetilde{K}_{\boldsymbol{a}}} v\, \psi_{\boldsymbol{a}}\, dx}{\int_{\widetilde{K}_{\boldsymbol{a}}} \varphi_{\boldsymbol{a}}\, dx}, \qquad \boldsymbol{a} \in \mathcal{N}_{v,0}. \tag{3.10}$$

Then define the interpolation operator $\Pi_h : V \to V_h$ as follows:

$$\Pi_h v = \sum_{\boldsymbol{a} \in \mathcal{N}_{v,0}} v_{\boldsymbol{a}} \varphi_{\boldsymbol{a}}. \tag{3.11}$$

The next result summarizes some basic estimates for Π_h. Its proof can be found in [21].

Theorem 3.2. *There exists an h-independent constant $C > 0$ such that for all $v \in V$ and $f \in L^2(\Omega)$,*

$$|v - \Pi_h v|^2_{1;\Omega} \le C |v|^2_{1;\Omega}, \tag{3.12}$$

$$\int_\Omega f(v - \Pi_h v)\,dx \le C|v|_{1;\Omega} \left(\sum_{a \in \mathcal{N}_{v,0}} h_a^2 \min_{f_a \in \mathbb{R}} \|f - f_a\|^2_{0;\widetilde{K}_a} \right)^{1/2}, \tag{3.13}$$

$$\sum_{K \in \mathcal{P}_h} \|h_K^{-1}(v - \Pi_h v)\|^2_{0;K} \le C|v|^2_{1;\Omega}, \tag{3.14}$$

$$\sum_{\gamma \in \mathcal{E}_h} \|h_\gamma^{-1/2}(v - \Pi_h v)\|^2_{0;\gamma} \le C|v|^2_{1;\Omega}. \tag{3.15}$$

The finite element method for the variational inequality (3.5) is

$$u_h \in V_h, \qquad a(u_h, v_h - u_h) + j(v_h) - j(u_h) \ge \ell(v_h - u_h) \qquad \forall\, v_h \in V_h. \tag{3.16}$$

The discrete problem has a unique solution $u_h \in V_h$ by the standard existence and uniqueness result on elliptic variational inequalities. We need the following characterization of the finite element solution, similar to that of the solution of the continuous problem.

Theorem 3.3. *The unique solution $u_h \in V_h$ of the discrete problem* (3.16) *is characterized by the existence of $\lambda_h \in L^\infty(\Gamma_2)$ such that*

$$a(u_h, v_h) + \int_{\Gamma_2} g\,\lambda_h v_h ds = \ell(v_h) \qquad \forall\, v_h \in V_h, \tag{3.17}$$

$$|\lambda_h| \le 1, \quad \lambda_h u_h = |u_h| \qquad \text{a.e. on } \Gamma_2. \tag{3.18}$$

Proof. Assuming (3.16), let us prove (3.17) and (3.18). Taking first $v_h = 0$ and then $v_h = 2u_h$ in (3.16), we obtain

$$a(u_h, u_h) + \int_{\Gamma_2} g\,|u_h|\,ds = \ell(u_h), \tag{3.19}$$

Together with (3.19) the relation (3.16) leads to

$$|\ell(v_h) - a(u_h, v_h)| \le \int_{\Gamma_2} g\,|v_h|\,ds \qquad \forall\, v_h \in V_h. \tag{3.20}$$

Write $V_h = V_h^0 \oplus V_h^\perp$, where $V_h^0 = V_h \cap H_0^1(\Omega)$ and $V_h^\perp$ is the orthogonal complement of V_h in $H^1_{\Gamma_1}(\Omega)$. It follows from (3.20) that $\ell(v_h) - a(u_h, v_h) = 0$ $\forall\, v_h \in V_h^0$. Notice that the trace operator from $V_h^\perp$ onto $V_h^\perp|_{\Gamma_2} \subset L^1(\Gamma_2)$ is an isomorphism. Therefore, the mapping $L(v_h) \equiv \ell(\widetilde{v}_h) - a(u_h, \widetilde{v}_h)$ can be viewed as a linear functional on $V_h^\perp|_{\Gamma_2}$, where $\widetilde{v}_h$ is any element from the space V_h whose trace on Γ_2 is v_h. It follows from (3.20) that

$$|L(v_h)| \le \int_{\Gamma_2} g\,|v_h|\,ds \qquad \forall\, v_h \in V_h^\perp|_{\Gamma_2}. \tag{3.21}$$

Thus, by the Hahn–Banach theorem the functional $L(v_h)$ can be extended to $L(v)$ on $L^1(\Gamma_2)$ and so there exists $\lambda_h \in L^\infty(\Gamma_2)$ such that

$$L(v) = \int_{\Gamma_2} \lambda_h g\, v\, ds \qquad \forall v \in L^1(\Gamma_2)$$

and $|\lambda_h| \leq 1$ a.e. on Γ_2, from which (3.17) follows. Taking now $v_h = u_h$ in relation (3.17), we have

$$a(u_h, u_h) + \int_{\Gamma_2} g\, \lambda_h u_h ds = \ell(u_h),$$

and using (3.19) we get

$$\int_{\Gamma_2} g\,(|u_h| - \lambda_h u_h)\, ds = 0.$$

Because $|\lambda_h| \leq 1$ a.e. on Γ_2, we must have $|u_h| = \lambda_h u_h$ a.e. on Γ_2. This completes the proof of (3.17) and (3.18).

Conversely, assume (3.17) and (3.18) hold. It follows from relation (3.17) that

$$a(u_h, v_h - u_h) + \int_{\Gamma_2} g\, \lambda_h (v_h - u_h)\, ds = \ell(v_h - u_h) \qquad \forall v_h \in V_h,$$

which can be rewritten as

$$a(u_h, v_h - u_h) + \int_{\Gamma_2} g\, \lambda_h v_h ds - \int_{\Gamma_2} g\, \lambda_h u_h ds = \ell(v_h - u_h) \qquad \forall v_h \in V_h.$$

Then, relation (3.18) implies that

$$a(u_h, v_h - u_h) + \int_{\Gamma_2} g\, \lambda_h v_h\, ds - \int_{\Gamma_2} g\, |u_h|\, ds = \ell(v_h - u_h) \qquad \forall v_h \in V_h.$$

Because $\lambda_h v_h \leq |v_h|$ a.e. on Γ_2, it follows immediately that u_h is the solution of the discrete problem (3.16). □

Convergence and a priori error estimates for the finite element method (3.16) can be found in the literature (e.g., [34, 35]). Here, we focus on the derivation and analysis of a posteriori error estimators that can be used in the adaptive finite element solution of variational inequalities.

In investigation of the efficiency of the a posteriori error estimators, we follow Verfürth [70], with special attention paid to the inequality feature of the problem. The argument makes use of the canonical bubble functions constructed for each element $K \in \mathcal{P}_h$ and each side $\gamma \in \mathcal{E}_h$.

Denote by P_K a polynomial space associated with the element K. The following two theorems provide some basic properties of the bubble functions used to derive lower bounds. For more details on bubble functions and proofs see [1].

Theorem 3.4. *Let $K \in \mathcal{P}_h$ and ψ_K be its corresponding bubble function. Then there exists a constant C, independent of h_K, such that for any $v \in P_K$,*

$$C^{-1}\|v\|_{0;K}^2 \le \int_K \psi_K v^2 dx \le C\|v\|_{0;K}^2,$$

$$C^{-1}\|v\|_{0;K} \le \|\psi_K v\|_{0;K} + h_K|\psi_K v|_{1;K} \le C\|v\|_{0;K}.$$

Theorem 3.5. *Let $K \in \mathcal{P}_h$ and $\gamma \in \mathcal{E}(K)$ be one of its sides. Let ψ_γ be the side bubble function corresponding to γ. Then there exists a constant C, independent of h_K, such that for any $v \in P_K$,*

$$C^{-1}\|v\|_{0;\gamma}^2 \le \int_\gamma \psi_\gamma v^2 ds \le C\|v\|_{0;\gamma}^2,$$

$$h_K^{-1/2}\|\psi_\gamma v\|_{0;K} + h_K^{1/2}|\psi_\gamma v|_{1;K} \le C\|v\|_{0;\gamma}.$$

3.3 Dual Formulation and A Posteriori Error Estimation

We now present a dual formulation for the model elliptic variational inequality within the framework of Theorem 3.1. The dual formulation is used in the derivation of a posteriori error estimators for approximate solutions.

Let us choose the spaces, functionals, and operators needed in applying Theorem 3.1. The space V is defined in (3.4). Let $Q = (L^2(\Omega))^d \times L^2(\Omega) \times L^2(\Gamma_2)$. Any element $\boldsymbol{q} \in Q$ is written as $\boldsymbol{q} = (\boldsymbol{q}_1, q_2, q_3)$, where $\boldsymbol{q}_1 \in (L^2(\Omega))^d$, $q_2 \in L^2(\Omega)$, and $q_3 \in L^2(\Gamma_2)$. Let V^* and $Q^* = (L^2(\Omega))^d \times L^2(\Omega) \times L^2(\Gamma_2)$ be the duals of V and Q placed in duality by the pairings $\langle\cdot,\cdot\rangle_V$ and $\langle\cdot,\cdot\rangle_Q$, respectively. Introduce a linear bounded operator $\Lambda : V \to Q$ by the relation

$$\Lambda v = (\nabla v, v, v|_{\Gamma_2}) \qquad \forall\, v \in V.$$

Define

$$F(v, \boldsymbol{q}) = \int_\Omega \left[\frac{1}{2}(|\boldsymbol{q}_1|^2 + |q_2|^2) - fv\right] dx + \int_{\Gamma_2} g\,|q_3|\, ds, \qquad v \in V,\ \boldsymbol{q} \in Q.$$

Then

$$J(v) = F(v, \Lambda v) \qquad \forall\, v \in V,$$

and we rewrite the minimization problem (3.6) as

$$u \in V, \qquad F(u, \Lambda u) = \inf_{v \in V} F(v, \Lambda v). \tag{3.22}$$

To apply Theorem 3.1, we need the conjugate function

$$F^*(\Lambda^* \boldsymbol{q}^*, -\boldsymbol{q}^*) \equiv \sup_{\substack{v \in V \\ \boldsymbol{q} \in Q}} \left\{ \langle \Lambda^* \boldsymbol{q}^*, v \rangle_V - \langle \boldsymbol{q}^*, \boldsymbol{q} \rangle_Q - F(v, \boldsymbol{q}) \right\},$$

where $\Lambda^* : Q^* \to V^*$ is the adjoint of Λ. We have

$$\begin{aligned} F^*(\Lambda^* \boldsymbol{q}^*, -\boldsymbol{q}^*) = \sup_{\substack{v \in V \\ \boldsymbol{q} \in Q}} \bigg\{ & \int_\Omega \left[\boldsymbol{q}_1^* \cdot \nabla v + (q_2^* + f)\, v \right] dx + \int_{\Gamma_2} q_3^* v \, ds \\ & - \int_\Omega \left(\frac{1}{2} |\boldsymbol{q}_1|^2 + \boldsymbol{q}_1^* \cdot \boldsymbol{q}_1 \right) dx - \int_\Omega \left(\frac{1}{2} |q_2|^2 + q_2^* q_2 \right) dx \\ & - \int_{\Gamma_2} \left(q_3^* q_3 + g\, |q_3| \right) ds \bigg\}. \end{aligned} \tag{3.23}$$

It can easily be verified that

$$\begin{aligned} \sup_{\boldsymbol{q}_1 \in (L^2(\Omega))^d} \left\{ - \int_\Omega \left(\frac{1}{2} |\boldsymbol{q}_1|^2 + \boldsymbol{q}_1^* \cdot \boldsymbol{q}_1 \right) dx \right\} &= \int_\Omega \frac{1}{2} |\boldsymbol{q}_1^*|^2 dx, \\ \sup_{q_2 \in L^2(\Omega)} \left\{ - \int_\Omega \left(\frac{1}{2} |q_2|^2 + q_2^* q_2 \right) dx \right\} &= \int_\Omega \frac{1}{2} |q_2^*|^2 dx, \\ \sup_{q_3 \in L^2(\Gamma_2)} \left\{ - \int_{\Gamma_2} \left(q_3^* q_3 + g\, |q_3| \right) ds \right\} &= \begin{cases} 0, & \text{if } |q_3^*| \le g \text{ a.e. on } \Gamma_2, \\ \infty, & \text{otherwise.} \end{cases} \end{aligned}$$

Note that the term

$$\sup_{v \in V} \left\{ \int_\Omega \left[\boldsymbol{q}_1^* \cdot \nabla v + (q_2^* + f)\, v \right] dx + \int_{\Gamma_2} q_3^* v \, ds \right\}$$

equals ∞, unless $\boldsymbol{q}^* \in Q^*$ satisfies

$$\int_\Omega \left[\boldsymbol{q}_1^* \cdot \nabla v + (q_2^* + f)\, v \right] dx + \int_{\Gamma_2} q_3^* v \, ds = 0 \qquad \forall v \in V;$$

and under this assumption, the above term equals 0. Thus, the conjugate function (3.23) is

$$F^*(\Lambda^* \boldsymbol{q}^*, -\boldsymbol{q}^*) = \begin{cases} \int_\Omega \frac{1}{2} \left(|\boldsymbol{q}_1^*|^2 + |q_2^*|^2 \right) dx, & \text{if } \boldsymbol{q}^* \in Q_{f,g}^*, \\ +\infty, & \text{otherwise,} \end{cases} \tag{3.24}$$

where the admissible dual function set

$$\begin{aligned} Q_{f,g}^* = \bigg\{ \boldsymbol{q}^* \in Q^* : & \int_\Omega \left[\boldsymbol{q}_1^* \cdot \nabla v + (q_2^* + f)\, v \right] dx + \int_{\Gamma_2} q_3^* v \, ds = 0 \qquad \forall v \in V, \\ & |q_3^*| \le g \text{ a.e. on } \Gamma_2 \bigg\}. \end{aligned} \tag{3.25}$$

Note that the classical form on the constraint $\boldsymbol{q}^* \in Q_{f,g}^*$ is

$$\begin{aligned} \operatorname{div} \boldsymbol{q}_1^* - q_2^* &= f \quad \text{in } \Omega, \\ \boldsymbol{q}_1^* \cdot \boldsymbol{\nu} + q_3^* &= 0 \quad \text{on } \Gamma_2, \\ |q_3^*| &\le g \quad \text{on } \Gamma_2. \end{aligned}$$

In conclusion, the dual problem of (3.22) is

$$\boldsymbol{p}^* \in Q_{f,g}^*, \quad -F^*(\Lambda^* \boldsymbol{p}^*, -\boldsymbol{p}^*) = \sup_{\boldsymbol{q}^* \in Q_{f,g}^*} \big\{ -F^*(\Lambda^* \boldsymbol{q}^*, -\boldsymbol{q}^*) \big\}. \tag{3.26}$$

Existence of solutions of the problems (3.22) and (3.26) is assured by Theorem 3.1, and moreover,

$$F(u, \Lambda u) = -F^*(\Lambda^* \boldsymbol{p}^*, -\boldsymbol{p}^*). \tag{3.27}$$

The function $\boldsymbol{q} \mapsto F^*(\Lambda^* \boldsymbol{q}^*, -\boldsymbol{q}^*)$ is strictly convex over $Q_{f,g}^*$, thus a solution of the dual problem (3.26) is unique.

Now let $w \in V$ be an (arbitrary) approximation of $u \in V$, the unique solution of (3.5). In the rest of the section, we present a general framework for a posteriori estimates of the error $u - w$. The error bounds are computable from the (known) approximant w. Later on, w is taken as the finite element solution of the variational inequality.

Consider

$$\frac{1}{2} a(u - w, u - w) = \frac{1}{2} a(w, w) - a(u, w) + \frac{1}{2} a(u, u).$$

By using (3.5) and (3.7),

$$\begin{aligned} \frac{1}{2} a(u - w, u - w) &= \frac{1}{2} a(w, w) - a(u, w - u) - \frac{1}{2} a(u, u) \\ &\le \frac{1}{2} a(w, w) + j(w) - j(u) - \ell(w - u) - \frac{1}{2} a(u, u) \\ &= J(w) - J(u). \end{aligned}$$

Relation (3.27) implies

$$J(u) = F(u, \Lambda u) = -F^*(\Lambda^* \boldsymbol{p}^*, -\boldsymbol{p}^*) \ge -F^*(\Lambda^* \boldsymbol{q}^*, -\boldsymbol{q}^*) \qquad \forall\, \boldsymbol{q}^* \in Q_{f,g}^*.$$

Therefore,

$$\frac{1}{2} a(u - w, u - w) \le J(w) + F^*(\Lambda^* \boldsymbol{q}^*, -\boldsymbol{q}^*) \qquad \forall\, \boldsymbol{q}^* = (\boldsymbol{q}_1^*, q_2^*, q_3^*) \in Q_{f,g}^*. \tag{3.28}$$

Introduce a function space

$$Q_r^* = (L^2(\Omega))^d \times L^2(\Omega),$$

and write $\boldsymbol{r}^* = (\boldsymbol{r}_1^*, r_2^*)$ for any $\boldsymbol{r}^* \in Q_r^*$. We have

$$\begin{aligned} J(w) + F^*(\Lambda^* \boldsymbol{q}^*, -\boldsymbol{q}^*) = & \int_\Omega \frac{1}{2}(|\nabla w + \boldsymbol{r}_1^*|^2 + |w + r_2^*|^2)\, dx \\ & - \int_\Omega (\nabla w \cdot \boldsymbol{r}_1^* + w r_2^* + f w)\, dx + \int_{\Gamma_2} g\, |w|\, ds \\ & + \int_\Omega \frac{1}{2}(|\boldsymbol{q}_1^*|^2 - |\boldsymbol{r}_1^*|^2 + |q_2^*|^2 - |r_2^*|^2)\, dx. \end{aligned} \tag{3.29}$$

Because $\boldsymbol{q}^* = (\boldsymbol{q}_1^*, q_2^*, q_3^*) \in Q^*_{f,g}$, from (3.25),

$$\int_\Omega (\boldsymbol{q}_1^* \cdot \nabla w + q_2^* w)\, dx + \int_{\Gamma_2} q_3^*\, w\, ds = -\int_\Omega f\, w\, dx.$$

Using this relation in (3.29) and recalling (3.28), we find that

$$\begin{aligned} & \frac{1}{2} a(u - w, u - w) \\ & \quad \le \int_\Omega \frac{1}{2}(|\nabla w + \boldsymbol{r}_1^*|^2 + |w + r_2^*|^2)\, dx + \int_\Omega \frac{1}{2}(|\boldsymbol{q}_1^* - \boldsymbol{r}_1^*|^2 + |q_2^* - r_2^*|^2)\, dx \\ & \quad + \int_\Omega [(\boldsymbol{q}_1^* - \boldsymbol{r}_1^*) \cdot (\nabla w + \boldsymbol{r}_1^*) + (q_2^* - r_2^*)(w + r_2^*)]\, dx \\ & \quad + \int_{\Gamma_2} (g\, |w| + q_3^* w)\, ds. \end{aligned}$$

So for any $\boldsymbol{q}^* \in Q^*_{f,g}$ and $\boldsymbol{r}^* \in Q^*_r$,

$$\begin{aligned} \frac{1}{2} a(u - w, u - w) \le & \int_\Omega (|\nabla w + \boldsymbol{r}_1^*|^2 + |w + r_2^*|^2)\, dx \\ & + \int_\Omega (|\boldsymbol{q}_1^* - \boldsymbol{r}_1^*|^2 + |q_2^* - r_2^*|^2)\, dx + \int_{\Gamma_2} (g\, |w| + q_3^* w)\, ds. \end{aligned}$$

Thus, we have established the following result.

Theorem 3.6. *Let $u \in V$ be the unique solution of (3.5), and $w \in V$ an approximation of u. Then the following estimate holds for any $\boldsymbol{r}^* \in Q^*_r$:*

$$\begin{aligned} \frac{1}{2} a(u - w, u - w) \le & \int_\Omega (|\nabla w + \boldsymbol{r}_1^*|^2 + |w + r_2^*|^2)\, dx \\ & + \inf_{\boldsymbol{q}^* \in Q^*_{f,g}} \left\{ \int_\Omega (|\boldsymbol{q}_1^* - \boldsymbol{r}_1^*|^2 + |q_2^* - r_2^*|^2)\, dx \right. \\ & \left. + \int_{\Gamma_2} (g\, |w| + q_3^* w)\, ds \right\}. \end{aligned} \tag{3.30}$$

Consider the second term

$$II \equiv \inf_{\boldsymbol{q}^* \in Q^*_{f,g}} \left\{ \int_\Omega (|\boldsymbol{q}_1^* - \boldsymbol{r}_1^*|^2 + |q_2^* - r_2^*|^2)\, dx + \int_{\Gamma_2} (g\, |w| + q_3^* w)\, ds \right\}$$

on the right side of the estimate (3.30). From the definition (3.25),

$$II = \inf_{\substack{\boldsymbol{q}^* \in Q^* \\ |q_3^*| \le g}} \sup_{v \in V} \left\{ \int_\Omega (|\boldsymbol{q}_1^* - \boldsymbol{r}_1^*|^2 + |q_2^* - r_2^*|^2)\, dx + \int_{\Gamma_2} (g\,|w| + q_3^* w)\, ds \right.$$
$$\left. + \int_\Omega [\boldsymbol{q}_1^* \cdot \nabla v + (q_2^* + f)\, v]\, dx + \int_{\Gamma_2} q_3^* v\, ds \right\}.$$

Here and below, the condition "$|q_3^*| \le g$" stands for "$|q_3^*| \le g$ a.e. on Γ_2." Substitute $\boldsymbol{q}_1^* - \boldsymbol{r}_1^*$ by $\boldsymbol{q}_1^*$, $q_2^* - r_2^*$ by q_2^*, and regroup the terms to get

$$II = \inf_{\substack{\boldsymbol{q}^* \in Q^* \\ |q_3^*| \le g}} \sup_{v \in V} \left\{ \int_\Omega (|\boldsymbol{q}_1^*|^2 + \boldsymbol{q}_1^* \cdot \nabla v + |q_2^*|^2 + q_2^* v)\, dx \right.$$
$$+ \int_\Omega [\boldsymbol{r}_1^* \cdot \nabla v + (r_2^* + f)\, v]\, dx$$
$$\left. + \int_{\Gamma_2} [g\,|w| + q_3^*(w + v)]\, ds \right\}$$
$$= \inf_{|q_3^*| \le g} \sup_{v \in V} \left\{ \int_\Omega \left[-\frac{1}{4}(|\nabla v|^2 + v^2) + \boldsymbol{r}_1^* \cdot \nabla v + (r_2^* + f) v \right] dx \right.$$
$$\left. + \int_{\Gamma_2} q_3^* v\, ds + \int_{\Gamma_2} (g\,|w| + q_3^* w)\, ds \right\}.$$

Define the residual

$$\mathcal{R}(q_3^*, \boldsymbol{r}^*) = \sup_{v \in V} \frac{1}{\|v\|_V} \left\{ \int_\Omega [\boldsymbol{r}_1^* \cdot \nabla v + (r_2^* + f)\, v]\, dx + \int_{\Gamma_2} q_3^* v\, ds \right\}. \quad (3.31)$$

Then

$$II \le \inf_{|q_3^*| \le g} \sup_{v \in V} \left\{ -\frac{1}{4} \|v\|_V^2 + \mathcal{R}(q_3^*, \boldsymbol{r}^*)\, \|v\|_V + \int_{\Gamma_2} (g\,|w| + q_3^* w)\, ds \right\}$$
$$= \inf_{|q_3^*| \le g} \left\{ \mathcal{R}(q_3^*, \boldsymbol{r}^*)^2 + \int_{\Gamma_2} (g\,|w| + q_3^* w)\, ds \right\}.$$

This last estimate is combined with Theorem 3.6, leading to the next result.

Theorem 3.7. *Let $u \in V$ be the unique solution of (3.5), and $w \in V$ an approximation of u. Then for any $\boldsymbol{r}^* = (\boldsymbol{r}_1^*, r_2^*) \in Q_r^*$,*

$$\frac{1}{2} a(u - w, u - w) \le \int_\Omega (|\nabla w + \boldsymbol{r}_1^*|^2 + |w + r_2^*|^2)\, dx$$
$$+ \inf_{|q_3^*| \le g} \left\{ \mathcal{R}(q_3^*, \boldsymbol{r}^*)^2 + \int_{\Gamma_2} (g\,|w| + q_3^* w)\, ds \right\}, \quad (3.32)$$

where the residual $\mathcal{R}(q_3^*, \boldsymbol{r}^*)$ is defined by (3.31).

In the next two sections, we make particular choices of $\boldsymbol{r}_1^*$ and r_2^* in (3.32) to obtain residual-based and recovery-based error estimates of finite element solutions of the model elliptic variational inequality.

3.4 Residual-Based Error Estimates for the Model Elliptic Variational Inequality

In (3.32), we let $\boldsymbol{r}_1^* = -\nabla w$ and $r_2^* = -w$. Then

$$\sqrt{\frac{1}{2}a(u-w,u-w)} \le \mathcal{R}, \tag{3.33}$$

where

$$\mathcal{R} = \inf_{|q_3^*|\le g} \left\{ \left[\sup_{v\in V} \frac{1}{\|v\|_V} \left(\int_\Omega [-\nabla w\cdot\nabla v + (-w+f)v]\,dx + \int_{\Gamma_2} q_3^* v\,ds \right) \right]^2 \right.$$
$$\left. + \int_{\Gamma_2} (g\,|w| + q_3^* w)\,ds \right\}^{1/2}. \tag{3.34}$$

Although it is possible to derive (3.33)–(3.34) through other approaches, we comment that Theorem 3.7 provides a general framework for various a posteriori error estimates with different choices of the auxiliary variable $\boldsymbol{r}^*$.

In the limiting case $g = 0$, the problem (3.5) reduces to the variational equation

$$u \in V, \qquad a(u,v) = \ell(v) \qquad \forall\, v \in V.$$

Correspondingly, the estimate (3.33)–(3.34) reduces to the familiar form

$$\sqrt{\frac{1}{2}a(u-w,u-w)} \le \sup_{v\in V} \frac{1}{\|v\|_V} \int_\Omega [(f-w)v - \nabla w\cdot\nabla v]\,dx,$$

which is a starting point in deriving some a posteriori error estimators for Galerkin approximations of linear elliptic partial differential equations (cf. [1]).

Now we focus on a posteriori analysis for the finite element solution error, that is, for the situation where the approximant $w = u_h$ is the finite element solution. By taking $q_3^* = -g\,\lambda_h$, and substituting v by $v_h - v$ in (3.34) we obtain

$$\mathcal{R} \le \sup_{v\in V} \frac{1}{\|v\|_V} \left\{ \int_\Omega [\nabla u_h\cdot\nabla(v-v_h) + (u_h-f)\,(v-v_h)]\,dx \right.$$
$$\left. + \int_{\Gamma_2} g\,\lambda_h\,(v-v_h)\,ds \right\}, \tag{3.35}$$

for any $v_h \in V_h$. Here we have used (3.17). For a given $v \in V$, take $v_h = \Pi_h v$ in (3.35), where $\Pi_h v$ is defined in (3.11).

$$\mathcal{R} \le \sup_{v \in V} \frac{1}{\|v\|_V} \left\{ \int_\Omega \left[\nabla u_h \cdot \nabla(v - \Pi_h v) + (u_h - f)\,(v - \Pi_h v) \right] dx \right.$$
$$\left. + \int_{\Gamma_2} g\,\lambda_h (v - \Pi_h v)\, ds \right\}.$$

Decompose the integrals into local contributions from each element $K \in \mathcal{P}_h$ and integrate by parts over K to obtain

$$\mathcal{R} \le \sup_{v \in V} \frac{1}{\|v\|_V} \sum_{K \in \mathcal{P}_h} \left\{ \int_{\partial K} \frac{\partial u_h}{\partial \nu_K} (v - \Pi_h v)\, ds \right.$$
$$+ \int_K (-\Delta u_h + u_h - f)(v - \Pi_h v)\, dx$$
$$\left. + \int_{\partial K \cap \Gamma_2} g\,\lambda_h (v - \Pi_h v)\, ds \right\}. \tag{3.36}$$

Define interior residuals for each element $K \in \mathcal{P}_h$ by

$$r_K = -\Delta u_h + u_h - f \quad \text{in } K, \tag{3.37}$$

and side residuals for each side $\gamma \in \mathcal{E}_{h,0,\Gamma_2} \equiv \mathcal{E}_{h,0} \cup \mathcal{E}_{h,\Gamma_2}$ by

$$R_\gamma = \begin{cases} \left[\frac{\partial u_h}{\partial \nu}\right]_\gamma & \text{if } \gamma \in \mathcal{E}_{h,0}, \\ \frac{\partial u_h}{\partial \nu} + g\,\lambda_h & \text{if } \gamma \in \mathcal{E}_{h,\Gamma_2}, \end{cases} \tag{3.38}$$

where the quantity

$$\left[\frac{\partial u_h}{\partial \nu}\right]_\gamma = \boldsymbol{\nu}_K \cdot (\nabla u_h)_K + \boldsymbol{\nu}_{K'} \cdot (\nabla u_h)_{K'}$$

represents the jump discontinuity in the approximation to the normal derivative on the side γ which separates the neighboring elements K and K'. By using definitions (3.37) and (3.38), relation (3.36) reduces to

$$\mathcal{R} \le \sup_{v \in V} \frac{1}{\|v\|_V} \left\{ \sum_{K \in \mathcal{P}_h} \int_K r_K (v - \Pi_h v)\, dx \right.$$
$$\left. + \sum_{\gamma \in \mathcal{E}_{h,0,\Gamma_2}} \int_\gamma R_\gamma (v - \Pi_h v)\, ds \right\}. \tag{3.39}$$

Using the estimates (3.14) and (3.15) in (3.39), and applying the Cauchy–Schwarz inequality, we have

$$\mathcal{R} \le \sup_{v \in V} \frac{1}{\|v\|_V} \left\{ C\, |v|_{1;\Omega} \Big(\sum_{K \in \mathcal{P}_h} h_K^2 \|r_K\|_{0;K}^2 + \sum_{\gamma \in \mathcal{E}_{h,0,\Gamma_2}} h_\gamma \|R_\gamma\|_{0;\gamma}^2 \Big)^{1/2} \right\}$$

$$\le C \left(\sum_{K \in \mathcal{P}_h} h_K^2 \|r_K\|_{0;K}^2 + \sum_{\gamma \in \mathcal{E}_{h,0,\Gamma_2}} h_\gamma \|R_\gamma\|_{0;\gamma}^2 \right)^{1/2}. \tag{3.40}$$

We summarize the above results in the form of a theorem.

Theorem 3.8. *Let u and u_h be the unique solutions of* (3.5) *and* (3.16), *respectively. Then the error $e_h = u - u_h$ satisfies the a posteriori estimate*

$$\|e_h\|_V^2 \le C \left(\sum_{K \in \mathcal{P}_h} h_K^2 \|r_K\|_{0;K}^2 + \sum_{\gamma \in \mathcal{E}_{h,0,\Gamma_2}} h_\gamma \|R_\gamma\|_{0;\gamma}^2 \right), \tag{3.41}$$

where r_K and R_γ are interior and side residuals, defined by (3.37) *and* (3.38), *respectively.*

In practical computations, the terms on the right side of (3.41) are regrouped by writing

$$\|e_h\|_V^2 \le C\, \eta_R^2, \qquad \eta_R^2 = \sum_{K \in \mathcal{P}_h} \eta_{R,K}^2, \tag{3.42}$$

where the local error indicator $\eta_{R,K}$ on each element K, defined by

$$\eta_{R,K}^2 = h_K^2 \|r_K\|_{0;K}^2 + \frac{1}{2} h_K \sum_{\gamma \in \mathcal{E}(K) \cap \mathcal{E}_{h,0}} \|R_\gamma\|_{0;\gamma}^2 + h_K \sum_{\gamma \in \mathcal{E}(K) \cap \mathcal{E}_{h,\Gamma_2}} \|R_\gamma\|_{0;\gamma}^2, \tag{3.43}$$

identifies contributions from each of the elements to the global error.

In the last part of the section, we explore the efficiency of the error bound from (3.41) or (3.42). We derive an upper bound for the error estimator. Integrating by parts over each element and using (3.8) and Theorem 3.18 we have, for any $v \in V$,

$$\begin{aligned} a(e_h, v) &= a(u, v) - a(u_h, v) \\ &= \int_\Omega f v\, dx - \int_{\Gamma_2} g\, \lambda\, v\, ds - \int_\Omega (\nabla u_h \cdot \nabla v + u_h v)\, dx \\ &= \sum_{K \in \mathcal{P}_h} \int_K (\Delta u_h - u_h + f) v\, dx + \sum_{\gamma \in \mathcal{E}_{h,0}} \int_\gamma - \left[\frac{\partial u_h}{\partial \nu} \right]_\gamma v\, ds \\ &\quad + \sum_{\gamma \in \mathcal{E}_{h,\Gamma_2}} \int_\gamma \left(-\frac{\partial u_h}{\partial \nu} - g\lambda \right) v\, ds. \end{aligned}$$

Thus,

$$a(e_h, v) = -\sum_{K\in\mathcal{P}_h}\int_K r_K v\,dx - \sum_{\gamma\in\mathcal{E}_{h,0,\Gamma_2}}\int_\gamma R_\gamma v\,ds + \sum_{\gamma\in\mathcal{E}_{h,\Gamma_2}}\int_\gamma g(\lambda_h-\lambda)v\,ds, \tag{3.44}$$

where r_K and R_γ are interior and side residuals defined for each element $K \in \mathcal{P}_h$ and each side $\gamma \in \mathcal{E}_{h,0,\Gamma_2}$ by (3.37) and (3.38), respectively. In order to simplify notation, we omit the subscripts K and γ. In the following, we apply Theorems 3.4 and 3.5 choosing P_K as the space of polynomials of degree less than or equal to l, and l is any integer larger than or equal to the local polynomial degree of the finite element functions. Let $\overline{r}$ be a discontinuous piecewise polynomial approximation to the residual r; that is, $\overline{r}|_K \in P_K$. Applying Theorem 3.4, we get

$$\|\overline{r}\|_{0;K}^2 \le C\int_K \psi_K \overline{r}^2 dx. \tag{3.45}$$

Because the function $v = \psi_K\overline{r}$ vanishes on the boundary ∂K, it can be extended to a function in V by 0 to the rest of the domain Ω. Inserting this extended function v in the residual equation (3.44), one obtains

$$a(e_h, \psi_K\overline{r}) = -\int_K r\psi_K\overline{r}\,dx.$$

Using this relation, we obtain

$$\int_K \psi_K\overline{r}^2 dx = \int_K \psi_K\overline{r}(\overline{r}-r)\,dx - a(e_h,\psi_K\overline{r}). \tag{3.46}$$

The terms on the right side of (3.46) are bounded by making use of the Cauchy–Schwarz inequality and the second part of Theorem 3.4,

$$\int_K \psi_K\overline{r}^2 dx \le C\|\overline{r}\|_{0;K}\|\overline{r}-r\|_{0;K} + C\,h_K^{-1}\|e_h\|_{1;K}\|\overline{r}\|_{0;K}.$$

Combined with (3.45) we have

$$\|\overline{r}\|_{0;K} \le C\left(\|\overline{r}-r\|_{0;K} + h_K^{-1}\|e_h\|_{1;K}\right).$$

With the aid of the triangle inequality, finally we get

$$\|r\|_{0;K} \le \|\overline{r}-r\|_{0;K} + \|\overline{r}\|_{0;K} \le C\left(\|\overline{r}-r\|_{0;K} + h_K^{-1}\|e_h\|_{1;K}\right). \tag{3.47}$$

Consider now an interior side $\gamma \in \mathcal{E}_{h,0}$. From the first part of Theorem 3.5 it follows that

$$\|R\|_{0;\gamma}^2 \le C\int_\gamma \psi_\gamma R^2 ds. \tag{3.48}$$

Let $\widetilde{\gamma}$ denote the subdomain of Ω consisting of the side γ and the two neighbouring elements. The function $v = \psi_\gamma R$ vanishes on $\partial\widetilde{\gamma}$ and as before it can be extended continuously to the whole domain Ω by 0 outside $\widetilde{\gamma}$. With this choice of v the residual equation (3.44) reduces to

$$a(e_h, \psi_\gamma R) = -\int_{\widetilde{\gamma}} r\psi_\gamma R\, dx - \int_\gamma \psi_\gamma R^2 ds.$$

Therefore,

$$\int_\gamma \psi_\gamma R^2 ds = -a(e_h, \psi_\gamma R) - \int_{\widetilde{\gamma}} r\psi_\gamma R\, dx. \tag{3.49}$$

Applying the Cauchy–Schwarz inequality and the second part of Theorem 3.5 to the terms on the right side of (3.49), we obtain

$$\int_\gamma \psi_\gamma R^2 ds \le C\, h_\gamma^{-1/2} \|e_h\|_{1;\widetilde{\gamma}} \|R\|_{0;\gamma} + C\, h_\gamma^{1/2} \|r\|_{0;\widetilde{\gamma}} \|R\|_{0;\gamma},$$

which, combined with (3.48) and (3.47), implies that for every interior side $\gamma \in \mathcal{E}_{h,0}$,

$$\|R\|_{0;\gamma} \le C\left(h_\gamma^{-1/2} \|e_h\|_{1;\widetilde{\gamma}} + h_\gamma^{1/2} \|\overline{r} - r\|_{0;\widetilde{\gamma}}\right). \tag{3.50}$$

Finally, consider those sides γ lying on $\overline{\Gamma_2}$. Denote $\overline{R} \in P_K$ an approximation to the residual $R = (\partial u_h/\partial\nu) + g\lambda_h$ on γ, $\gamma \in \mathcal{E}(K)$. The first part of Theorem 3.5 implies

$$\|\overline{R}\|^2_{0;\gamma} \le C \int_\gamma \psi_\gamma \overline{R}^2 ds. \tag{3.51}$$

Define the function $v = \psi_\gamma \overline{R}$ and let $\widetilde{\gamma}$ be the element whose boundary contains the side γ. Then $v|_{\partial\widetilde{\gamma}\setminus\gamma} = 0$. Extend this function to the whole domain by zero value outside $\widetilde{\gamma}$. The residual equation (3.44), with this choice of v, becomes

$$a(e_h, \psi_\gamma \overline{R}) = -\int_{\widetilde{\gamma}} r\psi_\gamma \overline{R}\, dx - \int_\gamma R\psi_\gamma \overline{R}\, ds + \int_\gamma g(\lambda_h - \lambda)\psi_\gamma \overline{R}\, ds,$$

which leads to

$$\int_\gamma \psi_\gamma \overline{R}^2 ds = \int_\gamma \psi_\gamma \overline{R}(\overline{R}-R)\, ds - a(e_h, \psi_\gamma \overline{R}) - \int_{\widetilde{\gamma}} r\psi_\gamma \overline{R}\, dx + \int_\gamma g(\lambda_h-\lambda)\psi_\gamma \overline{R}\, ds. \tag{3.52}$$

As before, the first three terms on the right side of (3.52) can be bounded by applying Theorem 3.5 and the Cauchy–Schwarz inequality. Using (3.51), we then obtain for each side $\gamma \in \mathcal{E}_{h,\Gamma_2}$,

$$\|\overline{R}\|^2_{0;\gamma} \le C\left(\|\overline{R}\|_{0;\gamma} \|\overline{R} - R\|_{0;\gamma} + h_\gamma^{-1/2} \|\overline{R}\|_{0;\gamma} \|e_h\|_{1;\widetilde{\gamma}} + h_\gamma^{1/2} \|\overline{R}\|_{0;\gamma} \|r\|_{0;\widetilde{\gamma}}\right) + \int_\gamma g(\lambda_h - \lambda)\psi_\gamma \overline{R}\, ds.$$

Multiplying this inequality by h_γ and summing over all sides $\gamma \in \mathcal{E}_{h,\Gamma_2}$, we get

$$\begin{aligned}\sum_{\gamma\in\mathcal{E}_{h,\Gamma_2}} h_\gamma \|\overline{R_\gamma}\|_{0;\gamma}^2 \le C &\sum_{\gamma\in\mathcal{E}_{h,\Gamma_2}} h_\gamma^{1/2}\|\overline{R_\gamma}\|_{0;\gamma} h_\gamma^{1/2}\|\overline{R_\gamma} - R_\gamma\|_{0;\gamma} \\ &+ C \sum_{\gamma\in\mathcal{E}_{h,\Gamma_2}} h_\gamma^{1/2}\|\overline{R_\gamma}\|_{0;\gamma}\|e_h\|_{1;\widetilde{\gamma}} \\ &+ C \sum_{\gamma\in\mathcal{E}_{h,\Gamma_2}} h_\gamma^{1/2}\|\overline{R_\gamma}\|_{0;\gamma} h_\gamma \|r_{\widetilde{\gamma}}\|_{0;\widetilde{\gamma}} + |R_{h,\Gamma_2}|, \qquad (3.53)\end{aligned}$$

where

$$R_{h,\Gamma_2} = \sum_{\gamma\in\mathcal{E}_{h,\Gamma_2}} \int_\gamma g\,(\lambda-\lambda_h)\, h_\gamma \psi_\gamma \overline{R_\gamma}\, ds.$$

We can bound R_{h,Γ_2} as follows,

$$\begin{aligned}|R_{h,\Gamma_2}| &\le \sum_{\gamma\in\mathcal{E}_{h,\Gamma_2}} h_\gamma^{1/2}\|g\,(\lambda-\lambda_h)\|_{0;\gamma} h_\gamma^{1/2}\|\psi_\gamma \overline{R_\gamma}\|_{0;\gamma} \\ &\le C \left(\sum_{\gamma\in\mathcal{E}_{h,\Gamma_2}} h_\gamma \|\lambda-\lambda_h\|_{0;\gamma}^2\right)^{1/2} \left(\sum_{\gamma\in\mathcal{E}_{h,\Gamma_2}} h_\gamma \|\overline{R_\gamma}\|_{0;\gamma}^2\right)^{1/2}.\end{aligned}$$

Use this bound in (3.53) and apply the Cauchy–Schwarz inequality to get

$$\begin{aligned}\sum_{\gamma\in\mathcal{E}_{h,\Gamma_2}} h_\gamma \|\overline{R_\gamma}\|_{0;\gamma}^2 \le C \Bigg(&\sum_{\gamma\in\mathcal{E}_{h,\Gamma_2}} h_\gamma \|\overline{R_\gamma} - R_\gamma\|_{0;\gamma}^2 + \sum_{\gamma\in\mathcal{E}_{h,\Gamma_2}} h_\gamma^2 \|r_{\widetilde{\gamma}}\|_{0;\widetilde{\gamma}}^2 \\ &+ \|e_h\|_V^2 + \sum_{\gamma\in\mathcal{E}_{h,\Gamma_2}} h_\gamma \|\lambda-\lambda_h\|_{0;\gamma}^2 \Bigg). \qquad (3.54)\end{aligned}$$

Combining (3.47) and (3.54), we finally conclude that

$$\begin{aligned}\sum_{\gamma\in\mathcal{E}_{h,\Gamma_2}} h_\gamma \|R_\gamma\|_{0;\gamma}^2 \le C \Bigg(&\sum_{\gamma\in\mathcal{E}_{h,\Gamma_2}} h_\gamma \|\overline{R_\gamma} - R_\gamma\|_{0;\gamma}^2 + \sum_{\gamma\in\mathcal{E}_{h,\Gamma_2}} h_\gamma^2 \|\overline{r_{\widetilde{\gamma}}} - r_{\widetilde{\gamma}}\|_{0;\widetilde{\gamma}}^2 \\ &+ \|e_h\|_V^2 + \sum_{\gamma\in\mathcal{E}_{h,\Gamma_2}} h_\gamma \|\lambda-\lambda_h\|_{0;\gamma}^2 \Bigg). \qquad (3.55)\end{aligned}$$

Summarizing (3.47), (3.50), and (3.55), we have

$$\eta_R^2 \le C \Bigg(\|e_h\|_V^2 + \sum_{\gamma \in \mathcal{E}_{h,\Gamma_2}} h_\gamma \|\lambda - \lambda_h\|_{0;\gamma}^2 + \sum_{K \in \mathcal{P}_h} h_K^2 \|r_K - \overline{r_K}\|_{0;K}^2 + \sum_{\gamma \in \mathcal{E}_{h,\Gamma_2}} h_\gamma \|R_\gamma - \overline{R_\gamma}\|_{0;\gamma}^2 \Bigg). \tag{3.56}$$

For our model problem, the element residual r_K of (3.37) and the side residual R_γ of (3.38), $-\Delta u_h + u_h$ in K and $\partial u_h/\partial \nu$ on γ are polynomials. Therefore, the terms $\|r_K - \overline{r_K}\|_{0;K}$ and $\|R_\gamma - \overline{R_\gamma}\|_{0;\gamma}$ in the right-hand side of (3.56) can be replaced by $\|f - f_K\|_{0;K}$ and $\|\lambda_h - \lambda_{h,\gamma}\|_{0;\gamma}$, with discontinuous piecewise polynomial approximations f_K and $\lambda_{h,\gamma}$.

Theorem 3.9. *Let η_R be defined as in (3.42). Then*

$$\eta_R^2 \le C \Bigg(\|e_h\|_V^2 + \sum_{\gamma \in \mathcal{E}_{h,\Gamma_2}} h_\gamma \|\lambda - \lambda_h\|_{0;\gamma}^2 + \sum_{K \in \mathcal{P}_h} h_K^2 \|f - f_K\|_{0;K}^2 + \sum_{\gamma \in \mathcal{E}_{h,\Gamma_2}} h_\gamma \|\lambda_h - \lambda_{h,\gamma}\|_{0;\gamma}^2 \Bigg) \tag{3.57}$$

with discontinuous piecewise polynomial approximations f_K, $\lambda_{h,\gamma}$ of f, λ_h.

Let us comment on the three summation terms in (3.57). As long as f has a suitable degree of smoothness, the approximation error $\sum_{K\in\mathcal{P}_h} h_K^2 \|f - f_K\|_{0;K}^2$ will be of higher order than $\|e_h\|_V^2$. Due to the inequality nature of the variational problem, in the efficiency bound (3.57) of the error estimator, there are extra terms involving λ and λ_h. A sharp bound of the term $\sum_{\gamma\in\mathcal{E}_{h,\Gamma_2}} h_\gamma \|\lambda - \lambda_h\|_{0;\gamma}^2$ is currently an open problem. Nevertheless, in Section 3.6, we present numerical results showing that the presence of this term in (3.57) does not have an effect on the efficiency of the error estimator. Similar numerical results can also be used as an evidence that the term $\sum_{\gamma\in\mathcal{E}_{h,\Gamma_2}} h_\gamma \|\lambda_h - \lambda_{h,\gamma}\|_{0;\gamma}^2$ does not have an effect on the efficiency of the error estimator.

3.5 Recovery-Based Error Estimates for the Model Elliptic Variational Inequality

An important class of a posteriori error estimates is based on local or global averaging of the gradient, for example, in the form of Zienkiewicz–Zhu gradient recovery technique [78, 79, 80]. It is known that in the case of structured grids and higher regularity solutions, such estimators are both efficient and reliable. Some work has been done for unstructured meshes as well (e.g., [74, 72]). In [8, 23], Carstensen and Bartels proved that all averaging techniques provide us with a reliable a posteriori error control for the Laplace equation with mixed boundary conditions on unstructured grids also. In the context of solving variational inequalities, gradient recovery type error estimates for elliptic obstacle problem have been derived recently in [9, 71].

In this section, we study a gradient recovery type error estimator for the finite element solution of the model problem, and we restrict our discussion to linear elements. Then all the nodes are also the vertices. To formulate the error estimator we need a gradient recovery operator. There are many types of gradient recovery operators. In order to have a "good" approximation of the true gradient ∇u, a set of conditions to be satisfied by the recovery operator was identified in [1]. These conditions lead to a more precise characterization of the form of the gradient recovery operator, summarized in Lemma 4.5 in [1]. In particular, the recovered gradient at a node $\boldsymbol{a}$ is a linear combination of the values of ∇u_h in a patch surrounding $\boldsymbol{a}$.

We define the gradient recovery operator $G_h : V_h \to (V_h)^d$ as follows:

$$G_h v_h(\boldsymbol{x}) = \sum_{\boldsymbol{a}\in\mathcal{N}_h} G_h v_h(\boldsymbol{a}) \varphi_{\boldsymbol{a}}(\boldsymbol{x}), \qquad G_h v_h(\boldsymbol{a}) = \frac{1}{|\widetilde{K}_{\boldsymbol{a}}|} \int_{\widetilde{K}_{\boldsymbol{a}}} \nabla v_h \, dx. \quad (3.58)$$

Linear elements are used, therefore

$$G_h v_h(\boldsymbol{a}) = \sum_{i=1}^{N_{\boldsymbol{a}}} \alpha_{\boldsymbol{a}}^i (\nabla v_h)_{K_{\boldsymbol{a}}^i}, \quad (3.59)$$

where $(\nabla v_h)_{K_{\boldsymbol{a}}^i}$ denotes the vector value of the gradient ∇v_h on the element $K_{\boldsymbol{a}}^i$, $\widetilde{K}_{\boldsymbol{a}} = \bigcup_{i=1}^{N_{\boldsymbol{a}}} K_{\boldsymbol{a}}^i$, $\alpha_{\boldsymbol{a}}^i = |K_{\boldsymbol{a}}^i|/|\widetilde{K}_{\boldsymbol{a}}|$, $i = 1, \ldots, N_{\boldsymbol{a}}$.

Recall from Section 3.4 that residual-type error estimates are derived by applying Theorem 3.7 with $\boldsymbol{r}^* = -(\nabla u_h, u_h)$, where u_h is the finite element solution. In this section, we consider a different choice.

Theorem 3.10. *Let u and u_h be the unique solutions of* (3.5) *and* (3.16), *respectively. Then*

$$\|u - u_h\|_V^2 \le C\eta_G^2 + C \sum_{\boldsymbol{a}\in\mathcal{N}_{h,0}} \left(h_{\boldsymbol{a}}^4 \|\nabla u_h\|_{0;\widetilde{K}_{\boldsymbol{a}}}^2 + h_{\boldsymbol{a}}^2 \min_{f_{\boldsymbol{a}}\in\mathbb{R}} \|f - f_{\boldsymbol{a}}\|_{0;\widetilde{K}_{\boldsymbol{a}}}^2 \right), \quad (3.60)$$

where

$$\eta_G^2 = \sum_{K\in\mathcal{P}_h} \eta_{G,K}^2, \quad (3.61)$$

$$\eta_{G,K}^2 = \|\nabla u_h - G_h u_h\|_{0;K}^2 + \sum_{\gamma\in\mathcal{E}(K)\cap\mathcal{E}_{h,\Gamma_2}} h_\gamma \|G_h u_h \cdot \boldsymbol{\nu}_\gamma + g\,\lambda_h\|_{0;\gamma}^2. \quad (3.62)$$

Proof. Let $\lambda_h \in L^\infty(\Gamma_2)$ be provided by Theorem 3.3. Apply Theorem 3.7 with $w = u_h$ and $\boldsymbol{r}^* = -(G_h u_h, u_h)$ to obtain

$$\frac{1}{2}a(u-u_h,u-u_h) \le \int_\Omega |\nabla u_h - G_h u_h|^2\,dx + \mathcal{R}^2, \tag{3.63}$$

where

$$\mathcal{R} = \sup_{v\in V} \frac{1}{\|v\|_V}\left\{\int_\Omega [G_h u_h\cdot\nabla v + (u_h - f)v]\,dx + \int_{\Gamma_2} g\lambda_h v\,ds\right\}. \tag{3.64}$$

Let Π_h be the interpolation operator defined by (3.11). Use (3.17) with $v_h = \Pi_h v$:

$$\int_\Omega [\nabla u_h\cdot\nabla \Pi_h v + (u_h - f)\,\Pi_h v]\,dx + \int_{\Gamma_2} g\,\lambda_h \Pi_h v\,ds = 0.$$

Therefore, we can rewrite (3.64) as

$$\begin{aligned}\mathcal{R} = \sup_{v\in V}\frac{1}{\|v\|_V}\Bigg\{ &\int_\Omega [G_h u_h\cdot\nabla(v-\Pi_h v) + (u_h - f)(v - \Pi_h v)]\,dx \\ &+ \int_{\Gamma_2} g\lambda_h(v-\Pi_h v)\,ds + \int_\Omega (G_h u_h - \nabla u_h)\cdot\nabla\Pi_h v\,dx\Bigg\}.\end{aligned}$$

By (3.12), we have

$$\|\nabla \Pi_h v\|_{0;\Omega} \le C\,\|v\|_V,$$

and so,

$$\sup_{v\in V}\frac{1}{\|v\|_V}\int_\Omega (G_h u_h - \nabla u_h)\cdot\nabla\Pi_h v\,dx \le C\,\|\nabla u_h - G_h u_h\|_{0;\Omega}.$$

Thus,

$$\mathcal{R} \le C\,\|\nabla u_h - G_h u_h\|_{0;\Omega} + \mathcal{R}_1, \tag{3.65}$$

where

$$\begin{aligned}\mathcal{R}_1 = \sup_{v\in V}\frac{1}{\|v\|_V}\Bigg\{ &\int_\Omega [G_h u_h\cdot\nabla(v-\Pi_h v) + (u_h - f)(v - \Pi_h v)]\,dx \\ &+ \int_{\Gamma_2} g\lambda_h(v-\Pi_h v)\,ds\Bigg\}.\end{aligned}$$

Integrate by parts over each element $K \in \mathcal{P}_h$ to get

$$\begin{aligned}\mathcal{R}_1 = \sup_{v\in V}\frac{1}{\|v\|_V}\sum_{K\in\mathcal{P}_h}\Bigg\{ &\int_K (-\mathrm{div}(G_h u_h) + u_h - f)(v - \Pi_h v)\,dx \\ &+ \sum_{\gamma\in\mathcal{E}(K)}\int_\gamma G_h u_h\cdot\boldsymbol{\nu}_\gamma(v-\Pi_h v)\,ds \\ &+ \sum_{\gamma\in\mathcal{E}(K)\cap\mathcal{E}_{h,\Gamma_2}}\int_\gamma g\,\lambda_h(v-\Pi_h v)\,ds\Bigg\}. \end{aligned} \tag{3.66}$$

Because $G_h u_h$ is continuous, the integrals on the interior sides $\gamma \in \mathcal{E}_{h,0}$ cancel each other. Write

$$-\mathrm{div}(G_h u_h) + u_h - f = \mathrm{div}(\nabla u_h - G_h u_h) + (-\Delta u_h + u_h - f)$$

and rearrange the terms in (3.66) to obtain

$$\begin{aligned}
\mathcal{R}_1 = \sup_{v \in V} \frac{1}{\|v\|_V} \Bigg\{ & \sum_{K \in \mathcal{P}_h} \int_K \mathrm{div}(\nabla u_h - G_h u_h)(v - \Pi_h v)\, dx \\
& + \sum_{K \in \mathcal{P}_h} \int_K r_K (v - \Pi_h v)\, dx \\
& + \sum_{\gamma \in \mathcal{E}_{h,\Gamma_2}} \int_\gamma (G_h u_h \cdot \boldsymbol{\nu}_\gamma + g\, \lambda_h)(v - \Pi_h v)\, ds \Bigg\}, \\
= \sup_{v \in V} \frac{1}{\|v\|_V} & \{ I + II + III \},
\end{aligned} \tag{3.67}$$

where $r_K = -\Delta u_h + u_h - f = u_h - f$ denotes the interior residual on element $K \in \mathcal{P}_h$. We use r to denote the piecewise interior residual; that is, $r|_K = r_K$ for $K \in \mathcal{P}_h$. To estimate the first summand on the right side of (3.67), we use an elementwise inverse inequality of the form

$$\|\mathrm{div}(\nabla u_h - G_h u_h)\|_{0;K} \le C h_K^{-1} \|\nabla u_h - G_h u_h\|_{0;K}. \tag{3.68}$$

Apply the Cauchy–Schwarz inequality, the inverse inequality (3.68), and the estimate (3.14) to get

$$\begin{aligned}
I &\le C \sum_{K \in \mathcal{P}_h} \|\mathrm{div}(\nabla u_h - G_h u_h)\|_{0;K} \|v - \Pi_h v\|_{0;K} \\
&\le C \sum_{K \in \mathcal{P}_h} \|\nabla u_h - G_h u_h\|_{0;K} \|h_K^{-1}(v - \Pi_h v)\|_{0;K} \\
&\le C \left(\sum_{K \in \mathcal{P}_h} \|\nabla u_h - G_h u_h\|_{0;K}^2 \right)^{1/2} \left(\sum_{K \in \mathcal{P}_h} \|h_K^{-1}(v - \Pi_h v)\|_{0;K}^2 \right)^{1/2} \\
&\le C\, |v|_{1;\Omega} \left(\sum_{K \in \mathcal{P}_h} \|\nabla u_h - G_h u_h\|_{0;K}^2 \right)^{1/2}.
\end{aligned} \tag{3.69}$$

For the second summand, we apply the estimate (3.13) to obtain

$$II \le C |v|_{1;\Omega} \left(\sum_{\boldsymbol{a} \in \mathcal{N}_{h,0}} h_{\boldsymbol{a}}^2 \min_{r_{\boldsymbol{a}} \in \mathbb{R}} \|r - r_{\boldsymbol{a}}\|_{0;\widetilde{K}_{\boldsymbol{a}}}^2 \right)^{1/2}.$$

Now

$$\sum_{\boldsymbol{a}\in\mathcal{N}_{h,0}} h_{\boldsymbol{a}}^2 \min_{r_{\boldsymbol{a}}\in\mathbb{R}} \|r - r_{\boldsymbol{a}}\|_{0;\widetilde{K}_{\boldsymbol{a}}}^2 \le 2 \sum_{\boldsymbol{a}\in\mathcal{N}_{h,0}} h_{\boldsymbol{a}}^2 \|u_h - \overline{u_h}\|_{0;\widetilde{K}_{\boldsymbol{a}}}^2 + 2 \sum_{\boldsymbol{a}\in\mathcal{N}_{h,0}} h_{\boldsymbol{a}}^2 \min_{f_{\boldsymbol{a}}\in\mathbb{R}} \|f - f_{\boldsymbol{a}}\|_{0;\widetilde{K}_{\boldsymbol{a}}}^2,$$

where $\overline{u_h}$ denotes the integral mean of u_h over $\widetilde{K}_{\boldsymbol{a}}$. Use the Poincaré inequality and an inverse inequality of the form (3.68) to get

$$\sum_{\boldsymbol{a}\in\mathcal{N}_{h,0}} h_{\boldsymbol{a}}^2 \min_{r_{\boldsymbol{a}}\in\mathbb{R}} \|r - r_{\boldsymbol{a}}\|_{0;\widetilde{K}_{\boldsymbol{a}}}^2 \le C \sum_{\boldsymbol{a}\in\mathcal{N}_{h,0}} h_{\boldsymbol{a}}^4 \|\nabla u_h\|_{0;\widetilde{K}_{\boldsymbol{a}}}^2 + 2 \sum_{\boldsymbol{a}\in\mathcal{N}_{h,0}} h_{\boldsymbol{a}}^2 \min_{f_{\boldsymbol{a}}\in\mathbb{R}} \|f - f_{\boldsymbol{a}}\|_{0;\widetilde{K}_{\boldsymbol{a}}}^2.$$

Therefore,

$$I\!I \le C|v|_{1;\Omega} \sum_{\boldsymbol{a}\in\mathcal{N}_{h,0}} \left(h_{\boldsymbol{a}}^4 \|\nabla u_h\|_{0;\widetilde{K}_{\boldsymbol{a}}}^2 + h_{\boldsymbol{a}}^2 \min_{f_{\boldsymbol{a}}\in\mathbb{R}} \|f - f_{\boldsymbol{a}}\|_{0;\widetilde{K}_{\boldsymbol{a}}}^2 \right). \tag{3.70}$$

Finally, with the aid of the Cauchy–Schwarz inequality and the estimate (3.15), the third summand on the right side of (3.67) can be bounded by

$$I\!I\!I \le C\,|v|_{1;\Omega} \left(\sum_{\gamma\in\mathcal{E}_{h,\Gamma_2}} h_\gamma \|G_h u_h \cdot \boldsymbol{\nu}_\gamma + g\,\lambda_h\|_{0;\gamma}^2 \right)^{1/2}. \tag{3.71}$$

Inserting (3.69) through (3.71) into (3.67), and using the Cauchy–Schwarz inequality and (3.65), we deduce that

$$\mathcal{R} \le C \left\{ \sum_{K\in\mathcal{P}_h} \|\nabla u_h - G_h u_h\|_{0;K}^2 + \sum_{\gamma\in\mathcal{E}_{h,\Gamma_2}} h_\gamma \|G_h u_h \cdot \boldsymbol{\nu}_\gamma + g\,\lambda_h\|_{0;\gamma}^2 + \sum_{\boldsymbol{a}\in\mathcal{N}_{h,0}} \left(h_{\boldsymbol{a}}^4 \|\nabla u_h\|_{0;\widetilde{K}_{\boldsymbol{a}}}^2 + h_{\boldsymbol{a}}^2 \min_{f_{\boldsymbol{a}}\in\mathbb{R}} \|f - f_{\boldsymbol{a}}\|_{0;\widetilde{K}_{\boldsymbol{a}}}^2 \right) \right\}^{1/2}. \tag{3.72}$$

Split the first term on the right side of estimate (3.63) into local contributions from each $K \in \mathcal{P}_h$ and insert (3.72) to conclude the proof. □

We can write (3.60) as

$$\|u - u_h\|_{1;\Omega} \le C\eta_G + R_h, \tag{3.73}$$

where

$$R_h = C\left[\sum_{\boldsymbol{a}\in\mathcal{N}_{h,0}} h_{\boldsymbol{a}}^4 \|\nabla u_h\|_{0;\widetilde{K}_{\boldsymbol{a}}}^2\right]^{1/2} + C\left[\sum_{\boldsymbol{a}\in\mathcal{N}_{h,0}} h_{\boldsymbol{a}}^2 \min_{f_{\boldsymbol{a}}\in\mathbb{R}} \|f - f_{\boldsymbol{a}}\|_{0;\widetilde{K}_{\boldsymbol{a}}}^2\right]^{1/2}.$$

We observe that usually the term R_h is of higher order compared to $\|u-u_h\|_{1;\Omega}$ which is of order $O(h)$ in the nondegenerate situations. This observation is argued as follows. First, it is easy to show from the definition of the finite element solutions that there is a constant C such that $\|u_h\|_{1;\Omega} \le C$ for any h. So the term

$$\left[\sum_{\boldsymbol{a}\in\mathcal{N}_{h,0}} h_{\boldsymbol{a}}^4 \|\nabla u_h\|_{0;\widetilde{K}_{\boldsymbol{a}}}^2\right]^{1/2}$$

is bounded by $O(h^2)$. Next, for $f \in L^2(\Omega)$,

$$\left(\sum_{\boldsymbol{a}\in\mathcal{N}_{h,0}} h_{\boldsymbol{a}}^2 \min_{f_{\boldsymbol{a}}\in\mathbb{R}} \|f - f_{\boldsymbol{a}}\|_{0;\widetilde{K}_{\boldsymbol{a}}}^2\right)^{1/2} = o(h);$$

and if $f \in H^1(\Omega)$, then

$$\left(\sum_{\boldsymbol{a}\in\mathcal{N}_{h,0}} h_{\boldsymbol{a}}^2 \min_{f_{\boldsymbol{a}}\in\mathbb{R}} \|f - f_{\boldsymbol{a}}\|_{0;\widetilde{K}_{\boldsymbol{a}}}^2\right)^{1/2} = O(h^2).$$

Thus, (3.73) illustrates the reliability of the error estimator η_G.

We now turn to the efficiency of the estimator. We relate the gradient recovery-based estimator η_G to the residual type estimator η_R. Recall that for the residual-type estimator $\eta_{R,K}$ is defined in (3.43) with the interior residual $r_K = -\Delta u_h + u_h - f$ in K and the side residual

$$R_\gamma = \begin{cases} [\frac{\partial u_h}{\partial \nu}]_\gamma & \text{if } \gamma \in \mathcal{E}_{h,0}, \\ \frac{\partial u_h}{\partial \nu} + g\,\lambda_h & \text{if } \gamma \in \mathcal{E}_{h,\Gamma_2}. \end{cases} \tag{3.74}$$

For the error estimator $\eta_R^2 = \sum_{K\in\mathcal{P}_h} \eta_{R,K}^2$, we have the inequality (3.57) for its efficiency.

Lemma 3.1. *Let $\eta_{G,K}$ be defined in (3.62). Then the following bound holds:*

$$\eta_{G,K}^2 \le C\left(\sum_{\gamma\in\mathcal{E}(K)\cap\mathcal{E}_{h,\Gamma_2}} h_\gamma \|R_\gamma\|_{0;\gamma}^2 + \sum_{\gamma'\in\mathcal{E}_{\widetilde{K}}} h_{\gamma'} \|R_{\gamma'}\|_{0;\gamma'}^2\right), \tag{3.75}$$

where $\mathcal{E}_{\widetilde{K}}$ denotes the set of inner sides of the patch $\widetilde{K}$ corresponding to the element K.

Proof. It follows from the definition of G_h that we have on each element K,

$$|\nabla u_h - G_h u_h|^2 = \left| \sum_{\boldsymbol{a}\in\mathcal{N}(K)} \varphi_{\boldsymbol{a}} \Big(\sum_{i=1}^{N_{\boldsymbol{a}}} \alpha_{\boldsymbol{a}}^i ((\nabla u_h)_K - (\nabla u_h)_{K_{\boldsymbol{a}}^i}) \Big) \right|^2$$
$$\le C \sum_{K'\subset\widetilde{K}} |(\nabla u_h)_K - (\nabla u_h)_{K'}|^2. \quad (3.76)$$

For any $K' \subset \widetilde{K}$, there is a sequence of inner edges $\gamma_1, \ldots, \gamma_m$, such that $\widetilde{\gamma_j} \cap \widetilde{\gamma_{j+1}} \ne \varnothing$ and $K \subset \widetilde{\gamma_1}$, $K' \subset \widetilde{\gamma_m}$. Hence,

$$|(\nabla u_h)_K - (\nabla u_h)_{K'}| \le \sum_{j=1}^{m} |[\nabla u_h]_{\gamma_j}| \le \sum_{\gamma\in\mathcal{E}_{\widetilde{K}}} |[\nabla u_h]_\gamma|. \quad (3.77)$$

Because u_h is continuous on Ω, $[\partial u_h/\partial t]_\gamma = 0$ for all $\gamma \in \mathcal{E}_{h,0}$, where $\partial u_h/\partial t$ is the tangential derivative of u_h. Therefore, $|[\nabla u_h]_\gamma| = |[\partial u_h/\partial \nu]_\gamma|$ if $\gamma \in \mathcal{E}_{h,0}$. The estimates (3.76) and (3.77) together with the shape regularity of the partition $\mathcal{P}_h$ imply

$$\|\nabla u_h - G_h u_h\|_{0;K}^2 \le C h_K^2 \sum_{\gamma\in\mathcal{E}_{\widetilde{K}}} \left[\frac{\partial u_h}{\partial \nu}\right]_\gamma^2 \le C \sum_{\gamma\in\mathcal{E}_{\widetilde{K}}} h_\gamma \int_\gamma \left[\frac{\partial u_h}{\partial \nu}\right]_\gamma^2 ds. \quad (3.78)$$

Let $K \in \mathcal{P}_h$ be such that $\mathcal{E}(K) \cap \mathcal{E}_{h,\Gamma_2} = \varnothing$. It follows from (3.78) and definitions of $\eta_{G,K}$ and R_γ that

$$\eta_{G,K}^2 \le C \sum_{\gamma\in\mathcal{E}_{\widetilde{K}}} h_\gamma \|R_\gamma\|_{0;\gamma}^2. \quad (3.79)$$

Consider now the case when the element K has at least one side lying on the boundary Γ_2. Let $\gamma \in \mathcal{E}(K) \cap \mathcal{E}_{h,\Gamma_2}$. Apply the triangle inequality to get

$$\begin{aligned}
\eta_{G,K}^2 &= \|\nabla u_h - G_h u_h\|_{0;K}^2 + \sum_{\gamma\in\mathcal{E}(K)\cap\mathcal{E}_{h,\Gamma_2}} h_\gamma \|G_h u_h \cdot \boldsymbol{\nu}_\gamma + g\lambda_h\|_{0;\gamma}^2 \\
&\le \|\nabla u_h - G_h u_h\|_{0;K}^2 \\
&\quad + \sum_{\gamma\in\mathcal{E}(K)\cap\mathcal{E}_{h,\Gamma_2}} h_\gamma \left(\|(\nabla u_h - G_h u_h)\cdot\boldsymbol{\nu}_\gamma\|_{0;\gamma} + \|R_\gamma\|_{0;\gamma}\right)^2 \\
&\le \|\nabla u_h - G_h u_h\|_{0;K}^2 \\
&\quad + \sum_{\gamma\in\mathcal{E}(K)\cap\mathcal{E}_{h,\Gamma_2}} 2h_\gamma \left(\|(\nabla u_h - G_h u_h)\cdot\boldsymbol{\nu}_\gamma\|_{0;\gamma}^2 + \|R_\gamma\|_{0;\gamma}^2\right). \quad (3.80)
\end{aligned}$$

From an inverse inequality and (3.78), we have

$$\begin{aligned}
h_\gamma \|(\nabla u_h - G_h u_h)\cdot\boldsymbol{\nu}_\gamma\|_{0;\gamma}^2 &\le h_K \|\nabla u_h - G_h u_h\|_{0;\partial K}^2 \\
&\le C\|\nabla u_h - G_h u_h\|_{0;K}^2 \le C \sum_{\gamma'\in\mathcal{E}_{\widetilde{K}}} h_{\gamma'} \|R_{\gamma'}\|_{0;\gamma'}^2. \quad (3.81)
\end{aligned}$$

Inserting (3.78) and (3.81) into (3.80) concludes the proof. □

From Lemma 3.1 and the inequality (3.57) we obtain

$$\eta_G^2 \le C \Bigg(\|u - u_h\|_V^2 + \sum_{\gamma \in \mathcal{E}_{h,\Gamma_2}} h_\gamma \|\lambda - \lambda_h\|_{0;\gamma}^2$$

$$+ \sum_{K \in \mathcal{P}_h} h_K^2 \|f - f_K\|_{0;K}^2 + \sum_{\gamma \in \mathcal{E}_{h,\Gamma_2}} h_\gamma \|\lambda_h - \lambda_{h,\gamma}\|_{0;\gamma}^2 \Bigg) \tag{3.82}$$

with discontinuous piecewise polynomial approximations f_K and $\lambda_{h,\gamma}$. The comments at the end of Section 3.4 apply to the three summation terms in (3.82).

3.6 Numerical Example on the Model Elliptic Variational Inequality

In this section, we provide some numerical results on a two-dimensional elliptic variational inequality to illustrate the effectiveness of the error estimators η_R of (3.42)–(3.43) and η_G of (3.61)–(3.62). We use triangular partitioning and linear elements for discretization, and a seven-point Gauss–Legendre quadrature to compute the load vector on each triangle. Numerical integration over a general triangle is done by the reference element technique. On the reference element

$$\hat{K} = \{(\xi, \eta) : \xi \ge 0,\ \eta \ge 0,\ 1 - \xi - \eta \ge 0\},$$

the seven-point Gauss–Legendre quadrature formula is defined by

$$\int_{\hat{K}} F(\xi, \eta)\, d\xi d\eta \approx \sum_{i=1}^{7} \omega_i F(\xi_i, \eta_i),$$

where the nodes $\{(\xi_i, \eta_i)\}_{i=1}^7$ and weights $\{\omega_i\}_{i=1}^7$ are given in Table 3.1. The discretized solution is computed by solving the equivalent minimization problem using an overrelaxation method with a relative error tolerance, in the maximum norm, of 10^{-6} (see [34, 35]).

In order to show the effectiveness of the adaptive procedure we compare numerical convergence orders of the approximate solutions. We compute these orders by considering families of uniform and adaptively refined partitions. Consider a sequence of finite element solutions u_h^{un} based on uniform partitions of the domain Ω. Starting with an initial coarse partition $\mathcal{P}_1$, we construct a family of nested meshes by subdividing each element into four congruent elements for the two-dimensional case. The solution from the most

Table 3.1 Nodes and weights of a quadrature over reference triangle

i	ξ_i	η_i	ω_i
1	$1/3$	$1/3$	$9/80$
2	$(6+\sqrt{15})/21$	$(6+\sqrt{15})/21$	$(155+\sqrt{15})/2400$
3	$(9-2\sqrt{15})/21$	$(6+\sqrt{15})/21$	$(155+\sqrt{15})/2400$
4	$(6+\sqrt{15})/21$	$(9-2\sqrt{15})/21$	$(155+\sqrt{15})/2400$
5	$(6-\sqrt{15})/21$	$(6-\sqrt{15})/21$	$(155-\sqrt{15})/2400$
6	$(9+2\sqrt{15})/21$	$(6-\sqrt{15})/21$	$(155-\sqrt{15})/2400$
7	$(6-\sqrt{15})/21$	$(9+2\sqrt{15})/21$	$(155-\sqrt{15})/2400$

refined mesh is taken as the "true" solution u used to compute the errors of the approximate solutions obtained on the other meshes.

Adaptive finite element solutions are obtained by the following algorithm.

1. Start with the initial partition $\mathcal{P}_h$ and corresponding finite element subspace V_h.
2. Compute the finite element solution $u_h^{ad} \in V_h$.
3. For each element $K \in \mathcal{P}_h$ compute the error estimator η_K, defined in (3.43) for the residual type and (3.62) for the gradient recovery type.
4. Let $\eta = \frac{1}{N}\sum_{K\in\mathcal{P}_h}\eta_K$ with N being the number of elements in partition $\mathcal{P}_h$. An element K is marked for refinement if $\eta_K > \mu\,\eta$, where μ is a prescribed threshold. In the example of this section, $\mu = 0.5$.
5. Perform refinement and obtain a new triangulation $\mathcal{P}_h$.
6. Return to step 2.

In the computation of the error indicator η_K we make use of the multiplier λ_h defined on $\Gamma_2 \subset \Gamma$. In what follows we describe how λ_h can be (approximately) recovered from the solution u_h using characterization (3.17). We compute the piecewise constant and the piecewise linear approximations to the Lagrange multiplier.

Denote by $\{\boldsymbol{a}^i\}_{i=1}^m$ the nodes of the partition $\mathcal{P}_h$ belonging to $\overline{\Gamma_2}$. Let $\{\varphi_i\}_{i=1}^m$ be the basis functions corresponding to the nodes $\{\boldsymbol{a}^i\}$. Let $\{\chi_i\}_{i=1}^m$ be the characteristic functions of the intervals $\{K_i\}$ belonging to $\overline{\Gamma_2}$ and defined as follows: K_i is the intersection of Γ_2 and the segment joining the midpoints of edges sharing $\boldsymbol{a}^i$ as a common point. We first determine a piecewise constant function

$$\lambda_{h,1}^{(0)} = \sum_{i=1}^{m} \lambda_{h,1}^{0,i}\chi_i$$

or a piecewise linear function

$$\lambda_{h,1}^{(1)} = \sum_{i=1}^{m} \lambda_{h,1}^{1,i}\varphi_i$$

by requiring an analogue of (3.17):

$$a(u_h, v_h) + \int_{\Gamma_2} g\,\lambda_{h,1}^{(k)} v_h ds = \ell(v_h) \qquad \forall\, v_h \in V_h \tag{3.83}$$

with $k = 0$ (piecewise constant) or $k = 1$ (piecewise linear). Denote $\boldsymbol{\lambda}_{h,1}^{(k)} = (\lambda_{h,1}^{k,1}, \ldots, \lambda_{h,1}^{k,m})^T$, $k = 0, 1$. We then project the components of $\boldsymbol{\lambda}_{h,1}^{(k)}$ onto the interval $[-1, 1]$ to get $\boldsymbol{\lambda}_{h,2}^{(k)} = (\lambda_{h,2}^{k,1}, \ldots, \lambda_{h,2}^{k,m})^T$:

$$\lambda_{h,2}^{k,i} = \max\{\min\{\lambda_{h,1}^{k,i}, 1\}, -1\}, \qquad i = 1, \ldots, m.$$

The piecewise constant approximation $\lambda_{h,2}^{(0)}$ and the piecewise linear approximation $\lambda_{h,2}^{(1)}$ of the multiplier λ_h on $\Gamma_2 \subset \Gamma$ can be computed as

$$\lambda_{h,2}^{(0)} = \sum_{i=1}^{m} \lambda_{h,2}^{0,i} \chi_i \qquad \text{and} \qquad \lambda_{h,2}^{(1)} = \sum_{i=1}^{m} \lambda_{h,2}^{1,i} \varphi_i. \tag{3.84}$$

We briefly comment on the method for finding $\lambda_{h,1}^{(k)}$, $k = 0, 1$. Let $n = \dim V_h$. Denote by $\boldsymbol{K}$ the standard $(n \times n)$ stiffness matrix and by $\boldsymbol{l} \in \mathbb{R}^n$ the standard load vector. Let $\boldsymbol{u} \in \mathbb{R}^n$ be the nodal value vector of the finite element solution u_h. Then the algebraic representation of (3.83) becomes

$$(\boldsymbol{K}\boldsymbol{u}, \boldsymbol{v})_{\mathbb{R}^n} + (g\boldsymbol{M}\boldsymbol{\lambda}_{h,1}^{(k)}, \boldsymbol{v}_c)_{\mathbb{R}^m} = (\boldsymbol{l}, \boldsymbol{v})_{\mathbb{R}^n} \qquad \forall\, \boldsymbol{v} \in \mathbb{R}^n, \tag{3.85}$$

where $\boldsymbol{v}_c$ denotes the subvector of $\boldsymbol{v}$, containing the nodal values of v_h at the nodes $\{\boldsymbol{a}^i\}_{i=1}^m \subset \overline{\Gamma_2}$ and $\boldsymbol{M}$ is a sparse $(m \times m)$ matrix. We can write $\boldsymbol{v} = (\boldsymbol{v}_i^T, \boldsymbol{v}_c^T)^T \in \mathbb{R}^{n-m} \times \mathbb{R}^m$ by assuming that the components of $\boldsymbol{v}_c$ are listed last. We similarly split $\boldsymbol{l}$ to $\boldsymbol{l}_i$ and $\boldsymbol{l}_c$. This decomposition yields a block structure for $\boldsymbol{K}$,

$$\boldsymbol{K} = \begin{pmatrix} \boldsymbol{K}_{ii} & \boldsymbol{K}_{ic} \\ \boldsymbol{K}_{ci} & \boldsymbol{K}_{cc} \end{pmatrix}.$$

Then (3.85) is equivalent to the following two relations:

$$\begin{aligned} \boldsymbol{K}_{ii}\boldsymbol{u}_i + \boldsymbol{K}_{ic}\boldsymbol{u}_c &= \boldsymbol{l}_i, \\ \boldsymbol{K}_{ci}\boldsymbol{u}_i + \boldsymbol{K}_{cc}\boldsymbol{u}_c + g\boldsymbol{M}\boldsymbol{\lambda}_{h,1}^{(k)} &= \boldsymbol{l}_c. \end{aligned}$$

Once the approximate solution u_h is computed, we can obtain from the second relation that

$$\boldsymbol{\lambda}_{h,1}^{(k)} = g^{-1}\boldsymbol{M}^{-1}(\boldsymbol{l}_c - \boldsymbol{K}_{ci}\boldsymbol{u}_i - \boldsymbol{K}_{cc}\boldsymbol{u}_c), \qquad k = 0, 1.$$

We use u_h^{un} for finite element solutions on uniform meshes, and u_h^{ad} for finite element solutions on adaptive meshes. We find that uses of piecewise linears and piecewise constants for the Lagrange multiplier lead to negligible differences in the adaptive meshes. For instance, in Example 3.1, we get identical adaptive meshes for the first three adaptive steps, and there is a difference of only one or two nodes for the fourth and fifth adaptive steps. Thus, in all the numerical examples in this chapter, u_h^{ad} refers to the adaptive finite element solution where piecewise linear functions are used for approximating the Lagrange multiplier in generating the adaptive mesh. Because adaptive solutions are involved, numerical solution errors are plotted against the number of degrees of freedom, rather than the meshsize.

Example 3.1. Let $\Omega = [0,1] \times [0,1]$ and $\Gamma_2 = \Gamma$. We solve the problem of finding $u \in H^1(\Omega)$ such that $\forall\, v \in H^1(\Omega)$,

$$\int_\Omega [\nabla u \cdot \nabla(v-u) + u\,(v-u)]\,dx + g\int_\Gamma |v|\,ds - g\int_\Gamma |u|\,ds \geq \int_\Omega f\,(v-u)\,dx$$

where

$$f = -\Delta w + w,$$
$$w = w_1 - w_2,$$

and for $i = 1, 2$,

$$w_i(\boldsymbol{x}) = \begin{cases} \exp(1/(r_i^2-1)) & \text{if } r_i < 1, \\ 0 & \text{otherwise} \end{cases}$$

with $r_i = [(x_1 - x_{1,0}^{(i)})^2 + (x_2 - x_{2,0}^{(i)})^2]^{1/2}/\varepsilon_i$. For the numerical results reported below, we let $g = 1$, $x_{1,0}^{(1)} = 0.8$, $x_{2,0}^{(1)} = 0.1$, $x_{1,0}^{(2)} = 0.3$, $x_{2,0}^{(2)} = 0.1$, $\varepsilon_1 = 0.25$, and $\varepsilon_2 = 0.2$.

We start with a coarse uniform triangulation shown on the left plot in Figure 3.2. Here, the interval $[0,1]$ is divided into $1/h$ equal parts with $h = 1/4$ which is successively halved. The numerical solution corresponding to $h = 1/256$ (66,049 nodes) is taken as the "true" solution u, shown in Figure 3.1.

We use the regular refinement technique (*red-blue-green* refinement), in which the triangle is divided into four triangles by joining the midpoints of edges and adjacent triangles are refined in order to avoid hanging nodes. For a detailed description of this and other refinement techniques currently used see, for example, [70] and references therein. Also, in order to improve the quality of triangulation, a smoothing procedure is used after each refinement. For each triangle K of the triangulation we compute the triangle quality measure defined by

$$Q(K) = \frac{4\sqrt{3}\,\text{area}(K)}{h_1^2 + h_2^2 + h_3^2},$$

where h_i, $i = 1, 2, 3$, are the side lengths of the triangle K. Note that $Q(K) = 1$ if $h_1 = h_2 = h_3$. A triangle is viewed to be of acceptable quality if $Q > 0.6$, otherwise we modify the mesh by moving the interior nodes toward the center

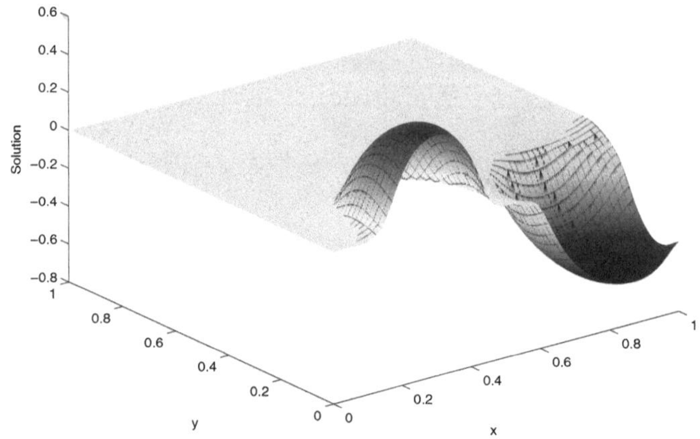

Fig. 3.1 "True" solution.

of mass of the polygon formed by the adjacent triangles. The adaptively refined triangulation after five iterations is shown on the right plot of Figure 3.2.

To have an idea of the convergence behaviour of the discrete Lagrange multipliers, we analyze the errors $\|\lambda^{(j)} - \lambda_h^{(k)}\|_{0;\Gamma}$, $j, k = 0, 1$ corresponding to the sequence of uniform refinements. Here, $\lambda^{(0)}$ and $\lambda^{(1)}$ are the piecewise constant and the piecewise linear approximations to the Lagrange multiplier corresponding to the parameter $h = 1/256$. Graphs of $\lambda_{h,1}^{(1)}$ and $\lambda_{h,2}^{(1)}$ with

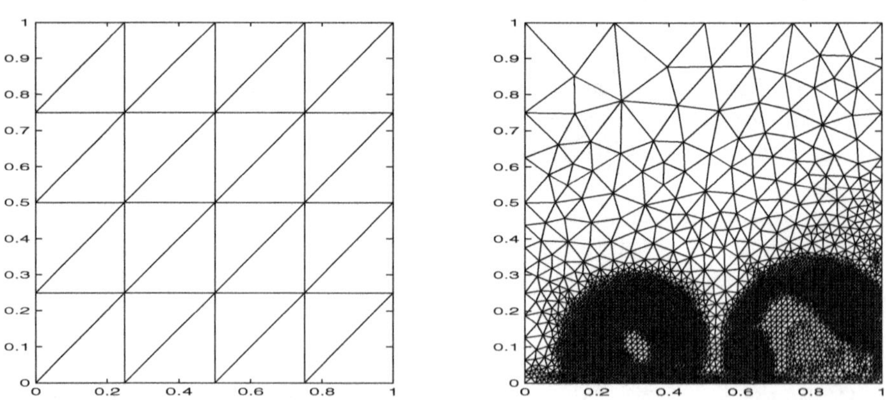

Fig. 3.2 Initial and adaptively refined partitions.

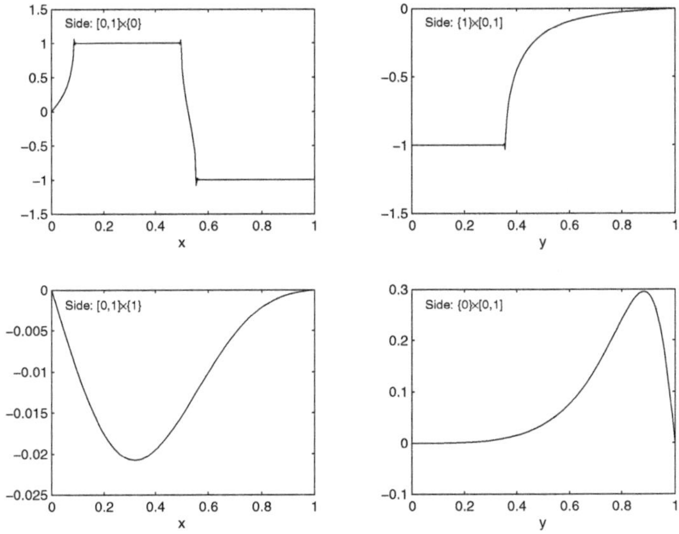

Fig. 3.3 Plots of $\lambda_{h,1}^{(1)}$.

$h = 1/256$ are provided in Figures 3.3 and 3.4. Note the peaks of $\lambda_{h,1}^{(1)}$ around $x_1 = 0.1$ and $x_1 = 0.5$ (upper left plot), and $x_2 = 0.3$ (upper right plot), which are eliminated by the projection of $\lambda_{h,1}^{(1)}$ onto the interval $[-1, 1]$. Figures 3.5

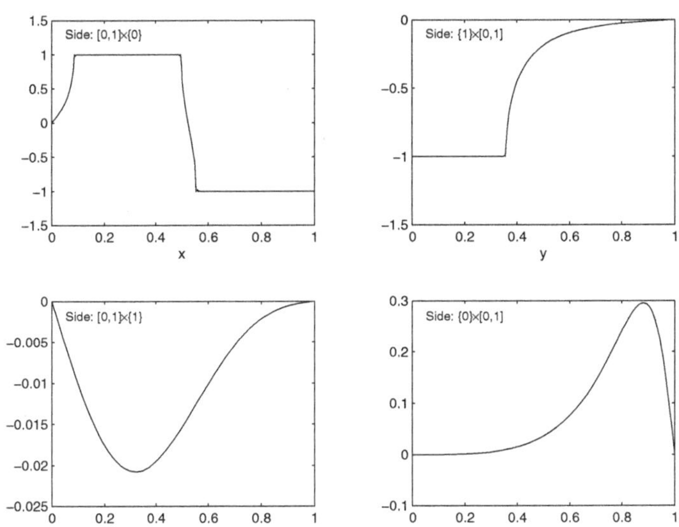

Fig. 3.4 Plots of $\lambda_{h,2}^{(1)}$.

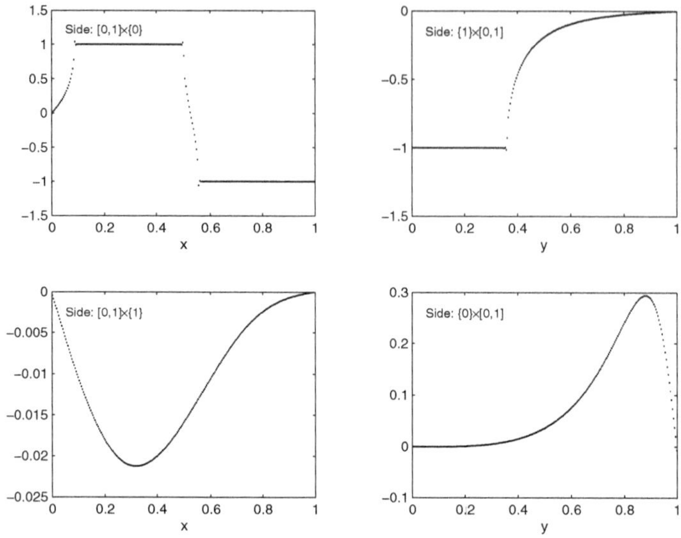

Fig. 3.5 Plots of $\lambda_{h,1}^{(0)}$.

and 3.6 contain the graphs of $\lambda_{h,1}^{(0)}$ and $\lambda_{h,2}^{(0)}$ corresponding to $h = 1/256$.

Figure 3.7 provides the error values $\|u - u_h^{un}\|_{1;\Omega}$ and $h^{1/2}\|\lambda^{(j)} - \lambda_h^{(k)}\|_{0;\Gamma}$, $j, k = 0, 1$. The numerical convergence orders of $h^{1/2}\|\lambda^{(j)} - \lambda_h^{(k)}\|_{0;\Gamma}$ are

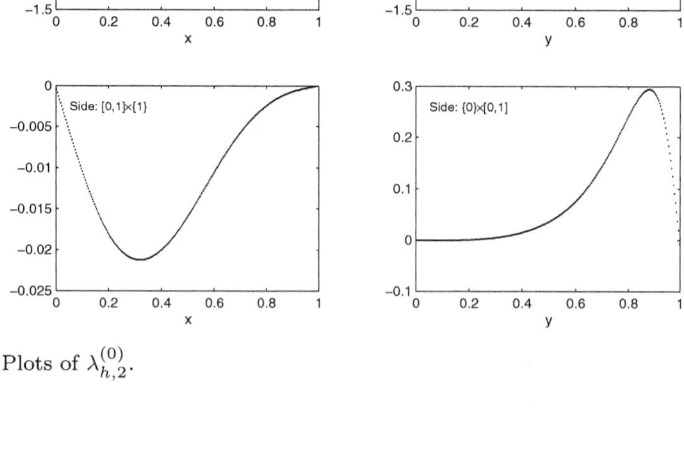

Fig. 3.6 Plots of $\lambda_{h,2}^{(0)}$.

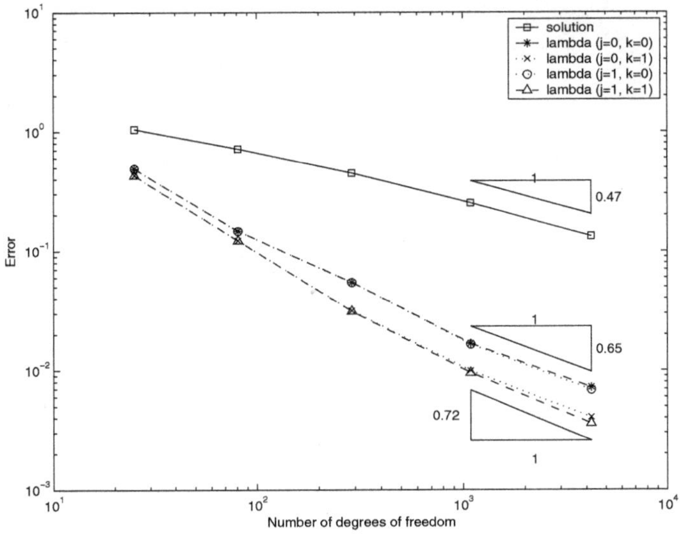

Fig. 3.7 Errors $\|u - u_h^{un}\|_{1;\Omega}$ ($\square$) versus $h^{1/2}\|\lambda^{(j)} - \lambda_h^{(k)}\|_{0;\Gamma}$ $(j,k = 0,1)$.

obviously higher than that of $\|u - u_h^{un}\|_{1;\Omega}$, indicating that the second term within the parentheses in the efficiency bounds (3.57) and (3.82) is expected to be of higher order compared to the first term $\|e_h\|^2_{1;\Omega}$.

We use an adaptive procedure based on both residual type and gradient recovery type estimates to obtain a sequence of approximate solutions u_h^{ad}. The adaptive finite element mesh after five adaptive iterations is shown on the right plot in Figure 3.2. Figures 3.8 and 3.9 contain the error values $\|u - u_h^{un}\|_{1;\Omega}$ and $\|u - u_h^{ad}\|_{1;\Omega}$. We observe a substantial improvement of the efficiency using adaptively refined meshes. Figures 3.10 and 3.11 provide the values of $\eta_I = (\sum_K \eta_{I,K}^2)^{1/2}$, $I \in \{R, G\}$, where $\eta_{I,K}$ are computed using either residual type estimator ($I = R$) or gradient recovery type estimator ($I = G$) on both uniform and adapted meshes. Table 3.2 contains the values of C_I computed for uniform and adaptive solutions:

$$C_I = \frac{\|u - u_h\|_{1;\Omega}}{\eta_I}, \qquad I \in \{R, G\}.$$

Table 3.2 Numerical values of C_R and C_G

h	1/4	1/8	1/16	1/32	1/64	1/128
C_R^{un}	1.01e-01	1.36e-01	1.60e-01	1.66e-01	1.64e-01	1.61e-01
C_R^{ad}	1.01e-01	1.36e-01	1.60e-01	1.67e-01	1.64e-01	1.65e-01
C_G^{un}	1.52e+00	1.15e+00	1.05e+00	9.50e-01	9.24e-01	9.09e-01
C_G^{ad}	1.52e+00	1.05e+00	1.04e+00	9.69e-01	8.30e-01	8.33e-01

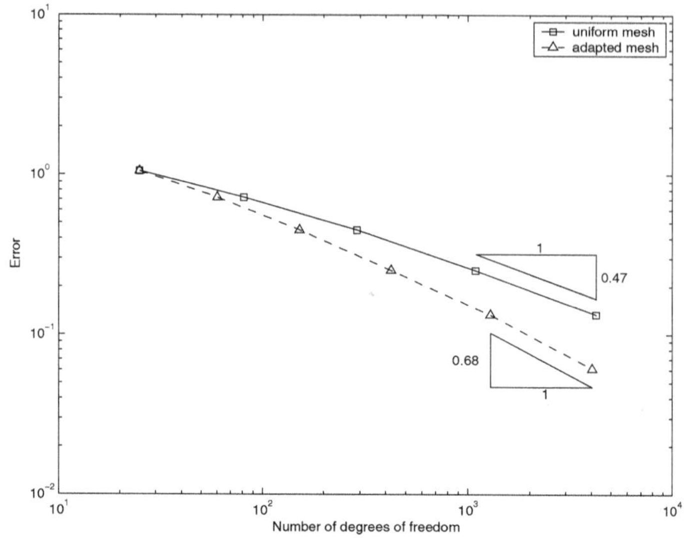

Fig. 3.8 Results based on residual type estimator.

It is seen from Table 3.2 that for this numerical example, the gradient recovery type error estimator provides a better prediction of the true error than the residual type error estimator, a phenomenon observed in numerous

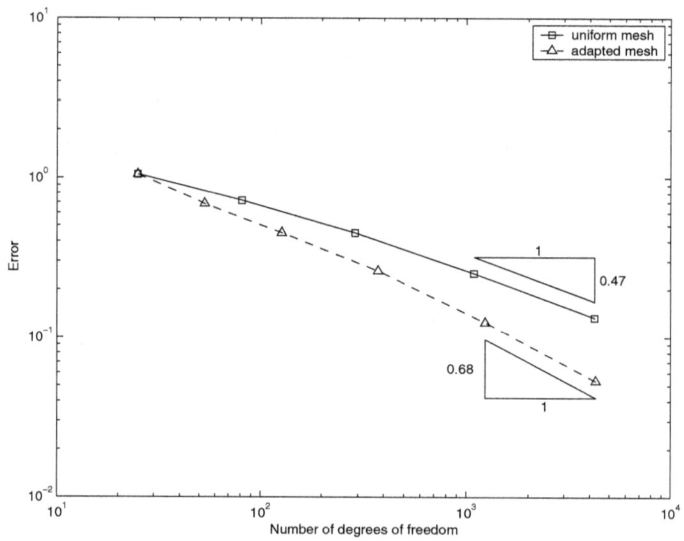

Fig. 3.9 Results based on recovery type estimator.

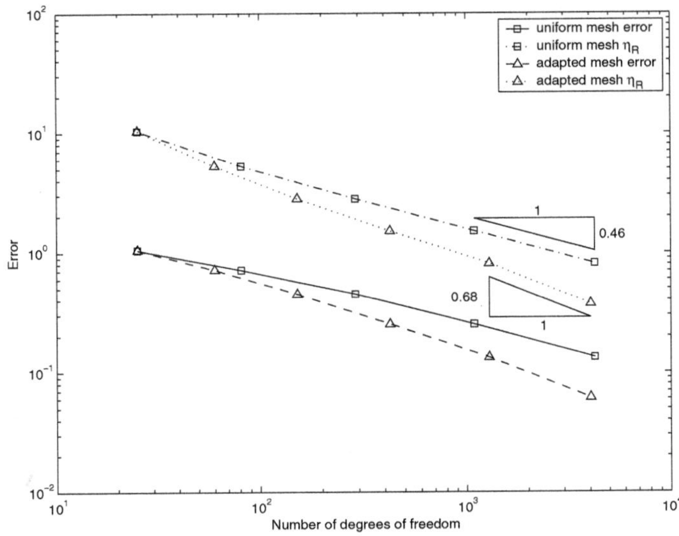

Fig. 3.10 Results based on residual type estimator.

references. In general, we use the a posteriori error estimates only for the purpose of designing adaptive meshes, due to the presence of the unknown constants C_R and C_G.

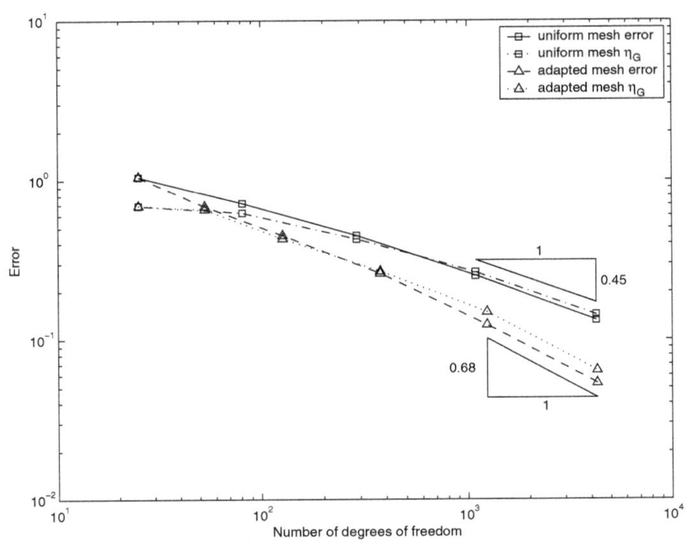

Fig. 3.11 Results based on recovery type estimator.

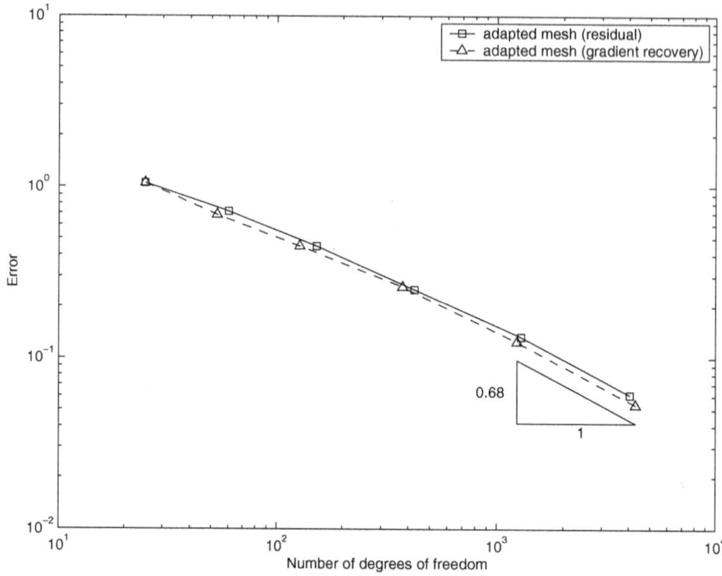

Fig. 3.12 Performance comparison of the two error estimators.

For comparison of the performance between the two error estimators, we show in Figure 3.12 the errors of the adaptive solutions corresponding to the two error estimators. We observe that the two error estimators lead to very similar solution accuracy for same amount of degrees of freedom. □

More numerical results can be found in [14].

3.7 Application to a Frictional Contact Problem

In this section, we take a frictional contact problem as an example to show that similar a posteriori error estimates can be derived for more complicated variational inequalities. Again, we denote by $\Omega \subset \mathbb{R}^d$ ($d \le 3$ in applications) a Lipschitz domain with boundary Γ. The outward unit normal exists a.e. on Γ and is denoted by $\boldsymbol{\nu}$. We use $\mathbb{S}^d$ for the space of second-order symmetric tensors on $\mathbb{R}^d$. The canonical inner products and corresponding norms on $\mathbb{R}^d$ and $\mathbb{S}^d$ are

$$\begin{aligned} \boldsymbol{u} \cdot \boldsymbol{v} &= u_i v_i, & |\boldsymbol{v}| &= (\boldsymbol{v} \cdot \boldsymbol{v})^{1/2} & \forall\, \boldsymbol{u},\ \boldsymbol{v} \in \mathbb{R}^d, \\ \boldsymbol{\sigma} : \boldsymbol{\varepsilon} &= \sigma_{ij} \varepsilon_{ij}, & |\boldsymbol{\sigma}| &= (\boldsymbol{\sigma} : \boldsymbol{\sigma})^{1/2} & \forall\, \boldsymbol{\sigma},\ \boldsymbol{\varepsilon} \in \mathbb{S}^d. \end{aligned}$$

Here, the indices i and j run between 1 and d, and the summation convention over repeated indices is used. We define the product spaces $\boldsymbol{L}^2(\omega) := (L^2(\omega))^d$

and $\boldsymbol{H}^1(\omega) := (H^1(\omega))^d$ equipped with the norms $\|\boldsymbol{v}\|_{k;\omega}^2 := \sum_{i=1}^d \|v_i\|_{k;\omega}^2$, $k = 0, 1$. When no ambiguity occurs, we use the same notation $\boldsymbol{v}$ to denote the function and its trace on the boundary. For a vector $\boldsymbol{v}$, we use its normal component $v_\nu = \boldsymbol{v}\cdot\boldsymbol{\nu}$ and tangential component $\boldsymbol{v}_\tau = \boldsymbol{v} - v_\nu \boldsymbol{\nu}$ at a point on the boundary. Similarly for a tensor $\boldsymbol{\sigma} \in \mathbb{S}^d$, we define its normal component $\sigma_\nu = \boldsymbol{\sigma}\boldsymbol{\nu}\cdot\boldsymbol{\nu}$ and tangential component $\boldsymbol{\sigma}_\tau = \boldsymbol{\sigma}\boldsymbol{\nu} - \sigma_\nu\boldsymbol{\nu}$. For a detailed treatment of traces for vector and tensor fields in contact problems and related spaces see [53] or [46].

The material occupying Ω is assumed linearly elastic. We denote by $\mathbb{C} : \Omega \times \mathbb{S}^d \to \mathbb{S}^d$ the elasticity tensor of the material. The fourth-order tensor $\mathbb{C}$ is assumed to be bounded, symmetric, and positive definite in Ω.

We briefly describe the physical setting of the frictional contact problem. Details and other related problems can be found in [53, 46]. Consider an elastic body occupying a bounded domain Ω in $\mathbb{R}^d$, $d \le 3$, with a Lipschitz boundary. The boundary Γ is partitioned as follows: $\Gamma = \overline{\Gamma_D} \cup \overline{\Gamma_N} \cup \overline{\Gamma_C}$ with Γ_D, Γ_N, and Γ_C relatively open and mutually disjoint, and $\operatorname{meas}(\Gamma_D) > 0$. The subscripts "$D$", "$N$", and "$C$" are intended as shorthand indications for Dirichlet, Neumann, and contact boundary conditions. We assume that the body is clamped on Γ_D; on the boundary part Γ_N surface tractions of density $\boldsymbol{f}_2 \in (L^2(\Gamma_N))^d$ are applied and on Γ_C the body is in bilateral contact with a rigid foundation. The contact is frictional and is modeled by Tresca's law. Volume forces of density $\boldsymbol{f}_1 \in (L^2(\Omega))^d$ act in Ω.

In classical formulation, the problem is to find a displacement field $\boldsymbol{u} : \Omega \to \mathbb{R}^d$ and a stress field $\boldsymbol{\sigma} : \Omega \to \mathbb{S}^d$ such that

$$\boldsymbol{\sigma} = \mathbb{C}\boldsymbol{\varepsilon}(\boldsymbol{u}) \quad \text{in } \Omega, \tag{3.86}$$
$$\boldsymbol{\varepsilon}(\boldsymbol{u}) = \tfrac{1}{2}(\nabla\boldsymbol{u} + (\nabla\boldsymbol{u})^T) \quad \text{in } \Omega, \tag{3.87}$$
$$\operatorname{Div}\boldsymbol{\sigma} + \boldsymbol{f}_1 = \boldsymbol{0} \quad \text{in } \Omega, \tag{3.88}$$
$$\boldsymbol{u} = \boldsymbol{0} \quad \text{on } \Gamma_D, \tag{3.89}$$
$$\boldsymbol{\sigma}\boldsymbol{\nu} = \boldsymbol{f}_2 \quad \text{on } \Gamma_N, \tag{3.90}$$
$$u_\nu = 0 \quad \text{on } \Gamma_C, \tag{3.91}$$
$$|\boldsymbol{\sigma}_\tau| \le g \quad \text{on } \Gamma_C, \tag{3.92}$$
$$|\boldsymbol{\sigma}_\tau| < g \ \Rightarrow\ \boldsymbol{u}_\tau = \boldsymbol{0} \quad \text{on } \Gamma_C, \tag{3.93}$$
$$|\boldsymbol{\sigma}_\tau| = g \ \Rightarrow\ \boldsymbol{u}_\tau = -\kappa\boldsymbol{\sigma}_\tau \text{ for some } \kappa \ge 0 \quad \text{on } \Gamma_C, \tag{3.94}$$

where the friction bound $g > 0$ on Γ_C. We comment that (3.86) is the constitutive relation of the linearized elasticity material, (3.87) defines the linearized strain and the displacement, and (3.88) is the equilibrium equation. The classical displacement and traction boundary conditions are given in (3.89) and (3.90). Contact conditions are described in (3.91)–(3.94). The bilateral contact feature is reflected by (3.91). The relations (3.92)–(3.94) represent Tresca's friction law.

In certain situations, the frictional contact problem stated above describes the material deformation quite accurately. In more complicated situations, such as when the contact zone is not prescribed a priori or when more realistic frictional contact laws are used, the frictional contact problem here can be viewed as an intermediate problem for a typical step in an iterative solution procedure for solving the more complicated contact problem.

To introduce a variational formulation of the problem, we let

$$\boldsymbol{V} = \{\boldsymbol{v} \in \boldsymbol{H}^1(\Omega) : \boldsymbol{v}|_{\Gamma_D} = \boldsymbol{0},\ v_\nu|_{\Gamma_C} = 0\}$$

with its inner product and norm defined by

$$(\boldsymbol{u}, \boldsymbol{v})_{\boldsymbol{V}} = \int_\Omega \boldsymbol{\varepsilon}(\boldsymbol{u}) : \boldsymbol{\varepsilon}(\boldsymbol{v})\, dx, \qquad \|\boldsymbol{v}\|_{\boldsymbol{V}} = (\boldsymbol{v}, \boldsymbol{v})_{\boldsymbol{V}}^{1/2}.$$

Because $\mathrm{meas}(\Gamma_D) > 0$, the Korn inequality holds and we see that $\|\boldsymbol{v}\|_{\boldsymbol{V}}$ is a norm over $\boldsymbol{V}$ which is equivalent to the canonical norm $\|\boldsymbol{v}\|_{1;\Omega}$. Over the space $\boldsymbol{V}$, we define

$$a(\boldsymbol{u}, \boldsymbol{v}) = \int_\Omega \mathbb{C}\boldsymbol{\varepsilon}(\boldsymbol{u}) : \boldsymbol{\varepsilon}(\boldsymbol{v})\, dx, \tag{3.95}$$

$$\ell(\boldsymbol{v}) = \int_\Omega \boldsymbol{f}_1 \cdot \boldsymbol{v}\, dx + \int_{\Gamma_N} \boldsymbol{f}_2 \cdot \boldsymbol{v}\, ds, \tag{3.96}$$

$$j(\boldsymbol{v}) = \int_{\Gamma_C} g|\boldsymbol{v}_\tau|\, ds. \tag{3.97}$$

A standard procedure leads to the following variational formulation of the problem (3.86)–(3.94):

$$\boldsymbol{u} \in \boldsymbol{V}, \qquad a(\boldsymbol{u}, \boldsymbol{v} - \boldsymbol{u}) + j(\boldsymbol{v}) - j(\boldsymbol{u}) \geq \ell(\boldsymbol{v} - \boldsymbol{u}) \qquad \forall\, \boldsymbol{v} \in \boldsymbol{V}. \tag{3.98}$$

The bilinear form $a(\cdot,\cdot) : \boldsymbol{V} \times \boldsymbol{V} \to \mathbb{R}$ is obviously continuous. Due to the assumption $\mathrm{meas}(\Gamma_D) > 0$, it is $\boldsymbol{V}$-elliptic. The functional $\ell : \boldsymbol{V} \to \mathbb{R}$ is linear and continuous, and $j : \boldsymbol{V} \to \overline{\mathbb{R}}$ is proper and lower semicontinuous convex. Therefore, the variational inequality (3.98) has a unique solution $\boldsymbol{u}$ in $\boldsymbol{V}$. Moreover, because the bilinear form $a(\cdot,\cdot)$ is symmetric and positive definite, solving the variational inequality (3.98) is equivalent to minimizing the energy functional

$$J(\boldsymbol{v}) = \frac{1}{2} a(\boldsymbol{v}, \boldsymbol{v}) - \ell(\boldsymbol{v}) + j(\boldsymbol{v})$$

over the space $\boldsymbol{V}$. The unique solution $\boldsymbol{u} \in \boldsymbol{V}$ of the problem (3.98) is characterized by the existence of a Lagrange multiplier $\boldsymbol{\lambda}_\tau \in (L^\infty(\Gamma_C))^d$ such that

$$a(\boldsymbol{u},\boldsymbol{v})+\int_{\Gamma_C} g\,\boldsymbol{\lambda}_\tau\cdot\boldsymbol{v}_\tau\,ds=\ell(\boldsymbol{v})\qquad\forall\,\boldsymbol{v}\in\boldsymbol{V},\tag{3.99}$$

$$|\boldsymbol{\lambda}_\tau|\le 1,\quad \boldsymbol{\lambda}_\tau\cdot\boldsymbol{u}_\tau=|\boldsymbol{u}_\tau|\qquad\text{a.e. on }\Gamma_C.\tag{3.100}$$

Turn now to a finite element approximation of the problem. Assume Ω has a polyhedral boundary Γ. Let $\{\mathcal{P}_h\}$ be a family of partitions of the domain Ω into straight-sided elements, and let $\boldsymbol{V}_h\subset\boldsymbol{V}$ be the associated standard finite element spaces of continuous piecewise polynomials of certain degree. Corresponding to the partition $\mathcal{P}_h$, we use the symbols h_K, h_γ, $\mathcal{E}(K)$, $\mathcal{E}_h$, $\mathcal{E}_{h,0}$ as introduced in Section 3.2. In addition, we use $\mathcal{E}_{h,\Gamma_N}$, $\mathcal{E}_{h,\Gamma_C}$ with obvious meanings. The discrete formulation of the variational inequality (3.98) reads: find $\boldsymbol{u}_h\in\boldsymbol{V}_h$ such that

$$a(\boldsymbol{u}_h,\boldsymbol{v}_h-\boldsymbol{u}_h)+j(\boldsymbol{v}_h)-j(\boldsymbol{u}_h)\ge\ell(\boldsymbol{v}_h-\boldsymbol{u}_h)\qquad\forall\,\boldsymbol{v}_h\in\boldsymbol{V}_h.\tag{3.101}$$

Like the continuous variational inequality (3.98), the discrete problem (3.101) has a unique solution $\boldsymbol{u}_h\in\boldsymbol{V}_h$ and it is characterized by the existence of $\boldsymbol{\lambda}_{h\tau}\in(L^\infty(\Gamma_C))^d$ such that

$$a(\boldsymbol{u}_h,\boldsymbol{v}_h)+\int_{\Gamma_C} g\,\boldsymbol{\lambda}_{h\tau}\cdot\boldsymbol{v}_{h\tau}\,ds=\ell(\boldsymbol{v}_h)\qquad\forall\,\boldsymbol{v}_h\in\boldsymbol{V}_h,\tag{3.102}$$

$$|\boldsymbol{\lambda}_{h\tau}|\le 1,\quad \boldsymbol{\lambda}_{h\tau}\cdot\boldsymbol{u}_{h\tau}=|\boldsymbol{u}_{h\tau}|\qquad\text{a.e. on }\Gamma_C,\tag{3.103}$$

where $\boldsymbol{v}_{h\tau}$ and $\boldsymbol{u}_{h\tau}$ denote the tangential components of $\boldsymbol{v}_h$ and $\boldsymbol{u}_h$, respectively. Analysis of the finite element approximation of such problems in the general context of variational inequalities is extensively discussed in [34, 35]. In the context of finite element approximations of a problem more general than the one considered in this chapter, one can find in [46, Section 8.2] an optimal-order a priori error estimate under additional solution regularity, and a convergence result without any additional solution regularity assumption.

Similar to the result in Section 3.4, we have the residual type error estimate

$$\|\boldsymbol{e}_h\|_{\boldsymbol{V}}^2\le C\,\eta_R^2,\qquad \eta_R^2=\sum_{K\in\mathcal{P}_h}\eta_{R,K}^2,$$

where

$$\eta_{R,K}^2=h_K^2\|\boldsymbol{r}_K\|_{0;K}^2+\frac{1}{2}h_K\sum_{\gamma\in\mathcal{E}(K)\cap\mathcal{E}_{h,0}}\|\boldsymbol{R}_\gamma\|_{0;\gamma}^2$$
$$+h_K\sum_{\gamma\in\mathcal{E}(K)\cap\mathcal{E}_{h,\Gamma_N}}\|\boldsymbol{R}_\gamma\|_{0;\gamma}^2+h_K\sum_{\gamma\in\mathcal{E}(K)\cap\mathcal{E}_{h,\Gamma_C}}\|\boldsymbol{R}_\gamma\|_{0;\gamma}^2,$$

and $\boldsymbol{r}_K$ and $\boldsymbol{R}_\gamma$ are the interior and side residuals, respectively, defined by ($\boldsymbol{\sigma}_h=\mathbb{C}\boldsymbol{\varepsilon}(\boldsymbol{u}_h)$):

$$\boldsymbol{r}_K = \operatorname{Div} \boldsymbol{\sigma}_h + \boldsymbol{f}_1 \qquad \text{in } K \in \mathcal{P}_h,$$

$$\boldsymbol{R}_\gamma = \begin{cases} -\left[\boldsymbol{\sigma}_h \boldsymbol{\nu}\right]_\gamma & \text{if } \gamma \in \mathcal{E}_{h,0}, \\ \boldsymbol{f}_2 - \boldsymbol{\sigma}_h \boldsymbol{\nu} & \text{if } \gamma \in \mathcal{E}_{h,\Gamma_N}, \\ -g\boldsymbol{\lambda}_{h\tau} - \boldsymbol{\sigma}_{h\tau} & \text{if } \gamma \in \mathcal{E}_{h,\Gamma_C}. \end{cases}$$

Moreover,

$$\eta_R^2 \le C \left(\|\boldsymbol{e}_h\|_V^2 + \sum_{\gamma \in \mathcal{E}_{h,\Gamma_C}} h_\gamma \|\boldsymbol{\lambda}_\tau - \boldsymbol{\lambda}_{h\tau}\|_{0;\gamma}^2 \right.$$

$$\left. + \sum_{K \in \mathcal{P}_h} h_K^2 \|\boldsymbol{f}_1 - \boldsymbol{f}_{1,K}\|_{0;K}^2 + \sum_{\gamma \in \mathcal{E}_{h,\Gamma_C}} h_\gamma \|\boldsymbol{\lambda}_{h\tau} - \boldsymbol{\lambda}_{h\tau,\gamma}\|_{0;\gamma}^2 \right)$$

with discontinuous piecewise polynomial approximations $\boldsymbol{f}_{1,K}$, $\boldsymbol{\lambda}_{h\tau,\gamma}$ of residuals $\boldsymbol{f}_1$, $\boldsymbol{\lambda}_{h\tau}$, respectively.

The results in Section 3.5 on recovery-based error estimate can be similarly extended to the finite element solution of the frictional contact problem (3.98).

We now present numerical results on two two-dimensional problems. In both examples, body forces are assumed to be negligible and the body is in bilateral frictional contact with a rigid foundation on the part Γ_C. The friction is modeled by Tresca's law with a given slip bound g. We use the adaptive algorithm stated at the beginning of Section 3.6, with $\mu = 1$.

Example 3.2. The physical setting of this example shown in Figure 3.13. The domain $\Omega = (0,4) \times (0,4)$ is the cross-section of a three-dimensional linearly elastic body and plane stress condition is assumed. On the part $\Gamma_D = \{4\} \times (0,4)$ the body is clamped. Oblique tractions act on the part $\{0\} \times (0,4)$ and the part $(0,4) \times \{4\}$ is traction free. Thus $\Gamma_N = (\{0\} \times (0,4)) \cup ((0,4) \times \{4\})$. The contact part of the boundary is $\Gamma_C = (0,4) \times \{0\}$.

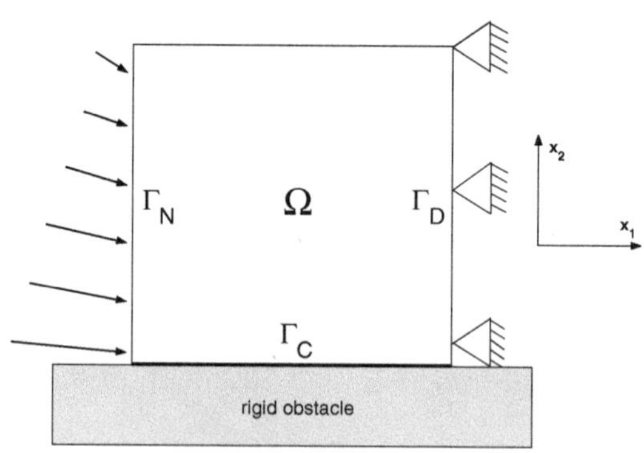

Fig. 3.13 Problem setting for Example 3.2.

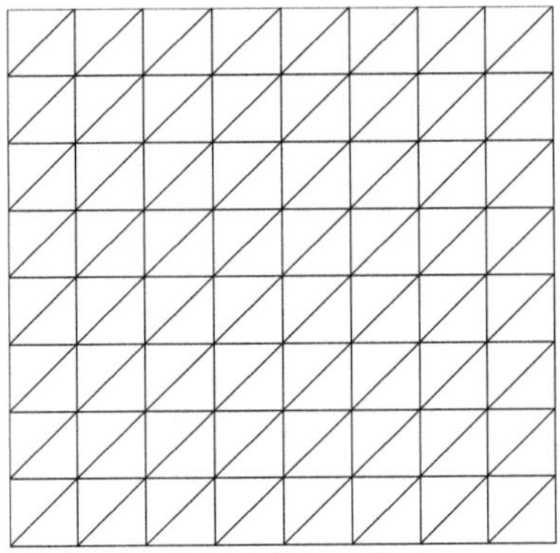

Fig. 3.14 Initial mesh.

The elasticity tensor $\mathbb{C}$ satisfies

$$(\mathbb{C}\boldsymbol{\varepsilon})_{ij} = \frac{E\nu}{1-\nu^2}(\varepsilon_{11}+\varepsilon_{22})\delta_{ij} + \frac{E}{1+\nu}\varepsilon_{ij}, \qquad 1 \le i,j \le 2,$$

where E is the Young's modulus, ν is the Poisson's ratio of the material, and δ_{ij} is the Kronecker symbol. We use the following data (the unit daN/mm^2 stands for "decanewtons per square millimeter"):

$$\begin{aligned} E &= 1500 \text{ daN/mm}^2, \\ \nu &= 0.4, \\ \boldsymbol{f}_1 &= (0,0) \text{ daN/mm}^2, \\ \boldsymbol{f}_2(x_1,x_2) &= (150(5-x_2), -75) \text{ daN/mm}^2, \\ g &= 450 \text{ daN/mm}^2. \end{aligned}$$

The initial uniform triangulation $\mathcal{P}_1$ (128 elements, 81 nodes) is shown in Figure 3.14 with the interval $[0,4]$ being divided into $4/h$ equal parts, $h = 1/2$. Then the triangulation is successively halved, and the numerical solution corresponding to $h = 1/64$ is taken as the "true" solution $\boldsymbol{u}$. To have an idea of the convergence behavior of the discrete Lagrange multipliers, we compute the errors $\|\boldsymbol{\lambda}_\tau - \boldsymbol{\lambda}_{h\tau}\|_{0;\Gamma_C}$ corresponding to the sequence of uniform refinements. Here, $\boldsymbol{\lambda}_\tau$ is the Lagrange multiplier corresponding to the parameter $h = 1/64$. Figure 3.15 provides a comparison of the errors $\|\boldsymbol{u} - \boldsymbol{u}_h^{un}\|_V$ and $h^{1/2}\|\boldsymbol{\lambda}_\tau - \boldsymbol{\lambda}_{h\tau}\|_{0;\Gamma_C}$. The numerical convergence order of $h^{1/2}\|\boldsymbol{\lambda}_\tau - \boldsymbol{\lambda}_{h\tau}\|_{0;\Gamma_C}$ is obviously higher than that of $\|\boldsymbol{u} - \boldsymbol{u}_h^{un}\|_V$, indicating that the second term in the efficiency bound is expected to be of higher order compared to the first term $\|\boldsymbol{u} - \boldsymbol{u}_h\|_V^2$. Note that in the two-dimensional case, $\boldsymbol{\lambda}_\tau$ and $\boldsymbol{\lambda}_{h\tau}$ are scalar-valued functions, also denoted by λ and λ_h. The (approximate) calculation of λ_h follows the procedure described in Section 3.6. Graphs of $\lambda_{h,1}$ and $\lambda_{h,2}$ with $h = 1/64$ are provided in Figures 3.16 and 3.17.

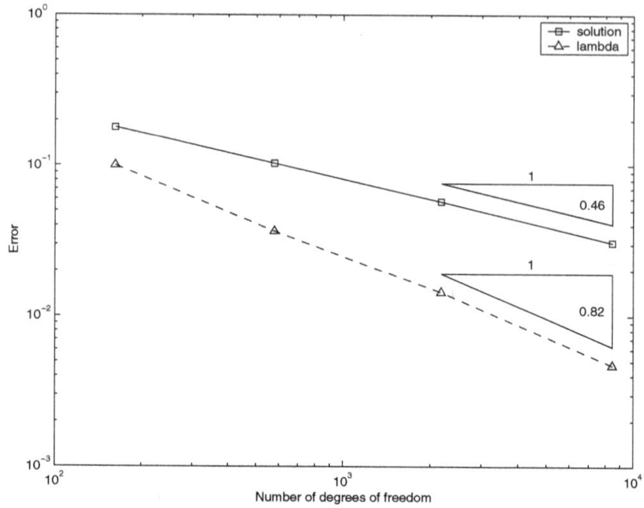

Fig. 3.15 Errors $\|\boldsymbol{u}-\boldsymbol{u}_h^{un}\|_{\boldsymbol{V}}$ ($\Box$) versus $h^{1/2}\|\boldsymbol{\lambda}_{h\boldsymbol{\tau}}-\boldsymbol{\lambda}_{h\boldsymbol{\tau}}^{un}\|_{0,\Gamma_C}$ ($\triangle$).

We use an adaptive procedure based on both residual type and recovery type estimates to obtain a sequence of approximate solutions $\boldsymbol{u}_h^{ad}$. The deformed configuration and the adaptive finite element mesh after four adaptive iterations are shown in Figures 3.18 (based on the residual type estimator, 5583 elements, 2921 nodes) and 3.19 (based on the recovery type estimator, 5437 elements, 2832 nodes). Figures 3.20 and 3.21 contain the error values $\|\boldsymbol{u}-\boldsymbol{u}_h^{un}\|_{\boldsymbol{V}}$ and $\|\boldsymbol{u}-\boldsymbol{u}_h^{ad}\|_{\boldsymbol{V}}$. We observe a substantial improvement of the efficiency using adaptively refined meshes. Figures 3.22 and 3.23 provide the values of $\eta_I = (\sum_K \eta_{I,K}^2)^{1/2}$, $I \in \{R, G\}$, where $\eta_{I,K}$ are computed using either residual type estimator ($I = R$) or recovery type estimator ($I = G$) on both uniform and adapted meshes. Table 3.3 contains the values of C_I computed for uniform and adaptive solutions:

$$C_I = \frac{\|\boldsymbol{u}-\boldsymbol{u}_h\|_{\boldsymbol{V}}}{\eta_I}, \qquad I\in\{R,G\}. \tag{3.104}$$

Table 3.3 Numerical values of C_R and C_G

h	1/2	1/4	1/8	1/16
C_R^{un}	2.63e-04	2.39e-04	2.18e-04	1.93e-04
C_R^{ad}	2.63e-04	2.39e-04	2.22e-04	2.01e-04
C_G^{un}	8.73e-04	8.05e-04	7.46e-04	6.90e-04
C_G^{ad}	8.73e-04	8.07e-04	7.46e-04	6.83e-04

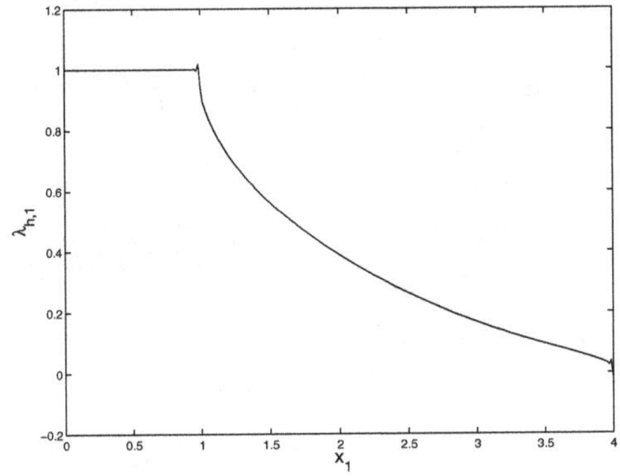

Fig. 3.16 Plot of $\lambda_{h,1}$.

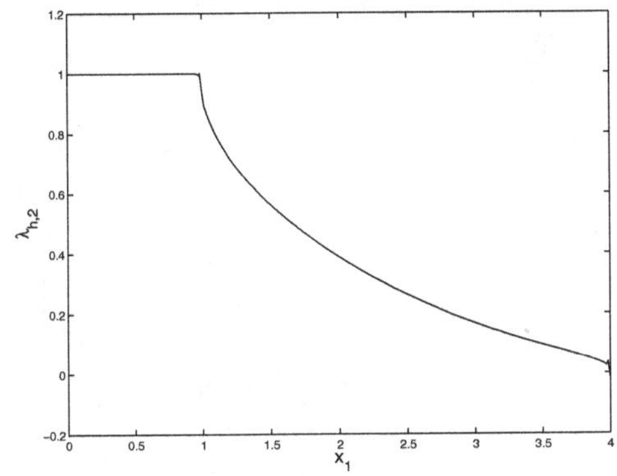

Fig. 3.17 Plot of $\lambda_{h,2}$.

Additional numerical experiments have been carried out in order to show the influence of the discrete Lagrange multipliers λ_h on the adaptive solution. To this end, we associate the residuals $\boldsymbol{R}_\gamma$ corresponding to the sides $\gamma \in \mathcal{E}_{h,\Gamma_C}$ with a weighting parameter Θ:

$$\begin{aligned}
\eta_{R,\Theta}^2 &= \sum_{K\in\mathcal{P}_h} \eta_{R,K,\Theta}^2, \\
\eta_{R,K,\Theta}^2 &= h_K^2 \|\boldsymbol{r}_K\|_{0;K}^2 + \frac{1}{2}\, h_K \sum_{\gamma\in\mathcal{E}(K)\cap\mathcal{E}_{h,0}} \|\boldsymbol{R}_\gamma\|_{0;\gamma}^2 \\
&\quad + h_K \sum_{\gamma\in\mathcal{E}(K)\cap\mathcal{E}_{h,\Gamma_N}} \|\boldsymbol{R}_\gamma\|_{0;\gamma}^2 \\
&\quad + \Theta\, h_K \sum_{\gamma\in\mathcal{E}(K)\cap\mathcal{E}_{h,\Gamma_C}} \|\boldsymbol{R}_\gamma\|_{0;\gamma}^2.
\end{aligned}$$

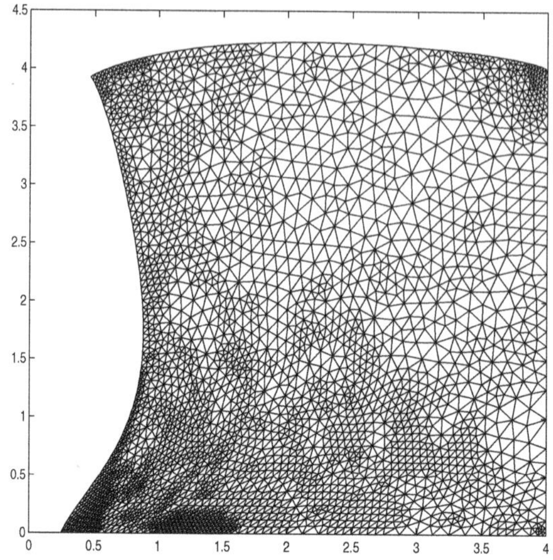

Fig. 3.18 Deformed configuration based on residual type estimator.

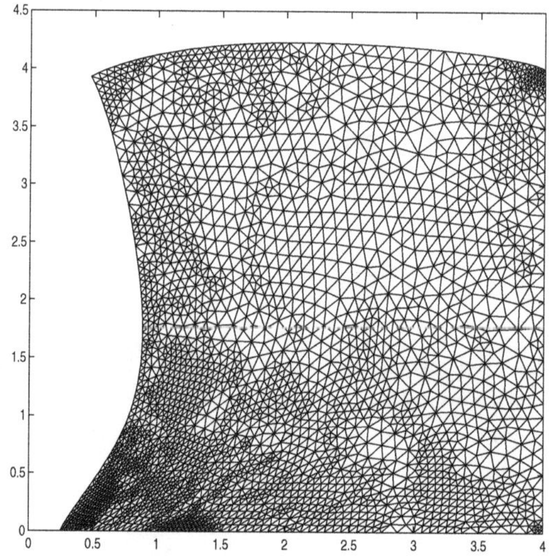

Fig. 3.19 Deformed configuration based on recovery type estimator.

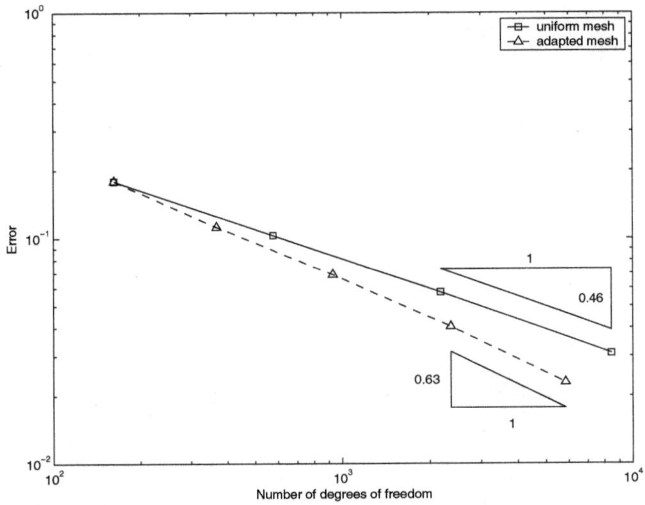

Fig. 3.20 Residual type estimator: $\|\boldsymbol{u} - \boldsymbol{u}_h^{un}\|_{\mathbf{V}}$ (□) versus $\|\boldsymbol{u} - \boldsymbol{u}_h^{ad}\|_{\mathbf{V}}$ (△).

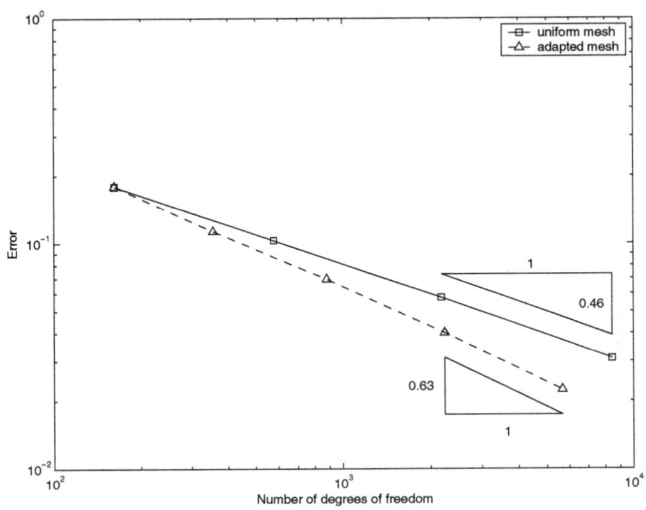

Fig. 3.21 Recovery type estimator: $\|\boldsymbol{u} - \boldsymbol{u}_h^{un}\|_{\mathbf{V}}$ (□) versus $\|\boldsymbol{u} - \boldsymbol{u}_h^{ad}\|_{\mathbf{V}}$ (△).

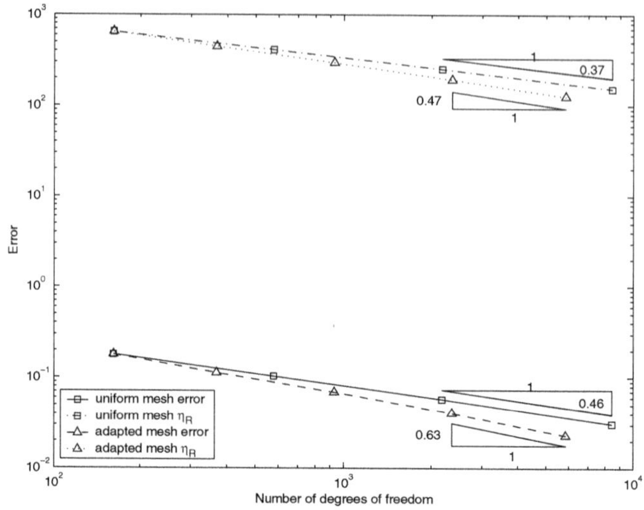

Fig. 3.22 Results based on residual type estimator.

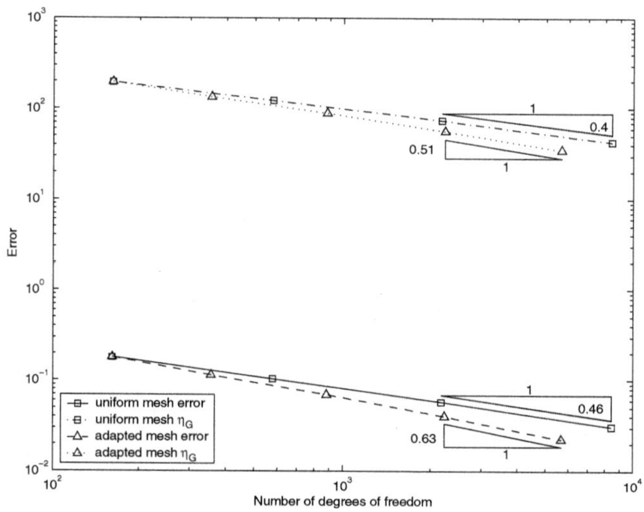

Fig. 3.23 Results based on recovery type estimator.

Table 3.4 Residual type estimator. Influence of Lagrange multiplier part on adaptive solution

$\Theta = 1$

adaptive level	0	1	2	3	4
d.o.f.	162	368	930	2356	5842
$\eta_{R,\Theta}$	6.49e+02	4.48e+02	2.96e+02	1.93e+02	1.26e+02
$\|\boldsymbol{u}-\boldsymbol{u}_h^{ad}\|_{\mathbf{V}}$	1.71e-01	1.07e-01	6.57e-02	3.87e-01	2.19e-02

$\Theta = 0.1$

d.o.f.	162	366	924	2342	5836
$\eta_{R,\Theta}$	6.41e+02	4.47e+02	2.96e+02	1.93e+02	1.25e+02
$\|\boldsymbol{u}-\boldsymbol{u}_h^{ad}\|_{\mathbf{V}}$	1.71e-01	1.08e-01	6.58e-02	3.86e-01	2.18e-02

$\Theta = 0.01$

d.o.f.	162	366	920	2318	5716
$\eta_{R,\Theta}$	6.40e+02	4.47e+02	2.96e+02	1.93e+02	1.26e+02
$\|\boldsymbol{u}-\boldsymbol{u}_h^{ad}\|_{\mathbf{V}}$	1.71e-01	1.08e-01	6.60e-02	3.86e-01	2.21e-02

We observe that the smaller the parameter Θ is, the less influence there is from the Lagrange multiplier part on the error estimator and hence on the adaptive mesh. An adaptive procedure is performed based on the error indicators $\eta_{R,\Theta}$ with $\Theta = 1, 0.1, 0.01$. Numerical results are summarized in Table 3.4. Similar numerical results are obtained for the gradient recovery type estimator. □

Example 3.3. The physical setting of this example is shown in Figure 3.24. The domain $\Omega = (0, 10) \times (0, 2)$ is the cross-section of a three-dimensional linearly elastic body with the plane stress condition assumed. On the part $\Gamma_D = (0, 10) \times \{2\}$ the body is clamped. Horizontal tractions act on the part $\{0\} \times (0, 2)$ and oblique tractions on $\{10\} \times (0, 2)$. Here $\Gamma_C = (0, 10) \times \{0\}$.

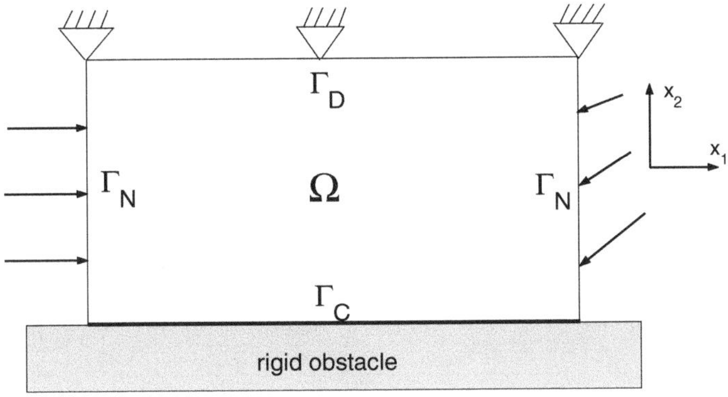

Fig. 3.24 Problem setting for Example 3.3.

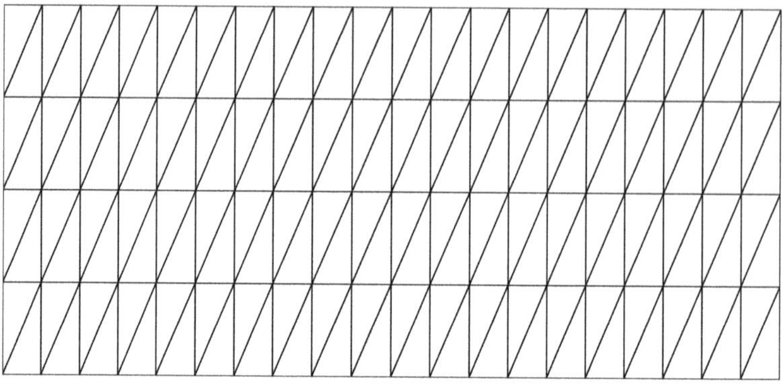

Fig. 3.25 Initial mesh.

The following data are used:

$$\begin{aligned} E &= 1000 \text{ daN/mm}^2, \\ \nu &= 0.3, \\ \boldsymbol{f}_1 &= (0,0) \text{ daN/mm}^2, \\ \boldsymbol{f}_2(x_1,x_2) &= \begin{cases} (500,0) \text{ daN/mm}^2 & \text{on } \{0\}\times(0,2), \\ (250x_2-750,-100) \text{ daN/mm}^2 & \text{on } \{10\}\times(0,2), \end{cases} \\ g &= 175 \text{ daN/mm}^2. \end{aligned}$$

We start with a coarse uniform triangulation shown in Figure 3.25, with 160 triangular elements and 105 nodes. For uniform triangulations, we divide the interval $[0,10]$ into $10/h$ equal parts and interval $[0,2]$ into $4/h$ parts. The numerical solution corresponding to $h=1/64$ is taken as the "true" solution $\boldsymbol{u}$. Figures 3.26 (based on the residual type estimator, 5222 elements, 2755 nodes) and 3.27 (based on the recovery type estimator, 4964 elements, 2624 nodes) show the approximate solution and refined mesh after four consecutive refinements. Graphs of $\lambda_{h,1}$ and $\lambda_{h,2}$ with $h=1/64$ are provided in Figures 3.28 and 3.29. Again, we compute the errors $\|\boldsymbol{u}-\boldsymbol{u}_h^{un}\|_{\boldsymbol{V}}$, $\|\boldsymbol{u}-\boldsymbol{u}_h^{ad}\|_{\boldsymbol{V}}$, $h^{1/2}\|\boldsymbol{\lambda}_\tau-\boldsymbol{\lambda}_{h\tau}\|_{0;\Gamma_C}$, and η_I, $I\in\{R,G\}$, whose values are provided in Figures 3.30–3.34. Table 3.5 contains the values of C_I, $I\in\{R,G\}$ defined in (3.104). □

Table 3.5 Numerical values of C_R and C_G

h	1/2	1/4	1/8	1/16
C_R^{un}	4.60e-04	4.27e-04	3.86e-04	3.24e-04
C_R^{ad}	4.60e-04	4.21e-04	3.83e-04	3.27e-04
C_G^{un}	1.39e-03	1.29e-03	1.22e-03	1.15e-03
C_G^{ad}	1.39e-03	1.29e-03	1.20e-03	1.12e-03

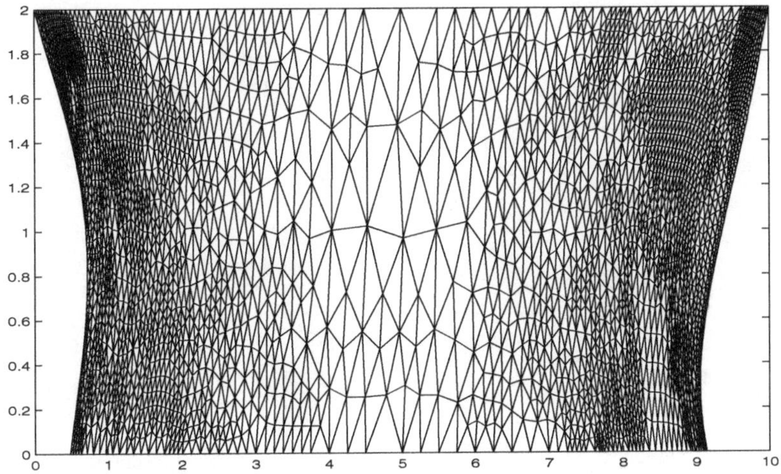

Fig. 3.26 Deformed configuration based on residual type estimator.

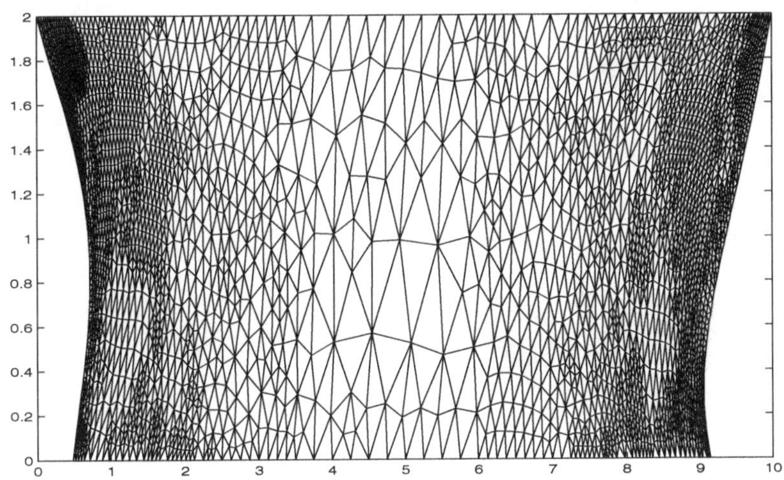

Fig. 3.27 Deformed configuration based on recovery type estimator.

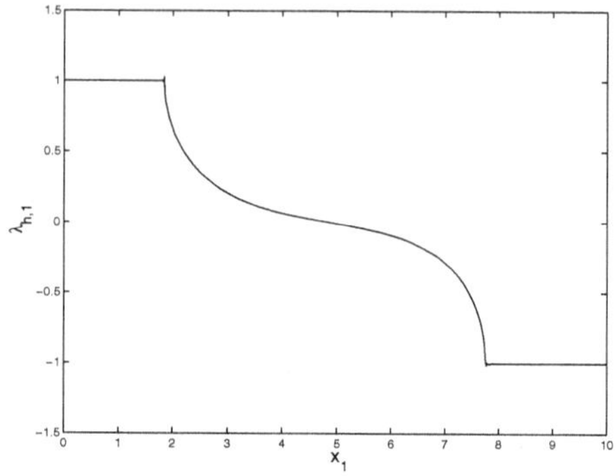

Fig. 3.28 Plot of $\lambda_{h,1}$.

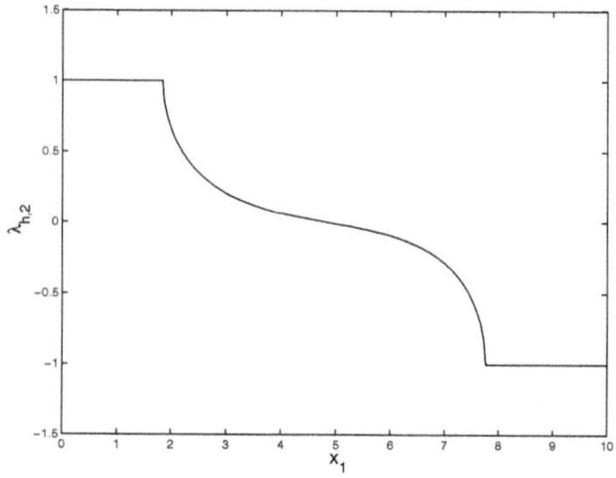

Fig. 3.29 Plot of $\lambda_{h,2}$.

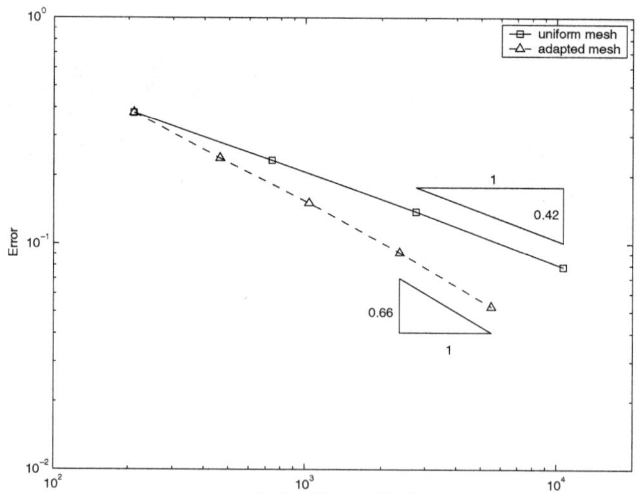

Fig. 3.30 Residual type estimator: $\|\boldsymbol{u} - \boldsymbol{u}_h^{un}\|_{\boldsymbol{V}}$ ($\square$) versus $\|\boldsymbol{u} - \boldsymbol{u}_h^{ad}\|_{\boldsymbol{V}}$ ($\triangle$).

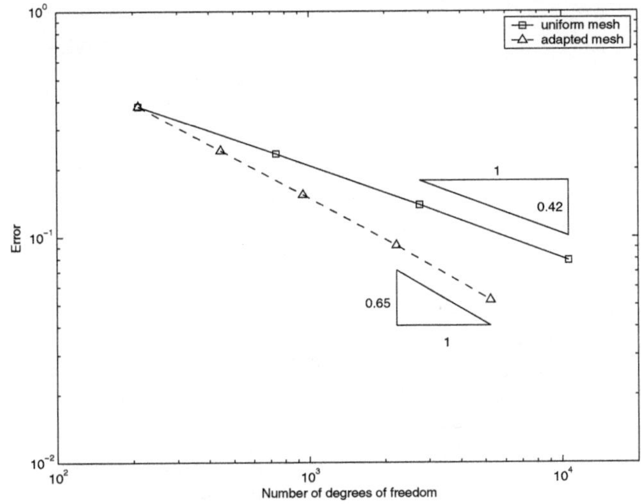

Fig. 3.31 Recovery type estimator: $\|\boldsymbol{u} - \boldsymbol{u}_h^{un}\|_{\mathbf{V}}$ ($\square$) versus $\|\boldsymbol{u} - \boldsymbol{u}_h^{ad}\|_{\mathbf{V}}$ ($\triangle$).

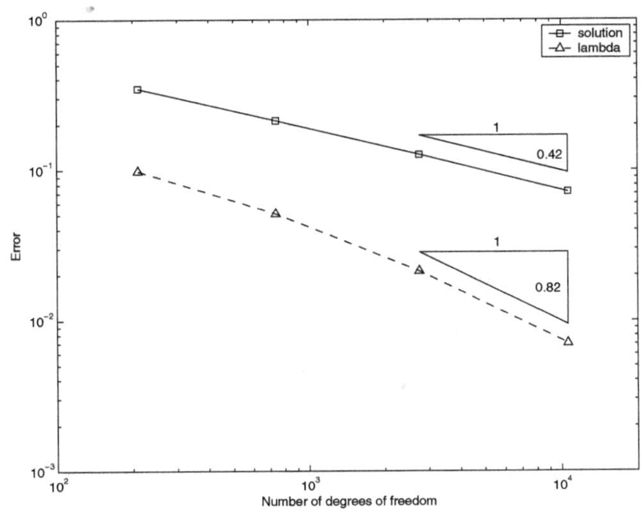

Fig. 3.32 $\|\boldsymbol{u} - \boldsymbol{u}_h^{un}\|_{\mathbf{V}}$ ($\square$) versus $h^{1/2}\|\boldsymbol{\lambda}_{h\mathcal{T}} - \boldsymbol{\lambda}_{h\mathcal{T}}^{un}\|_{0,\Gamma_C}$ ($\triangle$).

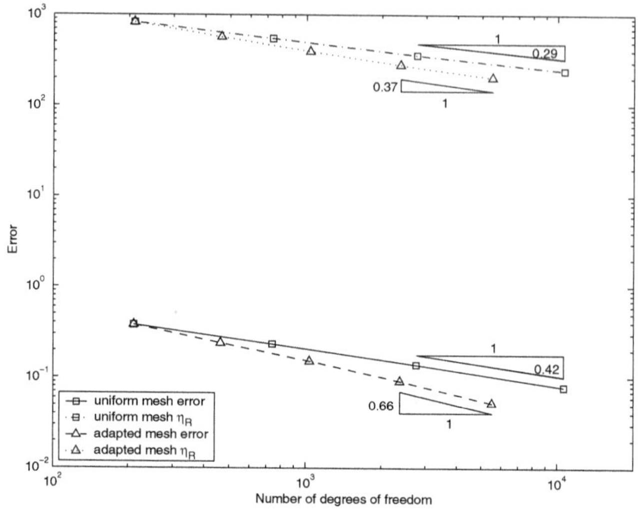

Fig. 3.33 Results based on residual type estimator.

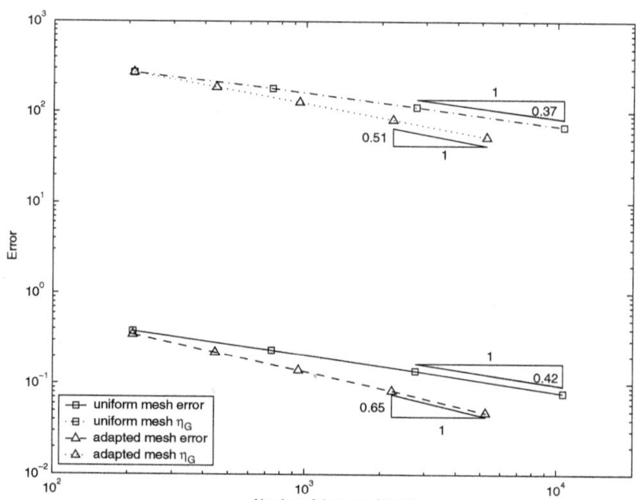

Fig. 3.34 Results based on recovery type estimator.

3.8 Quasistatic Variational Inequalities and Their Discretizations

The discussions in the previous sections on steady-state problems can be extended in adaptive solution of time-dependent variational inequalities. We illustrate this on some quasistatic variational inequalities. We first introduce an abstract quasistatic variational inequality. As is seen below, this variational inequality covers several quasistatic contact problems. A similar quasistatic variational inequality also arises in the study of a primal formulation of some plasticity problems [42].

Let X be a real Hilbert space with inner product $(\cdot,\cdot)_X$ and norm $\|\cdot\|_X$, $T>0$, and $1<p\le\infty$. Assume $a: X\times X\to\mathbb{R}$ is a symmetric, continuous, coercive bilinear form, $j: X\to\mathbb{R}$ is a continuous semi-norm X. Let there be given $f\in W^{1,p}(0,T;X)$ and $u_0\in X$ with the condition

$$a(u_0,v)+j(v)\ge(f(0),v)_X \qquad \forall\, v\in X.$$

We consider the following variational inequality.

Problem 3.1. Find $u:[0,T]\to X$ such that for a.e. $t\in(0,T)$,

$$a(u(t),v-\dot u(t))+j(v)-j(\dot u(t))\ge(f(t),v-\dot u(t))_X \qquad \forall\, v\in X, \tag{3.105}$$

and

$$u(0)=u_0. \tag{3.106}$$

A proof of the following well-posedness result can be found in [46].

Theorem 3.11. *Under the stated assumptions on the data, Problem 3.1 has a unique solution $u\in W^{1,p}(0,T;X)$.*

Next, we consider a semidiscrete scheme for Problem 3.1. We need a partition of the time interval: $[0,T]=\bigcup_{n=1}^N[t_{n-1},t_n]$ with $0=t_0<t_1<\cdots<t_N=T$. Denote $k_n=t_n-t_{n-1}$ for the length of the subinterval $[t_{n-1},t_n]$, and $k=\max_n k_n$ for the maximal step-size. For the given linear functional $f\in W^{1,p}(0,T;X)$ and the solution $u\in W^{1,p}(0,T;X)$, we use the notations $f_n=f(t_n)$ and $u_n=u(t_n)$, which are well-defined by the Sobolev embedding $W^{1,p}(0,T;X)\hookrightarrow C([0,T];X)$. The symbol $\delta_n u_n=(u_n-u_{n-1})/k_n$ is used to denote the backward divided difference.

Then a semidiscrete approximation of Problem 3.1 is the following.

Problem 3.2. Find $u^k=\{u_n^k\}_{n=0}^N\subset X$ such that

$$u_0^k=u_0 \tag{3.107}$$

and for $n=1,2,\ldots,N$,

$$a(u_n^k,v-\delta_n u_n^k)+j(v)-j(\delta_n u_n^k)\ge(f_n,v-\delta_n u_n^k)_X \qquad \forall\, v\in X. \tag{3.108}$$

We notice that under the assumptions of Theorem 3.11, Problem 3.2 has a unique solution. Let us briefly indicate how an error estimate can be derived for the semidiscrete solution defined in Problem 3.2. For this purpose, we assume $u \in H^2(0,T;X)$. Note that this assumption in particular implies that the solution u satisfies the inequality (3.105) for all $t \in (0,T)$.

Denote $e_n = u_n - u_n^k$ for the solution error, and let

$$\|v\|_a = a(v,v)^{1/2}$$

be the "energy" norm over X. We observe that from the assumptions on the bilinear form a, the quantity $\|v\|_a$ defines an equivalent norm on X. We take $v = \delta_n u_n^k$ in (3.105) at $t = t_n$, $v = \dot{u}_n$ in (3.108), and add the two inequalities to obtain

$$a(u_n, \delta_n u_n^k - \dot{u}_n) + a(u_n^k, \dot{u}_n - \delta_n u_n^k) \geq 0.$$

This relation can be rewritten as

$$a(e_n, e_n - e_{n-1}) \leq k_n a(e_n, \delta_n u_n - \dot{u}_n).$$

Now,

$$a(e_n, e_n - e_{n-1}) \geq \frac{1}{2}\left(\|e_n\|_a^2 - \|e_{n-1}\|_a^2\right),$$
$$a(e_n, \delta_n u_n - \dot{u}_n) \leq \frac{1}{2}\left(\|e_n\|_a^2 + \|\delta_n u_n - \dot{u}_n\|_a^2\right).$$

Hence,

$$\|e_n\|_a^2 \leq (1-k_n)^{-1}\|e_{n-1}\|_a^2 + k_n(1-k_n)^{-1}\|\delta_n u_n - \dot{u}_n\|_a^2.$$

It is easy to verify that when $k \leq 1/2$,

$$(1-k_n)^{-1} \leq e^{2\,k_n}.$$

Therefore,

$$\|e_n\|_a^2 \leq e^{2\,k_n}\|e_{n-1}\|_a^2 + k_n e^{2\,k_n}\|\delta_n u_n - \dot{u}_n\|_a^2.$$

An inductive argument leads to

$$\|e_n\|_a^2 \leq e^{2\,(k_n+\cdots+k_1)}\|e_0\|_a^2 + \sum_{j=0}^{n-1} k_{n-j} e^{2\,(k_n+\cdots+k_{n-j})}\|\delta_{n-j}u_{n-j} - \dot{u}_{n-j}\|_a^2.$$

Because $e_0 = 0$ and $k_n + \cdots + k_{n-j} \leq T$, we have

$$\|e_n\|_a^2 \leq c\sum_{j=0}^{n-1} k_{n-j}\|\delta_{n-j}u_{n-j} - \dot{u}_{n-j}\|_a^2.$$

Under the smoothness assumption $u \in H^2(0,T;X)$, we have

$$\|\delta_{n-j}u_{n-j} - \dot{u}_{n-j}\|_a^2 \le k_{n-j}\|\ddot{u}\|^2_{L^2(t_{n-j-1},t_{n-j};X)}.$$

Thus, we have shown the error estimate

$$\max_{1\le n\le N} \|u_n - u_n^k\|_X^2 \le c\sum_{j=1}^{N} k_j^2\|\ddot{u}\|^2_{L^2(t_{j-1},t_j;X)}. \tag{3.109}$$

It then follows that

$$\max_{1\le n\le N} \|u_n - u_n^k\|_X \le c\,k\,\|\ddot{u}\|_{L^2(0,T;X)}.$$

Note that the estimate (3.109) identifies the local contribution from each subinterval to the solution error, and therefore could be used to adjust the values of k_j to achieve an efficient time discretization.

Now we describe some contact problems that lead to a variational inequality of the type stated in Problem 3.1. We consider a quasistatic contact process of a linearly elastic body occupying a Lipschitz domain Ω in $\mathbb{R}^d$ ($d \le 3$ in applications). The boundary $\Gamma = \partial\Omega$ is partitioned as follows:$\Gamma = \overline{\Gamma_D} \cup \overline{\Gamma_N} \cup \overline{\Gamma_C}$ with Γ_D, Γ_N, and Γ_C relatively open and mutually disjoint, and meas$(\Gamma_D) > 0$. The time interval of interest is $(0,T)$. We assume that the body is clamped on $\Gamma_D \times (0,T)$, meaning that the body is clamped on Γ_D during the time interval $(0,T)$; a volume force of density $\boldsymbol{f}_1$ acts in $\Omega \times (0,T)$ and a surface traction of density $\boldsymbol{f}_2$ is applied on $\Gamma_N \times (0,T)$. Then the displacement field $\boldsymbol{u} : \Omega \times [0,T] \to \mathbb{R}^d$ and the stress field $\boldsymbol{\sigma} : \Omega \times [0,T] \to \mathbb{S}^d$ satisfy the relations:

$$\boldsymbol{\sigma} = \mathbb{C}\boldsymbol{\varepsilon}(\boldsymbol{u}) \quad \text{in } \Omega \times (0,T), \tag{3.110}$$

$$\boldsymbol{\varepsilon}(\boldsymbol{u}) = \tfrac{1}{2}(\nabla\boldsymbol{u} + (\nabla\boldsymbol{u})^T) \quad \text{in } \Omega \times (0,T), \tag{3.111}$$

$$\operatorname{Div}\boldsymbol{\sigma} + \boldsymbol{f}_1 = \boldsymbol{0} \quad \text{in } \Omega \times (0,T), \tag{3.112}$$

$$\boldsymbol{u} = \boldsymbol{0} \quad \text{on } \Gamma_D \times (0,T), \tag{3.113}$$

$$\boldsymbol{\sigma\nu} = \boldsymbol{f}_2 \quad \text{on } \Gamma_N \times (0,T), \tag{3.114}$$

$$\boldsymbol{u}(0) = \boldsymbol{u}_0 \quad \text{in } \Omega. \tag{3.115}$$

In (3.110), $\mathbb{C} : \Omega \times \mathbb{S}^d \to \mathbb{S}^d$ denotes the fourth-order elasticity tensor of the material, and is assumed to be bounded, symmetric, and positive definite in Ω. In (3.115), $\boldsymbol{u}_0$ is the given initial displacement field. In (3.114), $\boldsymbol{\nu}$ is the unit outward normal vector on the boundary Γ, which exists a.e. because Γ is assumed to be Lipschitz continuous.

The relations (3.110)–(3.115) are to be supplemented by contact conditions on $\Gamma_C \times (0,T)$. We assume the contact is bilateral (no loss of contact during the process) and the friction is modeled with Tresca's friction law (see, e.g., [29, 59]):

$$\left.\begin{array}{l} u_\nu = 0,\ |\boldsymbol{\sigma}_\tau| \le g, \\ |\boldsymbol{\sigma}_\tau| < g \ \Rightarrow\ \dot{\boldsymbol{u}}_\tau = \boldsymbol{0}, \\ |\boldsymbol{\sigma}_\tau| = g \ \Rightarrow\ \exists\lambda \ge 0 \text{ s.t. } \boldsymbol{\sigma}_\tau = -\lambda\dot{\boldsymbol{u}}_\tau \end{array}\right\} \text{ on } \Gamma_C \times (0,T). \tag{3.116}$$

Here, $g \geq 0$ represents the friction bound function, u_ν and $\boldsymbol{u}_\tau$ are the normal and tangential components of $\boldsymbol{u}$ on Γ, and $\boldsymbol{\sigma}_\tau$ is the tangential component of $\boldsymbol{\sigma}$.

To introduce a variational formulation of the problem, we need the space

$$V = \{\boldsymbol{v} \in (H^1(\Omega))^d : \boldsymbol{v} = \boldsymbol{0} \text{ a.e. on } \Gamma_D,\ v_\nu = 0 \text{ a.e. on } \Gamma_C\}$$

with the inner product and norm defined by

$$(\boldsymbol{u}, \boldsymbol{v})_V = \int_\Omega \mathbb{C}\boldsymbol{\varepsilon}(\boldsymbol{u}) : \boldsymbol{\varepsilon}(\boldsymbol{v})\, dx, \qquad \|\boldsymbol{v}\|_V = \sqrt{(\boldsymbol{v}, \boldsymbol{v})_V}.$$

Assume that the volume force and traction densities satisfy

$$\boldsymbol{f}_1 \in W^{1,\infty}(0, T; (L^2(\Omega))^d), \qquad \boldsymbol{f}_2 \in W^{1,\infty}(0, T; (L^2(\Gamma_N))^d), \tag{3.117}$$

and let the constant $g > 0$ be given. We define over the space V,

$$a(\boldsymbol{u}, \boldsymbol{v}) = \int_\Omega \mathbb{C}\boldsymbol{\varepsilon}(\boldsymbol{u}) : \boldsymbol{\varepsilon}(\boldsymbol{v})\, dx, \tag{3.118}$$

$$j(\boldsymbol{v}) = \int_{\Gamma_C} g\, |\boldsymbol{v}_\tau|\, ds \tag{3.119}$$

and denote by $\boldsymbol{l}(t)$ the element of V given by

$$(\boldsymbol{l}(t), \boldsymbol{v})_V = \int_\Omega \boldsymbol{f}_1(t) \cdot \boldsymbol{v}\, dx + \int_{\Gamma_N} \boldsymbol{f}_2(t) \cdot \boldsymbol{v}\, ds \qquad \forall\, \boldsymbol{v} \in V,\ t \in [0, T]. \tag{3.120}$$

We assume that the initial data satisfy

$$\boldsymbol{u}_0 \in V, \tag{3.121}$$

$$a(\boldsymbol{u}_0, \boldsymbol{v}) + j(\boldsymbol{v}) \geq (\boldsymbol{l}(0), \boldsymbol{v})_V \qquad \forall\, \boldsymbol{v} \in V. \tag{3.122}$$

Then a standard procedure leads to the following variational formulation for the frictional contact problem (3.110)–(3.115) and (3.116).

Problem 3.3. Find $\boldsymbol{u} : [0, T] \to V$ such that for a.e. $t \in (0, T)$,

$$a(\boldsymbol{u}(t), \boldsymbol{v} - \dot{\boldsymbol{u}}(t)) + j(\boldsymbol{v}) - j(\dot{\boldsymbol{u}}(t)) \geq (\boldsymbol{l}(t), \boldsymbol{v} - \dot{\boldsymbol{u}}(t))_V \qquad \forall\, \boldsymbol{v} \in V, \tag{3.123}$$

and

$$\boldsymbol{u}(0) = \boldsymbol{u}_0. \tag{3.124}$$

We observe that this is a special case of Problem 3.1. By Theorem 3.11, we see that under assumptions (3.117), (3.121), and (3.122), Problem 3.3 has a unique solution $\boldsymbol{u} \in W^{1,\infty}(0, T; V)$.

Some other contact conditions lead to a similar variational inequality. For example, suppose the contact condition is described by a simplified version of Coulomb's law (see, e.g., [29, 59]):

$$\left.\begin{array}{l}\sigma_\nu = S, \quad |\boldsymbol{\sigma}_\tau| \le \mu\,|\sigma_\nu|,\\ |\boldsymbol{\sigma}_\tau| < \mu\,|\sigma_\nu| \;\Rightarrow\; \dot{\boldsymbol{u}}_\tau = \mathbf{0},\\ |\boldsymbol{\sigma}_\tau| = \mu\,|\sigma_\nu| \;\Rightarrow \exists\,\lambda \ge 0 \text{ s.t. } \boldsymbol{\sigma}_\tau = -\lambda\dot{\boldsymbol{u}}_\tau\end{array}\right\} \quad \text{on } \Gamma_C \times (0,T). \tag{3.125}$$

Here $S \in L^\infty(\Gamma_C)$ is a given function on Γ_C and $\mu \in L^\infty(\Gamma_C)$ is the given coefficient of friction, $\mu \ge 0$ a.e. on Γ_C. Then the variational formulation is still of the form (3.123)–(3.124) with the same bilinear form a and

$$V = \{\boldsymbol{v} \in (H^1(\Omega))^d : \boldsymbol{v} = \mathbf{0} \text{ a.e. on } \Gamma_D\},$$

$$j(\boldsymbol{v}) = \int_{\Gamma_C} \mu\,|S|\,|\boldsymbol{v}_\tau|\,ds,$$

$$(\boldsymbol{l}(t),\boldsymbol{v})_V = \int_\Omega \boldsymbol{f}_1(t)\cdot\boldsymbol{v}\,dx + \int_{\Gamma_N} \boldsymbol{f}_2(t)\cdot\boldsymbol{v}\,ds + \int_{\Gamma_C} S\,v_\nu\,ds.$$

Analysis of this variational inequality is similar to that for the problem (3.123)–(3.124).

We now show how to derive a posteriori error estimates for finite element solutions of the temporally semidiscrete problem. We choose Problem 3.3 as the model quasistatic contact problem. Its temporal semidiscrete approximation is the following.

Problem 3.4. Find $\boldsymbol{u}^k = \{\boldsymbol{u}_n^k\}_{n=0}^N$, where $\boldsymbol{u}_n^k \in V$, $0 \le n \le N$, $\boldsymbol{u}_0^k = \boldsymbol{u}_0$, such that for $n = 1, 2, \ldots, N$,

$$a(\boldsymbol{u}_n^k, \boldsymbol{v} - \delta_n\boldsymbol{u}_n^k) + j(\boldsymbol{v}) - j(\delta_n\boldsymbol{u}_n^k) \ge (\boldsymbol{l}_n, \boldsymbol{v} - \delta_n\boldsymbol{u}_n^k) \qquad \forall\,\boldsymbol{v} \in V, \tag{3.126}$$

where $\boldsymbol{l}_n = \boldsymbol{l}(t_n)$.

Denote $\boldsymbol{w}_n^k = \delta_n\boldsymbol{u}_n^k$. Then $\boldsymbol{u}_n^k = \boldsymbol{u}_{n-1}^k + k_n\boldsymbol{w}_n^k$ and we can express the inequality problem (3.126) in terms of $\boldsymbol{w}_n^k$: Find $\boldsymbol{w}_n^k \in V$ such that

$$\begin{aligned} k_n a(\boldsymbol{w}_n^k, \boldsymbol{v} - \boldsymbol{w}_n^k) &+ j(\boldsymbol{v}) - j(\boldsymbol{w}_n^k)\\ &\ge (\boldsymbol{l}_n, \boldsymbol{v} - \boldsymbol{w}_n^k) - a(\boldsymbol{u}_{n-1}^k, \boldsymbol{v} - \boldsymbol{w}_n^k) \qquad \forall\,\boldsymbol{v} \in V. \end{aligned} \tag{3.127}$$

This inequality is equivalent to the minimization problem:

$$\boldsymbol{w}_n^k \in V, \qquad J_n(\boldsymbol{w}_n^k) = \inf_{\boldsymbol{v}\in V} J_n(\boldsymbol{v}), \tag{3.128}$$

where J_n is the functional

$$J_n(\boldsymbol{v}) = \frac{k_n}{2} a(\boldsymbol{v},\boldsymbol{v}) + j(\boldsymbol{v}) - (\boldsymbol{l}_n, \boldsymbol{v}) + a(\boldsymbol{u}_{n-1}^k, \boldsymbol{v}), \qquad \boldsymbol{v} \in V. \tag{3.129}$$

We now turn to a finite element approximation of Problem 3.4. We use the finite element space setting discussed in Section 3.7. For a given $\boldsymbol{v} \in (L^1(\Omega))^d$, similar to (3.10)–(3.11), we define the interpolation operator $\Pi_h : V \to V^h$ as follows:

$$\Pi_h \boldsymbol{v} = \sum_{\boldsymbol{a}\in\mathcal{N}_{v,0}} \boldsymbol{v}_{\boldsymbol{a}}\varphi_{\boldsymbol{a}}, \qquad \text{where } \boldsymbol{v}_{\boldsymbol{a}} = \frac{\int_{\widetilde{K}_{\boldsymbol{a}}} \boldsymbol{v}\,\psi_{\boldsymbol{a}}\,dx}{\int_{\widetilde{K}_{\boldsymbol{a}}} \varphi_{\boldsymbol{a}}\,dx}. \tag{3.130}$$

Applying Theorem 3.2, we have an h-independent constant $C > 0$ such that for all $\boldsymbol{v} \in V$ and $\boldsymbol{f} \in (L^2(\Omega))^d$,

$$|\boldsymbol{v} - \Pi_h \boldsymbol{v}|_{1;\Omega}^2 \le C|\boldsymbol{v}|_{1;\Omega}^2, \tag{3.131}$$

$$\int_\Omega \boldsymbol{f}\cdot(\boldsymbol{v} - \Pi_h \boldsymbol{v})\,dx \le C|\boldsymbol{v}|_{1;\Omega}\left(\sum_{\boldsymbol{a}\in\mathcal{N}_{h,0}} h_{\boldsymbol{a}}^2 \min_{\boldsymbol{f}_{\boldsymbol{a}}\in\mathbb{R}^d} \|\boldsymbol{f} - \boldsymbol{f}_{\boldsymbol{a}}\|_{0;\widetilde{K}_{\boldsymbol{a}}}^2\right)^{1/2}, \tag{3.132}$$

$$\sum_{K\in\mathcal{P}_h} \|h_K^{-1}(\boldsymbol{v} - \Pi_h \boldsymbol{v})\|_{0;K}^2 \le C|\boldsymbol{v}|_{1;\Omega}^2, \tag{3.133}$$

$$\sum_{\gamma\in\mathcal{E}_h} \|h_\gamma^{-1/2}(\boldsymbol{v} - \Pi_h \boldsymbol{v})\|_{0;\gamma}^2 \le C|\boldsymbol{v}|_{1;\Omega}^2. \tag{3.134}$$

The finite element approximation of the inequality problem (3.126) is to find $\boldsymbol{u}_n^{hk} \in V^h$ such that

$$a(\boldsymbol{u}_n^{hk}, \boldsymbol{v}^h - \delta_n \boldsymbol{u}_n^{hk}) + j(\boldsymbol{v}^h) - j(\delta_n \boldsymbol{u}_n^{hk}) \ge (\boldsymbol{l}_n, \boldsymbol{v}^h - \delta_n \boldsymbol{u}_n^{hk}) \qquad \forall\, \boldsymbol{v}^h \in V^h. \tag{3.135}$$

Corresponding to the formulation (3.127), the finite element method is to find $\boldsymbol{w}_n^{hk} \in V^h$ such that

$$\begin{aligned} &k_n a(\boldsymbol{w}_n^{hk}, \boldsymbol{v}^h - \boldsymbol{w}_n^{hk}) + j(\boldsymbol{v}^h) - j(\boldsymbol{w}_n^{hk}) \\ &\qquad \ge (\boldsymbol{l}_n, \boldsymbol{v}^h - \boldsymbol{w}_n^{hk}) - a(\boldsymbol{u}_{n-1}^{hk}, \boldsymbol{v}^h - \boldsymbol{w}_n^{hk}) \qquad \forall\, \boldsymbol{v}^h \in V^h. \end{aligned} \tag{3.136}$$

As in previous sections, $\boldsymbol{w}_n^k$ is characterized by the existence of a unique $\boldsymbol{\lambda}_{n\tau}^k \in L^\infty(\Gamma_C)$ such that

$$k_n a(\boldsymbol{w}_n^k, \boldsymbol{v}) + \int_{\Gamma_C} g\boldsymbol{\lambda}_{n\tau}^k \cdot \boldsymbol{v}_\tau\, ds = (\boldsymbol{l}_n, \boldsymbol{v}) - a(\boldsymbol{u}_{n-1}^k, \boldsymbol{v}) \qquad \forall\, \boldsymbol{v} \in V, \tag{3.137}$$

$$|\boldsymbol{\lambda}_{n\tau}^k| \le 1 \text{ a.e. on } \Gamma_C \qquad \text{and} \qquad \boldsymbol{\lambda}_{n\tau}^k \cdot \boldsymbol{w}_{n\tau}^k = |\boldsymbol{w}_{n\tau}^k| \text{ a.e. on } \Gamma_C; \tag{3.138}$$

$\boldsymbol{w}_n^{hk}$ is characterized by the existence of a unique $\boldsymbol{\lambda}_{n\tau}^{hk} \in L^\infty(\Gamma_C)$ such that

$$\begin{aligned} &k_n a(\boldsymbol{w}_n^{hk}, \boldsymbol{v}^h) + \int_{\Gamma_C} g\boldsymbol{\lambda}_{n\tau}^{hk} \cdot \boldsymbol{v}_\tau^h\, ds \\ &\qquad = (\boldsymbol{l}_n, \boldsymbol{v}^h) - a(\boldsymbol{u}_{n-1}^{hk}, \boldsymbol{v}^h) \qquad \forall\, \boldsymbol{v}^h \in V^h, \end{aligned} \tag{3.139}$$

$$|\boldsymbol{\lambda}_{n\tau}^{hk}| \le 1 \text{ a.e. on } \Gamma_C \qquad \text{and} \qquad \boldsymbol{\lambda}_{n\tau}^{hk} \cdot \boldsymbol{w}_{n\tau}^{hk} = |\boldsymbol{w}_{n\tau}^{hk}| \text{ a.e. on } \Gamma_C. \tag{3.140}$$

3.9 A Posteriori Error Estimates for the Quasistatic Contact Problem

We first provide a dual formulation for the problem (3.127). Let

$$Q = (L^2(\Omega))^{d\times d} \times (L^2(\Gamma_C))^d.$$

For $n = 1, 2, \dots, N$, define $F_n : V \times Q \to \overline{\mathbb{R}}$ by the formula

$$\begin{aligned} F_n(\boldsymbol{v}, \boldsymbol{q}) = &\int_\Omega \left[\frac{k_n}{2}\mathbb{C}\boldsymbol{q}_1 : \boldsymbol{q}_1 - \boldsymbol{f}_{1n}\cdot\boldsymbol{v} + \mathbb{C}\boldsymbol{\varepsilon}(\boldsymbol{u}_{n-1}^k) : \boldsymbol{\varepsilon}(\boldsymbol{v})\right] dx \\ &+ \int_{\Gamma_C} g\,|\boldsymbol{q}_2|\, ds - \int_{\Gamma_N} \boldsymbol{f}_{2n}\cdot\boldsymbol{v}\, ds, \end{aligned}$$

where $\boldsymbol{q} = (\boldsymbol{q}_1, \boldsymbol{q}_2) \in Q$, $\boldsymbol{f}_{1n} = \boldsymbol{f}_1(t_n)$, and $\boldsymbol{f}_{2n} = \boldsymbol{f}_2(t_n)$. Introduce a linear bounded operator $\Lambda : V \to Q$ by the relation

$$\Lambda\boldsymbol{v} = (\boldsymbol{\varepsilon}(\boldsymbol{v}), \boldsymbol{v}_\tau) \qquad \forall\, \boldsymbol{v} \in V.$$

Then for any $\boldsymbol{v} \in V$,

$$J_n(\boldsymbol{v}) = F_n(\boldsymbol{v}, \Lambda\boldsymbol{v}),$$

and the minimization problem (3.6) can be rewritten as

$$\boldsymbol{w}_n^k \in V, \qquad F_n(\boldsymbol{w}_n^k, \Lambda\boldsymbol{w}_n^k) = \inf_{\boldsymbol{v}\in V} F_n(\boldsymbol{v}, \Lambda\boldsymbol{v}). \tag{3.141}$$

Let V^* and $Q^* = (L^2(\Omega))^{d\times d} \times (L^2(\Gamma_C))^d$ be the duals of V and Q placed in duality by the pairings $\langle\cdot,\cdot\rangle_V$ and $\langle\cdot,\cdot\rangle_Q$, respectively. We need to compute the conjugate function F_n^* of F_n:

$$F_n^*(\Lambda^*\boldsymbol{q}^*, -\boldsymbol{q}^*) \equiv \sup_{\boldsymbol{v}\in V, \boldsymbol{q}\in Q} \left\{\langle\Lambda^*\boldsymbol{q}^*, \boldsymbol{v}\rangle_V - \langle\boldsymbol{q}^*, \boldsymbol{q}\rangle_Q - F_n(\boldsymbol{v}, \boldsymbol{q})\right\},$$

where $\Lambda^* : Q^* \to V^*$ is the adjoint of Λ. We have

$$\begin{aligned} F_n^*(\Lambda^*\boldsymbol{q}^*, -\boldsymbol{q}^*) = \sup_{\boldsymbol{v}\in V, \boldsymbol{q}\in Q} \Bigg\{ &\int_\Omega \Bigg[(\boldsymbol{q}_1^* - \mathbb{C}\boldsymbol{\varepsilon}(\boldsymbol{u}_{n-1}^k)) : \boldsymbol{\varepsilon}(\boldsymbol{v}) + \boldsymbol{f}_{1n}\cdot\boldsymbol{v} \\ &- \left(\frac{k_n}{2}\mathbb{C}\boldsymbol{q}_1 : \boldsymbol{q}_1 + \boldsymbol{q}_1^* : \boldsymbol{q}_1\right)\Bigg] dx \\ &+ \int_{\Gamma_N} \boldsymbol{f}_{2n}\cdot\boldsymbol{v}\, ds + \int_{\Gamma_C} [\boldsymbol{q}_2^*\cdot\boldsymbol{v}_\tau - (\boldsymbol{q}_2^*\cdot\boldsymbol{q}_2 + g\,|\boldsymbol{q}_2|)]\, ds\Bigg\}. \end{aligned} \tag{3.142}$$

Let $\mathbb{A} : \Omega \times \mathbb{S}^d \to \mathbb{S}^d$ be the fourth-order tensor, inverse to $\mathbb{C}$. Like $\mathbb{C}$, $\mathbb{A}$ is also bounded, symmetric, and positive definite in Ω. Then we can show that

$$F_n^*(\Lambda^* \boldsymbol{q}^*, -\boldsymbol{q}^*) = \begin{cases} \int_\Omega \frac{1}{2k_n} \mathbb{A} \boldsymbol{q}_1^* : \boldsymbol{q}_1 \, dx & \text{if } \boldsymbol{q}^* \in Q_{f,g}^*, \\ +\infty & \text{otherwise,} \end{cases}$$

where the admissible function set $Q_{f,g}^*$ consists of all $\boldsymbol{q}^* = (\boldsymbol{q}_1, \boldsymbol{q}_2) \in Q^*$ such that $|\boldsymbol{q}_2^*| \le g$ a.e. on Γ_C and

$$\int_\Omega \left[(\boldsymbol{q}_1^* - \mathbb{C}\boldsymbol{\varepsilon}(\boldsymbol{u}_{n-1}^k)) : \boldsymbol{\varepsilon}(\boldsymbol{v}) + \boldsymbol{f}_{1n} \cdot \boldsymbol{v} \right] dx + \int_{\Gamma_N} \boldsymbol{f}_{2n} \cdot \boldsymbol{v} \, ds + \int_{\Gamma_C} \boldsymbol{q}_2^* \cdot \boldsymbol{v}_\tau \, ds$$
$$= 0 \qquad \forall \, \boldsymbol{v} \in V. \tag{3.143}$$

The dual problem of (3.141) is to find $\boldsymbol{p}^* \in Q_{f,g}^*$ such that

$$-F_n^*(\Lambda^* \boldsymbol{p}^*, -\boldsymbol{p}^*) = \sup_{\boldsymbol{q}^* \in Q_{f,g}^*} \left\{ -F_n^*(\Lambda^* \boldsymbol{q}^*, -\boldsymbol{q}^*) \right\}. \tag{3.144}$$

Applying Theorem 3.1, we know that both problems (3.141) and (3.144) have unique solutions and the following duality relation holds:

$$F_n(\boldsymbol{w}_n^k, \Lambda \boldsymbol{w}_n^k) = -F_n^*(\Lambda^* \boldsymbol{p}^*, -\boldsymbol{p}^*). \tag{3.145}$$

As in Section 3.3, we first let $\boldsymbol{w}_n^{hk} \in V$ be any approximation of $\boldsymbol{w}_n^k$. By using (3.127) and (3.129) we obtain

$$\frac{k_n}{2} \|\boldsymbol{w}_n^k - \boldsymbol{w}_n^{hk}\|_V^2 = J_n(\boldsymbol{w}_n^{hk}) - J_n(\boldsymbol{w}_n^k).$$

Let $\boldsymbol{p}^*$ be the solution of the dual problem (3.144). Relation (3.145) implies

$$\begin{aligned} J_n(\boldsymbol{w}_n^k) &= F(\boldsymbol{w}_n^k, \Lambda \boldsymbol{w}_n^k) \\ &= -F_n^*(\Lambda^* \boldsymbol{p}^*, -\boldsymbol{p}^*) \ge -F_n^*(\Lambda^* \boldsymbol{q}^*, -\boldsymbol{q}^*) \qquad \forall \, \boldsymbol{q}^* \in Q_{f,g}^*. \end{aligned}$$

Therefore, for any $\boldsymbol{q}^* = (\boldsymbol{q}_1^*, \boldsymbol{q}_2^*) \in Q_{f,g}^*$ and $\boldsymbol{r}^* \in (L^2(\Omega))^{d \times d}$,

$$\begin{aligned} &\frac{k_n}{2} \|\boldsymbol{w}_n^k - \boldsymbol{w}_n^{hk}\|_V^2 \\ &\quad \le J_n(\boldsymbol{w}_n^{hk}) + F_n^*(\Lambda^* \boldsymbol{q}^*, -\boldsymbol{q}^*) + \int_\Omega \frac{1}{2k_n} \mathbb{A} \boldsymbol{r}^* : \boldsymbol{r}^* dx - \int_\Omega \frac{1}{2k_n} \mathbb{A} \boldsymbol{r}^* : \boldsymbol{r}^* dx \\ &\quad = \int_\Omega \frac{1}{2k_n} \mathbb{C}(k_n \boldsymbol{\varepsilon}(\boldsymbol{w}_n^{hk}) + \mathbb{A}\boldsymbol{r}^*) : (k_n \boldsymbol{\varepsilon}(\boldsymbol{w}_n^{hk}) + \mathbb{A}\boldsymbol{r}^*) \, dx \\ &\qquad + \int_\Omega \left[\mathbb{C}\boldsymbol{\varepsilon}(\boldsymbol{u}_{n-1}^k) : \boldsymbol{\varepsilon}(\boldsymbol{w}_n^{hk}) - \boldsymbol{f}_{1n} \cdot \boldsymbol{w}_n^{hk} \right] dx + \int_{\Gamma_N} \boldsymbol{f}_{2n} \cdot \boldsymbol{w}_n^{hk} \, ds \\ &\qquad - \int_\Omega \boldsymbol{\varepsilon}(\boldsymbol{w}_n^{hk}) : \boldsymbol{r}^* dx + \int_{\Gamma_C} g \, |\boldsymbol{w}_{n\tau}^{hk}| \, ds + \int_\Omega \frac{1}{2k_n} (\mathbb{A} \boldsymbol{q}_1^* : \boldsymbol{q}_1^* - \mathbb{A} \boldsymbol{r}^* : \boldsymbol{r}^*) \, dx. \end{aligned} \tag{3.146}$$

It follows immediately from (3.143) that for any $\boldsymbol{q}^* = (\boldsymbol{q}_1^*, \boldsymbol{q}_2^*) \in Q^*_{f,g}$,

$$\int_\Omega \boldsymbol{q}_1^* : \boldsymbol{\varepsilon}(\boldsymbol{v})\, dx + \int_{\Gamma_C} \boldsymbol{q}_2^* \cdot \boldsymbol{v}_\tau\, ds = \int_\Omega \left[\mathbb{C}\boldsymbol{\varepsilon}(\boldsymbol{u}_{n-1}^k) : \boldsymbol{\varepsilon}(\boldsymbol{v}) - \boldsymbol{f}_{1n} \cdot \boldsymbol{v}\right] dx \\ - \int_{\Gamma_N} \boldsymbol{f}_{2n} \cdot \boldsymbol{v}\, ds \qquad \forall\, \boldsymbol{v} \in V. \qquad (3.147)$$

Using (3.147) and regrouping terms in (3.146) we find that for any $\boldsymbol{q}^* \in Q^*_{f,g}$ and $\boldsymbol{r}^* \in (L^2(\Omega))^{d\times d}$,

$$\frac{k_n}{2}\|\boldsymbol{w}_n^k - \boldsymbol{w}_n^{hk}\|_V^2 \le \int_\Omega \frac{1}{k_n}\mathbb{C}(k_n\boldsymbol{\varepsilon}(\boldsymbol{w}_n^{hk}) + \mathbb{A}\boldsymbol{r}^*) : (k_n\boldsymbol{\varepsilon}(\boldsymbol{w}_n^{hk}) + \mathbb{A}\boldsymbol{r}^*)\, dx \\ + \int_\Omega \frac{1}{k_n}\mathbb{A}(\boldsymbol{q}_1^* - \boldsymbol{r}^*) : (\boldsymbol{q}_1^* - \boldsymbol{r}^*)\, dx \\ + \int_{\Gamma_C} (g\,|\boldsymbol{w}_{n\tau}^{hk}| + \boldsymbol{q}_2^* \cdot \boldsymbol{w}_{n\tau}^{hk})\, ds.$$

Thus, for any $\boldsymbol{r}^* \in (L^2(\Omega))^{d\times d}$,

$$\frac{k_n}{2}\|\boldsymbol{w}_n^k - \boldsymbol{w}_n^{hk}\|_V^2 \le \int_\Omega \frac{1}{k_n}\mathbb{C}k_n(\boldsymbol{\varepsilon}(\boldsymbol{w}_n^{hk}) + \mathbb{A}\boldsymbol{r}^*) : (k_n\boldsymbol{\varepsilon}(\boldsymbol{w}_n^{hk}) + \mathbb{A}\boldsymbol{r}^*)\, dx \\ + \inf_{\boldsymbol{q}^* \in Q^*_{f,g}} \left\{\int_\Omega \frac{1}{k_n}\mathbb{A}(\boldsymbol{q}_1^* - \boldsymbol{r}^*) : (\boldsymbol{q}_1^* - \boldsymbol{r}^*)\, dx \right. \\ \left. + \int_{\Gamma_C} (g\,|\boldsymbol{w}_{n\tau}^{hk}| + \boldsymbol{q}_2^* \cdot \boldsymbol{w}_{n\tau}^{hk})\, ds\right\}. \qquad (3.148)$$

The second term

$$II \equiv \inf_{\boldsymbol{q}^* \in Q^*_{f,g}} \left\{\int_\Omega \frac{1}{k_n}\mathbb{A}(\boldsymbol{q}_1^* - \boldsymbol{r}^*) : (\boldsymbol{q}_1^* - \boldsymbol{r}^*)\, dx + \int_{\Gamma_C} (g\,|\boldsymbol{w}_{n\tau}^{hk}| + \boldsymbol{q}_2^* \cdot \boldsymbol{w}_{n\tau}^{hk})\, ds\right\}$$

on the right-hand side of estimate (3.148) is bounded as follows. First, from the definition (3.143), the term II equals

$$\inf_{\substack{\boldsymbol{q}^* \in Q^* \\ |\boldsymbol{q}_2^*| \le g}} \sup_{\boldsymbol{v} \in V} \left\{\int_\Omega \frac{1}{k_n}\mathbb{A}(\boldsymbol{q}_1^* - \boldsymbol{r}^*) : (\boldsymbol{q}_1^* - \boldsymbol{r}^*)\, dx + \int_{\Gamma_C} (g\,|\boldsymbol{w}_{n\tau}^{hk}| + \boldsymbol{q}_2^* \cdot \boldsymbol{w}_{n\tau}^{hk})\, ds \right. \\ + \int_\Omega \left[(\boldsymbol{q}_1^* - \mathbb{C}\boldsymbol{\varepsilon}(\boldsymbol{u}_{n-1}^k)) : \boldsymbol{\varepsilon}(\boldsymbol{v}) + \boldsymbol{f}_{1n} \cdot \boldsymbol{v}\right] dx \\ \left. + \int_{\Gamma_N} \boldsymbol{f}_{2n} \cdot \boldsymbol{v}\, ds + \int_{\Gamma_C} \boldsymbol{q}_2^* \cdot \boldsymbol{v}_\tau\, ds\right\}.$$

Here and below, the condition "$|\boldsymbol{q}_2^*| \le g$" stands for "$|\boldsymbol{q}_2^*| \le g$ a.e. on Γ_C." Substitute $\boldsymbol{q}_1^* - \boldsymbol{r}_1^*$ by $\boldsymbol{q}_1^*$ and regroup the terms to see that II equals

$$\inf_{|q_2^*|\le g}\sup_{v\in V}\left\{\int_\Omega\left[-\frac{k_n}{4}\mathbb{C}\varepsilon(v):\varepsilon(v)+(r^*-\mathbb{C}\varepsilon(u_{n-1}^k)):\varepsilon(v)+f_{1n}\cdot v\right]dx\right.$$
$$\left.+\int_{\Gamma_N}f_{2n}\cdot v\,ds+\int_{\Gamma_C}q_2^*\cdot v_\tau\,ds+\int_{\Gamma_C}(g\,|w_{n\tau}^{hk}|+q_2^*\cdot w_{n\tau}^{hk})\,ds\right\}.$$

Define the residual

$$\mathcal{R}_n(q_2^*,r^*)=\sup_{v\in V}\frac{1}{\|v\|_V}\left\{\int_\Omega\left[(r^*-\mathbb{C}\varepsilon(u_{n-1}^{hk})):\varepsilon(v)+f_{1n}\cdot v\right]dx\right.$$
$$\left.+\int_{\Gamma_N}f_{2n}\cdot v\,ds+\int_{\Gamma_C}q_2^*\cdot v_\tau\,ds\right\}. \tag{3.149}$$

Then

$$II\le\inf_{|q_2^*|\le g}\sup_{v\in V}\left\{-\frac{k_n}{4}\|v\|_V^2+(\mathcal{R}_n(q_2^*,r^*)+\|u_{n-1}^k-u_{n-1}^{hk}\|_V)\|v\|_V\right.$$
$$\left.+\int_{\Gamma_C}(g\,|w_{n\tau}^{hk}|+q_2^*\cdot w_{n\tau}^{hk})\,ds\right\}$$
$$=\inf_{|q_2^*|\le g}\left\{\frac{1}{k_n}(\mathcal{R}_n(q_2^*,r^*)+\|u_{n-1}^k-u_{n-1}^{hk}\|_V)^2\right.$$
$$\left.+\int_{\Gamma_C}(g\,|w_{n\tau}^{hk}|+q_2^*\cdot w_{n\tau}^{hk})\,ds\right\}.$$

Therefore, we have the following result.

Theorem 3.12. *Let $w_n^k\in V$ be the unique solution of (3.127), and $w_n^{hk}\in V$ an approximation. Then for any $r^*\in(L^2(\Omega))^{d\times d}$, the following error bound*

$$\frac{k_n}{2}\|w_n^k-w_n^{hk}\|_V^2\le\int_\Omega\frac{1}{k_n}\mathbb{C}(k_n\varepsilon(w_n^{hk})+\mathbb{A}r^*):(k_n\varepsilon(w_n^{hk})+\mathbb{A}r^*)\,dx$$
$$+\inf_{|q_2^*|\le g}\left\{\frac{1}{k_n}(\mathcal{R}_n(q_2^*,r^*)+\|u_{n-1}^k-u_{n-1}^{hk}\|_V)^2\right.$$
$$\left.+\int_{\Gamma_C}(g\,|w_{n\tau}^{hk}|+q_2^*\cdot w_{n\tau}^{hk})\,ds\right\}$$

is valid, where the residual $\mathcal{R}_n(q_2^*,r^*)$ is defined by (3.31).

Theorem 3.12 provides a general framework for various a posteriori error estimates with different choices of the auxiliary variable r^*. From now on, w_n^{hk} is the finite element solution defined by (3.136). We choose $r^*=-k_n\mathbb{C}\varepsilon(w_n^{hk})$, and then replace the infimum over q_2^* by taking $q_2^*=-g\lambda_{n\tau}^{hk}$. Then Theorem 3.12 leads to the following corollary.

Corollary 3.1. *Let $u_n^k\in V$ be the unique solution of (3.126) and $u_n^{hk}\in V^h$ its finite element approximation defined by (3.135). Then there exists a constant C such that the following a posteriori error estimate holds:*

$$\|\boldsymbol{u}_n^k - \boldsymbol{u}_n^{hk}\|_V \le C\mathcal{R}_n + C\|\boldsymbol{u}_{n-1}^k - \boldsymbol{u}_{n-1}^{hk}\|_V, \tag{3.150}$$

where

$$\mathcal{R}_n \equiv \sup_{\boldsymbol{v}\in V} \frac{1}{\|\boldsymbol{v}\|_V} \left\{ \int_\Omega \left[\mathbb{C}\boldsymbol{\varepsilon}(\boldsymbol{u}_n^{hk}) : \boldsymbol{\varepsilon}(\boldsymbol{v}) - \boldsymbol{f}_{1n}\cdot\boldsymbol{v}\right] dx - \int_{\Gamma_N} \boldsymbol{f}_{2n}\cdot\boldsymbol{v}\, ds \right.$$
$$\left. + \int_{\Gamma_C} g\boldsymbol{\lambda}_{n\tau}^{hk}\cdot\boldsymbol{v}_\tau\, ds \right\}. \tag{3.151}$$

Let Π_h be the interpolation operator defined by (3.130). Substituting $\boldsymbol{v}$ by $\boldsymbol{v} - \Pi_h\boldsymbol{v}$ in (3.151) we obtain

$$\mathcal{R}_n \equiv \sup_{\boldsymbol{v}\in V} \frac{1}{\|\boldsymbol{v}\|_V} \left\{ \int_\Omega \left[\mathbb{C}\boldsymbol{\varepsilon}(\boldsymbol{u}_n^{hk}) : \boldsymbol{\varepsilon}(\boldsymbol{v}-\Pi_h\boldsymbol{v}) - \boldsymbol{f}_{1n}\cdot(\boldsymbol{v}-\Pi_h\boldsymbol{v})\right] dx \right.$$
$$\left. - \int_{\Gamma_N} \boldsymbol{f}_{2n}\cdot(\boldsymbol{v}-\Pi_h\boldsymbol{v})\, ds + \int_{\Gamma_C} g\boldsymbol{\lambda}_{n\tau}^{hk}\cdot(\boldsymbol{v}-\Pi_h\boldsymbol{v})_\tau\, ds \right\}.$$

Decompose the integrals into local contributions from each element $K \in \mathcal{P}_h$ and apply Green's formula over K to find

$$\mathcal{R}_n = \sup_{\boldsymbol{v}\in V} \frac{1}{\|\boldsymbol{v}\|_V} \left\{ \sum_{K\in\mathcal{P}_h} \int_K (-\operatorname{Div}\boldsymbol{\sigma}(\boldsymbol{u}_n^{hk}) - \boldsymbol{f}_{1n})\cdot(\boldsymbol{v}-\Pi_h\boldsymbol{v})\, dx \right.$$
$$+ \sum_{\gamma\in\mathcal{E}_{h,0}} \int_\gamma [\boldsymbol{\sigma}(\boldsymbol{u}_n^{hk})\boldsymbol{\nu}]\cdot(\boldsymbol{v}-\Pi_h\boldsymbol{v})\, ds$$
$$+ \sum_{\gamma\in\mathcal{E}_{h,\Gamma_N}} \int_\gamma (\boldsymbol{\sigma}(\boldsymbol{u}_n^{hk})\boldsymbol{\nu} - \boldsymbol{f}_{2n})\cdot(\boldsymbol{v}-\Pi_h\boldsymbol{v})\, ds$$
$$\left. + \sum_{\gamma\in\mathcal{E}_{h,\Gamma_C}} \int_\gamma \left(\boldsymbol{\sigma}(\boldsymbol{u}_n^{hk})_\tau + g\boldsymbol{\lambda}_{n\tau}^{hk}\right)\cdot(\boldsymbol{v}-\Pi_h\boldsymbol{v})_\tau\, ds \right\}, \tag{3.152}$$

where we used decomposition $\boldsymbol{\sigma\nu}\cdot\boldsymbol{v} = \sigma_\nu v_\nu + \boldsymbol{\sigma}_\tau\cdot\boldsymbol{v}_\tau = \boldsymbol{\sigma}_\tau\cdot\boldsymbol{v}_\tau$ a.e. on Γ_C. Define the interior residuals for each element $K \in \mathcal{P}_h$ by

$$\boldsymbol{r}_K = -\operatorname{Div}\boldsymbol{\sigma}(\boldsymbol{u}_n^{hk}) - \boldsymbol{f}_{1n} \qquad \text{in } K, \tag{3.153}$$

and side residuals for each side $\gamma \in \mathcal{E}_h$ by

$$\boldsymbol{R}_\gamma = \begin{cases} [\boldsymbol{\sigma}(\boldsymbol{u}_n^{hk})\boldsymbol{\nu}] & \text{if } \gamma\in\mathcal{E}_{h,0}, \\ \boldsymbol{\sigma}(\boldsymbol{u}_n^{hk})\boldsymbol{\nu} - \boldsymbol{f}_{2n} & \text{if } \gamma\in\mathcal{E}_{h,\Gamma_N}, \\ \boldsymbol{\sigma}(\boldsymbol{u}_n^{hk})_\tau + g\boldsymbol{\lambda}_{n\tau}^{hk} & \text{if } \gamma\in\mathcal{E}_{h,\Gamma_C}. \end{cases} \tag{3.154}$$

Note that residuals corresponding to the sides lying on $\overline{\Gamma_D}$ are considered to be 0. By using definitions (3.153) and (3.154), relation (3.152) reduces to

$$\mathcal{R}_n = \sup_{\boldsymbol{v}\in V} \frac{1}{\|\boldsymbol{v}\|_V} \left\{ \sum_{K\in\mathcal{P}_h} \int_K \boldsymbol{r}_K \cdot (\boldsymbol{v} - \Pi_h \boldsymbol{v})\, dx + \sum_{\gamma\in\mathcal{E}_h} \int_\gamma \boldsymbol{R}_\gamma \cdot (\boldsymbol{v} - \Pi_h \boldsymbol{v})\, ds \right\}. \tag{3.155}$$

Using the estimates (3.133) and (3.134) in (3.155), and applying the Cauchy–Schwarz inequality, we get

$$\begin{aligned} \mathcal{R}_n &\le \sup_{\boldsymbol{v}\in V} \frac{1}{\|\boldsymbol{v}\|_V} \left\{ C\, \|\boldsymbol{v}\|_V \Big(\sum_{K\in\mathcal{P}_h} h_K^2 \|\boldsymbol{r}_K\|_{0;K}^2 + \sum_{\gamma\in\mathcal{E}_h} h_\gamma \|\boldsymbol{R}_\gamma\|_{0;\gamma}^2 \Big)^{\frac{1}{2}} \right\} \\ &\le C \left(\sum_{K\in\mathcal{P}_h} h_K^2 \|\boldsymbol{r}_K\|_{0;K}^2 + \sum_{\gamma\in\mathcal{E}_h} h_\gamma \|\boldsymbol{R}_\gamma\|_{0;\gamma}^2 \right)^{\frac{1}{2}}. \end{aligned} \tag{3.156}$$

We summarize the above results in the form of a theorem.

Theorem 3.13. *Let $\boldsymbol{u}_n^k \in V$ and $\boldsymbol{u}_n^{hk} \in V^h$ be the unique solutions of* (3.126) *and* (3.135)*, respectively. Then the following a posteriori error estimate holds:*

$$\begin{aligned} \|\boldsymbol{u}_n^k - \boldsymbol{u}_n^{hk}\|_V \le{}& C \left(\sum_{K\in\mathcal{P}_h} h_K^2 \|\boldsymbol{r}_K\|_{0;K}^2 + \sum_{\gamma\in\mathcal{E}_h} h_\gamma \|\boldsymbol{R}_\gamma\|_{0;\gamma}^2 \right)^{\frac{1}{2}} \\ &+ C\, \|\boldsymbol{u}_{n-1}^k - \boldsymbol{u}_{n-1}^{hk}\|_V, \end{aligned} \tag{3.157}$$

where $\boldsymbol{r}_K$ and $\boldsymbol{R}_\gamma$ are interior and, respectively, side residuals, defined by (3.153) *and* (3.154).

In practical computations, the terms on the right side of (3.157) are regrouped by writing

$$\|\boldsymbol{u}_n^k - \boldsymbol{u}_n^{hk}\|_V \le C \left(\sum_{K\in\mathcal{P}_h} \eta_{K,R}^2 \right)^{\frac{1}{2}} + C \|\boldsymbol{u}_{n-1}^k - \boldsymbol{u}_{n-1}^{hk}\|_V, \tag{3.158}$$

where the local error indicator $\eta_{K,R}$ on each element K, defined by

$$\eta_{K,R}^2 = h_K^2 \|\boldsymbol{r}_K\|_{0;K}^2 + \frac{1}{2} \sum_{\gamma\in\mathcal{E}(K)\cap\mathcal{E}_{h,0}} h_\gamma \|\boldsymbol{R}_\gamma\|_{0;\gamma}^2 + \sum_{\gamma\in\mathcal{E}(K)\cap\mathcal{E}_{h,\Gamma}} h_\gamma \|\boldsymbol{R}_\gamma\|_{0;\gamma}^2, \tag{3.159}$$

identifies contributions from each of the elements to the global error.

Similar to the derivation given in the second half of Section 3.4, we have the following inequality on the efficiency of the error estimate:

$$\sum_{K\in\mathcal{P}_h}\eta_{K,R}^2 \le C\Bigg(\|\boldsymbol{u}_n^k-\boldsymbol{u}_n^{hk}\|_V^2+\sum_{\gamma\in\mathcal{E}_{h,\Gamma_C}} h_\gamma\|\boldsymbol{\lambda}_{n\tau}^k-\boldsymbol{\lambda}_{n\tau}^{hk}\|_{0;\gamma}^2$$
$$+\sum_{K\in\mathcal{P}_h} h_K^2\|\boldsymbol{f}_{1n}-\boldsymbol{f}_{1n,K}\|_{0,K}^2+\sum_{\gamma\in\mathcal{E}_{h,\Gamma_N}} h_\gamma\|\boldsymbol{f}_{2n}-\boldsymbol{f}_{2n,\gamma}\|_{0,\gamma}^2$$
$$+\sum_{\gamma\in\mathcal{E}_{h,\Gamma_C}} h_\gamma\|\boldsymbol{\lambda}_{n\tau}^{hk}-\boldsymbol{\lambda}_{n\tau,\gamma}^{hk}\|_{0,\gamma}^2\Bigg) \tag{3.160}$$

with discontinuous piecewise polynomial approximations $\boldsymbol{f}_{1n,K}$, $\boldsymbol{f}_{2n,\gamma}$, and $\boldsymbol{\lambda}_{n\tau,\gamma}^{hk}$ of $\boldsymbol{f}_{1n}$, $\boldsymbol{f}_{2n}$, and $\boldsymbol{\lambda}_{n\tau}^{hk}$.

We can apply Theorem 3.12 to derive a stress recovery type a posteriori error estimate, and we restrict our discussion to linear elements. Similar to (3.58)–(3.59), we define the stress recovery operator $G_h : V^h \to (V^h)^d$ as follows:

$$G_h\boldsymbol{\sigma}(\boldsymbol{v}^h)=\boldsymbol{\sigma}^*(\boldsymbol{v}^h)=\sum_{\boldsymbol{a}\in\mathcal{N}_h}\boldsymbol{\sigma}_{\boldsymbol{a}}^*\varphi_{\boldsymbol{a}},\qquad \boldsymbol{\sigma}_{\boldsymbol{a}}^*=\frac{1}{|\widetilde{K}_{\boldsymbol{a}}|}\int_{\widetilde{K}_{\boldsymbol{a}}}\boldsymbol{\sigma}(\boldsymbol{v}^h)\,dx. \tag{3.161}$$

In the case of linear elements

$$\boldsymbol{\sigma}_{\boldsymbol{a}}^*=\sum_{i=1}^{N\boldsymbol{a}}\alpha_{\boldsymbol{a}}^i(\boldsymbol{\sigma}(\boldsymbol{v}^h))_{K_{\boldsymbol{a}}^i}, \tag{3.162}$$

where $(\boldsymbol{\sigma}(\boldsymbol{v}^h))_{K_{\boldsymbol{a}}^i}$ denotes the tensor value of the stress $\boldsymbol{\sigma}(\boldsymbol{v}^h)$ on the element $K_{\boldsymbol{a}}^i$, $\widetilde{K}_{\boldsymbol{a}}=\bigcup_{i=1}^{N_{\boldsymbol{a}}}K_{\boldsymbol{a}}^i$, $\alpha_{\boldsymbol{a}}^i=|K_{\boldsymbol{a}}^i|/|\widetilde{K}_{\boldsymbol{a}}|$, $i=1,\dots,N_{\boldsymbol{a}}$.

Let $\boldsymbol{u}_n^{hk}$ and $\boldsymbol{u}_{n-1}^{hk}$ be the finite element solutions defined by (3.135) at the time steps n and $n-1$, respectively. Let $\boldsymbol{r}^*=-\boldsymbol{\sigma}^*(\boldsymbol{u}_n^{hk})+\boldsymbol{\sigma}(\boldsymbol{u}_{n-1}^{hk})$ and replace the infimum over $\boldsymbol{q}_2^*$ by taking $\boldsymbol{q}_2^*=-g\boldsymbol{\lambda}_{n\tau}^{hk}$, where $\boldsymbol{\sigma}^*(\boldsymbol{u}_n^{hk})$ is the recovered stress defined by (3.161) and $\boldsymbol{\lambda}_{n\tau}^{hk}$ is provided by Theorem 3.17. With these choices, estimate (3.32) becomes

$$\frac{k_n^2}{2}\|\boldsymbol{w}_n^k-\boldsymbol{w}_n^{hk}\|_V^2 \le \int_\Omega(\boldsymbol{\sigma}(\boldsymbol{u}_n^{hk})-\boldsymbol{\sigma}^*(\boldsymbol{u}_n^{hk})):(\boldsymbol{\varepsilon}(\boldsymbol{u}_n^{hk})-\boldsymbol{\varepsilon}^*(\boldsymbol{u}_n^{hk}))\,dx$$
$$+\left(\mathcal{R}_n+\|\boldsymbol{u}_{n-1}^k-\boldsymbol{u}_{n-1}^{hk}\|_V\right)^2$$
$$\le C\|\boldsymbol{\sigma}(\boldsymbol{u}_n^{hk})-\boldsymbol{\sigma}^*(\boldsymbol{u}_n^{hk})\|_{0;\Omega}^2+\left(\mathcal{R}_n+\|\boldsymbol{u}_{n-1}^k-\boldsymbol{u}_{n-1}^{hk}\|_V\right)^2, \tag{3.163}$$

where $\boldsymbol{\varepsilon}^*(\boldsymbol{u}_n^{hk})=\mathbb{A}\boldsymbol{\sigma}^*(\boldsymbol{u}_n^{hk})$ and the residual $\mathcal{R}_n$ is given by

$$\mathcal{R}_n\equiv\sup_{\boldsymbol{v}\in V}\frac{1}{\|\boldsymbol{v}\|_V}\Bigg\{\int_\Omega\left[\boldsymbol{\sigma}^*(\boldsymbol{u}_n^{hk}):\boldsymbol{\varepsilon}(\boldsymbol{v})-\boldsymbol{f}_{1n}\cdot\boldsymbol{v}\right]dx-\int_{\Gamma_N}\boldsymbol{f}_{2n}\cdot\boldsymbol{v}\,ds$$
$$+\int_{\Gamma_C}g\,\boldsymbol{\lambda}_{n\tau}^{hk}\cdot\boldsymbol{v}_\tau\,ds\Bigg\}. \tag{3.164}$$

We can then obtain the following result.

Theorem 3.14. *Let $\boldsymbol{u}_n^k \in V$ and $\boldsymbol{u}_n^{hk} \in V^h$ be the unique solutions of* (3.126) *and* (3.135), *respectively. Then the following a posteriori error estimate holds:*

$$\|\boldsymbol{u}_n^k - \boldsymbol{u}_n^{hk}\|_V \le C\left(\sum_{K\in\mathcal{P}_h} \eta_{K,G}^2\right)^{\frac{1}{2}} + C\left(\sum_{\boldsymbol{a}\in\mathcal{N}_{h,0}} h_{\boldsymbol{a}}^2 \min_{\boldsymbol{f}_{\boldsymbol{a}}\in\mathbb{R}^d} \|\boldsymbol{f}_{1n} - \boldsymbol{f}_{\boldsymbol{a}}\|_{0;\widetilde{K}_{\boldsymbol{a}}}^2\right)^{\frac{1}{2}}$$
$$+ C\|\boldsymbol{u}_{n-1}^k - \boldsymbol{u}_{n-1}^{hk}\|_V, \qquad (3.165)$$

where local indicators $\eta_{K,G}$ are computed for every element $K \in \mathcal{P}_h$ by

$$\eta_{K,G}^2 = \|\boldsymbol{\sigma}(\boldsymbol{u}_n^{hk}) - \boldsymbol{\sigma}^*(\boldsymbol{u}_n^{hk})\|_{0;K}^2 + \sum_{\gamma\in\mathcal{E}(K)\cap\mathcal{E}_{h,\Gamma_N}} h_\gamma \|\boldsymbol{\sigma}^*(\boldsymbol{u}_n^{hk})\boldsymbol{\nu} - \boldsymbol{f}_{2n}\|_{0;\gamma}^2$$
$$+ \sum_{\gamma\in\mathcal{E}(K)\cap\mathcal{E}_{h,\Gamma_C}} h_\gamma \|\boldsymbol{\sigma}^*(\boldsymbol{u}_n^{hk})_{\boldsymbol{\tau}} + g\boldsymbol{\lambda}_{n\boldsymbol{\tau}}^{hk}\|_{0;\gamma}^2. \qquad (3.166)$$

We observe that if $\boldsymbol{f}_{1n} \in (L^2(\Omega))^d$ then

$$\left(\sum_{\boldsymbol{a}\in\mathcal{N}_h} h_{\boldsymbol{a}}^2 \min_{\boldsymbol{f}_{\boldsymbol{a}}\in\mathbb{R}^d} \|\boldsymbol{f}_{1n} - \boldsymbol{f}_{\boldsymbol{a}}\|_{0;\widetilde{K}_{\boldsymbol{a}}}^2\right)^{1/2} = o(h),$$

and if $\boldsymbol{f}_{1n} \in (H^1(\Omega))^d$, then

$$\left(\sum_{\boldsymbol{a}\in\mathcal{N}_h} h_{\boldsymbol{a}}^2 \min_{\boldsymbol{f}_{\boldsymbol{a}}\in\mathbb{R}^d} \|\boldsymbol{f}_{1n} - \boldsymbol{f}_{\boldsymbol{a}}\|_{0;\widetilde{K}_{\boldsymbol{a}}}^2\right)^{1/2} = O(h^2).$$

Theorem 3.14 asserts that the estimator $\eta_G = (\sum_{K\in\mathcal{P}_h} \eta_{K,G}^2)^{1/2}$ is a reliable upper bound of a constant multiple of the error $\|\boldsymbol{u}_n^k - \boldsymbol{u}_n^{hk}\|_V$. We can also show the following inequality:

$$\sum_{K\in\mathcal{P}_h} \eta_{K,G}^2 \le C\Bigg(\|\boldsymbol{u}_n^k - \boldsymbol{u}_n^{hk}\|_V^2 + \sum_{\gamma\in\mathcal{E}_{h,\Gamma_C}} h_\gamma \|\boldsymbol{\lambda}_{n\boldsymbol{\tau}}^k - \boldsymbol{\lambda}_{n\boldsymbol{\tau}}^{hk}\|_{0;\gamma}^2$$
$$+ \sum_{K\in\mathcal{P}_h} h_K^2 \|\boldsymbol{f}_{1n} - \boldsymbol{f}_{1n,K}\|_{0,K}^2 + \sum_{\gamma\in\mathcal{E}_{h,\Gamma_N}} h_\gamma \|\boldsymbol{f}_{2n} - \boldsymbol{f}_{2n,\gamma}\|_{0,\gamma}^2$$
$$+ \sum_{\gamma\in\mathcal{E}_{h,\Gamma_C}} h_\gamma \|\boldsymbol{\lambda}_{n\boldsymbol{\tau}}^{hk} - \boldsymbol{\lambda}_{n\boldsymbol{\tau},\gamma}^{hk}\|_{0,\gamma}^2\Bigg), \qquad (3.167)$$

with discontinuous piecewise polynomial approximations $\boldsymbol{f}_{1n,K}$, $\boldsymbol{f}_{2n,\gamma}$, and $\boldsymbol{\lambda}_{n\boldsymbol{\tau},\gamma}^{hk}$ of $\boldsymbol{f}_{1n}$, $\boldsymbol{f}_{2n}$, and $\boldsymbol{\lambda}_{n\boldsymbol{\tau}}^{hk}$.

3.10 Numerical Example on the Quasistatic Contact Problem

We follow the procedure described in Section 3.6 for the finite element approximation.

In order to show the effectiveness of the adaptive procedure we compare numerical convergence orders of the approximate solutions. We compute these orders by considering families of uniform and locally refined triangulations. Consider a sequence of finite element solutions $\{\boldsymbol{u}^{hk}_{n,un}\}_{n=0}^{N}$ based on uniform triangulations of the domain Ω and the same uniform partition of the time interval. Starting with an initial coarse triangulation $\mathcal{P}_1$, we construct a family of nested meshes by the subdivision of each triangle into four triangles. The solution from the most refined mesh is taken as the "true" solution $\{\boldsymbol{u}^{k}_{n}\}_{n=0}^{N}$ that is used to compute the errors of the approximate solutions obtained on the other meshes.

The finite element solution $\{\boldsymbol{u}^{hk}_{n,ad}\}_{n=0}^{N}$ is obtained using the following adaptive algorithm:

1. Start with the initial triangulation $\mathcal{P}_h$ and corresponding finite element subspace V^h.
2. Compute the finite element solution $\{\boldsymbol{u}^{hk}_{n,ad}\}_{n=1}^{N}$, where $\boldsymbol{u}^{hk}_{n,ad} \in V^h$, $0 \le n \le N$.
3. At the time step t_N and for each element $K \in \mathcal{P}_h$ compute the error estimator $\eta_{K,I}$ of residual type ($I = R$) or gradient recovery type ($I = G$).
4. Let $\eta = (\sum_{K \in \mathcal{P}_h} \eta_{K,I})/N_e$, where N_e is the total number of elements. An element K is marked for refinement if $\eta_K > \mu\,\eta$, where μ is a prescribed threshold. In our example, $\mu = 1$.
5. Perform refinement and obtain a new triangulation $\mathcal{P}_h$.
6. Return to step 2.

Example 3.4. We consider the physical setting shown in Figure 3.35. The domain $\Omega = (0, 2) \times (0, 10)$ is the cross-section of a three-dimensional linearly

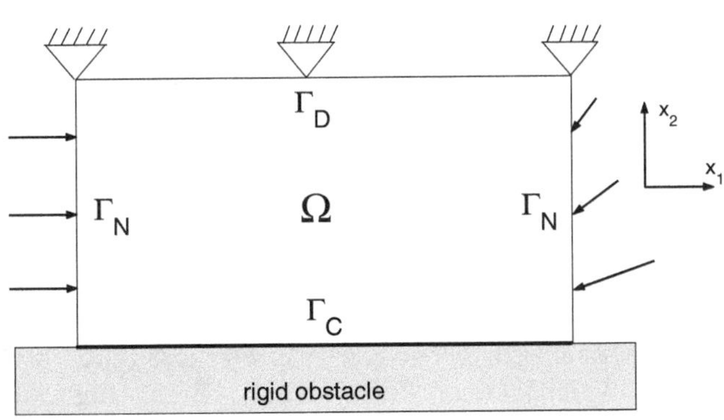

Fig. 3.35 Setting of the problem.

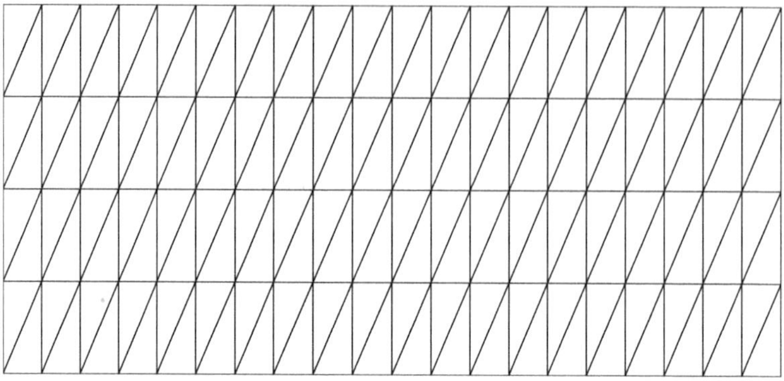

Fig. 3.36 Initial mesh.

elastic body and plane stress condition is assumed. On the part $\Gamma_D = (0,10)\times\{2\}$ the body is clamped. Oblique tractions act on the parts $\{0\}\times(0,2)$ and $\{10\}\times(0,2)$. Thus $\Gamma_N = (\{0\}\times(0,2))\cup(\{10\}\times(0,2))$. The contact part of the boundary is $\Gamma_C = (0,10)\times\{0\}$.

The elasticity tensor $\mathbb{C}$ satisfies

$$(\mathbb{C}\varepsilon)_{ij} = \frac{E\nu}{1-\nu^2}(\epsilon_{11}+\epsilon_{22})\delta_{ij} + \frac{E}{1+\nu}\epsilon_{ij}, \qquad 1\le i,j\le 2,$$

where E is the Young's modulus, ν is the Poisson's ratio of the material, and δ_{ij} is the Kronecker symbol. We use the following data:

$$\begin{aligned}
&E = 1000 \text{ daN/mm}^2, \qquad \nu = 0.3,\\
&\boldsymbol{f}_1 = (0,0) \text{ daN/mm}^2,\\
&\boldsymbol{f}_2(x_1,x_2,t) = (2.5t, 0) \text{ daN/mm}^2 \qquad \text{if } (x_1,x_2)\in\{0\}\times(0,2),\\
&\boldsymbol{f}_2(x_1,x_2,t) = (2(x_2-2)t, -t) \text{ daN/mm}^2 \qquad \text{if } (x_1,x_2)\in\{10\}\times(0,2),\\
&g = 1 \text{ daN/mm}^2, \qquad \boldsymbol{u}_0 = \boldsymbol{0}m, \qquad T = 1 \text{ sec.}
\end{aligned}$$

The time step considered is $t = 0.2$ sec. We start with the initial triangulation $\mathcal{P}_1$ (160 elements, 105 nodes) shown in Figure 3.36. Here, the interval $[0,1]$ is divided into $1/h$ equal parts with $h = 1/2$ which is successively halved. The numerical solution corresponding to $h = 1/64$ is taken as the "true" solution $\{\boldsymbol{u}_n^k\}_{n=0}^N$. To have an idea of the convergence behaviour of the discrete Lagrange multipliers, we compute the errors $\max_n \|\boldsymbol{\lambda}_{n\tau}^k - \boldsymbol{\lambda}_{n\tau}^{hk}\|_{0;\Gamma_C}$ corresponding to the sequence of uniform refinements. Here, $\{\boldsymbol{\lambda}_{n\tau}^k\}_{n=0}^N$ is the Lagrange multiplier corresponding to the parameter $h = 1/64$. Figure 3.43 provides the relative error values $\max_n \|\boldsymbol{u}_n^k - \boldsymbol{u}_{n,un}^{hk}\|_V$ and $h^{1/2}\max_n \|\boldsymbol{\lambda}_{n\tau}^k - \boldsymbol{\lambda}_{n\tau}^{hk}\|_{0;\Gamma_C}$. The numerical convergence order of $h^{1/2}\max_n \|\boldsymbol{\lambda}_{n\tau}^k - \boldsymbol{\lambda}_{n\tau}^{hk}\|_{0;\Gamma_C}$ is obviously higher than that of $\max_n \|\boldsymbol{u}_n^k - \boldsymbol{u}_{n,un}^{hk}\|_V$, indicating that the second term in the efficiency bounds (3.160) and (3.167) is expected to be of higher order compared to the first term $\|\boldsymbol{u}_n^k - \boldsymbol{u}_n^{hk}\|_V^2$. Graphs of $\lambda_{h,1}$ and $\lambda_{h,2}$ with $h = 1/64$ are provided in Figures 3.39 and 3.40.

We use an adaptive procedure based on both residual type and gradient recovery type estimates to obtain a sequence of approximate solutions $\{\boldsymbol{u}_{n,ad}^{hk}\}_{n=0}^N$. The deformed configuration and the adaptive finite element mesh after four adaptive iterations are shown in Figure 3.37 (based on the residual type estimator, 4792 elements, 2531 nodes) and Figure 3.38 (based on the recovery type estimator, 4894 elements, 2583 nodes). Figures 3.41 and 3.42 contain the relative error values $\max_n \|\boldsymbol{u}_n^k - \boldsymbol{u}_{n,un}^{hk}\|_V$ and $\max_n \|\boldsymbol{u}_n^k - \boldsymbol{u}_{n,ad}^{hk}\|_V$. We observe a substantial improvement of the efficiency using adaptively refined meshes. Figures 3.44 and 3.45 provide the values of $\eta_I = (\sum_K \eta_{K,I}^2)^{1/2}$, $I \in \{R, G\}$, where $\eta_{K,I}$ are computed using either residual type estimator ($I = R$) or recovery type estimator ($I = G$) on both uniform and adapted meshes. Table 3.6 contains the values of C_I computed for uniform and adaptive solutions:

$$C_I = \max_n \frac{\|\boldsymbol{u}_n^k - \boldsymbol{u}_n^{hk}\|_V}{\|\boldsymbol{u}_n^k\|_V \ \eta_I}, \qquad I \in \{R, G\}. \qquad \square$$

Table 3.6 Numerical values of C_R and C_G

h	1/2	1/4	1/8	1/16
C_R^{un}	9.29e-02	8.48e-02	8.27e-02	8.12e-02
C_R^{ad}	9.29e-02	8.37e-02	7.99e-02	7.56e-02
C_G^{un}	3.18e-01	2.99e-01	2.99e-01	2.99e-01
C_G^{ad}	3.18e-01	2.94e-01	2.92e-01	2.77e-01

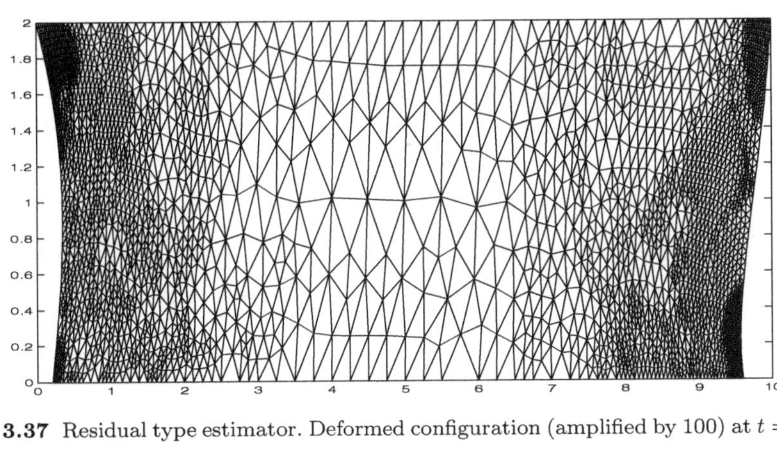

Fig. 3.37 Residual type estimator. Deformed configuration (amplified by 100) at $t = 1$ sec.

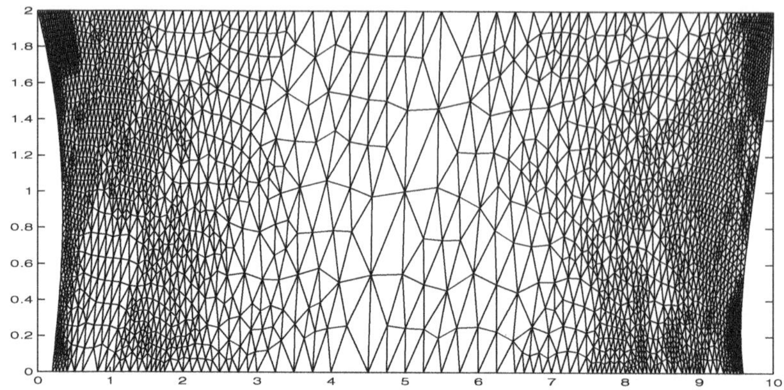

Fig. 3.38 Recovery type estimator. Deformed configuration (amplified by 100) at $t = 1$ sec.

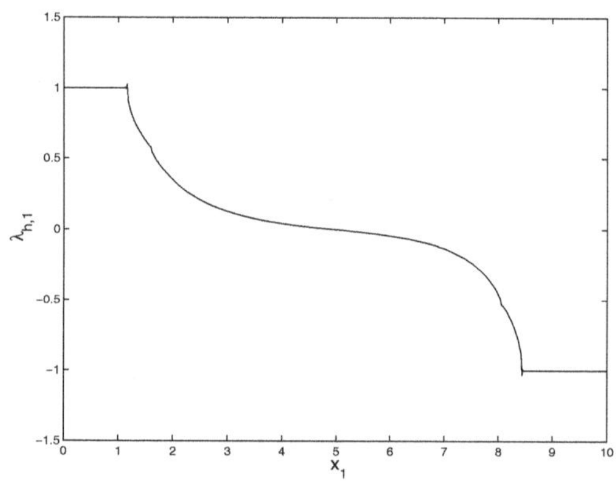

Fig. 3.39 Plot of λ_1^h at $t = 1$ sec.

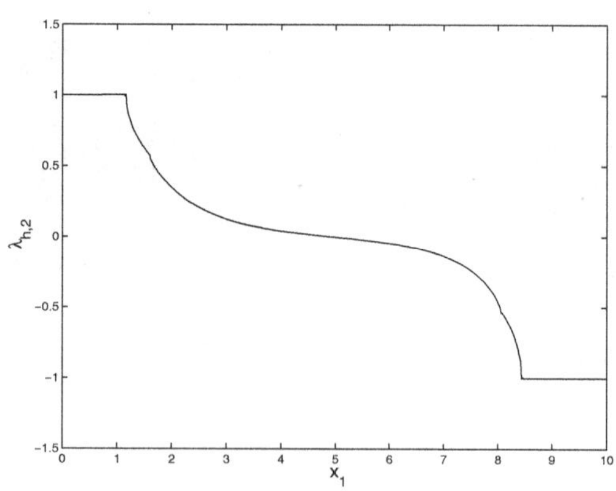

Fig. 3.40 Plot of λ_2^h at $t = 1$ sec.

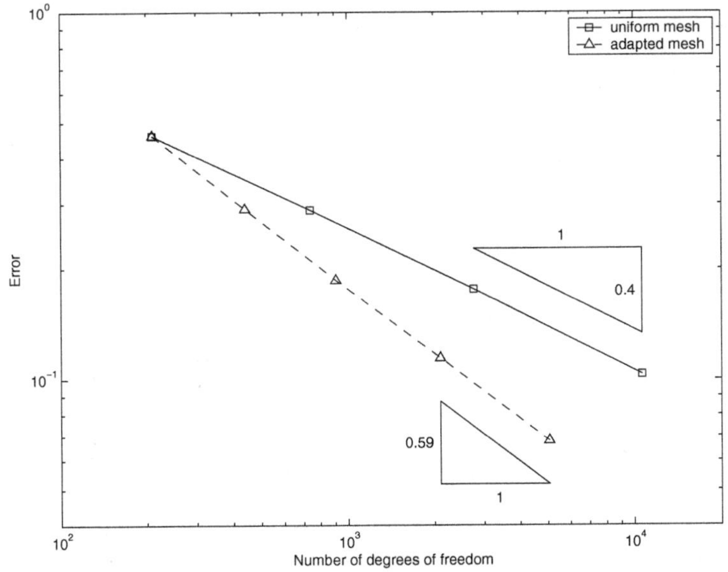

Fig. 3.41 Residual type estimator. $\max_n \|\boldsymbol{u}_n^k - \boldsymbol{u}_{n,un}^{hk}\|_V$ ($\square$) versus $\max_n \|\boldsymbol{u}_n^k - \boldsymbol{u}_{n,ad}^{hk}\|_V$ ($\triangle$).

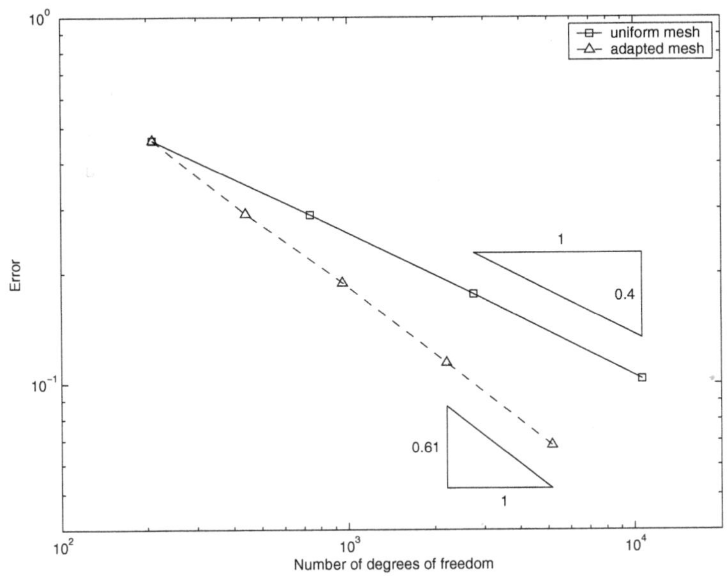

Fig. 3.42 Recovery type estimator. $\max_n \|\boldsymbol{u}_n^k - \boldsymbol{u}_{n,un}^{hk}\|_V$ ($\square$) versus $\max_n \|\boldsymbol{u}_n^k - \boldsymbol{u}_{n,ad}^{hk}\|_V$ ($\triangle$).

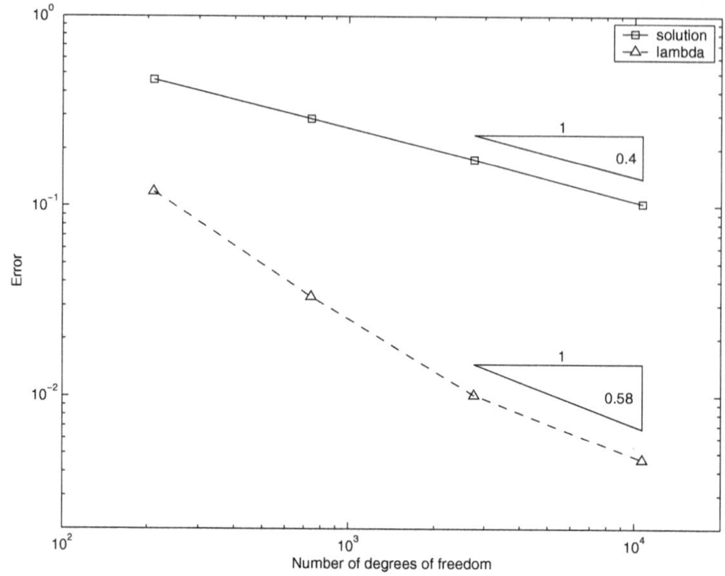

Fig. 3.43 $\max_n \|\boldsymbol{u}_n^k - \boldsymbol{u}_{n,un}^{hk}\|_V$ (□) versus $h^{1/2} \max_n \|\boldsymbol{\lambda}_{n\boldsymbol{\tau}}^k - \boldsymbol{\lambda}_{n\boldsymbol{\tau}}^{hk}\|_{0,\Gamma_C}$ (△).

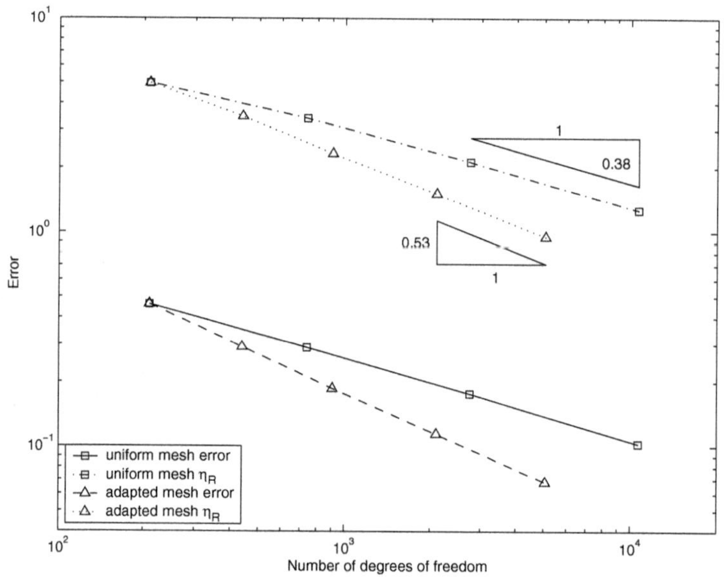

Fig. 3.44 Residual type estimator. $\max_n \|\boldsymbol{u}_n^k - \boldsymbol{u}_{n,un}^{hk}\|_V$ and η_R on uniform mesh (□) versus $\max_n \|\boldsymbol{u}_n^k - \boldsymbol{u}_{n,ad}^{hk}\|_V$ and η_R on adapted mesh (△).

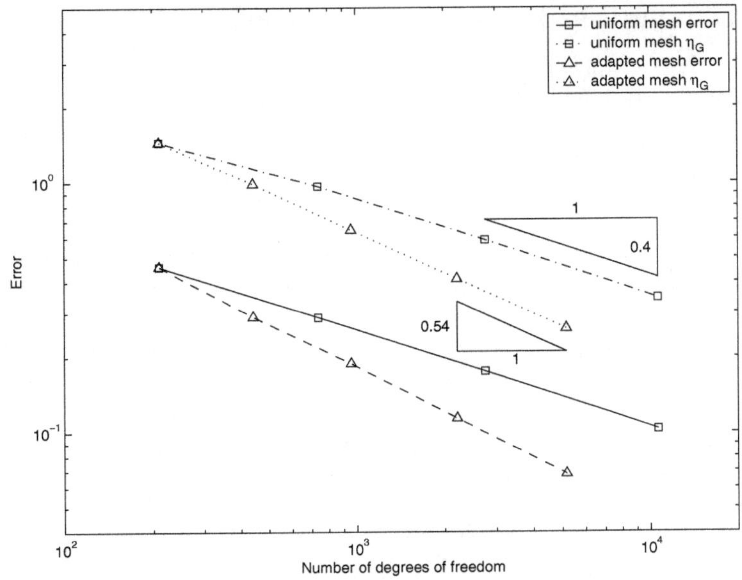

Fig. 3.45 Recovery type estimator. $\max_n \|\boldsymbol{u}_n^k - \boldsymbol{u}_{n,un}^{hk}\|_V$ and η_G on uniform mesh (□) versus $\max_n \|\boldsymbol{u}_n^k - \boldsymbol{u}_{n,ad}^{hk}\|_V$ and η_G on adapted mesh (△).

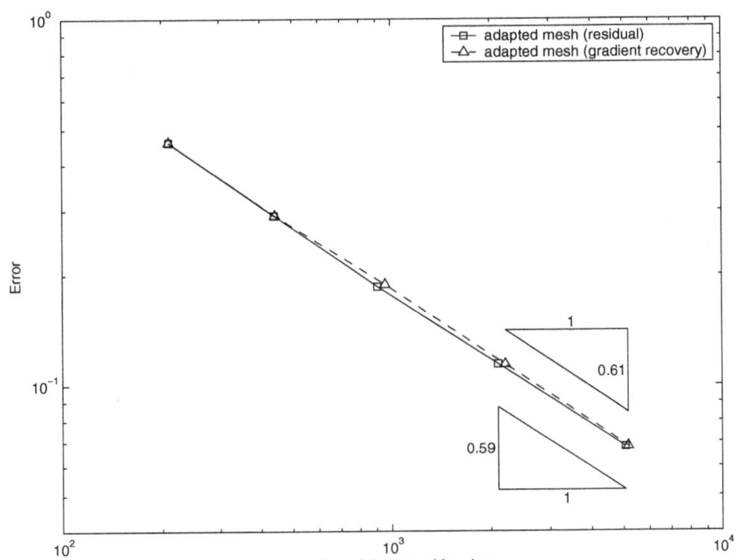

Fig. 3.46 Residual type estimator versus recovery type estimator.

3.11 Concluding Remarks

This chapter presents a general framework on a posteriori error estimation for finite element solutions of some variational inequalities, through the employment of the duality theory in convex analysis. The general error estimate contains an auxiliary variable. Different choices of the auxiliary variable leads to different a posteriori error bounds; in particular, residual type and recovery type error estimates are shown and analyzed in this chapter. The a posteriori error estimates are used in developing adaptive finite element algorithms for the variational inequalities. The effectiveness of the adaptive finite element algorithms is demonstrated in several numerical examples.

Due to the inequality nature of the problems, efficiency bounds for the error estimators contain terms related to approximations of Lagrange multipliers; see, for example, (3.57) and (3.82). Currently, sharp bounds for such terms are unknown. Ideally, it is hoped that terms such as

$$\sum_{\gamma \in \mathcal{E}_{h,\Gamma_2}} h_\gamma \|\lambda - \lambda_h\|_{0;\gamma}^2$$

can be bounded by $o(h^2)$, and then efficiency of the error estimators can be rigorously deduced from (3.57) and (3.82). Numerical results reported in the chapter strongly suggest the terms involving the Lagrange multiplier approximations are of higher order, and performance of the adaptive finite element algorithms is not sensitive with respect to the calculation of the Lagrange multipliers.

The derivation and analysis of the a posteriori error estimates are done on some model problems in this chapter, and can be extended to other or more general variational inequalities, including those arising in frictional contact mechanics. Another direction for future research is to develop adaptive algorithms for simultaneous time and space discretizations of time-dependent variational inequalities, and their applications in solving evolutionary problems in contact mechanics.

Acknowledgments We thank Professor David Y. Gao for various helpful suggestions on this chapter. This work was supported by NSF under grant DMS-0106781.

References

1. M. Ainsworth and J.T. Oden, *A Posteriori Error Estimation in Finite Element Analysis*, John Wiley, New York, 2000.
2. M. Ainsworth, J.T. Oden, and C.Y. Lee, Local *a posteriori* error estimators for variational inequalities, *Numer. Meth. PDE* **9** (1993), 23–33.

3. J. Alberty, C. Carstensen, and D. Zarrabi, Adaptive numerical analysis in primal elastoplasticity with hardening, *Comput. Meth. Appl. Mech. Eng.* **171** (1999), 175–204.
4. I. Babuška and A.K. Aziz, Survey lectures on the mathematical foundations of the finite element method, in A.K. Aziz, ed., *The Mathematical Foundations of the Finite Element Method with Applications to Partial Differential Equations*, Academic Press, New York, 1972, pp. 3–359.
5. I. Babuška and W.C. Rheinboldt, Error estimates for adaptive finite element computations, *SIAM J. Numer. Anal.* **15** (1978), 736–754.
6. I. Babuška and W.C. Rheinboldt, *A posteriori* error estimates for the finite element method, *Int. J. Numer. Meth. Eng.* **12** (1978), 1597–1615.
7. I. Babuška and T. Strouboulis, *The Finite Element Method and Its Reliability*, Oxford University Press, Oxford, 2001.
8. S. Bartels and C. Carstensen, Each averaging technique yields reliable a posteriori error control in FEM on unstructured grids. Part II: Higher order FEM, *Math. Comp.* **71** (2002), 971–994.
9. S. Bartels and C. Carstensen, Averaging techniques yield reliable a posteriori finite element error control for obstacle problems. Preprint Nr. 2/2001. Publications of the Max-Planck-Institute for Mathematics in the Sciences, Leipzig, Germany.
10. R. Becker and R. Rannacher, An optimal control approach to a posteriori error estimation in finite element methods, in A. Iserles, ed., *Acta Numerica*, Vol. 10, Cambridge University Press, Cambridge, 2001, pp. 1–102.
11. T. Belytschko, W.K. Liu, and B. Moran, *Nonlinear Finite Elements for Continua and Structures*, Wiley, Chichester, England, 2000.
12. H. Blum and F.-T. Suttmeier, An adaptive finite element discretisation for a simplified Signorini problem, *CALCOLO* **37** (2000), 65–77.
13. H. Blum and F.-T. Suttmeier, Weighted error estimates for finite element solutions of variational inequalities, *Computing* **65** (2000), 119–134.
14. V. Bostan and W. Han, Recovery-based error estimation and adaptive solution of elliptic variational inequalities of the second kind, *Commun. Math. Sci.* **2** (2004), 1–18.
15. V. Bostan, W. Han, and B.D. Reddy, A posteriori error analysis for elliptic variational inequalities of the second kind, in K.J. Bathe, ed., *Computational Fluid and Solid Mechanics 2003*, Proceedings of Second MIT Conference on Computational Fluid and Solid Mechanics, June 17–20, Elsevier Science, Oxford, 2003, pp. 1867–1870.
16. V. Bostan, W. Han, and B.D. Reddy, A posteriori error estimation and adaptive solution of elliptic variational inequalities of the second kind, *Appl. Numer. Math.* **52** (2004), 13–38.
17. D. Braess, *Finite Elements: Theory, Fast Solvers, and Applications in Solid Mechanics*, third edition, Cambridge University Press, Cambridge, 2007.
18. S.C. Brenner and L.R. Scott, *The Mathematical Theory of Finite Element Methods*, third edition, Springer-Verlag, New York, 2008.
19. F. Brezzi and M. Fortin, *Mixed and Hybrid Finite Element Methods*, Springer-Verlag, Berlin, 1991.
20. C. Carstensen, Numerical analysis of the primal problem of elastoplasticity with hardening, *Numer. Math.* **82** (1999), 577–597.
21. C. Carstensen, Quasi-interpolation and a posteriori analysis in finite element methods, *RAIRO Math. Model. Num.* **33** (1999), 1187–1202.
22. C. Carstensen and J. Alberty, Averaging techniques for reliable a posteriori FE-error control in elastoplasticity with hardening, *Comput. Meth. Appl. Mech. Eng.* **192** (2003), 1435–1450.
23. C. Carstensen and S. Bartels, Each averaging technique yields reliable a posteriori error control in FEM on unstructured grids. Part I: Low order conforming, nonconforming, and mixed FEM, *Math. Comp.* **71** (2002), 945–969.

24. C. Carstensen and R. Verfürth, Edge residuals dominate a posteriori error estimates for low order finite element methods, *SIAM J. Numer. Anal.* **36** (1999), 1571–1587.
25. Z. Chen and R.H. Nochetto, Residual type a posteriori error estimates for elliptic obstacle problems, *Numer. Math.* **84** (2000), 527–548.
26. P.G. Ciarlet, *The Finite Element Method for Elliptic Problems*, North-Holland, Amsterdam, 1978.
27. P.G. Ciarlet, Basic error estimates for elliptic problems, in P.G. Ciarlet and J.-L. Lions, eds., *Handbook of Numerical Analysis*, Vol. II, North-Holland, Amsterdam, 1991, pp. 17–351.
28. Ph. Clément, Approximation by finite element functions using local regularization, *RAIRO Numer. Anal.* **R-2** (1975), 77–84.
29. G. Duvaut and J.-L. Lions, *Inequalities in Mechanics and Physics*, Springer-Verlag, Berlin, 1976.
30. I. Ekeland and R. Temam, *Convex Analysis and Variational Problems*, North-Holland, Amsterdam, 1976.
31. D. French, S. Larsson, and R.H. Nochetto, Pointwise a posteriori error analysis for an adaptive penalty finite element method for the obstacle problem, *Comput. Meth. Appl. Math.* **1** (2001), 18–38.
32. M.B. Giles and E. Süli, Adjoint methods for PDEs: A posteriori error analysis and postprocessing by duality, in A. Iserles, ed., *Acta Numerica*, Vol. 11, Cambridge University Press, Cambridge, 2002, pp. 145–236.
33. V. Girault and P.-A. Raviart, *Finite Element Methods for Navier-Stokes Equations, Theory and Algorithms*, Springer-Verlag, Berlin, 1986.
34. R. Glowinski, *Numerical Methods for Nonlinear Variational Problems*, Springer-Verlag, New York, 1984.
35. R. Glowinski, J.-L. Lions, and R. Trémolières, *Numerical Analysis of Variational Inequalities*, North-Holland, Amsterdam, 1981.
36. W. Han, Finite element analysis of a holonomic elastic-plastic problem, *Numer. Math.* **60** (1992), 493–508.
37. W. Han, *A posteriori* error analysis for linearizations of nonlinear elliptic problems and their discretizations, *Math. Meth. Appl. Sci.* **17** (1994), 487–508.
38. W. Han, Quantitative error estimates for idealizations in linear elliptic problems, *Math. Meth. Appl. Sci.* **17** (1994), 971–987.
39. W. Han, *A Posteriori Error Analysis via Duality Theory, with Applications in Modeling and Numerical Approximations*, Springer Science+Business Media, 2005.
40. W. Han and S. Jensen, The Kačanov method for a nonlinear variational inequality of the second kind arising in elastoplasticity, *Chinese Ann. Math. Ser. B* **17** (1996), 129–138.
41. W. Han, S. Jensen, and I. Shimansky, The Kačanov method for some nonlinear problems, *Appl. Numer. Math.* **24** (1997), 57–79.
42. W. Han and B.D. Reddy, On the finite element method for mixed variational inequalities arising in elastoplasticity, *SIAM J. Numer. Anal.* **32** (1995), 1778–1807.
43. W. Han and B.D. Reddy, Computational plasticity: The variational basis and numerical analysis, *Comput. Mech. Adv.* **2** (1995), 283–400.
44. W. Han and B.D. Reddy, *Plasticity: Mathematical Theory and Numerical Analysis*, Springer-Verlag, New York, 1999.
45. W. Han, B.D. Reddy, and G.C. Schroeder, Qualitative and numerical analysis of quasistatic problems in elastoplasticity, *SIAM J. Numer. Anal.* **34** (1997), 143–177.
46. W. Han and M. Sofonea, *Quasistatic Contact Problems in Viscoelasticity and Viscoplasticity*, American Mathematical Society, Providence, RI, International Press, Somerville, MA, 2002.
47. J. Haslinger, I. Hlaváček, and J. Nečas, Numerical methods for unilateral problems in solid mechanics, in P.G. Ciarlet and J.-L. Lions, eds., *Handbook of Numerical Analysis*, Vol. IV, North-Holland, Amsterdam, 1996, pp. 313–485.

48. I. Hlaváček, J. Haslinger, J. Nečas, and J. Lovíšek, *Solution of Variational Inequalities in Mechanics*, Springer-Verlag, New York, 1988.
49. R.H.W. Hoppe and R. Kornhuber, Adaptive multilevel methods for obstacle problems, *SIAM J. Numer. Anal.* **31** (1994), 301–323.
50. T.J.R. Hughes, *The Finite Element Method*, Prentice-Hall, Englewood Cliffs, NJ, 1987.
51. C. Johnson, *Numerical Solutions of Partial Differential Equations by the Finite Element Method*, Cambridge University Press, Cambridge, 1987.
52. C. Johnson, Adaptive finite element methods for the obstacle problem, *Math. Models Meth. Appl. Sci.* **2** (1992), 483–487.
53. N. Kikuchi and J.T. Oden, *Contact Problems in Elasticity: A Study of Variational Inequalities and Finite Element Methods*, SIAM, Philadelphia, 1988.
54. D. Kinderlehrer and G. Stampacchia, *An Introduction to Variational Inequalities and Their Applications*, Academic Press, New York, 1980.
55. R. Kornhuber, A posteriori error estimates for elliptic variational inequalities, *Comput. Math. Appl.* **31** (1996), 49–60.
56. R.H. Nochetto, K.G. Siebert, and A. Veeser, Pointwise a posteriori error control for elliptic obstacle problems, *Numer. Math.* **95** (2003), 163–195.
57. J.T. Oden, Finite elements: An introduction, in P.G. Ciarlet and J.L. Lions, eds., *Handbook of Numerical Analysis*, Vol. II, North-Holland, Amsterdam, 1991, pp. 3–15.
58. J.T. Oden and J.N. Reddy, *An Introduction to the Mathematical Theory of Finite Elements*, John Wiley, New York, 1976.
59. P.D. Panagiotopoulos, *Inequality Problems in Mechanics and Applications*, Birkhäuser, Boston, 1985.
60. S.I. Repin, A posteriori error estimation for variational problems with uniformly convex functionals, *Math. Comp.* **69** (2000), 481–500.
61. S.I. Repin and L.S. Xanthis, A posteriori error estimation for elasto-plastic problems based on duality theory, *Comput. Meth. Appl. Mech. Eng.* **138** (1996), 317–339.
62. J.E. Roberts and J.-M. Thomas, Mixed and hybrid methods, in P.G. Ciarlet and J.-L. Lions, eds., *Handbook of Numerical Analysis*, Vol. II, North-Holland, Amsterdam, 1991, pp. 523–639.
63. Ch. Schwab, *p- and hp-Finite Element Methods*, Oxford University Press, 1998.
64. G. Strang and G. Fix, *An Analysis of the Finite Element Method*, Prentice-Hall, Englewood Cliffs, NJ, 1973.
65. F.-T. Suttmeier, General approach for a posteriori error estimates for finite element solutions of variational inequalities, *Comput. Mech.* **27** (2001), 317–323.
66. B. Szabó and I. Babuška, *Finite Element Analysis*, John Wiley, New York, 1991.
67. V. Thomée, *Galerkin Finite Element Methods for Parabolic Problems*, Lecture Notes in Mathematics, No. 1054, Springer-Verlag, New York, 1984.
68. V. Thomée, *Galerkin Finite Element Methods for Parabolic Problems*, second edition, Springer, New York, 2006.
69. A. Veeser, Efficient and reliable a posteriori error estimators for elliptic obstacle problems, *SIAM J. Numer. Anal.* **39** (2001), 146–167.
70. R. Verfürth, *A Review of A Posteriori Error Estimation and Adaptive Mesh Refinement Techniques*, Wiley and Teubner, New York, 1996.
71. N. Yan, A posteriori error estimators of gradient recovery type for elliptic obstacle problems, *Adv. Comput. Math.* **15** (2001), 333–362.
72. N. Yan and A. Zhou, Gradient recovery type a posteriori error estimates for finite element approximations on irregular meshes, *Comput. Meth. Appl. Mech. Eng.* **190** (2001), 4289–4299.
73. E. Zeidler, *Nonlinear Functional Analysis and Its Applications, III: Variational Methods and Optimization*, Springer-Verlag, New York, 1986.
74. Z. Zhang, A posteriori error estimates on irregular grids based on gradient recovery, *Adv. Comput. Math.* **15** (2001), 363–374.

75. O.C. Zienkiewicz, Origins, milestones and directions of the finite element method—A personal view, in P.G. Ciarlet and J.-L. Lions, eds., *Handbook of Numerical Analysis*, Vol. IV, North-Holland, Amsterdam, 1996, pp. 3–67.
76. O.C. Zienkiewicz and R.L. Taylor, *The Finite Element Method*, Vol. 1 (Basic Formulation and Linear Problems), McGraw-Hill, New York, 1989.
77. O.C. Zienkiewicz and R.L. Taylor, *The Finite Element Method*, Vol. 2 (Solid and Fluid Mechanics, Dynamics and Nonlinearity), McGraw-Hill, New York, 1991.
78. O.C. Zienkiewicz and J.Z. Zhu, A simple error estimator and adaptive procedure for practical engineering analysis, *Int. J. Numer. Meth. Eng.* **24** (1987), 337–357.
79. O.C. Zienkiewicz and J.Z. Zhu, The superconvergent patch recovery and a posteriori error estimates. Part 1: The recovery technique, *Int. J. Numer. Meth. Eng.* **33** (1992), 1331–1364.
80. O.C. Zienkiewicz and J.Z. Zhu, The superconvergent patch recovery and a posteriori error estimates. Part 2: Error estimates and adaptivity, *Int. J. Numer. Meth. Eng.* **33** (1992), 1365–1382.

Chapter 4
Time–Frequency Analysis of Brain Neurodynamics

W. Art Chaovalitwongse, W. Suharitdamrong, and P.M. Pardalos

Summary. The characteristics of neurodynamics of intracranial electroencephalogram (EEG) at different frequency bands were investigated in a sample of two patients with epilepsy. The results indicate a tendency for the gamma, theta, and alpha frequency bands in EEG signals to have a higher dimensional complexity than the beta and gamma frequency bands. We also investigate the time–frequency component decomposition of EEG signals and observe very different perceptual complexity and a difference in evoked spectral responses, which could be a reflection of neuronal recruitment that triggers the epileptogenic process. The results of this study may provide insights to the brain network's mechanism by which local and regional circuits can continuously form and reform with different regions functionally disconnected from other brain areas.

Key words: EEG, brain dynamics, time–frequency distribution, chaos theory, optimization

4.1 Introduction

The electroencephalogram (EEG) measures brainwaves of different frequencies within specific areas of the brain. Electrodes are implanted and placed

W. Art Chaovalitwongse
Department of Industrial and Systems Engineering, Rutgers, The State University of New Jersey, Piscataway, NJ, e-mail: wchaoval@rci.rutgers.edu

W. Suharitdamrong
Department of Industrial and Systems Engineering, University of Florida, Gainesville, FL, e-mail: wichais@ufl.edu

P.M. Pardalos
Departments of Industrial and Systems Engineering, Computer Science, and Biomedical Engineering, University of Florida, Gainesville, FL, e-mail: pardalos@cao.ise.ufl.edu

D.Y. Gao, H.D. Sherali, (eds.), *Advances in Applied Mathematics and Global Optimization*, Advances in Mechanics and Mathematics 17, DOI 10.1007/978-0-387-75714-8_4, on specific sites in the temporal lobe or the scalp to monitor and record the electrical impulses and neuronal activities in the brain. The extent of the noise effect depends on whether the additive noise caused by the remote electrodes is sufficient to induce important variations of the dynamics across the electrodes. It is well known that electrocortical activity exhibits microchaos while macroscopically behaving as a linear, near-equilibrium system. For this reason, EEG time series recorded at macroscopic scale may exhibit chaos, and it is common for physical phenomena to exhibit different dynamic features at different scales. Characterizing EEG signals by using measures of chaos derived from the theory of nonlinear dynamics has been proven to accurately reflect underlying attractor properties. The study of nonlinear dynamics involves the statistical investigation of molecular motion, sound waves, the weather, and other dynamical systems. As EEG exhibits particular dynamic properties at particular scales, which can be called far-from-equilibrium cellular and small cellular group interactions, the properties of EEG contrast with near-equilibrium dynamics and linear wave superposition at the macroscopic level. In this research, we focus on a question of whether the macroscopic EEG time series preserves chaotic dynamical features carried through from the underlying neural events driving the wave motion and how to apply a statistical analysis to study neurodynamics of EEG time series. Before one could claim to understand the neurodynamics of cortical neurons, one needs to understand how their dynamics differ with scale and how they interact across scale associated with the brain electrical activity. In this research, we direct our application to the epilepsy research. We investigate the neurodynamics of the EEG and its sensitivity to different frequency bands linked to epileptic waveforms in the brain of patients with epilepsy.

4.1.1 The Fact About Epilepsy and EEG

Epilepsy is the second most common serious brain disorder after stroke. Worldwide, at least 40 million people or 1% of the population currently suffer from epilepsy. Epilepsy is a chronic condition of diverse etiologies with the common symptom of spontaneous recurrent seizures, which is characterized by intermittent paroxysmal and highly organized rhythmic neuronal discharges in the cerebral cortex. Seizures can temporarily disrupt normal brain functions such as motor control, responsiveness, and recall, which typically last from seconds to a few minutes. There is a localized structural change in neuronal circuitry within the cerebrum which produces organized quasirhythmic discharges in some types of epilepsy (i.e., focal or partial epilepsy). These discharges then spread from the region of origin (epileptogenic zone) to activate other areas of the cerebral hemisphere. Nonetheless, the mechanism by which these fixed disturbances in local circuitry produce intermittent disturbances of brain function is not well comprehended. The development of the epileptic state can be considered as changes in network circuitry of neurons in the brain. When neuronal networks are activated, they produce changes in voltage potential, which can be captured by EEG. These changes are reflected by wriggling lines along the time axis in a typical EEG recording. A typical electrode montage for EEG recordings in our study is shown in Figure 4.1.

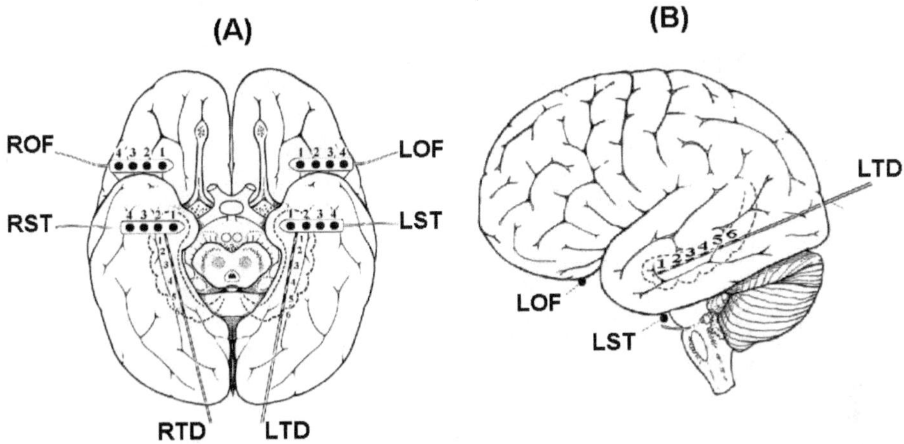

Fig. 4.1 Inferior transverse views of the brain, illustrating approximate depth and subdural electrode placement for EEG recordings. Subdural electrode strips are placed over the left orbitofrontal (LOF), right orbitofrontal (ROF), left subtemporal (LST), and right subtemporal (RST) cortex. Depth electrodes are placed in the left temporal depth (LTD) and right temporal depth (RTD) to record hippocampal activity.

The model for macroscopic neurodynamics depends on the epileptiforms in EEG time series (with the presence of noise) at different frequency bands. The effect of electrical activities at different frequency bands upon microscopic neural activity of high chaoticity level of the signal can be used to portray the mutual information between macroscopic pools of neurons in the brain. As instantaneous natural frequencies, damping factors, and coupling coefficients describing the dynamics of small pools of coupled neurons are stochastically independent, macroscopic neuronal wave motions are assumed to obey superposition and position near-equilibrium. This assumption leads to the intriguing possibility that local coherence among pools of neuronal cells might be generated from subsets of cells currently engaged in high-amplitude microscopic chaos, the local pool then providing a driving signal to the linear and near-equilibrium macroscopic dynamics. This mechanism appears to offer a further means of bridging dynamic interactions across scale. For this reason, the research is motivated to apply the nonlinear measures based on the theory of nonlinear dynamics as they have been proved capable of capturing the microscopic and macroscopic chaos in EEG time series.

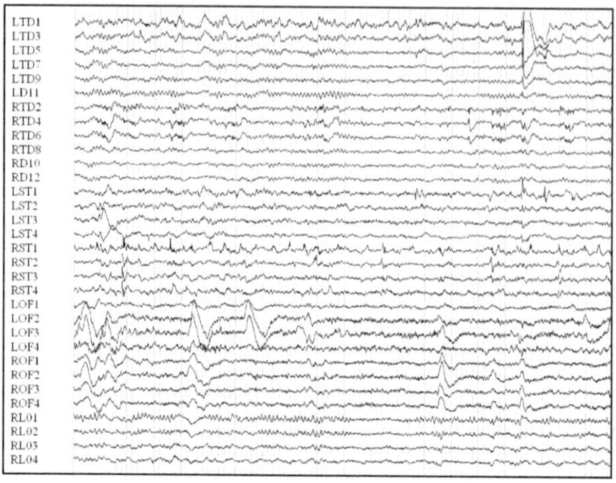

Fig. 4.2 Twenty second EEG recording of the pre-ictal state of a typical epileptic seizure obtained from 32 electrodes. Each horizontal trace represents the voltage recorded from electrode sites listed in the left column (see Figure 4.1 for anatomical location of electrodes).

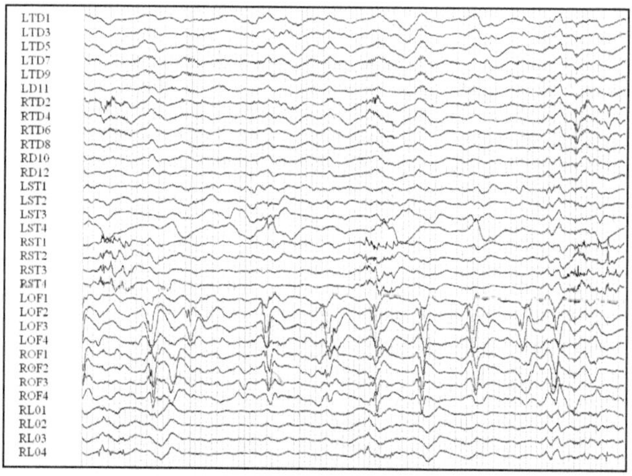

Fig. 4.3 Twenty second EEG recording of the post-ictal state of a typical epileptic seizure obtained from 32 electrodes. Each horizontal trace represents the voltage recorded from electrode sites listed in the left column (see Figure 4.1 for anatomical location of electrodes).

4.1.2 EEG Frequency Bands

The EEG measures brainwaves of different frequencies within the brain. Electrodes are placed on specific sites on the scalp to detect and record the electrical impulses within the brain as shown in Figures 4.1 to 4.3.

The raw EEG has usually been described in terms of frequency bands: Delta (0.1–3 Hz), Theta (4–8 Hz), Alpha (8–12 Hz), Beta (13–30 Hz), and Gamma (above 30 Hz).

Delta (0.1–3 Hz)

The EEG Delta waves are less than 4 Hz and occur in deep sleep and in some abnormal processes also during experiences of "empathy state." Delta waves are involved with our ability to integrate and release. As it reflects the unconscious mind, the information in our unconscious mind can be accessed through Delta. It is the dominant rhythm in infants up to one year of age and it is present in stages 3 and 4 of sleep. The EEG Delta waves tend to be the highest in amplitude and the slowest waves. However, most individuals diagnosed with attention deficit disorder, naturally increase rather than decrease Delta activity when trying to focus. The inappropriate Delta response often severely restricts the ability to focus and maintain attention. It is as if the brain is locked into a perpetual drowsy state [8].

Theta (4–8 Hz)

Theta activity has a frequency of 4 to 8 Hz and is classified as "slow" activity. It can be seen in connection with creativity, intuition, daydreaming, and fantasizing and is a repository for memories, emotions, and sensations. Theta waves are strong during internal focus, meditation, prayer, and spiritual awareness. It reflects the state between wakefulness and sleep. Theta is believed to reflect activity from the limbic system and hippocampal regions. Theta is observed in anxiety, behavioral activation, and behavioral inhibition [8].

Alpha (8–12 Hz)

Alpha waves are those between 8 and 12 Hz. Good healthy alpha production promotes mental resourcefulness, aids in the ability to mentally coordinate, and enhances an overall sense of relaxation and fatigue. When Alpha predominates most people feel at ease and calm. Alpha also appears to bridge the conscious to the subconscious. Alpha rhythms are reported to be derived from the white matter of the brain. The white matter can be considered the part of the brain that connects all parts with each other. Alpha is a common state for the brain and occurs whenever a person is alert, but not actively processing information. It can also be used as a marker for alertness and sleep. Alpha has been linked to extroversion (introverts show less), creativity (creative subjects show alpha when listening and coming to a solution for creative problems), and mental work. Alpha is one of the brain's most important frequencies for learning and using information taught in the classroom and on the job [8].

Beta (13–30 Hz)

Beta activity is considered to be "fast" activity. It reflects desynchronized active brain tissue. It is usually seen on both sides in symmetrical distribution and is most evident frontally; however, it may be absent or distorted in areas of cortical damage. Beta activity is generally regarded as a normal rhythm and is the dominant rhythm in those who are alert or anxious or who have their eyes open. It is the state that most of brain is in when we have our eyes open and are listening and thinking during analytical problem solving, judgment, decision making, or processing information about the world around us [8].

Gamma (above 30 Hz)

Gamma activity is the only frequency group found in every part of the brain. When the brain needs to simultaneously process information from different areas, it is hypothesized that the 40 Hz activity consolidates the required areas for simultaneous processing. A good memory is associated with well-regulated and efficient 40 Hz activity, whereas a 40 Hz deficiency creates learning disabilities [8].

4.1.3 Chaos in Brain

The term "chaos" in the theory of nonlinear dynamics is associated with exponential divergence of trajectories in phase space, which can reflect sensitivity to initial conditions. The evolution of chaos theory over the past two decades has mostly dealt with apparently simple systems with few degrees of freedom that can exhibit chaotic behavior. The brain is certainly a complex high-dimensional system, which has been proven to exhibit high-dimensional chaos associated with many brain state variables. The theory of chaos in brain appears that phenomena characteristic of many complex nonlinear systems (e.g., self-organization, interactions at multiple temporal and spatial scales, stable spatial structure in the presence of temporal chaos, etc.) occur in neocortex and may be closely aligned with cognitive processing. One of the most important concepts in the brain theories is that linear or quasilinear phenomena at one scale can coexist with highly nonlinear phenomena at another scale. This belief lies in communicating several general concepts, which are apparently critical to brain dynamic function, to disparate fields.

A proof of concept about the nonstationarity of EEG time series comes from that fact that although locally chaotic EEG dynamics with globally linear dynamics does not address evidence for nonstationary temporal signals in limit set dimension, in this study we deal with multichannel EEG time series and embed the time series in higher dimension, which makes globally chaotic EEG to be nonlinear dynamics in spatiotemporal aspects. This concept suggests that the dynamical properties of EEG time series in our model demonstrate significant large-scale temporal and spatial heterogeneity in cortical function, which can produce macroscopic chaos. Our approaches based on the theory of nonlinear dynamics seem to address the evidence for nonstationarity of EEG time series as a function of spatiotemporal correlation of coupled oscillators. A macroscopic model of EEG should address some issues that arise specifically at the transition from locally chaotic dynamics to the macroscopic scale. In this study, we consider the system of nonlinear EEG dynamics, including standing waves with frequency in the 10 Hz range (within a factor of about two or three), which will give us insights on the increasing frequency with maturation of the alpha rhythm and negative correlation between brain size and alpha frequency during seizure episodes.

The organization of the succeeding sections of this chapter is as follows. The background of previous studies in neocortical dynamics, neurodynamics of the brain, and the measures of chaoticity are described in Section 4.2. In Section 4.3, the design of experiment in this study is described as well as the computational methods to test the hypothesis. The results are described in Section 4.4. The conclusions and discussion are addressed in Section 4.5.

4.2 Neurodynamics of the Brain

In the last decade, time series analysis based on chaos theory and the theory of nonlinear dynamics, which are among the most interesting and growing research topics, has been applied to time series data with some degree of success. The concepts of chaos theory and the theory of nonlinear dynamics have not only been useful to analyze specific systems of ordinary differential equations or iterated maps, but have also offered new techniques for time series analysis. Moreover, a variety of experiments has shown that a recorded time series is driven by a deterministic dynamical system with a low-dimensional chaotic attractor, which is defined as the phase space point or set of points representing the various possible steady-state conditions of a system, an equilibrium state or group of states to which a dynamical system converges. Thus, the theories of chaos and nonlinear dynamics have provided new theoretical and conceptual tools that allow us to capture, understand, and link the complex behaviors of simple systems. Characterization and quantification of the dynamics of nonlinear time series are also important steps toward understanding the nature of random behavior and may enable us to predict the occurrences of some specific events that follow temporal dynamical patterns in the time series.

Several quantitative system approaches incorporating statistical technique nonlinear methods based on chaos theory have been successfully used to study epilepsy because the aperiodic and unstable behavior of the epileptic brain is suitable to nonlinear techniques that allow precise tracking of the temporal evolution. Our previous studies have shown that seizures are deterministic rather than random. Consequently, studies of the spatiotemporal dynamics in long-term intracranial EEGs, from patients with temporal lobe epilepsy, demonstrated the predictability of epileptic seizures; that is, seizures develop minutes to hours before clinical onset. The period of a seizure's development is called a pre-ictal transition period, which is characterized by gradual dynamical changes in EEG signals of critical electrode sites approximately 1/2 to 1 hour duration before the ictal onset [4, 22, 17, 20, 14, 18, 15, 26, 29, 34, 33]. During a pre-ictal transition period, gradual dynamical changes can be exposed by a progressive convergence (entrainment) of dynamical measures (e.g., short-term maximum Lyapunov exponents, $STL_{\max}$) at specific anatomical areas and cortical sites, in the neocortex and hippocampus. Another measure we have used in the state space created from the EEG at individual electrode sites in the brain, average angular frequency ($\bar{\Omega}$), has produced promising results too. The value of $\bar{\Omega}$ quantifies the average rate of the temporal change in the state of a system and is measured in rads/sec. Although the existence of the pre-ictal transition period has recently been confirmed and further defined by other investigators [9, 10, 23, 31, 24, 29], the characterization of this spatiotemporal transition is still far from complete. For instance, even in the same patient, a different set of cortical sites may exhibit a pre-ictal transition from one seizure to the next. In addition, this convergence of the normal sites with the epileptogenic focus (critical cortical sites) is reset after each seizure [4, 21, 18, 29]. Therefore, complete or partial post-ictal resetting of pre-ictal transition of the epileptic brain, affects the route to the subsequent seizure, contributing to the apparently nonstationary nature of the entrainment process. In those studies, however, the critical site selections are not trivial but extremely important because most groups of brain sites are irrelevant to the occurrences of the seizures and only certain groups of sites have dynamical convergence in the pre-ictal transition.

Because the brain is a nonstationary system, algorithms used to estimate measures of the brain dynamics should be capable of automatically identifying and appropriately weighing existing transients in the data. In a chaotic system, orbits originating from similar initial conditions (nearby points in the state space) diverge exponentially (expansion process). The rate of divergence is an important aspect of the system dynamics and is reflected in the value of Lyapunov exponents and dynamical phase. Estimates of Lyapunov exponents and phase velocity of EEG time series are shown to be consistent with a global theory of EEG in which waves are partly transmitted along corticocortical fibers in the brain. General features of neocortical dynamics and their implications for theoretical descriptions are considered. Such features include multiple scales of interaction, multiple connection lengths, local and global time constants, dominance of collective interactions at most scales, and periodic boundary conditions. Epileptiforms and other epileptic activities in EEG time series may be directly related to the continuous forming and reforming of local and regional circuits that are functionally disconnected from tissue involved in global operation.

4.2.1 Estimation of Lyapunov Exponents

The method we developed for estimation of short term largest Lyapunov exponents ($STL_{\max}$), an estimate of $L_{\max}$ for nonstationary data, is explained in detail elsewhere [13, 16, 36]. Herein we present only a short description of our method. Construction of the embedding phase space from a data segment $x(t)$ of duration T is made with the method of delays. The vectors X_i in the phase space (see Figure 4.4) are constructed as

$$X_i = (x(t_i), x(t_i + \tau) \ldots x(t_i + (p-1) * \tau)), \tag{4.1}$$

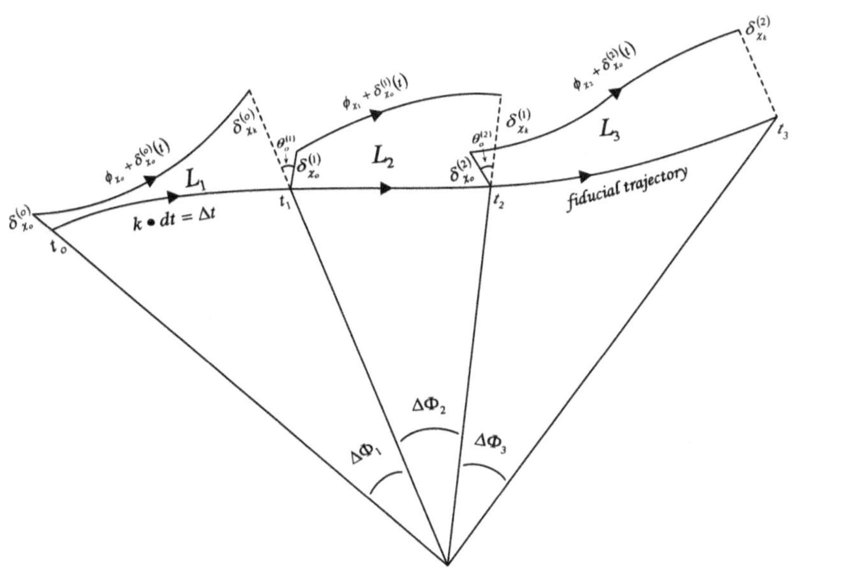

Fig. 4.4 Diagram illustrating the estimation of $STL_{\max}$ measures in the state space. The fiducial trajectory, the first three local Lyapunov exponents (L_1, L_2, L_3), is shown.

where τ is the selected time lag between the components of each vector in the phase space, p is the selected dimension of the embedding phase space, and $t_i \in [1, T-(p-1)\tau]$. If we denote by L the estimate of the short-term largest Lyapunov exponent $STL_{\max}$ then:

$$L = \frac{1}{N_a \Delta t} \sum_{i=1}^{N_a} \log_2 \frac{|\delta X_{i,j}(\Delta t)|}{|\delta X_{i,j}(0)|} \tag{4.2}$$

with

$$\delta X_{i,j}(0) = X(t_i) - X(t_j), \tag{4.3}$$

$$\delta X_{i,j}(\Delta t) = X(t_i + \Delta t) - X(t_j + \Delta t), \tag{4.4}$$

where

- $X(t_i)$ is the point of the fiducial trajectory $\phi_t(X(t_0))$ with $t = t_i$, $X(t_0) = (x(t_0), \ldots, x(t_0+(p-1)*\tau))$, and $X(t_j)$ is a properly chosen vector adjacent to $X(t_i)$ in the phase space (see below).
- $\delta X_{i,j}(0) = X(t_i) - X(t_j)$ is the displacement vector at t_i, that is, a perturbation of the fiducial orbit at t_i, and $\delta X_{i,j}(\Delta t) = X(t_i + \Delta t) - X(t_j + \Delta t)$ is the evolution of this perturbation after time Δt.
- $t_i = t_0 + (i-1) * \Delta t$ and $t_j = t_0 + (j-1) * \Delta t$, where $i \in [1, N_a]$ and $j \in [1, N]$ with $j \neq i$.
- Δt is the evolution time for $\delta X_{i,j}$, that is, the time one allows $\delta X_{i,j}$ to evolve in the phase space. If the evolution time Δt is given in sec, then L is in bits per second.
- t_0 is the initial time point of the fiducial trajectory and coincides with the time point of the first data in the data segment of analysis. In the estimation of L, for a complete scan of the attractor, t_0 should move within $[0, \Delta t]$.
- N_a is the number of local $L_{\max}$s estimated within a duration T data segment. Therefore, if D_t is the sampling period of the time domain data, $T = (N-1)D_t = N_a \Delta t + (p-1)\tau$.

We computed the $STL_{\max}$ profiles using the method proposed by Iasemidis and coworkers [13], which is a modification of the method by Wolf et al. [36]. We call the measure short term to distinguish it from those used to study autonomous dynamical systems studies. Modification of Wolf's algorithm is necessary to better estimate $STL_{\max}$ in small data segments that include transients, such as inter-ictal spikes. The modification is primarily in the search procedure for a replacement vector at each point of a fiducial trajectory. For example, in our analysis of the EEG, we found that the crucial parameter of the $L_{\max}$ estimation procedure, in order to distinguish between the pre-ictal, the ictal, and the post-ictal stages, was not the evolution time Δt nor the angular separation $V_{i,j}$ between the evolved displacement vector $\delta X_{i-1,j}(\Delta t)$ and the candidate displacement vector $\delta X_{i,j}(0)$ (as was claimed in Frank et al. [11]). The crucial parameter is the adaptive estimation in time and

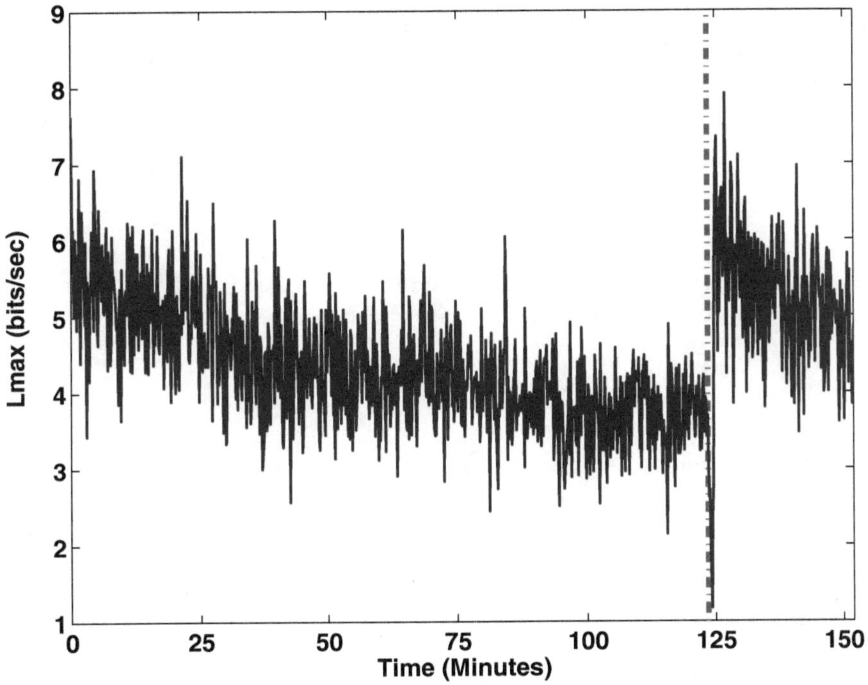

Fig. 4.5 Smoothed $STL_{\max}$ profiles over 2 hours derived from an EEG signal recorded at RTD2 (patient 1). A seizure (SZ ♯ 10) started and ended between the two vertical dashed lines. The estimation of the $L_{\max}$ values was made by dividing the signal into nonoverlapping segments of 10.24 sec each, using $p = 7$ and $\tau = 20$ msec for the phase space reconstruction. The smoothing was performed by a 10 point (1.6 min) moving average window over the generated $STL_{\max}$ profiles.

phase space of the magnitude bounds of the candidate displacement vector to avoid catastrophic replacements. Results from simulation data of known attractors have shown the improvement in the estimates of L achieved by using the proposed modifications [13]. In the pre-ictal state, depicted in Figure 4.5, one can see a trend of $STL_{\max}$ toward lower values over the whole pre-ictal period, with one prominent drop in the value of $STL_{\max}$ approximately 24 minutes prior to the seizure (denoted by an asterisk in the figure). This pre-ictal drop in $STL_{\max}$ can be explained as an attempt of the system toward a new state of less degrees of freedom long before the actual seizure [17].

4.2.2 Estimation of EEG Phase Velocity

Motivated by the representation of a state as a vector in the state space, another chaoticity measure employed in this research is a use of frequency-wave spectra to obtain phase velocity estimates. To estimate the phase velocity, we define the difference in phase between two evolved states $X(t_i)$ and $X(t_i+\Delta t)$ as $\Delta\Phi_i$ [19]. Then, denoting with $(\Delta\Phi)$ the average of the local phase differences $\Delta\Phi_i$ between the vectors in the state space, we have:

$$\Delta\Phi = \frac{1}{N_\alpha} \cdot \sum_{i=1}^{N_\alpha} \Delta\Phi_i, \tag{4.5}$$

where N_α is the total number of phase differences estimated from the evolution of $X(t_i)$ to $X(t_i + \Delta t)$ in the state space, according to:

$$\Delta\Phi_i = \left| \arccos \frac{X(t_i) \cdot X(t_i + \Delta t)}{\|X(t_i)\| \cdot \|X(t_i + \Delta t)\|} \right|. \tag{4.6}$$

Then, the average angular frequency $\bar{\Omega}$ is:

$$\bar{\Omega} = \frac{1}{\Delta t} \cdot \Delta\Phi. \tag{4.7}$$

If Δt is given in sec, then $\bar{\Omega}$ is given in rad/sec. Thus, whereas $STL_{\max}$ measures the local stability of the state of the system on average, $\bar{\Omega}$ measures how fast a local state of the system changes on average (e.g., dividing $\bar{\Omega}$ by 2π, the rate of change of the state of the system is expressed in $\sec^{-1}$ = Hz).

An example of a typical $\bar{\Omega}$ profile over time is given in Figure 4.6. The values are estimated from a 60 minute long EEG sample recorded from an electrode located in the epileptogenic hippocampus. The EEG sample includes a 2 minute seizure that occurs in the middle of the recording. The state space was reconstructed from sequential, nonoverlapping EEG data segments of 2048 points (sampling frequency 200 Hz, hence each segment of 10.24 sec in duration) with $p = 7$ and $\tau = 4$, as for the estimation of $STL_{\max}$ profiles [19]. The pre-ictal, ictal, and post-ictal states correspond to medium, high, and lower values of $\bar{\Omega}$, respectively. The highest $\bar{\Omega}$ values were observed during the ictal period, and higher $\bar{\Omega}$ values were observed during the pre-ictal period than during the post-ictal period. This pattern roughly corresponds to the typical observation of higher frequencies in the original EEG signal ictally, and lower EEG frequencies post-ictally. However, these observations can hardly denote a long-term warning of an impending seizure. The estimates of phase velocity of EEG data exhibit evidence for an underlying characteristic velocity of the EEG time series along the cortical surface. These results support the general theoretical idea that an EEG is composed of traveling waves, which are partly combined to form standing waves and propagated along corticocortical fibers. However, the results demonstrate that the existence of waves at such large scales does not preclude the simultaneous existence of waves at several smaller scales in which propagation can be adequately described in terms of exclusively intracortical interactions.

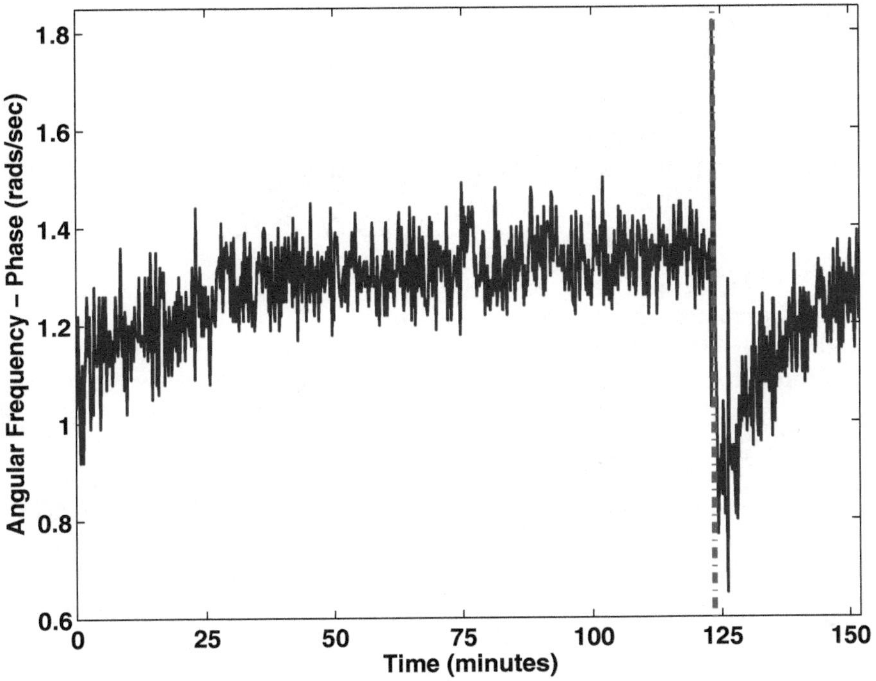

Fig. 4.6 A typical $\bar{\Omega}$ profile before, during and after an epileptic seizure, estimated from the EEG recorded from a site in the epileptogenic hippocampus; the seizure occurred between the vertical lines.

4.2.3 Spatiotemporal Dynamics

Based on the estimated $STL_{\max}$ and $\bar{\Omega}$ profiles at individual cortical sites, the temporal evolution of the stability of each cortical site is quantified. However, the system under consideration (brain) has a spatial extent and, as such, information about the transition of the system towards the ictal state should also be included in the interactions of its spatial components. The spatial dynamics of this transition are captured by consideration of the relations of the $STL_{\max}$ (and $\bar{\Omega}$) between different cortical sites. For example, if a similar transition occurs at different cortical sites, the $STL_{\max}$ of the involved sites are expected to converge to similar values prior to the transition. We have called such participating sites "critical electrode sites." We have used periods of 10 minutes (i.e., moving windows including approximately 60 $STL_{\max}$ values over time at each electrode site) to test the convergence at the 0.01 statistical significance level. We employed the T-index (from the well-known paired T-statistics for comparisons of means) as a measure of distance between the mean values of pairs of $STL_{\max}$ profiles over time. The T-index at time t between electrode sites i and j is defined as

$$T_{i,j}(t) = \sqrt{N} \times |E\{STL_{\max,i} - STL_{\max,j}\}| / \sigma_{i,j}(t), \tag{4.8}$$

where $E\{\cdot\}$ is the sample average difference for the $STL_{\max,i} - STL_{\max,j}$ estimated over a moving window $w_t(\lambda)$ defined as

$$w_t(\lambda) = \begin{cases} 1 & \text{if } \lambda \in [t-N-1, t], \\ 0 & \text{if } \lambda \notin [t-N-1, t], \end{cases}$$

where N is the length of the moving window. Then, $\sigma_{i,j}(t)$ is the sample standard deviation of the $STL_{\max}$ differences between electrode sites i and j within the moving window $w_t(\lambda)$. The thus defined T-index follows a t-distribution with $N-1$ degrees of freedom. For the estimation of the $T_{i,j}(t)$ indices in our data we used $N = 60$ (i.e., average of 60 differences of $STL_{\max}$ exponents between sites i and j per moving window of approximately 10 minute duration). Therefore, a two-sided t-test with $N-1$ ($= 59$) degrees of freedom, at a statistical significance level α should be used to test the null hypothesis, H_o: "brain sites i and j acquire identical $STL_{\max}$ values at time t." In this experiment, we set $\alpha = 0.01$, the probability of a type I error, or better, the probability of falsely rejecting H_o if H_o is true, is 1%. For the T-index to pass this test, the $T_{i,j}(t)$ value should be within the interval [0,2.662].

4.2.4 Optimization in the Brain Neurodynamics

Having quantified the spatiotemporal dynamics of the brain, we propose optimization techniques to identify the critical electrode sites. This problem can be naturally modeled as a quadratic 0–1 program, which has been extensively used to study Ising spin glass models [1, 2, 3, 12, 25]. Specifically, the critical electrode selection problem is formulated as a quadratic 0–1 knapsack problem with the objective function to minimize the average T-index (a measure of statistical distance between the mean values of $STL_{\max}$) among electrode sites and the knapsack constraint to identify the number of critical cortical sites. The problem is formally formulated as follows.

Let A be an $n \times n$ matrix, whose each element $a_{i,j}$ represents the T-index between electrode i and j within a 10 minute window before the onset of a seizure. Define $x = (x_1, \ldots, x_n)$, where each x_i represents the cortical electrode site i. If the cortical site i is selected to be one of the critical electrode sites, then $x_i = 1$; otherwise, $x_i = 0$.

A quadratic function is defined on R^n by

$$\min f(x) = x^T A x, \qquad \text{s.t. } x_i \in \{0,1\}, i = 1, \ldots, n, \tag{4.9}$$

where A is an $n \times n$ matrix [27, 28]. Next, we add a linear constraint, $\sum_{i=1}^{n} x_i = k$, where k is the number of critical electrode sites that we want to select. We now consider the following linearly constrained quadratic 0–1 problem:

$$\bar{P}: \min f(x) = x^T Ax, \quad \text{s.t.} \sum_{i=1}^{n} x_i = k \quad \text{for some } k, x \in \{0,1\}^n, \ A \in R^{n\times n}.$$

Problem $\bar{P}$ can be formulated as a quadratic 0–1 problem of the form as in (4.9) by using an exact penalty. If $A = (a_{ij})$ then let

$$M = 2\left[\sum_{j=1}^{n}\sum_{i=1}^{n} |a_{ij}|\right] + 1.$$

Then, we have the following equivalent problem P.

$$P: \min g(x) = x^T Ax + M\left(\sum_{i=1}^{n} x_i - k\right)^2, \qquad \text{s.t. } x \in \{0,1\}^n, A \in R^{n\times n}.$$

Such a problem can be solved by applying a branch and bound algorithm with a dynamic rule for fixing variables [27, 28].

Previous studies by our group have shown the existence of resetting of the brain after seizure onset [35, 21, 32], that is, divergence of $STL_{\max}$ profiles after seizures. Therefore, to ensure that the optimal group of critical sites shows this divergence, we reformulate this optimization problem by adding one more quadratic constraint. The quadratically constrained quadratic zero–one problem is given by:

$$\min \ x^T Ax \tag{4.10}$$

$$\text{s.t. } \sum_{i=1}^{n} x_i = k \tag{4.11}$$

$$x^T Bx \geq T_\alpha k(k-1), \tag{4.12}$$

where $x_i \in \{0,1\} \ \forall i \in \{1,\ldots,n\}$.

Note that the matrix $B = (b_{ij})$ is the T-index matrix of brain sites i and j within 10 minute windows after the onset of a seizure. T_α is the critical value of the T-index, as previously defined, to reject H_o: "two brain sites acquire identical $\bar{\Omega}$ values within time window $w_t(\lambda)$."

With one more quadratic constraint, the quadratic 0–1 problem becomes much harder to solve. Note that in the approach, a branch and bound algorithm with a dynamic rule for fixing variables cannot be applied to solve this problem because of the additional quadratic constraint [27, 28]. A conventional linearization has been proposed to solve this problem by introducing a new variable for each product of two variables and adding some additional constraints, and then formulating this problem as a mixed-integer linear (MILP) problem. Specifically, for each product $x_i x_j$, we introduce a new 0–1 variable, $x_{ij} = x_i x_j$ $(i \neq j)$. Note that $x_{ii} = x_i^2 = x_i$ for $x_i \in \{0, 1\}$. The equivalent MILP problem is given by

$$\min \sum_i \sum_j a_{ij} x_{ij} \tag{4.13}$$

$$\text{s.t.} \sum_{i=1}^{n} x_i = k, \tag{4.14}$$

$$x_{ij} \leq x_i, \qquad \text{for } i, j = 1, \ldots, n \ (i \neq j) \tag{4.15}$$

$$x_{ij} \leq x_j, \qquad \text{for } i, j = 1, \ldots, n \ (i \neq j) \tag{4.16}$$

$$x_i + x_j - 1 \leq x_{ij}, \qquad \text{for } i, j = 1, \ldots, n \ (i \neq j) \tag{4.17}$$

$$\sum_i \sum_j b_{ij} x_{ij} \geq T_\alpha k(k-1), \tag{4.18}$$

where $x_i \in \{0, 1\}$ and $0 \leq x_{ij} \leq 1$, $i, j = 1, \ldots, n$.

Although the above-mentioned linearization technique can be used to solve the quadratically constrained quadratic integer program, a better linearization technique has been proposed [5]. The reason is that the above formulation is computationally inefficient as n increases. A more efficient MILP for electrode selection problem proposed in [5, 26] is given by

$$\min \sum_{i=1}^{n} s_i \tag{4.19}$$

$$\text{s.t.} \sum_{i=1}^{n} x_i - k = 0 \tag{4.20}$$

$$-\sum_{j=1}^{n} a_{ij} x_j + s_i + y_i = 0, \qquad \text{for } i = 1, \ldots, n \tag{4.21}$$

$$y_i - M(1 - x_i) \leq 0, \qquad \text{for } i = 1, \ldots, n \tag{4.22}$$

$$h_i - M' x_i \leq 0, \qquad \text{for } i = 1, \ldots, n \tag{4.23}$$

$$-\sum_{j=1}^{n} d_{ij} x_j + h_i \leq 0, \qquad \text{for } i = 1, \ldots, n \tag{4.24}$$

$$\sum_{i=1}^{n} h_i \geq T_\alpha k(k-1), \tag{4.25}$$

where $x_i \in \{0, 1\}$ and $s_i, y_i, h_i \geq 0$, for $i, j = 1, \ldots, n$, and $M' = \|A\|_\infty$ and $M = \|B\|_\infty$.

4.3 Design of Experiments

In this study, we investigate the properties of the brain neurodynamics from the continuous long-term multichannel intracranial EEG recordings that had been acquired from two patients with medically intractable temporal lobe epilepsy. The recordings were obtained as part of a pre-surgical clinical evaluation. They had been obtained using Nicolet BMSI 4000 and 5000 recording systems, using a 0.1 Hz high-pass and a 70 Hz low-pass filter. Each record included a total of 28 to 32 intracranial electrodes (8 subdural and 6 hippocampal depth electrodes for each cerebral hemisphere).

In this framework, we filter the EEG data and then estimate $STL_{\max}$ and phase velocity, which measure the order or disorder of EEG signals recorded from individual electrode sites. Because the EEG consists of several frequency components (as described in the introduction) embedded in a single time series, bandpass filters are used to extract a particular frequency band from the EEG. In this experiment Butterworth filters are used to filter specified frequency bands (Delta, Theta, Alpha, Beta, and Gamma). In this study, we use the 10th order of Butterworth filters for each frequency band. Filter parameters were generated by using the butter command in MATLAB. Then the EEG from each channel was passed to each filter as shown in Figure 4.7.

In these experiments, we aim to investigate if the mechanism of epileptogenesis implies a specific function for each frequency band response. Stimuli by epileptogenesis processes are likely to elicit dynamical changes in intracortical interactions. Such stimulation might trigger neurodynamical processes and we investigate whether additional processes will follow.

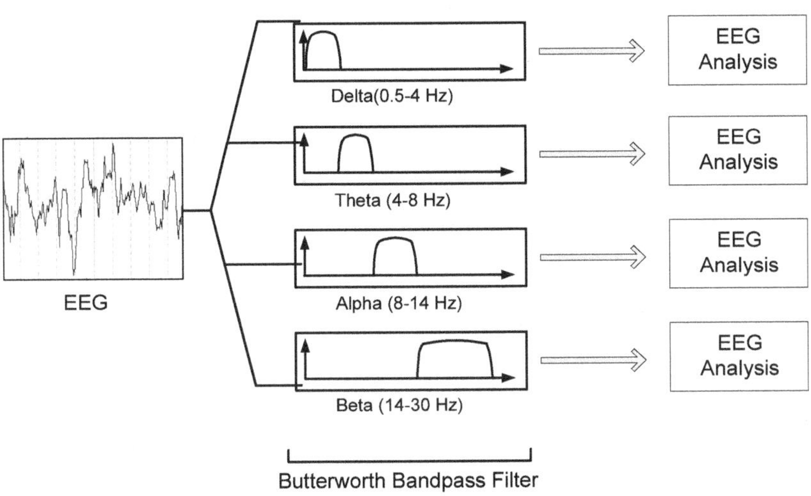

Fig. 4.7 Filter design of the proposed analysis for each frequency band.

4.3.1 Hypothesis I

To determine whether synchronized and/or repetitive activity of neurons serves a specific epileptogenesis process, one must compare brain responses in two states (normal and abnormal) only one of which induces changes in neurodynamics. In this study, we speculate that the chaoticity level of different frequency bands of EEG data moving in the same direction lead to synchronous and fast oscillatory activity of neurons activated by the epileptogenesis stimuli. We hypothesize that filtering EEG data by using a different cutoff frequency might give different effects/synchronous and fast oscillatory activity of the EEG, which can be reflected from the average value of $STL_{\max}$ profiles. To test this hypothesis, we have to implement the narrowband low-pass filter (LPF) using an IIR filter. Note that designing this filter is a very difficult task (but very straightforward to implement). One of the reasons is that it is extremely hard for a filter to get such a sharp response. Another reason is that a perfectly linear phase (constant group delay) cannot be realized using IIR filtering. There are several approaches to designing an approximately linear phase IIR filter. A more widely used approach is to iteratively design filters that simultaneously minimize errors in magnitude and group delay. Another widely used approach is to design an IIR filter that approximates the desired magnitude response (e.g., an elliptic filter) and then design an IIR all-pass filter which compensates for the nonlinear phase.

In this study, we implement IIR filtering procedures using the FD toolbox in MATLAB, which makes it possible to design a very good LPF magnitude response using an elliptic filter. This technique allows us to keep the poles within a circle of a specified radius. This approach employs a least-pth algorithm that attempts to minimize an L_p-norm error. The signal's magnitude response of the L_p-norm error is given by:

$$|H|_p = \frac{1}{2\pi} \int_{-\pi}^{\pi} |H(\omega) - H_d(\omega)|^p \, W(\omega) \, d\omega^{\frac{1}{p}}, \tag{4.26}$$

where H is the actual frequency response, H_d is the desired response, and W is some weighting function. In practice, the weighting function is normally equal to 1 over the passband and stopband and 0 in the transition band. Note that minimizing the L_2-norm is equivalent to minimizing the root mean square (RMS) error in the magnitude. In contrast, the L_∞-norm is equivalent to minimizing the maximum error over the frequencies of interest. Once the magnitude response in our experiment has been set, we need to perform group delay equalization to yield approximately constant group delay using a least-pth algorithm to constrain the radius of the poles.

To construct a LPF, we employ a frequency response function of the Butterworth LPF. This type of filter is especially useful because the random errors involved in the raw position EEG data obtained through reconstruction are characterized by relatively high frequency contents. The Butterworth

LPF can be expressed by

$$|H_c(j\omega)|^2 = \frac{1}{1 + (j\omega/j\omega_c)^{2n}}, \tag{4.27}$$

where $j = \sqrt{-1}$, ω is the frequency (rad/s), ω_c is the cutoff frequency (rad/s), and $n =$ the order of the filter. When $\omega = 0$, the magnitude-squared function (H_c^2) equals 1 and the frequency component is completely passed. When $\omega = \infty$, H_c^2 equals 0 and the frequency component is completely stopped. Between the passband and the stopband, there is the transition band ($0 < H_c^2 < 1$) in which the frequency component will be partially passed but partially stopped at the same time. When $\omega = \omega_c$, H_c^2 always equals 0.5 (half-power) regardless of the order of the filter. The Butterworth LPF is usually represented by the transfer function of its normalized form, $|H(j\omega)|^2 = 1/\left(1 + \omega^{2n}\right)$. The frequency response function of the Butterworth filter involves complex numbers ($j\omega$). Thus, the magnitude-squared function is the product of the response function pairs $H_c(j\omega)$ and $H_c(-j\omega)$ given by

$$|H_c(j\omega)|^2 = H_c(j\omega) \cdot H_c(-j\omega) = \frac{1}{1 + (j\omega/j\omega_c)^{2n}}.$$

The Butterworth LPF gives transfer functions that are rational functions but finding its roots results in a transfer function of the form:

$$\frac{1}{(s - s_1)(s - s_2)\cdots(s - s_n)} = \frac{1}{s_n + a_{n-1}s^{n-1} + \cdots + a_1 s + 1},$$

where a_i is the root of $1 + (-1)^n s^{2n} = 0$.

4.3.2 Hypothesis II

Before the ictal period, there are frequency components that have frequency responses that can be used to verify that the brain is fundamentally parallel. In this study, neuropsychological and neuroanatomical evidence might suggest that both its function and its organization are modular. To what extent can parallels be drawn between modularity of epileptogenesis processes as exemplified by cerebral localization of function which might be revealed by an epileptogenic zone? In addition, we speculate that this phenomenon might be a result of quadratic phase couplings. We hypothesize that this phenomenon is a result of quadratic phase couplings, which can be manifested through the bispectrum of the EEG signal by decomposing the Wigner–Ville time–frequency components of the EEG signal. The frequency components appearing in the spectrogram should form a nonperfect sine wave that creates a higher-order harmonic of the signals. By employing Choi–Williams time–frequency distribution, we should be able to capture the frequency responses of the EEG that show only the main component of EEG signals (not the harmonic component), which is speculated to be reflected from the epileptogenesis processes.

Wigner–Ville Distribution

The Wigner–Ville distribution (WVD) is employed to capture the temporal development of epileptogenesis, which might be reflected from the time-dependent variation in amplitude changes of each frequency band. The WVD is defined as follows:

$$W(t,f) = \frac{1}{2\pi}\int_{-\infty}^{\infty} h\left(t-\frac{\tau}{2}\right) h^*\left(t+\frac{\tau}{2}\right) e^{2\pi i f\tau}\, d\tau, \tag{4.28}$$

where $h(t)$ is the time series data [7]. In practice, we normally use the discrete analogue of the previous equation, which is represented by $W_{jk} = \sum_{\ell=-N/2}^{N/2} h_{(j-\ell/2)}\ h_{(j+\ell/2)}\ e^{2\pi ik\ell/N}$. The WVD is the most suitable and promising time–frequency distribution map for our study because it satisfies a large number of desirable mathematical properties including: (a) energy conservation (the energy of h can be obtained by integrating the WVD of h all over the time–frequency), (b) marginal properties (the energy spectral density and the instantaneous power can be obtained as marginal distributions of $W(t,f)$), and (c) compatibility with filterings (the WVD expresses the fact that if a signal y is the convolution of x and z, the WVD of y is the time-convolution between the WVD of z and the WVD of x).

Choi–Williams Distribution

The Choi–Williams distribution (CWD) expresses the EEG time series from the spectral density point of view. The CWD, which is a time–frequency distribution of Cohen's class, introduces an exponential function of the time series (also called exponential distribution). The exponential kernel is used to control the cross-terms as represented in the generalized ambiguity function domain. It is also considered to be the Fourier transform of the time-indexed autocorrelation function $K(t,\tau)$ estimated at a given time t. The CWD is given by

$$C(t,f) = \frac{1}{2\pi}\int_{-\infty}^{\infty} e^{j\tau f} K(t,\tau)\, d\tau, \tag{4.29}$$

where

$$K(t,\tau) = \int_{-\infty}^{\infty} \frac{1}{\sqrt{4\pi\tau^2/\sigma}} \cdot \exp\left[-\frac{(u-t)^2}{4\tau^2/\sigma}\right] \cdot h\left(u-\frac{\tau}{2}\right) \cdot h^*\left(u+\frac{\tau}{2}\right) du, \tag{4.30}$$

where σ is a factor controlling the suppression of cross-terms and the frequency resolution. Note that $C(t,f)$ becomes the Wigner–Ville distribution when $\sigma \to \infty$. It also satisfies the marginal conditions for all the values of σ. To give a reasonable accuracy to the estimate $K(t, \tau)$, the range of time average should be increased for the large value of τ [6]. It is worth mentioning that the Choi–Williams transform is quite expensive in terms of computational complexity.

4.4 Results

Based on the experimental design, the results of these hypothesis testings will give more insights in complex physical components in different frequency bands of EEG signals, which are essential features of neocortical dynamics at multiple scales of interaction with different frequency bands and dominance of collective interactions at most scales, and periodic boundary conditions. In addition, the time–frequency component decomposition may provide insights to the brain network's mechanism by which local and regional circuits can continuously form and reform with different regions functionally disconnected from other tissue (a form of self-organization). In the neurodynamics theory, the switching between more local and more global operation in time–frequency components is governed by local and global control parameters speculated to change due to the influences of various neuromodulators in epileptogenic processes. The following two sections explain the results of hypothesis testings and resultant observations in this study.

4.4.1 Hypothesis I

Figure 4.8 illustrates $STL_{\max}$ profiles of original EEG data and EEG data after passing a lowpass filter (LPF) of 10, 20, 30, and 50 Hz. Obviously, the brain activity at Delta and Theta bands (under 10 Hz) is more chaotic than the activity at higher frequency bands. However, during a seizure, the drop in the chaoticity at Delta and Theta bands is much less prominent than that of higher frequency bands. After a seizure, we observe an activity at the high Gamma band (over 50 Hz), which makes the EEG signal become less chaotic (more ordered) 10 minutes after a seizure. It is worth noting that there is no significant activity in the EEG data between 20–50 Hz, which considered to be in high Beta and low Gamma bands. This result verifies the theory used to categorize frequency bands in EEG data based on neuroanatomical connections in the brain at higher frequency bands. If cortical synchrony of EEG data at different frequency bands were an indicator of epileptogenesis processes, it could be expected that ongoing activity between 10–50 Hz bands would invoke more complex integration processes during the seizure, which is reflected by more pronounced synchronized activity.

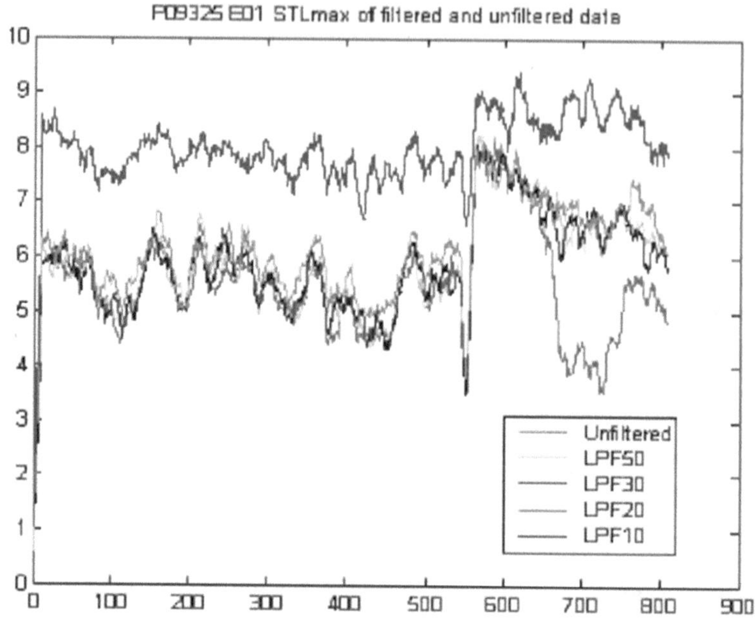

Fig. 4.8 $STL_{\max}$ profiles of EEG data from Electrode 1: Unfiltered and after passing LPF 10, 20, 30, 40, and 50 Hz.

Based on the implication of our previous observation, the $STL_{\max}$ profile of filtered (LPF-10) appears to contain characteristics different from those at higher frequency bands. Also we observed that the EEG activity at 20–50 Hz bands does not contain pronounced characteristics associated with epileptogenic processes. Figure 4.9 illustrates $STL_{\max}$ profiles of original EEG data and EEG data from Electrode 1 (LTD1) after passing lowpass filter of 10 and 50 Hz, or bandpass filter (BPF) of 10 and 20 Hz. Surprisingly, the EEG activity at the frequency range of 10–20 Hz is much less chaotic than that at other frequency bands and the EEG activity at the frequency range lower than 10 Hz is much more chaotic than that at other frequency bands. This implies that EEG activity at Alpha and low Beta bands is much less chaotic than EEG data at all frequency bands. In addition, EEG activity at the Delta and Theta bands is much more chaotic than EEG data at all frequency bands. This phenomenon can be explained by the fact that more and less complex manipulative activity is also correlated with different patterns of synchronized oscillatory brain activity in different frequency ranges and synchronized Delta–Theta activity can be recorded from the motor and somatosensory cortices of the brain. Furthermore, Delta activity is related with attention deficit disorder which could occur in patients with epilepsy as they sometimes lose their ability to focus or maintain attention. Figure 4.10 illustrates $STL_{\max}$ profiles of original EEG data and EEG data from Electrode

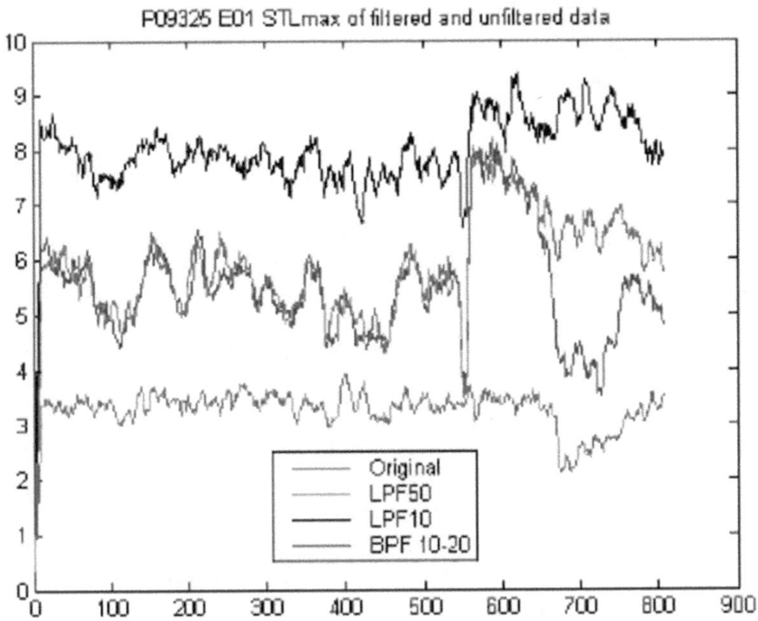

Fig. 4.9 $STL_{\max}$ profiles of EEG data from Electrode 1 after passing LPF 10 Hz, BPF 10–20 Hz, and LPF 50 Hz.

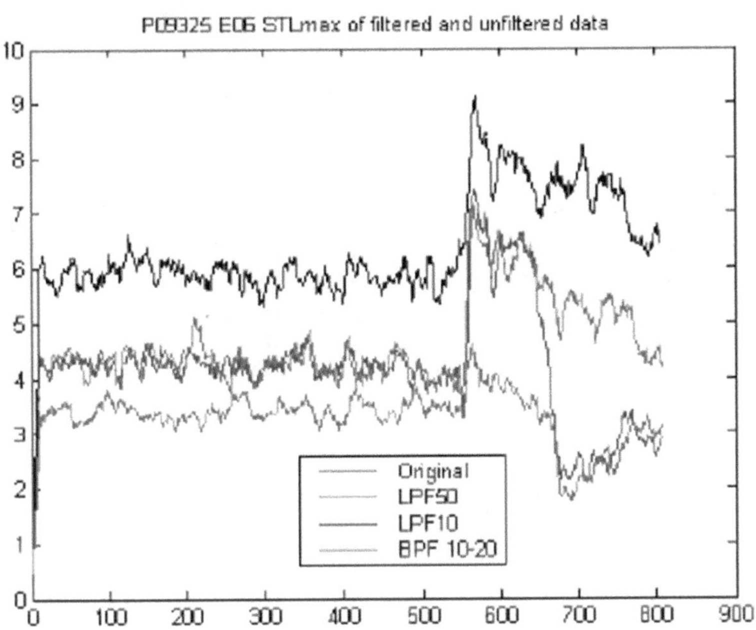

Fig. 4.10 Electrode 6 frequency band using LPF 10 BPF10–20 LPF 50.

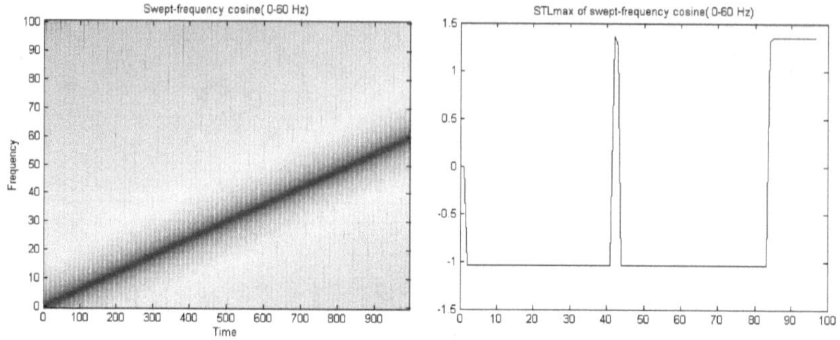

Fig. 4.11 Sweep frequency and $STL_{\max}$ profile of EEG data.

6 (RTD2) after passing a lowpass filter of 10 and 50 Hz, or bandpass filter of 10 and 20 Hz. Note that Figure 4.10 shows consistent findings as illustrated in Figure 4.9, which confirms our hypothesis.

We also observed a drop of $STL_{\max}$ profiles of original EEG about 20 minutes after the seizure. This drop cannot be observed if the EEG data are filtered at LPF 50 Hz, which is postulated to be the enhancement of gamma-band responses over 50 Hz, which might be attributable to the level of sensorimotor integration required by complex movements or activity of the brain to recover from seizure episodes. These results are also consistent with the ones confirmed by an MEG investigation of human brain responses during the brain recovery period. In addition, the complexity of the brain activity could also be critical for the gamma-band response to occur as EEG recordings normally are used to investigate changes of gamma-band activity associated with tasks such as verbal and visual–spatial problem solving. Although these results are consistent with the assumption that cognitive processes, that is, selective attention to sensorimotor integration, underlie stronger gamma-band responses, they might also arise from the complexity of the movement to be performed. If no differences occur in these bands, the changes of the brain activity during seizure episodes cannot be the cause of an effect visible in lower frequencies. Figure 4.11 illustrates the sweep frequency and $STL_{\max}$ profile of EEG data.

4.4.2 Hypothesis II

An example of 5 second EEG data before a seizure onset analyzed in this study for hypothesis II and its spectrogram are illustrated in Figure 4.12. We postulate that before the ictal period, there are frequency components that demonstrate neuropsychological and neuroanatomical evidence of the modularity of the brain network. In addition, epileptogenic processes are believed to be a result of quadratic phase couplings in the epileptogenic zone. Figure 4.13

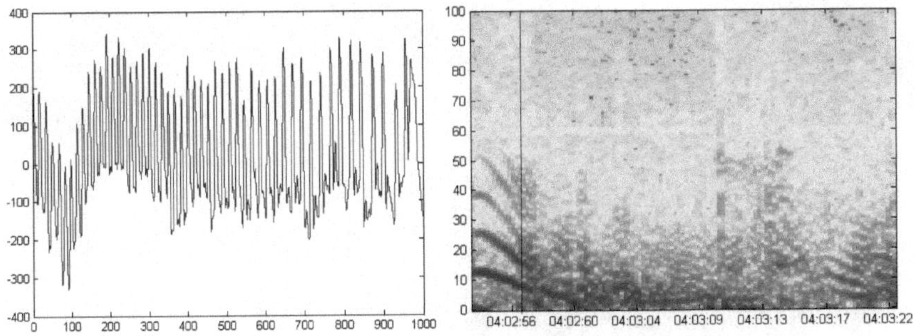

Fig. 4.12 An example of 5 sec data of EEG before a seizure onset and its spectrogram.

illustrates the bispectrum of the pre-ictal EEG signal by decomposing the Wigner time–frequency components. As shown, the decomposed frequency components form an imperfect sine wave, which is a result of higher-order harmonic of the pre-ictal EEG signal. In Figure 4.14, we employ the Choi–Williams time–frequency distribution to capture the main component of the frequency responses of the pre-ictal EEG signal, which can reflect a formation of the epileptogenic process.

In this study, we have demonstrated that epileptogenic processes incur specific changes in high-frequency brain responses when two time–frequency components are compared. As observed in Figures 4.13 and 4.14, the time–frequency components have very different perceptual complexity and a difference in evoked spectral responses, which could be a reflection of neuronal recruitment that triggers the epileptogenic process. On the cortical level, the epileptogenesis should lead to cell assembly ignition, which gamma-band responses to the post-ictal period should be stronger than responses to the other stages (see Figure 4.8).

4.5 Conclusions and Discussion of Future Research

This research has given critical contributions to neocortical dynamics and the nonlinear dynamics methods to study the brain dynamics and epileptogenic processes, which contain almost none of these features to significant degree, and are likely to have a wide spectrum of applications to successful theories of neocortical function. We have also demonstrated that successful theories must either approximate interactions between neuronal masses at microscopic and macroscopic levels consistent with the scale of the experiment (e.g., electrode size and location) by using modern statistical methods and methods in chaos theory to express variables at experimentally interesting scales in terms of degrees of chaoticity over microscopic and macroscopic variables and their distribution functions at different scales. In the future,

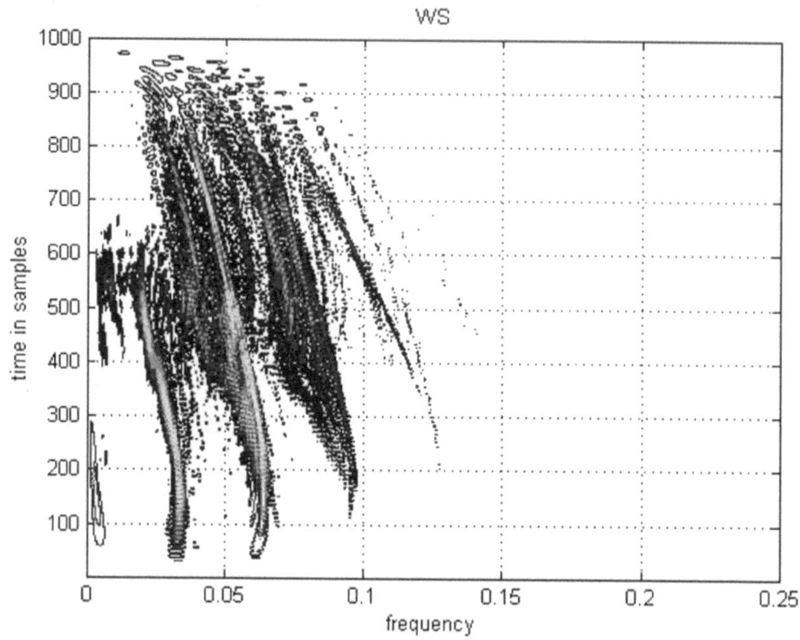

Fig. 4.13 Wigner–Ville time–frequency of 5 sec EEG data before a seizure onset.

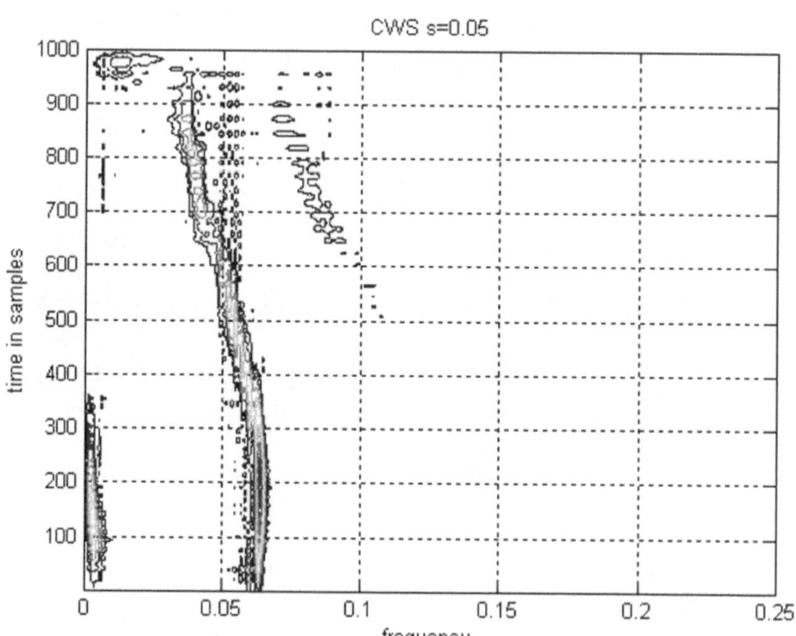

Fig. 4.14 Choi–Williams time–frequency of 5 sec EEG data before a seizure onset.

we also plan to expand our research to take into account more from genuine neocortical theory, which is based on real anatomy and physiology (as understood at the time) and contains no free (arbitrary) parameters. Although this might be an indication that cortical spatiotemporal responses change with the chaoticity of the EEG, such as continuity, there is still much research to be done to study how these responses relate to perception of complex forms. Nevertheless, we can broadly characterize the brain network in terms of the dynamics underlying EEG information processing. The brain network modules are directly mapped onto the spatiotemporal properties of the EEG, which are governed by attractor dynamics extending over time. In attractor networks, the changes in the dynamics of the brain network can be reflected from epileptogenic stimuli as sets of initial points corresponding to those stimuli converge onto an attractor. During the pre-ictal and ictal periods, the complete convergence of the brain dynamics onto an attractor is postulated to be such a sequence of the ideal network's response to epileptogenesis, which will be an infinite transient whose path is governed by the input and the attractor structure of the brain network.

The results from this study suggested that different EEG frequency bands operate in different functions and have different architectures. This separation of different EEG frequency bands by function and "anatomy" is worth study because because it may explain correspondences between deficits in the brain network's performance when specific brain functioning is damaged, which might be attributed to the localization of epileptogenesis. In the future, in order to address issues such as epileptogenesis localization, it is crucially important to understand representation and processing in connectivity models of the brain network as a tool for understanding neuropathologies. We postulated that there are crucial connections during epileptogenic processes governed by the pyramidal cells in which the probability of a synapse occurring between two cells varies inversely with the distance between them independent of their spatial separation. In addition, those connections can be driven by the long-range apical system of the brain connectivity, coordinated in its projections between cortical areas in a space-independent fashion.

If one confines the time scale over which interactions occur to be short, it might be possible to regard these long-range connections as forming a number of feedforward systems of epileptogenesis. Ideally, if one can project back any signals, in theory, it is possible that the process would take too long to arrive to modify the initial feedforward stimulus and the epileptogenic processes be intervened. In addition, one possibility is to regard spatially localized collections of cells in the neocortex as components of a system. To monitor the time scale interactions of the brain network, we assume some randomness in the strength of individual connections, therefore, the strength of connections between groups of cells is approximately symmetric over a short range on average. Although the probability of connection between individual cells is space-dependent, this does not imply multiple connections between pairs of nearby cells.

As fundamentally unstable brain cell units, their behavior is subject to change independent of their interactions with other neuronal groups. Furthermore, if the group of neuronal interactions to a symmetrically connected attractor is conceivable, it will also be highly constrained in the time scale to operate effectively. Another possible assumption for epileptogenic localization is that neuronal networks with spatially localized connections behave as selective dynamic patterns, forming systems in parallel to reaction–diffusion systems of brain connectivity. If this assumption holds, then the time and space constants for the decay of activity depending on the relative strengths and distributions of excitatory and inhibitory neurons and the level of "background" neuronal potential action activity in the system become critical. For this reason, in the future, we need to find an explanation of how the brain functions associated with epileptogenesis may be separated by time scales rather than anatomy theory. One may explain neuropsychological dissociations of function as neurochemical systems operating over different time courses as anatomical epileptogenic localizations. Sometimes it may be appropriate to equate "lesions" of connectivity models with localized epileptogenic zones. Nevertheless, brain dysfunction may be attributable to epileptogenesis to remote or diffuse modulatory systems. It also requires greater understanding in drawing inferences from neuropsychology to support modular connectivity models of epileptogenesis and to explain the basis of neuropathologies of epileptogenic zones.

The studies of the connections of neocortical dynamics and neuronal networks also need to be addressed in order for us to have greater understanding about the brain. Our group has done some preliminary studies focusing on the idea that connections between assumed functional units at different scales (e.g., neurons, minicolumns, macrocolumns) can be symmetric or asymmetric, depending on the spatial scale of such units [30]. In a realm of network connectivity studies, the nature of connections at different scales is a critical theoretical issue. Also, features other than symmetry of such interactions (e.g., the density of the network's connections) are important and addressed in that study [30]. Because neocortical dynamic variables may behave quite differently at different scales, the "interconnectivities" are studied. Furthermore, interactions across scales are also important, as emphasized in [30]. Any theory of EEG constructed at smaller scales must be coarse grained before comparisons are made with scalp data. In the future, we plan to experiment on the same analyses with some of these ideas on scalp EEG data. We expect chaotic or quasiperiodic behavior may be observed depending on the spatial filter implicit in the experimental methods.

Acknowledgments The authors would like to thank members of the Bioengineering Research Partnership at the brain dynamics laboratory, Brain Institute, University of Florida, for their fruitful comments and discussion. Research was partially supported by Rutgers Research Council Grant-202018 and NIH grant R01-NS-39687-01A1.

References

1. G.G. Athanasiou, C.P. Bachas, and W.F. Wolf. Invariant geometry of spin-glass states. *Phys. Rev. B*, 35:1965–1968, 1987.
2. F. Barahona. On the computational complexity of spin glass models. *J. Phys. A: Math. Gen.*, 15:3241–3253, 1982.
3. F. Barahona. On the exact ground states of three-dimensional Ising spin glasses. *J. Phys. A: Math. Gen.*, 15:L611–L615, 1982.
4. W. Chaovalitwongse, P.M. Pardalos, L.D. Iasemidis, D.-S. Shiau, and J.C. Sackellares. Applications of global optimization and dynamical systems to prediction of epileptic seizures. In P.M. Pardalos, J.C. Sackellares, L.D. Iasemidis, and P.R. Carney, editors, *Quantitative Neurosciences*, pages 1–36. Kluwer Academic, 2004.
5. W.A. Chaovalitwongse, P.M. Pardalos, and O.A. Prokoyev. A new linearization technique for multi-quadratic 0–1 programming problems. *Oper. Res. Lett.*, 32(6):517–522, 2004.
6. H.I. Choi and W.J. Williams. Improved time-frequency representation of multicomponent signals using exponential kernels. *IEEE Trans. Acoustics, Speech Signal*, 37:862–871, 1989.
7. L. Cohen. Time-frequency distribution—a review. *Proc. IEEE*, 77:941–981, 1989.
8. C.T. Cripe. Brainwave and EEG: The language of the brain. http://www.crossroads institute.org/eeg.html, 2004.
9. L. Diambra, J.C. Bastos de Figueiredo, and C.P. Malta. Epileptic activity recognition in EEG recording. *Physica A*, 273:495–505, 1999.
10. C.E. Elger and K. Lehnertz. Seizure prediction by non-linear time series analysis of brain electrical activity. *Europ. J. Neurosci.*, 10:786–789, 1998.
11. W.G. Frank, T. Lookman, M.A. Nerenberg, C. Essex, J. Lemieux, and W. Blume. Chaotic time series analyses of epileptic seizures. *Physica D*, 46:427–438, 1990.
12. R. Horst, P.M. Pardalos, and N.V. Thoai. *Introduction to Global Optimization*. Kluwer Academic, 1995.
13. L.D. Iasemidis. *On the dynamics of the human brain in temporal lobe epilepsy*. PhD thesis, University of Michigan, Ann Arbor, 1991.
14. L.D. Iasemidis, P.M. Pardalos, J.C. Sackellares, and D.-S. Shiau. Quadratic binary programming and dynamical system approach to determine the predictability of epileptic seizures. *J. Combin. Optim.*, 5:9–26, 2001.
15. L.D. Iasemidis, P.M. Pardalos, D.-S. Shiau, W. Chaovalitwongse, K. Narayanan, A. Prasad, K. Tsakalis, P.R. Carney, and J.C. Sackellares. Long term prospective on-line real-time seizure prediction. *Journal of Clinical Neurophysiology*, 116(3):532–544, 2005.
16. L.D. Iasemidis, J.C. Principe, and J.C. Sackellares. Measurement and quantification of spatiotemporal dynamics of human epileptic seizures. In M. Akay, editor, *Nonlinear Biomedical Signal Processing*, pages 294–318. Wiley–IEEE Press, vol. II, 2000.
17. L.D. Iasemidis and J.C. Sackellares. The evolution with time of the spatial distribution of the largest Lyapunov exponent on the human epileptic cortex. In D.W. Duke and W.S. Pritchard, editors, *Measuring Chaos in the Human Brain*, pages 49–82. World Scientific, 1991.
18. L.D. Iasemidis, D.-S. Shiau, W. Chaovalitwongse, J.C. Sackellares, P.M. Pardalos, P.R. Carney, J.C. Principe, A. Prasad, B. Veeramani, and K. Tsakalis. Adaptive epileptic seizure prediction system. *IEEE Trans. Biomed. Eng.*, 50(5):616–627, 2003.
19. L.D. Iasemidis, D.-S. Shiau, P.M. Pardalos, and J.C. Sackellares. Phase entrainment and predictability of epileptic seizures. In P.M. Pardalos and J.C. Principe, editors, *Biocomputing*, pages 59–84. Kluwer Academic, 2001.
20. L.D. Iasemidis, D.-S. Shiau, J.C. Sackellares, and P.M. Pardalos. Transition to epileptic seizures: Optimization. In D.Z. Du, P.M. Pardalos, and J. Wang, editors, *DIMACS series in Discrete Mathematics and Theoretical Computer Science*, pages 55–74. American Mathematical Society, 1999.

21. L.D. Iasemidis, D.-S. Shiau, J.C. Sackellares, P.M. Pardalos, and A. Prasad. Dynamical resetting of the human brain at epileptic seizures: Application of nonlinear dynamics and global optimization tecniques. *IEEE Trans. Biomed. Eng.*, 51(3):493–506, 2004.
22. L.D. Iasemidis, H.P. Zaveri, J.C. Sackellares, and W.J. Williams. Phase space topography of the electrocorticogram and the Lyapunov exponent in partial seizures. *Brain Topog.*, 2:187–201, 1990.
23. K. Lehnertz and C.E. Elger. Can epileptic seizures be predicted? Evidence from nonlinear time series analysis of brain electrical activity. *Phys. Rev. Lett.*, 80:5019–5022, 1998.
24. B. Litt, R. Esteller, J. Echauz, D.A. Maryann, R. Shor, T. Henry, P. Pennell, C. Epstein, R. Bakay, M. Dichter, and G. Vachtservanos. Epileptic seizures may begin hours in advance of clinical onset: A report of five patients. *Neuron*, 30:51–64, 2001.
25. M. Mezard, G. Parisi, and M.A. Virasoro. *Spin Glass Theory and Beyond.* World Scientific, 1987.
26. P.M. Pardalos, W. Chaovalitwongse, L.D. Iasemidis, J.C. Sackellares, D.-S. Shiau, P.R. Carney, O.A. Prokopyev, and V.A. Yatsenko. Seizure warning algorithm based on optimization and nonlinear dynamics. *Math. Program.*, 101(2):365–355, 2004.
27. P.M. Pardalos and G. Rodgers. Parallel branch and bound algorithms for unconstrained quadratic zero-one programming. In R. Sharda et al., editor, *Impact of Recent Computer Advances on Operations Research.* North-Holland, 1989.
28. P.M. Pardalos and G. Rodgers. Computational aspects of a branch and bound algorithm for quadratic zero-one programming. *Computing*, 45:131–144, 1990.
29. P.M. Pardalos, J.C. Sackellares, L.D. Iasemidis, and P.R. Carney. *Quantitative Neurosciences.* Kluwer Academic, 2004.
30. O.A. Prokopyev, V. Boginski, W. Chaovalitwongse, P.M. Pardalos, J.C. Sackellares, and P.R. Carney. Network-based techniques in EEG data analysis and epileptic brain modeling. In P.M. Pardalos and A. Vazacopoulos, editors, *Data Mining in Biomedicine.* Kluwer Academic, 2005.
31. M. Le Van Quyen, J. Martinerie, M. Baulac, and F. Varela. Anticipating epileptic seizures in real time by non-linear analysis of similarity between EEG recordings. *NeuroReport*, 10:2149–2155, 1999.
32. J.C. Sackellares, L.D. Iasemidis, R.L. Gilmore, and S.N. Roper. Epileptic seizures as neural resetting mechanisms. *Epilepsia*, 38(S3):189, 1997.
33. J.C. Sackellares, L.D. Iasemidis, R.L. Gilmore, and S.N. Roper. Epilepsy - When chaos fails. In K. Lehnertz, J. Arnhold, P. Grassberger, and C.E. Elger, editors, *Chaos in the Brain?* World Scientific, 2002.
34. J.C. Sackellares, L.D. Iasemidis, and D.-S. Shiau. Detection of the preictal transition in scalp EEG. *Epilepsia*, 40(S7):176, 1999.
35. D.-S. Shiau, Q. Luo, S.L. Gilmore, S.N. Roper, P.M. Pardalos, J.C. Sackellares, and L.D. Iasemidis. Epileptic seizures resetting revisited. *Epilepsia*, 41(S7):208–209, 2000.
36. A. Wolf, J.B. Swift, H.L. Swinney, and J.A. Vastano. Determining Lyapunov exponents from a time series. *Physica D*, 16:285–317, 1985.

Chapter 5
Nonconvex Optimization for Communication Networks

Mung Chiang

Summary. Nonlinear convex optimization has provided both an insightful modeling language and a powerful solution tool to the analysis and design of communication systems over the last decade. A main challenge today is on nonconvex problems in these applications. This chapter presents an overview on some of the important nonconvex optimization problems in communication networks. Four typical applications are covered: Internet congestion control through nonconcave network utility maximization, wireless network power control through geometric and sigmoidal programming, DSL spectrum management through distributed nonconvex optimization, and Internet intra-domain routing through nonconvex, nonsmooth optimization. A variety of nonconvex optimization techniques are showcased: sum-of-squares programming through successive SDP relaxation, signomial programming through successive GP relaxation, leveraging specific structures in these engineering problems for efficient and distributed heuristics, and changing the underlying protocol to enable a different problem formulation in the first place. Collectively, they illustrate three alternatives of tackling nonconvex optimization for communication networks: going "through" nonconvexity, "around" nonconvexity, and "above" nonconvexity.

Key words: Digital subscriber line, duality, geometric programming, Internet, network utility maximization, nonconvex optimization, power control, routing, semidefinite programming, sum of squares, TCP/IP, wireless network

---

Mung Chiang
Electrical Engineering Department, Princeton University, Princeton, NJ 08544, U.S.A.
e-mail: chiangm@princeton.edu

D.Y. Gao, H.D. Sherali, (eds.), *Advances in Applied Mathematics and Global Optimization*
Advances in Mechanics and Mathematics 17, DOI 10.1007/978-0-387-75714-8_5,

5.1 Introduction

There have been two major "waves" in the history of optimization theory and its applications: the first started with linear programming (LP) and the simplex method in the late 1940s, and the second with convex optimization and the interior point method in the late 1980s. Each has been followed by a transforming period of "appreciation-application cycle": as more people appreciate the use of LP/convex optimization, more look for their formulations in various applications; then more work on its theory, efficient algorithms, and software; the more powerful the tools become, and in turn more people appreciate its usage. Communication systems benefit significantly from both waves; the vast array of many success stories includes multicommodity flow solutions (e.g., Bellman–Ford algorithm) from LP, and network utility maximization and robust transceiver design from convex optimization.

Much of the current research is about the potential of the third wave, on nonconvex optimization. If one word is used to differentiate between easy and hard problems, convexity is probably the "watershed." But if a longer description length is allowed, useful conclusions can be drawn even for nonconvex optimization. Indeed, convexity is a very disturbing watershed, because it is not a topological invariant under change of variable (e.g., see geometric programming) or higher-dimension embedding (e.g., see sum of squares method). A variety of approaches has been proposed to tackle nonconvex optimization problems: from successive convex approximation to dualization, from nonlinear transformation to turn an apparently nonconvex problem into a convex problem to characterization of attraction regions and systematically jumping out of a local optimum, and from leveraging the specific structures of the problems (e.g., difference of convex functions, concave minimization, low rank nonconvexity) to developing more efficient branch-and-bound procedures.

Researchers in communications and networking have been examining nonconvex optimization using domain-specific structures in important problems in the areas of wireless networking, Internet engineering, and communication theory. Perhaps four typical topics best illustrate the variety of challenging issues arising from nonconvex optimization in communication systems:

- Nonconvex objective to be minimized. An example is congestion control for inelastic application traffic, where a nonconcave utility function needs to be maximized.
- Nonconvex constraint set. An example is power control in the low SIR regime.
- Integer constraints. Two important examples are single path routing and multiuser detection.
- Constraint sets that are convex but require an exponential number of inequalities to explicitly describe. An example is optimal scheduling in multihop wireless networks under certain interference models. The problem of wireless scheduling will not be discussed in this chapter. Interested readers can refer to [73] for a unifying framework of the problem.

This chapter overviews the latest results in recent publications about the first two topics, with a particular focus on showing the connections between the engineering intuitions about important problems in communication networks and the state-of-the-art algorithms in nonconvex optimization theory. Most of the results surveyed here were obtained in 2005–2006, and the problems driven by fundamental issues in the Internet, wireless, and broadband access networks. As this chapter illustrates, even after much progress made in recent years, there are still many challenging mysteries to be resolved on these important nonconvex optimization problems.

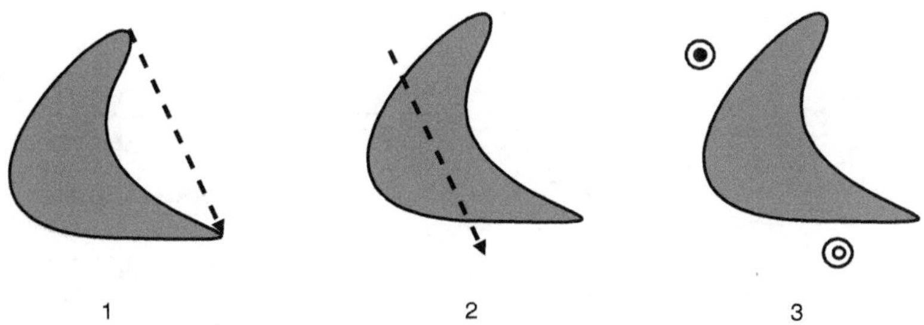

Fig. 5.1 Three major types of approaches when tackling nonconvex optimization problems in communication networks: Go (1) through, (2) around, or (3) above nonconvexity.

It is interesting to point out that, as illustrated in Figure 5.1, there are at least three very different approaches to tackle the difficult issue of nonconvexity.

- *Go "through" nonconvexity.* In this approach, we try to solve the difficult nonconvex problem; for example, we may use successive convex relaxations (e.g., sum-of-squares, signomial programming), utilize special structures in the problem (e.g., difference of convex functions, generalized quasiconcavity), or leverage smarter branch and bound methods.
- *Go "around" nonconvexity.* In this approach, we try to avoid solving the convex problem; for example, we may discover a change of variables that turns the seemingly nonconvex problem into a convex one, determine conditions under which the problem is convex or the KKT point is unique, or make approximations to make the problem convex.
- *Go "above" nonconvexity.* In this approach, we try to reformulate the nonconvex problem in the first place to make it more "solvable" or "approximately solvable." We observe that optimization problem formulations are induced by some underlying assumptions on what the network architectures and protocols should look like. By changing these assumptions, a different, much easier-to-solve or easier-to-approximate formulations may result. We refer to this approach as *design for optimizability*, which is concerned with redrawing architectures to make the resulting optimization

problem easier to solve. This approach of changing a hard problem into an easier one is in contrast to *optimization*, which tries to solve a given, possibly difficult, problem.

The four topics chosen in this chapter span a range of application contexts and tasks in communication networks. The sources of difficulty in these nonconvex optimization problems are summarized in Table 5.1, together with the key ideas in solving them and the type of approaches used. For more details beyond this brief overview chapter, please refer to the related publications [29, 19, 14, 35, 7, 70, 71] by the author and coworkers and the references therein.

Table 5.1 Summary of four nonconvex optimization problems in this chapter

Section	*Application*	*Task*	*Difficulty*	*Solution*	*Approach*
5.2	Internet	Congestion control	Nonconcave U	Sum of squares	"Through"
5.3	Wireless	Power control	Posynomial ratio	Geometric program	"Around"
5.4	DSL	Spectrum management	Posynomial ratio	Problem structure	"Around"
5.5	Internet	Routing	Nonconvex constraint	Approximation	"Above"

5.2 Internet Congestion Control

5.2.1 Introduction

Basic Network Utility Maximization

Since the publication of the seminal paper [37] by Kelly, Maulloo, and Tan in 1998, the framework of network utility maximization (NUM) has found many applications in network rate allocation algorithms and Internet congestion control protocols (e.g., surveyed in [45, 60]). It has also led to a systematic understanding of the entire network protocol stack in the unifying framework of "layering as optimization decomposition" (e.g., surveyed in [13, 49, 44]). By allowing nonlinear concave utility objective functions, NUM substantially expands the scope of the classical LP-based network flow problems.

Consider a communication network with L links, each with a fixed capacity of c_l bps, and S sources (i.e., end-users), each transmitting at a source rate of x_s bps. Each source s emits one flow, using a fixed set $L(s)$ of links in its path, and has a utility function $U_s(x_s)$. Each link l is shared by a set $S(l)$ of sources. Network utility maximization, in its basic version, is the following problem of maximizing the total utility of the network $\sum_s U_s(x_s)$, over the source rates $\mathbf{x}$, subject to linear flow constraints $\sum_{s:l \in L(s)} x_s \leq c_l$ for all links l:

$$\begin{array}{ll} \text{maximize} & \sum_s U_s(x_s) \\ \text{subject to} & \sum_{s \in S(l)} x_s \leq c_l, \ \ \forall l, \\ & \mathbf{x} \succeq 0, \end{array} \tag{5.1}$$

where the variables are $\mathbf{x} \in \mathbf{R}^S$.

There are many nice properties of the basic NUM model due to several simplifying assumptions of the utility functions and flow constraints, which provide the mathematical tractability of problem (5.1) but also limit its applicability. In particular, the utility functions $\{U_s\}$ are often assumed to be increasing and strictly concave functions.

Assuming that $U_s(x_s)$ becomes concave for large enough x_s is reasonable, because the law of diminishing marginal utility eventually will be effective. However, U_s may not be concave throughout its domain. In his seminal paper in 1995, Shenker [57] differentiated inelastic network traffic from elastic traffic. Utility functions for elastic traffic were modeled as strictly concave functions. Although inelastic flows with nonconcave utility functions represent important applications in practice, they have received little attention and rate allocation among them has only a limited mathematical foundation. There have been three recent publications [41, 29, 19] (see also earlier work in [69, 42, 43] related to the approach in [41]) on this topic.

In this section, we investigate the extension of the basic NUM to maximization of nonconcave utilities, as in the approach of [19]. We provide a centralized algorithm for offline analysis and establishment of a performance benchmark for nonconcave utility maximization when the utility function is a polynomial or signomial. Based on the semialgebraic approach to polynomial optimization, we employ convex sum-of-squares (SOS) relaxations solved by a sequence of semidefinite programs (SDP), to obtain increasingly tighter upper bounds on total achievable utility for polynomial utilities. Surprisingly, in all our experiments, a very low-order and often a minimal-order relaxation yields not just a bound on attainable network utility, but the globally maximized network utility. When the bound is exact, which can be proved using a sufficient test, we can also recover a globally optimal rate allocation.

Canonical Distributed Algorithm

A reason that the assumption of a utility function's concavity is upheld in many papers on NUM is that it leads to three highly desirable mathematical properties of the basic NUM:

- It is a convex optimization problem, therefore the global minimum can be computed (at least in centralized algorithms) in worst-case polynomial-time complexity [4].
- Strong duality holds for (5.1) and its Lagrange dual problem. A zero duality gap enables a dual approach to solve (5.1).
- Minimization of a separable objective function over linear constraints can be conducted by distributed algorithms based on the dual approach.

Indeed, the basic NUM (5.1) is such a "nice" optimization problem that its theoretical and computational properties have been well studied since the 1960s in the field of monotropic programming (e.g., as summarized in [54]). For network rate allocation problems, a dual-decomposition-based distributed algorithm has been widely studied (e.g., in [37, 45]), and is summarized below.

Zero duality gap for (5.1) states that solving the Lagrange dual problem is equivalent to solving the primal problem (5.1). The Lagrange dual problem is readily derived. We first form the Lagrangian of (5.1):

$$L(\mathbf{x}, \boldsymbol{\lambda}) = \sum_s U_s(x_s) + \sum_l \lambda_l \left(c_l - \sum_{s \in S(l)} x_s \right),$$

where $\lambda_l \geq 0$ is the Lagrange multiplier (can be interpreted as the link congestion price) associated with the linear flow constraint on link l. Additivity of total utility and linearity of flow constraints lead to a Lagrangian dual decomposition into individual source terms:

$$\begin{aligned} L(\mathbf{x}, \boldsymbol{\lambda}) &= \sum_s \left[U_s(x_s) - \left(\sum_{l \in L(s)} \lambda_l \right) x_s \right] + \sum_l c_l \lambda_l \\ &= \sum_s L_s(x_s, \lambda^s) + \sum_l c_l \lambda_l, \end{aligned}$$

where $\lambda^s = \sum_{l \in L(s)} \lambda_l$. For each source s, $L_s(x_s, \lambda^s) = U_s(x_s) - \lambda^s x_s$ only depends on local x_s and the link prices λ_l on those links used by source s.

The Lagrange dual function $g(\boldsymbol{\lambda})$ is defined as the maximized $L(\mathbf{x}, \boldsymbol{\lambda})$ over $\mathbf{x}$. This "net utility" maximization obviously can be conducted distributively by each source, as long as the aggregate link price $\lambda^s = \sum_{l \in L(s)} \lambda_l$ is available to source s, where source s maximizes a strictly concave function $L_s(x_s, \lambda^s)$ over x_s for a given λ^s:

$$x_s^*(\lambda^s) = \text{argmax} \left[U_s(x_s) - \lambda^s x_s \right], \ \forall s. \tag{5.2}$$

The Lagrange dual problem is

$$\begin{array}{ll} \text{minimize} & g(\boldsymbol{\lambda}) = L(\mathbf{x}^*(\boldsymbol{\lambda}), \boldsymbol{\lambda}) \\ \text{subject to} & \boldsymbol{\lambda} \succeq 0, \end{array} \tag{5.3}$$

where the optimization variable is $\boldsymbol{\lambda}$. Any algorithms that find a pair of primal–dual variables $(\mathbf{x}, \boldsymbol{\lambda})$ that satisfy the KKT optimality condition would solve (5.1) and its dual problem (5.3). One possibility is a distributed, iterative subgradient method, which updates the dual variables $\boldsymbol{\lambda}$ to solve the dual problem (5.3):

$$\lambda_l(t+1) = \left[\lambda_l(t) - \alpha(t)\left(c_l - \sum_{s \in S(l)} x_s(\lambda^s(t))\right)\right]^+, \quad \forall l, \tag{5.4}$$

where t is the iteration number and $\alpha(t) > 0$ are step sizes. Certain choices of step sizes, such as $\alpha(t) = \alpha_0/t$, $\alpha_0 > 0$, guarantee that the sequence of dual variables $\boldsymbol{\lambda}(t)$ will converge to the dual optimal $\boldsymbol{\lambda}^*$ as $t \to \infty$. The primal variable $\mathbf{x}(\boldsymbol{\lambda}(t))$ will also converge to the primal optimal variable $\mathbf{x}^*$. For a primal problem that is a convex optimization, the convergence is towards the global optimum.

The sequence of the pair of algorithmic steps (5.2, 5.4) forms what we refer to as the *canonical distributed algorithm*, which solves the network utility optimization problem (5.1) and the dual (5.3) and computes the optimal rates $\mathbf{x}^*$ and link prices $\boldsymbol{\lambda}^*$.

Nonconcave Network Utility Maximization

It is known that for many multimedia applications, user satisfaction may assume a nonconcave shape as a function of the allocated rate. For example, the utility for voice applications is better described by a sigmoidal function: with a convex part at low rate and a concave part at high rate, and a single inflexion point x^0 (with $U_s''(x^0) = 0$) separating the two parts. Furthermore, in some other models of utility functions, the concavity assumption on U_s is also related to the elasticity assumption on rate demands by users. When demands for x_s are not perfectly elastic, $U_s(x_s)$ may not be concave.

Suppose we remove the critical assumption that $\{U_s\}$ are concave functions, and allow them to be any nonlinear functions. The resulting NUM becomes nonconvex optimization and significantly harder to be analyzed and solved, even by centralized computational methods. In particular, a local optimum may not be a global optimum and the duality gap can be strictly positive. The standard distributive algorithms that solve the dual problem may produce infeasible or suboptimal rate allocation.

There have been several recent publications on distributed algorithms for nonconcave utility maximization. In [41], a "self-regulation" heuristic is proposed to avoid the resulting oscillation in rate allocation and is shown to converge to an optimal rate allocation asymptotically when the proportion of nonconcave utility sources vanishes. In [29], a set of sufficient conditions and necessary conditions is presented under which the canonical distributed algo-

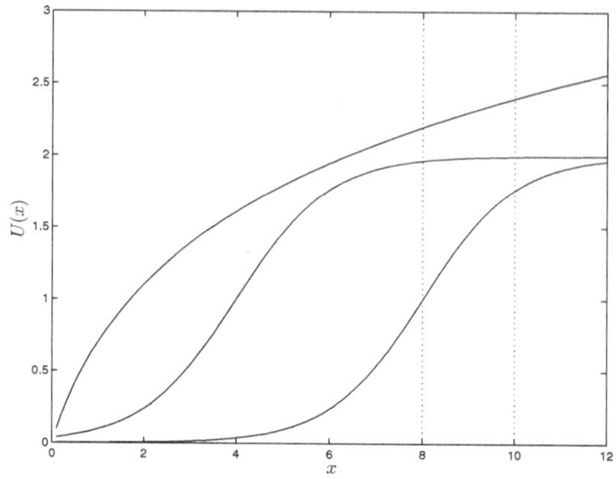

Fig. 5.2 Some examples of utility functions $U_s(x_s)$: it can be concave or sigmoidal as shown in the graph, or any general nonconcave function. If the bottleneck link capacity used by the source is small enough, that is, if the dotted vertical line is pushed to the left, a sigmoidal utility function effectively becomes a convex utility function.

rithm still converges to the globally optimal solution. However, these conditions may not hold in many cases. These two approaches illustrate the choice between admission control and capacity planning to deal with nonconvexity (see also the discussion in [36]). But neither approach provides a theoretically polynomial-time and practically efficient algorithm (distributed or centralized) for nonconcave utility maximization.

In [19], using a family of convex semidefinite programming (SDP) relaxations based on the sum-of-squares (SOS) relaxation and the positivstellensatz theorem in real algebraic geometry, we apply a centralized computational method to bound the total network utility in polynomial time. A surprising result is that for all the examples we have tried, wherever we could verify the result, the tightest possible bound (i.e., the globally optimal solution) of NUM with nonconcave utilities is computed with a very low-order relaxation. This efficient numerical method for offline analysis also provides the benchmark for distributed heuristics.

These three different approaches: proposing distributed but suboptimal heuristics (for sigmoidal utilities) in [41], determining optimality conditions for the canonical distributed algorithm to converge globally (for all nonlinear utilities) in [29], and proposing an efficient but centralized method to compute the global optimum (for a wide class of utilities that can be transformed into polynomial utilities) in [19] (and this section), are complementary in the study of distributed rate allocation by nonconcave NUM.

5.2.2 *Global Maximization of Nonconcave Network Utility*

Sum-of-Squares Method

We would like to bound the maximum network utility by γ in polynomial time and search for a tight bound. Had there been no link capacity constraints, maximizing a polynomial is already an NP-hard problem, but can be relaxed into an SDP [58]. This is because testing if the following bounding inequality holds $\gamma \geq p(\mathbf{x})$, where $p(\mathbf{x})$ is a polynomial of degree d in n variables, is equivalent to testing the positivity of $\gamma - p(\mathbf{x})$, which can be relaxed into testing if $\gamma - p(\mathbf{x})$ can be written as a sum of squares (SOS): $p(\mathbf{x}) = \sum_{i=1}^{r} q_i(\mathbf{x})^2$ for some polynomials q_i, where the degree of q_i is less than or equal to $d/2$. This is referred to as the SOS relaxation. If a polynomial can be written as a sum of squares, it must be nonnegative, but not vice versa. Conditions under which this relaxation is tight have been studied since Hilbert. Determining if a sum of squares decomposition exists can be formulated as an SDP feasibility problem, thus polynomial-time solvable.

Constrained nonconcave NUM can be relaxed by a generalization of the Lagrange duality theory, which involves nonlinear combinations of the constraints instead of linear combinations in the standard duality theory. The key result is the positivstellensatz, due to Stengle [62], in real algebraic geometry, which states that for a system of polynomial inequalities, either there exists a solution in $\mathbf{R}^n$ or there exists a polynomial which is a certificate that no solution exists. This infeasibility certificate has recently been shown to be also computable by an SDP of sufficient size [51, 50], a process that is referred to as the sum-of-squares method and automated by the software SOSTOOLS [52] initiated by Parrilo in 2000. For a complete theory and many applications of SOS methods, see [51] and references therein.

Furthermore, the bound γ itself can become an optimization variable in the SDP and can be directly minimized. A nested family of SDP relaxations, each indexed by the degree of the certificate polynomial, is guaranteed to produce the exact global maximum. Of course, given the problem is NP-hard, it is not surprising that the worst-case degree of certificate (thus the number of SDP relaxations needed) is exponential in the number of variables. What is interesting is the observation that in applying SOSTOOLS to nonconcave utility maximization, a very low-order, often the minimum-order relaxation already produces the globally optimal solution.

Application of SOS Method to Nonconcave NUM

Using sum-of-squares and the positivstellensatz, we set up the following problem whose objective value converges to the optimal value of problem (5.1), where $\{U_i\}$ are now general polynomials, as the degree of the polynomials involved is increased.

$$\begin{array}{l} \text{minimize} \quad \gamma \\ \text{subject to} \\ \gamma - \sum_s U_s(x_s) - \sum_l \lambda_l(\mathbf{x})(c_l - \sum_{s \in S(l)} x_s) \\ - \sum_{j,k} \lambda_{jk}(\mathbf{x})(c_j - \sum_{s \in S(j)} x_s)(c_k - \sum_{s \in S(k)} x_s) - \\ \ldots - \lambda_{12\ldots n}(\mathbf{x})(c_1 - \sum_{s \in S(1)} x_s) \ldots (c_n - \sum_{s \in S(n)} x_s) \\ \text{ is SOS,} \\ \lambda_l(\mathbf{x}), \lambda_{jk}(\mathbf{x}), \ldots, \lambda_{12\ldots n}(\mathbf{x}) \text{ are SOS.} \end{array} \quad (5.5)$$

The optimization variables are γ and all of the coefficients in polynomials $\lambda_l(\mathbf{x})$, $\lambda_{jk}(\mathbf{x})$, ..., $\lambda_{12\ldots n}(\mathbf{x})$. Note that $\mathbf{x}$ is not an optimization variable; the constraints hold for all $\mathbf{x}$, therefore imposing constraints on the coefficients. This formulation uses Schmüdgen's representation of positive polynomials over compact sets [56].

Let D be the degree of the expression in the first constraint in (5.5). We refer to problem (5.5) as the SOS relaxation of order D for the constrained NUM. For a fixed D, the problem can be solved via SDP. As D is increased, the expression includes more terms, the corresponding SDP becomes larger, and the relaxation gives tighter bounds. An important property of this nested family of relaxations is guaranteed convergence of the bound to the global maximum.

Regarding the choice of degree D for each level of relaxation, clearly a polynomial of odd degree cannot be SOS, so we need to consider only the cases where the expression has even degree. Therefore, the degree of the first nontrivial relaxation is the largest even number greater than or equal to degree $\sum_s U_s(x_s)$, and the degree is increased by 2 for the next level.

A key question now becomes: how do we find out, after solving an SOS relaxation, if the bound happens to be exact? Fortunately, there is a sufficient test that can reveal this, using the properties of the SDP and its dual solution. In [31, 39], a parallel set of relaxations, equivalent to the SOS ones, is developed in the dual framework. The dual of checking the nonnegativity of a polynomial over a semialgebraic set turns out to be finding a sequence of moments that represent a probability measure with support in that set. To be a valid set of moments, the sequence should form a positive semidefinite moment matrix. Then, each level of relaxation fixes the size of this matrix (i.e., considers moments up a certain order) and therefore solves an SDP. This is equivalent to fixing the order of the polynomials appearing in SOS relaxations. The sufficient rank test checks a rank condition on this moment matrix and recovers (one or several) optimal $\mathbf{x}^*$, as discussed in [31].

In summary, we have the following algorithm for centralized computation of a globally optimal rate allocation to nonconcave utility maximization, where the utility functions can be written as or converted into polynomials.

Algorithm 1. Sum-of-squares for nonconcave utility maximization.

1. Formulate the relaxed problem (5.5) for a given degree D.
2. Use SDP to solve the Dth order relaxation, which can be conducted using SOSTOOLS [52].
3. If the resulting dual SDP solution satisfies the sufficient rank condition, the Dth-order optimizer $\gamma^*(D)$ is the globally optimal network utility, and a corresponding $\mathbf{x}^*$ can be obtained.[1]
4. Increase D to $D+2$, that is, the next higher-order relaxation, and repeat.

In the following section, we give examples of the application of SOS relaxation to the nonconcave NUM. We also apply the above sufficient test to check if the bound is exact, and if so, we recover the optimum rate allocation $\mathbf{x}^*$ that achieve this tightest bound.

5.2.3 Numerical Examples and Sigmoidal Utilities

Polynomial Utility Examples

First, consider quadratic utilities (i.e., $U_s(x_s) = x_s^2$) as a simple case to start with (this can be useful, for example, when the bottleneck link capacity limits sources to their convex region of a sigmoidal utility). We present examples that are typical, in our experience, of the performance of the relaxations.

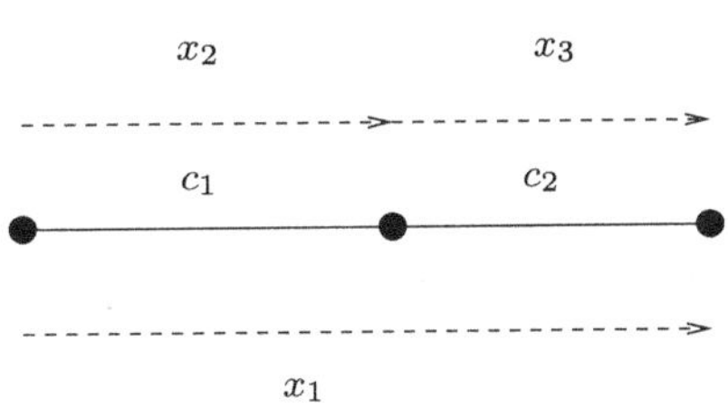

Fig. 5.3 Network topology for Example 5.1.

Example 5.1. *A small illustrative example.* Consider the simple 2-link, 3-user network shown in Figure 5.3, with $\mathbf{c} = [1,\ 2]$. The optimization problem is

$$\begin{array}{ll} \text{maximize} & \sum_s x_s^2 \\ \text{subject to} & x_1 + x_2 \leq 1 \\ & x_1 + x_3 \leq 2 \\ & x_1, x_2, x_3 \geq 0. \end{array} \tag{5.6}$$

[1] Otherwise, $\gamma^*(D)$ may still be the globally optimal network utility but is only provably an upper bound.

The first level relaxation with $D = 2$ is

$$\begin{array}{l} \text{minimize } \gamma \\ \text{subject to} \\ \gamma - (x_1^2 + x_2^2 + x_3^2) - \lambda_1(-x_1 - x_2 + 1) - \lambda_2(-x_1 \\ -x_3 + 2) - \lambda_3 x_1 - \lambda_4 x_2 - \lambda_5 x_3 - \lambda_6(-x_1 - x_2 + 1) \\ (-x_1 - x_3 + 2) - \lambda_7 x_1(-x_1 - x_2 + 1) - \lambda_8 x_2(-x_1 \\ -x_2 + 1) - \lambda_9 x_3(-x_1 - x_2 + 1) - \lambda_{10} x_1(-x_1 - x_3 + 2) \\ -\lambda_{11} x_2(-x_1 - x_3 + 2) - \lambda_{12} x_3(-x_1 - x_3 + 2) - \\ \lambda_{13} x_1 x_2 - \lambda_{14} x_1 x_3 - \lambda_{15} x_2 x_3 \text{ is SOS,} \\ \lambda_i \geq 0,\ i = 1, \ldots, 15. \end{array} \tag{5.7}$$

The first constraint above can be written as $x^T Q x$ for $x = [1,\ x_1,\ x_2,\ x_3]^T$ and an appropriate Q. For example, the (1,1) entry which is the constant term reads $\gamma - \lambda_1 - 2\lambda_2 - 2\lambda_6$, the (2,1) entry, coefficient of x_1, reads $\lambda_1 + \lambda_2 - \lambda_3 + 3\lambda_6 - \lambda_7 - 2\lambda_{10}$, and so on. The expression is SOS if and only if $Q \geq 0$. The optimal γ is 5, which is achieved by, for example, $\lambda_1 = 1$, $\lambda_2 = 2$, $\lambda_3 = 1$, $\lambda_8 = 1$, $\lambda_{10} = 1$, $\lambda_{12} = 1$, $\lambda_{13} = 1$, $\lambda_{14} = 2$ and the rest of the λ_i equal to zero. Using the sufficient test (or, in this example, by inspection) we find the optimal rates $\mathbf{x}_0 = [0,\ 1,\ 2]$.

In this example, many of the λ_i could be chosen to be zero. This means not all product terms appearing in (5.7) are needed in constructing the SOS polynomial. Such information is valuable from the decentralization point of view, and can help determine to what extent our bound can be calculated in a distributed manner. This is a challenging topic for future work.

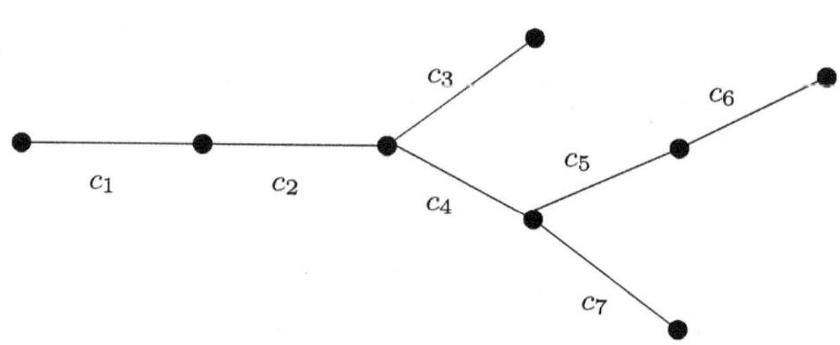

Fig. 5.4 Network topology for Example 5.2.

Example 5.2. *Larger tree topology.* As a larger example, consider the network shown in Figure 5.4 with seven links. There are nine users, with the following routing table that lists the links on each user's path.

x_1	x_2	x_3	x_4	x_5	x_6	x_7	x_8	x_9
1,2	1,2,4	2,3	4,5	2,4	6,5,7	5,6	7	5

For $\mathbf{c} = [5,\ 10,\ 4,\ 3,\ 7,\ 3,\ 5]$, we obtain the bound $\gamma = 116$ with $D = 2$, which turns out to be globally optimal, and the globally optimal rate vector can be recovered: $\mathbf{x}_0 = [5,\ 0,\ 4,\ 0,\ 1,\ 0,\ 0,\ 5,\ 7]$. In this example, exhaustive search is too computationally intensive, and the sufficient condition test plays an important role in proving the bound is exact and in recovering $\mathbf{x}_0$.

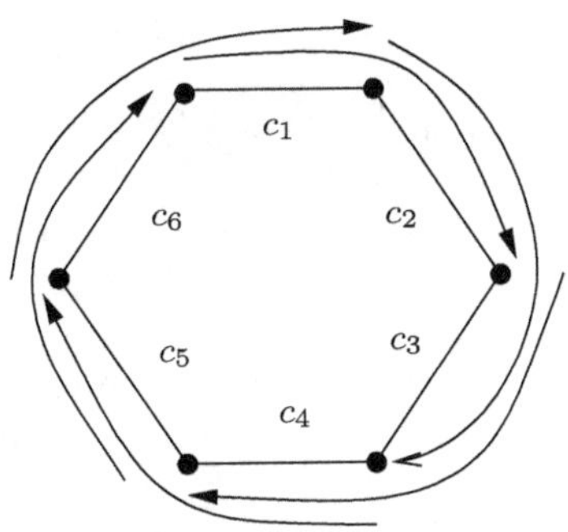

Fig. 5.5 Network topology for Example 5.3.

Example 5.3. *Large m-hop ring topology.* Consider a ring network with n nodes, n users, and n links where each user's flow starts from a node and goes clockwise through the next m links, as shown in Figure 5.5 for $n = 6$, $m = 2$. As a large example, with $n = 25$, $m = 2$, and capacities chosen randomly for a uniform distribution on $[0,\ 10]$, using relaxation of order $D = 2$ we obtain the exact bound $\gamma = 321.11$ and recover an optimal rate allocation. For $n = 30$, $m = 2$, and capacities randomly chosen from $[0,\ 15]$, it turns out that $D = 2$ relaxation yields the exact bound 816.95 and a globally optimal rate allocation.

Sigmoidal Utility Examples

Now consider sigmoidal utilities in a standard form:

$$U_s(x_s) = \frac{1}{1 + e^{-(a_s x_s + b_s)}},$$

where $\{a_s, b_s\}$ are constant integers. Even though these sigmoidal functions are not polynomials, we show the problem can be cast as one with polynomial cost and constraints, with a change of variables.

Example 5.4. *Sigmoidal utility.* Consider the simple 2-link, 3-user example shown in Figure 5.3 for $a_s = 1$ and $b_s = -5$.

The NUM problem is to

$$\begin{array}{ll} \text{maximize} & \sum_s \frac{1}{1+e^{-(x_s-5)}} \\ \text{subject to} & x_1 + x_2 \leq c_1 \\ & x_1 + x_3 \leq c_2 \\ & \mathbf{x} \geq 0. \end{array} \tag{5.8}$$

Let $y_s = 1/\left(1 + e^{-(x_s-5)}\right)$, then $x_s = -\log((1/y_s) - 1) + 5$. Substituting for x_1, x_2 in the first constraint, arranging terms and taking exponentials, then multiplying the sides by $y_1 y_2$ (note that $y_1, y_2 > 0$), we get

$$(1 - y_1)(1 - y_2) \geq e^{(10-c_1)} y_1 y_2,$$

which is polynomial in the new variables $\mathbf{y}$. This applies to all capacity constraints, and the nonnegativity constraints for x_s translate to $y_s \geq 1/\left(1 + e^5\right)$. Therefore the whole problem can be written in polynomial form, and SOS methods apply. This transformation renders the problem polynomial for general sigmoidal utility functions, with any a_s and b_s.

We present some numerical results, using a small illustrative example. Here SOS relaxations of order 4 ($D = 4$) were used. For $c_1 = 4$, $c_2 = 8$, we find $\gamma = 1.228$, which turns out to be a global optimum, with $\mathbf{x}_0 = [0,\ 4,\ 8]$ as the optimal rate vector. For $c_1 = 9$, $c_2 = 10$, we find $\gamma = 1.982$ and $\mathbf{x}_0 = [0,\ 9,\ 10]$. Now place a weight of 2 on y_1, and the other y_s have weight one; we obtain $\gamma = 1.982$ and $\mathbf{x}_0 = [9,\ 0,\ 1]$.

In general, if $a_s \neq 1$ for some s, however, the degree of the polynomials in the transformed problem may be very high. If we write the general problem as

$$\begin{array}{ll} \text{maximize} & \sum_s \frac{1}{1+e^{-(a_s x_s + b_s)}} \\ \text{subject to} & \sum_{s \in S(l)} x_s \leq c_l, \ \ \forall l, \\ & \mathbf{x} \geq 0, \end{array} \tag{5.9}$$

each capacity constraint after transformation will be

$$\prod_s (1 - y_s)^{r_{ls} \Pi_{k \neq s} a_k} \geq$$
$$\exp\left(-\prod_s a_s \left(c_l + \sum_s r_{ls}/a_s b_s\right)\right) \prod_s y_s^{r_{ls} \prod_{k \neq s} a_k},$$

where $r_{ls} = 1$ if $l \in L(s)$ and equals 0 otherwise. Because the product of the a_s appears in the exponents, $a_s > 1$ significantly increases the degree of the polynomials appearing in the problem and hence the dimension of the SDP in the SOS method.

It is therefore also useful to consider alternative representations of sigmoidal functions such as the following rational function:

$$U_s(x_s) = \frac{x_s^n}{a + x_s^n},$$

where the inflection point is $x^0 = \left(\left(a(n-1)\right) / (n+1)\right)^{1/n}$ and the slope at the inflection point is $U_s(x^0) = \left((n-1)/4n\right)\left((n+1) / (a(n-1))\right)^{1/n}$. Let

$y_s = U_s(x_s)$; the NUM problem in this case is equivalent to

$$\begin{array}{ll} \text{maximize} & \sum_s y_s \\ \text{subject to} & x_s^n - y_s x_s^n - a y_s = 0 \\ & \sum_{s \in S(l)} x_s \leq c_l, \ \forall l \\ & \mathbf{x} \geq 0 \end{array} \tag{5.10}$$

which again can be accommodated in the SOS method and be solved by Algorithm 1.

The benefit of this choice of utility function is that the largest degree of the polynomials in the problem is $n+1$, therefore growing linearly with n. The disadvantage compared to the exponential form for sigmoidal functions is that the location of the inflection point and the slope at that point cannot be set independently.

5.2.4 Alternative Representations for Convex Relaxations to Nonconcave NUM

The SOS relaxation we used in the last two sections is based on Schmüdgen's representation for positive polynomials over compact sets described by other polynomials. We now briefly discuss two other representations of relevance to the NUM, that are interesting from both theoretical (e.g., interpretation) and computational points of view.

LP Relaxation

Exploiting linearity of the constraints in NUM and with the additional assumption of nonempty interior for the feasible set (which holds for NUM), we can use Handelman's representation [30] and refine the positivstellensatz condition to obtain the following convex relaxation of nonconcave NUM problem.

$$\begin{array}{l} \text{maximize } \gamma \\ \text{subject to} \\ \gamma - \sum_s U_s(x_s) = \sum_{\alpha \in N^L} \lambda_\alpha \prod_{l=1}^{L} (c_l - \sum_{s \in S(l)} x_s)^{\alpha_l}, \ \forall \mathbf{x} \\ \lambda_\alpha \geq 0, \ \forall \alpha, \end{array} \tag{5.11}$$

where the optimization variables are γ and λ_α, and α denotes an ordered set of integers $\{\alpha_l\}$.

Fixing D where $\sum_l \alpha_l \leq D$, and equating the coefficients on the two sides of the equality in (5.11), yields a linear program (LP). There are no

SOS terms, therefore no semidefiniteness conditions. As before, increasing the degree D gives higher-order relaxations and a tighter bound.

We provide a pricing interpretation for problem (5.11). First, normalize each capacity constraint as $1 - u_l(x) \geq 0$, where $u_l(x) = \sum_{s \in S(l)} x_s / c_l$. We can interpret $u_l(x)$ as *link usage*, or the probability that link l is used at any given point in time. Then, in (5.11), we have terms linear in u such as $\lambda_l(1 - u_l(x))$, in which λ_l has a similar interpretation as in concave NUM, as the price of using link l. We also have product terms such as $\lambda_{jk}(1-u_j(x))(1-u_k(x))$, where $\lambda_{jk} u_j(x) u_k(x)$ indicates the probability of simultaneous usage of links j and k, for links whose usage probabilities are independent (e.g., they do not share any flows). Products of more terms can be interpreted similarly.

Although the above price interpretation is not complete and does not justify all the terms appearing in (5.11) (e.g., powers of the constraints, product terms for links with shared flows), it does provide some useful intuition: this relaxation results in a pricing scheme that provides better incentives for the users to observe the constraints, by giving an additional reward (because the corresponding term adds positively to the utility) for simultaneously keeping two links free. Such incentive helps tighten the upper bound and eventually achieve a feasible (and optimal) allocation.

This relaxation is computationally attractive because we need to solve an LPs instead of the previous SDPs at each level. However, significantly more levels may be required [40].

Relaxation with No Product Terms

Putinar [53] showed that a polynomial positive over a compact set[2] can be represented as an SOS-combination of the constraints. This yields the following convex relaxation for nonconcave NUM problem.

$$\begin{array}{ll} \text{maximize} & \gamma \\ \text{subject to} & \\ \gamma - \sum_s U_s(x_s) = \sum_{l=1}^{L} \lambda_l(\mathbf{x})(c_l - \sum_{s \in S(l)} x_s), \ \forall \mathbf{x} & \\ \lambda(\mathbf{x}) \text{ is SOS}, & \end{array} \tag{5.12}$$

where the optimization variables are the coefficients in $\lambda_l(\mathbf{x})$. Similar to the SOS relaxation (5.5), fixing the order D of the expression in (5.12) results in an SDP. This relaxation has the nice property that no product terms appear: the relaxation becomes exact with a high enough D without the need of product terms. However, this degree might be much higher than what the previous SOS method requires.

---

[2] With an extra assumption that always holds for linear constraints as in NUM problems.

5.2.5 Concluding Remarks and Future Directions

We consider the NUM problem in the presence of inelastic flows, that is, flows with nonconcave utilities. Despite its practical importance, this problem has not been studied widely, mainly due to the fact it is a nonconvex problem. There has been no effective mechanism, centralized or distributed, to compute the globally optimal rate allocation for nonconcave utility maximization problems in networks. This limitation has made performance assessment and design of networks that include inelastic flows very difficult.

In one of the recent works on this topic [19], we employed convex SOS relaxations, solved by a sequence of SDPs, to obtain high-quality, increasingly tighter upper bounds on total achievable utility. In practice, the performance of our SOSTOOLS-based algorithm was surprisingly good, and bounds obtained using a polynomial-time (and indeed a low-order and often minimal-order) relaxation were found to be exact, achieving the global optimum of nonconcave NUM problems. Furthermore, a dual-based sufficient test, if successful, detects the exactness of the bound, in which case the optimal rate allocation can also be recovered. This performance of the proposed algorithm brings up a fundamental question on whether there is any particular property or structure in nonconcave NUM that makes it especially suitable for SOS relaxations.

We further examined the use of two more specialized polynomial representations, one that uses products of constraints with constant multipliers, resulting in LP relaxations; and at the other end of spectrum, one that uses a linear combination of constraints with SOS multipliers. We expect these relaxations to give higher-order certificates, thus their potential computational benefits need to be examined further. We also show they admit economics interpretations (e.g., prices, incentives) that provide some insight on how the SOS relaxations work in the framework of link congestion pricing for the simultaneous usage of multiple links.

An important research issue to be further investigated is decentralization methods for rate allocation among sources with nonconcave utilities. The proposed algorithm here is not easy to decentralize, given the products of the constraints or polynomial multipliers that destroy the separable structure of the problem. However, when relaxations become exact, the sparsity pattern of the coefficients can provide information about partially decentralized computation of optimal rates. For example, if after solving the NUM offline, we obtain an exact bound, then if the coefficient of the cross-term $x_i x_j$ turns out to be zero, it means users i and j do not need to communicate to each other to find their optimal rates. An interesting next step in this area of research is to investigate a distributed version of the proposed algorithm through limited message passing among clusters of network nodes and links.

5.3 Wireless Network Power Control

5.3.1 Introduction

Due to the broadcast nature of radio transmission, data rates and other quality of service (QoS) issues in a wireless network are affected by interference. This is particularly important in CDMA systems where users transmit at the same time over the same frequency bands and their spreading codes are not perfectly orthogonal. Transmit power control is often used to tackle this problem of signal interference [12]. We study how to optimize over the transmit powers to create the optimal set of signal-to-interference ratios (SIR) on wireless links. Optimality here can be with respect to a variety of objectives, such as maximizing a systemwide efficiency metric (e.g., the total system throughput), or maximizing a QoS metric for a user in the highest QoS class, or maximizing a QoS metric for the user with the minimum QoS metric value (i.e., a maxmin optimization).

The objective represents a systemwide goal to be optimized; however, individual users' QoS requirements also need to be satisfied. Any power allocation must therefore be constrained by a feasible set formed by these minimum requirements from the users. Such a constrained optimization captures the tradeoff between user-centric constraints and some network-centric objective. Because a higher power level from one transmitter increases the interference levels at other receivers, there may not be any feasible power allocation to satisfy the requirements from all the users. Sometimes an existing set of requirements can be satisfied, but when a new user is admitted into the system, there exist no more feasible power control solutions, or the maximized objective is reduced due to the tightening of the constraint set, leading to the need for admission control and admission pricing, respectively.

Because many QoS metrics are nonlinear functions of SIR, which is in turn a nonlinear (and neither convex nor concave) function of transmit powers, in general, power control optimization or feasibility problems are difficult nonlinear optimization problems that may appear to be NP-hard problems. Following [14, 35], this section shows that, when SIR is much larger than 0 dB, a class of nonlinear optimization called geometric programming (GP) can be used to efficiently compute the globally optimal power control in many of these problems, and efficiently determine the feasibility of user requirements by returning either a feasible (and indeed optimal) set of powers or a certificate of infeasibility. This also leads to an effective admission control and admission pricing method.

The key observation is that despite the apparent nonconvexity, through log change of variable the GP technique turns these constrained optimizations of power control into convex optimization, which is intrinsically tractable despite its nonlinearity in objective and constraints. However, when SIR is comparable to or below 0 dB, the power control problems are truly nonconvex with no efficient and global solution methods. In this case, we present a heuristic that is provably convergent and empirically almost always computes the globally optimal power allocation by solving a sequence of GPs through the approach of successive convex approximations.

The GP approach reveals the hidden convexity structure, which implies efficient solution methods and the global optimality of any local optimum in power control problems with nonlinear objective functions. It clearly differentiates the tractable formulations in a high-SIR regime from the intractable ones in a low-SIR regime. Power control by GP is applicable to formulations in both cellular networks with single-hop transmission between mobile users and base stations, and ad hoc networks with multihop transmission among the nodes, as illustrated through several numerical examples in this section. Traditionally, GP is solved by centralized computation through the highly efficient interior point methods. In this section we present a new result on how GP can be solved distributively with message passing, which has independent value to general maximization of coupled objective, and applies it to power control problems with a further reduction of message-passing overhead by leveraging the specific structures of power control problems.

More generally, the technique of nonlinear change of variables, including the log change of variables, to reveal "hidden" convexity in optimization formulations has recently become quite popular in the communication network research community.

5.3.2 Geometric Programming

GP is a class of nonlinear, nonconvex optimization problems with many useful theoretical and computational properties. It was invented in 1967 by Duffin, Peterson, and Zener [17], and much of the development by the early 1980s was summarized in [1]. Because a GP can be turned into a convex optimization problem, a local optimum is also a global optimum, the Lagrange duality gap is zero under mild conditions, and a global optimum can be computed very efficiently. Numerical efficiency holds both in theory and in practice: interior point methods applied to GP have provably polynomial-time complexity [48], and are very fast in practice with high-quality software downloadable from the Internet (e.g., the MOSEK package). Convexity and duality properties of GP are well understood, and large-scale, robust numerical solvers for GP are available. Furthermore, special structures in GP and its Lagrange dual problem lead to distributed algorithms, physical interpretations, and computational acceleration beyond the generic results for convex optimization. A detailed tutorial of GP and comprehensive survey of its recent applications to communication systems and to circuit design can be found in [11] and [3], respectively. This section contains a brief introduction of GP terminology.

There are two equivalent forms of GP: standard form and convex form. The first is a constrained optimization of a type of function called posynomial, and the second form is obtained from the first through a logarithmic change of variable.

We first define a monomial as a function $f : \mathbf{R}^n_{++} \rightarrow \mathbf{R}$:

$$f(\mathbf{x}) = d x_1^{a^{(1)}} x_2^{a^{(2)}} \dots x_n^{a^{(n)}},$$

where the multiplicative constant $d \geq 0$ and the exponential constants $a^{(j)} \in \mathbf{R}$, $j = 1, 2, \dots, n$. A sum of monomials, indexed by k below, is called a posynomial:

$$f(\mathbf{x}) = \sum_{k=1}^{K} d_k x_1^{a_k^{(1)}} x_2^{a_k^{(2)}} \dots x_n^{a_k^{(n)}},$$

where $d_k \geq 0$, $k = 1, 2, \dots, K$, and $a_k^{(j)} \in \mathbf{R}$, $j = 1, 2, \dots, n$, $k = 1, 2, \dots, K$. For example, $2x_1^{-\pi} x_2^{0.5} + 3x_1 x_3^{100}$ is a posynomial in $\mathbf{x}$, $x_1 - x_2$ is not a posynomial, and x_1/x_2 is a monomial, thus also a posynomial.

Minimizing a posynomial subject to posynomial upper bound inequality constraints and monomial equality constraints is called GP in *standard form*:

$$\begin{array}{ll} \text{minimize} & f_0(\mathbf{x}) \\ \text{subject to} & f_i(\mathbf{x}) \leq 1, \quad i = 1, 2, \dots, m, \\ & h_l(\mathbf{x}) = 1, \quad l = 1, 2, \dots, M, \end{array} \tag{5.13}$$

where f_i, $i = 0, 1, \dots, m$, are posynomials: $f_i(\mathbf{x}) = \sum_{k=1}^{K_i} d_{ik} x_1^{a_{ik}^{(1)}} x_2^{a_{ik}^{(2)}} \dots x_n^{a_{ik}^{(n)}}$, and h_l, $l = 1, 2, \dots, M$, are monomials: $h_l(\mathbf{x}) = d_l x_1^{a_l^{(1)}} x_2^{a_l^{(2)}} \dots x_n^{a_l^{(n)}}$.

GP in standard form is not a convex optimization problem, because posynomials are not convex functions. However, with a logarithmic change of the variables and multiplicative constants: $y_i = \log x_i$, $b_{ik} = \log d_{ik}$, $b_l = \log d_l$, and a logarithmic change of the functions' values, we can turn it into the following equivalent problem in $\mathbf{y}$.

$$\begin{array}{ll} \text{minimize} & p_0(\mathbf{y}) = \log \sum_{k=1}^{K_0} \exp(\mathbf{a}_{0k}^T \mathbf{y} + b_{0k}) \\ \text{subject to} & p_i(\mathbf{y}) = \log \sum_{k=1}^{K_i} \exp(\mathbf{a}_{ik}^T \mathbf{y} + b_{ik}) \leq 0, \quad i = 1, 2, \dots, m, \\ & q_l(\mathbf{y}) = \mathbf{a}_l^T \mathbf{y} + b_l = 0, \quad l = 1, 2, \dots, M. \end{array} \tag{5.14}$$

This is referred to as GP in *convex form*, which is a convex optimization problem because it can be verified that the log-sum-exp function is convex [4].

In summary, GP is a nonlinear, nonconvex optimization problem that can be transformed into a nonlinear convex problem. GP in standard form can be used to formulate network resource allocation problems with nonlinear objectives under nonlinear QoS constraints. The basic idea is that resources are often allocated proportional to some parameters, and when resource allo-

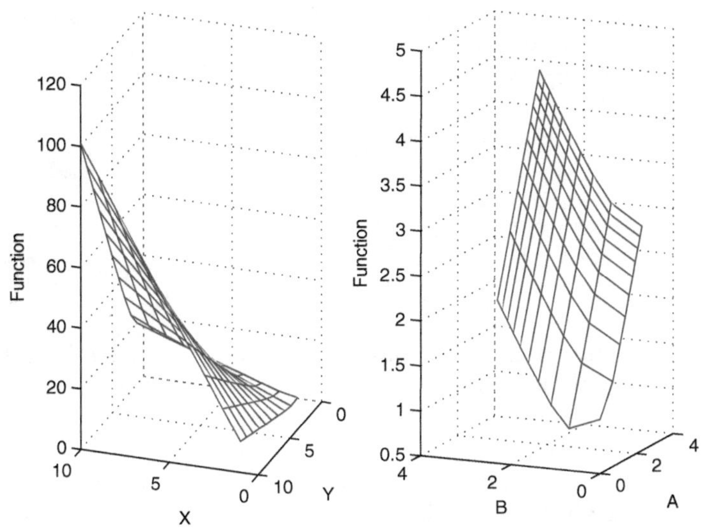

Fig. 5.6 A bivariate posynomial before (left graph) and after (right graph) the log transformation. A nonconvex function is turned into a convex one.

cations are optimized over these parameters, we are maximizing an inverted posynomial subject to lower bounds on other inverted posynomials, which are equivalent to GP in standard form.

SP/GP, SOS/SDP

Note that, although the posynomial seems to be a nonconvex function, it becomes a convex function after the log transformation, as shown in an example in Figure 5.6. Compared to the (constrained or unconstrained) minimization of a polynomial, the minimization of a posynomial in GP relaxes the integer constraint on the exponential constants but imposes a positivity constraint on the multiplicative constants and variables. There is a sharp contrast between these two problems: polynomial minimization is NP-hard, but GP can be turned into convex optimization with provably polynomial-time algorithms for a global optimum.

In an extension of GP called signomial programming discussed later in this section, the restriction of nonnegative multiplicative constants is removed. This results in a general class of nonlinear and truly nonconvex problems that is simultaneously a generalization of GP and polynomial minimization over the positive quadrant, as summarized in the comparison Table 5.2.

Table 5.2 Comparison of GP, constrained polynomial minimization over the positive quadrant (PMoP), and signomial programming (SP). All three types of problems minimize a sum of monomials subject to upper bound inequality constraints on sums of monomials, but have different definitions of monomial: $c\prod_j x_j^{a^{(j)}}$, as shown in the table. GP is known to be polynomial-time solvable, but PMoP and SP are not.

	GP	$PMoP$	SP
c	$\mathbf{R}_+$	$\mathbf{R}$	$\mathbf{R}$
$a^{(j)}$	$\mathbf{R}$	$\mathcal{Z}_+$	$\mathbf{R}$
x_j	$\mathbf{R}_{++}$	$\mathbf{R}_{++}$	$\mathbf{R}_{++}$

The objective function of signomial programming can be formulated as minimizing a ratio between two posynomials, which is not a posynomial (because posynomials are closed under positive multiplication and addition but not division). As shown in Figure 5.7, a ratio between two posynomials is a nonconvex function both before and after the log transformation. Although it does not seem likely that signomial programming can be turned into a convex optimization problem, there are heuristics to solve it through a sequence of GP relaxations. However, due to the absence of algebraic structures found in polynomials, such methods for signomial programming currently lack a theoretical foundation of convergence to global optimality. This is in contrast to the sum-of-squares method [51], which uses a nested family of SDP relax-

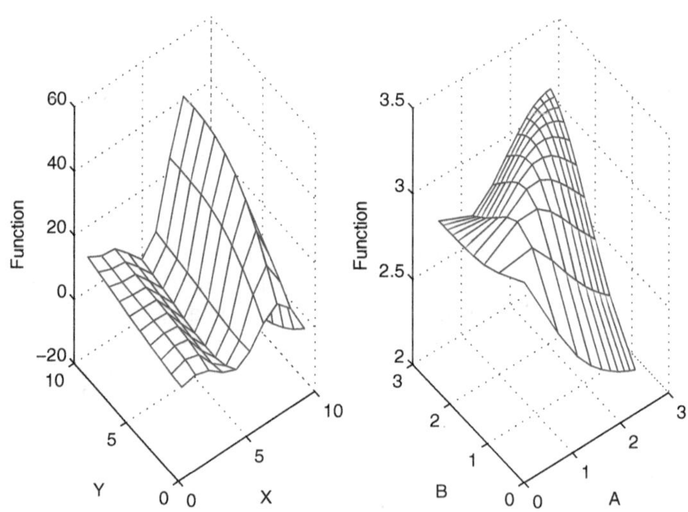

Fig. 5.7 Ratio between two bivariate posynomials before (left graph) and after (right graph) the log transformation. It is a nonconvex function in both cases.

ations to solve constrained polynomial minimization problems as explained in the last section.

5.3.3 *Power Control by Geometric Programming: Convex Case*

Various schemes for power control, centralized or distributed, have been extensively studied since the 1990s based on different transmission models and application needs (e.g., in [2, 26, 47, 55, 63, 72]). This section summarizes the new approach of formulating power control problems through GP. The key advantage is that globally optimal power allocations can be efficiently computed for a variety of nonlinear systemwide objectives and user QoS constraints, even when these nonlinear problems appear to be nonconvex optimization.

Basic Model

Consider a wireless (cellular or multihop) network with n logical transmitter/receiver pairs. Transmit powers are denoted as $P_1, \dots, P_n$. In the cellular uplink case, all logical receivers may reside in the same physical receiver, that is, the base station. In the multihop case, because the transmission environment can be different on the links comprising an end-to-end path, power control schemes must consider each link along a flow's path.

Under Rayleigh fading, the power received from transmitter j at receiver i is given by $G_{ij}F_{ij}P_j$ where $G_{ij} \geq 0$ represents the path gain (it may also encompass antenna gain and coding gain) that is often modeled as proportional to $d_{ij}^{-\gamma}$, where d_{ij} denotes distance, γ is the power fall-off factor, and F_{ij} model Rayleigh fading and are independent and exponentially distributed with unit mean. The distribution of the received power from transmitter j at receiver i is then exponential with mean value $\mathbf{E}\left[G_{ij}F_{ij}P_j\right] = G_{ij}P_j$. The SIR for the receiver on logical link i is:

$$\mathsf{SIR}_i = \frac{P_i G_{ii} F_{ii}}{\sum_{j \neq i}^{N} P_j G_{ij} F_{ij} + n_i} \tag{5.15}$$

where n_i is the noise power for receiver i.

The constellation size M used by a link can be closely approximated for MQAM modulations as follows. $M = 1 + (-\phi_1 / (\ln(\phi_2 \mathsf{BER}))) \mathsf{SIR}$, where BER is the bit error rate and ϕ_1, ϕ_2 are constants that depend on the modulation type. Defining $K = -\phi_1 / (\ln(\phi_2 \mathsf{BER}))$ leads to an expression of the data rate R_i on the ith link as a function of the SIR: $R_i = (1/T) \log_2(1 + K\mathsf{SIR}_i)$, which can be approximated as

$$R_i = \frac{1}{T}\log_2(K\mathsf{SIR}_i) \tag{5.16}$$

when $K\mathsf{SIR}$ is much larger than 1. This approximation is reasonable either when the signal level is much higher than the interference level or, in CDMA systems, when the spreading gain is large. For notational simplicity in the rest of this section, we redefine G_{ii} as K times the original G_{ii}, thus absorbing constant K into the definition of SIR.

The aggregate data rate for the system can then be written as

$$R_{\text{system}} = \sum_i R_i = \frac{1}{T}\log_2\left[\prod_i \mathsf{SIR}_i\right].$$

So in the high SIR regime, aggregate data rate maximization is equivalent to maximizing a product of SIR. The system throughput is the aggregate data rate supportable by the system given a set of users with specified QoS requirements.

Outage probability is another important QoS parameter for reliable communication in wireless networks. A channel outage is declared and packets lost when the received SIR falls below a given threshold SIR_{th}, often computed from the BER requirement. Most systems are interference-dominated and the thermal noise is relatively small, thus the ith link outage probability is

$$\begin{aligned} P_{\text{o},i} &= \textbf{Prob}\{\mathsf{SIR}_i \le \mathsf{SIR}_{\text{th}}\} \\ &= \textbf{Prob}\{G_{ii}F_{ii}P_i \le \mathsf{SIR}_{\text{th}}\sum_{j\ne i} G_{ij}F_{ij}P_j\}. \end{aligned}$$

The outage probability can be expressed as [38]

$$P_{\text{o},i} = 1 - \prod_{j\ne i}\frac{1}{1+\frac{\mathsf{SIR}_{\text{th}}G_{ij}P_j}{G_{ii}P_i}},$$

which means that the upper bound $P_{\text{o},i} \le P_{\text{o},i,\max}$ can be written as an upper bound on a posynomial in $\mathbf{P}$:

$$\prod_{j\ne i}\left(1+\frac{\mathsf{SIR}_{\text{th}}G_{ij}P_j}{G_{ii}P_i}\right) \le \frac{1}{1-P_{\text{o},i,\max}}. \tag{5.17}$$

Cellular Wireless Networks

We first present how GP-based power control applies to cellular wireless networks with one-hop transmission from N users to a base station. These results extend the scope of power control by the classical solution in CDMA systems that equalizes SIRs, and those by the iterative algorithms (e.g., in [2, 26, 47]) that minimize total power (a linear objective function) subject to SIR constraints.

We start the discussion on the suite of power control problem formulations with a simple objective function and basic constraints. The following constrained problem of maximizing the SIR of a particular user i^* is a GP.

$$\begin{array}{ll} \text{maximize} & R_{i^*}(\mathbf{P}) \\ \text{subject to} & R_i(\mathbf{P}) \geq R_{i,\min}, \ \ \forall i, \\ & P_{i1}G_{i1} = P_{i2}G_{i2}, \\ & 0 \leq P_i \leq P_{i,\max}, \ \ \forall i. \end{array}$$

The first constraint, equivalent to $\mathsf{SIR}_i \geq \mathsf{SIR}_{i,\min}$, sets a floor on the SIR of other users and protects these users from user i^* increasing her transmit power excessively. The second constraint reflects the classical power control criterion in solving the near–far problem in CDMA systems: the expected received power from one transmitter $i1$ must equal that from another $i2$. The third constraint is regulatory or system limitations on transmit powers. All constraints can be verified to be inequality upper bounds on posynomials in transmit power vector $\mathbf{P}$.

Alternatively, we can use GP to maximize the minimum rate among all users. The maxmin fairness objective

$$\text{maximize}_{\mathbf{P}} \ \ \min_i \{R_i\}$$

can be accommodated in GP-based power control because it can be turned into equivalently maximizing an auxiliary variable t such that $\mathsf{SIR}_i(\mathbf{P}) \geq \exp(t)$, $\forall i$, which has a posynomial objective and constraints in $(\mathbf{P}, t)$.

Example 5.5. *A small illustrative example.* A simple system comprised of five users is used for a numerical example. The five users are spaced at distances d of $1, 5, 10, 15$, and 20 units from the base station. The power fall-off factor $\gamma = 4$. Each user has a maximum power constraint of $P_{\max} = 0.5$ mW. The noise power is 0.5 μW for all users. The SIR of all users, other than the user we are optimizing for, must be greater than a common threshold SIR level β. In different experiments, β is varied to observe the effect on the optimized user's SIR. This is done independently for the near user at $d = 1$, a medium distance user at $d = 15$, and the far user at $d = 20$. The results are plotted in Figure 5.8.

Several interesting effects are illustrated. First, when the required threshold SIR in the constraints is sufficiently high, there is no feasible power control solution. At moderate threshold SIR, as β is decreased, the optimized SIR initially increases rapidly. This is because it is allowed to increase its own power by the sum of the power reductions in the four other users, and the noise is relatively insignificant. At low threshold SIR, the noise becomes more significant and the power tradeoff from the other users less significant,

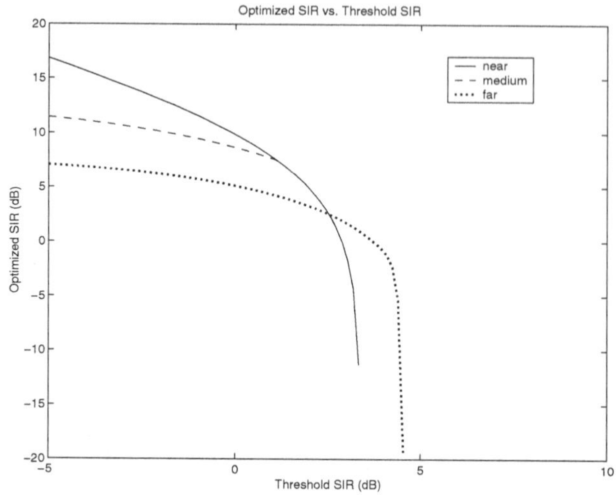

Fig. 5.8 Constrained optimization of power control in a cellular network (Example 5.5).

so the curve starts to bend over. Eventually, the optimized user reaches its upper bound on power and cannot utilize the excess power allowed by the lower threshold SIR for other users. This is exhibited by the transition from a sharp bend in the curve to a much shallower sloped curve.

We now proceed to show that GP can also be applied to the problem formulations with an overall system objective of total system throughput, under both user data rate constraints and outage probability constraints.

The following constrained problem of maximizing system throughput is a GP.

$$\begin{array}{ll} \text{maximize} & R_{\text{system}}(\mathbf{P}) \\ \text{subject to} & R_i(\mathbf{P}) \geq R_{i,\min}, \ \ \forall i, \\ & P_{\text{o},i}(\mathbf{P}) \leq P_{\text{o},i,\max}, \ \ \forall i, \\ & 0 \leq P_i \leq P_{i,\max}, \ \ \forall i \end{array} \tag{5.18}$$

where the optimization variables are the transmit powers $\mathbf{P}$. The objective is equivalent to minimizing the posynomial $\prod_i \text{ISR}_i$, where ISR is 1/SIR. Each ISR is a posynomial in $\mathbf{P}$ and the product of posynomials is again a posynomial. The first constraint is from the data rate demand $R_{i,\min}$ by each user. The second constraint represents the outage probability upper bounds $P_{\text{o},i,\max}$. These inequality constraints put upper bounds on posynomials of $\mathbf{P}$, as can be readily verified through (5.16) and (5.17). Thus (5.18) is indeed a GP, and efficiently solvable for global optimality.

There are several obvious variations of problem (5.18) that can be solved by GP; for example, we can lower bound R_{system} as a constraint and maximize R_{i^*} for a particular user i^*, or have a total power $\sum_i P_i$ constraint or objective function.

Table 5.3 Suite of power control optimization solvable by GP

Objective Function	*Constraints*
(A) Max R_{i^*} (specific user)	(a) $R_i \geq R_{i,\min}$ (rate constraint)
(B) Max $\min_i R_i$ (worst-case user)	(b) $P_{i1}G_{i1} = P_{i2}G_{i2}$ (near–far constraint)
(C) Max $\sum_i R_i$ (total throughput)	(c) $\sum_i R_i \geq R_{\text{system},\min}$ (sum rate constraint)
(D) Max $\sum_i w_i R_i$ (weighted rate sum)	(d) $P_{\text{o},i} \leq P_{\text{o},i,\max}$ (outage probability constraint)
(E) Min $\sum_i P_i$ (total power)	(e) $0 \leq P_i \leq P_{i,\max}$ (power constraint)

The objective function to be maximized can also be generalized to a weighted sum of data rates, $\sum_i w_i R_i$, where $\mathbf{w} \succeq 0$ is a given weight vector. This is still a GP because maximizing $\sum_i w_i \log \mathsf{SIR}_i$ is equivalent to maximizing $\log \prod_i \mathsf{SIR}_i^{w_i}$, which is in turn equivalent to minimizing $\prod_i \text{ISR}_i^{w_i}$. Now use auxiliary variables $\{t_i\}$, and minimize $\prod_i t_i^{w_i}$ over the original constraints in (5.18) plus the additional constraints $\text{ISR}_i \leq t_i$ for all i. This is readily verified to be a GP in $(\mathbf{x}, \mathbf{t})$, and is equivalent to the original problem.

Generalizing the above discussions and observing that high-SIR assumption is needed for GP formulation only when there are sums of $\log(1 + \mathsf{SIR})$ in the optimization problem, we have the following summary.

Proposition 5.1. *In the high-SIR regime, any combination of objectives* (A)–(E) *and constraints* (a)–(e) *in Table* 5.3 (*pick any one of the objectives and any subset of the constraints*) *is a power control optimization problem that can be solved by GP, that is, can be transformed into a convex optimization with efficient algorithms to compute the globally optimal power vector. When objectives* (C)–(D) *or constraints* (c)–(d) *do not appear, the power control optimization problem can be solved by GP in any SIR regime.*

In addition to efficient computation of the globally optimal power allocation with nonlinear objectives and constraints, GP can also be used for admission control based on feasibility study described in [11], and for determining which QoS constraint is a performance bottleneck, that is, met tightly at the optimal power allocation.[3]

[3] This is because most GP solution algorithms solve both the primal GP and its Lagrange dual problem, and by the complementary slackness condition, a resource constraint is tight at optimal power allocation when the corresponding optimal dual variable is nonzero.

Extensions

In wireless multihop networks, system throughput may be measured either by end-to-end transport layer utilities or by link layer aggregate throughput. GP application to the first approach has appeared in [10], and those to the second approach in [11]. Furthermore, delay and buffer overflow properties can also be accommodated in the constraints or objective function of GP-based power control.

5.3.4 Power Control by Geometric Programming: Nonconvex Case

If we maximize the total throughput R_{system} in the medium to low SIR case (i.e., when SIR is not much larger than 0 dB), the approximation of $\log(1+\mathsf{SIR})$ as $\log \mathsf{SIR}$ does not hold. Unlike SIR, which is an inverted posynomial, $1+\mathsf{SIR}$ is not an inverted posynomial. Instead, $1/(1+\mathsf{SIR})$ is a ratio between two posynomials:

$$\frac{f(\mathbf{P})}{g(\mathbf{P})} = \frac{\sum_{j \neq i} G_{ij} P_j + n_i}{\sum_j G_{ij} P_j + n_i}. \tag{5.19}$$

Minimizing, or upper bounding, a ratio between two posynomials belongs to a truly nonconvex class of problems known as complementary GP [1, 11] that is in general an NP-hard problem. An equivalent generalization of GP is signomial programming [1, 11]: minimizing a signomial subject to upper bound inequality constraints on signomials, where a signomial $s(\mathbf{x})$ is a sum of monomials, possibly with negative multiplicative coefficients: $s(\mathbf{x}) = \sum_{i=1}^{N} c_i g_i(\mathbf{x})$ where $\mathbf{c} \in \mathbf{R}^N$ and $g_i(\mathbf{x})$ are monomials.[4]

Successive Convex Approximation Method

Consider the following nonconvex problem,

$$\begin{array}{ll} \text{minimize} & f_0(\mathbf{x}) \\ \text{subject to} & f_i(\mathbf{x}) \leq 1, \quad i = 1, 2, \ldots, m, \end{array} \tag{5.20}$$

where f_0 is convex without loss of generality,[5] but the $f_i(\mathbf{x})$s, $\forall i$ are nonconvex. Because directly solving this problem is NP-hard, we want to solve it by

[4] An SP can always be converted into a complementary GP, because an inequality in SP, which can be written as $f_{i1}(\mathbf{x}) - f_{i2}(\mathbf{x}) \leq 1$, where f_{i1}, f_{i2} are posynomials, is equivalent to an inequality $f_{i1}(\mathbf{x})/(1 + f_{i2}(\mathbf{x})) \leq 1$ in complementary GP.

[5] If f_0 is nonconvex, we can move the objective function to the constraint by introducing auxiliary scalar variable t and writing minimize t subject to the additional constraint $f_0(\mathbf{x}) - t \leq 0$.

a series of approximations $\tilde{f}_i(\mathbf{x}) \approx f_i(\mathbf{x}), \forall \mathbf{x}$, each of which can be optimally solved in an easy way. It is known [46] that if the approximations satisfy the following three properties, then the solutions of this series of approximations converge to a point satisfying the necessary optimality Karush–Kuhn–Tucker (KKT) conditions of the original problem.

(1) $f_i(\mathbf{x}) \le \tilde{f}_i(\mathbf{x})$ for all $\mathbf{x}$.
(2) $f_i(\mathbf{x}_0) = \tilde{f}_i(\mathbf{x}_0)$ where x_0 is the optimal solution of the approximated problem in the previous iteration.
(3) $\nabla f_i(\mathbf{x}_0) = \nabla \tilde{f}_i(\mathbf{x}_0)$.

The following algorithm describes the generic successive approximation approach. Given a method to approximate $f_i(\mathbf{x})$ with $\tilde{f}_i(\mathbf{x})$, $\forall i$, around some point of interest $\mathbf{x}_0$, the following algorithm provides the output of a vector that satisfies the KKT conditions of the original problem.

Algorithm 2. Successive approximation to a nonconvex problem.

1. Choose an initial feasible point $\mathbf{x}^{(0)}$ and set $k = 1$.
2. Form an approximated problem of (5.20) based on the previous point $\mathbf{x}^{(k-1)}$.
3. Solve the kth approximated problem to obtain $\mathbf{x}^{(k)}$.
4. Increment k and go to step 2 until convergence to a stationary point.

Single condensation method. Complementary GPs involve upper bounds on the ratio of posynomials as in (5.19); they can be turned into GPs by approximating the denominator of the ratio of posynomials, $g(\mathbf{x})$, with a monomial $\tilde{g}(\mathbf{x})$, but leaving the numerator $f(\mathbf{x})$ as a posynomial.

The following basic result can be readily proved using the arithmetic-mean–geometric-mean inequality.

Lemma 5.1. *Let $g(\mathbf{x}) = \sum_i u_i(\mathbf{x})$ be a posynomial. Then*

$$g(\mathbf{x}) \ge \tilde{g}(\mathbf{x}) = \prod_i \left(\frac{u_i(\mathbf{x})}{\alpha_i} \right)^{\alpha_i}. \tag{5.21}$$

If, in addition, $\alpha_i = u_i(\mathbf{x}_0)/g(\mathbf{x}_0)$, $\forall i$, for any fixed positive $\mathbf{x}_0$, then $\tilde{g}(\mathbf{x}_0) = g(\mathbf{x}_0)$, and $\tilde{g}(\mathbf{x})$ is the best local monomial approximation to $g(\mathbf{x})$ near $\mathbf{x}_0$ in the sense of first-order Taylor approximation.

The above lemma easily leads to the following

Proposition 5.2. *The approximation of a ratio of posynomials $f(\mathbf{x})/g(\mathbf{x})$ with $f(\mathbf{x})/\tilde{g}(\mathbf{x})$ where $\tilde{g}(\mathbf{x})$ is the monomial approximation of $g(\mathbf{x})$ using the arithmetic-geometric mean approximation of Lemma 5.1 satisfies the three conditions for the convergence of the successive approximation method.*

Double condensation method. Another choice of approximation is to make a double monomial approximation for both the denominator and numerator in (5.19). However, in order to satisfy the three conditions for the convergence of the successive approximation method, a monomial approximation for the numerator $f(\mathbf{x})$ should satisfy $f(\mathbf{x}) \le \tilde{f}(\mathbf{x})$.

Applications to Power Control

Figure 5.9 shows a block diagram of the approach of GP-based power control for a general SIR regime [64]. In the high SIR regime, we need to solve only one GP. In the medium to low SIR regimes, we solve truly nonconvex power control problems that cannot be turned into convex formulation through a series of GPs.

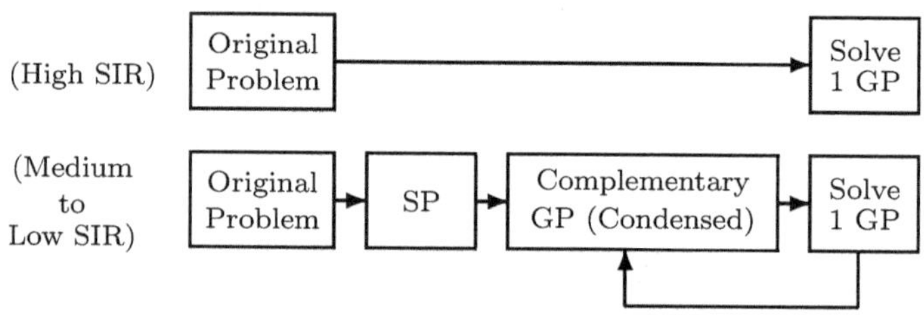

Fig. 5.9 GP-based power control in different SIR regimes.

GP-based power control problems in the medium to low SIR regimes become SP (or, equivalently, complementary GP), which can be solved by the single or double condensation method. We focus on the single condensation method here. Consider a representative problem formulation of maximizing total system throughput in a cellular wireless network subject to user rate and outage probability constraints in problem (5.18), which can be explicitly written out as

$$\begin{array}{ll} \text{minimize} & \prod_{i=1}^{N} \frac{1}{1+\mathsf{SIR}_i} \\ \text{subject to} & (2^{TR_{i,\min}} - 1)\frac{1}{\mathsf{SIR}_i} \le 1, \;\; i = 1, \dots, N, \\ & (\mathsf{SIR}_{\text{th}})^{N-1}(1 - P_{\text{o},i,\max}) \prod_{j\ne i}^{N} \frac{G_{ij}P_j}{G_{ii}P_i} \le 1, \;\; i = 1, \dots, N, \\ & P_i(P_{i,\max})^{-1} \le 1, \;\; i = 1, \dots, N. \end{array} \tag{5.22}$$

All the constraints are posynomials. However, the objective is not a posynomial, but a ratio between two posynomials as in (5.19). This power control problem can be solved by the condensation method by solving a series of GPs. Specifically, we have the following single-condensation algorithm.

Algorithm 3. Single condensation GP power control.

1. Evaluate the denominator posynomial of the objective function in (5.22) with the given $\mathbf{P}$.

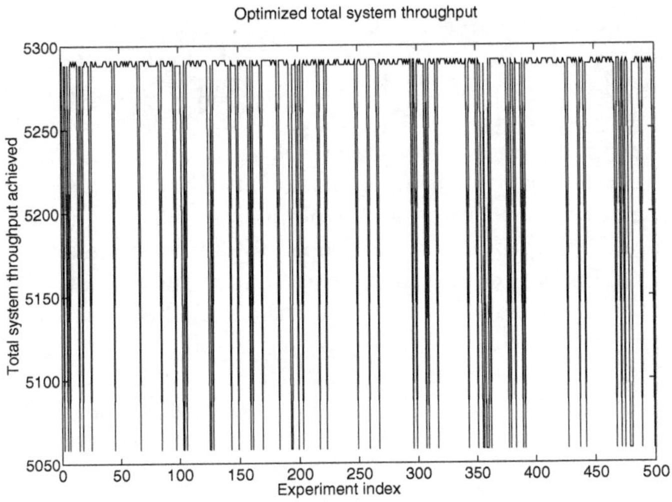

Fig. 5.10 Maximized total system throughput achieved by the (single) condensation method for 500 different initial feasible vectors (Example 5.6). Each point represents a different experiment with a different initial power vector.

2. Compute for each term i in this posynomial,

$$\alpha_i = \frac{\text{value of } i\text{th term in posynomial}}{\text{value of posynomial}}.$$

3. Condense the denominator posynomial of the (5.22) objective function into a monomial using (5.21) with weights α_i.

4. Solve the resulting GP using an interior point method.

5. Go to step 1 using $\mathbf{P}$ of step 4.

6. Terminate the kth loop if $\left\|\mathbf{P}^{(k)} - \mathbf{P}^{(k-1)}\right\| \le \epsilon$ where ϵ is the error tolerance for exit condition.

As condensing the objective in the above problem gives us an underestimate of the objective value, each GP in the condensation iteration loop tries to improve the accuracy of the approximation to a particular minimum in the original feasible region. All three conditions for convergence are satisfied, and the algorithm is convergent. Empirically through extensive numerical experiments, we observe that it almost always computes the globally optimal power allocation.

Example 5.6. *Single condensation example.* We consider a wireless cellular network with three users. Let $T = 10^{-6}$ s, $G_{ii} = 1.5$, and generate G_{ij}, $i \neq j$, as independent random variables uniformly distributed between 0 and 0.3. Threshold SIR is $\mathsf{SIR}_{\text{th}} = -10$ dB, and minimal data rate requirements are 100 kbps, 600 kbps, and 1000 kbps for logical links 1, 2, and 3, respectively.

Maximal outage probabilities are 0.01 for all links, and maximal transmit powers are 3 mW, 4 mW, and 5 mW for links 1, 2, and 3, respectively. For each instance of problem (5.22), we pick a random initial feasible power vector $\mathbf{P}$ uniformly between 0 and $\mathbf{P}_{\text{max}}$. Figure 5.10 compares the maximized total network throughput achieved over 500 sets of experiments with different initial vectors. With the (single) condensation method, SP converges to different optima over the entire set of experiments, achieving (or coming very close to) the global optimum at 5290 bps 96% of the time and a local optimum at 5060 bps 4% of the time. The average number of GP iterations required by the condensation method over the same set of experiments is 15 if an extremely tight exit condition is picked for SP condensation iteration: $\epsilon = 1 \times 10^{-10}$. This average can be substantially reduced by using a larger ϵ; for example, increasing ϵ to 1×10^{-2} requires on average only 4 GPs.

We have thus far discussed a power control problem (5.22) where the objective function needs to be condensed. The method is also applicable if some constraint functions are signomials and need to be condensed [14, 35].

5.3.5 Distributed Algorithm

A limitation for GP-based power control in ad hoc networks (without base stations) is the need for centralized computation (e.g., by interior point methods). The GP formulations of power control problems can also be solved by a new method of distributed algorithm for GP. The basic idea is that each user solves its own local optimization problem and the coupling among users is taken care of by message passing among the users. Interestingly, the special structure of coupling for the problem at hand (all coupling among the logical links can be lumped together using interference terms) allows one to further reduce the amount of message passing among the users. Specifically, we use a dual decomposition method to decompose a GP into smaller subproblems whose solutions are jointly and iteratively coordinated by the use of dual variables. The key step is to introduce auxiliary variables and to add extra equality constraints, thus transferring the coupling in the objective to coupling in the constraints, which can be solved by introducing "consistency pricing" (in contrast to "congestion pricing"). We illustrate this idea through an unconstrained GP followed by an application of the technique to power control.

Distributed Algorithm for GP

Suppose we have the following unconstrained standard form GP in $\mathbf{x} \succ 0$,

$$\text{minimize } \textstyle\sum_i f_i(x_i, \{x_j\}_{j \in I(i)}), \tag{5.23}$$

where x_i denotes the local variable of the ith user, $\{x_j\}_{j\in I(i)}$ denotes the coupled variables from other users, and f_i is either a monomial or posynomial. Making a change of variable $y_i = \log x_i, \forall i$, in the original problem, we obtain

$$\text{minimize } \textstyle\sum_i f_i(e^{y_i}, \{e^{y_j}\}_{j\in I(i)}).$$

We now rewrite the problem by introducing auxiliary variables y_{ij} for the coupled arguments and additional equality constraints to enforce consistency:

$$\begin{array}{ll} \text{minimize} & \sum_i f_i(e^{y_i}, \{e^{y_{ij}}\}_{j\in I(i)}) \\ \text{subject to} & y_{ij} = y_j, \forall j \in I(i), \forall i. \end{array} \tag{5.24}$$

Each ith user controls the local variables $(y_i, \{y_{ij}\}_{j\in I(i)})$. Next, the Lagrangian of (5.24) is formed as

$$\begin{aligned} L(\{y_i\}, \{y_{ij}\}; \{\gamma_{ij}\}) &= \sum_i f_i(e^{y_i}, \{e^{y_{ij}}\}_{j\in I(i)}) + \sum_i \sum_{j\in I(i)} \gamma_{ij}(y_j - y_{ij}) \\ &= \sum_i L_i(y_i, \{y_{ij}\}; \{\gamma_{ij}\}), \end{aligned}$$

where

$$L_i(y_i, \{y_{ij}\}; \{\gamma_{ij}\}) = f_i(e^{y_i}, \{e^{y_{ij}}\}_{j\in I(i)}) + \left(\sum_{j:i\in I(j)} \gamma_{ji}\right) y_i - \sum_{j\in I(i)} \gamma_{ij} y_{ij}. \tag{5.25}$$

The minimization of the Lagrangian with respect to the primal variables $(\{y_i\}, \{y_{ij}\})$ can be done simultaneously and distributively by each user in parallel. In the more general case where the original problem (5.23) is constrained, the additional constraints can be included in the minimization at each L_i.

In addition, the following master Lagrange dual problem has to be solved to obtain the optimal dual variables or consistency prices $\{\gamma_{ij}\}$,

$$\max_{\{\gamma_{ij}\}} g(\{\gamma_{ij}\}), \tag{5.26}$$

where

$$g(\{\gamma_{ij}\}) = \sum_i \min_{y_i, \{y_{ij}\}} L_i(y_i, \{y_{ij}\}; \{\gamma_{ij}\}).$$

Note that the transformed primal problem (5.24) is convex with zero duality gap; hence the Lagrange dual problem indeed solves the original standard GP problem. A simple way to solve the maximization in (5.26) is with the following subgradient update for the consistency prices,

$$\gamma_{ij}(t+1) = \gamma_{ij}(t) + \delta(t)(y_j(t) - y_{ij}(t)). \tag{5.27}$$

Appropriate choice of the stepsize $\delta(t) > 0$, for example, $\delta(t) = \delta_0/t$ for some constant $\delta_0 > 0$, leads to convergence of the dual algorithm.

Summarizing, the ith user has to: (i) minimize the function L_i in (5.25) involving only local variables, upon receiving the updated dual variables $\{\gamma_{ji}, j : i \in I(j)\}$, and (ii) update the local consistency prices $\{\gamma_{ij}, j \in I(i)\}$ with (5.27), and broadcast the updated prices to the coupled users.

Applications to Power Control

As an illustrative example, we maximize the total system throughput in the high SIR regime with constraints local to each user. If we directly applied the distributed approach described in the last section, the resulting algorithm would require knowledge by each user of the interfering channels and interfering transmit powers, which would translate into a large amount of message passing. To obtain a practical distributed solution, we can leverage the structures of power control problems at hand, and instead keep a local copy of each of the effective received powers $P_{ij}^R = G_{ij}P_j$. Again using problem (5.18) as an example formulation and assuming high SIR, we can write the problem as (after the log change of variable)

$$\begin{array}{ll} \text{minimize} & \sum_i \log\left(G_{ii}^{-1}\exp(-\tilde{P}_i)\left(\sum_{j\neq i}\exp(\tilde{P}_{ij}^R) + \sigma^2\right)\right) \\ \text{subject to} & \tilde{P}_{ij}^R = \tilde{G}_{ij} + \tilde{P}_j, \end{array} \tag{5.28}$$

Constraints are local to each user, for example, (a), (d), and (e) in Table 5.3. The partial Lagrangian is

$$\begin{aligned} L = & \sum_i \log\left(G_{ii}^{-1}\exp(-\tilde{P}_i)\left(\sum_{j\neq i}\exp(\tilde{P}_{ij}^R) + \sigma^2\right)\right) \\ & + \sum_i\sum_{j\neq i}\gamma_{ij}\left(\tilde{P}_{ij}^R - \left(\tilde{G}_{ij} + \tilde{P}_j\right)\right), \end{aligned} \tag{5.29}$$

and the local ith Lagrangian function in (5.29) is distributed to the ith user, from which the dual decomposition method can be used to determine the optimal power allocation $\mathbf{P}^*$. The distributed power control algorithm is summarized as follows.

Algorithm 4. Distributed power allocation update to maximize R_{system}.

At each iteration t:

1. The ith user receives the term $\left(\sum_{j\neq i}\gamma_{ji}(t)\right)$ involving the dual variables from the interfering users by message-passing and minimizes the following local Lagrangian with respect to $\tilde{P}_i(t)$, $\left\{\tilde{P}_{ij}^R(t)\right\}_j$ subject to the local constraints.

$$L_i\left(\tilde{P}_i(t), \left\{\tilde{P}_{ij}^R(t)\right\}_j ; \{\gamma_{ij}(t)\}_j\right)$$
$$= \log\left(G_{ii}^{-1}\exp(-\tilde{P}_i(t))\left(\sum_{j\neq i}\exp(\tilde{P}_{ij}^R(t)) + \sigma^2\right)\right)$$
$$+\sum_{j\neq i}\gamma_{ij}\tilde{P}_{ij}^R(t) - \left(\sum_{j\neq i}\gamma_{ji}(t)\right)\tilde{P}_i(t).$$

2. The ith user estimates the effective received power from each of the interfering users $P_{ij}^R(t) = G_{ij}P_j(t)$ for $j \neq i$, updates the dual variable by

$$\gamma_{ij}(t+1) = \gamma_{ij}(t) + (\delta_0/t)\left(\tilde{P}_{ij}^R(t) - \log G_{ij}P_j(t)\right), \tag{5.30}$$

and then broadcasts them by message passing to all interfering users in the system.

Example 5.7. *Distributed GP power control.* We apply the distributed algorithm to solve the above power control problem for three logical links with $G_{ij} = 0.2$, $i \neq j$, $G_{ii} = 1$, $\forall i$, maximal transmit powers of 6 mW, 7 mW, and 7 mW for links 1, 2, and 3 respectively. Figure 5.11 shows the convergence of the dual objective function towards the globally optimal total throughput of the network. Figure 5.12 shows the convergence of the two auxiliary variables in links 1 and 3 towards the optimal solutions.

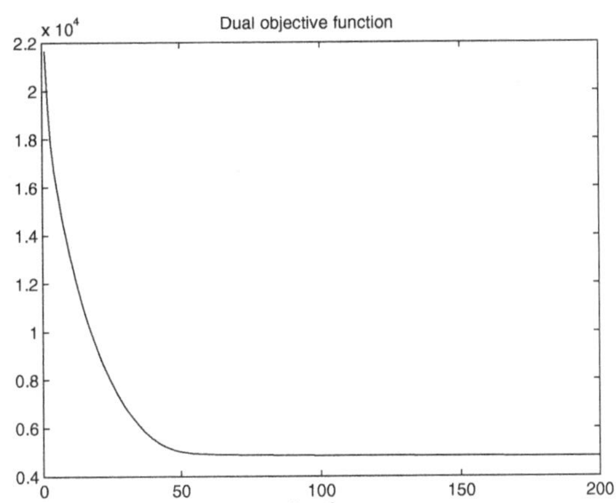

Fig. 5.11 Convergence of the dual objective function through distributed algorithm (Example 5.7).

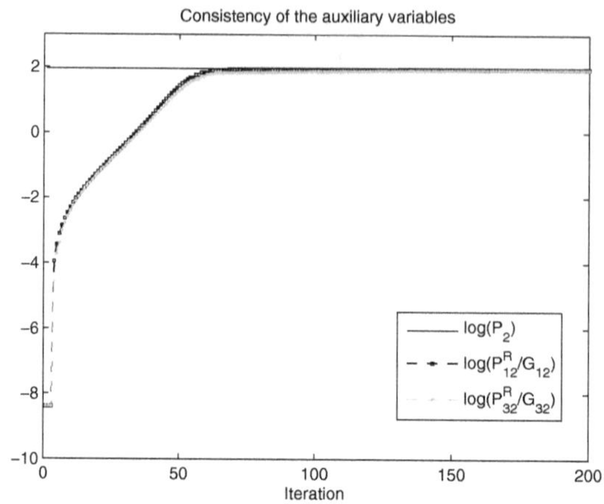

Fig. 5.12 Convergence of the consistency constraints through distributed algorithm (Example 5.7).

5.3.6 Concluding Remarks and Future Directions

Power control problems with nonlinear objective and constraints may seem to be difficult, NP-hard problems to solve for global optimality. However, when SIR is much larger than 0 dB, GP can be used to turn these problems into intrinsically tractable convex formulations, accommodating a variety of possible combinations of objective and constraint functions involving data rate, delay, and outage probability. Then interior point algorithms can efficiently compute the globally optimal power allocation even for a large network. Feasibility analysis of GP naturally leads to admission control and pricing schemes. When the high SIR approximation cannot be made, these power control problems become SP and may be solved by the heuristic of the condensation method through a series of GPs. Distributed optimal algorithms for GP-based power control in multihop networks can also be carried out through message passing.

Several challenging research issues in the low-SIR regime remain to be further explored. These include, for example, the reduction of SP solution complexity (e.g., by using high-SIR approximation to obtain the initial power vector and by solving the series of GPs only approximately except the last GP), and the combination of SP solution and distributed algorithm for distributed power control in low SIR regime. We also note that other approaches to tackle nonconvex power control problems have been studied, for example, the use of a particular utility function of rate to turn the problem into a convex one [28].

5.4 DSL Spectrum Management

5.4.1 Introduction

Digital subscriber line (DSL) technologies transform traditional voice-band copper channels into high-bandwidth data pipes, which are currently capable of delivering data rates up to several Mbps per twisted-pair over a distance of about 10 kft. The major obstacle for performance improvement in today's DSL systems (e.g., ADSL and VDSL) is *crosstalk*, which is the interference generated between different lines in the same binder. The crosstalk is typically 10–20 dB larger than the background noise, and direct crosstalk cancellation (e.g., [6, 27]) may not be feasible in many cases due to complexity issues or as a result of unbundling. To mitigate the detriments caused by crosstalk, *static spectrum management* which mandates spectrum mask or flat power backoff across all frequencies (i.e., tones) has been implemented in the current system.

Dynamic spectrum management (DSM) techniques, on the other hand, can significantly improve data rates over the current practice of static spectrum management. Within the current capability of the DSL modems, each modem has the capability to shape its own power spectrum density (PSD) across different tones, but can only treat crosstalk as background noise (i.e., no signal level coordination, such as vector transmission or iterative decoding, is allowed), and each modem is inherently a single-input–single-output communication system. The objective would be to optimize the PSD of all users on all tones (i.e., continuous power loading or discrete bit loading), such that they are "compatible" with each other and the system performance (e.g., weighted rate sum as discussed below) is maximized.

Compared to power control in wireless networks treated in the last section, the channel gains are not time-varying in DSL systems, but the problem dimension increases tremendously because there are many "tones" (or frequency carriers) over which transmission takes place. Nonconvexity still remains a major technical challenge, and high SIR approximation in general cannot be made. However, utilizing the specific structures of the problem (e.g., the interference channel gain values), an efficient and distributed heuristic is shown to perform close to the optimum in many realistic DSL network scenarios.

Following [7], this section presents a new algorithm for spectrum management in frequency selective interference channels for DSL, called autonomous spectrum balancing (ASB). It is the first DSL spectrum management algorithm that satisfies all of the following requirements for performance and complexity. It is autonomous (distributed algorithm across the users without explicit information exchange) with linear-complexity, while provably convergent, and comes close to the globally optimal rate region in practice. ASB overcomes the bottlenecks in the state-of-the-art algorithms in DSM, including IW, OSB, and ISB summarized below.

Let K be the number of tones and N the number of users (lines). The *iterative waterfilling* (IW) algorithm [74] is among one of the first DSM algorithms proposed. In IW, each user views any crosstalk experienced as additive Gaussian noise, and seeks to maximize its data rate by "waterfilling" over the aggregated noise plus interference. No information exchange is needed among users, and all the actions are completely autonomous. IW leads to a great performance increase over the static approach, and enjoys a low complexity that is linear in N. However, the greedy nature of IW leads to a performance far from optimal in the near–far scenarios such as mixed CO/RT deployment and upstream VDSL.

To address this, an *optimal spectrum balancing* (OSB) algorithm [9] has been proposed, which finds the best possible spectrum management solution under the current capabilities of the DSL modems. OSB avoids the selfish behaviors of individual users by aiming at the maximization of a total weighted sum of user rates, which corresponds to a boundary point of the achievable rate region. On the other hand, OSB has a high computational complexity that is exponential in N, which quickly leads to intractability when N is larger than 6. Moreover, it is a completely centralized algorithm where a spectrum management center at the central office needs to know the *global information* (i.e., all the noise PSDs and crosstalk channel gains in the same binder) to perform the algorithm.

As an improvement to the OSB algorithm, an *iterative spectrum balancing* (ISB) algorithm [8] has been proposed, which is based on a weighted sum rate maximization similar to OSB. Different from OSB, ISB performs the optimization iteratively through users, which leads to a quadratic complexity in N. Close to optimal performance can be achieved by the ISB algorithm in most cases. However, each user still needs to know the global information as in OSB, thus ISB is still a centralized algorithm and is considered to be impractical in many cases.

This section presents the ASB algorithm [7], which attains near-optimal performance in an implementable way. The basic idea is to use the concept of a reference line to mimic a "typical" victim line in the current binder. By setting the power spectrum level to protect the reference line, a good balance between selfish and global maximizations can be achieved. The ASB algorithm enjoys a linear complexity in N and K, and can be implemented in a completely autonomous way. We prove the convergence of ASB under both sequential and parallel updates.

Table 5.4 compares various aspects of different DSM algorithms. Utilizing the structures of the DSL problem, in particular, the lack of channel variation and user mobility, is the key to provide a linear complexity, distributed, convergent, and almost optimal solution to this coupled, nonconvex optimization problem.

Table 5.4 Comparison of different DSM algorithms

Algorithm	Operation	Complexity	Performance	Reference
IW	Autonomous	$O(KN)$	Suboptimal	[74]
OSB	Centralized	$O(Ke^N)$	Optimal	[9]
ISB	Centralized	$O(KN^2)$	Near optimal	[8]
ASB	Autonomous	$O(KN)$	Near optimal	[7]

5.4.2 System Model

Using the notation as in [9, 8], we consider a DSL bundle with $\mathcal{N} = \{1, \ldots, N\}$ modems (i.e., lines, users) and $\mathcal{K} = \{1, \ldots, K\}$ tones. Assume discrete multitone (DMT) modulation is employed by all modems; transmission can be modeled independently on each tone as

$$\mathbf{y}_k = \mathbf{H}_k \mathbf{x}_k + \mathbf{z}_k.$$

The vector $\mathbf{x}_k = \{x_k^n, n \in \mathcal{N}\}$ contains transmitted signals on tone k, where x_k^n is the signal transmitted onto line n at tone k. $\mathbf{y}_k$ and $\mathbf{z}_k$ have similar structures. $\mathbf{y}_k$ is the vector of received signals on tone k. $\mathbf{z}_k$ is the vector of additive noise on tone k and contains thermal noise, alien crosstalk, single-carrier modems, radio frequency interference, and so on. $\mathbf{H}_k = [h_k^{n,m}]_{n,m \in \mathcal{N}}$ is the $N \times N$ channel transfer matrix on tone k, where $h_k^{n,m}$ is the channel from TX m to RX n on tone k. The diagonal elements of $\mathbf{H}_k$ contain the direct-channels whereas the off-diagonal elements contain the crosstalk channels. We denote the transmit power spectrum density (PSD) $s_k^n = \mathcal{E}\left\{|x_k^n|^2\right\}$. In the last section's notation for single-carrier systems, we would have $s_k^n = P_n$, $\forall k$. For convenience we denote the vector containing the PSD of user n on all tones as $\mathbf{s}^n = \{s_k^n, k \in \mathcal{K}\}$. We denote the DMT symbol rate as f_s.

Assume that each modem treats interference from other modems as noise. When the number of interfering modems is large, the interference can be well approximated by a Gaussian distribution. Under this assumption the achievable bit loading of user n on tone k is

$$b_k^n = \log\left(1 + \frac{1}{\Gamma} \frac{s_k^n}{\sum_{m \neq n} \alpha_k^{n,m} s_k^m + \sigma_k^n}\right), \tag{5.31}$$

where $\alpha_k^{n,m} = |h_k^{n,m}|^2 / |h_k^{n,n}|^2$ is the normalized crosstalk channel gain, and σ_k^n is the noise power density normalized by the direct channel gain $|h_k^{n,n}|^2$. Here Γ denotes the SINR-gap to capacity, which is a function of the desired BER, coding gain, and noise margin [61]. Without loss of generality, we assume $\Gamma = 1$. The data rate on line n is thus

$$R^n = f_s \sum_{k \in \mathcal{K}} b_k^n. \tag{5.32}$$

Each modem n is typically subject to a total power constraint P^n, due to the limitations on each modem's analogue frontend.

$$\sum_{k \in \mathcal{K}} s_k^n \le P^n. \tag{5.33}$$

5.4.3 Spectrum Management Problem Formulation

One way to define the spectrum management problem is start with the following optimization problem.

$$\begin{array}{ll} \text{maximize} & R^1 \\ \text{subject to} & R^n \ge R^{n,\text{target}}, \quad \forall n > 1 \\ & \sum_{k \in \mathcal{K}} s_k^n \le P^n, \forall n. \end{array} \tag{5.34}$$

Here $R^{n,\text{target}}$ is the target rate constraint of user n. In other words, we try to maximize the achievable rate of user 1, under the condition that all other users achieve their target rates $R^{n,\text{target}}$. The mutual interference in (5.31) causes Problem (5.34) to be coupled across users on each tone, and the individual total power constraint causes Problem (5.34) to be coupled across tones as well.

Moreover, the objective function in Problem (5.34) is nonconvex due to the coupling of interference, and the convexity of the rate region cannot be guaranteed in general.

However, it has been shown in [75] that the duality gap between the dual gap of Problem (5.34) goes to zero when the number of tones K gets large (e.g., for VDSL), thus Problem (5.34) can be solved by the dual decomposition method, which brings the complexity as a function of K down to linear. Moreover, a frequency-sharing property ensures the rate region is convex with large enough K, and each boundary point of the boundary point of the rate region can be achieved by a weighted rate maximization as (following [9]),

$$\begin{array}{ll} \text{maximize} & R^1 + \sum_{n>1} w^n R^n \\ \text{subject to} & \sum_{k \in \mathcal{K}} s_k^n \le P^n, \forall n \in \mathcal{N}, \end{array} \tag{5.35}$$

such that the nonnegative weight coefficient w^n is adjusted to ensure that the target rate constraint of user n is met. Without loss of generality, here we define $w^1 = 1$. By changing the rate constraints $R^{n,\text{target}}$ for users $n > 1$ (or equivalently, changing the weight coefficients, w^n for $n > 1$), every boundary point of the convex rate region can be traced.

We observe that at the optimal solutions of (5.34), each user chooses a PSD level that leads to a good balance of maximization of its own rate and minimization of the damages it causes to the other users. To accurately calculate the latter, the user needs to know the global information of the noise PSDs and crosstalk channel gains. However, if we aim at a less aggressive objective and only require each user give enough protection to the other users in the binder while maximizing her own rate, then global information may not be needed. Indeed, we can introduce the concept of a "reference line", a virtual line that represents a "typical" victim in the current binder. Then instead of solving (5.34), each user tries to maximize the achievable data rate on the reference line, subject to its own data rate and total power constraint. Define the rate of the reference line to user n as

$$R^{n,\text{ref}} = \sum_{k\in\mathcal{K}} \tilde{b}_k^n = \sum_{k\in\mathcal{K}} \log\left(1 + \frac{\tilde{s}_k}{\tilde{\alpha}_k^n s_k^n + \tilde{\sigma}_k}\right).$$

The coefficients $\{\tilde{s}_k, \tilde{\sigma}_k, \tilde{\alpha}_k^n, \forall k, n\}$ are parameters of the reference line and can be obtained from field measurement. They represent the conditions of a "typical" victim user in an interference channel (here a binder of DSL lines), and are known to the users a priori. They can be further updated on a much slower time scale through channel measurement data. User n then wants to solve the following problem local to itself,

$$\begin{array}{ll} \text{maximize} & R^{n,\text{ref}} \\ \text{subject to} & R^n \geq R^{n,\text{target}}, \\ & \sum_{k\in\mathcal{K}} s_k^n \leq P^n. \end{array} \tag{5.36}$$

By using Lagrangian relaxation on the rate target constraint in Problem (5.36) with a weight coefficient (dual variable) w^n, the relaxed version of (5.36) is

$$\begin{array}{ll} \text{maximize} & w^n R^n + R^{n,\text{ref}} \\ \text{subject to} & \sum_{k\in\mathcal{K}} s_k^n \leq P^n. \end{array} \tag{5.37}$$

The weight coefficient w^n needs to be adjusted to enforce the rate constraint.

5.4.4 ASB Algorithms

We first introduce the basic version of the ASB algorithm (ASB-I), where each user n chooses the PSD $\mathbf{s}^n$ to solve (5.36), and updates the weight coefficient w^n to enforce the target rate constraint. Then we introduce a variation of the ASB algorithm (ASB-II) that enjoys even lower computational complexity and provable convergence.

ASB-I

For each user n, replacing the original optimization (5.36) with the Lagrange dual problem

$$\max_{\lambda^n \geq 0,\, \sum_{k\in\mathcal{K}} s_k^n \leq P^n} \sum_{k\in\mathcal{K}} \max_{s_k^n} J_k^n\left(w^n, \lambda^n, s_k^n, s_k^{-n}\right), \tag{5.38}$$

where

$$J_k^n\left(w^n, \lambda^n, s_k^n, s_k^{-n}\right) = w^n b_k^n + \tilde{b}_k^n - \lambda^n s_k^n. \tag{5.39}$$

By introducing the dual variable λ^n, we decouple (5.36) into several smaller subproblems, one for each tone. And define J_k^n as user n's objective function on tone k. The optimal PSD that maximizes J_k^n for given w^n and λ^n is

$$s_k^{n,I}\left(w^n, \lambda^n, s_k^{-n}\right) = \arg \max_{s_k^n \in [0, P^n]} J_k^n\left(w^n, \lambda^n, s_k^n, s_k^{-n}\right), \tag{5.40}$$

which can be found by solving the first-order condition,

$$\partial J_k^n\left(w^n, \lambda^n, s_k^n, s_k^{-n}\right) / \partial s_k^n = 0,$$

which leads to

$$\frac{w^n}{s_k^{n,I} + \sum_{m\neq n} \alpha_k^{n,m} s_k^m + \sigma_k^n} - \frac{\tilde{\alpha}_k^n \tilde{s}_k}{\left(\tilde{s}_k + \tilde{\alpha}_k^n s_k^{n,I} + \tilde{\sigma}_k\right)\left(\tilde{\alpha}_k^n s_k^{n,I} + \tilde{\sigma}_k\right)} - \lambda^n = 0. \tag{5.41}$$

Note that (5.41) can be simplified into a cubic equation that has three solutions. The optimal PSD can be found by substituting these three solutions back to the objective function $J_k^n\left(w^n, \lambda^n, s_k^n, s_k^{-n}\right)$, as well as checking the boundary solutions $s_k^n = 0$ and $s_k^n = P^n$, and picking the one that yields the largest value of J_k^n.

The user then updates λ^n to enforce the power constraint, and updates w^n to enforce the target rate constraint. The complete algorithm is given as follows, where ε_λ and ε_w are small stepsizes for updating λ^n and w^n.

Algorithm 5. Autonomous Spectrum Balancing.

repeat
  for each user $n = 1, \ldots, N$
    repeat
      for each tone $k = 1, \ldots, K$, find
        $s_k^{n,I} = \arg\max_{s_k^n \geq 0} J_k^n$
      $\lambda^n = \left[\lambda^n + \varepsilon_\lambda\left(\sum_k s_k^{n,I} - P^n\right)\right]^+;$
      $w^n = \left[w^n + \varepsilon_w\left(R^{n,\text{target}} - \sum_k b_k^n\right)\right]^+;$
    until convergence
  end
until convergence

ASB-II with Frequency-Selective Waterfilling

To obtain the optimal PSD in ASB-I (for fixed λ^n and w^n), we have to solve the roots of a cubic equation. To reduce the computational complexity and gain more insights of the solution structure, we assume that the reference line operates in the high SIR regime whenever it is active: If $\tilde{s}_k > 0$, then $\tilde{s}_k \gg \tilde{\sigma}_k \gg \alpha_k^{n,m} s_k^n$ for any feasible s_k^n, $n \in \mathcal{N}$, and $k \in \mathcal{K}$. This assumption is motivated by our observations on optimal solutions in the DSL type of interference channels. It means that the reference PSD is much larger than the reference noise, which is in turn much larger than the interference from user n. Then on any tone $k \in \bar{\mathcal{K}} = \{k \mid \tilde{s}_k > 0, k \in \mathcal{K}\}$, the reference line's achievable rate is

$$\log\left(1 + \frac{\tilde{s}_k}{\tilde{\alpha}_k^n s_k^n + \tilde{\sigma}_k}\right) \approx \log\left(\frac{\tilde{s}_k}{\tilde{\sigma}_k}\right) - \frac{\tilde{\alpha}_k^n s_k^n}{\tilde{\sigma}_k},$$

and user n's objective function on tone k can be approximated by

$$J_k^{n,II,1}\left(w^n, \lambda^n, s_k^n, s_k^{-n}\right) = w^n b_k^n - \frac{\tilde{\alpha}_k^n s_k^n}{\tilde{\sigma}_k} - \lambda^n s_k^n + \log\left(\frac{\tilde{s}_k}{\tilde{\sigma}_k}\right).$$

The corresponding optimal PSD is

$$s_k^{n,II,1}\left(w^n, \lambda^n, s_k^{-n}\right) = \left(\frac{w_n}{\lambda^n + \tilde{\alpha}_k^n/\tilde{\sigma}_k} - \sum_{m \neq n} \alpha_k^{n,m} s_k^m - \sigma_k^n\right)^+. \quad (5.42)$$

This is a waterfilling type of solution and is intuitively satisfying: the PSD should be smaller when the power constraint is tighter (i.e., λ_n is larger), or the interference coefficient to the reference line $\tilde{\alpha}_k^n$ is higher, or the noise level on the reference line $\tilde{\sigma}_k$ is smaller, or there is more interference plus noise $\sum_{m \neq n} \alpha_k^{n,m} s_k^m + \sigma_k^n$ on the current tone. It is different from the conventional waterfilling in that the water level in each tone is not only determined by the dual variables w^n and λ^n, but also by the parameters of the reference line, $\tilde{\alpha}_k^n/\tilde{\sigma}_k$.

On the other hand, on any tone where the reference line is inactive, that is, $k \in \bar{\mathcal{K}}^C = \{k \mid \tilde{s}_k = 0, k \in \mathcal{K}\}$, the objective function is

$$J_k^{n,II,2}\left(w^n, \lambda^n, s_k^n, s_k^{-n}\right) = w^n b_k^n - \lambda^n s_k^n,$$

and the corresponding optimal PSD is

$$s_k^{n,II,2}\left(w^n, \lambda^n, s_k^{-n}\right) = \left(\frac{w^n}{\lambda^n} - \sum_{m \neq n} \alpha_k^{n,m} s_k^m - \sigma_k^n\right)^+. \quad (5.43)$$

This is the same solution as the iterative waterfilling.

The choice of optimal PSD in ASB-II can be summarized as the following.

$$s_k^{n,II}\left(w^n,\lambda^n,s_k^{-n}\right)=\begin{cases}\left(\frac{w^n}{\lambda^n+\tilde{\alpha}_k^n/\tilde{\sigma}_k}-\sum_{m\neq n}\alpha_k^{n,m}s_k^m-\sigma_k^n\right)^+, & k\in\bar{\mathcal{K}},\\ \left(\frac{w^n}{\lambda^n}-\sum_{m\neq n}\alpha_k^{n,m}s_k^m-\sigma_k^n\right)^+, & k\in\bar{\mathcal{K}}^C.\end{cases} \tag{5.44}$$

This is essentially a waterfilling type of solution, with different water levels for different tones (frequencies). We call it *frequency selective waterfilling.*

5.4.5 Convergence Analysis

In this section, we show the convergence for both ASB-I and ASB-II, for the case where users fix their weight coefficients w^n, which is also called *rate adaptive* (RA) spectrum balancing [61] that aims at maximizing users' rates subject to power constraint.

Convergence in the Two-User Case

The first result is on the convergence of ASB-I algorithm, with fixed $\mathbf{w}=\left(w^1,w^2\right)$ and $\boldsymbol{\lambda}=\left(\lambda^1,\lambda^2\right)$.

Proposition 5.3. *The ASB-I algorithm converges in a two-user case under fixed* $\mathbf{w}$ *and* $\boldsymbol{\lambda}$*, if users start from initial PSD values* $\left(s_k^1,s_k^2\right)=\left(0,P^2\right)$ *or* $\left(s_k^1,s_k^2\right)=\left(P^1,0\right)$ *on all tones.*

The proof of Proposition 5.3 uses supermodular game theory [65] and strategy transformation similar to [32].

Now consider the ASB-II algorithm where two users sequentially optimize their PSD levels under fixed values of $\mathbf{w}$, but adjust $\boldsymbol{\lambda}$ to enforce the power constraint. Denote $s_k^{n,t}$ as the PSD of user n in tone k after iteration t, where $\sum_k s_k^{n,t}=P^n$ is satisfied for any n and t. One iteration is defined as one round of updates of all users. We can show that

Proposition 5.4. *The ASB-II algorithm globally converges to the unique fixed point in a two-user system under fixed* $\mathbf{w}$*, if* $\max_k\alpha_k^{2,1}\max_k\alpha_k^{1,2}<1$.

The convergence result of iterative waterfilling in the two-user case [74] is a special case of Proposition 5.4 by setting $\tilde{s}_k=0,\forall k$.

We further extend the convergence results to a system with an arbitrary $N>2$ of users. We consider both sequential and parallel PSD updates of the users. In the more realistic but harder-to-analyze parallel updates, time is divided into slots, and each user n updates the PSD simultaneously in each time slot according to (5.44) based on the PSDs in the previous slot, where the λ^n is adjusted such that the power constraint is satisfied.

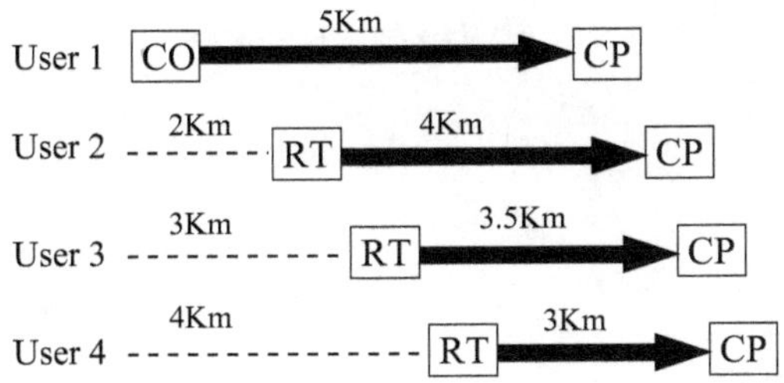

Fig. 5.13 An example of mixed CO/RT deployment topology (Example 5.8).

Proposition 5.5. *Assume* $\max_{n,m,k} \alpha_k^{n,m} < 1/(N-1)$*; then the ASB-II algorithm globally converges (to the unique fixed point) in an N-user system under fixed* $\mathbf{w}$*, with either sequential or parallel updates.*

Proposition 5.5 contains the convergence of iterative waterfilling in an N-user case with sequential updates (proved in [15]) as a special case of ASB convergence with sequential or parallel updates. Moreover, the convergence proof for the parallel updates turns out to be simpler than the one for sequential updates. The proof extends that of Proposition 5.4, and can be found in [7].

5.4.6 Simulation Results

Example 5.8. *Mixed CO-RT DSL.* Here we summarize a typical numerical example comparing the performances of ASB algorithms with IW, OSB, and ISB. We consider a standard mixed central office (CO) and remote terminal (RT) deployment. A four-user scenario has been selected to make a comparison with the highly complex OSB algorithm possible. As depicted in Figure 5.13 the scenario consists of one CO distributed line, and three RT distributed lines. The target rates on RT1 and RT2 have both been set to 2 Mbps. For a variety of different target rates on RT3, the CO line attempts to maximize its own data rate either by transmitting at full power in IW, or by setting its corresponding weight w_{co} to unity in OSB, ISB, and ASB.

This produces the rate regions shown in Figure 5.14, which shows that ASB achieves near optimal performance similar to OSB and ISB, and significant gain over IW even though both ASB and IW are autonomous. For example, with a target rate of 1 Mbps on CO, the rate on RT3 reaches 7.3 Mbps under the ASB algorithm, which is a 121% increase compared with the 3.3 Mbps achieved by IW. We have also performed extensive simulations (more than 10,000 scenarios) with different CO and RT positions, line lengths, and reference line parameters. We found that the performance of ASB is

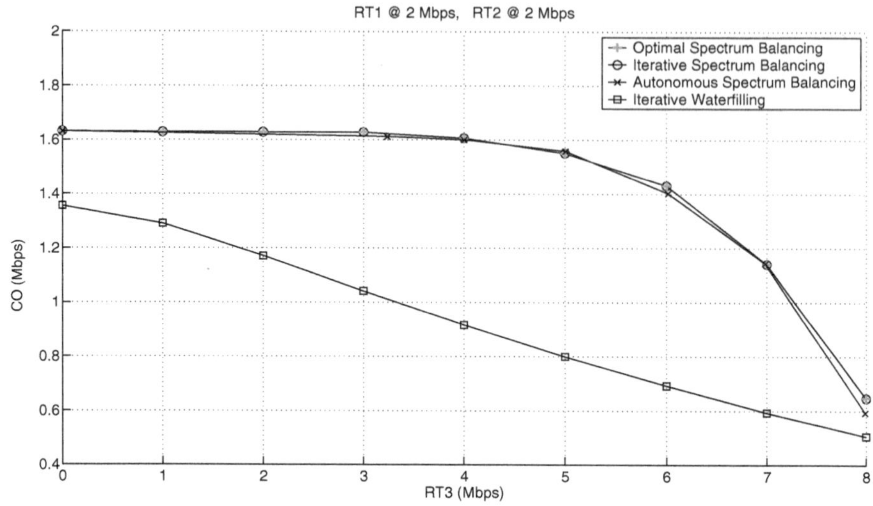

Fig. 5.14 Rate regions obtained by ASB, IW, OSB, and ISB (Example 5.8).

very insensitive to definition of the reference line: with a single choice of the reference line we observe good performance in a broad range of scenarios.

5.4.7 Concluding Remarks and Future Directions

Dynamic spectrum management techniques can greatly improve the performance of DSL lines by inducing cooperation among interfering users in the same binder. For example, the iterative waterfiling algorithm is a completely autonomous DSM algorithm with linear complexity in the number of users and number of tones, but the performance could be far from optimal in the mixed CO/RT deployment scenario. The optimal spectrum balancing and iterative spectrum balancing algorithms achieve optimal and close to optimal performances, respectively, but have high complexities in terms of the number of users and are completely centralized.

This section surveys an autonomous dynamic spectrum management algorithm called autonomous spectrum balancing. ASB utilizes the concept of "reference line", which mimics a typical victim line in the binder. By setting the power spectrum level to protect the reference line, a good balance between selfish and global maximizations can be achieved. Compared with IW, OSB, and ISB, the ASB algorithm enjoys completely autonomous operations, low (linear) complexity in both the number of users and number of tones. Simulation shows that the ASB algorithm achieves close to optimal performance and is robust to the choice of reference line parameters.

We conclude this section by highlighting the key ideas behind ASB. The reference line represents the statistical average of all victims within a typical network, and can be thought as a "static pricing". This differentiates the ASB algorithm with power control algorithms in the wireless setting, where pricing mechanisms have to be adaptive to the change of channel fading states and network topology, or Internet congestion control, where time-varying congestion pricing signals are used to align selfish interests for social welfare maximization. By using static pricing, no explicit message passing among the users is needed and the algorithm becomes autonomous across the users. This is possible because of the static nature of the channel gains in DSL networks.

Mathematically, the surprisingly good rate region results by ASB means that the specific engineering problem structures in this nonconvex and coupled optimization problem can be leveraged to provide a very effective approximation solution algorithm. Furthermore, robustness of the attained rate region with respect to perturbation of the reference line parameters has been verified to be very strong. This means that the dependence of the values of the local maxima of this nonconvex optimization problem on crosstalk channel coefficients is sufficiently insensitive for the observed robustness to hold.

There are several exciting further directions to pursue with ASB, for example, convergence conditions for ASB-I, extensions to intercarrier-interference cases, and bounds on optimality gap that are empirically verified to be very small. Interactions of ASB with link layer scheduling have resulted in further improvement of throughput in DSL networks [33, 67].

5.5 Internet Routing

5.5.1 Introduction

Most large IP (Internet protocol) networks run interior gateway protocols (IGPs) such as OSPF (open shortest path first) or IS-IS (intermediate system–intermediate system) that select paths based on link weights. Routers use these protocols to exchange link weights and construct a complete view of the topology inside the autonomous system (AS). Then, each router computes shortest paths (where the length of a path is the sum of the weights on the links) and creates a table that controls the forwarding of each IP packet to the next hop in its route. To handle the presence of multiple shortest paths, in practice, a router typically splits traffic roughly evenly over each of the outgoing links along a shortest path to the destination. The link weights are typically configured by the network operators or automated management systems, through centralized computation, to satisfy traffic-engineering goals, such as minimizing the maximum link utilization or the sum of link cost [24]. Following common practice, we use the the sum of some increasing and convex link cost functions as the primary comparison metric and the optimization objective in this section.

Setting link weights under OSPF and IS-IS can be categorized as *link-weight-based* traffic engineering, where a set of link weights can uniquely and distributively determine the flow of traffic within the network for any given traffic matrix. The traffic matrix can be computed based on traffic measurements (e.g., [20]) or may represent explicit subscriptions or reservations from users. Link-weight-based traffic engineering has two key components: a *centralized* approach for setting the routing parameters (i.e., link weights) and a *distributed* way of using these link weights to decide the routes to forward packets. Setting the routing parameters based on a networkwide view of the topology and traffic, rather than the local views at each router, can achieve better performance [22].

Evaluation of various traffic engineering schemes, in terms of total link cost minimization, can be made against the performance benchmark of optimal routing (OPT), which can direct traffic along any paths in any proportion. The formulation can be found, for example, in [70]. OPT models an idealized routing scheme that can establish one or more explicit paths between every pair of nodes, and distribute an arbitrary amount of traffic on each of the paths.

It is easy to construct examples where OSPF, one of the most prevalent IP routing protocols today, with the best link weighting performs substantially (5000 times) worse than OPT in terms of minimizing sum of link cost. In addition, finding the best link weights under OSPF is NP-hard [24]. Although the best OSPF link weights can be found by solving an integer linear program (ILP) formulation, such an approach is impractical even for a midsize network. Many heuristics, including local search [24] and simulated annealing [5, 18] have been proposed to search for the best link weights under OSPF. Among them, local-search technique is the most attractive method in finding a good setting of the link weights for large-scale networks. Even though OSPF with a good setting of the weights performs within a few percent of OPT for some practical scenarios [24, 18, 5], there are still many realistic situations where the performance gap between OSPF and OPT could be significant even at low utilization.

There are two main reasons for the difficulty in tuning OSPF for good performance. First, the routing mechanism restricts the traffic to be routed only on shortest paths (and evenly split across shortest paths, an issue that has been addressed in [59]). Second, link weights and the traffic matrix are not integrated into the optimization formulation.

Both bottlenecks are overcome in the distributed exponentially weighted flow splitting (DEFT) protocol developed in [70]:

1. Traffic is allowed to be routed on nonshortest paths, with exponential penalty on path lengths.

2. An innovative optimization formulation is proposed, where both link weights and flows are variables. It leads to an effective two-stage iterative method.

As a result, DEFT, discussed in this section, has the following desirable properties.

- It determines a unique flow of traffic for a given link weight setting in polynomial time.
- It is provably always better than OSPF in terms of minimizing the maximum link utilization or the sum of link cost.
- It is readily implemented as an extension to the existing IGP (e.g., OSPF).
- The traffic engineering under DEFT with the two-stage iterative method realizes near-optimal flow of traffic even for large-scale network topologies.
- The optimizing procedure for DEFT converges much faster than that for OSPF.

In summary, DEFT provides a new way to compute link weights for OSPF that exceeds the current benchmark based on local search methods while reducing computational complexity at the same time. Furthermore, the performance turns out to be very close to the much more complicated and difficult to implement family of MPLS-type routing protocols, which allows arbitrary flow splitting.

More recently in [71], we have proved that a variation of DEFT, called PEFT, can provably achieve the optimal traffic engineering as a link-state routing protocol with hop-by-hop forwarding, with the optimal link weights computed in polynomial time and much faster than local search methods for link weight computation for OSPF. This answers the question on optimal traffic engineering by link state routing conclusively and positively.

5.5.2 DEFT: Framework and Properties

Given a directed graph $\mathbf{G} = (\mathbf{V}, \mathbf{E})$ with capacity $c_{u,v}$ for each link (u, v), let $D(s, t)$ denote the traffic demand originated from node s and destined for node t. $\Phi(f_{u,v}, c_{u,v})$ is a strictly increasing convex function of flow $f_{u,v}$ on link (u, v), typically a piecewise linear cost [24, 59] as shown in equation (5.45). The networkwide objective is to minimize $\sum_{(u,v)\in\mathbf{E}} \Phi(f_{u,v}, c_{u,v})$.

$$\Phi(f_{u,v}, c_{u,v}) = \begin{cases} f_{u,v} & f_{u,v}/c_{u,v} \le 1/3 \\ 3f_{u,v} - 2/3\, c_{u,v} & 1/3 \le f_{u,v}/c_{u,v} \le 2/3 \\ 10f_{u,v} - 16/3\, c_{u,v} & 2/3 \le f_{u,v}/c_{u,v} \le 9/10 \\ 70f_{u,v} - 178/3\, c_{u,v} & 9/10 \le f_{u,v}/c_{u,v} \le 1 \\ 500f_{u,v} - 1468/3\, c_{u,v} & 1 \le f_{u,v}/c_{u,v} \le 11/10 \\ 5000f_{u,v} - 16318/3\, c_{u,v} & 11/10 \le f_{u,v}/c_{u,v}. \end{cases} \tag{5.45}$$

In link-weight-based traffic engineering, each router u needs to make an independent decision on how to split the traffic destined for node t among its outgoing links only using link weights. Therefore, it calls for a function ($\Gamma(\cdot) \geq 0$) to represent the traffic allocation.

Shortest path routing (e.g., OSPF) evenly splits flow across all the outgoing links as long as they are on the shortest paths. First of all, we need a variable to indicate whether link (u, v) is on the shortest path to t. Denote $w_{u,v}$ as the weight for link (u, v), and d_u^t as the shortest distance from node u to node t; then $d_v^t + w_{u,v}$ is the distance from u to t when routed through v. The gap of the two above distances, $h_{u,v}^t = d_v^t + w_{u,v} - d_u^t$ is always larger than or equal to 0. Then (u, v) is on the shortest path to t if and only if $h_{u,v}^t = 0$. Accordingly, we can use a unit step function of $h_{u,v}^t$ to represent the traffic allocation for OSPF as follows.

$$\Gamma(h_{u,v}^t) = \begin{cases} 1, & \text{if } h_{u,v}^t = 0 \\ 0, & \text{if } h_{u,v}^t > 0. \end{cases} \tag{5.46}$$

The flow proportion on the outgoing link (u, v) destined for t at u is

$$\Gamma(h_{u,v}^t) / \sum_{(u,j)\in\mathbf{E}} \Gamma(h_{u,j}^t).$$

Denote $f_{u,v}^t$ as the flow on link (u, v) destined for node t and f_u^t as the flow sent along the shortest path of node u destined for t; then

$$f_{u,v}^t = f_u^t \, \Gamma(h_{u,v}^t). \tag{5.47}$$

The $\Gamma(h_{u,v}^t)$ function (5.46) (i.e., evenly splitting) results in intractability in searching for the best link weights under OSPF. In part inspired by Fong et al.'s work in [21], we can define a new $\Gamma(h_{u,v}^t)$ function to allow for flow on nonshortest paths. Intuitively, we may want to send more traffic on the shortest path than on a nonshortest path. Moreover, the traffic on a nonshortest path should be 0 if the distance gap between the nonshortest path and the shortest path is infinitely large. Based on the above intuition, $\Gamma(h_{u,v}^t)$ should be a strictly decreasing continuous function of $h_{u,v}^t$ bounded within $[0, 1]$. The exponential function is one of the natural choices, and the performance of using such function turns out to be excellent.

In [70], we propose an IGP with distributed exponentially weighted flow splitting:

$$\Gamma(h_{u,v}^t) = \begin{cases} e^{-h_{u,v}^t}, & \text{if } d_u^t > d_v^t \\ 0, & \text{otherwise;} \end{cases} \tag{5.48}$$

that is, the routers can direct traffic on nonshortest paths, with an exponential penalty on longer paths.

The following properties of DEFT can be proved [70].

Proposition 5.6. *DEFT can realize any acyclic flow for a single-destination demand within polynomial time. It can also achieve optimal routing with a single destination within polynomial time. For any traffic matrix, it can determine a unique flow for a given link weighting within polynomial time.*

Proposition 5.7. *DEFT is always better than OSPF in terms of minimizing total link cost or the maximum link utilization.*

5.5.3 DEFT: Optimization Formulation and Solutions

Note that it is still difficult to directly integrate the exponentially weighted flow splitting of DEFT into an optimization formulation because of its discrete feature; that is, the traffic destined for node t can be sent through link (u, v) if and only if $d_u^t > d_v^t$. Instead of introducing some binary variables, we relax (5.48) into (5.49) first, and then, by properly setting the lower bound of all link weights, a constant parameter $w_{\min}$, make such relaxation as tight as we want:

$$\Gamma(h_{u,v}^t) = e^{-h_{u,v}^t}. \tag{5.49}$$

Indeed, consider a flow solution satisfying (5.49); there is a link (u, v) where $d_v^t \geq d_u^t$ and $f_{u,v}^t > 0$, then $f_{u,v}^t \leq f_u^t e^{-h_{u,v}^t} = f_u^t e^{-(d_v^t + w_{u,v} - d_u^t)} \leq f_u^t e^{-w_{\min}}$. If $w_{\min}$ is large enough, this flow portion, which is infeasible to DEFT on link (u, v), could be neglected.

Therefore, we present the following optimization problem, called ORIG, using the relaxed rule of flow splitting as the approximation for the traffic engineering under DEFT.

$$\begin{array}{lll}
\text{minimize} & \sum_{(u,v)\in\mathbf{E}} \Phi(f_{u,v}, c_{u,v}) & (5.50) \\
\text{subject to} & \sum_{z:(y,z)\in\mathbf{E}} f_{y,z}^t - \sum_{x:(x,y)\in\mathbf{E}} f_{x,y}^t = D(y,t), \forall y \neq t & (5.51) \\
& f_{u,v} = \sum_{t\in\mathbf{V}} f_{u,v}^t, & (5.52) \\
& h_{u,v}^t = d_v^t + w_{u,v} - d_u^t, & (5.53) \\
& f_{u,v}^t = f_u^t e^{-h_{u,v}^t}, & (5.54) \\
& f_u^t = \max_{(u,v)\in\mathbf{E}} f_{u,v}^t, & (5.55) \\
\text{variables} & w_{u,v} \geq w_{\min}, f_u^t, d_u^t, h_{u,v}^t, f_{u,v}^t, f_{u,v} \geq 0. & (5.56)
\end{array}$$

Note that both the flow splittings and the link weights are incorporated as optimization variables in one problem, with further constraints relating them. Constraint (5.51) is to ensure flow conservation at an intermediate node y. Constraint (5.52) is for flow aggregation on each link. Constraint (5.53) is from the definition of gap of shortest distance. Constraints (5.54) and (5.55) come from (5.47) and (5.49). In addition, (5.54) and (5.55) also imply that

$f^t_{u,v} \leq f^t_u$, and that $h^t_{u,v}$ of at least one of an outgoing links (u, v) of node u destined for node t should be 0; that is, the link (u, v) is on the shortest path from node u to node t.

Problem ORIG is nonsmooth and nonconvex due to nonsmooth constraint (5.55) and nonlinear equality (5.54). In [70], we propose a two-stage iterative relaxation to solve problem ORIG.

First, we relax constraint (5.55) into (5.57) below:

$$f^t_u \leq \sum\nolimits_{(u,v)\in\mathbf{E}} f^t_{u,v}, \quad \forall t \in \mathbf{V}, \quad \forall u \in \mathbf{V}. \tag{5.57}$$

Equations (5.50)–(5.54), (5.56), and (5.57) constitute problem APPROX.

We only need to obtain a "reasonably" accurate solution (link weighting $\mathbf{W}$) to problem APPROX because the inaccuracy caused by the relaxation (5.57) will be compensated by a successive refinery process later. From the $\mathbf{W}$, we can derive the shortest path tree $\mathbf{T}(\mathbf{W}, t)$[6] for each destination t, and all other dependent variables $(d^t_u, h^t_{u,v}, f^t_u, f^t_{u,v}, f_{u,v})$ within DEFT.

We then use these values as the initial point (which is also strictly feasible) for a new problem REFINE, which consists of equations (5.50)–(5.54), (5.56), and (5.58) below:

$$f^t_u = f^t_{u,v}, \quad \forall t \in \mathbf{V} \cap \forall u \in \mathbf{V} \cap (u, v) \in \mathbf{T}(\mathbf{W}, t). \tag{5.58}$$

With the two-stage iterative method, we are left with two optimization problems, APPROX and REFINE, both of which have convex objective functions and twice continuously differentiable constraints. To solve the large-scale nonlinear problems APPROX and REFINE (with $O(|V||E|)$ variables and constraints), we extend the primal–dual interior point filter line search algorithm, IPOPT [68], by solving a set of barrier problems for a decreasing sequences of barrier parameters μ converging to 0.

In summary, in solving problem APPROX, we mainly want to determine the shortest path tree for each destination (i.e., deciding which outgoing link should be chosen on the shortest path). Then in solving problem REFINE, we can tune the link weights (and the corresponding flow) with the same shortest path trees as in APPROX.

The pseudocode of the proposed two-stage iterative method for DEFT is shown in Algorithms 6A and 6B. Most instructions are self-explanatory. Function DEFT_FLOW($\mathbf{W}$) is used to derive a flow from a set of link weights $\mathbf{W}$. Given the initial and ending values for barrier parameter μ, maximum iteration number, with/without initial link weighting/flow, function DEFT_IPOPT() returns a new set of link weights as well as a new flow. Note that, as shown in Algorithm 6B, when DEFT_IPOPT() is used for problem APPROX, it returns with the last iteration rather than the iteration with the best $\mathbf{Flow}_i$ in terms of the objective value as in problem REFINE. This

[6] To keep $\mathbf{T}(\mathbf{W}, t)$ as a tree, only one downstream node is chosen if a node can reach the destination through several downstream nodes with the same distance.

is because problem APPROX has different constraints from problem ORIG and a too greedy method may leave small search freedom for the successive REFINE problem. Finally, we need to specify initial and terminative μ values, ($\mu_{\text{init}} \geq \mu_{\text{end\_approx}} \geq \mu_{\text{end\_refine}}$), and maximum iteration number $\text{Iter}_{\text{approx}} \geq \text{Iter}_{\text{refine}}$. As shown in the next section, it is straightforward to specify these parameters.

Algorithm 6A. DEFT Solution.

1. $(\mu, \mathbf{W}) \leftarrow \text{DEFT\_IPOPT}(\mu_{\text{init}}, \mu_{\text{end\_approx}}, \text{Iter}_{\text{approx}}, \textbf{nil})$
2. $\text{Initial\_Point} \leftarrow (\mathbf{W}, \text{DEFT\_FLOW}(\mathbf{W}))$
3. $(\mu, \mathbf{W}) \leftarrow \text{DEFT\_IPOPT}(\mu, \mu_{\text{end\_refine}}, \text{Iter}_{\text{refine}}, \text{Initial\_Point})$
4. Return $(\mathbf{W}, \text{DEFT\_FLOW}(\mathbf{W}))$

Algorithm 6B. DEFT IPOPT.

If Initial_Point $\neq$ **nil** Then
Initiate the problem with Initial_Point /*REFINE*/
End if
For **each** iteration $i \leq \text{Iter}_{\max}$ with $\mu_{\text{start}} \geq \mu \geq \mu_{\text{end}}$ do
  $\mu_i \leftarrow$ current value for μ
  $\mathbf{W}_i \leftarrow$ current values for all $w_{u,v}$
  $\mathbf{Flow}_i \leftarrow \text{DEFT\_FLOW}(\mathbf{W}_i)$
end for
If Initial_Point = **nil** then
return $(\mu_i, \mathbf{W}_i)$ of the last iteration /*APPROX*/
else
return $(\mu_i, \mathbf{W}_i)$ of the iteration with the best $\mathbf{Flow}_i$ in terms of objective value /*REFINE*/
end if

5.5.4 Numerical Examples

We summarize some of the numerical results in [70] on various schemes under many practical scenarios. We employ the same cost function (5.45) as in [23]. The primary metric used is the optimality gap, in terms of total link cost, compared against the value achieved by optimal routing using CPLEX 9.1 [16] via AMPL [25]. The secondary metric used is the maximum link utilization. We do not reproduce the performance of some obvious link-weight-based traffic engineering approaches for OSPF, for example, UnitOSPF (setting all link weights to 1), RandomOSPF (choosing the weights randomly), InvCapOSPF (setting the weight of an link inversely proportional to its capacity as recommended by Cisco), or L2OSPF (setting the weight proportional to its physical Euclidean distance) [23], because none of them performs as well as the state-of-the-art local search method proposed in [23]. In addition, because DEFT is always better than OSPF in terms of minimizing the maximum link utilization or the sum of link cost, we bypass the scenarios where OSPF can achieve near-optimal solution. Instead, we are particularly interested in those scenarios where OSPF does not perform well.

For fair comparisons, we use the same topology and traffic matrix as those in [23]. The 2-level hierarchical networks were generated using GT-ITM, which consists of two kinds of links: local access links with 200-unit capacity and long distance links with 1000-unit capacity. In the second type of topology, the random topologies, the probability of having a link between two nodes is a constant parameter and all link capacities are 1000 units.

Although AT&T's proprietary code of local search used in [23] is not publicly available, there is an open source software project with IGP weight optimization, TOTEM 1.1 [66]. It follows the same lines as [23], and has similar quality of the results. It is slightly slower due to the lack of implementation of the dynamic Dijkstra algorithm. We use the same parameter setting for local search as in [24, 23] where link weight is restricted as an integer from 1 to 20, initial link weights are chosen randomly, and the best result is collected after 5000 iterations.

To implement the proposed two-stage iterative method for DEFT, we modify another open source software, IPOPT 3.1 [34], and adjust its AMPL interface to integrate it into our test environment. We choose $\mu_{\text{init}} = 0.1$ for most cases except for $\mu_{\text{init}} = 10$ for the 100-node network with heavy traffic load. We also choose $\mu_{\text{end\_approx}} = 10^{-4}$, $\mu_{\text{end\_refine}} = 10^{-9}$, and maximum iteration number $\text{Iter}_{\text{approx}} = 1000$, $\text{Iter}_{\text{refine}} = 400$. The code terminates earlier if the optimality gap has been less than 0.1%.

Example 5.9. *DEFT and OSPF on 2-level topology.* The results for a 2-level topology with 50 nodes and 212 links with seven different traffic matrices are shown in Table 5.5. The results are also depicted graphically in Figure 5.15.

Table 5.5 Results of 2-level topology with 50 nodes and 212 links

Total Traffic Demand	1700	2000	2200	2500	2800	3100	3400
Ave Link Load-OPT	0.128	0.148	0.17	0.192	0.216	0.242	0.267
Max Link Load-OPT	0.667	0.667	0.667	0.9	0.9	0.9	0.9
Opt. Gap-OSPF	2.8%	4.4%	7.2%	9.4%	20.7%	64.2%	222.8%
Opt. Gap-DEFT	0.1%	0.1%	0.1%	0.1%	0.1%	0.1%	0.1%

In addition to the two metrics, optimality gap in terms of total link cost and maximum link utilization,[7] we also show the average link utilization under optimal routing as an indication of network load. From the results, we

[7] Note, however, maximum link utilization is not a metric as comprehensive as total link cost because it cannot indicate whether there are multiple overcongested links.

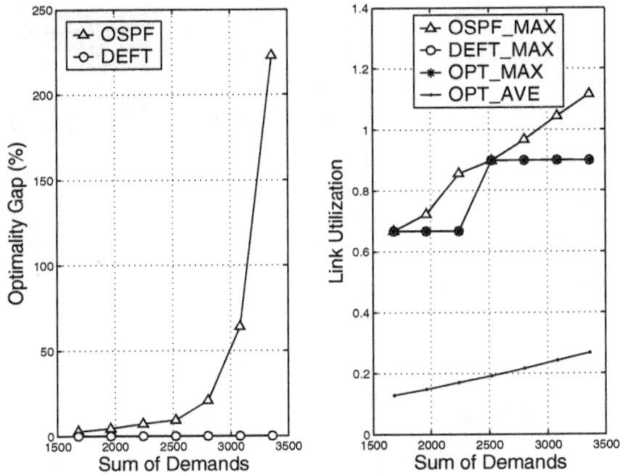

Fig. 5.15 Comparison of DEFT and local search OSPF in terms of optimality gap and maximum link utilization for a 2-level topology with 50 nodes and 212 links (Example 5.9).

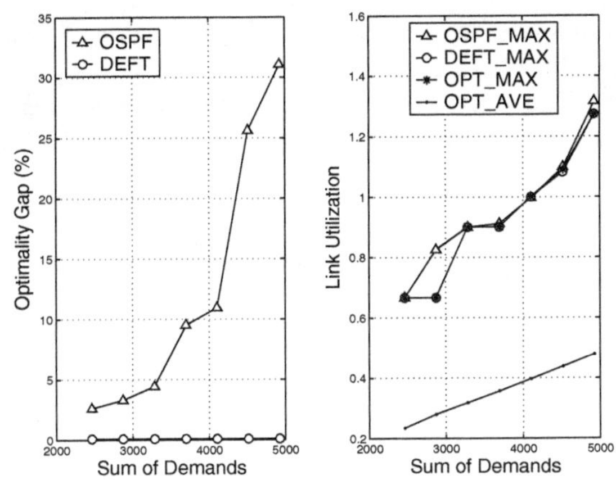

Fig. 5.16 2-level topology with 50 nodes and 148 links (Example 5.9).

can observe that the gap between OSPF and optimal routing can be very significant (up to 222.8%) for a practical network scenario, even when the average link utilization is low ($\leq$27%). In contrast, DEFT can achieve almost the same performance as the optimal routing in terms of both total link cost and maximum link utilization.

Example 5.10. *DEFT and OSPF on random topology.* Similar observations can be found for other scenarios, for example, as shown in Figure 5.17. Without exception, the curves of the DEFT scheme (the horizontal lines almost coinciding with x-axes) almost completely overlap those of optimal routing,

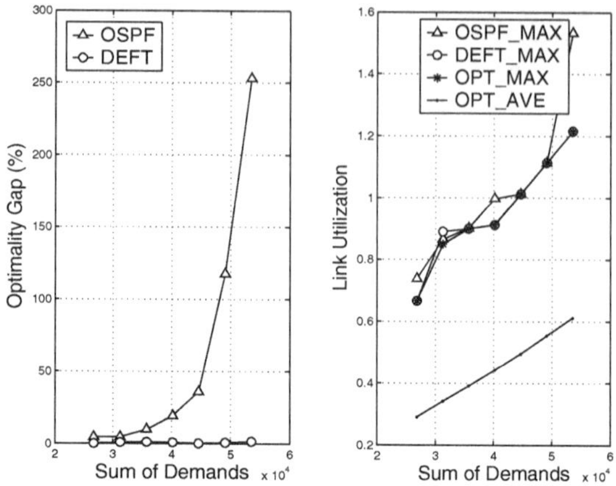

Fig. 5.17 Random topology with 50 nodes and 245 links (Example 5.10).

in terms of total link cost and maximum link utilization. Among these numerical experiments, the maximum optimality gap of OSPF is as high as 252% and that of DEFT is only at worst 1.5%. In addition, DEFT reduces the maximum link utilization compared to OSPF on all tests, and substantially on some tests.

Simulations on rate of convergence, as well as comparisons of computation and implementation complexity, can be found in [70].

5.5.5 Concluding Remarks and Future Directions

Network operators today try to alleviate congestion in their own network by tuning the parameters in IGP. Unfortunately, traffic engineering under OSPF or IS-IS to avoid networkwide congestion is computationally intractable, forcing the use of local-search techniques. While staying within the context of link-weight-based traffic engineering, we propose a new protocol called [70] distributed exponentially weighted flow splitting. DEFT significantly outperforms the state-of-the-art OSPF local search mechanisms in minimizing networkwide congestion. The success of DEFT can be attributed to two additional features. First, DEFT can put traffic on nonshortest paths, with an exponential penalty on longer paths. Second, DEFT solves the resulting optimization problem by integrating link weights and the corresponding traffic distribution together in the formulation. The novel formulation leads to a much more efficient way of tuning link-weight than the existing local search heuristic for OSPF.

DEFT is readily implementable as an extension to existing IGPs. It is provably always better than OSPF in minimizing the sum of link cost. DEFT retains the simplicity of having routers compute paths based on configurable link weights, while approaching the performance of the much more complex routing protocols that can split traffic arbitrarily over any paths. In summary, in terms of minimizing total link cost, performance of OSPF by local search heuristics is at best what is attained by solving the ILP, which is substantially outperformed by DEFT that comes very close to the optimal routing. In terms of a performance–complexity tradeoff, DEFT clearly exceeds OSPF.

In this section, we only address the link weighting under DEFT for a given traffic matrix. The next challenge would be to explore robust optimization under DEFT, optimizing to select a single weight setting that works for a range of traffic matrices and/or a range of link/node failure scenarios. Extension of the ideas behind DEFT to routing across different autonomous systems managed by different network operators is another interesting future direction.

In the larger picture of "design for optimizability", DEFT shows one case where by changing the underlying protocol, the resulting new optimization formulation becomes much more readily solvable or approximable. We expect this new approach to tackle nonconvex problems to bring many new results and insights to the engineering of communication networks. Indeed, in an extension of DEFT work [71], we have developed the first provably optimal link state routing protocol with hop by hop forwarding, called PEFT, which achieves optimal traffic engineering with polynomial time (and very fast in practice) computation of optimal link weights.

Acknowledgments The author would like to acknowledge collaborations with Raphael Cendrillon, Maryam Fazel, Prashanth Hande, Jianwei Huang, Daniel Palomar, Jennifer Rexford, Chee Wei Tan, and Dahai Xu while working on the five publications related to this survey [29, 19, 14, 7, 70], as well as very helpful discussions on these topics with Stephen Boyd, Rob Calderbank, John Doyle, David Gao, Jiayue He, David Julian, Jang-Won Lee, Ying Li, Steven Low, Marc Moonen, Daniel O'Neill, Asuman Ozdaglar, Pablo Parrilo, Ness Shroff, R. Srikant, Ao Tang, and Shengyu Zheng. The work reported in this chapter has been supported in part by the following grants: AFOSR FA9550-06-1-0297, DARPA W911NF-07-1-0057, NSF CNS-0519880 and CNS-0720570, and ONR N00014-07-1-0864.

References

1. M. Avriel, Ed., *Advances in Geometric Programming,* Plenum Press, New York, 1980.
2. N. Bambos, "Toward power-sensitive network architectures in wireless communications: Concepts, issues, and design aspects," *IEEE Pers. Commun. Mag.*, vol. 5, no. 3, pp. 50–59, 1998.
3. S. Boyd, S. J. Kim, L. Vandenberghe, and A. Hassibi, "A tutorial on geometric programming," *Optim. Eng.*, vol. 8, no. 1, pp. 67–127, 2007.
4. S. Boyd and L. Vandenberghe, *Convex Optimization,* Cambridge University Press, 2004.

5. L. Buriol, M. Resende, C. Ribeiro, and M. Thorup, "A memetic algorithm for OSPF routing," *Proc. 6th INFORMS Telecom*, 2002, pp. 187–188.
6. R. Cendrillon, G. Ginis, and M. Moonen, "Improved linear crosstalk precompensation for downstream VDSL," in *Proc. IEEE International Conference on Acoustics, Speech, and Signal Processing*, 2004.
7. R. Cendrillon, J. Huang, M. Chiang, and M. Moonen, "Autonomous Spectrum Balancing (ASB) for digital subscriber loop," *IEEE Trans. Signal Process.*, vol. 55, no. 8, pp. 4241–4257, August 2007.
8. R. Cendrillon and M. Moonen, "Iterative spectrum balancing for digital subscriber lines," in *Proc. IEEE International Communications Conference*, 2005.
9. R. Cendrillon, W. Yu, M. Moonen, J. Verlinden, and T. Bostoen, "Optimal multi-user spectrum management for digital subscriber lines," *IEEE Trans. Commun.*, July 2006.
10. M. Chiang, "Balancing transport and physical layers in wireless multihop networks: Jointly optimal congestion control and power control," *IEEE J. Selected Areas Commun.*, vol. 23, no. 1, pp. 104–116, January 2005.
11. M. Chiang, "Geometric programming for communication systems," *Foundations Trends Commun. Inf. Theor.*, vol. 2, no. 1–2, pp. 1–156, August 2005.
12. M. Chiang, P. Hande, T. Lan, and C. W. Tan, "Power control in wireless cellular networks," *Foundations and Trends in Networking*, vol. 2, no. 4, pp. 381–533, July 2008.
13. M. Chiang, S. H. Low, R. A. Calderbank, and J. C. Doyle, "Layering as optimization decomposition," *Proceedings of the IEEE*, vol. 95, no. 1, pp. 255–312, January 2007.
14. M. Chiang, C. W. Tan, D. Palomar, D. O'Neill, and D. Julian, "Power control by geometric programming," *IEEE Trans. Wireless Commun.*, vol. 6, no. 7, pp. 2640–2651, July 2007.
15. S. T. Chung, "Transmission schemes for frequency selective Gaussian interference channels," Ph.D. dissertation, Stanford University, 2003.
16. ILOG CPLEX, http://www.ilog.com/products/cplex/.
17. R. J. Duffin, E. L. Peterson, and C. Zener, *Geometric Programming: Theory and Applications*, Wiley, 1967.
18. M. Ericsson, M. Resende, and P. Pardalos, "A genetic algorithm for the weight setting problem in OSPF routing," *J. Combin. Optim.*, vol. 6, pp. 299–333, 2002.
19. M. Fazel and M. Chiang, "Nonconcave network utility maximization by sum of squares programming," *Proc. IEEE Conference on Decision and Control*, December 2005.
20. A. Feldmann, A. Greenberg, C. Lund, N. Reingold, J. Rexford, and F. True, "Deriving traffic demands for operational IP networks: Methodology and experience," *IEEE/ACM Trans. Netw.*, June 2001.
21. J. H. Fong, A. C. Gilbert, S. Kannan, and M. J. Strauss, "Better alternatives to OSPF routing," *Algorithmica*, vol. 43, no. 1–2, pp. 113–131, 2005.
22. B. Fortz, J. Rexford, and M. Thorup, "Traffic engineering with traditional IP routing protocols," *IEEE Commun. Mag.*, October 2002.
23. B. Fortz and M. Thorup, "Increasing internet capacity using local search," *Comput. Optim. Appl.*, vol. 29, no. 1, pp. 13–48, 2004.
24. B. Fortz and M. Thorup, "Internet traffic engineering by optimizing OSPF weights," *Proc. IEEE INFOCOM*, May 2000.
25. R. Fourer, D. M. Gay, and B. W. Kernighan, *AMPL: A Modeling Language for Mathematical Programming*, Thomson, Danvers, MA, 1993.
26. G. Foschini and Z. Miljanic, "A simple distributed autonomous power control algorithm and its convergence," *IEEE Trans. Vehicular Technol.*, vol. 42, no. 4, 1993.
27. G. Ginis and J. Cioffi, "Vectored transmission for digital subscriber line systems," *IEEE J. Selected Areas Commun.*, vol. 20, no. 5, pp. 1085–1104, 2002.
28. P. Hande, S. Rangan, M. Chiang, and X. Wu, "Distributed uplink power control for optimal SIR assignment in cellular data networks," To appear in *IEEE/ACM Trans. Netw.*, 2008.

29. P. Hande, S. Zhang, and M. Chiang, "Distributed rate allocation for inelastic flows," *IEEE/ACM Trans. Netw.*, vol. 15, no. 6, pp. 1240–1253, December 2007.
30. D. Handelman, "Representing polynomials by positive linear functions on compact convex polyhedra," *Pacific J. Math.*, vol. 132, pp. 35–62, 1988.
31. D. Henrion and J. B. Lasserre, "Detecting global optimality and extracting solutions in GloptiPoly," Research report, LAAS-CNRS, 2003.
32. J. Huang, R. Berry, and M. L. Honig, "A game theoretic analysis of distributed power control for spread spectrum ad hoc networks," *Proc. IEEE International Symposium of Information Theory*, July 2005.
33. J. Huang, C. W. Tan, M. Chiang, and R. Cendrillon, "Statistical multiplexing over DSL networks," *Proc. IEEE INFOCOM*, May 2007.
34. IPOPT, http://projects.coin-or.org/Ipopt.
35. D. Julian, M. Chiang, D. ONeill, and S. Boyd, "QoS and fairness constrained convex optimization of resource allocation for wireless cellular and ad hoc networks," *Proc. IEEE INFOCOM*, June 2002.
36. F. P. Kelly, "Models for a self-managed Internet," *Philosoph. Trans. Royal Soc.*, A358, 2335–2348, 2000.
37. F. P. Kelly, A. Maulloo, and D. Tan, "Rate control for communication networks: shadow prices, proportional fairness and stability," *J. Oper. Res. Soc.,* vol. 49, no. 3, pp. 237–252, March 1998.
38. S. Kandukuri and S. Boyd, "Optimal power control in interference limited fading wireless channels with outage probability specifications," *IEEE Trans. Wireless Commun.*, vol. 1, no. 1, pp. 46–55, January 2002.
39. J. B. Lasserre, "Global optimization with polynomials and the problem of moments," *SIAM J. Optim.*, vol. 11, no. 3, pp. 796–817, 2001.
40. J. B. Lasserre, "Polynomial programming: LP-relaxations also converge," *SIAM J. Optim.*, vol. 15, no. 2, pp. 383–393, 2004.
41. J. W. Lee, R. Mazumdar, and N. B. Shroff, "Non-convex optimization and rate control for multi-class services in the Internet," *IEEE/ACM Trans. Netw.*, vol. 13, no. 4, pp. 827–840, August 2005.
42. J. W. Lee, R. Mazumdar, and N. B. Shroff, "Downlink power allocation for multi-class CDMA wireless networks", *IEEE/ACM Trans. Netw.*, vol. 13, no. 4, pp. 854–867, August 2005.
43. J. W. Lee, R. Mazumdar, and N. B. Shroff, "Opportunistic power scheduling for multi-server wireless systems with minimum performance constraints," *IEEE Trans. Wireless Commun.*, vol. 5, no. 5, May 2006.
44. X. Lin, N. B. Shroff, and R. Srikant, "A tutorial on cross-layer optimization in wireless networks," *IEEE J. Selected Areas Commun.*, August 2006.
45. S. H. Low, "A duality model of TCP and queue management algorithms," *IEEE/ACM Trans. Netw.*, vol. 11, no. 4, pp. 525–536, August 2003.
46. B. R. Marks and G. P. Wright, "A general inner approximation algorithm for nonconvex mathematical program," *Oper. Res.*, 1978.
47. D. Mitra, "An asynchronous distributed algorithm for power control in cellular radio systems," *Proc. 4th WINLAB Workshop*, Rutgers University, NJ, 1993.
48. Yu. Nesterov and A. Nemirovsky, *Interior Point Polynomial Methods in Convex Programming,* SIAM Press, 1994.
49. D. Palomar and M. Chiang, "Alternative distributed algorithms for network utility maximization: Framework and applications," *IEEE Trans. Autom. Control*, vol. 52, no. 12, pp. 2254–2269, December 2007.
50. P. A. Parrilo, Structured semidefinite programs and semi-algebraic geometry methods in robustness and optimization," PhD thesis, Caltech, May 2002.
51. P. A. Parrilo, "Semidefinite programming relaxations for semi-algebraic problems," *Math. Program.*, vol. 96, pp. 293–320, 2003.

52. S. Prajna, A. Papachristodoulou, and P. A. Parrilo, "SOSTOOLS: Sum of squares optimization toolbox for Matlab," available from `http://www.cds.caltech.edu/sostools`, 2002–04.
53. M. Putinar, "Positive polynomials on compact semi-algebraic sets," *Indiana Univ. Math. J.*, vol. 42, no. 3, pp. 969–984, 1993.
54. R. T. Rockafellar, *Network Flows and Monotropic Programming*, Athena Scientific, 1998.
55. C. Saraydar, N. Mandayam, and D. Goodman, "Pricing and power control in a multicell wireless data network", *IEEE J. Selected Areas Commun.*, vol. 19, no. 10, pp. 1883–1892, October 2001.
56. K. Schmüdgen, "The K-moment problem for compact semi-algebraic sets," *Math. Ann.*, vol. 289, no. 2, pp. 203–206, 1991.
57. S. Shenker, "Fundamental design issues for the future Internet," *IEEE J. Selected Areas Commun.*, vol. 13, no. 7, pp. 1176–1188, September 1995.
58. N. Z. Shor, "Quadratic optimization problems," *Soviet J. Computat. Syst. Sci.*, vol. 25, pp. 1–11, 1987.
59. A. Sridharan, R. Guérin, and C. Diot, "Achieving near-optimal traffic engineering solutions for current OSPF/IS-IS networks," *IEEE/ACM Trans. Netw.*, vol. 13, no. 2, pp. 234–247, 2005.
60. R. Srikant, *The Mathematics of Internet Congestion Control*, Birkhäuser 2004.
61. T. Starr, J. Cioffi, and P. Silverman, *Understanding Digital Subscriber Line Technology*, Prentice Hall, 1999.
62. G. Stengle, "A Nullstellensatz and a Positivstellensatz in semialgebraic geometry," *Math. Ann.*, vol. 207, no. 2, pp. 87–97, 1974.
63. C. Sung and W. Wong, "Power control and rate management for wireless multimedia CDMA systems," *IEEE Trans. Commun.*, vol. 49, no. 7, pp. 1215–1226, 2001.
64. C. W. Tan, D. Palomar and M. Chiang, "Distributed Optimization of Coupled Systems with Applications to Network Utility Maximization," *Proc. IEEE International Conference of Acoustic, Speech, and Signal Processing*, May 2006.
65. D. M. Topkis, *Supermodularity and Complementarity*, Princeton University Press, 1998.
66. TOTEM, http://totem.info.ucl.ac.be.
67. P. Tsiaflakis, Y. Yi, M. Chiang, and M. Moonen, "Throughput and delay of DSL dynamic spectrum management," *Proc. IEEE GLOBECOM*, December 2008.
68. A. Wächter and L. T. Biegler, "On the implementation of a primal-dual interior point filter line search algorithm for large-scale nonlinear programming," *Math. Program.*, vol. 106, no. 1, pp. 25–57, 2006.
69. M. Xiao, N. B. Shroff, and E. K. P. Chong, "Utility based power control in cellular wireless systems," *IEEE/ACM Trans. Netw.*, vol. 11, no. 10, pp. 210–221, March 2003.
70. D. Xu, M. Chiang, and J. Rexford, "DEFT: Distributed exponentially-weighted flow splitting," *Proc. IEEE INFOCOM*, May 2007.
71. D. Xu, M. Chiang, and J. Rexford, "Link state routing with hop by hop forwarding can achieve optimal traffic engineering," *Proc. IEEE INFOCOM*, April 2008.
72. R. Yates, "A framework for uplink power control in cellular radio systems," *IEEE J. Selected Areas Commun.*, vol. 13, no. 7, pp. 1341–1347, 1995.
73. Y. Yi, A. Proutiere, and M. Chiang, "Complexity of wireless scheduling: Impact and tradeoffs," *Proc. ACM Mobihoc*, May 2008.
74. W. Yu, G. Ginis, and J. Cioffi, "Distributed multiuser power control for digital subscriber lines," *IEEE J. Selected Areas Commun.*, vol. 20, no. 5, pp. 1105–1115, June 2002.
75. W. Yu, R. Lui, and R. Cendrillon, "Dual optimization methods for multiuser orthogonal frequency division multiplex systems," *Proc. IEEE GLOBECOM*, November 2004.

Chapter 6
Multilevel (Hierarchical) Optimization: Complexity Issues, Optimality Conditions, Algorithms

Altannar Chinchuluun, Panos M. Pardalos, and Hong-Xuan Huang

Summary. In this chapter we discuss some algorithmic and theoretical results on multilevel programming including complexity issues, optimality conditions, and algorithmic methods for solving multilevel programming problems. We also discuss an approach, which is called the multivariate partition approach, for solving a single-level mathematical programming problem based on its equivalent multilevel programming formulation.

Key words: Hierarchy, multilevel programming, multivariate partition approach

6.1 Introduction

The word *hierarchy* comes from the Greek word "$\iota\epsilon\rho\alpha\rho\chi\iota\alpha$," a system of graded (religious) authority. Hierarchical (multilevel) structures are found in many complex systems and in particular in biology. Biological systems are characterized by hierarchical architectural designs in which organization is controlled on length scales ranging from the molecular to macroscopic. These hierarchical architectures rely on critical interfaces that link structural elements of disparate scale.

Nature makes very different systems (that have specific hierarchical composite structures) out of very similar molecular constituents. First, the structures are organized in discrete levels. Second, the levels of structural organiza-

Altannar Chinchuluun · Panos M. Pardalos
Department of Industrial and Systems Engineering, University of Florida, Gainesville, FL 32611, U.S.A., e-mail: altannar@ufl.edu, pardalos@ufl.edu

Hong-Xuan Huang
Department of Industrial Engineering, Tsinghua University
e-mail: huanghongxuan@tsinghua.org.cn

D.Y. Gao, H.D. Sherali, (eds.), *Advances in Applied Mathematics and Global Optimization*
Advances in Mechanics and Mathematics 17, DOI 10.1007/978-0-387-75714-8_6,

tion are held together by specific interactions between components. Finally, these interacting levels are organized into an oriented distinct hierarchical composite system of specific function.

The mathematical study of hierarchical structures can be found in diverse scientific disciplines including environment, ecology, biology, chemical engineering, classification theory, databases, network design, game theory, and economics. The study of hierarchy occurring in biological structures reveals interesting properties as well as limitations due to different properties of molecules. Understanding the complexity of hierarchical designs requires systems methodologies that are amenable to modeling, analyzing, and optimizing these structures.

Hierarchical optimization (or multilevel) can be used to study properties of these hierarchical designs. In hierarchical optimization, the constraint domain is implicitly determined by a series of optimization problems that must be solved in a predetermined sequence. Hierarchical optimization is a generalization of mathematical programming. The simplest two-level (or bilevel) programming problem describes a hierarchical system composed of two levels of decision makers.

A bilevel programming problem consists of two optimization problems where the constraint set of the upper-level problem is implicitly determined by the lower-level problem.

$$\begin{aligned} &\min_x \; F(x,y) \\ &\text{s.t. } G(x,y) \leq 0, \end{aligned} \tag{6.1}$$

where y solves

$$\begin{aligned} &\min_z \;\; f(x,z) \\ &\text{s.t.} \;\; g(x,z) \leq 0. \\ &x \in X \subset \mathbb{R}^n, \;\; y,z \in Y \subset \mathbb{R}^m, \end{aligned}$$

where $X \subseteq \mathbb{R}^n$ and $Y \subseteq \mathbb{R}^m$ are compact sets, and $G : \mathbb{R}^n \times \mathbb{R}^m \to \mathbb{R}^p$, $g : \mathbb{R}^n \times \mathbb{R}^m \to \mathbb{R}^q$, and $F, f : \mathbb{R}^n \times \mathbb{R}^m \to \mathbb{R}$ are scalar functions. Let $\Omega = \{(x,y) \mid G(x,y) \leq 0, \; g(x,y) \leq 0, \; (x,y) \in \mathbb{R}^n \times \mathbb{R}^m\}$ be the constraint set of the problem. We also denote the set of solutions of the lower-level programming problem by $S(x)$ for any $x \in X$.

When the objective functions and constraint functions are linear, Problem (6.1) is called a bilevel linear programming problem. For a bilevel linear programming problem, it is known that the solution to the problem occurs at an extreme point of the feasible set [16, 9]. If at least one of the objective functions or constraint functions is nonlinear, then Problem (6.1) is called a nonlinear bilevel programming problem. In general, bilevel optimization problems are nonconvex, and therefore, it is not easy to find globally optimal solutions.

Bilevel programming has a wide range of applications in many areas including economics [8], network design [82, 15], transport planning and modeling [69, 71], credit allocation [74], and electric power pricing [51]. The reader is referred to Du and Pardalos [38] and Migdalas et al. [72] for a more comprehensive survey of applications. An excellent bibliographical survey of bilevel programming can also be found in Vicente and Calamai [88].

It seems that hierarchical structures are harder to be managed than completely centralized systems. Then what are the rationalities for hierarchical structures to exist? Answers to such questions may help us to understand the reason behind hierarchical structures in biology.

Multilevel programming problems have been studied extensively in their general setting for the last decade. Many algorithmic developments are based on the properties of special cases of the bilevel problem (and the more general problem) and reformulations to equivalent or approximating models, presumably more tractable. Most of the exact methods are based on branch-and-bound or cutting plane techniques and can handle only moderately sized problems.

The present chapter is organized as follows. In Section 6.2, we consider complexity issues of bilevel programming, and Section 6.3 covers optimality conditions of the problem. We discuss some of the algorithmic methods for solving the bilevel programming problem in Section 6.4. We present a multivariate partitional approach for solving single-level mathematical programming problem based on its equivalent bilevel programming formulation in Section 6.5. We also discuss two real-world applications of bilevel programming in transportation systems in Section 6.6, and some open problems in multilevel programming in Section 6.7.

6.2 Complexity Issues

Complexity analyzes the intrinsic difficulty of optimization problems and reveals surprising connections among many other optimization problems and their solutions.

6.2.1 Complexity of Finding a Global Solution

Calamai and Vicente [22] proposed a technique to generate bilevel programming problems. Using their technique, a linear bilevel programming problem, which has an exponential number of local minima, can be generated. Jeroslow [56] has shown that the linear bilevel program is $\mathcal{NP}$-hard. His results were subsequently confirmed and simplified by Ben-Ayed and Blair [14] and Blair [19], and strengthened by Hansen et al. [46]. Hansen et al. demonstrated that the linear bilevel problem without the upper-level constraints, which is a more restricted max-min problem, is strongly $\mathcal{NP}$-hard by considering the complexity of a special case of a linear max-min optimization problem. They reduced KERNEL to this problem whereas KERNEL is known to be $\mathcal{NP}$-hard [43]. Max-min linear programs can be formulated as convex maximization (or concave minimization) problems. The computational complexity of min-max optimization problems has been extensively studied in Ko and Lin [61] These problems are naturally characterized by Π_2^P, the second level of polynomial-time hierarchy.

For higher-level hierarchies, Jeroslow has shown that the optimal value of a $(k+1)$-level linear program is Σ_k-hard.

6.2.2 Complexity of Local Search Methods

Computing locally optimal solutions is presumably easier than finding globally optimal solutions in practice. However, from the complexity point of view it has been shown that the problem of checking local optimality for a feasible point and the problem of checking whether a local minimum is strict, are NP-hard even for instances of quadratic problems with a simple structure in the constraints and the objective. These results have been proved using the classical 3-satisfiability problem.

Vicente et al. [90] proved that checking local optimality in bilevel programming is an $\mathcal{NP}$-hard problem. The proof uses a similar idea as in Pardalos and Schnitger [77].

6.2.3 Approximation Algorithms

Jeroslow [56] observed that for any constant factor c, it remains $\mathcal{NP}$-hard to find a solution within a multiplicative factor of c of the optimum.

Deng et al. [37] have extended this result to disallow even an additive constant for a sufficiently small multiplicative factor. That is, for sufficiently small $c_1 > 0$ and for some $c_2 > 0$, there is no algorithm that can guarantee a solution within $(1 + c_1) \cdot \mathit{optimum} + c_2$ unless $\mathcal{NP} = \mathcal{P}$.

6.2.4 Polynomially Solvable Problems

Liu and Spencer [65] have introduced a polynomial-time algorithm for a bilevel linear problem when there are a constant number of lower-level control variables.

Deng et al. [37] have presented a much simpler proof for the above result which also allows for an extension to a k-level linear programming problem when the total number of variables controlled by lower-level linear programs is a constant.

6.3 Optimality Conditions

Bard [9] first derived optimality conditions for the bilevel programming problem based on an equivalent single-level programming problem of the bilevel programming problem. However, Clark and Westerberg [30] showed a counterexample of these conditions.

Ye et al. [95] presented necessary optimality conditions for the generalized bilevel programming problem, which is a bilevel programming problem with a variational inequality as the lower-level problem. They proved that the generalized bilevel programming problem is equivalent to single-level problems under some assumptions on the objective function. Then, Karush–Kuhn–Tucker optimality conditions are applied to the single-level problems to derive optimality conditions for the generalized bilevel programming problem.

Let us define the concept of global and local optimality of the bilevel problem:

$$\begin{aligned} &\min_x \; F(x, y) \\ &\text{s.t. } G(x) \leq 0, \end{aligned} \tag{6.2}$$

y solves

$$\begin{aligned} &\min_z \; f(x, z) \\ &\text{s.t. } g(x, z) \leq 0. \\ &x \in \mathbb{R}^n, \; y, z \in \mathbb{R}^m, \end{aligned}$$

where $X = \{x \mid G(x) \leq 0\}$ is a compact set.

Definition 6.1. A point (x^*, y^*) is called a local optimal solution of Problem (6.2), if and only if, $x^* \in X$, $y^* \in S(x^*)$ with

$$F(x^*, y^*) \leq F(x^*, y), \quad \text{for all } y \in S(x^*),$$

and there exists $\delta > 0$ such that

$$\phi(x^*) \leq \phi(x) \quad \text{for all } x \in X \cap B(x^*, \delta),$$

where $\phi(x) = \min_y \{F(x, y) \mid y \in S(x)\}$.

The point (x^*, y^*) is called a global optimal solution of Problem (6.2) if we can select as $\delta = \infty$.

Here, $B(x^*, \delta)$ is the open ball of radius δ centered at point $x^* \in X$; that is, $B(x^*, \delta) = \{x \in \mathbb{R}^n \mid \|x - x^*\| < \delta\}$. As we mentioned before, the bilevel problem can be formulated as a single-level problem by replacing the lower-level problem by its Karush–Kuhn–Tucker conditions under some constraint qualification if the lower level problem is convex. Then, the resulting problem is a smooth optimization problem. However, Scheel and Scholtes [80] showed that the equivalent single-level formulation of the bilevel problem violates most of the usual constraint qualifications. Therefore, they introduced a nonsmooth version of Karush–Kuhn–Tucker conditions:

$$\begin{aligned} \min_{x,y,\lambda} \ & F(x,y) \\ \text{s.t.} \ & G(x) \leq 0, \\ & \nabla_y L(x,y,\lambda) = 0, \\ & \min\{-g_j(x,y), \lambda_j\} = 0, \ j = 1, \ldots, q, \end{aligned} \tag{6.3}$$

where $L(x, y, \lambda) = f(x, y) + \lambda^T g(x, y)$ is the Lagrangian of the lower-level problem.

Then, every local solution of Problem (6.2) is a local solution of Problem (6.3). It is not difficult to see that every local solution (x^*, y^*, λ^*) of Problem (6.3) is also a local solution of the problem

$$\begin{aligned} \min_{x,y,\lambda} \ & F(x,y) \\ \text{s.t.} \ & G(x) \leq 0, \\ & \nabla_y L(x,y,\lambda) = 0, \\ & g_j(x,y) = 0 \text{ for } g_j(x^*,y^*) = 0, \\ & \lambda_j = 0 \text{ for } \lambda_j^* = 0, \\ & g_j(x,y) \leq 0 \text{ for } g_j(x^*,y^*) < 0, \\ & \lambda_j \geq 0 \text{ for } \lambda_j^* > 0. \end{aligned} \tag{6.4}$$

Let us consider the Mangasarian–Fromowitz constraint qualification [68]:

Definition 6.2. We say that Mangasarian–Fromowitz constraint qualification (MFCQ) is satisfied at point (x, y) if there exists a vector $d \in \mathbb{R}^m$ satisfying

$$d^T \nabla_y g_j(x,y) < 0 \quad \text{for all } j \in I(x,y) = \{i \mid g_i(x,y) = 0, \ i = 1, \ldots, q\}.$$

Now, we are ready to present a theorem [35], which is an application of the results in [80], for Problem (6.3).

Theorem 6.1. *Let (x^*, y^*, λ^*) be a local minimum of Problem* (6.3)*. Suppose that (MFCQ) is satisfied for Problem* (6.4) *at (x^*, y^*, λ^*). Then, there exist multipliers (u, v, w, r) such that*

$$
\begin{aligned}
&\nabla F(\chi^*) + u^T(\nabla_x G(x^*), 0) + \nabla(\nabla_y L(\chi^*, \lambda^*)v) + w^T \nabla g(\chi^*) = 0,\\
&\nabla_y g(\chi^*)v - r = 0,\\
&g_j(\chi^*)w_j = 0, \; j = 1, \dots, q,\\
&\lambda_j^* r_j = 0, \; j = 1, \dots, q,\\
&w_j r_j \geq 0, \; j \in \{i \mid g_i(x^*, y^*) = \lambda_i^* = 0\},\\
&u^T G(x^*) = 0,\\
&u \geq 0,
\end{aligned}
$$

where $\chi^* = (x^*, y^*)$.

Vicente and Calamai [89] presented necessary and sufficient optimality conditions for bilevel programming problems with quadratic strictly convex lower-level problems using on the local geometry of the problems. Based on the geometrical property of the problems, they observed that the set of feasible directions at a point is a finite union of some convex sets, and extended the first- and second-order optimality conditions of single-level programming to bilevel programming.

6.4 Algorithms

Algorithmic approaches for the bilevel problem may be grouped as follows: (1) extreme point ranking methods, (2) branch-and-bound algorithms, (3) complementarity pivot algorithms, (4) descent methods, (5) penalty function methods, and (6) trust region methods. In this section, we discuss some of these algorithmic approaches for solving the bilevel programming problem. In particularly, extreme point algorithms, branch-and-bound algorithms, and a multicriteria approach are covered. Complementarity pivoting is a method based on replacing the lower-level problem with its Karush–Kuhn–Tucker conditions. Many authors suggested different pivot algorithms for the modified problem including [18, 59, 73]. Descent methods [79, 90, 62, 40] look for a local solution of the bilevel problem based on feasible descent directions with respect to the upper-level function. Penalty function methods [4, 5, 60, 20, 93] are also local search methods for solving the bilevel problem.

6.4.1 Extreme Point Algorithms

As we mentioned before, the linear bilevel programming problem has a nice property that the solution of the problem occurs at an extreme point of the feasible set. Therefore, the problem can be solved using a vertex enumeration technique. Candler and Townsley [24] proposed the first enumeration algorithm for the linear bilevel programming problem, which has no upper-level constraints and the lower-level problem with unique solution. Their algorithm enumerates the basis of the lower-level problem.

One of the well-known vertex enumeration algorithms is the kth best method introduced by Bialas and Karwan [17]. The kth best method enumerates the basis of the resulting problem of the bilevel problem after relaxing the objective function of the lower-level problem.

Tuy et al. [86] proposed a global optimization approach for the linear two-level problem:

$$\begin{aligned} &\min_{x \geq 0} c_u^T x + d_u^T y \\ &\text{s.t. } A_u x + B_u y \leq r_u, \end{aligned} \tag{6.5}$$

where y solves

$$\begin{aligned} &\min_{z \geq 0} c_l^T x + d_l^T z \\ &\text{s.t. } A_l x + B_l z \leq r_l \leq 0. \\ &x \in \mathbb{R}^n, \ y, z \in \mathbb{R}^m. \end{aligned}$$

We can choose $c_l = 0$ because the value of this parameter does not affect the solution of the linear bilevel problem. Therefore, Problem (6.5) is equivalent to the following reverse convex programming problem.

$$\begin{aligned} &\min c_u^T x + d_u^T y \\ &\text{s.t. } Ax + By \leq r \\ &\quad\ d_l^T y \leq S(x) \\ &\quad\ x, y \geq 0, \end{aligned}$$

where $A = (A_u, A_l)^T$, $B = (B_u, B_l)^T$, $r = (r_u, r_l)^T$, and $S(x)$ is the optimal objective function value of the lower-level problem

$$\begin{aligned} &\min d_l^T z \\ &\text{s.t. } A_l x + B_l z \leq r_l \\ &\quad\ y \geq 0. \end{aligned}$$

The function $S(x)$ is a convex polyhedral function and the problem is a linear program with an additional reverse convex constraint. Several algorithms [49, 83] are available for these types of problems. They applied the approach proposed by Tuy [84] to reduce the dimension of the above optimization problem. Then, a vertex enumeration technique is applied to solve the resulting problem.

Zhang and Liu [96] proposed an extreme point algorithm to solve a mathematical programming problem with variational inequalities. As we mentioned before, this problem generalizes the bilevel programming problem in which the lower-level problem is convex. Some other vertex enumeration methods can be found in [32, 26].

6.4.2 Branch-and-Bound Algorithms

One way to tackle the bilevel problem is to replace the lower-level problem by its corresponding Karush–Kuhn–Tucker conditions and obtain a single-level mathematical programming problem. However, if the lower-level problem is nonconvex, the corresponding single-level problem is no longer equivalent to the bilevel problem. Most of the branch-and-bound algorithms for the bilevel programming problem are based on the single-level programming problem

$$
\begin{aligned}
\min\ & F(x,y) \\
\text{s.t.}\ & G(x,y) \le 0,\ \ g(x,y) \le 0, \\
& \nabla_y f(x,y) + \sum_{j=1}^{q} \lambda_j \nabla_y g(x,y) = 0, \\
& \lambda_j g_j(x,y) = 0, \\
& \lambda \ge 0,\ \ x \in X,\ \ y \in Y.
\end{aligned}
$$

Due to the stationary and complementarity conditions, the above problem is nonconvex even if the bilevel problem is linear.

Bard and Moore [12] implemented a branch-and-bound algorithm, which was initially suggested by Fortuny-Amat and McCarl [41], for solving the bilevel problem where the constraints and the objective function of the upper-level problem are linear and, the objective function of the lower-level problem is quadratic.

Hansen et al. [46] derived necessary optimality conditions for linear bilevel programming, expressed in terms of active constraints of the lower-level problem. Based on the optimality conditions, they established a branch-and-bound algorithm for linear bilevel programming. Their computational results showed that this approach was favorable for the linear case compared to the methods by Bard and Moore [12] and Júdice and Faustino [58]. Al-Khayyal et al. [6] and Bard [10] also introduced branch-and-bound algorithms for the quadratic bilevel problem.

Gümüs and Floudas [45] proposed a branch-and-bound algorithm for the bilevel programming problem. Their approach is based on a relaxation, made by Karush–Kuhn–Tucker conditions of the feasible region of the bilevel problem. The relaxed problem is solved using the global optimization technique described in [2, 3]. When the objective and constraint functions are twice differentiable and the linear constraint qualification holds for the lower-level program constraints, this approach guarantees global optimality. Bard and

Moore [13] and Wen and Yang [92] have also proposed branch-and-bound algorithms for integer and mixed integer linear bilevel programming problems.

6.4.3 A Multicriteria Approach

Several authors have been interested in the relationship between bicriteria programming and the bilevel programming problem including Bard [9] and Ünlü [87] for the linear case. They claimed that an optimal solution of the linear bilevel problem is a nondominated point for the objective functions of the upper-level and the lower-level programming problems. However, Candler [23], Clarke and Westerberg [30], and Haurie et al. [48] reported counterexamples of this argument. Fülöp [42] pointed out that more than two criteria are needed to establish the relationship between multicriteria programming and bilevel programming. Fülöp showed that, for each linear bilevel programming problem, there exists some linear multicriteria problem such that the global solution of the first problem is an optimal solution for minimizing the upper-level objective function over the Pareto optimal set of the second problem. In order to discuss a multicriteria approach, we need to define the concept of domination.

Definition 6.3. A point a in a set M is said to be nondominated with respect to an order "$\prec$", if there does not exist any point $b \in M$ such that $b \prec a$ and $a \neq b$.

Recently, Fliege and Vicente [39] established the link between these two programs. Let us consider the bilevel programming problem (6.1) with $X = \mathbb{R}^n$ and $Y = \mathbb{R}^m$.

They defined an order such that every nondominated point of $\mathbb{R}^n \times \mathbb{R}^m$ is a solution to the bilevel problem. Let $(x_1, y_1), (x_2, y_2) \in \mathbb{R}^n \times \mathbb{R}^m$. Then the order "$\prec$" can be defined as follows.

Definition 6.4. $x^1 = (x_1, y_1) \prec x^2 = (x_2, y_2)$ is equivalent to the following conditions.

- $x_1 = x_2$ and $f(x_1, y_1) < f(x_2, y_2)$

or

- $\|\nabla_y f(x_1, y_1)\|_2 = 0$ and $F(x_1, y_1) < F(x_2, y_2)$.

Then, they proved that a nondominated point of $\mathbb{R}^n \times \mathbb{R}^m$, with respect to the order defined above, is a solution of the bilevel problem.

Let us define the following function,

$$\varphi : (x, y) \rightarrow (x, F(x, y), f(x, y), \|\nabla_y f(x, y)\|_2).$$

A cone K can be defined in the image space of φ, $\mathbb{R}^n \times \mathbb{R} \times \mathbb{R} \times \mathbb{R}$, as

$$K = \{(x, f_1, f_2, d) \mid (x = 0 \text{ and } f_2 > 0) \text{ or } (f_1 > 0 \text{ and } d \geq 0)\}.$$

Then, we can define the usual cone order "$<_K$" as

$$a <_K b \Leftrightarrow b - a \in K.$$

Therefore the relation between these two orders is given by the following theorem [39].

Theorem 6.2. *If $\varphi(x, y)$ is a nondominated point with respect to "$<_K$", then (x, y) is nondominated with respect to "$\prec$".*

This theorem says that if we find a nondominated point with repect to "$<_K$", then the point should be an optimal solution of the bilevel program. Fliege and Vicente proposed a multicriteria approach to compute nondominated points with respect to the order.

6.5 Multivariate Partition Approach

In this section we discuss an approach introduced by Huang and Pardalos [52] for solving a nonlinear single-level mathematical programming problem based on its equivalent multilevel programming formulation. Let us consider the following multivariate optimization problem

$$\begin{aligned} &\min\ f(x) \\ &\text{s.t. } x \in D, \end{aligned} \tag{6.6}$$

where $D \subset \mathbb{R}^n$ $(n > 1)$ is a robust set, and $f(x)$ is a continuous function in D.

Huang and Pardalos [52] proposed a general approach, which was called the multivariate partition approach, for finding a stationary point or a local minimum of Problem (6.6). This method is based on a multilevel optimization formulation of Problem (6.6) and uses the following idea. We partition all the independent variables of Problem (6.6) into a number of groups and find an improved solution of Problem (6.6) based on solutions of some small-sized problems with respect to each group of variables.

In order to present an equivalent multilevel optimization problem of Problem (6.6), we need to define the concept of a partition of a set.

Definition 6.5. For a given set S, a set $\triangle_1, \ldots, \triangle_p$ is called a partition of the set S if the following conditions are satisfied.

(i) $\triangle_i \subseteq S$ is nonempty for $i = 1, \ldots, p$.
(ii) $\triangle_i \cap \triangle_j = \varnothing$ for $i \neq j$ and $i, j = 1, \ldots, p$.
(iii) $\bigcup_{i=1}^{p} \triangle_i = S$.

Let $S = \{x_1, \dots, x_n\}$, and let $\bar{\triangle}_i = S \setminus \triangle_i$ for $i = 1, \dots, p$.

Then a multilevel optimization formulation of Problem (6.6) can be written as

$$\min_{y_{\sigma_1} \in D_{\sigma_1}} \{ \min_{y_{\sigma_2} \in D_{\sigma_2}} \cdots \{ \min_{y_{\sigma_p} \in D_{\sigma_p}} f(\triangle_1, \dots, \triangle_p)\}\}, \tag{6.7}$$

where $\sigma = (\sigma_1, \dots, \sigma_p)$ is a permutation of the set $\{1, \dots, p\}$ and D_{σ_i} is the feasible domain of the variable y_{σ_i} corresponding to the set $\triangle_i$ for each $i = 1, \dots, p$.

A special case of Problem (6.7) is the bilevel optimization problem. Let $\{I^-, I^+\}$ be a partition of the index set $I = \{1, \dots, n\}$ of Problem (6.6). Let us denote the corresponding subsets of S with respect to I^- and I^+ by S^- and S^+, respectively. We also denote the feasible domains of the variables y^- corresponding to the set S^- and y^+ corresponding to the set S^+ by D_{S^-} and D_{S^+}, respectively. Then the bilevel optimization formulation of Problem (6.6) can be written in the form

$$\min_{y^- \in D_{S^-}} g(S^-), \tag{6.8}$$

where $g(S^-)$ is the optimal value of the problem

$$\min_{y^+ \in D_{S^+}} f(y^-, S^+). \tag{6.9}$$

The multivariate partition approach consists of the following steps.

Strategy I. For each $i = 1, \dots, p$, we obtain an improved point $\tilde{y}_{\sigma_i}$ corresponding to $S^+ = \triangle_{\sigma_i}$ by solving Problem (6.9) with $S^- = \bar{\triangle}_i = S \setminus \triangle_{\sigma_i}$.
or
Strategy II. For each $i = 1, \dots, p$, we denote $\bar{\triangle}_i^- = \{\tilde{y}_{\sigma_j} \mid j < i,\ j = 1, \dots, p\}$, $\bar{\triangle}_i^+ = \{y_{\sigma_j} \mid j > i,\ j = 1, \dots, p\}$, and $S^- = \bar{\triangle}_i^- \cup \bar{\triangle}_i^+$. We obtain an improved point $\tilde{y}_{\sigma_i}$ corresponding to $S^+ = \triangle_{\sigma_i}$ by solving Problem (6.9) with S^-.
Improved Point. Based on the improved exploratory point $\tilde{y} = (\tilde{y}_{\sigma_1}, \dots, \tilde{y}_{\sigma_p})$, we obtain an improved feasible point y_σ^* using a search and decision scheme.

The convergence of the multivariate partitional approach was proved in [52] under some assumptions. We note that it is sufficient to find local solutions of Problems (6.9) in order to find a local solution of Problem (6.6) in the algorithm. Therefore, even if we find global solutions of Problems (6.9) in Strategies I and II, there is no guarantee that we can find a global solution of Problem (6.6). Huang and Pardalos showed that the multivariate partition approach could be regarded as an extension of the Jacobi and Gauss–Seidel iterative algorithms. Next, we show some other examples of the multivariate partition approach.

6.5.1 *Examples*

K-Means Type Algorithms

Clustering is the unsupervised classification of the patterns. The K-means [66] method is one of the most popular clustering methods that have been applied in a variety of fields including pattern recognition, information retrieval, document extraction, microbiology analysis, and so on. We now show that the K-means algorithm can be seen as a special case of the multivariate partition approach. The goal of this method is to classify a given data set through a certain number of clusters, such that some metric relative to the centroids of the clusters is minimized. We can define our problem mathematically as follows.

Suppose that we are given a set X of a finite number of points in d-dimensional Euclidean space $\mathbb{R}^d$, that is,

$$X = (x^1, x^2, \ldots, x^n) \quad \text{where } x^i \in \mathbb{R}^d,\ i = 1, 2, \ldots, n.$$

We aim at finding a partition $C_j \neq \varnothing$, $j = 1, 2, \ldots, k$:

$$\begin{aligned} X = \bigcup_{j=1}^{k} C_j,\ C_j \cap C_l \\ = \varnothing \quad \text{for all } j \neq l, \end{aligned}$$

of X, which minimizes the squared error function

$$\begin{aligned} f(C_1, C_2, \ldots, C_k) &= g(c^1, c^2, \ldots, c^k) \\ &= \sum_{j=1}^{k} \sum_{x^i \in C_j} \|c^j - x^i\|^2, \end{aligned}$$

where $\|\cdot\|$ denotes the Euclidian norm, c^j is the center of the cluster C_j,

$$c^j = \frac{1}{|C_j|} \sum_{x^i \in C_j} x^i, \quad j = 1, 2, \ldots, k, \tag{6.10}$$

and $|C_j|$ is the cardinality of C_j, $j = 1, 2, \ldots, k$.

We note that, in optimal solutions, a cluster can be represented by its centroid because each point should be assigned to the cluster that has the closest centroid. Thus, we can think of variables of our optimization problem as centroids of the clusters.

The algorithm is composed of the following steps.

Algorithm 1. (`K-means`) An algorithm for solving the clustering problem.

Step 1. Initialize the centroids c_0^j, $j = 1, 2, \ldots, k$. Set $q = 0$.
Step 2. Assign each point x^i $(i = 1, 2, \ldots, n)$ to the cluster that has the closest centroid c^j $(j \in \{1, 2, \ldots, k\})$, that is,

$$j = \operatorname{argmin}_{1 \le l \le k} \|x^i - c_q^l\|^2.$$

Step 3. When all points have been assigned, for $j = 1, 2, \ldots, k$, calculate the new position c_{q+1}^j of the centroid j using equation (6.10).
Step 4. If $c_q^j = c_{q+1}^j$ for all $j = 1, 2, \ldots, k$, then stop, otherwise set $q = q + 1$ and go to Step 2.

From the algorithm, we can see that Step 3 of the algorithm uses Strategy I of the multivariate partitional approach. Equation (6.10) finds a local solution of problem

$$\begin{aligned} &\min\ g(c^1, \ldots, c^k) \\ &\text{s.t. } c^j \in \mathbb{R}^d \end{aligned}$$

whereas all other centroids (except centroid j) are fixed. One way to find a solution, which is close to a global minimum, is to locate centroid j at $x^1, \ldots, x^n$ and choose the one with the least value of g among them after reassigning the points to the centroids. This idea is closely related to the j-means clustering algorithm [47].

Coordinate Descent Method

If $n = p$ or $\Delta_i = \{x_i\}$ for $i = 1, \ldots, n$ in the partition of the multivariate partition approach, then many existing optimization algorithms designed for solving Problem (6.6) are closely related to this partition including the coordinate descent method. Basically, each coordinate axis is searched, and a descent is only made along a unit vector. The cyclic coordinate descent method minimizes a function $f(x_1, \ldots, x_n)$ cyclically with respect to the coordinate variables. That is, first x_1 is searched, then x_2, and so on. Various variations are possible. One advantage of these methods is their easy implementation.

Hirsch et al. (2007) proposed a GRASP (greedy randomized search procedure [50]), which is a special case of the coordinate descent method, for solving the following nonlinear programming problem with box constraint.

$$\begin{aligned} &\min\ f(x) \\ &\text{s.t. } l \le x \le u, \end{aligned}$$

where $f(x) : \mathbb{R}^n \to \mathbb{R}$, and l, u are lower and upper bounds for the values of the variable x.

GRASP starts with an initial point $x = (l+u)/2$. Then, for each $i = 1, \ldots, n$, we solve a problem of one variable x_i and other coordinate values of the current solution x are fixed. These problems are solved by discretizing the solution space using the grid density h. Based on solutions to these problems, the method finds an improved feasible point. Then the above procedure is repeated. If the number of iterations with no improvements achieves value N, the value of h is reduced to its half, and the process is restarted. We now present the algorithm.

Algorithm 2. (GRASP) An algorithm for solving nonlinear programming problems

Input: K, N, f, l, u.
Initialization. Set $x = (l+u)/2$, $h = 1$, $Val = f(x)$, $k = 1$.
Step 1. If $k > K$, terminate the algorithm and x is an optimal (not necessary to be global optimal) solution. If $j > N$, set $h := h/2$ and $j = 0$. We find a feasible point $y \in \mathbb{R}^n$ using the procedure *GreedyRandomSol*(x,h).
Step 2. If $f(y) < f(x)$, set $j := 0$, $x := y$ and $Val := f(x)$. Otherwise, set $j = j+1$.
Step 3. Set $k := k+1$, and go to Step 1.

Procedure *GreedyRandomSol* can be interpreted as follows.

Algorithm 3. *GreedyRandomSol*

Input: f, l, b, h, $\alpha \in (0,1)$, x.
Step 1. For each $i = 1, \ldots, n$, find a solution x_i^0 of the problem

$$\begin{aligned} &\min f(x_1, \ldots, x_n) \\ &\text{s.t. } l_i \le x_i \le u_i. \end{aligned} \tag{6.11}$$

Let $g_i = f(x_1, \ldots, x_{i-1}, x_i^0, x_{i+1}, \ldots, x_n)$ for $i = 1, \ldots, n$.
Step 2. Let min and max be the minimum and the maximum values of the set $\{g_1, \ldots, g_n\}$, respectively. Let $S = \varnothing$.
For each $i = 1, \ldots, n$, $S := S \cup \{i\}$ if $g_i \le min + \alpha(max - min)$.
Step 3. Select an index l from the set S randomly, and set $y := x$ and $y_l := x_l^0$.

Problem (6.11) can be solved by discretizing the solution space. The mesh points can be chosen as $l_i, l_i + h, l_i + 2h, \ldots, l_i + mh$, where m is the largest integer number such that $l_i + mh \le u_i$. Then we choose the minimum of the objective function values at the mesh points. It is not difficult to see that GRASP is a special case of the multivariate partition approach. In this case, the partition is $\{\triangle_1, \ldots, \triangle_n\}$, where $\triangle_i = x_i$ for all $i = 1, \ldots, n$, and $D_i = [l_i, u_i]$ for all $i = 1, \ldots, n$.

6.5.2 Applications of the Multivariate Partition Approach

Lennard-Jones Problem

The Lennard-Jones problem is one of the most challenging global optimization problems in molecular biophysics. The problem is to find such a structure of a protein, a cluster of N atoms, interacting via the Lennard-Jones (dimensionless nonquantized pair) potential : $v(r) = r^{-12} - 2r^{-6}$, that its energy E is (globally) minimal. Let the code $P_N = \{x_1, \ldots, x_N\}$ be the collection of centers of N atoms. Then the potential energy E is defined as follows.

$$E(x_1, \ldots, x_N) = \sum_{1 \le i < j \le N} v(\|x_i - x_j\|),$$

where $v(r) = r^{-12} - 2r^{-6}$.

Therefore, our optimization problem can be formulated as

$$\begin{aligned} \min\ E(x_1, \ldots, x_N) &= \sum_{1 \le i < j \le N} v(\|x_i - x_j\|), \\ \text{s.t. } x_1, \ldots, x_N &\in \mathbb{R}^3. \end{aligned}$$

Huang et al. [53] presented necessary optimality conditions for the Lennard–Jones problem and applied the multivariate partitional approach to the Lennard-Jones problem. The partition can be described by a center of each atom; that is, $\triangle_i = x_i$ for $i = 1, \ldots, N$. They obtained all the expected global minima of the problem with $2 \le N \le 56$ using a quasi-Newton accelerating approach for the auxiliary problem (6.9).

Spherical Code Problem

The spherical code problem has many applications including physics, molecular biology, and chemistry. Tammes' problem is one of the well-known cases of the spherical code problem which is referred to distribute points on a unit sphere such that they maximize the minimum distance between any pair of points. In general, we distribute points on a unit sphere according to a certain generalized energy. Let a code $P_N = \{x_1, \ldots, x_N\}$ be a set of N points on a unit sphere in $\mathbb{R}^n$. Then the *s-energy* associated with the spherical code P_N is

$$w(s, P_N) = \begin{cases} \sum\limits_{i<j} \|x_i - x_j\|^{-s} & \text{if } s \neq 0 \\ \sum\limits_{i<j} \ln\left(\frac{1}{\|x_i - x_j\|}\right) & \text{if } s = 0, \end{cases}$$

and the spherical code problem can be formulated mathematically as

$$\begin{aligned} &\min f_s(x_1,\ldots,x_N) \\ &\text{s.t. } x_i \in S^n = \{x \mid \|x\| = 1,\ x \in \mathbb{R}^n\}\ \mathit{for}\ i = 1,\ldots,n, \end{aligned}$$

where

$$f_s(P_N) = \begin{cases} w(s,P_N) & \text{if } s \geq 0, \\ -w(s,P_N) & \text{if } s < 0. \end{cases}$$

In [54], Huang et al. used a multivariate partition approach for the spherical code problems. The partition of the variables is similar to that of the Lennard-Jones problem.

6.6 Applications

Multilevel optimization has many applications in different fields including economics, transportation systems, engineering system design, environmental engineering, and mechanics. Here, we discuss two of these applications in transportation systems.

The concentration of human activities in urban areas has given rise to congestion problems that create negative environmental or economical effects. Therefore, many researchers have developed efficient methods for road network design in order to improve transportation systems. The network design problem (NDP) determines a set of parameters that optimizes the road network. The model includes traffic signal control, traffic information provision, congestion charge, new transportation modes, and road expansion or deletion. The NDP is usually formulated as a bilevel programming problem. The upper-level part defines system design and the lower-level part defines travellers' behavior. NDPs are classified into two categories: discrete and continuous variations. LeBlanc [63] formulated a bilevel programming problem for discrete NDP. Discrete models usually consider link or lane additions. Abdulaal and LeBlanc [1] later described continuous models of the NDP. Continuous models are concerned with network improvements that can be modeled as continuous variables including lane and lateral clearance changes and other enhancements. The NDP has also been investigated by Janson and Husaini [55], Magnanti and Wong [67], and Xiong and Schneider [94]. For a more comprehensive survey about the NDP, the reader is referred to Migdalas [71] and Cascetta [25].

Another application of bilevel programming is the signal setting problem (SSP) which maximizes network performance by optimizing traffic signals. The main difficulty of the problem arises from the existing interaction between network performance and users' choices of the routes. Gartner et al. [44] formulated the SSP as a bilevel programming problem, where the upper-level part represents the public network manager and the lower-level part represents users' behavior.

Some other applications of multilevel optimization can be found in Migdalas [71], Migdalas et al. [72], and Bard [11].

6.7 Some Interesting Open Problems

6.7.1 Polynomially Solvable Problems

As we mentioned before, the linear bilevel programming problem is known to be $\mathcal{NP}$-hard. However, if the lower-level problem has a constant number of variables, the linear bilevel problem is polynomially solvable. Hence, a similar question may arise: whether we can find a polynomial-time algorithm for the linear bilevel problem in which the upper-level problem has a constant number of variables. Therefore, it would be interesting if we can find polynomially solvable special cases of the nonlinear bilevel programming problem.

6.7.2 The Relation Between Multilevel Programming and Multicriteria Programming

In Section 6.4, we discussed a bicriteria approach by Fliege and Vicente [39] for solving bilevel problems. They introduced an order in the Euclidean space and showed that an optimal solution of the bilevel problem is a nondominated point with respect to the order. However, they did not exactly define any multicriteria problem. The problem of finding equivalent multicriteria programs of multilevel programs is still an open question. As we mentioned before, several attempts, including Bard [7] and Ünlü [87], have been made to discover the relation between bilevel linear programming and multicriteria programming. Later, Wen and Hsu [91] pointed out that the theorem in [87], which illustrates the relation between multilevel programming and multicriteria programming, is valid only under an additional constraint. Let us consider the linear bilevel programming problem

$$\min_x F(x,y) = c_u^T x + d_u^T y \tag{6.12}$$

where y solves

$$\begin{aligned} &\min_z f(x,z) = c_l^T x + d_l^T z \\ &\text{s.t. } Ax + Bz \le r \le 0. \\ &x \in \mathbb{R}^n,\ y,z \in \mathbb{R}^m \end{aligned}$$

and the multicriteria programming problem

$$\begin{aligned} &\min\ (c_u^T x + d_u^T y, d_l^T z) \\ &\text{s.t. } Ax + Bz \leq r. \end{aligned} \tag{6.13}$$

Then, Wen and Hsu stated the following theorem.

Theorem 6.3. *If $\nabla F^T \nabla \bar{f} \geq 0$, where $\bar{f} = f(0, y)$, then the optimal solution to the bilevel programming problem* (6.12) *is an efficient solution to the bicriteria programming problem* (6.13).

The condition in the above theorem restricts the use of bicriteria approaches to the linear bilevel problem. Thus, it would be useful if we can derive an equivalent multicriteria programming problem, which may have three or more objective functions, for the billevel problem (6.12) without any strong conditions. Therefore, it might be interesting if we state special cases of the nonlinear multilevel programming problem, which are equivalent to multicriteria programming problems.

6.7.3 Multilevel Multicriteria Programming Problems

Most of the real-world optimization problems require more than one objective function. Therefore, one of the challenging problems in mathematical programming is the multilevel multicriteria programming problem; that is, each decision maker has several objectives. The general bilevel multicriteria programming problem has the form

$$\begin{aligned} &\min_x\ F(x, y) = (F_1(x, y), \ldots, F_l(x, y)) \\ &\text{s.t. } G(x, y) \leq 0, \end{aligned} \tag{6.14}$$

where y is Pareto optimal for

$$\begin{aligned} &\min_z\ f(x, z) = (f(x, z), \ldots, f_k(x, z)) \\ &\text{s.t. } g(x, z) \leq 0. \\ &x \in X \subset \mathbb{R}^n,\ y, z \in Y \subset \mathbb{R}^m, \end{aligned}$$

where $X \subseteq \mathbb{R}^n$ and $Y \subseteq \mathbb{R}^m$ are compact sets, $G : \mathbb{R}^n \times \mathbb{R}^m \to \mathbb{R}^p$, $g : \mathbb{R}^n \times \mathbb{R}^m \to \mathbb{R}^q$, and $F_i, f_j : \mathbb{R}^n \times \mathbb{R}^m \to \mathbb{R}$, $i = 1, \ldots, l$ and $j = 1, \ldots, k$, are scalar functions.

Definition 6.6. A point $y^* \in Y$ with $f(x, y)$ is called Pareto optimal for the lower-level multicriteria problem, if and only if there exists no point $z \in Y$ such that $f_i(x, z) \leq f_i(y)$ for all $i = 1, 2, \ldots, k$ and $f_j(x, z) < f_j(x, y)$ for at least one index $j \in \{1, 2, \ldots, k\}$.

Solving the bilevel multicriteria problem (6.14) means we look for Pareto optimal soltutions of the problem (6.14):

Definition 6.7. A point $(x^*, y^*) \in X \times Y$ with $F(x^*, y^*)$, where y^* is Pareto optimal for the lower-level problem, is called Pareto optimal for (6.14), if and only if there exists no point $(x, y) \in X \times Y$, where y is Pareto optimal for the lower-level problem, such that $F_i(x) \leq F_i(x^*)$ for all $i = 1, 2, \ldots, l$ and $F_j(x, y) < F_j(x^*, y^*)$ for at least one index $j \in \{1, 2, \ldots, l\}$.

To our knowledge, this problem has not been studied by many researchers. Recently, Sinha and Sinha [81] studied multilevel decentralized programming problems in which decision makers have absolute control over certain decision variables but some variables may be controlled by two or more decision makers. In this case, the problem has conflicting objectives, but the decision makers are placed in hierarchical order.

We have briefly discussed optimality conditions for multilevel programming in Section 6.3. Many researchers also considered optimality conditions and duality results for multicriteria programming problems including Preda [78], Jeyakumar and Mond [57], Liang et al. [64], and Chinchuluun et al. [28]. Based on the optimality conditions and duality results of both multicriteria programming and multilevel programming, optimality duality of multilevel multiobjective programming can also be studied.

6.8 Concluding Remarks

In this brief survey, we have shown a number of theoretical results for deterministic multilevel programming. These results include complexity issues, optimality conditions, and algorithmic approaches for solving multilevel programming problems. We have also presented a method, which is called the multivariate partition approach, for solving single-level mathematical programming problems based on their equivalent bilevel programming formulations. Some open questions have also been included at the end of the chapter. This survey is not comprehensive; we have not focused on stochastic issues and connection with decomposition methods.

Acknowledgments The research of the first two authors is partially supported by NSF, Air Force, and CRDF grants. The research of the third author is partially supported by National Science Foundation of China under Project 10201017.

References

1. Abdulaal, M.S., LeBlanc, L.J.: Continuous Equilibrium Network Design Models. *Transportation Research*, **13B**, 19–32 (1979)
2. Adjiman, C.S., Androulakis, I.P., Floudas, C.A.: Global Optimization of MINLP Problems in Process Synthesis. *Computers and Chemical Engineering*, **21**, S445–S450 (1997)

3. Adjiman, C.S., Androulakis, I.P., Floudas, C.A.: Global Optimization of Mixed Integer Nonlinear Problems. *AIChE Journal*, **46**, 1769–1797 (2000)
4. Aiyoshi, E., Shimizu, K.: Hierarchical Decentralized Systems and Its New Solution by a Barrier Method. *IEEE Transactions on Systems, Management and Cybernetics*, 11, 444–449 (1981)
5. Aiyoshi, E., Shimizu, K.: A Solution Method for the Static Constrained Stackelberg Problem via Penalty Method. *IEEE Transactions on Automatic Control*, **29**, 1111–1114 (1984)
6. Al-Khayyal, F., Horst, R., Pardalos, P.: Global Optimization of Concave Functions Subject to Quadratic Constraints: An Application in Nonlinear Bilevel Programming. *Annals of Operations Research*, **34**, 125–147 (1992)
7. Bard, J.F.: An Efficient Point Algorithm for a Linear Two-Stage Optimization Problem. *Operations Research*, **31**, 670–684 (1983)
8. Bard, J.F.: Regulating Nonnuclear Industrial Waste by Hazard Classification. *Journal of Environmental Systems*, **13**, 21–41 (1983–84)
9. Bard, J.F.: Optimality Conditions for the Bilevel Programming Problem. *Naval Research Logistics Quarterly*, **31**, 13–26 (1984)
10. Bard, J.F.: Convex Two-Level Optimization. *Mathematical Programming*, **40**, 15–27 (1988)
11. Bard, J.F.: *Practical Bilevel Optimization: Algorithms and Applications.* Kluwer Academic, Boston (1998)
12. Bard, J., Moore, J.: A Branch and Bound Algorithm for the Bilevel Programming Problem. *SIAM Journal on Scientific and Statistical Computing*, **11**, 281–292 (1990)
13. Bard, J., Moore, J.: An Algorithm for the Discrete Bilevel Programming Problem. *Naval Research Logistics*, **39**, 419–435 (1992)
14. Ben-Ayed, O., Blair, C.E.: Computational Difficulties of Bilevel Linear Programming. *Operations Research*, **38**, 556–559 (1990)
15. Ben-Ayed, O., Boyce, D.E., Blair, C.E., III: A General Bilevel Linear Programming Problem Formulation of the Network Design Problem. *Transportation Research*, **22**, 311–318 (1988)
16. Bialas, W.F., Karwan, M.H.: On Two Level Optimization. *IEEE Transactions on Automatic Control*, **27**, 211–214 (1982)
17. Bialas, W.F., Karwan, M.H.: Two-Level Linear Programming. *Management Science*, **30**, 1004–1020 (1984)
18. Bialas, W., Karwan, M., Shaw, J.: A Parametric Complementary Pivot Approach for Two-Level Linear Programming. Technical Report 80-2, Operations Research Program, State University of New York, Buffalo (1980)
19. Blair, C.E.: The Computational Complexity of Multi-Level Linear Programs. *Annals of Operations Research*, **34**, 13–19 (1992)
20. Calamai, Z.B., Conn, A.: An Exact Penalty Function Approach for the Linear Bilevel Programming Problem. Technical Report 180-O-170591, Department of Systems Design Engineering, University of Waterloo (1991)
21. Calamai, P.H., Moré, J.J.: Projected Gradient Methods for Linearly Constrained Problems. *Mathematical Programming*, **39**, 93–116 (1987)
22. Calamai, P., Vicente, L.: Generating Linear and Linear-Quadratic Bilevel Programming Problems. *SIAM Journal on Scientific and Statistical Computing*, **14**, 770–782 (1993)
23. Candler, W.: A Linear Bilevel Programming Algorithm: A Comment. *Computers and Operations Research*, **15**, 297–298 (1988)
24. Candler, W., Townsley, R.: A Linear Two-Level Programming Problem. *Computers and Operations Research*, **9**, 59–76 (1982)
25. Cascetta, E.: *Transportation Systems Engineering: Theory and Methods.* Kluwer Academic Press, Boston (2001)

26. Chen, Y., Florian, M.: On the Geometry Structure of Linear Bilevel Programs: A Dual Approach. Technical Report CRT-867, Centre de Recherche sur les Transports (1992)
27. Chen, Y., Florian, M.: The Nonlinear Bilevel Programming Problem: Formulations, Regularity and Optimality Conditions. *Optimization*, **32**, 193–209 (1995)
28. Chinchuluun, A., Yuan, D.H., Pardalos, P.M.: Optimality Conditions and Duality for Nondifferentiable Multiobjective Fractional Programming with Generalized Convexity. *Annals of Operations Research*, **154**, 133–147 (2007)
29. Clarke, F.H.: *Optimization and Nonsmooth Analysis.* Wiley & Sons, New York (1983)
30. Clarke, P., Westerberg, A.: A Note on the Optimality Conditions for the Bilevel Programming Problem. *Naval Research Logistics*, **35**, 413–418 (1988)
31. Colson, B., Marcotte, P., Savard, G.: Bilevel Programming: A Survey. *4OR*, **3**, 87–107 (2005)
32. Dempe, S.: A Simple Algorithm for the Linear Bilevel Programming Problem. *Optimization*, **18**, 373–385 (1987)
33. Dempe, S.: A Necessary and a Sufficient Optimality Condition for Bilevel Programming Problems. *Optimization*, **25**, 341–354 (1992)
34. Dempe, S.: Annotated Bibliography on Bilevel Programming and Mathematical Programs with Equilibrium Constraints. *Optimization*, **52**, 333–359 (2003)
35. Dempe, S.: *Foundations of Bilevel Programming.* NOIA vol 61, Kluwer Academic, Boston (2002)
36. Deng, X.: Complexity Issues in Bilevel Linear Programming. In: Migdalas, A., Pardalos, P.M., Värbrand, P. (eds) *Multilevel Optimization: Algorithms and Applications.* Kluwer Academic, Boston, 149–164 (1997)
37. Deng, X., Wang, Q., Wang, S.: On the Complexity of Linear Bilevel Programming. In: *Proceeding of the 1st International Symposium on Operations Research and Applications*, 205–212 (1995)
38. Du, D.Z., Pardalos, P.M. (eds): *Minimax and Applications.* Kluwer Academic, Boston (1995)
39. Fliege, J., Vicente, L.N.: A Multicriteria Approach to Bilevel Optimization. *Journal of Optimization Theory and Applications*, **131**, 209–225 (2006)
40. Florian, M., Chen, Y.: A Bilevel Programming Approach to Estimating O-D Matrix by Traffic Counts. Technical Report CRT-750, Centre de Recherche sur les Transports (1991)
41. Fortuny-Amat, J., McCarl, B.: A Representation and Economic Interpretation of a Two Level Programming Problem. *Journal of Operations Research Society*, **32**, 783–792 (1981)
42. Fülöp, J.: On the Equivalence Between a Linear Bilevel Programming Problem and Linear Optimization Over the Efficient Set. Technical Report WP 93-1, Laboratory of Operations Research and Decision Systems, Computer and Automation Institute, Hungarian Academy of Sciences (1993)
43. Garey, M.R., Johnson, D.S.: *Computers and Intractability: A Guide to the Theory of $\mathcal{NP}$-Completeness.* W.H. Freeman, San Francisco (1979)
44. Gartner, N.H., Gershwin, S.B., Little, J.D.C., Ross, P.: Pilot Study of Computer Based Urban Traffic Management. *Transportation Research*, **14B**, 203–217 (1980)
45. Gümüs, Z.H., Floudas, C.A.: Global Optimization of Nonlinear Bilevel Programming Problems. *Journal of Global Optimization*, **20**, 1–31 (2001)
46. Hansen, P., Jaumard, B., Savard, G.: New Branch and Bound Rules for Linear Bilevel Programming. *SIAM Journal on Scientific and Statistical Computing*, **13**, 1194–1217 (1992)
47. Hansen, P., Mladenović, N.: J-Means: A New Local Search Heuristic for Minimum Sum of Squares Clustering. *Pattern Recognition*, **34**, 405–413 (2001)
48. Haurie, A., Savard, G., White, D.: A Note On: An Efficient Point Algorithm for a Linear Two-Stage Optimization Problem. *Operations Research*, **38**, 553–555 (1990)

49. Hillestad, R.J., Jacobsen, S.E.: Linear Programs with an Additional Reverse Convex Constraint. *Applied Mathematics and Optimization*, **6**, 257–269 (1980)
50. Hirsch, M.J., Meneses, C.N., Pardalos, P.M., Resende, M.G.C.: Global Optimization by Continuous Grasp, *Optimization Letters*, **1**, 201–212 (2007)
51. Hobbs, B.F., Nelson, S.K.: A Nonlinear Bilevel Model for Analysis of Electric Utility Demand-Side Planning Issues. *Annals of Operations Research*, **34**, 255–274 (1992)
52. Huang, H.X., Pardalos, P.M.: A Multivariate Partition Approach to Optimization Problems. *Cybernetics and Systems Analysis*, **38**, 265–275 (2002)
53. Huang, H.X., Pardalos, P.M., Shen Z.J.: Equivalent Formulations and Necessary Optimality Conditions for the Lennard-Jones Problem. *Journal of Global Optimization*, **22**, 97–118 (2002)
54. Huang, H.X., Pardalos, P.M., Shen Z.J.: A Point Balance Algorithm for the Spherical Code Problem. *Journal of Global Optimization*, **19**, 329–344 (2001)
55. Janson, B.N., Husaini, A.: Heuristic Ranking and Selection Procedures for Network Design Problems. *Journal of Advanced Transportation*, **21**, 17–46 (1987)
56. Jeroslow, R.G.: The Polynomial Hierarchy and a Simple Model for Competetive Analysis. *Mathematical Programming*, **32**, 146–164 (1985)
57. Jeyakumar, V., Mond, B.: On Generalized Convex Mathematical Programming. *Journal of the Australian Mathematical Society Series B*, **34**, 43–53 (1992)
58. Júdice, J.J., Faustino, A.M.: The Solution of the Linear Bilevel Programming Problem by Using the Linear Complementarity Problem. *Investigação Operacional*, **8**, 77–95 (1988)
59. Júdice, J., Faustino, A.: A Sequential LCP Method for Bilevel Linear Programming. *Annals of Operations Research*, **34**, 80–106 (1992)
60. Ishizuka, Y., Aiyosi, E.: Double Penalty Method for Bilevel Optimization Problems. *Annals of Operations Research*, **34**, 73–88 (1992)
61. Ko, K.I., Lin, C.L.: On the Complexity of Min-Max Optimization Problems and Their Approximation. In: Du, D.Z., Pardalos, P.M. (eds) *Minimax and Applications*. Kluwer Academic, Boston, 219–239 (1995)
62. Kolstad, C., Lasdon, L.: Derivative Evaluation and Computational Experience with Large Bilevel Mathematical Programs. *Journal of Optimization Theory and Applications*, **65**, 485–499 (1990)
63. Leblanc, L.J.: An Algorithm for the Discrete Network Design Problem. *Transportation Science*, **9**, 183–199 (1975)
64. Liang, Z.A., Huang, H.X., Pardalos, P.M.: Efficiency Conditions and Duality for a Class of Multiobjective Fractional Programming Problems. *Journal Global Optimization*, **27**, 447–471 (2003)
65. Liu, Y., Spencer, T.H.: Solving a Bilevel Linear Program When the Inner Decision Maker Controls Few Variables. *European Journal of Operational Research*, **81**, 644–651 (1995)
66. MacQueen, J.B.: Some Methods for Classification and Analysis of Multivariate Observations. In: *Proceedings of the Fifth Berkeley Symposium on Mathematical Statistics and Probability*. University of California Press, Berkeley, vol 1, 281–297 (1967)
67. Magnanti, T.L., Wong, R.T.: Network Design and Transportation Planning: Models and Algorithms. *Transportation Science*, **18**, 1–55 (1984)
68. Mangasarian, O.L., Fromowitz, S.: The Fritz John Necessary Optimality Condition in the Presence of Equality and Inequality Constraints. *Journal of Mathematical Analysis and Applications*, **17**, 37–47 (1967)
69. Marcotte, P.: Network Optimization with Continuous Control Parameters. *Transportation Science*, **17**, 181–197 (1983)
70. Marcotte, P., Zhu, D.L.: Exact and Inexact Penalty Methods for the Generalized Bilevel Programming Problem. *Mathematical Programming*, **74**, 141–157 (1996)
71. Migdalas, A.: Bilevel Programming in Traffic Planning: Models, Methods and Challenge. *Journal of Global Optimization*, **7**, 381–405 (1995)

72. Migdalas, A., Pardalos, P.M., Värbrand, P. (eds): *Multilevel Optimization: Algorithms and Applications.* NOIA vol 20, Kluwer Academic, Boston(1998)
73. Önal, H.: A Modified Simplex Approach for Solving Bilevel Linear Programming Problems. *European Journal of Operational Research*, **67**, 126–135 (1993)
74. Önal, H., Darmawan, D.H., Johnson, S.H., III: A Multilevel Analysis of Agricultural Credit Distribution in East Java, Indonesia, *Computers and Operations Research*, **22**, 227–236 (1995)
75. Outrata, J.: Necessary Optimality Conditions for Stackelberg Problems. *Journal of Optimization Theory and Applications*, **76**, 305–320 (1993)
76. Pardalos, P.M., Deng, X.: Complexity Issues in Hierarchical Optimization. In: Mirkin, B., McMorris, F.R., Roberts, F.S., Rzhetsky, A. (eds) *Mathematical Hierarchies and Biology.* DIMACS Series, vol 37, American Mathematical Society, Providence, 219–224 (1997)
77. Pardalos, P.M., Schnitger, G.: Checking Local Optimality in Constrained Quadratic Programming is NP-Hard. *Operations Research Letters*, **7**, 33–35 (1988)
78. Preda, V.: On Sufficiency and Duality for Multiobjective Programs. *Journal of Mathematical Analysis and Applications*, **166**, 365–377 (1992)
79. Savard, G., Gauvin, J.: The Steepest Descent Direction for the Nonlinear Bilevel Programming Problem. *Operations Research Letters*, **15**, 275–282 (1994)
80. Scheel, H., Scholtes, S.: Mathematical Programs with Complementarity Constraints: Stationarity, Optimality, and Sensitivity. *Mathematics of Operations Research*, **25**, 1–22 (2000)
81. Sinha, S., Sinha, S.B.: KKT Transformation Approach for Multi-Objective Multi-Level Linear Programming Problems. *European Journal of Operational Research*, **143**, 19–31 (2002)
82. Suh, S., Kim, T.: Solving Nonlinear Bilevel Programming Models of the Equilibrium Network Design Problem: A Comparative Review. *Annals of Operations Research*, **34**, 203–218 (1992)
83. Thuong, N.V., Tuy, H.: A Finite Algorithm for Solving Linear Programs with an Additional Reverse Convex Constraint. *Lecture Notes in Economics and Management Systems*, **225**, 291–302 (1984)
84. Tuy, H.: Polyhedral Annexation, Dualization and Dimension Reduction Technique in Global Optimization. *Journal of Global Optimization*, **1**, 229–244 (1991)
85. Tuy, H., Migdalas, A., Hoai-Phuong, N.T.: A Novel Approach to Bilevel Nonlinear Programming, *Journal of Global Optimization*, http://www.ingentaconnect.com/content/klu/jogo, **38**, 527–554 (2007)
86. Tuy, H., Migdalas, A., Värbrand, P.: A Global Optimization Approach for the Linear Two-Level Program. *Journal of Global Optimization*, **3**, 1–23 (1993)
87. Ünlü, G.: A Linear Bilevel Programming Algorithm Based on Bicriteria Programming. *Computers and Operations Research*, **14**, 173–179 (1987)
88. Vicente, L.N., Calamai, P.H.: Bilevel and Multilevel Programming: A Bibliography Review. *Journal of Global Optimization*, **5**, 291–306 (1994)
89. Vicente, L.N., Calamai, P.H.: Geometry and Local Optimality Conditions for Bilevel Programs with Quadratic Strictly Convex Lower Levels. In: Du, D.Z., Pardalos, P.M. (eds) *Minimax and Applications.* Kluwer Academic, Boston, 141–151 (1995)
90. Vicente, L., Savard, G., Júdice, J.: Descent Approaches for Quadratic Bilevel Programming. *Journal of Optimization Theory and Applications*, **81**, 379–399 (1994)
91. Wen, U., Hsu, S.: A Note on a Linear Bilevel Programming Algorithm Based on Bicriteria Programming. *Computers and Operations Research*, **16**, 79–83 (1989)
92. Wen, U., Yang, Y.: Algorithms for Solving the Mixed Integer Two-Level Linear Programming Problem. *Computers and Operations Research*, **17**, 133–142 (1990)
93. White, D., Anandalingam, G.: A Penalty Function Approach for Solving Bi-Level Linear Programs. *Journal of Global Optimization*, **3**, 397–419 (1993)

94. Xiong, Y., Schneider, J.B.: Transportation Network Design Using a Cumulative Genetic Algorithm and Neural Network. *Transportation Research Record*, **1364**, 37–44 (1995)
95. Ye, J.J., Zhu, D.L., Zhu, Q.J.: Exact Penalization and Necessary Optimality Conditions for Generalized Bilevel Programming. *SIAM Journal on Optimization*, **7**, 481–507 (1997)
96. Zhang, J., Liu, G.: A New Extreme Point Algorithm and Its Application in PSQP Algorithms for Solving Mathematical Programs with Linear Complementarity Constraints. *Journal of Global Optimization*, **19**, 345–361 (2001)

Chapter 7
Central Path Curvature and Iteration-Complexity for Redundant Klee–Minty Cubes

Antoine Deza, Tamás Terlaky, and Yuriy Zinchenko

Summary. We consider a family of linear optimization problems over the n-dimensional Klee–Minty cube and show that the central path may visit all of its vertices in the same order as simplex methods do. This is achieved by carefully adding an exponential number of redundant constraints that forces the central path to take at least $2^n - 2$ sharp turns. This fact suggests that any feasible path-following interior-point method will take at least $O(2^n)$ iterations to solve this problem, whereas in practice typically only a few iterations (e.g., 50) suffices to obtain a high-quality solution. Thus, the construction potentially exhibits the worst-case iteration-complexity known to date which almost matches the theoretical iteration-complexity bound for this type of methods. In addition, this construction gives a counterexample to a conjecture that the total central path curvature is $O(n)$.

Key words: Linear programming, central path, interior-point methods, total curvature

7.1 Introduction

Consider the following linear programming problem: $\min c^T x$ such that $Ax \geq b$ where $A \in \mathbb{R}^{m \times n}$, $b \in \mathbb{R}^m$, and $c, x \in \mathbb{R}^n$.

In theory, the so-called feasible path-following interior-point methods exhibit polynomial iteration-complexity: starting at a point on the central path they take at most $O(\sqrt{m} \ln \nu)$ iterations to attain a ν-relative decrease in the duality gap. Moreover, if L is the bit-length of the input data, it takes

Antoine Deza · Tamás Terlaky · Yuriy Zinchenko
Advanced Optimization Laboratory, Department of Computing and Software, McMaster University, Hamilton, ON L8S 4K1, Canada
e-mail: zinchen@mcmaster.ca

D.Y. Gao, H.D. Sherali, (eds.), *Advances in Applied Mathematics and Global Optimization*
Advances in Mechanics and Mathematics 17, DOI 10.1007/978-0-387-75714-8_7,

at most $O(\sqrt{m}L)$ iterations to solve the problem exactly; see, for instance, [11]. However, in practice typically only a few iterations, usually less than 50, suffices to obtain a high-quality solution. This remarkable difference stands behind the tremendous success of interior-point methods in applications.

Let $\psi : [\alpha, \beta] \to \mathbb{R}^n$ be a C^2 map with nonzero derivative $\forall t \in [\alpha, \beta]$. Denote its arc length by

$$l(t) := \int_{\alpha}^{t} \|\dot{\psi}(\tau)\| d\tau,$$

its parametrization by the arc length by $\psi_{\text{arc}}(l) : [0, l(\beta)] \to \mathbb{R}^n$, and its curvature at the point l,

$$\kappa(l) := \frac{d}{dl}\ddot{\psi}_{\text{arc}}(l).$$

The total curvature K is defined as

$$K := \int_{0}^{l(\beta)} \|\kappa(l)\| dl.$$

Intuitively, the total curvature is a measure of how far off a certain curve is from being a straight line. Thus, it has been hypothesized that the total curvature of the central path is positively correlated with the number of iterations that any Newton-like path following method will take to traverse this curve, in particular, the number of iterations for feasible path-following interior-point methods, for example, long-step or predictor-corrector.

The worst-case behavior for path-following interior-point methods has already been under investigation, for example, Todd and Ye [13] gave a lower iteration-complexity bound of order $\sqrt[3]{m}$ necessary to guarantee a fixed decrease in the central path parameter and consequently in the duality gap. At the same time, different notions for the curvature of the central path have been examined. The relationship between the number of approximately straight segments of the central path introduced by Vavasis and Ye [14] and a certain curvature measure of the central path introduced by Sonnevend, Stoer, and Zhao [12] and further analyzed in [15], was further studied by Monteiro and Tsuchiya in [9]. Dedieu, Malajovich, and Shub [1] investigated a properly averaged total curvature of the central path. Nesterov and Todd [10] studied the Riemannian curvature of the central path in particular relevant to the so-called short-step methods. We follow a constructive approach originated in [4, 5] which is driven by the geometrical properties of the central path to address these questions.

We consider a family of linear optimization problems over the n-dimensional Klee–Minty cube and show that the central path may visit all of its vertices in the same order as simplex methods do. This is achieved by carefully adding an exponential number of redundant constraints that forces the central path to take at least $2^n - 2$ sharp turns. We derive explicit formulae for the number of the redundant constraints needed. In particular, we give a bound of

$O(n2^{3n})$ on the number of redundant constraints when the distances to those are chosen uniformly. When these distances are chosen to decay geometrically, we give a slightly tighter bound of the same order $n^3 2^{2n}$ as in [5].

The behavior of the central path suggests that any feasible path-following interior-point method will take at least order 2^n iterations to solve this problem. Thus, the construction potentially exhibits the worst-case iteration-complexity known to date which almost matches the theoretical iteration-complexity bound for this type of methods. However, state-of-the art linear optimization solvers that include preprocessing of the problem as described in [6, 7] are expected to recognize and remove the redundant constraints in no more than two passes. This underlines the importance of the implementation of efficient preprocessing algorithms.

We show that the total curvature of the central path for the construction is at least exponential in n and, therefore, provides a counterexample to a conjecture of Dedieu and Shub [2] that it can be bounded by $O(n)$. Also, the construction may serve as an example where one can relate the total curvature and the number of iterations almost exactly.

The chapter is organized as follows. In Section 7.2 we introduce a family of linear programming problems studied along with a set of sufficient conditions that ensure the desired behavior for the central path and give a lower bound on the total curvature of the central path, in Section 7.3 we outline the approach to determine the number of the redundant constraints required, and Sections 7.4 and 7.5 contain a detailed analysis of the two distinct models for the distances to the redundant constraints. We give a brief conclusion in Section 7.6.

7.2 Sufficient Conditions for Bending the Central Path and the Total Curvature

Let $x \in \mathbb{R}^n$. Consider the following optimization problem.

$$\left\{\begin{array}{rclll}
\min & & x_n & & \\
0 & \leq & x_1 & \leq 1 & \\
\varepsilon x_{k-1} & \leq & x_k & \leq 1 - \varepsilon x_{k-1} & k = 2, \ldots, n \\
0 & \leq & d_1 + x_1 & & \text{repeated } h_1 \text{ times} \\
\varepsilon x_1 & \leq & d_2 + x_2 & & \text{repeated } h_2 \text{ times} \\
 & & \vdots & & \\
\varepsilon x_{n-1} & \leq & d_n + x_n & & \text{repeated } h_n \text{ times.}
\end{array}\right.$$

The feasible region is the Klee–Minty n-cube and is denoted by $\mathcal{C} \subset \mathbb{R}^n$. Denote $d := (d_1, \ldots, d_n) \in \mathbb{R}^n_+$ – the vector containing the distances to the redundant constraints from $\mathcal{C}$, $h := (h_1, \ldots, h_n) \in \mathbb{N}^n$ – the vector containing the number of the redundant constraints.

By analogy with the unit cube $[0,1]^n$, we denote the vertices of $\mathcal{C}$ as follows. For $S \subset \{1,\dots,n\}$, a vertex v^S of $\mathcal{C}$ satisfies

$$v_1^S = \begin{cases} 1 & \text{if } 1 \in S \\ 0 & \text{otherwise} \end{cases}$$
$$v_k^S = \begin{cases} 1 - \varepsilon v_{k-1}^S & \text{if } k \in S \\ \varepsilon v_{k-1}^S & \text{otherwise} \end{cases} \qquad k = 2,\dots,n.$$

Define δ-neighborhood $\mathcal{N}_\delta(v^S)$ of a vertex v^S, with the convention $x_0 = 0$, by

$$\mathcal{N}_\delta(v^S) := \left\{ x \in \mathcal{C} : \begin{cases} 1 - x_k - \varepsilon x_{k-1} \le \varepsilon^{k-1}\delta & \text{if } k \in S \\ x_k - \varepsilon x_{k-1} \le \varepsilon^{k-1}\delta & \text{otherwise} \end{cases} \quad k = 1,\dots,n \right\}.$$

Remark 7.1. Observe that $\forall S \subseteq \{1,\dots,n\}$ for $\mathcal{N}_\delta(v^S)$ to be pairwise-disjoint it suffices $\varepsilon + \delta < 1/2$: given $\varepsilon, \delta > 0$, the shortest amongst all n coordinates' distance between the neighborhoods, equal to $(1 - 2\varepsilon - 2\varepsilon\delta)$, is attained along the second coordinate and must be positive, which is readily implied.

For brevity of the notation we introduce slack variables corresponding to the constraints in the problem above as follows:

$$\begin{aligned}
s_1 &= x_1 & & \\
s_k &= x_k - \varepsilon x_{k-1} & & k = 2,\dots,n \\
\bar{s}_1 &= 1 - x_1 & & \\
\bar{s}_k &= 1 - \varepsilon x_{k-1} - x_k & & k = 2,\dots,n \\
\tilde{s}_1 &= d_1 + x_1 & & \\
\tilde{s}_k &= d_k + (x_n - \varepsilon x_{n-1}) & & k = 2,\dots,n.
\end{aligned}$$

Recall that the analytic center χ corresponds to the unique maximizer

$$\arg\max_x \sum_{i=1}^n (\ln s_i + \ln \bar{s}_i + h_i \ln \tilde{s}_i).$$

Also, recall that the primal central path $\mathcal{P}$ can be characterized as the closure of the set of maximizers

$$\left\{ x \in \mathbb{R}^n : x = \arg\max_{x : x_n = \alpha} \sum_{i=1}^n (\ln s_i + \ln \bar{s}_i + h_i \ln \tilde{s}_i), \text{ for some } \alpha \in (0, \chi_n) \right\}.$$

Therefore, setting to 0 the derivatives of $\sum_{i=1}^n (\ln s_i + \ln \bar{s}_i + h_i \ln \tilde{s}_i)$ with respect to x_n,

$$\frac{1}{s_n} - \frac{1}{\bar{s}_n} + \frac{h_n}{\tilde{s}_n} = 0, \tag{7.1}$$

and with respect to x_k,

$$\frac{1}{s_k} - \frac{\varepsilon}{s_{k+1}} - \frac{1}{\bar{s}_k} - \frac{\varepsilon}{\bar{s}_{k+1}} + \frac{h_k}{\tilde{s}_k} - \frac{\varepsilon h_{k+1}}{\tilde{s}_{k+1}} = 0, \qquad k = 1, \ldots, n-1, \tag{7.2}$$

combined give us necessary and sufficient conditions for $x = \chi$. Furthermore, (7.2) combined with $x_n = \alpha \in (0, \chi_n)$ gives us necessary and sufficient conditions for $x \in \mathcal{P} \setminus (\{\mathbf{0}\} \cup \{\chi\})$ where $\mathbf{0} \in \mathbb{R}^n$ denotes the origin.

Given $\varepsilon, \delta > 0$, the sufficient conditions for $h = h(d, \varepsilon, \delta)$ to guarantee that the central path $\mathcal{P}$ visits the (disjoint) δ-neighborhoods of each vertex of $\mathcal{C}$ may be summarized in the following proposition. We write $\mathbf{1}$ for the vector of all ones in $\mathbb{R}^n$.

Proposition 7.1. *Fix $\varepsilon, \delta > 0$. Denote for $k = 2, \ldots, n$*

$$\mathcal{I}_\delta^k := \{x \in \mathcal{C} : \bar{s}_k \geq \varepsilon^{k-1}\delta, s_k \geq \varepsilon^{k-1}\delta\}$$

and

$$\mathcal{B}_\delta^k := \{x \in \mathcal{C} : \bar{s}_{k-1} \leq \varepsilon^{k-2}\delta, s_{k-2} \leq \varepsilon^{k-3}\delta, \ldots, s_1 \leq \delta\}.$$

If $h = h(d, \varepsilon, \delta) \in \mathbb{N}^n$ satisfies

$$Ah \geq \frac{3}{\delta}\mathbf{1} \tag{7.3}$$

and

$$\frac{h_k}{d_k + 1}\varepsilon^{k-1} \geq \frac{h_{k+1}}{d_{k+1}}\varepsilon^k + \frac{3}{\delta}, \qquad k = 1, \ldots, n-1, \tag{7.4}$$

where

$$A = \begin{pmatrix} \frac{1}{d_1+1} & \frac{-\varepsilon}{d_2} & 0 & 0 & \cdots & 0 & 0 \\ \frac{-1}{d_1} & \frac{2\varepsilon}{d_2+1} & \frac{-\varepsilon^2}{d_3} & 0 & \cdots & 0 & 0 \\ \vdots & \vdots & \ddots & \ddots & \vdots & \vdots & \vdots \\ \frac{-1}{d_1} & 0 & \cdots & \frac{2\varepsilon^{k-1}}{d_k+1} & \frac{-\varepsilon^k}{d_{k+1}} & \cdots & 0 \\ \vdots & \vdots & \vdots & \vdots & \ddots & \ddots & \vdots \\ \frac{-1}{d_1} & 0 & 0 & 0 & \cdots & \frac{2\varepsilon^{n-2}}{d_{n-1}+1} & \frac{-\varepsilon^{n-1}}{d_n} \\ \frac{-1}{d_1} & 0 & 0 & 0 & \cdots & 0 & \frac{2\varepsilon^{n-1}}{d_n+1} \end{pmatrix}$$

then

$$\mathcal{I}_\delta^k \cap \mathcal{P} \subset \mathcal{B}_\delta^k.$$

Proof. Fix $k \geq 2$ and let $x \in \mathcal{I}_\delta^k \cap \mathcal{P}$.

Let $j \leq k-2$. Summing up all of the ith equations of (7.2) over $i = j, \ldots, (k-2)$, each multiplied by ε^{i-1}, and then subtracting the $(k-1)$st equation multiplied by ε^{k-2}, we have

$$-\frac{2h_{k-1}\varepsilon^{k-2}}{\tilde{s}_{k-1}}+\frac{h_j\varepsilon^{j-1}}{\tilde{s}_j}+\frac{h_k\varepsilon^{k-1}}{\tilde{s}_k}+\frac{\varepsilon^{j-1}}{s_j}+\frac{\varepsilon^{k-1}}{s_k}+\frac{\varepsilon^{k-1}}{\bar{s}_k}$$
$$=\frac{2\varepsilon^{k-2}}{s_{k-1}}+\frac{\varepsilon^{j-1}}{\bar{s}_j}+2\sum_{i=j}^{k-3}\frac{\varepsilon^i}{\bar{s}_{i+1}}.$$

Because $\tilde{s}_{k-1} < d_{k-1}+1$, $\tilde{s}_k > d_k$, $\tilde{s}_j > d_j$, and $s_k \geq \varepsilon^{k-1}\delta$, $\bar{s}_k \geq \varepsilon^{k-1}\delta$ as $x \in \mathcal{I}_\delta^k$, from the above we get

$$\frac{2h_{k-1}\varepsilon^{k-2}}{d_{k-1}+1}-\frac{h_j\varepsilon^{j-1}}{d_j}-\frac{h_k\varepsilon^{k-1}}{d_k}\leq\frac{\varepsilon^{j-1}}{s_j}+\frac{2}{\delta}.$$

From (7.4) it follows that $(h_1/d_1) \geq \left(h_j\varepsilon^{j-1}/d_j\right)$, thus we can write

$$-\frac{h_1}{d_1}+\frac{2h_{k-1}\varepsilon^{k-2}}{d_{k-1}+1}-\frac{h_k\varepsilon^{k-1}}{d_k}\leq\frac{\varepsilon^{j-1}}{s_j}+\frac{2}{\delta};$$

that is, as $(3/\delta) \leq -(h_1/d_1) + \left(\left(2h_{k-1}\varepsilon^{k-2}\right)/(d_{k-1}+1)\right) - \left(h_k\varepsilon^{k-1}/d_k\right)$ by (7.3), we have

$$s_j^n \leq \varepsilon^{j-1}\delta, \qquad \forall j \leq k-2.$$

In turn, the $(k-1)$ st equation of (7.2),

$$\frac{h_{k-1}\varepsilon^{k-2}}{\tilde{s}_{k-1}}-\frac{h_k\varepsilon^{k-1}}{\tilde{s}_k}=\frac{\varepsilon^{k-2}}{\bar{s}_{k-1}}+\frac{\varepsilon^{k-1}}{s_k}+\frac{\varepsilon^{k-1}}{\bar{s}_k}-\frac{\varepsilon^{k-2}}{s_{k-1}}$$

implies

$$\frac{h_{k-1}\varepsilon^{k-2}}{d_{k-1}+1}-\frac{h_k\varepsilon^{k-1}}{d_k}\leq\frac{\varepsilon^{k-2}}{\bar{s}_{k-1}}+\frac{2}{\delta}$$

and since $(3/\delta) \leq \left(\left(h_{k-1}\varepsilon^{k-2}\right)/(d_{k-1}+1)\right) - \left(h_k\varepsilon^{k-1}/d_k\right)$ by (7.4), we have $\bar{s}_{k-1} \leq \varepsilon^{k-2}\delta$. □

Proposition 7.2. *Fix* $\varepsilon, \delta > 0$. *If* $h \in \mathbb{N}^n$ *satisfies* (7.3) *and* (7.4), *then* $\chi \in \mathcal{N}_\delta(v^{\{n\}})$.

Proof. Summing up all of the ith equations of (7.2) over $i = k, \ldots, (n-1)$, each multiplied by ε^{i-1}, and then subtracting (7.1) multiplied by ε^{n-1}, we have

$$\frac{\varepsilon^{k-1}}{s_k}-\frac{\varepsilon^{k-1}}{\bar{s}_k}+\frac{h_k\varepsilon^{k-1}}{\tilde{s}_k}-\frac{2\varepsilon^{n-1}}{s_n}-2\sum_{i=k}^{n-2}\frac{\varepsilon^i}{\bar{s}_{i+1}}-\frac{2h_n\varepsilon^{n-1}}{\tilde{s}_n}=0$$

implying

$$\frac{2h_n\varepsilon^{n-1}}{\tilde{s}_n}-\frac{h_k\varepsilon^{k-1}}{\tilde{s}_k}\leq\frac{\varepsilon^{k-1}}{s_k}.$$

Because $\tilde{s}_n \leq d_n + 1$, $\tilde{s}_k \geq d_k$ and by (7.4), $(h_1/d_1) \geq \left(h_k \varepsilon^{k-1}/d_k\right)$, from the above we get

$$\frac{2h_n\varepsilon^{n-1}}{d_n+1} - \frac{h_1}{d_1} \leq \frac{\varepsilon^{k-1}}{s_k},$$

combined with $\left(\left(2h_n\varepsilon^{n-1}\right)/\left(d_n+1\right)\right) - (h_1/d_1) \geq (3/\delta)$ (from (7.3)) this leads to

$$s_k \leq \frac{\varepsilon^{k-1}}{3}\delta \qquad k = 1, \ldots, n-1.$$

In turn, (7.1) implies $\left(h_n\varepsilon^{n-1}/\tilde{s}_n\right) \leq \left(\varepsilon^{n-1}/\bar{s}_n\right)$. And because $\tilde{s}_n < d_n + 1$ and by (7.4), $\left(h_n\varepsilon^{n-1}/\left(d_n+1\right)\right) \geq (3/\delta)$, we have $(3/\delta) \leq \left(\varepsilon^{n-1}/\bar{s}_n\right)$; that is, $\bar{s}_n \leq \left(\varepsilon^{n-1}/3\right)\delta$. □

Corollary 7.1. *Fix $\varepsilon, \delta > 0$ such that $\varepsilon + \delta < 1/2$. If $h \in \mathbb{N}^n$ satisfies (7.3) and (7.4), then the central path $\mathcal{P}$ intersects the disjoint δ-neighborhoods of all the vertices of $\mathcal{C}$. Moreover, $\mathcal{P}$ is confined to a polyhedral tube defined by $\mathcal{T}_\delta := \bigcup_{k=1}^n \left(\left(\bigcap_{j=k+1}^n (\mathcal{I}_\delta^j)^c\right) \cap \mathcal{B}_\delta^k\right)$ with the convention $\mathcal{B}_\delta^1 = \mathcal{C}$.*

Remark 7.2. Observe that $\mathcal{T}_0$ is the sequence of connected edges of $\mathcal{C}$ starting from $v^{\{n\}}$ and terminating at $v^{\varnothing}$, and is precisely the path followed by the simplex method on the original Klee–Minty problem as it pivots along the edges of $\mathcal{C}$.

For simplicity of the notation we write $\mathcal{T}$ instead of $\mathcal{T}_\delta$ when the choice of δ is clear. For a fixed δ, we define a turn of $\mathcal{T}$ adjacent to a vertex v^S, or corresponding to $\mathcal{N}_\delta(v^S)$ if the δ-neighborhoods are disjoint because in the latter case $\mathcal{N}_\delta(v^S)$ determines v^S uniquely, to be the angle between the two edges of $\mathcal{C}$ that belong to $\mathcal{T}_0$ and connect at this vertex.

Intuitively, if a smooth curve is confined to a narrow tube that makes a sharp turn, then the curve itself must at least make a similar turn and thus have a total curvature bounded away from zero. It might be worthwhile to substantiate this intuition with a proposition.

Proposition 7.3. *Let $\Psi : [0, T] \to \mathbb{R}^2$ be C^2, parameterized by its arc length t, such that $\Psi([0,T]) \subset \{(x,y) : 0 \leq x \leq a+b, 0 \leq y \leq b\} \cup \{(x,y) : a \leq x \leq a+b, -a \leq y \leq b\}$ and $\Psi(0) \in \{0\} \times [0,b]$, $\Psi(T) \in [a, a+b] \times \{-a\}$. Then the total curvature K of Ψ satisfies $K \geq \arcsin\left(1 - 2b^2/a^2\right)$.*

Proof. By the mean-value theorem, for any τ such that $\Psi_1(\tau) = a$ we have $\tau \geq a$; recall that $\|\dot{\Psi}\| = 1$. Thus, by the same theorem, $\exists t_1$ such that $|\Psi_2(t_1)| \leq b/a$. Similarly, $\exists t_2$ such that $|\Psi_1(t_2)| \leq b/a$. Now map the values of the derivative of Ψ at t_1 and t_2 onto a sphere and recall that the total curvature K between these two points corresponds to the length of a connecting curve on the sphere, thus bounded below by the length of the geodesic (which in this case is the same as the angular distance).

A simple calculation completes the proof. □

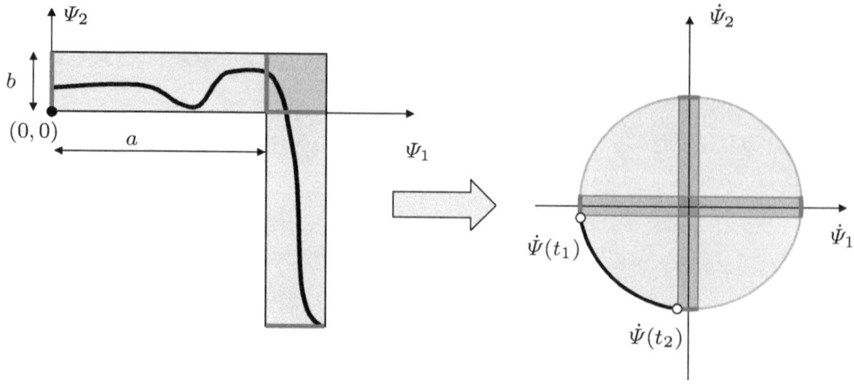

Fig. 7.1 Total curvature and geodesics.

Remark 7.3. Note that if $b/a \to 0$, then the corresponding lower bound on the total curvature K approaches $\pi/2$.

Next we construct a simple bound on the total curvature of $\mathcal{P}$ by picking suitable d, ε, and finally δ small enough, together with h, that results in a "narrow" polyhedral tube $\mathcal{T}$.

For $X \subseteq \mathbb{R}^n$ denote its orthogonal projection onto a linear subspace spanned by a subset $S \subseteq \{1, \ldots, n\}$ of coordinates, with coordinates corresponding to S^c suppressed, by X_S. For $x, z \in \mathbb{R}^n$ we denote (x, z) the straight line segment connecting the point x and z.

Corollary 7.2. *Fix $n \geq 2$. If $d_i = (n-1)2^{n-i+2}$, $i = 1, \ldots, n$,*

$$\varepsilon = \frac{n-1}{2n},$$

$$\delta = \frac{1}{32n^2}\left(\frac{4}{5}\right)^{n-2},$$

and h satisfies

$$h = \left\lfloor \left(1 + \frac{\delta}{3} \max_i \sum_{j=1}^{n} |a_{ij}|\right) \frac{3}{\delta}\widetilde{h} \right\rfloor,$$

where $A\widetilde{h} = \mathbf{1}$, then the total curvature of the central path $\mathcal{P}$ satisfies

$$K \geq \frac{1}{2n}\left(\frac{8}{5}\right)^{n-2}.$$

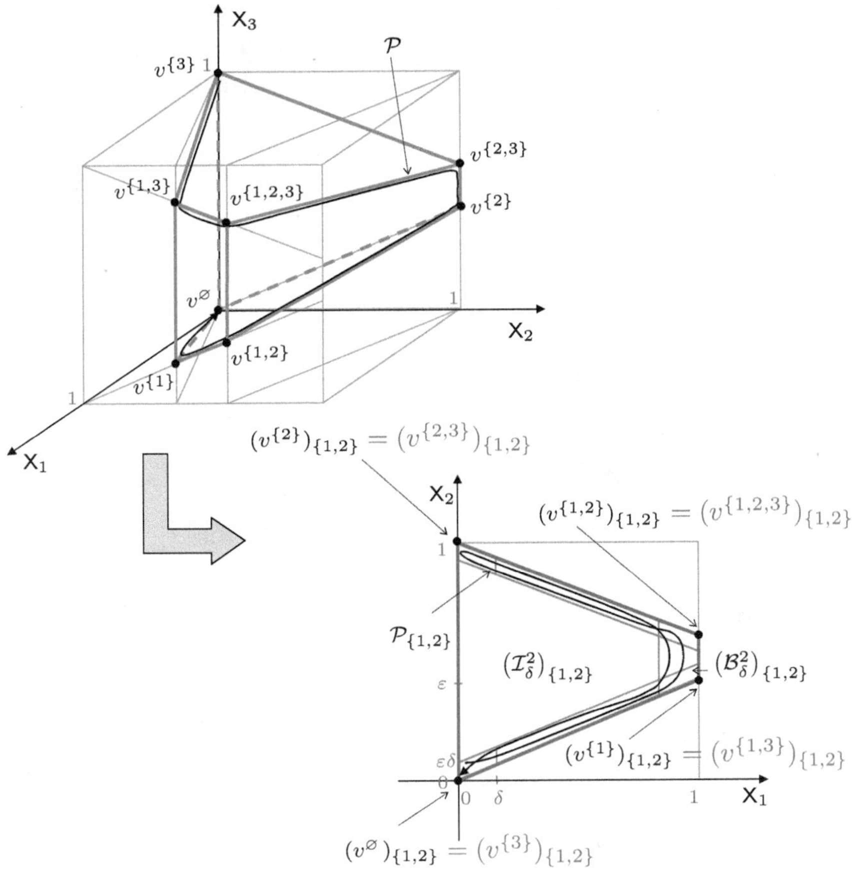

Fig. 7.2 Planar projection of the central path for $n = 3$.

Proof. That $\varepsilon, \delta, h = h(d, \varepsilon, \delta)$ above satisfies the conditions of Corollary 7.1 and thus $\mathcal{P}$ is confined to the polyhedral tube $\mathcal{T}$ is established in Section 7.5.

Instead of analyzing $\mathcal{P} \in \mathbb{R}^n$ directly we derive the lower bound on the total curvature of $\mathcal{P}$ based on its planar projection $\mathcal{P}_{\{1,2\}}$.

From $\mathcal{I}_\delta^k \cap \mathcal{P} \subset \mathcal{B}_\delta^k$, $k = 2, \ldots, n$, it follows that $\mathcal{P}_{\{1,2\}}$ will traverse the two-dimensional Klee–Minty cube $\mathcal{C}_{\{1,2\}}$ at least 2^{n-2} times, every time originating in either $\mathcal{N}_\delta(v^{\varnothing})_{\{1,2\}}$ or $\mathcal{N}_\delta(v^{\{2\}})_{\{1,2\}}$ and terminating in the other neighborhood, while confined to the polyhedral tube $\mathcal{T}_{\{1,2\}} = (\{s_2 \le \varepsilon\delta\} \cup \{\bar{s}_1 \le \delta\} \cup \{\bar{s}_2 \le \varepsilon\delta\}) \cap \mathcal{C}_{\{1,2\}}$. Thus, $\mathcal{P}_{\{1,2\}}$ will make at least 2^{n-1} "sharp turns", each corresponding to a turn in $\mathcal{N}_\delta(v^{\{1,2\}})_{\{1,2\}}$ or $\mathcal{N}_\delta(v^{\{1\}})_{\{1,2\}}$.

In order to understand how the turns of $\mathcal{P}_{\{1,2\}}$ contribute to the total curvature of $\mathcal{P}$ we need the following lemma.

Lemma 7.1. *Let* $\widehat{u}, \widehat{v} \in \mathbb{R}^3$ *and* $u = (\widehat{u}_{\{1,2\}}, 0), v = (\widehat{v}_{\{1,2\}}, 0)$. *If the angle*

$$\epsilon := \pi - \arccos \arg \min_{\substack{\widehat{w} \in \mathrm{span}\{\widehat{u},\widehat{v}\}, \\ w \in \mathrm{span}\{u,v\}, \\ \|\widehat{w}\|=\|w\|=1}} \widehat{w}^T w$$

between the hyperplane spanned by $\widehat{u}, \widehat{v}$ *and the hyperplane spanned by* u, v *does not exceed* $\arcsin \varepsilon$, *then the angle* $\widehat{\alpha}$ *between* $\widehat{u}$ *and* $\widehat{v}$ *satisfies*

$$\frac{\cos\alpha - \varepsilon^2\left(\frac{1-\cos\alpha}{2}\right)}{1+\varepsilon^2\left(\frac{1-\cos\alpha}{2}\right)} \le \cos\widehat{\alpha} \le \frac{\cos\alpha + \varepsilon^2\left(\frac{1+\cos\alpha}{2}\right)}{1+\varepsilon^2\left(\frac{1+\cos\alpha}{2}\right)}$$

where α *is the angle between* u *and* v.

Proof. Without loss of generality we may assume $\|u\| = \|v\| = 1$ with

$$u_1 = \sin\tfrac{\alpha}{2} = \sqrt{\tfrac{1-\cos\alpha}{2}},\ v_1 = -\sin\tfrac{\alpha}{2} = -\sqrt{\tfrac{1-\cos\alpha}{2}},$$
$$u_2 = \cos\tfrac{\alpha}{2} = \sqrt{\tfrac{1+\cos\alpha}{2}},\ v_2 = \cos\tfrac{\alpha}{2} = \sqrt{\tfrac{1+\cos\alpha}{2}},$$

and, assuming that the angle ϵ is precisely $\arcsin\varepsilon$, parameterize $\mathrm{span}\{\widehat{u},\widehat{v}\}$ by $\mathrm{span}\{u, v\}$ and $z = (z_1, z_2, 0)$ such that $\|z\| = 1$, writing $x \in \mathrm{span}\{\widehat{u},\widehat{v}\}$ as $x = (x_1, x_2, x_{\{1,2\}}^T z_{\{1,2\}}\varepsilon)$.

Introducing β such that $z_1 = \cos\beta$ and $z_2 = \sin\beta$ we have

$$\widehat{u} = \left(\sqrt{\frac{1-\cos\alpha}{2}}, \sqrt{\frac{1+\cos\alpha}{2}}, \varepsilon\cos\left(\beta - \frac{\pi}{2} + \frac{\alpha}{2}\right)\right),$$

$$\widehat{v} = \left(-\sqrt{\frac{1-\cos\alpha}{2}}, \sqrt{\frac{1+\cos\alpha}{2}}, \varepsilon\cos\left(\beta - \frac{\pi}{2} + \frac{\alpha}{2}\right)\right),$$

and, therefore,

$$\cos\widehat{\alpha} = \frac{\widehat{u}^T\widehat{v}}{\|\widehat{u}\|\|\widehat{v}\|} = \frac{\cos\alpha + \varepsilon^2\cos\left(\beta - \frac{\pi}{2} + \frac{\alpha}{2}\right)\cos\left(\beta - \frac{\pi}{2} - \frac{\alpha}{2}\right)}{\sqrt{1+\varepsilon^2\cos^2\left(\beta - \frac{\pi}{2} + \frac{\alpha}{2}\right)}\sqrt{1+\varepsilon^2\cos^2\left(\beta - \frac{\pi}{2} - \frac{\alpha}{2}\right)}}.$$

Denoting $\gamma := \beta - (\pi/2)$ and differentiating the above with respect to γ we get

$$(\cos\widehat{\alpha})'_\gamma = \frac{(1+\varepsilon^2)(-32\varepsilon^2\sin 2\gamma + 16\varepsilon^2\sin(2\gamma+2\alpha) + 16\varepsilon^2\sin(2\gamma-2\alpha))}{D},$$

where

$$D = \left(16 + 8\varepsilon^2 \cos(2\gamma - \alpha) + 16\varepsilon^2 + 8\varepsilon^2 \cos(2\gamma + \alpha) + 2\varepsilon^4 \cos 2\alpha \right.$$
$$\left. + 2\varepsilon^4 \cos 4\gamma + 4\varepsilon^4 \cos(2\gamma + \alpha) + 4\varepsilon^4 \cos(2\gamma - \alpha) + 4\varepsilon^4\right)^{3/2}.$$

Setting the derivative to 0 and simplifying the numerator we obtain the necessary condition for the extremum of $\cos\widehat{\alpha}$,

$$32\varepsilon^2(1+\varepsilon^2)\sin 2\gamma(\cos 2\alpha - 1) = 0.$$

That is, $\gamma = k\,(\pi/2)$ for $k = 0, \pm 1, \pm 2$, and so on. In particular, it follows that the minimum of $\cos\widehat{\alpha}$ is attained at $\beta_{\min} = 0$ and the maximum is attained at $\beta_{\max} = \pi/2$. The bounds are obtained by further substituting the critical values of β into the expression for $\cos\widehat{\alpha}$ and observing the monotonicity with respect to ε. □

Although the full-dimensional tube $\mathcal{T}$ might make quite wide turns, the projected tube $\mathcal{T}_{\{1,2\}}$ is bound to make the same sharp turn equal to $\left((\pi/2) + \arcsin\left(\varepsilon/\sqrt{1+\varepsilon^2}\right)\right)$ each time $\mathcal{T}$ passes through the δ-neighborhood of a vertex v^S, $1 \in S$ (e.g., consider the turn adjacent to $v^{\{1,3\}}$ for $n = 3$).

For a moment, equip $\mathcal{C}$ and $\mathcal{T}$ with a superscript $\bar{n}$ to indicate the dimension of the cube, that is, the largest number of linearly independent vectors in $\operatorname{span}(\{v^S : v^S \in \mathcal{C}^{\bar{n}}\})$. Recalling the $\mathcal{C}^n$ defining constraints, namely $\varepsilon x_{n-1} \le x_n \le 1 - \varepsilon x_{n-1}$, we note that by construction of the Klee–Minty cube, whenever we increase the dimension from $\bar{n}$ to $\bar{n}+1$, $\mathcal{C}^{\bar{n}}$ is affinely transformed into "top" and "bottom" $\bar{n}$-dimensional faces $F^{\bar{n}+1}_{\text{top}}$ and $F^{\bar{n}+1}_{\text{bottom}}$ of $\mathcal{C}^{\bar{n}+1}$; that is,

$$F^{\bar{n}+1}_{\text{top}} = \begin{pmatrix} I \\ (0,\ldots,0,-\varepsilon) \end{pmatrix} \mathcal{C}^{\bar{n}} + \begin{pmatrix} 0 \\ \vdots \\ 0 \\ 1 \end{pmatrix},$$

$$F^{\bar{n}+1}_{\text{bottom}} = \begin{pmatrix} I \\ (0,\ldots,0,\varepsilon) \end{pmatrix} \mathcal{C}^{\bar{n}},$$

where I is the identity $\bar{n}\times\bar{n}$ matrix, and $\mathcal{C}^{\bar{n}+1}$ is the convex hull of $F^{\bar{n}+1}_{\text{top}}$ and $F^{\bar{n}+1}_{\text{bottom}}$. Consequently, any two-dimensional space spanned by two connected edges of $\mathcal{C}^{\bar{n}+1}$ from $\mathcal{T}_0^{\bar{n}+1} \cap F^{\bar{n}+1}_{\text{top}}$ or $\mathcal{T}_0^{\bar{n}+1} \cap F^{\bar{n}+1}_{\text{bottom}}$ is obtained by tilting the two-dimensional space spanned by the two corresponding edges of $\mathcal{C}^{\bar{n}}$ from $\mathcal{T}_0^{\bar{n}}$, lifted to $\mathbb{R}^{\bar{n}+1}$ by setting the $(\bar{n}+1)$st coordinate to zero, by an angle not exceeding $\arcsin\left(\varepsilon/\sqrt{1+\varepsilon^2}\right)$, and moreover, not exceeding $\arcsin\varepsilon$. Therefore, we are in position to apply Lemma 7.1 to bound how fast the cosine of a turn α^S of $\mathcal{T}^n$ adjacent to any $v^S \in \mathcal{C}^n$ with $1 \in S$ may approach its two boundary values of 1 or -1 by induction on the dimension n.

Fixing $n = 3$, $S \subseteq \{1,2,3\}$ such that $1 \in S$, adding and subtracting 1 to $\cos\alpha^S$ we get

$$1+\cos\alpha^S \geq \frac{1+\cos\alpha^{S_{\{1,2\}}}}{1+\varepsilon^2\left(\frac{1-\cos\alpha^{S_{\{1,2\}}}}{2}\right)}$$

and

$$1-\cos\alpha^S \geq \frac{1-\cos\alpha^{S_{\{1,2\}}}}{1+\varepsilon^2\left(\frac{1+\cos\alpha^{S_{\{1,2\}}}}{2}\right)}.$$

Furthermore, for any $n \geq 3$ and v^S with $1 \in S$ we can write

$$1+\cos\alpha^S \geq \frac{1+\cos\alpha^{S_{\{1,2\}}}}{(1+\varepsilon^2)^{n-2}} \geq \frac{1-\varepsilon}{(1+\varepsilon^2)^{n-2}},$$
$$1-\cos\alpha^S \geq \frac{1-\cos\alpha^{S_{\{1,2\}}}}{(1+\varepsilon^2)^{n-2}} \geq \frac{1+(2\varepsilon/\sqrt{5})}{(1+\varepsilon^2)^{n-2}},$$

recalling $-2\varepsilon/\sqrt{5} \geq \cos\alpha^{S_{\{1,2\}}} = -\varepsilon/\sqrt{1+\varepsilon^2} \geq -\varepsilon$ because $\varepsilon \geq 2$.

Observe that by construction of a polyhedral tube $\mathcal{T}$, a single linearly connected component of $\mathcal{T} \setminus \left(\bigcup_{S\subseteq\{1,\ldots,n\}} N_\delta(v^S)\right)$ may be uniquely identified with an edge (v^R, v^S), $R, S \subseteq \{1,\ldots,n\}$, of $\mathcal{C}$ from $\mathcal{T}_0$ by having a nonempty intersection with this component and thus we denote such a component by $L_{(v^R,v^S)}$ and refer to it as a section of $\mathcal{T}$ corresponding to (v^R, v^S). Moreover, recalling the definition of $N_\delta(v^S)$ and $\mathcal{T}$, and noting that $\sqrt{\delta^2+(\varepsilon\delta)^2+(\varepsilon\delta)^3+\cdots} \leq \delta+\varepsilon\delta+\cdots \leq 2\delta$ because $\varepsilon \leq 1/2$, we get that within a given section of a tube $L_{(v^R,v^S)}$ the Euclidean distance from $\forall x \in L_{(v^R,v^S)}$ to the compact closure of $(v^R, v^S) \cap L_{(v^R,v^S)}$ is bounded from above by 2δ.

Let us consider what happens to the central path in the proximity of a vertex $v^S \in \mathcal{C}$ such that $1 \in S$. We do so by manufacturing a surrogate for a part of $\mathcal{T}$ that is easier to analyze.

Fix $v^S \in \mathcal{C}$ with $1 \in S$ and denote the two adjacent vertices to which v^S is connected by the two edges from $\mathcal{T}_0$ by v^R and v^Q. Without loss of generality we may assume that

$$v^R_{\{1,2\}} = (0,1),$$
$$v^S_{\{1,2\}} = (1,1-\varepsilon),$$
$$v^Q_{\{1,2\}} = (1,\varepsilon),$$

and $v^R_n > v^S_n > v^Q_n$, so that the central path $\mathcal{P}$ enters the part of the polyhedral tube $\mathcal{T}$ sectioned between these three vertices via $N_\delta(v^R)$ and exits via $N_\delta(v^Q)$.

Define four auxiliary points $\overline{x}, \overline{z} \in (v^R, v^S)$ and $\underline{x}, \underline{z} \in (v^S, v^Q)$ satisfying

$$\overline{x}_{\{1,2\}} = (1-3\delta, 1-\varepsilon+3\varepsilon\delta) + \tfrac{1/2-\varepsilon-3\delta}{\sqrt{1+\varepsilon^2}}(-1,\varepsilon),$$
$$\overline{z}_{\{1,2\}} = (1-3\delta, 1-\varepsilon+3\varepsilon\delta),$$
$$\underline{x}_{\{1,2\}} = (1, 1-\varepsilon-3\delta),$$
$$\underline{z}_{\{1,2\}} = (1, 1/2).$$

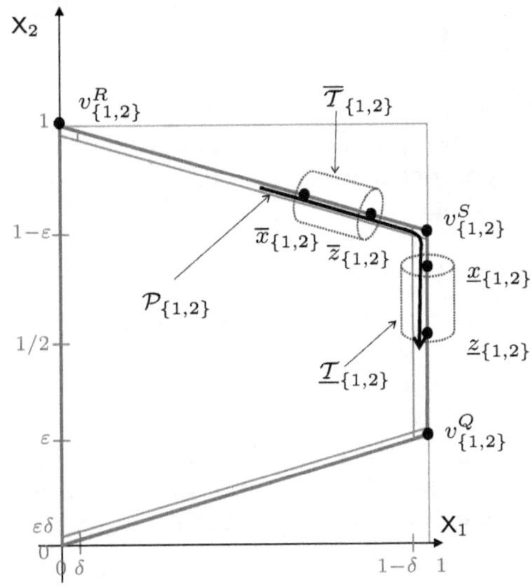

Fig. 7.3 Schematic drawing for the cylindrical tube segments.

Because the distance from any point to the (part of the) identifying edge of $L_{(v^R,v^S)}$ or $L_{(v^S,v^Q)}$ is no greater than 2δ and because $(\cdot)_{\{1,2\}}$ corresponds to the orthogonal projection from $\mathbb{R}^n$ onto its first two coordinates, we can define two cylindrical tube segments:

$$\begin{aligned}\overline{\mathcal{T}} := &\{x \in \mathbb{R}^n : \min_{z\in(\overline{x},\overline{z})} \|x - z\| \le 2\delta\}\\ &\cap \{x \in \mathbb{R}^n : (\overline{x} - \overline{z})^T x \le (\overline{x} - \overline{z})^T \overline{x}\}\\ &\cap \{x \in \mathbb{R}^n : (\overline{z} - \overline{x})^T x \le (\overline{z} - \overline{x})^T \overline{z}\}\end{aligned}$$

and

$$\begin{aligned}\underline{\mathcal{T}} := &\{x \in \mathbb{R}^n : \min_{z\in(\underline{x},\underline{z})} \|x - z\| \le 2\delta\}\\ &\cap \{x \in \mathbb{R}^n : (\underline{x} - \underline{z})^T x \le (\underline{x} - \underline{z})^T \underline{x}\}\\ &\cap \{x \in \mathbb{R}^n : (\underline{z} - \underline{x})^T x \le (\underline{z} - \underline{x})^T \underline{z}\}\end{aligned}$$

such that

$$\overline{\mathcal{T}} \supset L_{(v^R,v^S)} \cap \{x \in \mathbb{R}^n : (\overline{x} - \overline{z})^T x \le (\overline{x} - \overline{z})^T \overline{x}, (\overline{z} - \overline{x})^T x \le (\overline{z} - \overline{x})^T \overline{z}\},$$

$$\underline{\mathcal{T}} \supset L_{(v^S,v^Q)} \cap \{x \in \mathbb{R}^n : (\underline{x} - \underline{z})^T x \le (\underline{x} - \underline{z})^T \underline{x}, (\underline{z} - \underline{x})^T x \le (\underline{z} - \underline{x})^T \underline{z}\},$$

and

$$\overline{\mathcal{T}} \cap \left(N_\delta(v^R) \cup N_\delta(v^S)\right) = \underline{\mathcal{T}} \cap \left(N_\delta(v^S) \cup N_\delta(v^Q)\right) = \varnothing.$$

Therefore, $\mathcal{P}$ will traverse $\overline{\mathcal{T}}$ and $\underline{\mathcal{T}}$, first entering $\overline{\mathcal{T}}$ through its face corresponding to $(\overline{x} - \overline{z})^T x = (\overline{x} - \overline{z})^T \overline{x}$ and exiting through the face corresponding to $(\overline{z}-\overline{x})^T x = (\overline{z}-\overline{z})^T \overline{z}$, and then entering $\underline{\mathcal{T}}$ at a point with $(\underline{x}-\underline{z})^T x = (\underline{x}-\underline{z})^T \underline{x}$ and exiting through a point with $(\underline{z}-\underline{x})^T x = (\underline{z}-\underline{x})^T \underline{z}$.

Now we choose a new system of orthogonal coordinates in $\mathbb{R}^n$ that allows us apply the argument similar to that of Proposition 7.3 as follows. Let the first two coordinates correspond to the linear subspace spanned by $(\overline{x}, \overline{z})$ and $(\underline{x}, \underline{z})$; align the second coordinate axis with the vector $(\underline{z}, \underline{x})$, so that the vector $(\overline{x}, \overline{z})$ forms the same angle equal to α^S with the second coordinate axis as with $(\underline{x}, \underline{z})$. Choose the rest $(n-2)$ coordinates so that they form an orthogonal basis for $\mathbb{R}^n$.

Consider parameterization of $\mathcal{P}$ by its arc length, $\mathcal{P}_{\text{arc}}$. Because the shortest distance between the two parallel faces of $\overline{\mathcal{T}}$ that correspond to $\{x \in \mathbb{R}^n : (\overline{x}-\overline{z})^T x = (\overline{x}-\overline{z})^T \overline{x}\}$ and $\{x \in \mathbb{R}^n : (\overline{z}-\overline{x})^T x = (\overline{z}-\overline{x})^T \overline{z}\}$ is equal to $\|(\overline{x}, \overline{z})\| = 1/2 - \varepsilon - 3\delta$, by the mean-value theorem it takes at least $(1/2-\varepsilon-3\delta)$ change of the arc length parameter for $\mathcal{P}_{\text{arc}}$ to traverse $\overline{\mathcal{T}}$. Noting that while traversing the tube $\overline{\mathcal{T}}$ the second coordinate of $\mathcal{P}_{\text{arc}}$ might change at most by $2 \cdot |2\delta \sin\alpha^S| + |(1/2-\varepsilon-3\delta)\cos\alpha^S|$, by the same theorem we deduce that $\exists t_1$ such that

$$
\begin{aligned}
\left|\left(\dot{\mathcal{P}}_{\text{arc}}(t_1)\right)_2\right| &\le \frac{2|2\delta \sin\alpha^S| + |(1/2-\varepsilon-3\delta)\cos\alpha^S|}{1/2-\varepsilon-3\delta} \\
&\le |\cos\alpha^S| + \frac{4\delta}{1/2-\varepsilon-3\delta}.
\end{aligned}
$$

Analogously, considering $\underline{\mathcal{T}}$ along the ith coordinate with $i \neq 2$ we conclude that $\forall i \neq 2$, $\exists t_i$ such that

$$
\left|\left(\dot{\mathcal{P}}_{\text{arc}}(t_i)\right)_i\right| \le \frac{4\delta}{1/2-\varepsilon-3\delta}.
$$

We use the points $t_1, t_2, \ldots, t_n$ to compute a lower bound on the total curvature contribution of a turn of $\mathcal{P}$ next to v^S: recalling $\|\dot{\mathcal{P}}_{\text{arc}}\| = 1$, the total curvature of the part of $\mathcal{P}$ that passes through $\overline{\mathcal{T}}$ and $\underline{\mathcal{T}}$ (i.e., resulting from a turn of $\mathcal{T}$ adjacent to v^S) may be bounded below by the length of the shortest curve on a unit n-sphere that connects points $\dot{\mathcal{P}}_{\text{arc}}(t_1), \dot{\mathcal{P}}_{\text{arc}}(t_2), \ldots, \dot{\mathcal{P}}_{\text{arc}}(t_n)$ in any order. For simplicity, the latter length may be further bounded below by

$$
\begin{aligned}
\mathcal{K}^S := & \min_{\substack{x^i \in \mathbb{R}^n, \quad i=1,\ldots,n: \\ \|x^i\|=1, \quad \forall i, \\ |x_1^1| \le |\cos\alpha^S| + \frac{4\delta}{1/2-\varepsilon-3\delta}, \\ |x_j^j| \le \frac{4\delta}{1/2-\varepsilon-3\delta}, \quad j \ge 2}} \max_{j \ge 2} \operatorname{dist}(x^1, x^j) \\
\ge & \min_{\substack{x^i \in \mathbb{R}^n, \quad i=1,\ldots,n: \\ \|x^1\|=1, \\ |x_1^1| \le |\cos\alpha^S| + \frac{4\delta}{1/2-\varepsilon-3\delta}, \\ |x_j^j| \le \frac{4\delta}{1/2-\varepsilon-3\delta}, \quad j \ge 2}} \max_{j \ge 2} \|x^1 - x^j\|,
\end{aligned}
$$

where $\mathrm{dist}(x, z)$ is the length of the shortest curve on a unit sphere between points x and z, that is, the geodesic. Clearly, the critical value for the last expression is attained, in particular, at $x^i \in \mathbb{R}^n_+$, $\forall i$, when $\|x^1\| = 1$, $\|x^1 - x^j\| = \|x^1 - x^i\|$, $i, j \geq 2$, and

$$x^1_1 = |\cos\alpha^S| + \frac{4\delta}{1/2 - \varepsilon - 3\delta}, \qquad x^j_j = \frac{4\delta}{1/2 - \varepsilon - 3\delta}, \qquad j \geq 2.$$

It follows that

$$\sum_{j=2}^{n} (x^1_j)^2 = 1 - \left(|\cos\alpha^S| + \frac{4\delta}{1/2 - \varepsilon - 3\delta}\right)^2 \geq 1 - |\cos\alpha^S| - \frac{4\delta}{1/2 - \varepsilon - 3\delta}$$

and, because $|\cos\alpha^S| \leq 1 - \left((1-\varepsilon) \,/\, (1+\varepsilon^2)^{n-2}\right)$,

$$\sum_{j=2}^{n} (x^1_j)^2 \geq \frac{1-\varepsilon}{(1+\varepsilon^2)^{n-2}} - \frac{4\delta}{1/2 - \varepsilon - 3\delta},$$

resulting in

$$\begin{aligned} x^1_j \geq (x^1_j)^2 &\geq \frac{1}{n-1}\left(\frac{1-\varepsilon}{(1+\varepsilon^2)^{n-2}} - \frac{4\delta}{1/2 - \varepsilon - 3\delta}\right) \\ &\geq \frac{1}{n-1} \cdot \frac{1}{2(1+1/4)^{n-2}} - \frac{1}{n-1}\left(\frac{4\delta}{1/2 - \varepsilon - 3\delta}\right), \qquad j \geq 2. \end{aligned}$$

Therefore, recalling $\varepsilon = (n-1)/2n$ and $\delta = \left(1/32n^2\right)(4/5)^{n-2}$, we can write

$$\begin{aligned} \mathcal{K}^S &\geq \|x^1 - x^2\| \\ &\geq x^1_2 - x^2_2 \\ &\geq \frac{1}{2(n-1)}\left(\frac{4}{5}\right)^{n-2} - \left(1 + \frac{1}{n-1}\right)\frac{4\delta}{1/2 - \varepsilon - 3\delta} \\ &\geq \frac{1}{2(n-1)}\left(\frac{4}{5}\right)^{n-2} - \frac{n}{8n^2(n-1)}\left(\frac{4}{5}\right)^{n-2} \frac{1}{\frac{1}{2n} - \frac{3}{32n^2}\left(\frac{4}{5}\right)^{n-2}} \\ &\geq \frac{1}{2(n-1)}\left(\frac{4}{5}\right)^{n-2} - \frac{n}{8n^2(n-1)}\left(\frac{4}{5}\right)^{n-2} 2n \\ &\geq \frac{1}{4n}\left(\frac{4}{5}\right)^{n-2}. \end{aligned}$$

Finally, recalling that the polyhedral tube $\mathcal{T}$ makes 2^{n-1} such turns, we conclude that the total curvature of $\mathcal{P}$ indeed satisfies $K \geq (1/2n)\,(8/5)^{n-2}$. □

The bound on the total curvature K of $\mathcal{P}$ established above is obviously not tight. We expect the true order of K to be 2^n up to a multiplier, rational in n.

Remark 7.4. In $\mathbb{R}^2$, by combining the optimality conditions (7.1) and (7.2) for the analytic center χ with that of the central path $\mathcal{P}$ visiting the δ-neighborhoods of the vertices $v^{\{1\}}$ and $v^{\{1,2\}}$ one can show that for δ below a certain threshold both d_1 and d_2 are bounded away from 0 by a constant. In turn, this implies that for fixed feasible d_1, d_2, the necessary conditions (7.1) and (7.2) for h chosen such that the central path visits the δ-neighborhoods of all the vertices of $\mathcal{C}$ are "asymptotically equivalent" as $\delta \downarrow 0$ to the sufficient conditions (7.3) and (7.4), up to a constant multiplier. Here the term asymptotic equivalence refers to the convergence of the normalized extreme rays and the vertices of the unbounded polyhedra given by the set of necessary conditions for a fixed d to those of the polyhedra given by the set of sufficient conditions (7.3) and (7.4).

This suggests that the following might be true. In $\mathbb{R}^n$

$$\min_{i=1,\ldots,n} d_i \geq \hat{d} > 0,$$

where $\hat{d}$ is independent of n, δ, ε. Moreover, the necessary conditions for $\mathcal{P}$ to visit the δ-neighborhoods of all the vertices of $\mathcal{C}$ for a fixed d are asymptotically equivalent as $\delta \downarrow 0$ to the sufficient conditions (7.3) and (7.4). If, furthermore, we confine ourselves to only bounded subsets of all such feasible (d, h) corresponding to, say,

$$\sum_{i=1}^{n} h_i \leq H_\delta^* := 2 \min_{d,h} \sum_{i=1}^{n} h_i$$

then the conditions (7.3) and (7.4) are tight, in a sense that if we denote the set of all (d, h) satisfying the necessary conditions for $\mathcal{P}$ to visit all the δ-neighborhoods intersected with $\{h : \sum_i h \leq H_\delta^*\}$ as Necc_δ, the set of all (d, h) satisfying (7.3) and (7.4) intersected with $\{h : \sum_i h \leq H_\delta^*\}$ as Suff_δ, then for some small enough $\hat{\delta}$ there exists $M, m > 0$ independent of ε such that

$$\mathrm{Necc}_{m\delta} \subseteq \mathrm{Suff}_\delta \subseteq \mathrm{Necc}_{M\delta}, \qquad 0 < \delta \leq \hat{\delta}.$$

7.3 Finding $h \in \mathbb{N}^n$ Satisfying Conditions (7.3) and (7.4)

We write $f(n) \approx g(n)$ for $f(n), g(n) : \mathbb{N} \to \mathbb{R}$ if $\exists c, C > 0$ such that $cf(n) \leq g(n) \leq Cf(n)$, $\forall n$; the argument n is usually omitted from the notation.

Denote $b := \mathbf{1}\,(3/\delta)$. Let us first concentrate on finding $h \in \mathbb{N}^n$ such that (7.3) holds. If the integrality condition on h is relaxed, a solution to (7.3) can be found by simply solving $Ah = b$. Note that

$$\|A\|_{1,\infty} = \max_i \sum_{j=1}^{n} |a_{ij}|$$

is, in fact, small for d – large componentwise and $\varepsilon < 1/2$. So to find an integral h we can

- Solve $A\widehat{h} = (1+\gamma)b$ for some small $\gamma > 0$.
- Set $h = \lfloor \widehat{h} \rfloor$.

Observe that for h to satisfy (7.3), it is enough to require $\max_i(A(\widehat{h} - h) - \gamma b)_i \leq 0$. In turn, this can be satisfied by choosing $\gamma > 0$ such that

$$\gamma \frac{3}{\delta} \geq \max_i \sum_{j=1}^{n} |a_{ij}|.$$

In Section 7.3.1 we show how to solve this system of linear equations. In Section 7.3.2 we demonstrate that under some assumption on d, (7.4) is already implied by (7.3), and consequently the rounding of $\widehat{h}$ will not cause a problem for (7.4) either.

Remark 7.5. The choice of rounding down instead of rounding up is arbitrary.

7.3.1 Solving the Linear System

Because $(1+\gamma)b = (1+\gamma)\,(3/\delta)\,\mathbf{1}$, we can first solve $A\widetilde{h} = \mathbf{1}$ and then scale $\widetilde{h}$ by $(1+\gamma)\,(3/\delta)$. Our current goal is to find the solution to $A\widetilde{h} = \mathbf{1}$.

For an arbitrary invertible $B \in \mathbb{R}^{n\times n}$ and $y, z \in \mathbb{R}^n$ such that $1 + z^T B^{-1} y \neq 0$, the solution to

$$(B + yz^T)x = b$$

can be written as

$$x = B^{-1}(b - \alpha y),$$

where

$$\alpha = \frac{x^T B^{-1} b}{1 + z^T B^{-1} y}$$

(for writing $(B + yz^T)x = Bx + y(z^T x) = b$, denoting $\alpha := z^T x$, we can express $x = B^{-1}(b - \alpha y)$ and substitute this x into $(B + yz^T)x = b$ again to compute α).

Denoting

$$B := \begin{pmatrix} \frac{1}{d_1+1} & \frac{-\varepsilon}{d_2} & 0 & 0 & \cdots & 0 & 0 \\ 0 & \frac{2\varepsilon}{d_2+1} & \frac{-\varepsilon^2}{d_3} & 0 & \cdots & 0 & 0 \\ \vdots & \vdots & \ddots & \ddots & \vdots & \vdots & \vdots \\ 0 & 0 & \cdots & \frac{2\varepsilon^{k-1}}{d_k+1} & \frac{-\varepsilon^k}{d_{k+1}} & \cdots & 0 \\ \vdots & \vdots & \vdots & \vdots & \ddots & \ddots & \vdots \\ 0 & 0 & 0 & 0 & \cdots & \frac{2\varepsilon^{n-2}}{d_{n-1}+1} & \frac{-\varepsilon^{n-1}}{d_n} \\ 0 & 0 & 0 & 0 & \cdots & 0 & \frac{2\varepsilon^{n-1}}{d_n+1} \end{pmatrix},$$

$$y^T := \frac{-1}{d_1}(0, 1, 1, \ldots, 1),$$

$$z^T := (1, 0, 0, \ldots, 0),$$

we can compute the solution to $(B + yz^T)\widetilde{h} \equiv A\widetilde{h} = \mathbf{1}$ as

$$\widetilde{h} = B^{-1}\mathbf{1} + \frac{\alpha}{d_1} B^{-1} \begin{pmatrix} 0 \\ 1 \\ \vdots \\ 1 \end{pmatrix}, \tag{7.5}$$

where

$$\alpha = \frac{\sum_{j=1}^n B_{1j}^{-1}}{1 - \frac{1}{d_1}\sum_{j=2}^n B_{1j}^{-1}}. \tag{7.6}$$

So to get the explicit formula for $\widetilde{h}$ we need to compute B^{-1} and show that d can be chosen such that α is well defined; that is, $1 - (1/d_1)\sum_{j=2}^n B_{1j}^{-1} \neq 0$.

In order to invert B first note that it satisfies

$$B = \mathrm{Diag}\left(\frac{1}{d_1+1}, \frac{2\varepsilon}{d_2+1}, \frac{2\varepsilon^2}{d_3+1}, \ldots, \frac{2\varepsilon^{n-1}}{d_n+1}\right)(I + S),$$

where a superdiagonal matrix $S \in \mathbb{R}^{n\times n}$ is such that

$$S_{ij} = \begin{cases} \frac{-\varepsilon(d_i+1)}{d_{i+1}} & j = i+1, i = 1, \ldots, n-1 \\ 0 & \text{otherwise.} \end{cases}$$

Recall that $(I+Z)^{-1} = I - Z + Z^2 - Z^3 + \cdots$ for any $Z \in \mathbb{R}^{n\times n}$ such that these matrix-power series converge. In our case, the powers of S are easy to compute for $1 \leq k \leq n-1$,

$$S^k_{ij} = \begin{cases} \prod_{l=i}^{i+k-1} S_{l,l+1} & j = i+k, i = 1, \dots, n-k \\ 0 & \text{otherwise,} \end{cases}$$

and $S^m = 0$ for all $m \geq n$, so the inverse of $(I + S)$ can be computed as above and the inverse of B can be further computed by post-multiplying by the inverse of

$$\text{Diag}\left(\frac{1}{d_1+1}, \frac{2\varepsilon}{d_2+1}, \frac{2\varepsilon^2}{d_3+1} \dots, \frac{2\varepsilon^{n-1}}{d_n+1}\right).$$

Therefore, B^{-1} is equal to

$$\begin{pmatrix} d_1+1 & \frac{(d_1+1)(d_2+1)}{2d_2} & \frac{(d_1+1)(d_2+1)(d_3+1)}{4d_2d_3} & \frac{(d_1+1)\cdots(d_4+1)}{8d_2d_3d_4} & \cdots & \frac{\prod_{j=1}^n(d_j+1)}{2^{n-1}\prod_{j=2}^n d_j} \\ 0 & \frac{d_2+1}{2\varepsilon} & \frac{(d_2+1)(d_3+1)}{4d_3\varepsilon} & \frac{(d_2+1)\cdots(d_4+1)}{8d_3d_4\varepsilon} & \cdots & \frac{\prod_{j=2}^n(d_j+1)}{2^{n-1}\varepsilon\prod_{j=3}^n d_j} \\ 0 & 0 & \frac{d_3+1}{2\varepsilon^2} & \frac{(d_3+1)(d_4+1)}{4d_4\varepsilon^2} & \cdots & \frac{\prod_{j=3}^n(d_j+1)}{2^{n-2}\varepsilon^2\prod_{j=4}^n d_j} \\ 0 & 0 & 0 & \frac{d_4+1}{2\varepsilon^3} & \cdots & \frac{\prod_{j=4}^n(d_j+1)}{2^{n-3}\varepsilon^3\prod_{j=5}^n d_j} \\ \vdots & \vdots & \vdots & \vdots & \ddots & \vdots \\ 0 & 0 & 0 & 0 & \cdots & \frac{d_n+1}{2\varepsilon^{n-1}} \end{pmatrix}.$$

7.3.2 *Partial Implication for Sufficient Conditions*

Observe that in order for $\widetilde{h} \in \mathbb{R}^n_+$, we must have $\alpha > 0$ as in (7.6). For if not (i.e., if $\alpha < 0$), then denoting

$$\beta_1 := z^T B^{-1} \begin{pmatrix} 0 \\ 1 \\ \vdots \\ 1 \end{pmatrix} = \sum_{j=2}^n B^{-1}_{1j}$$

and writing

$$\alpha = \frac{\beta_1 + d_1 + 1}{1 - \frac{\beta_1}{d_1}} \tag{7.7}$$

we must have $\beta_1 > d_1 > 0$. So

$$\frac{-\alpha}{d_1} = \frac{\beta_1 + d_1 + 1}{\beta_1 - d_1} > 1$$

and from (7.5) it follows that if $(\alpha/d_1) < -1$, then $\widetilde{h}_2, \widetilde{h}_3, \ldots, \widetilde{h}_n < 0$.

From now on we assume $\alpha > 0$ (in Sections 7.4.1, 7.5.1 we show how to achieve this by choosing d appropriately). Note that in this case $(\alpha/d_1) > 1$.

Suppose $h \in \mathbb{N}^n$ is such that (7.3) holds. If, furthermore, $\left(h_i\varepsilon^{i-1}\right)/d_i$ is dominated by h_1/d_1 for $i = 1, \ldots, n$, then (7.3) already implies (7.4). Therefore, it is left to show that d can be chosen such that $h = \lfloor \widehat{h} \rfloor$ satisfies the domination condition above.

For this to hold it suffices

$$\frac{\frac{3}{\delta}(1+\gamma)\widetilde{h}_1 - 1}{d_1} \geq \frac{\frac{3}{\delta}(1+\gamma)\widetilde{h}_i + 1}{d_i}\varepsilon^{i-1}, \qquad i = 2, \ldots, n,$$

where $\widetilde{h}$ solves $A\widetilde{h} = \mathbf{1}$. The above is implied by

$$\frac{d_1 + 1 + \beta_1 + \frac{\alpha}{d_1}\beta_1 - \frac{1}{6}}{d_1} \geq \frac{(1+\frac{\alpha}{d_1})\frac{\beta_i}{\varepsilon^{i-1}} + \frac{1}{6}}{d_i}\varepsilon^{i-1}, \qquad i = 2, \ldots, n$$

because $\gamma > 0$, $\delta < 1/2$, where

$$\beta_i := \varepsilon^{i-1}(B^{-1}\mathbf{1})_i.$$

This can be written as

$$\left(1 + \frac{\alpha}{d_1}\right)\frac{\beta_1}{d_1} + 1 + \frac{5}{6d_1} \geq \left(1 + \frac{\alpha}{d_1}\right)\frac{\beta_i}{d_i} + \frac{\varepsilon^{i-1}}{6d_i}, \qquad i = 2, \ldots, n.$$

In particular, if we have

$$\frac{\beta_1}{d_1} > \frac{\beta_i}{d_i}, \qquad i = 2, \ldots, n \tag{7.8}$$

then the above inequality holds true if

$$1 \geq \frac{\varepsilon}{6d_1} - \frac{5}{6d_1},$$

that is, because $\varepsilon < 1/2$, for $d_i > 0$ for $i = 1, \ldots, n$.

Finally, observe that if $d_1 \geq d_i$, $i \geq 2$, and $d_1 = O(2^n)$, then the magnitude of $\widetilde{h}$ is primarily determined by α: recalling (7.5), (7.7), we write

$$\widetilde{h} = \alpha \left(\frac{B^{-1}}{d_1} \begin{pmatrix} 0 \\ 1 \\ \vdots \\ 1 \end{pmatrix} + \frac{B^{-1}}{d_1}\mathbf{1} \right)$$

$$\leq \alpha \left(\frac{B^{-1}}{d_1} \begin{pmatrix} 0 \\ 1 \\ \vdots \\ 1 \end{pmatrix} + \frac{\left(1 - \frac{\beta_1}{d_1}\right) B^{-1}}{d_1 + 1} \mathbf{1} \right)$$

$$= \alpha \left(\frac{B^{-1}}{d_1} \begin{pmatrix} 0 \\ 1 \\ \vdots \\ 1 \end{pmatrix} + \left(1 - \frac{\beta_1}{d_1}\right) \frac{B^{-1}}{d_1 + 1} \begin{pmatrix} 0 \\ 1 \\ \vdots \\ 1 \end{pmatrix} + \left(1 - \frac{\beta_1}{d_1}\right) \begin{pmatrix} d_1 + 1 \\ 0 \\ \vdots \\ 0 \end{pmatrix} \right).$$

Because $d_i > 0$ for $i = 1, \ldots, n$, we have $(d_i + 1)/2d_i > \frac{1}{2}$ and so $1 - (\beta_1/d_1) < 1/2^{n-1}$, implying

$$\widetilde{h} < \alpha \left(\frac{B^{-1}}{d_1} \begin{pmatrix} 0 \\ 1 \\ \vdots \\ 1 \end{pmatrix} + \frac{B^{-1}}{2^{n-1}(d_1 + 1)} \begin{pmatrix} 0 \\ 1 \\ \vdots \\ 1 \end{pmatrix} + \frac{1}{2^{n-1}} \begin{pmatrix} d_1 + 1 \\ 0 \\ \vdots \\ 0 \end{pmatrix} \right) \tag{7.9}$$

$$\approx \alpha \frac{B^{-1}}{d_1} \begin{pmatrix} 0 \\ 1 \\ \vdots \\ 1 \end{pmatrix} \quad \text{for large } n.$$

7.4 Uniform Distances to the Redundant Hyperplanes

Clearly, many different choices for d are possible. In this section we explore the uniform model for d; that is $d_1 = d_2 = \cdots = d_n$.

7.4.1 Bounding α

For $\alpha > 0$ we need

$$\beta_1 < d_1.$$

Denoting

$$\xi := \frac{d_1 + 1}{2d_1}$$

the above can be written as

$$(d_1 + 1)\xi + (d_1 + 1)\xi^2 + \cdots + (d_1 + 1)\xi^{n-1} < d_1$$

or, equivalently,

$$\xi^2(1+\xi+\xi^2+\cdots+\xi^{n-2}) < \frac{1}{2},$$

same as

$$\xi^2\left(\frac{1-\xi^{n-1}}{1-\xi}\right) < \frac{1}{2}.$$

In other words, we want

$$p(\xi) := 1-\xi-2\xi^2+2\xi^{n+1} > 0.$$

Note that as $d_1 \to \infty$, $\xi \to \frac{1}{2}$ and $p(\xi) \to 2\xi^{n+1} > 0$, so $p(\xi) > 0$ for d_1 large enough.

Observe that $\xi > \frac{1}{2}$ for any $d_1 > 0$, and

$$\begin{array}{rcll} p'\left(\frac{1}{2}\right) & = & -3+\frac{2(n+1)}{2^n} < 0 & \text{for } n=1,2,\ldots \\ p''\left(\frac{1}{2}\right) & = & -4+\frac{2(n+1)n}{2^{n-1}}\begin{cases} =0 \\ >0 \\ <0 \end{cases} & \begin{array}{l}\text{for } n=1 \\ \text{for } n=2,3,4 \\ \text{otherwise}\end{array} \\ p^{(k)}\left(\frac{1}{2}\right) & = & \frac{1}{2^{n-k}}\frac{(n+1)!}{(n-k+1)!} > 0 & \text{for } k \geq 3, \end{array}$$

so

$$p\left(\frac{1}{2}+\Delta\xi\right) \geq \widetilde{p}\left(\frac{1}{2}+\Delta\xi\right) := p\left(\frac{1}{2}\right)+p'\left(\frac{1}{2}\right)\Delta\xi+p''\left(\frac{1}{2}\right)\frac{\Delta\xi^2}{2}.$$

Thus, to guarantee $p(\xi) > 0$, it is enough to require

$$\widetilde{p}\left(\frac{1}{2}+\Delta\xi\right) \geq 0$$

letting $\xi = \frac{1}{2}+\Delta\xi$.

Denoting

$$\begin{aligned} \widetilde{a} &:= -2+\frac{n(n+1)}{2^{n-1}} \\ \widetilde{b} &:= -3+\frac{2(n+1)}{2^n} \\ \widetilde{c} &:= \frac{1}{2^n} \end{aligned}$$

and

$$\Delta\xi^* := \begin{cases} \frac{-\widetilde{b}+\sqrt{\widetilde{b}^2-4\widetilde{a}\widetilde{c}}}{2\widetilde{a}}, & 2 \leq n \leq 4 \\ \frac{-\widetilde{b}-\sqrt{\widetilde{b}^2-4\widetilde{a}\widetilde{c}}}{2\widetilde{a}}, & n > 4 \end{cases} \tag{7.10}$$

(the smallest positive root of $\widetilde{p}(\frac{1}{2} + \Delta\xi) = 0$), we conclude that $\beta_1 < d_1$ as long as

$$\xi = \frac{d_1 + 1}{2d_1} \leq \frac{1}{2} + \Delta\xi^*.$$

That is,

$$d_1 = \frac{1}{2\Delta\xi} \geq \frac{1}{2\Delta\xi^*}. \tag{7.11}$$

Note that for $d_1 = d_2 = \cdots = d_n$, (7.8) holds, so (7.4) is readily implied by (7.3).

It is left to demonstrate how to choose d_1 satisfying the above to guarantee a moderate growth in $\widetilde{h}$ as $n \to \infty$.

7.4.2 Picking a "Good" d_1

Note that as $n \to \infty$, we have

$$\Delta\xi^* \to \frac{1}{3 \cdot 2^n}$$

by expanding the square root in (7.10) as a first-order Taylor series. Also,

$$1 - \frac{\beta_1}{d_1} = p(\xi) \geq \widetilde{p}(\xi)$$

and hence

$$\frac{1}{1 - \frac{\beta_1}{d_1}} \leq \frac{1}{\widetilde{p}(\xi)}.$$

In fact, for large n, $p(\xi) \approx \widetilde{p}(\xi)$, for $\frac{1}{2} \leq \xi \leq \frac{1}{2} + \Delta\xi^*$. In turn, $\widetilde{p}(\xi)$ is almost linear on this interval because, as $n \to \infty$,

$$\widetilde{b} \to -3 \leftarrow \frac{\widetilde{c}}{\Delta\xi^*}$$

(we compare the slope of $\widetilde{p}(\frac{1}{2} + \Delta\xi)$ at $\Delta\xi = 0$ with its decrement as a function of $\Delta\xi$ over $[0, \Delta\xi^*]$).

Recalling that the growth of $\widetilde{h}$ is primarily determined by α for large n (see (7.9)), our goal becomes to minimize α. From (7.7), (7.11), noting that $\beta_1 \approx d_1$ for large n (also recall $\beta_1 < d_1$), we get

$$\alpha \leq \frac{\beta_1 + (d_1 + 1)}{\widetilde{p}(\xi)} \approx \frac{2d_1}{\widetilde{p}(\xi)} \approx \frac{2}{2\Delta\xi} \cdot \frac{1}{\widetilde{c} - \widetilde{c}\frac{\Delta\xi}{\Delta\xi^*}}$$

and, moreover the right-hand side of this expression approximates α fairly well. So, to approximately minimize $\widetilde{h}$, we maximize

$$\Delta\xi \cdot \left(\widetilde{c} - \widetilde{c}\frac{\Delta\xi}{\Delta\xi^*}\right)$$

for $0 \leq \Delta\xi \leq \Delta\xi^*$, which corresponds to setting

$$\Delta\xi = \frac{\Delta\xi^*}{2}$$

thus resulting in

$$\alpha \to 6 \cdot 2^{2n} \quad \text{as} \quad n \to \infty$$

and

$$\widetilde{h}_i = O\left(\frac{2^{2n}}{\varepsilon^{i-1}}\right), \qquad i = 1, \ldots, n.$$

7.4.3 An Explicit Formula

For a given n, ε, δ, compute $\Delta\xi^*$ according to (7.10), set $\Delta\xi = \Delta\xi^*/2$, d_1 as in (7.11). Compute the solution to $A\widetilde{h} = \mathbf{1}$ using (7.5), where α is computed according to (7.6). Set $\gamma = (\delta/3) \max_i \sum_{j=1}^n |a_{ij}|$ and, finally,

$$h = \left\lfloor (1+\gamma)\frac{3}{\delta}\widetilde{h} \right\rfloor .$$

From (7.9) it follows that (for large n)

$$\widetilde{h}_i \approx \frac{\alpha}{d_1} B^{-1} \begin{pmatrix} 0 \\ 1 \\ \vdots \\ 1 \end{pmatrix} \leq \frac{\alpha}{\varepsilon^{i-1}}, \qquad i = 1, \ldots, n.$$

We are interested in picking ε, δ, to minimize the total number of the redundant constraints, $\sum_{i=1}^n h_i$. Recalling $\varepsilon + \delta < \frac{1}{2}$, denoting

$$g(\varepsilon) := \frac{\varepsilon(\varepsilon^{-n} - 1)}{1 - 3\varepsilon + 2\varepsilon^2}$$

for large n we can write

$$\begin{aligned} \textstyle\sum_{i=1}^n h_i &\approx \tfrac{3}{\delta} \textstyle\sum_{i=1}^n \widetilde{h}_i \approx \tfrac{3}{\delta} \textstyle\sum_{i=1}^n \alpha \left(\tfrac{1}{\varepsilon}\right)^{i-1} \\ &\approx 6 \cdot 2^{2n} \tfrac{6}{1-2\varepsilon} \tfrac{1-\left(\frac{1}{\varepsilon}\right)^n}{1-\frac{1}{\varepsilon}} \\ &= 9 \cdot 2^{2n+2} g(\varepsilon) \end{aligned}$$

(it is natural to pick δ as close to $\frac{1}{2} - \varepsilon$ as possible, say $\delta = .999\left(\frac{1}{2} - \varepsilon\right)$).

In order to bound

$$g^* := \min_{0<\varepsilon<\frac{1}{2}} g(\varepsilon)$$

we first bound the minimizer ε^* of the above, next we bound the derivative of $g(\varepsilon)$, and finally using the mean-value theorem we bound g^* itself.

Observing

$$\begin{aligned} g'(\varepsilon) &= \frac{(1-3\varepsilon+2\varepsilon^2)((1-n)\varepsilon^{-n}-1)-(4\varepsilon-3)\varepsilon(\varepsilon^n-1)}{(1-3\varepsilon+2\varepsilon^2)^2} \\ &= \frac{(1-2\varepsilon)(1-n-\varepsilon^n)+\varepsilon(3-4\varepsilon)(\varepsilon^{n-1}+\varepsilon^{n-2}+\cdots+1)}{\varepsilon^n(1-\varepsilon)(1-2\varepsilon)^2} \\ &= \frac{3-4\varepsilon}{\varepsilon^n(1-\varepsilon)(1-2\varepsilon)^2}\left(\frac{1-n}{2}+\frac{n-1+\varepsilon^n}{2(3-4\varepsilon)}+\left(\frac{1}{2}\varepsilon^n+\varepsilon^{n-1}+\cdots+\varepsilon\right)\right) \end{aligned}$$

and noting that in the expression above the second and the third summands are monotone-increasing functions of ε (for $\varepsilon \in (0,1/2), n \geq 1$), for $n \geq 3$ we can write $\varepsilon^* \in (\varepsilon^L, \varepsilon^U)$ with

$$\varepsilon^L := \frac{n-5/4}{2n}$$

because for $n \geq 2$ we have

$$\begin{aligned} &(1-2\varepsilon)(1-n-\varepsilon^n)+\varepsilon(3-4\varepsilon)(\varepsilon^{n-1}+\varepsilon^{n-2}+\cdots+1)\big|_{\varepsilon=\varepsilon^L} \\ &\quad = \frac{(-4n^2+15n-25)+(-16n^2-40n+25)\left(\frac{n-5/4}{2n}\right)^n}{4n(4n+5)} < 0, \end{aligned}$$

where the first summand in brackets has no real roots with respect to n and the second summand is negative for $n \geq 2$, so $g'(\varepsilon^L) < 0$ and thus $g'(\varepsilon) < 0$ for $0 < \varepsilon < \varepsilon^L$, and

$$\varepsilon^U := \frac{n-1}{2n}$$

because for $n \geq 3$,

$$\begin{aligned} &(1-2\varepsilon)(1-n-\varepsilon^n)+\varepsilon(3-4\varepsilon)(\varepsilon^{n-1}+\varepsilon^{n-2}+\cdots+1)\big|_{\varepsilon=\varepsilon^U} \\ &\quad = \frac{n-1-n^2-2n+2\left(\frac{n-1}{2n}\right)^n}{n(n+1)} \\ &\quad = \frac{n-1-\left((n-1)(n+3)\left(\frac{n-1}{2n}\right)^n+\left(\frac{n-1}{2n}\right)^n\right)}{n(n+1)} \\ &\quad > \frac{n-1}{n(n+1)}\left(1-\frac{n+4}{2^n}\right) > 0, \end{aligned}$$

so $g'(\varepsilon^U) > 0$ and thus $g'(\varepsilon) > 0$ for $\varepsilon^U < \varepsilon < 1/2$. Consequently, for $\varepsilon \in (\varepsilon^L, \varepsilon^U)$ we have

$$g'(\varepsilon) \geq \min_{\varepsilon \in [\varepsilon^L, \varepsilon^U]} \frac{-n(1-2\varepsilon)}{\varepsilon^n (1-\varepsilon)(1-2\varepsilon)^2} = \frac{-2n^3}{n+1} \left(\frac{8n}{4n-5}\right)^n$$

and, therefore, by the mean-value theorem,

$$g(\varepsilon^L) + \min_{\varepsilon \in [\varepsilon^L, \varepsilon^U]} g'(\varepsilon)(\varepsilon^U - \varepsilon^L) \leq g^* \leq g(\varepsilon^L);$$

that is,

$$0 < \frac{4n}{5}\left(\frac{4n-5}{4n+5}\right)\left(\left(\frac{8n}{4n-5}\right)^n - 1\right) - \frac{n^2}{4(n+1)}\left(\frac{8n}{4n-5}\right)^n$$

$$\leq g^* \leq -\frac{4n}{5}\left(\frac{4n-5}{4n+5}\right)\left(\left(\frac{8n}{4n-5}\right)^n - 1\right)$$

for $n \geq 3$.

In turn, this results in $\sum_{i=1}^n h_i = O(n2^{3n})$, and, in fact, $\sum_{i=1}^n h_i$ has the same order lower bound as well by the above, noting that

$$\left(\frac{8n}{4n-5}\right)^n = 2^n \left(1 + \frac{5}{4n-5}\right)^n \to e^{5/4} \cdot 2^n$$

as $n \to \infty$, for a suitably chosen small $\varepsilon > 0$.

7.5 Geometrically Decaying Distances to the Redundant Hyperplanes

Next we explore the geometric model for d: $d_i = \omega\,(1/\tilde{\varepsilon})^{n-i+1}$, $i = 1, \ldots, n$.

7.5.1 Bounding α

As in Section 7.4.1, we need to guarantee $\beta_1 < d_1$.

Firstly, we give a lower bound on $(\Delta_k)_{k=1,\ldots,n}$ recursively defined by

$$1 - \Delta_{k+1} = \frac{\tilde{d}_{k+1} + 1}{2\tilde{d}_{k+1}}(2 - \Delta_k), \qquad k = 0, \ldots, n-1 \tag{7.12}$$

with $\Delta_0 = 1$, and where

$$\tilde{d}_k = \omega \left(\frac{1}{\tilde{\varepsilon}}\right)^k, \qquad k = 1, \ldots, n$$

with some constant ω.

We have $d_i = \tilde{d}_{n-i+1}$ for $i = 1, \ldots, n$, and

$$\frac{\beta_1}{d_1} = \frac{\tilde{d}_n + 1}{\tilde{d}_n}(1 - \Delta_{n-1}), \qquad \frac{\beta_i}{d_i} = 1 - \Delta_{n-i+1}, \qquad i = 2, \ldots, n.$$

Note that to satisfy $\beta_1 < d_1$ we necessarily must have $1 - \Delta_k < 1$ for $k = 1, \ldots, n-1$.

From (7.12) it follows that

$$\Delta_{k+1} = \frac{\Delta_k}{2} - \frac{\tilde{\varepsilon}^{k+1}}{\omega} + \frac{\Delta_k}{2}\frac{\tilde{\varepsilon}^{k+1}}{\omega} \geq \frac{\Delta_k}{2} - \frac{\tilde{\varepsilon}^{k+1}}{\omega}, \qquad k = 1, \ldots, n-1$$

and hence

$$\Delta_k \geq \frac{\Delta_1}{2^{k-1}} - \frac{\tilde{\varepsilon}^2}{\omega 2^{k-2}}(1 + (2\tilde{\varepsilon}) + (2\tilde{\varepsilon})^2 + \cdots + (2\tilde{\varepsilon})^{k-2}), \qquad k = 2, \ldots, n.$$

Observing $\Delta_1 = \frac{1}{2}(1 - (\tilde{\varepsilon}/\omega))$ we can write the above inequality as

$$\begin{aligned}\Delta_k &\geq \frac{1}{2^k}\left(1 - \frac{\tilde{\varepsilon}}{\omega}\right) - \frac{\tilde{\varepsilon}^2}{\omega 2^{k-2}}\sum_{i=0}^{k-2}(2\tilde{\varepsilon})^i \\ &= \frac{1}{2^k}\left(1 - \frac{\tilde{\varepsilon}}{\omega} - \frac{4\tilde{\varepsilon}^2}{\omega}\sum_{i=0}^{k-2}(2\tilde{\varepsilon})^i\right), \qquad k = 2, \ldots, n.\end{aligned}$$

Now, for α to be positive, that is, for

$$1 - \frac{\tilde{d}_n + 1}{\tilde{d}_n}(1 - \Delta_{n-1}) = 1 - \left(1 - \Delta_{n-1} + \frac{1}{\tilde{d}_n} - \frac{\Delta_{n-1}}{\tilde{d}_n}\right) > 0,$$

it suffices

$$\Delta_{n-1} - \frac{1}{\tilde{d}_n} > 0$$

which is implied by

$$\frac{1}{2^{n-1}}\left(1 - \frac{\tilde{\varepsilon}}{\omega} - \frac{4\tilde{\varepsilon}^2}{\omega}\sum_{i=0}^{n-3}(2\tilde{\varepsilon})^i\right) - \frac{\tilde{\varepsilon}^n}{\omega} > 0.$$

If $\tilde{\varepsilon} = \frac{1}{2}$, the above translates into

$$\frac{1}{2^{n-1}} - \frac{n-1}{\omega \cdot 2^{n-1}} > 0;$$

that is

$$\omega > n - 1$$

resulting in

$$d_i = \omega 2^{n-i+1} > (n-1)2^{n-i+1}, \qquad i = 1, \ldots, n.$$

It is left to verify that $h_i \varepsilon^{i-1}/d_i$ is indeed dominated by h_1/d_1 for $i = 1, \ldots, n$, to ensure (7.4) as in Section 7.3.2. In particular, we demonstrate (7.8), as in the case of uniform d. Recalling

$$\frac{\beta_1}{d_1} = \frac{\tilde{d}_n + 1}{\tilde{d}_n}(1 - \Delta_{n-1})$$

and

$$\frac{\beta_i}{d_i} = 1 - \Delta_{n-i+1} \qquad \text{for } i = 2, \ldots, n,$$

it immediately follows that

$$\frac{\beta_1}{d_1} > \frac{\beta_2}{d_2}.$$

Also observe

$$\frac{\beta_1}{d_1} > \frac{\beta_2}{d_2} > \frac{\beta_3}{d_3} > \cdots > \frac{\beta_n}{d_n}$$

because, recalling (7.12) and $0 < \Delta_k < 1, k = 1, \ldots, n-1$,

$$\begin{aligned}(1 - \Delta_{k+1}) - (1 - \Delta_k) &= \frac{1}{2}\left(2 - \Delta_k + \frac{2 - \Delta_k}{\tilde{d}_{k+1}}\right) - 1 + \Delta_k \\ &= \frac{\Delta_k}{2} + \frac{2 - \Delta_k}{2\tilde{d}_{k+1}} > 0.\end{aligned}$$

7.5.2 Picking a "Good" ω

As in Section 7.4.2, we would like to minimize α with respect to ω, which, in the case of $\tilde{\varepsilon} = \frac{1}{2}$, can be well approximated from above by

$$\left(2\tilde{d}_n + 1\right)\left(\Delta_{n-1} - \frac{1}{\tilde{d}_n}\right)^{-1} \approx 2\tilde{d}_n\left(\frac{1}{2^{n-1}} - \frac{n-1}{\omega \cdot 2^{n-1}}\right)^{-1}.$$

We look for

$$\min_{\omega>n-1} \omega \cdot 2^n \left(\frac{1}{2^{n-1}} - \frac{n-1}{\omega \cdot 2^{n-1}} \right)^{-1};$$

that is,

$$\min_{\omega>n-1} \frac{\omega^2}{\omega - n + 1}$$

or equivalently

$$\min_{\omega>n-1} (2 \ln \omega - \ln(\omega - n + 1)) .$$

Setting the gradient to 0, we obtain

$$\frac{2}{\omega} - \frac{1}{\omega - n + 1} = \frac{2(\omega - n + 1) - \omega}{\omega(\omega - n + 1)} = 0$$

which gives us the minimizer

$$\omega = 2(n-1)$$

with the corresponding value of $\alpha \approx (n-1)2^{2n+1}$. This results in

$$d_i = (n-1)2^{n-i+2}, \qquad i = 1, \ldots, n$$

and

$$\widetilde{h}_i = O\left(\frac{n 2^{2n}}{(2\varepsilon)^{i-1}} \right), \qquad i = 1, \ldots, n.$$

7.5.3 An Explicit Formula

For a given n, ε, δ, set $d_i = (n-1)2^{n-i+2}$ for $i = 1, \ldots, n$ and compute the solution to $A\widetilde{h} = \mathbf{1}$ using (7.5). Set

$$h = \left\lfloor \left(1 + \frac{\delta}{3} \max_i \sum_{j=1}^{n} |a_{ij}| \right) \frac{3}{\delta} \widetilde{h} \right\rfloor .$$

From (7.9) it follows that for large n

$$\widetilde{h}_i \approx \frac{\alpha}{d_1} B^{-1} \begin{pmatrix} 0 \\ 1 \\ \vdots \\ 1 \end{pmatrix} \le \alpha \left(\frac{1}{2\varepsilon} \right)^{i-1}, \qquad i = 1, \ldots, n.$$

We choose ε, δ, to minimize the total number of the redundant constraints, $\sum_{i=1}^{n} h_i$. Recalling $\varepsilon + \delta < 1/2$, for large n we can write

$$\begin{aligned}
\sum_{i=1}^{n} h_i &\approx \frac{3}{\delta}\sum_{i=1}^{n}\widetilde{h}_i \approx \frac{3}{\delta}\sum_{i=1}^{n}\alpha\left(\frac{1}{2\varepsilon}\right)^{i-1} \\
&\approx 3(n-1)2^{2n+1}\,\frac{2}{1-2\varepsilon}\,\frac{1-\left(\frac{1}{2\varepsilon}\right)^n}{1-\frac{1}{2\varepsilon}} \\
&= 3(n-1)2^{2n+2}\,2\varepsilon\frac{\left(\frac{1}{2\varepsilon}\right)^n-1}{(2\varepsilon-1)^2} \\
&\le 3(n-1)2^{2n+2}\,\frac{\left(\frac{1}{2\varepsilon}\right)^n-1}{(2\varepsilon-1)^2}.
\end{aligned} \tag{7.13}$$

In fact, we would expect ε to be close to $1/2$ in order for $\sum_{i=1}^{n} h_i$ to be minimized, so the last inequality also gives us a good approximation, namely

$$\sum_{i=1}^{n} h_i \approx 3(n-1)2^{2n+2}\frac{\left(\frac{1}{2\varepsilon}\right)^n-1}{(2\varepsilon-1)^2}. \tag{7.14}$$

Indeed, denoting

$$\zeta := 2\varepsilon$$

and introducing

$$f(\zeta) := \frac{\left(\frac{1}{\zeta}\right)^n-1}{(\zeta-1)^2}$$

we can write the last two lines in (7.13) as

$$3(n-1)2^{2n+2}\zeta f(\zeta) \le 3(n-1)2^{2n+2} f(\zeta).$$

Differentiating $\zeta f(\zeta)$ we get

$$(\zeta f(\zeta))' = \frac{(n+1-\zeta^n)(\zeta+1)-2n}{\zeta^n(1-\zeta)^3} < 0 \qquad \text{for } 0 < \zeta < \frac{n-1}{n+1}$$

because $(n+1-\zeta^n)(\zeta+1)-2n < (n+1)(\zeta+1)-2n < 0$ for such ζ, and therefore, the function is decreasing on this interval. So to maximize

$$\zeta f(\zeta) \approx \frac{\sum_{i=1}^{n} h_i}{3(n-1)2^{2n+2}}$$

we must take $\zeta > (n-1)\,/\,(n+1)$, thus justifying (7.14).

Next we demonstrate how to minimize $\sum_{i=1}^{n} h_i$ with respect to $0 < \varepsilon < 1/2$. We note that the approximate minimum of $\sum_{i=1}^{n} h_i$ corresponds to

$$f^* := \min_{0<\zeta<1} f(\zeta).$$

To analyze f^* we proceed as follows: first we demonstrate that for n large enough $f(\zeta)$ is convex on $(0,1)$; then we produce lower and upper bounds on the root of $f'(\zeta) = 0$ in this interval, thus bounding the minimizer; and finally we compute upper and lower bounds on f^* using Taylor expansions of $f(\zeta)$.

Observe

$$f'(\zeta) = \frac{(n+2)\zeta^{-n} - n\zeta^{-n-1} - 2}{(1-\zeta)^3}$$

and

$$f''(\zeta) = \frac{(n^2+6n+6)\zeta^2 - (2n^2+6n)\zeta + n^2 + n - 6\zeta^{n+2}}{\zeta^{n+2}(1-\zeta)^4}.$$

Denoting

$$\begin{aligned}
\hat{a} &:= n^2 + 6n + 6 \\
\hat{b} &:= -(2n^2 + 6n) \\
\hat{c} &:= n^2 + n
\end{aligned}$$

for $f''(\zeta)$ to be positive on $(0,1)$ it suffices to show that the minimum of

$$\hat{a}\zeta^2 + \hat{b}\zeta + \hat{c}$$

is greater or equal than 6 (recall $0 < \zeta < 1$, so that $6\zeta^{n+2} < 6$). In turn,

$$\min_{\zeta} \left(\hat{a}\zeta^2 + \hat{b}\zeta + \hat{c}\right) = \left(\hat{a}\zeta^2 + \hat{b}\zeta + \hat{c}\right)\Big|_{\zeta = -\hat{b}/2\hat{a}} = \frac{-\hat{b}^2}{4\hat{a}} + \hat{c}$$

so for $f''(\zeta) > 0$ on $(0,1)$ it is enough to have

$$-\frac{(2n^2+6n)^2}{4(n^2+6n+6)} + n^2 + n \geq 6.$$

The above can be rewritten as

$$\frac{4n^3 + 12n^2 + 24n}{4(n^2+6n+6)} = n - \frac{3n^2}{n^2+6n+6} \geq 6$$

and is clearly implied by

$$n - 3 \geq 6.$$

Therefore, $f(\zeta)$ is convex on $(0,1)$ as long as $n \geq 9$.

Furthermore, at

$$\zeta^L := \frac{n-1}{n+1}$$

we have

$$f'(\zeta^L) = \frac{\left(\frac{n+1}{n-1}\right)^n \frac{2}{n-1} - 2}{\left(1 - \frac{n-1}{n+1}\right)^3} < 0, \qquad \text{for } n \geq 2.$$

On the other hand, at

$$\zeta^U := \frac{n-1}{n}$$

we have

$$f'(\zeta^U) = \frac{\left(\frac{n}{n-1}\right)^n \left(\frac{n-2}{n-1}\right) - 2}{\left(1 - \frac{n-1}{n}\right)^3} < 0$$

and relying on all the derivatives of $(1+z)^n$ for $n \in \mathbb{N}$ being positive at $z = 0$, expanding

$$\left(\frac{n}{n-1}\right)^n = \left(1 + \frac{1}{n-1}\right)^n$$

into second-order Taylor series, we get

$$f'(\zeta^U) \geq \frac{\left(1 + \frac{n}{n-1} + \frac{n(n-1)}{2(n-1)^2}\right)\left(\frac{n-2}{n-1}\right) - 2}{\left(1 - \frac{n-1}{n}\right)^3} = \frac{n^2 - 4n}{2(n-1)\left(1 - \frac{n-1}{n}\right)^3} \geq 0$$

for $n \geq 4$.

Now we are in position to give bounds on f^* for $n \geq 9$: by convexity it follows that

$$f(\zeta^L) + f'(\zeta^L)(\zeta^U - \zeta^L) \leq f^* \leq f(\zeta^L);$$

that is,

$$\begin{aligned} 0 < \left(\frac{n+1}{n-1}\right)^n \left(\frac{(n+1)^2}{4} - \frac{(n+1)^2}{4n}\right) - \frac{(n+1)^2}{4n} \\ \leq f^* \leq \left(\left(\frac{n+1}{n-1}\right)^n - 1\right) \frac{(n+1)^2}{4}. \end{aligned}$$

Consequently $\sum_{i=1}^n h_i = O(n^3 2^{2n})$ with the same order lower bound in the best case for $\varepsilon \approx (n-1)\,/\,(2(n+1))$.

Remark 7.6. Similar analysis can be easily carried out for $\tilde{d}_k = \omega\,(1/\tilde{\varepsilon})^k$ for $k = 1, \ldots, n$, with $\tilde{\varepsilon} = \varepsilon$. Because

$$\begin{aligned} \alpha &\approx \frac{2\tilde{d}_n + 1}{1 - \frac{\tilde{d}_n + 1}{\tilde{d}_n}(1 - \Delta_{n-1})} \\ &\approx \frac{2\omega}{\varepsilon^n} \left(\frac{1}{2^{n-1}} \left(1 - \frac{\varepsilon}{\omega} - \frac{4\varepsilon^2}{\omega} \sum_{i=0}^{n-3} (2\varepsilon)^i \right) - \frac{\varepsilon^n}{\omega} \right)^{-1} \\ &= \frac{2^n}{\varepsilon^n} \frac{\omega^2}{\omega - \left(\varepsilon + (2\varepsilon)^2 \sum_{i=0}^{n-3} (2\varepsilon)^i + (2\varepsilon)^{n-1}\varepsilon\right)} \end{aligned}$$

and is approximately minimized at $\omega = 2\left(\varepsilon + (2\varepsilon)^2 \sum_{i=0}^{n-3}(2\varepsilon)^i + (2\varepsilon)^{n-1}\varepsilon\right)$, resulting in $\alpha \approx 4\,(2/\varepsilon)^n \left(\varepsilon + (2\varepsilon)^2 \sum_{i=0}^{n-3}(2\varepsilon)^i + (2\varepsilon)^{n-1}\varepsilon\right)$, we have

$$\sum_{i=1}^{n} h_i \approx \frac{3}{\delta}\sum_{i=1}^{n} \widetilde{h}_i \approx 3\frac{2}{1-2\varepsilon}4\left(\frac{2}{\varepsilon}\right)^n \left(\varepsilon + (2\varepsilon)^2 \sum_{i=0}^{n-3}(2\varepsilon)^i + (2\varepsilon)^{n-1}\varepsilon\right) n$$

$$= 12n2^{2n}\frac{1}{(1-2\varepsilon)(2\varepsilon)^n}\left(2\varepsilon + (2\varepsilon)^n + 2(2\varepsilon)^2\,\frac{1-(2\varepsilon)^{n-2}}{1-2\varepsilon}\right).$$

Noting that

$$\frac{1}{(1-2\varepsilon)(2\varepsilon)^n}\left(2\varepsilon + (2\varepsilon)^n + 2(2\varepsilon)^2\,\frac{1-(2\varepsilon)^{n-2}}{1-2\varepsilon}\right) \le \frac{2(n-1)}{(1-2\varepsilon)(2\varepsilon)^n}$$

for $\varepsilon \in (0, 1/2)$ and thus the latter may be bounded from above by the value of the right-hand side at $\varepsilon = n/\left(2(n+1)\right)$, that is, $2e(n+1)(n-1)$, we get

$$\sum_{i=1}^{n} h_i \le 24e(n+1)(n-1)n2^{2n} = O(n^3 2^{2n}).$$

The resulting estimate for the number of the redundant constraints is not much different from the case of $\tilde{\varepsilon} = 1/2$, so this model for d is not discussed here in any more details.

7.6 Conclusions

We provide sufficient conditions for the central path to intersect small neighborhoods of all the vertices of the Klee–Minty n-cube; see Propositions 7.1, 7.2, and Corollary 7.1. More precisely, we derive explicit formulae for the number of redundant constraints for the Klee–Minty n-cube example given in [4, 5]. We give a smaller number of redundant constraints of order $n2^{3n}$ when the distances to those are chosen uniformly, as opposed to the previously established $O(n^2 2^{6n})$. When these distances are chosen to decay geometrically, we give a slightly tighter bound of the same order $n^3 2^{2n}$ as in [5], that results in a provably smaller number of constraints in practice.

We argue that in $\mathbb{R}^2$ the sufficient conditions presented are tight and indicate that the same is likely true in higher dimensions.

Our construction potentially gives rise to linear programming instances that exhibit the worst case iteration-complexity for path-following interior-point methods, which almost matches its theoretical counterpart.

Considering the n-dimensional simplex, Megiddo and Shub [8] demonstrated that the total curvature of the central path can be as large as order

n. Combined with Corollary 7.2, it follows that the worst-case lower bound on the total curvature of the central path is at least exponential in n up to a rational multiplier. We conjecture that the total curvature of the central path is $O(m)$; see [3].

Acknowledgments We would like to thank Michael Shub for a stimulating and inspiring discussion that triggered the work on the total curvature of the central path. Research supported by an NSERC Discovery grant, by a MITACS grant, and by the Canada Research Chair program.

References

1. J-P. Dedieu, G. Malajovich, and M. Shub. On the curvature of the central path of linear programming theory. *Foundations of Computational Mathematics*, 5:145–171, 2005.
2. J-P. Dedieu and M. Shub. Newton flow and interior point methods in linear programming. *International Journal of Bifurcation and Chaos*, 15:827–839, 2005.
3. A. Deza, T. Terlaky, and Y. Zinchenko. Polytopes and arrangements: Diameter and curvature. *Operations Research Letters*, 36-2:215–222, 2008.
4. A. Deza, E. Nematollahi, R. Peyghami, and T. Terlaky. The central path visits all the vertices of the Klee–Minty cube. *Optimization Methods and Software*, 21-5:851–865, 2006.
5. A. Deza, E. Nematollahi, and T. Terlaky. How good are interior point methods? Klee–Minty cubes tighten iteration-complexity bounds. *Mathematical Programming*, 113-1:1–14, 2008.
6. J. Gondzio. Presolve analysis of linear programs prior to applying an interior point method. *INFORMS Journal on Computing*, 9-1:73–91, 1997.
7. I. Maros. *Computational Techniques of the Simplex Method.* Kluwer Academic, Boston, MA.
8. N. Megiddo and M. Shub. Boundary behavior of interior point algorithms in linear programming. *Mathematics of Operations Research*, 14-1:97–146, 1989.
9. R. Monteiro and T. Tsuchiya. A strong bound on the integral of the central path curvature and its relationship with the iteration complexity of primal-dual path-following LP algorithms. *Optimization Online*, September 2005.
10. Yu. Nesterov and M. Todd. On the Riemannian geometry defined by self-concordant barriers and interior-point methods. *Foundations of Computational Mathematics*, 2:333–361, 2002.
11. C. Roos, T. Terlaky, and J-Ph. Vial. *Theory and Algorithms for Linear Optimization: An Interior Point Approach.* Springer, New York, second edition, 2006.
12. G. Sonnevend, J. Stoer, and G. Zhao. On the complexity of following the central path of linear programs by linear extrapolation. II. *Mathematical Programming*, 52:527–553, 1991.
13. M. Todd and Y. Ye. A lower bound on the number of iterations of long-step and polynomial interior-point linear programming algorithms. *Annals of Operations Research*, 62:233–252, 1996.
14. S. Vavasis and Y. Ye. A primal-dual interior-point method whose running time depends only on the constraint matrix. *Mathematical Programming*, 74:79–120, 1996.
15. G. Zhao and J. Stoer. Estimating the complexity of a class of path-following methods for solving linear programs by curvature integrals. *Applied Mathematics and Optimization*, 27:85–103, 1993.

Chapter 8
Canonical Duality Theory: Connections between Nonconvex Mechanics and Global Optimization

David Y. Gao and Hanif D. Sherali

Dedicated to Professor Gilbert Strang on the occasion of his 70th birthday

Summary. This chapter presents a comprehensive review and some new developments on canonical duality theory for nonconvex systems. Based on a tricanonical form for quadratic minimization problems, an insightful relation between canonical dual transformations and nonlinear (or extended) Lagrange multiplier methods is presented. Connections between complementary variational principles in nonconvex mechanics and Lagrange duality in global optimization are also revealed within the framework of the canonical duality theory. Based on this framework, traditional saddle Lagrange duality and the so-called biduality theory, discovered in convex Hamiltonian systems and d.c. programming, are presented in a unified way; together, they serve as a foundation for the triality theory in nonconvex systems. Applications are illustrated by a class of nonconvex problems in continuum mechanics and global optimization. It is shown that by the use of the canonical dual transformation, these nonconvex constrained primal problems can be converted into certain simple canonical dual problems, which can be solved to obtain all extremal points. Optimality conditions (both local and global) for these extrema can be identified by the triality theory. Some new results on general nonconvex programming with nonlinear constraints are also presented as applications of this canonical duality theory. This review brings some fundamentally new insights into nonconvex mechanics, global optimization, and computational science.

Key words: Duality, triality, Lagrangian duality, nonconvex mechanics, global optimization, nonconvex variations, canonical dual transformations, critical point theory, semilinear equations, NP-hard problems, quadratic programming

David Y. Gao, Department of Mathematics, Virginia Tech, Blacksburg, VA 24061, U.S.A. e-mail: gao@vt.edu

Hanif D. Sherali, Grado Department of Industrial and Systems Engineering, Virginia Tech, Blacksburg, VA 24061, U.S.A., e-mail: hanifs@vt.edu

D.Y. Gao, H.D. Sherali, (eds.), *Advances in Applied Mathematics and Global Optimization*
Advances in Mechanics and Mathematics 17, DOI 10.1007/978-0-387-75714-8_8,

8.1 Introduction

Complementarity and duality are two inspiring, closely related concepts. Together they play fundamental roles in multidisciplinary fields of mathematical science, especially in engineering mechanics and optimization.

The study of complementarity and duality in mathematics and mechanics has had a long history since the well-known Legendre transformation was formally introduced in 1787. This elegant transformation plays a key role in complementary duality theory. In classical mechanical systems, each energy function defined in a configuration space is linked via the Legendre transformation with a complementary energy in the dual (source) space, through which the Lagrangian and Hamiltonian can be formulated. In static systems, the convex total potential energy leads to a saddle Lagrangian through which a beautiful saddle min-max duality theory can be constructed. This saddle Lagrangian plays a central role in classical duality theory in convex analysis and constrained optimization. In convex dynamic systems, however, the total action is usually a nonconvex d.c. function, that is, the difference of convex kinetic energy and total potential functions. In this case, the classical Lagrangian is no longer a saddle function, but the Hamiltonian is convex in each of its variables. It turns out that instead of the Lagrangian, the Hamiltonian has been extensively used in convex dynamics. From a geometrical point of view, Lagrangian and Hamiltonian structures in convex systems and d.c. programming display an appealing symmetry, which was widely studied by their founders. Unfortunately, such a symmetry in nonconvex systems breaks down. It turns out that in recent times, tremendous effort and attention have been focused on the role of symmetry and symmetry-breaking in Hamiltonian mechanics in order to gain a deeper understanding into nonlinear and nonconvex phenomena (see Marsden and Ratiu, 1995).

The earliest examples of the Lagrangian duality in engineering mechanics are probably the complementary energy principles proposed by Haar and von Kármán in 1909 for elastoperfectly plasticity and Hellinger in 1914 for continuum mechanics. Since the boundary conditions in Hellinger's principle were clarified by E. Reissner in 1953 (see Reissner, 1996), the complementary–dual variational principles and methods have been studied extensively for more than 50 years by applied mathematicians and engineers (see Arthurs, 1980, Noble and Sewell, 1972).[1] The development of mathematical duality theory in convex variational analysis and optimization has had a similar history since W. Fenchel proposed the well-known Fenchel transformation in 1949. After the revolutionary concepts of superpotential and subdifferentials introduced by J. J. Moreau in 1966 in the study of frictional mechanics,

[1] Eric Reissner (PhD 1938) was a professor in the Department of Mathematics at MIT from 1949 to 1969. According to Gil Strang, since Reissner moved to the Department of Mechanical and Aerospace Engineering at University of California, San Diego in 1969, many applied mathematicians in the field of continuum mechanics, especially solid mechanics, switched from mathematical departments to engineering schools in the United States.

the modern mathematical theory of duality has been well developed by celebrated mathematicians such as R. T. Rockafellar (1967, 1970, 1974), Moreau (1968), Ekeland (1977, 2003), I. Ekeland and R. Temam (1976), F. H. Clarke (1983, 1985), Auchmuty (1986, 2001), G. Strang (1979–1986), and Moreau, Panagiotopoulos, and Strang (1988). Mathematically speaking, in linear elasticity where the total potential energy is convex, the Hellinger–Reissner complementary variational principle in engineering mechanics is equivalent to a Fenchel–Moreau–Rockafellar type dual variational problem. The so-called generalized complementary variational principle is actually the saddle Lagrangian duality theory, which serves as the foundation for hybrid/mixed finite element methods, and has been subjected to extensive study during the past 40 years (see Strang and Fix (1973), Oden and Lee (1977), Pian and Tong (1980), Pian and Wu (2006), Han (2005), and the references cited therein).

Early in the beginning of the last century, Haar and von Kármán (1909) had already realized that in nonlinear variational problems of continuum mechanics, the direct approaches for solving minimum potential energy (primal problem) can only provide upper bounding solutions. However, the minimum complementary energy principle (i.e., the maximum Lagrangian dual problem) provides a lower bound (the mathematical proof of Haar–von Kármán's principle was given by Greenberg in 1949). In safety analysis of engineering structures, the upper and lower bounding approximations to the so-called collapse states of the elastoplastic structures are equally important to engineers. Therefore, the primal–dual variational methods have been studied extensively by engineers for solving nonsmooth nonlinear problems (see Gao, 1991, 1992, Maier, 1969, 1970, Temam and Strang, 1980, Casciaro and Cascini, 1982, Gao, 1986, Gao and Hwang, 1988, Gao and Cheung, 1989, Gao and Strang, 1989b, Gao and Wierzbicki, 1989, Gao and Onate, 1990, Tabarrok and Rimrott, 1994). The article by Maier et al. (2000) serves as an excellent survey on the developments for applications of the Lagrangian duality in engineering structural mechanics. In mathematical programming and computational science, the so-called primal–dual interior point methods are also based on the Lagrangian duality theory, which has emerged as a revolutionary technique during the last 15 years. Complementary to the interior-point methods, the so-called pan-penalty finite element programming developed by Gao in 1988 (1988a,b) is indeed a primal–dual exterior-point method. He proved that in rigid-perfectly plastic limit analysis, the exterior penalty functional and the associated perturbation method possess an elegant physical meaning, which led to an efficient dimension rescaling technique in large-scale nonlinear mixed finite element programming problems (Gao, 1988b).

In mathematical programming and analysis, the subject of complementarity is closely related to constrained optimization, variational inequality, and fixed point theory. Through the classical Lagrangian duality, the KKT conditions of constrained optimization problems lead to corresponding complementarity problems. The primal–dual schema has continued to evolve for linear and convex mathematical programming during the past 20 years (see Walk, 1989, Wright, 1998). However, for nonconvex systems, it is well known that the KKT conditions are only necessary under certain regularity conditions for global optimality. Moreover, the underlying nonlinear complementarity problems are fundamentally difficult due to the nonmonotonicity of the nonlinear operators, and also, many problems in global optimization are NP-hard. The well-developed Fenchel–Moreau–Rockafellar duality theory will produce a so-called duality gap between the primal problem and its Lagrangian dual. Therefore, how to formulate perfect dual problems (with a zero duality gap) is a challenging task in global optimization and nonconvex analysis. Extensions of the classical Lagrangian duality and the primal–dual schema to nonconvex systems are ongoing research endeavors (see Aubin and Ekeland, 1976, Ekeland, 1977, Thach, 1993, 1995, Thach, Konno, and Yokota, 1996, Singer, 1998, Gasimov, 2002). On the flip side, the Hellinger–Reissner complementary energy principle, emanating from large deformation mechanics, holds for both convex and nonconvex problems. It is very interesting to note that around the same time period of Reissner's work, the generalized potential variational principle in finite deformation elastoplasticity was proposed independently by Hu Hai-chang (1955) and K. Washizu (1955). These two variational principles are perfectly dual to each other (i.e., with zero duality gap) and play important roles in large deformation mechanics and computational methods. The inner relations between the Hellinger–Reissner and Hu–Washizu principles were discovered by Wei-Zang Chien in 1964 when he proposed a systematic method to construct generalized variational principles in solid mechanics (see Chien, 1980).

Mechanics and mathematics have been complementary partners since Newton's time, and the history of science shows much evidence of the beneficial influence of these disciplines on each other. However, the independent developments of complementary–duality theory in mathematics and mechanics for more than a half century have generated a "duality gap" between the two partners. In modern analysis, the mathematical theory of duality was mainly based on the Fenchel transformation. During the last three decades, many modified versions of the Fenchel–Moreau–Rockafellar duality have been proposed. One, the so-called relaxation method in nonconvex mechanics, can be used to solve the relaxed convex problems (see Atai and Steigmann, 1998, Dacorogna, 1989, Ye, 1992). However, due to the duality gap, these relaxed solutions do not directly yield real solutions to the nonconvex primal problems. Thus, tremendous efforts have been focused recently on finding the so-called perfect duality theory in global optimization. On the other hand, it seems that most engineers and scientists prefer the classical Legendre transformation. It turns out that their attention has been mainly focused on how to use traditional Lagrange multiplier methods and complementary constitutive laws to correctly formulate complementary variational principles for numerical computational and application purposes. Although the generalized Hellinger–Reissner principle leads to a perfect duality between the nonconvex potential variational problem and its complementary–dual, and has many important consequences in large deformation theory and computational mechanics, the extremality property of this well-known principle, as well as the Hu–Washizu principle, remained an open problem for more than 40 years, and this raised many arguments in large deformation theory and nonconvex mechanics (see Levinson, 1965, Veubeke, 1972, Koiter, 1976, Ogden, 1975, 1977, Lee and Shield, 1980a,b, Guo, 1980).

Actually, this open problem was partially solved in 1989 in the joint work of Gao and Strang (1989a) on nonconvex/nonsmooth variational problems. In order to recover the lost symmetry between the nonconvex primal problem and its dual, they introduced a so-called *complementary gap function*, which leads to a nonlinear Lagrangian duality theory in fully nonlinear variational problems. They proved that if this gap function is positive on a dual feasible space, the generalized Hellinger–Reissner energy is a saddle-Lagrangian. Therefore, this gap function provides a sufficient condition in nonconvex variational problems. However, the extremality conditions for negative gap function were ignored until 1997 when Gao (1997) got involved with a project on postbuckling problems in nonconvex mechanics. He discovered that if this gap function is negative, the generalized Hellinger–Reissner energy (the so-called super-Lagrangian) is concave in each of its variables, which led to a biduality theory. Therefore, a canonical duality theory has gradually developed, first in nonconvex mechanics, and then in global optimization (see Gao, 1990–2005). This new theory is composed mainly of a potentially useful canonical dual transformation and an associated *triality theory*, whose components comprise a saddle min-max duality and two pairs of double-min, double-max dualities. The canonical dual transformation can be used to formulate perfect dual problems without a duality gap, whereas the triality theory can be used to identify both global and local extrema.

The goal of this chapter is to present a comprehensive review on the canonical duality theory within a unified framework, and to expose its role in establishing connections between nonconvex mechanics and global optimization. Applications to constrained nonconvex optimization problems are shown to reveal some important new results that are fundamental to global optimization theory. This chapter should be of interest to both the operations research and applied mathematics communities. In order to make this presentation easy to follow by interdisciplinary readers, our attention here is mainly focused on smooth systems, although some concepts from nonsmooth analysis have been used in later sections.

8.2 Quadratic Minimization Problems

Let us begin with the simplest quadratic minimization problem (in short, the primal problem $(\mathcal{P}_q)$):

$$(\mathcal{P}_q): \quad \min \left\{ P(u) = \frac{1}{2}\langle u, Au\rangle - \langle u, f\rangle \ : \ u \in \mathcal{U}_k \right\}, \tag{8.1}$$

where $\mathcal{U}_k$ is an open subset of a linear space $\mathcal{U}$; A is a linear symmetrical operator, which maps each $u \in \mathcal{U}$ into its dual space $\mathcal{U}^*$; the bilinear form $\langle u, u^*\rangle : \mathcal{U} \times \mathcal{U}^* \to \mathbb{R}$ puts $\mathcal{U}$ and $\mathcal{U}^*$ in duality; $f \in \mathcal{U}^*$ is a given input, and $P : \mathcal{U} \to \mathbb{R}$ represents the total cost (action) of the system. The criticality condition $\delta P(u) = 0$ leads to a linear equation

$$Au = f, \tag{8.2}$$

which is called the *fundamental equation* (or equilibrium equation) in mathematical physics. By the fact that $A : \mathcal{U} \to \mathcal{U}^*$ is a symmetrical operator, we have the following canonical decomposition,

$$A = \Lambda^* D \Lambda, \tag{8.3}$$

where $\Lambda : \mathcal{U} \to \mathcal{V}$ is a so-called *geometrical operator*, which maps each $u \in \mathcal{U}$ into a so-called *intermediate space* $\mathcal{V}$, and the symmetrical operator D links $\mathcal{V}$ with its dual space $\mathcal{V}^*$. The bilinear form $\langle v \, ; v^*\rangle : \mathcal{V} \times \mathcal{V}^* \to \mathbb{R}$ puts $\mathcal{V}$ and $\mathcal{V}^*$ in duality. We distinguish between the notations $\langle \ , \ \rangle$ and $\langle \ ; \ \rangle$ according to the differences of the dual spaces $\mathcal{U} \times \mathcal{U}^*$ and $\mathcal{V} \times \mathcal{V}^*$ on which they are respectively defined. The mapping $v^* = Dv \in \mathcal{V}^*$ is called the *duality equation*. The adjoint operator $\Lambda^* : \mathcal{V}^* \to \mathcal{U}^*$, defined by

$$\langle \Lambda u \, ; v^*\rangle = \langle u, \Lambda^* v^*\rangle,$$

is also called the *balance operator*. Thus, by the use of the intermediate pair (v, v^*), the fundamental equation (8.2) can be split into the so-called *tri-canonical form*

$$\left.\begin{array}{ll} \text{(a) geometrical equation:} & \Lambda u = v \\ \text{(b) duality equation:} & Dv = v^* \\ \text{(c) balance equation:} & \Lambda^* v^* = f \end{array}\right\} \Rightarrow \ \Lambda^* D \Lambda u = f. \tag{8.4}$$

In mathematical physics, the duality equation $v^* = Dv$ is also recognized as the *constitutive law* and the operator D depends on the physical properties of the system considered.

The pair (v, v^*) is said to be a *canonical dual pair* on $\mathcal{V}_a \times \mathcal{V}_a^* \subset \mathcal{V} \times \mathcal{V}^*$ if the duality mapping $D : \mathcal{V}_a \subset \mathcal{V} \to \mathcal{V}_a^* \subset \mathcal{V}^*$ is one-to-one and onto. Generally speaking, most physical variables appear in dual pairs; that is, there exists a Gâteaux differentiable function $V : \mathcal{V}_a \to \mathbb{R}$ such that the duality relation $v^* = \delta V(v) : \mathcal{V}_a \to \mathcal{V}_a^*$ is revertible, where $\delta V(v)$ represents the Gâteaux derivative of V at v. In mathematical physics, such a function is called *free energy*. Its Legendre conjugate $V^*(v^*) : \mathcal{V}^* \to \mathbb{R}$, defined by the *Legendre transformation*

$$V^*(v^*) = \text{sta}\{\langle v; v^*\rangle - V(v) \ : \ v \in \mathcal{V}_a\}, \tag{8.5}$$

is called *complementary energy*, where sta{ } denotes finding stationary points of the statement in { }. In order to study the canonical duality theory, consider the following definition.

Definition 8.1. A real-valued function $V : \mathcal{V}_a \subset \mathcal{V} \to \mathbb{R}$ is called a canonical function on $\mathcal{V}_a$ if its Legendre conjugate $V^*(v^*)$ can be uniquely defined on $\mathcal{V}_a^* \subset \mathcal{V}^*$ such that the following relations hold on $\mathcal{V}_a \times \mathcal{V}_a^*$:

$$v^* = \delta V(v) \ \Leftrightarrow \ v = \delta V^*(v^*) \ \Leftrightarrow \ \langle v \, ; \, v^*\rangle = V(v) + V^*(v^*). \tag{8.6}$$

Clearly, if $D : \mathcal{V}_a \to \mathcal{V}_a^*$ is invertible, the quadratic function $V(v) = \frac{1}{2}\langle v; Dv\rangle$ is canonical on $\mathcal{V}_a$ and its Legendre conjugate $V^*(v^*) = \frac{1}{2}\langle D^{-1}v^*; v^*\rangle$ is a canonical function on $\mathcal{V}_a^*$. Generally speaking, if $V : \mathcal{V}_a \to \mathbb{R}$ is a canonical function and $v^* = \delta V(v)$, then (v, v^*) is a canonical dual pair on $\mathcal{V}_a \times \mathcal{V}_a^*$. The one-to-one canonical duality relation serves as a foundation for the canonical dual transformation method reviewed in the following sections. The definition of the canonical pairs and functions can be generalized to nonsmooth systems where the Fenchel transformation and subdifferential have to be applied (see Gao, 2000a,c). This is discussed in the context of constrained global optimization problems in Section 8.8 of this chapter.

In order to study general problems, we denote the linear function $\langle u, f\rangle$ by $U(u)$. If the feasible space $\mathcal{U}_k$ can be written in the form of

$$\mathcal{U}_k = \{u \in \mathcal{U}_a | \ \Lambda u \in \mathcal{V}_a\}, \tag{8.7}$$

then the problem $(\mathcal{P}_q)$ can be written in a general form

$$(\mathcal{P}): \quad \min\{P(u) = V(\Lambda u) - U(u) \ : \ u \in \mathcal{U}_k\}. \tag{8.8}$$

This general form covers many problems in applications. In continuum mechanics, the feasible set $\mathcal{U}_k$ is usually called the *kinetically admissible space*. In statics, where the function $V(v)$ is viewed as an internal (or stored) energy and $U(u)$ is considered as an external energy, the cost function $P(u)$ is the so-called *total potential* and $(\mathcal{P})$ represents a minimal potential variational problem. In dynamical systems if $V(v)$ is considered as a kinetic energy and $U(u)$ is the total potential, then $P(u)$ is called the *total action* of the system. In this case, the variational problem associated with the general form $(\mathcal{P})$ is the well-known *least action principle*. A diagrammatic representation of this tricanonical decomposition is shown in Figure 8.1.

The development of the $\Lambda^* D\Lambda$-operator theory was apparently initiated by von Neumann in 1932, and was subsequently extended and put into a more general setting in the studies of complementary variational principles in continuum mechanics by Rall (1969), Arthurs (1980), Tonti (1972a,b), Oden and Reddy (1983), and Sewell (1987). In mathematical analysis, the tricanonical form of $A = \Lambda^* D\Lambda$ has also been used to develop a mathematical theory

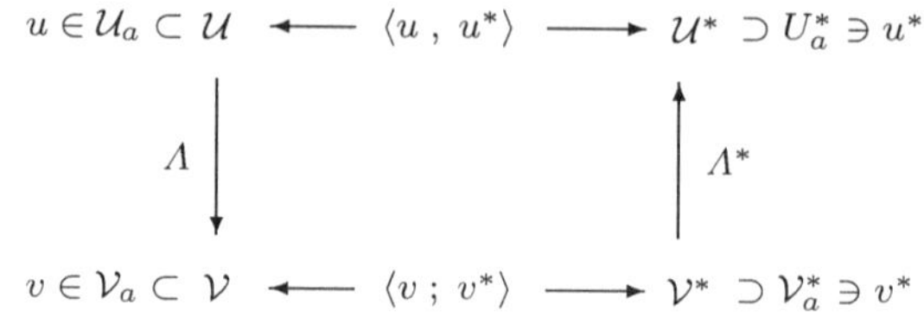

Fig. 8.1 Diagrammatic representation for quadratic systems.

of duality by Rockafellar (1970), Ekeland and Temam (1976), Toland (1978, 1979), Auchmuty (1983), Clarke (1985), and many others. In the excellent textbook by Strang (1986), the trifactorization $A = \Lambda^* D \Lambda$ for linear operators can be seen through an application of continuum theories to discrete systems. In what follows, we list some simple examples. More applications can be found in the monograph Gao (2000a).

8.2.1 Quadratic Optimization Problems in $\mathbb{R}^n$

First, we consider $\mathcal{U}$ as a finite-dimensional space such that $\mathcal{U} = \mathcal{U}^* = \mathbb{R}^n$. Thus $A : \mathcal{U} \to \mathcal{U}^*$ is a symmetric matrix in $\mathbb{R}^{n\times n}$ and the bilinear form $\langle u, u^* \rangle = u^T u^*$ is simply a dot-product in $\mathbb{R}^n$. By linear algebra, the canonical decomposition $A = \Lambda^* D \Lambda$ can be performed in many ways (see Strang, 1986), where $\Lambda : \mathbb{R}^n \to \mathbb{R}^m$ is a matrix, $D : \mathbb{R}^m \to \mathbb{R}^m$ is a symmetrical matrix, and $\Lambda^* = \Lambda^T$ maps $\mathcal{V}^* = \mathbb{R}^m$ back to $\mathcal{U}^* = \mathbb{R}^n$. The bilinear forms $\langle * \, , \, * \rangle$ and $\langle * \, ; \, * \rangle$ are simply dot products in $\mathbb{R}^n$ and $\mathbb{R}^m$, respectively, that is,

$$\langle \Lambda u; v^* \rangle = \sum_{i=1}^{m} \left(v_i^* \sum_{j=1}^{n} \Lambda_{ij} u_j \right) = \sum_{j=1}^{n} \left(u_j \sum_{i=1}^{m} \Lambda_{ij} v_i^* \right) = \langle u, \ \Lambda^T v^* \rangle.$$

If the matrix A is positive semidefinite, we can always choose a geometrical operator Λ to ensure that the matrix $D \in \mathbb{R}^{m\times m}$ is positive definite. In this case the problem $(\mathcal{P})$ is a convex program and any solution of the fundamental equation $Au = f$ also solves the minimization problem $(\mathcal{P})$.

If the matrix A is indefinite, the quadratic function $\frac{1}{2}\langle u, Au \rangle$ is nonconvex. From linear algebra, it follows then that by choosing a particular linear operator $\Lambda : \mathbb{R}^n \to \mathbb{R}^m$, the matrix A can be written in the tricanonical form:

$$A = (\Lambda^T, I) \begin{pmatrix} D & 0 \\ 0 & -C \end{pmatrix} \begin{pmatrix} \Lambda \\ I \end{pmatrix}, \tag{8.9}$$

where $D \in \mathbb{R}^{m\times m}$ is positive definite, $C \in \mathbb{R}^{n\times n}$ is positive semidefinite, and I is an identity in $\mathbb{R}^n$. In this case, both $V(v) = \frac{1}{2}\langle v; Dv\rangle$ and $U(u) = \frac{1}{2}\langle u, Cu\rangle + \langle u, f\rangle$ are convex quadratic functions, but

$$P(u) = V(\Lambda u) - U(u) = \frac{1}{2}\langle \Lambda u; D\Lambda u\rangle - \frac{1}{2}\langle u, Cu\rangle - \langle u, f\rangle$$

is a *nonconvex d.c. function*, that is, a difference of convex functions. In this case, the problem $(\mathcal{P})$ is a nonconvex quadratic minimization and the solution of $Au = f$ is only a critical point of $P(u)$.

Nonconvex quadratic programming and d.c. programming are important from both the mathematical and application viewpoints. Sahni (1974) first showed that for a negative definite matrix A, the problem $(\mathcal{P})$ is NP-hard. This result was also proved by Vavasis (1990, 1991) and by Pardalos (1991). During the last decade, several authors have shown that the general quadratic programming problem $(\mathcal{P})$ is an *NP-hard problem* in global optimization (cf. Murty and Kabadi, 1987, Horst et al., 2000). It was shown by Pardalos and Vavasis (1991) that even when the matrix A is of rank one with exactly one negative eigenvalue, the problem is NP-hard. In order to solve this difficult problem, much effort has been devoted during the last decade. Comprehensive surveys have been given by Floudas and Visweswaran (1995) for quadratic programming, and by Tuy (1995) for d.c. optimization.

8.2.2 Variational Problems in Continuum Mechanics

In continuous systems the linear space $\mathcal{U}$ is usually a function space over a time–space domain, and the linear mapping A is a differential operator. In classical Newtonian dynamics, for example, the fundamental equation (8.2) is a second-order differential equation

$$Au = -mu'' = f,$$

where f is an applied force field. In this case, $\Lambda = \mathrm{d}/\mathrm{d}t$ is a linear differential operator, $m > 0$ is a mass density, and $\Lambda^* = -\mathrm{d}/\mathrm{d}t$ can be defined by integrating by parts over a time domain $T \subset \mathbb{R}$ with boundary ∂T:

$$\langle \Lambda u; v^*\rangle = \int_T u'v^* \,\mathrm{d}t = \int_T u(-v^*)' \,\mathrm{d}t = \langle u, \Lambda^* v^*\rangle,$$

subject to the boundary conditions $u(t)v^*(t) = 0,\ \forall t \in \partial T$.

For Newton's law, $D = m$ is a constant and the tricanonical form $Au = \Lambda^* D\Lambda u = -mu'' = f$ is Newton's equilibrium equation. The quadratic form

$$V(\Lambda u) = \frac{1}{2}\langle u, Au\rangle = \frac{1}{2}\langle \Lambda u; D\Lambda u\rangle = \frac{1}{2}\int_T mu'^2 \,\mathrm{d}t$$

represents the internal (or kinetic) energy of the system, and the linear term

$$U(u) = \int_T uf \, dt$$

represents the external energy of the system. The function $P(u) = V(\Lambda u) - U(u)$ is called the total action, which is a convex functional.

For Einstein's law, however, $D = m(t) = m_o/\sqrt{1 - c^2/v^2}$ depends on the velocity $v = u'$, where $m_o > 0$ is a constant and c is the speed of light. In this case, the tricanonical form $Au = f$ leads to Einstein's theory of special relativity:

$$-\frac{d}{dt}\left(\frac{m_o}{\sqrt{1 - u'^2/c}}\frac{d}{dt}u\right) = f.$$

The kinetic energy

$$V(v) = \int_T -m_o\sqrt{1 - v^2/c^2} \, dt$$

is no longer quadratic, but is still a convex functional on $\mathcal{V}_a = \{v \in \mathcal{L}^\infty(T) \mid v(t) < c, \quad \forall t \in T\}$. By using the canonical dual transformation, the nonlinear minimization problem $(\mathcal{P})$ can be solved analytically (see Gao, 2000b).

In mass–spring systems, $A = -(m\partial_{tt} + k)$ and the fundamental equation (8.2) has the form:

$$Au = -mu'' - ku = f.$$

The additional term ku represents the spring force and $k > 0$ is a spring constant. In this case, if we let $\Lambda = (\partial_t, 1)^T$ be a vector-valued operator, the second-order linear differential operator A can still be written in the $\Lambda^* D \Lambda$ form as

$$A = -(m\frac{\partial^2}{\partial t^2} + k) = \left[-\frac{\partial}{\partial t}, \ 1\right]\begin{bmatrix} m & 0 \\ 0 & -k \end{bmatrix}\begin{bmatrix} \frac{\partial}{\partial t} \\ 1 \end{bmatrix}. \tag{8.10}$$

As evident here, if we let $\Lambda = (\partial_t, 1)^T$ be a vector-valued operator, the operator D is indefinite. However, if we let $\Lambda = \partial_t$, then similar to (8.9), we have $D = m$, which is positive definite. Thus in this dynamical system, we have.

$$V(v) = \int_T \frac{1}{2}mv^2 \, dt, \quad U(u) = \int_T \left(\frac{1}{2}ku^2 - uf\right) dt,$$

where the quadratic function $U(u)$ represents the total potential energy. The quadratic functional given by

$$P(u) = V(\Lambda u) - U(u) = \int_T \frac{1}{2}mu_{,t}^2 \, dt - \int_T [\frac{1}{2}ku^2 - uf] \, dt \tag{8.11}$$

is the well-known *total action*, which is again a d.c. functional.

Actually, every function $P(u) \in \mathcal{C}^2$ is d.c. on any compact convex set $\mathcal{U}_k$, and any d.c. optimization problem can be reduced to the canonical form (see Tuy, 1995):

$$\min\{V(\Lambda u) \; : \; U(u) \leq 0, \;\; G(u) \geq 0\}, \tag{8.12}$$

where V, U, and G are convex functions. In the next section, we demonstrate how the tricanonical $\Lambda^* D \Lambda$-operator theory serves as a framework for the Lagrangian duality theory.

8.3 Canonical Lagrangian Duality Theory

Classical Lagrangian duality was originally studied by Lagrange in analytical mechanics. In engineering mechanics it has been recognized as the *complementary variational principle,* and has been subjected to extensive study for more than several centuries. In this section, we show its connection to constrained optimization/variational problems. In addition to the well-known saddle Lagrangian duality theory, a so-called super-Lagrangian duality is presented within a unified framework, which leads to a biduality theorem in d.c. programming and convex Hamiltonian systems.

Recall the general primal problem (8.8)

$$(\mathcal{P}): \;\; \min\{P(u) = V(\Lambda u) - U(u) \; : \; u \in \mathcal{U}_k\}, \tag{8.13}$$

where $V : \mathcal{V}_a \subset \mathcal{V} \to \mathbb{R}$ is a canonical function, $U : \mathcal{U}_a \to \mathbb{R}$ is a Gâteaux differentiable function, either linear or canonical, and $\mathcal{U}_k = \{u \in \mathcal{U}_a \mid \Lambda u \in \mathcal{V}_a\}$ is a convex feasible set. Without loss of generality, we assume that the geometrical operator $\Lambda : \mathcal{U}_a \to \mathcal{V}$ can be chosen in a way such that the canonical function $V : \mathcal{V}_a \to \mathbb{R}$ is convex. By the definition of the canonical function, the duality relation $v^* = \delta V(v) : \mathcal{V}_a \to \mathcal{V}_a^*$ leads to the following Fenchel–Young equality on $\mathcal{V}_a \times \mathcal{V}_a^*$,

$$V(v) = \langle v; v^* \rangle - V^*(v^*).$$

Substituting this into equation (8.13), the Lagrangian $L(u, v^*) : \mathcal{U}_a \times \mathcal{V}_a^* \to \mathbb{R}$ associated with the canonical problem $(\mathcal{P})$ can be defined by

$$L(u, v^*) = \langle \Lambda u; v^* \rangle - V^*(v^*) - U(u). \tag{8.14}$$

Definition 8.2. (Canonical Lagrangian) A function $L : \mathcal{U}_a \times \mathcal{V}_a^* \to \mathbb{R}$ associated with the problem $(\mathcal{P})$ is called a canonical Lagrangian if it is a canonical function on $\mathcal{V}_a^*$ and a canonical or linear function on $\mathcal{U}_a$.

The criticality condition $\delta L(\bar{u}, \bar{v}^*) = 0$ leads to the well-known Lagrange equations:

$$\begin{aligned} \Lambda\bar{u} &= \delta V^*(\bar{v}^*) \\ \Lambda^*\bar{v}^* &= \delta U(\bar{u}). \end{aligned} \tag{8.15}$$

By the fact that $V : \mathcal{V}_a \to \mathcal{V}_a^*$ is a canonical function, the Lagrange equations (8.15) are equivalent to $\Lambda^*\delta V(\Lambda\bar{u}) = \delta U(\bar{u})$. If $(\bar{u}, \bar{v}^*)$ is a critical point of $L(u, v^*)$, then $\bar{u}$ is a critical point of $P(u)$ on $\mathcal{U}_k$.

Because the canonical function V is assumed to be convex on $\mathcal{V}_a$, the canonical Lagrangian $L(u, v^*)$ is concave on $\mathcal{V}_a^*$. Thus, the extremality conditions of the critical point of $L(u, v^*)$ depend on the convexity of the function $U(u)$. Two important duality theories are associated with the canonical Lagrangian, as shown in Sections 8.3.1 and 8.3.2 below.

8.3.1 Saddle-Lagrangian Duality

First, we assume that $U(u)$ is a concave function on $\mathcal{U}_a$. In this case, $L(u, v^*)$ is a saddle-Lagrangian; that is, $L(u, v^*)$ is convex on $\mathcal{U}_a$ and concave on $\mathcal{V}_a^*$. By the traditional definition, a pair $(\bar{u}, \bar{v}^*)$ is called a *saddle point* of $L(u, v^*)$ on $\mathcal{U}_a \times \mathcal{V}_a^*$ if

$$L(u, \bar{v}^*) \geq L(\bar{u}, \bar{v}^*) \geq L(\bar{u}, v^*), \qquad \forall (u, v^*) \in \mathcal{U}_a \times \mathcal{V}_a^*. \tag{8.16}$$

The classical saddle-Lagrangian duality theory can be presented precisely by the following theorem.

Theorem 8.1. (Saddle-Min-Max Theorem) *Suppose that the function $U : \mathcal{U}_a \to \mathbb{R}$ is concave and there exists a linear operator $\Lambda : \mathcal{U}_a \to \mathcal{V}_a$ such that the canonical Lagrangian $L : \mathcal{U}_a \times \mathcal{V}_a^* \to \mathbb{R}$ is a saddle function. If $(\bar{u}, \bar{v}^*) \in \mathcal{U}_a \times \mathcal{V}_a^*$ is a critical point of $L(u, v^*)$, then*

$$\min_{u \in \mathcal{U}_k} \max_{v^* \in \mathcal{V}_a^*} L(u, v^*) = L(\bar{u}, \bar{v}^*) = \max_{v^* \in \mathcal{V}_k^*} \min_{u \in \mathcal{U}_a} L(u, v^*). \tag{8.17}$$

By using this theorem, the dual function $P^d(v^*)$ can be defined as

$$P^d(v^*) = \min_{u \in \mathcal{U}_a} L(u, v^*) = U^\flat(\Lambda^* v^*) - V^*(v^*), \tag{8.18}$$

where $U^\flat : \mathcal{U}^* \to \mathbb{R}$ is a Fenchel conjugate function of U defined by the Fenchel transformation

$$U^\flat(u^*) = \min_{u \in \mathcal{U}_a} \{\langle u, u^* \rangle - U(u)\}. \tag{8.19}$$

Because $U(u)$ is a concave function on $\mathcal{U}_a$, the Fenchel conjugate $U^\flat$ is also a concave function on $\mathcal{U}_a^* \subset \mathcal{U}^*$. Thus, on the dual feasible space $\mathcal{V}_k^*$ defined by

$$\mathcal{V}_k^* = \{v^* \in \mathcal{V}_a^* |\ \Lambda^* v^* \in \mathcal{U}_a^*\}, \tag{8.20}$$

the problem, which is dual to $(\mathcal{P})$, can be proposed as the following,

$$(\mathcal{P}^d): \quad \max \left\{ P^d(v^*) \ : \ v^* \in \mathcal{V}_k^* \right\}. \tag{8.21}$$

The saddle min-max duality theory leads to the following well-known result.

Theorem 8.2. (Saddle-Lagrangian Duality Theorem) *Suppose that $L(u, v^*) : \mathcal{U}_a \times \mathcal{V}_a^* \to \mathbb{R}$ is a canonical saddle Lagrangian and $(\bar{u}, \bar{v}^*)$ is a critical point of $L(u, v^*)$. Then $\bar{u}$ is a global minimizer of $P(u)$, $\bar{v}^*$ is a global maximizer of $P^d(v^*)$, and*

$$\min_{u \in \mathcal{U}_k} P(u) = P(\bar{u}) = L(\bar{u}, \bar{v}^*) = P^d(\bar{v}^*) = \max_{v^* \in \mathcal{V}_k^*} P^d(v^*). \tag{8.22}$$

Particularly, for a given $f \in \mathcal{U}_a^*$ such that $U(u) = \langle u, f \rangle$ is a linear function on $\mathcal{U}_a$, the Fenchel-conjugate $U^\flat(u^*)$ can be computed as

$$U^\flat(u^*) = \min_{u \in \mathcal{U}_a} \{ \langle u, u^* \rangle - U(u) \} = \begin{cases} 0 & \text{if } u^* = f, \\ -\infty & \text{otherwise.} \end{cases} \tag{8.23}$$

Its effective domain is $\mathcal{U}_a^* = \{ u^* \in \mathcal{U}^* | \ u^* = f \}$. Thus, the dual feasible space can be well defined as $\mathcal{V}_k^* = \{ v^* \in \mathcal{V}_a^* | \ \Lambda^* v^* = f \}$, and the dual problem is a concave maximization problem with a linear constraint:

$$(\mathcal{P}^d): \quad \max \{ P^d(v^*) = -V^*(v^*) \ : \ \Lambda^* v^* = f, \ \ v^* \in \mathcal{V}_a^* \}. \tag{8.24}$$

By using the Lagrange multiplier $u \in \mathcal{U}_a$ to relax the linear constraint, we have

$$L(u, v^*) = -V^*(v^*) + \langle u, (\Lambda^* v^* - f) \rangle,$$

which is exactly the canonical Lagrangian (8.14) associated with the problem $(\mathcal{P})$ if the Lagrange multiplier u is in $\mathcal{U}_a$ such that $V(\Lambda u)$ is a canonical function on $\mathcal{V}_a$. This shows that the classical Lagrangian can be obtained in two ways:

1. Legendre transformation method (by choosing a proper linear operator Λ in $(\mathcal{P})$)
2. Classical Lagrange multiplier method (by relaxing the constraint $\Lambda^* v^* = f$ in $(\mathcal{P}^d)$)

In engineering mechanics, because V^* is called the complementary energy, the constrained problem

$$\min \{ V^*(v^*) \ : \ \Lambda^* v^* = f, \ \ v^* \in \mathcal{V}_a^* \}$$

is also called the *complementary variational problem* and the Lagrangian $L(u, v^*)$ is called the *generalized complementary energy*. In computational mechanics, the saddle-Lagrangian duality theory serves as a foundation for mixed and hybrid finite element methods.

8.3.2 Super-Lagrangian Duality

If the function $U : \mathcal{U}_a \to \mathbb{R}$ is convex, the canonical Lagrangian $L(u, v^*)$ is concave in each of its variables $u \in \mathcal{U}_a$ and $v^* \in \mathcal{V}_a^*$. However, $L(u, v^*)$ may not be concave in $(u, v^*) \in \mathcal{U}_a \times \mathcal{V}_a^*$ (see examples in Gao, 2000a). In this case, consider the following definition that was introduced in Gao (2000a).

Definition 8.3. A point $(\bar{u}, \bar{v}^*)$ is said to be a *supercritical* (or ∂^+-*critical*) *point* of L on $\mathcal{U}_a \times \mathcal{V}_a^*$ if

$$L(\bar{u}, v^*) \leq L(\bar{u}, \bar{v}^*) \geq L(u, \bar{v}^*), \qquad \forall (u, v^*) \in \mathcal{U}_a \times \mathcal{V}_a^*. \tag{8.25}$$

A function $L : \mathcal{U}_a \times \mathcal{V}_a^* \to \mathbb{R}$ is said to be a *supercritical* (or ∂^+) *function* on $\mathcal{U}_a \times \mathcal{V}_a^*$ if it is concave in each of its arguments; that is,

$$\begin{aligned} L : \mathcal{U}_a \to \mathbb{R} \text{ is concave}, &\quad \forall v^* \in \mathcal{V}_a^*, \\ L : \mathcal{V}_a^* \to \mathbb{R} \text{ is concave}, &\quad \forall u \in \mathcal{U}_a. \end{aligned}$$

In particular, if the supercritical function $L : \mathcal{U}_a \times \mathcal{V}_a^* \to \mathbb{R}$ is a Lagrange form, it is called a *super-Lagrangian.*

From a duality viewpoint, a point $(\bar{u}, \bar{v}^*)$ is said to be a *subcritical* (or ∂^--*critical*) *point* of L on $\mathcal{U}_a \times \mathcal{V}_a^*$ if

$$L(\bar{u}, v^*) \geq L(\bar{u}, \bar{v}^*) \leq L(u, \bar{v}^*), \qquad \forall (u, v^*) \in \mathcal{U}_a \times \mathcal{V}_a^*. \tag{8.26}$$

This definition comes from the subdifferential (see Gao, 2000a):

$$\bar{v}^* \in \partial^- V(v) = \{v^* \in \mathcal{V}_a^* \mid V(v) - V(\bar{v}) \geq \langle v - \bar{v}; \bar{v}^* \rangle, \ \ \forall v \in \mathcal{V}_a\}.$$

Clearly, $(\bar{u}, \bar{v}^*)$ is a supercritical point of L on $\mathcal{U}_a \times \mathcal{V}_a^*$ if and only if it is a subcritical point of $-L$ on $\mathcal{U}_a \times \mathcal{V}_a^*$.

Theorem 8.3. (Super-Lagrangian Duality Theorem (Gao, 2000a)) *Suppose that there exists a linear operator $\Lambda : \mathcal{U}_a \to \mathcal{V}_a$ such that $L : \mathcal{U}_a \times \mathcal{V}_a^* \to \mathbb{R}$ is a super-Lagrangian. If $(\bar{u}, \bar{v}^*) \in \mathcal{U}_a \times \mathcal{V}_a^*$ is a supercritical point of $L(u, v^*)$ on $\mathcal{U}_a \times \mathcal{V}_a^*$, then either the supermaximum theorem in the form*

$$\max_{u \in \mathcal{U}_k} \max_{v^* \in \mathcal{V}_a^*} L(u, v^*) = L(\bar{u}, \bar{v}^*) = \max_{v^* \in \mathcal{V}_k^*} \max_{u \in \mathcal{U}_a} L(u, v^*) \tag{8.27}$$

holds, or the supermin-max theorem in the form

$$\min_{u \in \mathcal{U}_k} \max_{v^* \in \mathcal{V}_a^*} L(u, v^*) = L(\bar{u}, \bar{v}^*) = \min_{v^* \in \mathcal{V}_k^*} \max_{u \in \mathcal{U}_a} L(u, v^*) \tag{8.28}$$

holds.

Based on this super-Lagrangian duality theorem, a dual function to the nonconvex d.c. function $P(u) = V(\Lambda u) - U(u)$ can be formulated as

$$P^d(v^*) = \max_{u \in \mathcal{U}_a} L(u, v^*) = U^\sharp(\Lambda^* v^*) - V^*(v^*), \tag{8.29}$$

where $U^\sharp : \mathcal{V}^* \to \mathbb{R}$ is defined by the super-Fenchel transformation

$$U^\sharp(u^*) = \max\{\langle u, u^* \rangle - U(u) \quad : \quad u \in \mathcal{U}_a\}. \tag{8.30}$$

Suppose that $\mathcal{U}_a^* \subset \mathcal{U}^*$ is an effective domain of $U^\sharp$. Then on the dual feasible space $\mathcal{V}_k^* = \{v^* \in \mathcal{V}_a^* | \ \Lambda^* v^* \in \mathcal{U}_a^*\}$, we have the following result.

Theorem 8.4. (Biduality Theory (Gao, 2000a)) *If $(\bar{u}, \bar{v}^*)$ is a super-critical point of $L(u, v^*)$, then either the double-min theorem in the form*

$$\min_{u \in \mathcal{U}_k} P(u) = P(\bar{u}) = L(\bar{u}, \bar{v}^*) = P^d(\bar{v}^*) = \min_{v^* \in \mathcal{V}_k^*} P^d(v^*) \tag{8.31}$$

holds, or the double-max theorem in the form

$$\max_{u \in \mathcal{U}_k} P(u) = P(\bar{u}) = L(\bar{u}, \bar{v}^*) = P^d(\bar{v}^*) = \max_{v^* \in \mathcal{V}_k^*} P^d(v^*) \tag{8.32}$$

holds.

The Hamiltonian $H : \mathcal{U}_a \times \mathcal{V}_a^* \to \mathbb{R}$ associated with the Lagrangian is defined by

$$H(u, v^*) = \langle \Lambda u; \ v^* \rangle - L(u, v^*) = V^*(v^*) + U(u). \tag{8.33}$$

Clearly, if $L(u, v^*)$ is a super-Lagrangian, the Hamiltonian $H(u, v^*)$ is convex in each of its variables and in terms of $H(u, v^*)$, the Lagrange equations (8.15) can be written in the so-called *Hamiltonian canonical form*:

$$\Lambda u = \delta_{v^*} H(u, v^*), \qquad \Lambda^* v^* = \delta_u H(u, v^*). \tag{8.34}$$

However, this nice symmetrical form and the convexity of the Hamiltonian do not afford new insights into understanding the extremality conditions of the nonconvex problem. The super-Lagrangian duality theory plays an important role in d.c. programming, convex Hamilton systems, and global optimization.

8.3.3 Applications in Quadratic Programming and Commentary

Now, let us consider the nonconvex quadratic programming problem $(\mathcal{P}_q)$ where the cost function is a d.c. function

$$P(u) = \frac{1}{2}\langle \Lambda u; D\Lambda u \rangle - \frac{1}{2}\langle u, Cu \rangle - \langle u, f \rangle$$

as discussed in (8.2.1), where D is a positive definite matrix in $\mathbb{R}^{m\times m}$, and $C \in \mathbb{R}^{n\times n}$ is positive semidefinite. Because $U(u) = \frac{1}{2}\langle u, Cu\rangle + \langle u, f\rangle$ in this case is convex, the Lagrangian

$$L(u, v^*) = \langle \Lambda u; v^*\rangle - \frac{1}{2}\langle D^{-1}v^*; v^*\rangle - \frac{1}{2}\langle u, Cu\rangle - \langle u, f\rangle$$

is a super-Lagrangian. By using the super-Fenchel transformation, we have

$$\begin{aligned} U^\sharp(u) &= \max_{u\in\mathcal{U}_a}\{\langle u, u^* - f\rangle - \frac{1}{2}\langle u, Cu\rangle\} \\ &= \frac{1}{2}\langle C^+(u^* - f), (u^* - f)\rangle, \end{aligned}$$

subject to $u^* - f \in \mathcal{C}(C)$, where C^+ is a pseudo-inverse of C and $\mathcal{C}(C)$ represents the column space of C. Thus, on the dual feasible space

$$\mathcal{V}_k^* = \{v^* \in \mathcal{V}_a \subset \mathbb{R}^m \mid \Lambda^T v^* - f \in \mathcal{C}(C)\}, \tag{8.35}$$

the dual function

$$P^d(v^*) = \frac{1}{2}\langle C^+(\Lambda^* v^* - f), \Lambda^* v^* - f\rangle - \frac{1}{2}\langle D^{-1}v^*; v^*\rangle \tag{8.36}$$

is also a d.c. function. The biduality theorem shows that the optimal values of the primal and dual problems are equal. If $\bar{u}$ solves the primal (either minimization or maximization) and $\Lambda^*\bar{v}^* - f \in \partial^- U(\bar{u})$, then $\bar{v}^*$ solves the dual.

One of the earliest and best known double-min duality schemes was formulated by Toland (1978) for the d.c. minimization problem

$$\min\{W(u) - U(u) \ : \ u \in \text{dom } W\}, \tag{8.37}$$

where $W(u)$ is an arbitrary function, $U(u)$ is a convex proper lsc function on $\mathbb{R}^n$, and dom W represents effective domain of W. The dual problem is

$$\min\{U^\sharp(u^*) - W^\sharp(u^*) \ : \ u^* \in \text{dom } U^\sharp\}, \tag{8.38}$$

which is also a d.c. minimization problem in $\mathbb{R}^n$. The generalizations were made by Auchmuty (1983) to general nonconvex functionals with a linear operator Λ. Since then, several important duality concepts have been developed and studied for nonconvex optimization and d.c. programming by Crouzeix (1981), Hiriart-Urruty (1985), Singer (1998), Penot and Volle (1990), Tuy (1995), Thach (1993, 1995), and many others. A detailed review on duality in d.c. programming appears in Tuy (1995). Much of the foregoing discussion is based on generalized nonconvex functionals, which are allowed to be extended-real-valued. In order to avoid difficulties such as $\infty - \infty$, a modified version of the double-min duality in optimization was presented in Rockafellar and Wets (1998). It is traditional in the calculus of variations and optimization that the primal problem is always taken to be a minimization problem. However, this tradition somewhat obscures our view of more general problems. In convex Hamiltonian systems where $V(v) = \frac{1}{2}\langle \Lambda u, D\Lambda u\rangle$ is a kinetic energy function and $U(u) = \frac{1}{2}\langle u, Cu\rangle + \langle u, f\rangle$ is a total potential energy function, the d.c. function $P(u) = V(\Lambda u) - U(u)$ represents a total action of the system. As pointed out in Ekeland (1990) and Gao (2000a), in the context of convex dynamical systems, the least action principle is somehow misleading because the action is a d.c. function that takes minimum and maximum values periodically over the time domain. Both the min- and the max-primal problems have to be considered simultaneously in a period. The biduality theorem reveals a periodic behavior of dynamical systems.

In two-person game theory, the biduality theory shows that the d.c. programming problem has two Nash equilibrium points.

The super-Lagrangian duality and the associated biduality theory were first proposed in the monograph Gao (2000a). Based on this theory and the tricanonical form $\Lambda^* D\Lambda$, we reformulated the nonconvex quadratic programming problem in a dual form of (8.36), which is well defined on the dual feasible space $\mathcal{V}_k^* \subset \mathbb{R}^m$ (8.35). Because $m \leq n$, we believe this new dual form will play an important role in nonconvex quadratic programming theory.

8.4 Complementary Variational Principles in Continuum Mechanics

This section presents two simple applications of the canonical Lagrange duality theory in continuum mechanics. The first application shows the connection between the mathematical theory of saddle-Lagrangian duality and the complementary energy variational principles in static linear elasticity, which are well known in solid mechanics and computational mechanics. Indeed, the application of the super-Lagrangian duality theory to convex Hamiltonian systems may bring some important insights into extremality conditions in dynamic systems.

8.4.1 Linear Elasticity

Let us consider an elastic material in $\mathbb{R}^3$ occupying a simple connected domain $\Omega \subset \mathbb{R}^3$ with boundary $\Gamma = \partial\Omega = \Gamma_u \cup \Gamma_t$ such that $\Gamma_u \cap \Gamma_t = \varnothing$. On Γ_u, the boundary displacement $\bar{u}$ is given, whereas on Γ_t, a surface traction $\bar{\mathbf{t}}$ is prescribed. Suppose that the elastic body is subjected to a distributed force field f. The equilibrium equation $Au = f$ has the following form,

$$-\frac{\partial}{\partial x_j}\left(D_{ijkl}\frac{\partial u_k(x)}{\partial x_l}\right) = f_i(x), \qquad \forall x \in \Omega, \tag{8.39}$$

where $D = \{D_{ijkl}\}$ $(i,j,k,l = 1,2,3)$ is a positive definite fourth-order elastic tensor, satisfying $D_{ijkl} = D_{jikl} = D_{klij}$, and Einstein's summation convention over the repeated subindices is used here. In this problem, $A = -\operatorname{div} D \operatorname{grad}$ is an elliptic operator, $\Lambda = \operatorname{grad}$ is a gradient, and $v = \operatorname{grad} u$ is called the deformation gradient. Its symmetrical part is an infinitesimal strain tensor, denoted as $\boldsymbol{\epsilon} = \frac{1}{2}(\nabla u + (\nabla u)^T)$. The dual variable $v^* = D\boldsymbol{\epsilon}$ is a stress tensor, usually denoted by $\boldsymbol{\sigma}$. In this infinite-dimensional system $\mathcal{U} = \mathcal{L}^2(\Omega; \mathbb{R}^3) = \mathcal{U}^*$ and $\mathcal{V} = \mathcal{L}^2(\Omega; \mathbb{R}^{3\times 3}) = \mathcal{V}^*$. The bilinear forms are defined by

$$\langle u, f\rangle = \int_\Omega u \cdot f \, \mathrm{d}\Omega, \qquad \langle \boldsymbol{\epsilon}, \boldsymbol{\sigma}\rangle = \int_\Omega \boldsymbol{\epsilon} : \boldsymbol{\sigma} \, \mathrm{d}\Omega,$$

where $\boldsymbol{\epsilon} : \boldsymbol{\sigma} = \operatorname{tr}(\boldsymbol{\epsilon} \cdot \boldsymbol{\sigma}) = \epsilon_{ij}\sigma_{ij}$. The adjoint operator Λ^* in this case is $\Lambda^* = \{-\operatorname{div} \text{ in } \Omega, \ \mathbf{n}\cdot \text{ on } \Gamma\}$, and $-\operatorname{div}$ is also called the formal adjoint of $\Lambda = \operatorname{grad}$. Let

$$\mathcal{U}_a = \{u \in \mathcal{U} \mid u(x) = \bar{u}(x), \ \forall x \in \Gamma_u\}$$
$$\mathcal{V}_a = \{\boldsymbol{\epsilon} \in \mathcal{V} \mid \boldsymbol{\epsilon}(x) = \boldsymbol{\epsilon}^T(x), \ \forall x \in \Omega\}.$$

Thus on the feasible space, that is, the so-called *statically admissible space* $\mathcal{U}_k = \{u \in \mathcal{U}_a \mid \Lambda u \in \mathcal{V}_a\}$, the quadratic form

$$P(u) = \int_\Omega \frac{1}{2}(\nabla u) : D : (\nabla u) \, \mathrm{d}\Omega - \int_{\Gamma_t} u \cdot f \, \mathrm{d}\Gamma \tag{8.40}$$

is the so-called *total potential* of the deformed elastic body. The *minimal potential principle* leads to the convex variational problem

$$\min\{P(u) \ : \ u \in \mathcal{U}_k\}. \tag{8.41}$$

The functional $V(\boldsymbol{\epsilon}) = \frac{1}{2}\langle \boldsymbol{\epsilon}; D\boldsymbol{\epsilon}\rangle$ is call the *internal (or stored) potential.* Its Legendre conjugate

$$V^*(\boldsymbol{\sigma}) = \{\langle \boldsymbol{\epsilon}, \boldsymbol{\sigma}\rangle - U(\boldsymbol{\epsilon}) | \ \boldsymbol{\sigma} = D : \boldsymbol{\epsilon}\} = \int_\Omega \frac{1}{2}\boldsymbol{\sigma} : D^{-1} : \boldsymbol{\sigma} \, \mathrm{d}\Omega$$

is known as the *complementary energy* in solid mechanics. Because

$$U(u) = \int_\Omega u \cdot f \, \mathrm{d}\Omega + \int_{\Gamma_u} u \cdot \bar{\mathbf{t}} \, \mathrm{d}\Gamma$$

is linear, which is also called the *external potential*, the Lagrangian associated with the total potential $P(u)$, as given by

$$L(u,\boldsymbol{\sigma}) = \int_\Omega [(\nabla u) : \boldsymbol{\sigma} - \frac{1}{2}\boldsymbol{\sigma} : D^{-1} : \boldsymbol{\sigma}]\, \mathrm{d}\Omega - \int_{\Gamma_u} u \cdot \bar{\mathbf{t}}\, \mathrm{d}\Gamma, \tag{8.42}$$

can be considered as a saddle Lagrangian, which is the well-known generalized *Hellinger–Reissner complementary energy.* Thus, by the saddle Lagrangian duality, the dual functional $P^d(\boldsymbol{\sigma})$ is defined by

$$P^d(\boldsymbol{\sigma}) = \min_{u \in \mathcal{U}_a} L(u,\boldsymbol{\sigma}) = U^\flat(\Lambda^*\boldsymbol{\sigma}) - V^*(\boldsymbol{\sigma}),$$

where

$$U^\flat(\Lambda^*\boldsymbol{\sigma}) = \min_u \left\{ \int_\Omega (\nabla u) : \boldsymbol{\sigma})\, \mathrm{d}\Omega - \int_\Omega u \cdot f\, \mathrm{d}\Omega - \int_{\Gamma_t} u \cdot \bar{\mathbf{t}}\, \mathrm{d}\Gamma \right\}$$

$$= \begin{cases} \int_{\Gamma_u} \bar{u} \cdot \boldsymbol{\sigma} \cdot \mathbf{n}\, \mathrm{d}\Gamma & \text{if } -\operatorname{div} \boldsymbol{\sigma} = 0 \quad \text{in } \Omega, \quad \boldsymbol{\sigma} \cdot \mathbf{n} = \bar{\mathbf{t}} \text{ on } \Gamma_t, \\ -\infty & \text{otherwise.} \end{cases}$$

Thus, on the dual feasible space, that is, the so-called *statically admissible space* defined by

$$\mathcal{V}_k^* = \{\boldsymbol{\sigma} \in \mathcal{V}_a^* \mid -\operatorname{div} \boldsymbol{\sigma} = 0 \quad \text{in } \Omega, \quad \boldsymbol{\sigma} \cdot \mathbf{n} = \bar{\mathbf{t}} \quad \text{on } \Gamma_t\},$$

the dual problem for this linear elasticity case is given by

$$\max \left\{ P^d(\boldsymbol{\sigma}) = \int_{\Gamma_u} \bar{u} \cdot \boldsymbol{\sigma} \cdot \mathbf{n}\, \mathrm{d}\Gamma - \int_\Omega \frac{1}{2}\boldsymbol{\sigma} : D^{-1} : \boldsymbol{\sigma}\, \mathrm{d}\Omega \ : \ \boldsymbol{\sigma} \in \mathcal{V}_k^* \right\}. \tag{8.43}$$

This is a concave maximization problem with linear constraints. The Lagrange multiplier u for the equilibrium constraints is the solution of the primal problem.

In continuum mechanics, the functional $-P^d$, denoted by

$$P^c(\boldsymbol{\sigma}) = \int_\Omega \frac{1}{2}\boldsymbol{\sigma} : D^{-1} : \boldsymbol{\sigma}\, \mathrm{d}\Omega - \int_{\Gamma_u} \bar{u} \cdot \boldsymbol{\sigma} \cdot \mathbf{n}\, \mathrm{d}\Gamma,$$

is called the total complementary energy. Thus, instead of the dual problem (8.43), the minimum complementary variational problem

$$\min \{P^c(\boldsymbol{\sigma}) \ : \ \boldsymbol{\sigma} \in \mathcal{V}_k^*\}$$

has been extensively studied by engineers, which serves as a foundation for the so-called *stress*, or *equilibrium, finite element methods.*

8.4.2 Convex Hamiltonian Systems

Recall the mass–spring dynamical system discussed in Section 8.2, where the total action is a d.c. function of the form

$$\begin{aligned} P(u) &= V(\Lambda u) - U(u) \\ &= \int_T \frac{1}{2} m(u_{,t})^2 \,\mathrm{d}t - \int_T [\frac{1}{2} k u^2 - uf] \,\mathrm{d}t. \end{aligned} \tag{8.44}$$

The Lagrangian

$$L(u,p) = \int_T [u_{,t} p - \frac{1}{2} m^{-1} p^2 - \frac{1}{2} k u^2] \,\mathrm{d}t - \int_T uf \,\mathrm{d}t$$

is not a saddle function, thus the Hamiltonian

$$\begin{aligned} H(u,p) &= \langle \Lambda u, p \rangle - L(u,p) \\ &= \int_T [\frac{1}{2} m^{-1} p^2 + \frac{1}{2} k u^2] \,\mathrm{d}t + \int_T uf \,\mathrm{d}t \end{aligned} \tag{8.45}$$

was extensively used in classical dynamical systems. One of the main reasons for this could be that $H(u,p)$ is convex. Thus, the original differential equation $Au = -mu_{,tt} - ku = f$ can be written in the well-known *Hamiltonian canonical form:*

$$\Lambda u = \delta_p H(u,p), \qquad \Lambda^* p = \delta_u H(u,p). \tag{8.46}$$

However, an important phenomenon has been hiding in the shadow of this convex Hamiltonian for centuries. Because $L(u,p)$ is a super-Lagrangian, the dual action can be formulated as

$$\begin{aligned} P^d(p) &= \max_u L(u,p) \\ &= \frac{1}{2} \int_T k^{-1} (p_{,t} - f)^2 \,\mathrm{d}t - \frac{1}{2} \int_T m^{-1} p^2 \,\mathrm{d}t, \end{aligned}$$

which is also a d.c. functional. The biduality theory

$$\min P(u) = \min P^d(p),$$

$$\max P(u) = \max P^d(p)$$

shows that the well-known least action principle in periodic dynamical systems is actually a misnomer; that is, the periodic solution $u(t)$ does not minimize the total action $P(u)$, which could be either a minimizer or a maximizer, depending on the time period (see Gao, 2000a).

8.5 Nonconvex Problems with Double-Well Energy

We now turn our attention to duality theory in nonconvex systems by considering a very simple problem in $\mathbb{R}^n$:

$$(\mathcal{P}_w): \ \min \left\{ P(u) = \frac{1}{2}\alpha \left(\frac{1}{2}|Bu|^2 - \lambda \right)^2 - \langle u, f \rangle \ : \ u \in \mathbb{R}^n \right\}, \tag{8.47}$$

where $B \in \mathbb{R}^{m \times n}$ is a matrix, $\alpha, \lambda > 0$ are positive constants, and $|v|$ denotes the Euclidean norm of v. The criticality condition $\delta P(u) = 0$ leads to a coupled nonlinear algebraic system in $\mathbb{R}^n$:

$$\alpha \left(\frac{1}{2}|Bu|^2 - \lambda \right) B^T B u = f. \tag{8.48}$$

Clearly, it is difficult to solve this nonlinear system by direct methods. Also, due to the nonconvexity of $P(u)$, any solution to this nonlinear system satisfies only a necessary condition. The nonconvex function $W(v) = \frac{1}{2}\alpha(\frac{1}{2}|v|^2 - \lambda)^2$ is a so-called *double-well energy*, which was first studied by van der Waals in fluid mechanics in 1895 (see Rowlinson, 1979). For each given parameter $\lambda > 0$, $W(v)$ has two minimizers and one local maximizer (see Figure 8.2a). The global and local minimizers depend on the input f (see Figure 8.2b). This double-well function has extensive applications in mathematical physics. In phase transitions of shape memory alloys, or in the mathematical theory of superconductivity, $W(v)$ is the well-known Landau second-order free energy, and each of its local minimizers represents a possible phase state of the material. In quantum mechanics, if v represents the Higgs' field strength, then $W(v)$ is the energy. It was discovered in the context of postbuckling analysis of large deformed beam models, that the total potential is also a double-well energy (see Gao, 2000d), and each potential well represents a possible buckled beam state. More examples can be found in a recent review article (Gao, 2003b).

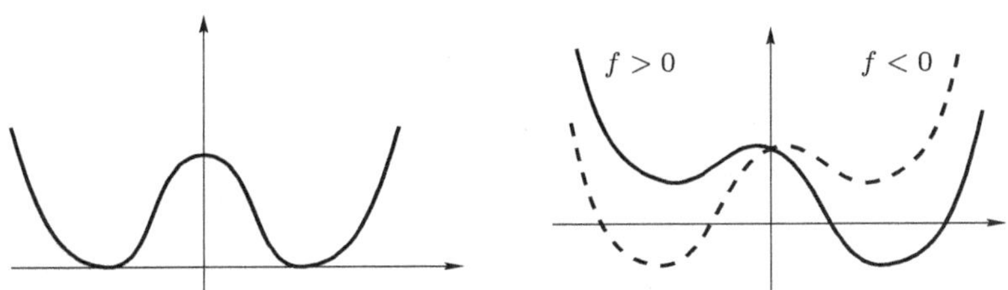

(a) Graph of $W(u) = \frac{1}{2}(\frac{1}{2}u^2 - \lambda)^2$ (b) Graphs of $P(u) = W(u) - fu$

Fig. 8.2 Double-well energy and nonconvex potential functions.

8.5.1 Classical Lagrangian and Duality Gap

If we choose $\Lambda = B$ as a linear operator, the primal function can be written in the traditional form $P(u) = W(Bu) - U(u)$, where $U(u) = \langle u, f \rangle$ is a linear function. Because the duality relation $v^* = \delta W(v) = \alpha(\frac{1}{2}|v|^2 - \lambda)v$ is not one-to-one, the Legendre conjugate

$$W^*(v^*) = \text{sta}\{\langle v, v^* \rangle - W(v) \ : \ v \in \mathbb{R}^m\}$$

is not uniquely defined. Thus, the entity (v, v^*) associated with the nonconvex function $W(v)$ is not a canonical dual pair. By using the Fenchel transformation

$$W^\sharp(v^*) = \max\{\langle v, v^* \rangle - W(v) \ : \ v \in \mathbb{R}^m\},$$

the traditional Lagrangian (associated with the linear operator $\Lambda = B$) can still be defined as

$$L(u, v^*) = \langle Bu, v^* \rangle - W^\sharp(v^*) - \langle u, f \rangle. \tag{8.49}$$

Thus, the classical Lagrangian duality theory $P^\sharp(v^*) = \max_u L(u, v^*)$ leads to the well-known Fenchel–Rockafellar dual problem

$$(\mathcal{P}^\sharp) : \ \max_{v^* \in \mathbb{R}^m} \{P^\sharp(v^*) = -W^\sharp(v^*) \ : \ B^T v^* = f\}. \tag{8.50}$$

This is a linearly constrained concave maximization problem. The Lagrange multiplier for the linear constraint set is u. However, due to the nonconvexity of $W(v)$, the Fenchel–Young inequality

$$W(v) + W^\sharp(v^*) \leq \langle v, v^* \rangle$$

leads to a weak duality relation

$$\min P \geq \max P^\sharp.$$

The nonzero value $\theta \equiv \min P(u) - \max P^\sharp(v^*)$ is called the *duality gap*. This duality gap shows that the classical Lagrange multiplier u may not be a solution to the primal problem. Thus, the Fenchel–Rockafellar duality theory can be used mainly for solving convex problems. In order to eliminate this duality gap, many modified Lagrangian dualities have been proposed during recent years (see, for examples, Aubin and Ekeland, 1976, Rubinov et al., 2001, 2003, Goh and Yang, 2002, Huang and Yang, 2003, Zhou and Yang, 2004). Most of these mathematical approaches are based on penalization of a class of augmented Lagrangian functions. On the other hand, the canonical duality theory addressed in the next section is based on a fundamental truth in physics; that is, physical variables appear in (canonical) pairs. The one-to-one canonical duality relation leads to a perfect duality theory in mathematical physics and global optimization.

8.5.2 Canonical Dual Transformation and Triality Theory

In order to recover the duality gap, a *canonical duality theory* was developed during the last 15 years: first in nonconvex mechanics and analysis (see Gao and Strang, 1989a,b, Gao, 1997, 1998a, 2000a), then in global optimization (see Gao, 2000a,c, 2003a, 2004b). The key idea of this theory is to choose a right operator (usually nonlinear) $\xi = \Lambda(u)$ such that the nonconvex function $W(u)$ can be written in the canonical form

$$W(u) = V(\Lambda(u)),$$

where $V(\xi)$ is a canonical function of $\xi = \Lambda(u)$. For the present nonconvex problem (8.47), instead of $\Lambda = B$, we choose

$$\xi = \Lambda(u) = \frac{1}{2}|Bu|^2, \tag{8.51}$$

which is a quadratic map from $\mathcal{U} = \mathbb{R}^n$ into $\mathcal{V}_a = \{\xi \in \mathbb{R} \mid \xi \geq 0\}$. Thus, the canonical function

$$V(\xi) = \frac{1}{2}\alpha(\xi - \lambda)^2$$

is simply a scale-valued quadratic function well defined on $\mathcal{V}_a$, which leads to a linear duality relation

$$\varsigma = \delta V(\xi) = \alpha(\xi - \lambda).$$

Let $\mathcal{V}_a^* = \{\varsigma \in \mathbb{R} \mid \varsigma \geq -\alpha\lambda\}$ be the range of this duality mapping. So (ξ, ς) forms a canonical duality pair on $\mathcal{V}_a \times \mathcal{V}_a^*$, and the Legendre conjugate V^* is also a quadratic function:

$$V^*(\varsigma) = \text{sta}\left\{ \langle \xi ; \varsigma \rangle - \frac{1}{2}\alpha(\xi - \lambda)^2 \ : \ \xi \in \mathcal{V}_a \right\} = \frac{1}{2}\alpha^{-1}\varsigma^2 + \lambda\varsigma.$$

Thus, replacing $W(u) = V(\Lambda(u)) = \langle \Lambda(u); \varsigma\rangle - V^*(\varsigma)$ in $P(u) = W(u) - U(u)$, the so-called *total complementary function* (Gao and Strang, 1989a, Gao, 2000a) can be defined by

$$\begin{aligned} \Xi(u, \varsigma) &= \langle \Lambda(u) \ ; \ \varsigma \rangle - V^*(\varsigma) - U(u) \\ &= \frac{1}{2}|Bu|^2\varsigma - \frac{1}{2}\alpha^{-1}\varsigma^2 - \lambda\varsigma - u^T f. \end{aligned} \tag{8.52}$$

The criticality condition $\delta \Xi(u, \varsigma) = 0$ leads to the following canonical equilibrium equations.

$$(\frac{1}{2}|Bu|^2 - \lambda) = \alpha^{-1}\varsigma, \tag{8.53}$$

$$\varsigma B^T B u = f. \tag{8.54}$$

Equation (8.53) is actually the inverse duality relation $\xi = \delta V^*(\varsigma)$, which is equivalent to $\varsigma = \alpha(\frac{1}{2}|Bu|^2 - \lambda)$. Thus, equation (8.54) is identical to the Euler equation (8.48). This shows that the critical point of the total complementary function is also a critical point of the primal problem. For a fixed $\varsigma \neq 0$, solving (8.54) for u gives

$$u = \frac{1}{\varsigma}(B^T B)^{-1} f. \tag{8.55}$$

Substituting this result into the total complementary function leads to the canonical dual function

$$P^d(\varsigma) = -\frac{1}{2\varsigma} f^T (B^T B)^{-1} f - \lambda\varsigma - \frac{1}{2}\alpha^{-1}\varsigma^2, \tag{8.56}$$

which is well defined on the dual feasible space given by

$$\mathcal{V}_k^* = \{\varsigma \in \mathcal{V}_a^* \mid \varsigma \neq 0\} = \{\varsigma \in \mathbb{R} \mid \varsigma \geq -\alpha\lambda, \ \ \varsigma \neq 0\}.$$

The criticality condition $\delta P^d(\varsigma) = 0$ gives the canonical dual algebraic equation:

$$2\varsigma^2(\alpha^{-1}\varsigma + \lambda) = f^T (B^T B)^{-1} f. \tag{8.57}$$

Theorem 8.5. (Gao, 2000c) *For any given parameters α, $\lambda > 0$, and vector $f \in \mathbb{R}^n$, the canonical dual function* (8.56) *has at most three critical points $\bar{\varsigma}_i$ $(i = 1, 2, 3)$ satisfying*

$$\bar{\varsigma}_1 > 0 > \bar{\varsigma}_2 \geq \bar{\varsigma}_3. \tag{8.58}$$

For each of these roots, the vector

$$\bar{u}_i = (B^T B)^{-1} f / \bar{\varsigma}_i, \qquad \text{for } i = 1, 2, 3, \tag{8.59}$$

is a critical point of the nonconvex function $P(u)$ in Problem (8.47), *and we have*

$$P(\bar{u}_i) = P^d(\bar{\varsigma}_i), \qquad \forall i = 1, 2, 3. \tag{8.60}$$

The original version of this theorem was first discovered in a postbifurcation problem of a large deformed beam model in 1997 (Gao, 1997), which shows that there is no duality gap between the nonconvex function $P(u)$ and its canonical dual $P^d(\varsigma)$. The dual algebraic equation (8.57) can be solved exactly to obtain all critical points, therefore the vector $\{\bar{u}_i\}$ defined by (8.59) yields a complete set of solutions to the nonlinear algebraic system (8.48).

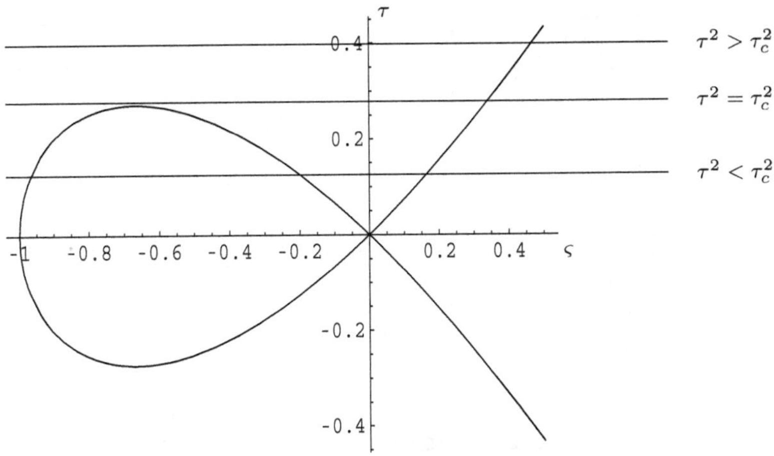

Fig. 8.3 Graph of the dual algebraic equation (8.57) and a geometrical proof of the triality theorem.

Let $\tau^2 = f^T(B^TB)^{-1}f$. In algebraic geometry, the graph of the algebraic equation $\tau^2 = 2\varsigma^2(\alpha^{-1}\varsigma + \lambda)$ is the so-called *singular algebraic curve* in (ς, τ)-space (i.e., the point $\varsigma = 0$ is on the curve; cf. Silverman and Tate, 1992). From this algebraic curve, we can see that there exists a constant τ_c such that if $\tau^2 > \tau_c^2$, the dual algebraic equation (8.57) has a unique solution $\varsigma > 0$. It has three real solutions if and only if $\tau^2 < \tau_c^2$.

It is interesting to note that for $\varsigma > 0$, the total complementary function $\Xi(u, \varsigma)$ is a saddle function and the well-known saddle min-max theory leads to

$$\min_u \max_{\varsigma>0} \Xi(u,\varsigma) = \Xi(\bar{u},\bar{\varsigma}) = \max_{\varsigma>0} \min_u \Xi(u,\varsigma). \tag{8.61}$$

This means that $\bar{u}_1$ is a global minimizer of $P(u)$ and $\bar{\varsigma}_1$ is a global maximizer on the open domain $\varsigma > 0$. However, for $\varsigma < 0$, the total complementary function $\Xi(u, \varsigma)$ is concave in both u and $\varsigma < 0$; that is, it is a supercritical function. Thus, by the biduality theory, we have that either

$$\min_u \max_{\varsigma<0} \Xi(u,\varsigma) = \Xi(\bar{u},\bar{\varsigma}) = \min_{\varsigma<0} \max_u \Xi(u,\varsigma) \tag{8.62}$$

holds on a neighborhood of $(\bar{u}, \bar{\varsigma})$, or

$$\max_u \max_{\varsigma<0} \Xi(u,\varsigma) = \Xi(\bar{u},\bar{\varsigma}) = \max_{\varsigma<0} \max_u \Xi(u,\varsigma). \tag{8.63}$$

Actually, the extremality conditions can be easily viewed through the graph of $P^d(\varsigma)$ (see Figure 8.4). To compare with this canonical dual function, the graph of $P(u)$ for $n = 1$ is also shown in Figure 8.4. Precisely, we have the following result (see Gao, 2000a,b).

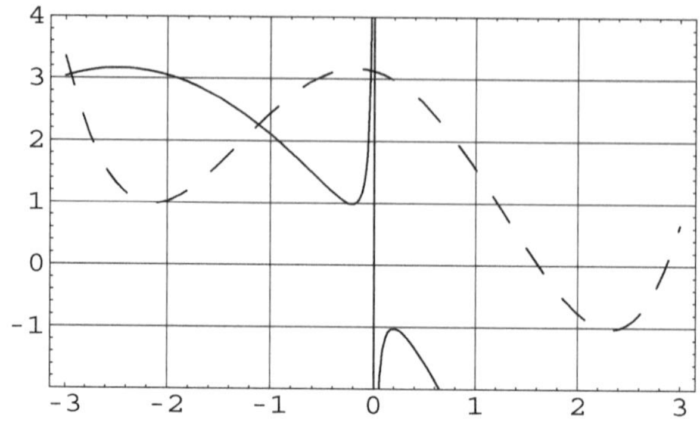

Fig. 8.4 Graphs of $P(x)$ (dashed) and $P^d(\varsigma)$ (solid) for $n = 1$.

Theorem 8.6. (Complete Solutions for Problem $(\mathcal{P}_w)$ (Gao, 1998a, 2000a)) *For certain given parameters α, $\lambda > 0$, and the vector $f \in \mathbb{R}^n$, if $\tau^2 > \tau_c^2 = 8\alpha^2\lambda^3/27$, then the canonical dual function $P^d(\varsigma)$ has only one critical point $\bar{\varsigma} > 0$, which is a global maximizer of $P^d(\varsigma)$, and $\bar{u} = (B^T B)^{-1} f/\bar{\varsigma}$ is a global minimizer of $P(u)$.*

If $\tau^2 < \tau_c^2$, the canonical dual function $P^d(\varsigma)$ has three critical points $\bar{\varsigma}_1 > 0 > \bar{\varsigma}_2 > \bar{\varsigma}_3$ such that $\bar{u}_1$ is a global minimizer, $\bar{u}_2$ is a local minimizer, and $\bar{u}_3$ is a local maximizer of $P(u)$.

8.5.3 Canonical Dual Solutions to Nonconvex Variational Problems

Similar to the nonconvex optimization problem (8.47) with the double-well function, let us now consider the following typical nonconvex variational problem,

$$(\mathcal{P}): \min_{u \in \mathcal{U}_k} \left\{ P(u) = \int_0^1 \frac{1}{2}\alpha \left(\frac{1}{2} u'^2 - \lambda \right)^2 \mathrm{d}x - \int_0^1 uf \,\mathrm{d}x \right\}, \tag{8.64}$$

where $f(x)$ is a given function, $\lambda > 0$ is a parameter, and

$$\mathcal{U}_k = \{ u \in \mathcal{L}^2[0,1] |\; u' \in \mathcal{L}^4[0,1],\;\; u(0) = 0 \}$$

is an admissible space. Compared with Problem (8.47), we see that the linear operator B in this case is a differential operator $\mathrm{d}/\mathrm{d}x$. This variational problem appears frequently in association with phase transitions in fluids and solids, and in postbuckling analysis of large deformed structures. The criticality condition $\delta P(u) = 0$ leads to a nonlinear differential equation in the domain $(0, 1)$ with the natural boundary condition at $x = 1$; that is,

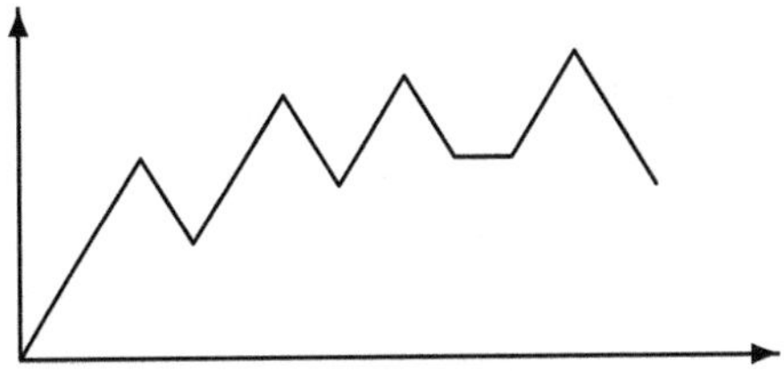

Fig. 8.5 Zigzag function: Solution to the nonlinear boundary value problem (8.65).

$$\left[\alpha u' \left(\frac{1}{2}u'^2 - \lambda\right)\right]' + f(x) = 0, \quad \forall x \in (0, 1), \tag{8.65}$$

$$\alpha u' \left(\frac{1}{2}u'^2 - \lambda\right) = 0 \quad \text{at } x = 1. \tag{8.66}$$

Due to its nonlinearity, a solution to this boundary value problem is not unique. Particularly, if we let $f(x) = 0$, the equation (8.65) could have three real roots $u'(x) = \{0, \pm\sqrt{2\lambda}\}$. Thus, any zigzag curve $u(x)$ with slope $\{0, \pm\sqrt{2\lambda}\}$ solves the boundary value problem, but may not be a global minimizer of the total energy $P(u)$. This problem shows an important fact that in nonconvex analysis the criticality condition is only necessary, but not sufficient for solving variational problems. Traditional direct approaches for solving nonconvex variational problems are very difficult, or impossible. However, by using the canonical dual transformation, this problem can be solved completely. To see this, we introduce a new "strain measure"

$$\xi = \Lambda(u) = \frac{1}{2}u'^2,$$

such that the canonical functional

$$V(\xi) = \int_0^1 \frac{1}{2}\alpha(\xi - \lambda)^2 \, dx$$

is convex on $\mathcal{V}_a = \{\xi \in \mathcal{L}^2[0, 1] \mid \xi(x) \geq 0 \; \forall x \in (0, 1)\}$, and the duality relation $\varsigma = \delta V(\xi) = \alpha(\xi - \lambda)$ is one-to-one. Thus, its Legendre conjugate can be simply obtained as

$$\begin{aligned} V^*(\varsigma) &= \text{sta}\left\{\int_0^1 \xi\varsigma \, dx - V(\xi) \;\; : \;\; \xi \in \mathcal{V}_a\right\} \\ &= \int_0^1 \left(\frac{1}{2}\alpha^{-1}\varsigma^2 + \lambda\varsigma\right) \, dx. \end{aligned}$$

Similar to (8.52), the total complementary function is

$$\Xi(u,\varsigma) = \int_0^1 \left(\frac{1}{2}u'^2\varsigma - \frac{1}{2}\alpha^{-1}\varsigma^2 - \lambda\varsigma\right)\, dx - \int_0^1 uf\, dx. \tag{8.67}$$

For a given $\varsigma \neq 0$, the canonical dual functional can be obtained as

$$P^d(\varsigma) = \text{sta}\{\Xi(u,\varsigma) \,:\, u \in \mathcal{U}_k\} = -\int_0^1 \left(\frac{\tau^2}{2\varsigma} + \lambda\varsigma + \frac{1}{2}\alpha^{-1}\varsigma^2\right)\, dx, \tag{8.68}$$

where $\tau(x)$ is defined by

$$\tau = -\int_0^x f(x)\, dx + c, \tag{8.69}$$

and the integral constant c depends on the boundary condition. The criticality condition $\delta P^d(\varsigma) = 0$ leads to the dual equilibrium equation

$$2\varsigma^2(\alpha^{-1}\varsigma + \lambda) = \tau^2. \tag{8.70}$$

This algebraic equation is the same as (8.57), which can be solved analytically as stated below.

Theorem 8.7. (Analytical Solutions and Triality Theorem (Gao, 1998a, 2000b)) *For any given input function $f(x)$ such that $\tau(x)$ is defined by* (8.69)*, the dual algebraic equation* (8.70) *has at most three real roots ς_i $(i = 1, 2, 3)$ satisfying*

$$\bar{\varsigma}_1(x) > 0 > \bar{\varsigma}_2(x) \geq \bar{\varsigma}_3(x).$$

For each $\bar{\varsigma}_i$, the function

$$\bar{u}_i(x) = \int_0^x \frac{\tau}{\bar{\varsigma}_i}\, dx \tag{8.71}$$

is a critical point of the variational problem (8.64)*. Moreover, $\bar{u}_1(x)$ is a global minimizer, $\bar{u}_2(x)$ is a local minimizer, and $\bar{u}_3(x)$ is a local maximizer; that is,*

$$P(\bar{u}_1) = \min_u \max_{\varsigma>0} \Xi(u,\varsigma) = \max_{\varsigma>0} \min_u \Xi(u,\varsigma) = P^d(\bar{\varsigma}_1); \tag{8.72}$$

$$P(\bar{u}_2) = \min_u \max_{\varsigma\in(\bar{\varsigma}_3,0)} \Xi(u,\varsigma) = \min_{\varsigma\in(\bar{\varsigma}_3,0)} \max_u \Xi(u,\varsigma) = P^d(\bar{\varsigma}_2); \tag{8.73}$$

$$P(\bar{u}_3) = \max_u \max_{\varsigma<\bar{\varsigma}_2} \Xi(u,\varsigma) = \max_{\varsigma<\bar{\varsigma}_2} \max_u \Xi(u,\varsigma) = P^d(\bar{\varsigma}_3). \tag{8.74}$$

As a complete theory, the triality theorem was first discovered in post-buckling analysis of large deformed elastic beam models (Gao, 1997). The biduality theory was developed two years later during the writing of the monograph Gao (2000a). However, the original idea of the canonical dual transformation and the saddle-min-max theorem (8.72) were from the joint work by Gao and Strang in the study of complementary variational problems in nonconvex/nonsmooth boundary value problems (Gao and Strang, 1989a,b). Theorems 8.5 and 8.7 were also first proposed in the context of continuum mechanics (see Section 8.7, and Gao, 1999a,c, 2000b, Li and Gupta, 2006).

8.6 Canonical Duality Theory in General Nonconvex Systems

In this section, we discuss the canonical dual transformation and its associated triality theory for solving the following general nonconvex problem

$$(\mathcal{P}): \quad \min\{P(u) = W(u) - U(u) \ : \ \forall u \in \mathcal{U}_k\}, \tag{8.75}$$

where $W(u)$ is a general nonconvex function on an open set $\mathcal{U}_a \subset \mathcal{U}$, $U : \mathcal{U}_a \to \mathbb{R}$ is a Gâteaux differentiable function, either linear or canonical, and $\mathcal{U}_k \subset \mathcal{U}_a$ is a feasible space. The canonical dual transformation for solving more general problems can be found in Gao (1998a, 2000a,c).

8.6.1 Canonical Dual Transformation and Framework

The key idea of the canonical dual transformation is to choose a Gâteaux differentiable *geometrical operator* $\xi = \Lambda(u) : \mathcal{U}_a \to \mathcal{V}_a$ and a canonical function $V(\xi) : \mathcal{V}_a \to \mathbb{R}$ such that the nonconvex function $W(u)$ can be written as

$$W(u) = V(\Lambda(u)). \tag{8.76}$$

Because $V(\xi)$ is a canonical function on $\mathcal{V}_a$, its Legendre conjugate can be defined uniquely on $\mathcal{V}_a^* \subset \mathcal{V}^*$ by

$$V^*(\varsigma) = \text{sta}\{\langle \xi, \varsigma\rangle - V(\xi) \ : \ \forall \xi \in \mathcal{V}_a\}, \tag{8.77}$$

and on $\mathcal{V}_a \times \mathcal{V}_a^*$, we have

$$\varsigma = \delta V(\xi) \ \Leftrightarrow \ \xi = \delta V^*(\varsigma) \ \Leftrightarrow \ \langle \xi \, ; \, \varsigma\rangle = V(\xi) + V^*(\varsigma). \tag{8.78}$$

Replacing $W(u)$ by $V(\Lambda(u))$ and letting $\mathcal{U}_k = \{u \in \mathcal{U}_a \mid \Lambda(u) \in \mathcal{V}_a\}$, the primal problem $(\mathcal{P})$ can be written in the canonical form:

$$(\mathcal{P}): \quad \min\{P(u) = V(\Lambda(u)) - U(u) \ : \ \forall u \in \mathcal{U}_k\}. \tag{8.79}$$

Because $\Lambda(u)$ is Gâteaux differentiable, by the chain rule we have $\delta V(\Lambda(u)) = \Lambda_t(u)\delta_\xi V(\Lambda(u))$, where $\Lambda_t(u)$ is the Gâteaux derivative of $\Lambda(u)$ and $\delta_\xi V(\Lambda(u))$ represents the Gâteaux derivative of V with respective to $\xi = \Lambda(u)$. Its adjoint $\Lambda_t^*(u)$ is defined by

$$\langle \Lambda_t(u)u \; ; \; \varsigma \rangle = \langle u \; , \; \Lambda_t^*(u)\varsigma \rangle.$$

Thus, the criticality condition $\delta P(u) = 0$ leads to the *canonical equilibrium equation*

$$\Lambda_t^*(u)\delta_\xi V(\Lambda(u)) - \delta U(u) = 0. \tag{8.80}$$

In terms of the canonical duality pair (ξ, ς), the canonical equilibrium equation (8.80) can be written in the tricanonical forms:

$$\begin{array}{ll} \text{(a) Geometrical equation:} & \Lambda(u) = \xi. \\ \text{(b) Constitutive equation:} & \delta V(\xi) = \varsigma. \\ \text{(c) Balance equation:} & \Lambda_t^*(u)\varsigma = \delta U(u). \end{array} \tag{8.81}$$

In many applications, where the function $U(u)$ is usually linear on $\mathcal{U}_a$, the nonlinearity of the problem $(\mathcal{P})$ mainly depends on Λ and V. In this case, the nonlinearities of the general nonconvex problem can be classified by the following definition (Gao, 2000a).

Definition 8.4. (Nonlinearity Classification) The problem $(\mathcal{P})$ is said to be *geometrically nonlinear* if the operator $\Lambda(u)$ is nonlinear, *physically nonlinear* if the constitutive relation $\varsigma = \delta V(\xi)$ is nonlinear, and *fully nonlinear* if it is both geometrically and physically nonlinear.

Generally speaking, the nonconvexity of $P(u)$ is mainly due to the geometrical nonlinearity. For a nonlinear operator $\Lambda(u)$, the following operator decomposition introduced by Gao and Strang (1989a) plays an important role in canonical duality theory,

$$\Lambda(u) = \Lambda_t(u)u + \Lambda_c(u), \tag{8.82}$$

where $\Lambda_c = \Lambda(u) - \Lambda_t(u)u$ is the so-called *complementary operator* of Λ_t. By this decomposition (8.82), Gao and Strang discovered in the case where $U(u)$ is a linear function, that the duality gap existing in classical Lagrangian duality theory can be naturally recovered by the so-called *complementary gap function* defined by

$$G_c(u, \varsigma) = -\langle \Lambda_c(u) \; ; \; \varsigma \rangle. \tag{8.83}$$

The diagrammatic representation for a *fully nonlinear canonical system* is given in Figure 8.6.

Based on the canonical form of the primal problem (8.79), the total complementary function $\Xi : \mathcal{U}_a \times \mathcal{V}_a^* \to \mathbb{R}$ can be formulated as

$$\Xi(u, \varsigma) = \langle \Lambda(u); \varsigma \rangle - V^*(\varsigma) - U(u), \tag{8.84}$$

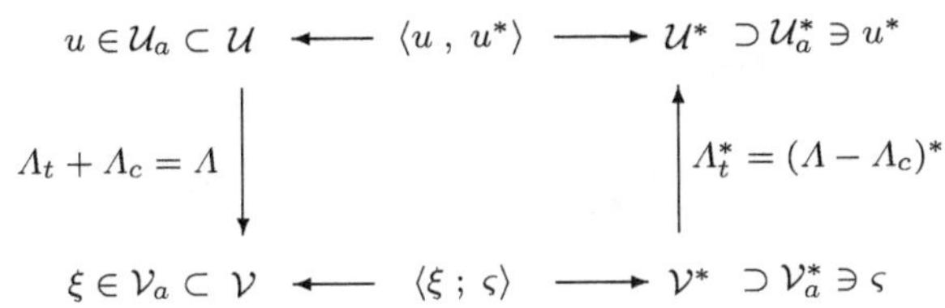

Fig. 8.6 Diagrammatic representation in fully nonlinear systems.

which is also called the *generalized complementary energy* in nonconvex variational problems and continuum mechanics (Gao and Strang, 1989a, Gao, 2000a), or the *nonlinear Lagrangian* in global optimization (Gao, 2000c). For each fixed $u \in \mathcal{U}_a$, the mapping $\Xi(u, \cdot) : \mathcal{V}^*_a \to \mathbb{R}$ is a canonical function. However, the property of the mapping $\Xi(\cdot, \varsigma) : \mathcal{U}_a \to \mathbb{R}$ will depend on the geometrical operator $\Lambda(u)$. Therefore, for a given $\varsigma \in \mathcal{V}^*_a$, we introduce a new (parametric) function

$$G_\varsigma(u) := \langle \Lambda(u) \, ; \, \varsigma \rangle - U(u), \qquad \forall \, u \in \mathcal{U}_a. \tag{8.85}$$

Clearly, for a fixed $\varsigma \in \mathcal{V}^*_a$, the criticality condition

$$\delta G_\varsigma(\bar{u}; u) = \langle \Lambda_t(\bar{u}) u \, ; \, \varsigma \rangle - \delta U(\bar{u}; u) = 0, \qquad \forall u \in \mathcal{U}_a$$

leads to the balance equation $\Lambda^*_t(\bar{u})\varsigma - \delta U(\bar{u}) = 0$. This function plays an important role in canonical duality theory. By introducing the so-called *canonical dual feasible space* $\mathcal{V}^*_k$ defined by

$$\mathcal{V}^*_k = \{\varsigma \in \mathcal{V}^*_a \mid \Lambda^*_t(u)\varsigma = \delta U(u), \quad \forall u \in \mathcal{U}_a\}, \tag{8.86}$$

the canonical dual function $P^d : \mathcal{V}^*_k \to \mathbb{R}$ can be formulated via $\Xi(u, \varsigma)$ as

$$P^d(\varsigma) = \text{sta}\{\Xi(u, \varsigma) \ : \ u \in \mathcal{U}_a\} = U^\Lambda(\varsigma) - V^*(\varsigma), \tag{8.87}$$

where $U^\Lambda : \mathcal{V}^*_k \to \mathbb{R}$ is called Λ-conjugate transformation of U, defined by (see Gao, 2000a)

$$U^\Lambda(\varsigma) = \text{sta}\{\langle \Lambda(u) \, ; \, \varsigma \rangle - U(u) \ : \ u \in \mathcal{U}_a\}. \tag{8.88}$$

Theorem 8.8. (Canonical Dual Transformation (Gao, 2000a)) *The function*

$$P^d(\varsigma) = U^\Lambda(\varsigma) - V^*(\varsigma) : \mathcal{V}^*_k \to \mathbb{R}$$

is canonically dual to $P(u) = V(\Lambda(u)) - U(u) : \mathcal{U}_k \to \mathbb{R}$ *in the sense that if* $(\bar{u}, \bar{\varsigma})$ *is a critical point of* $\Xi(u, \varsigma)$, *then* $\bar{u}$ *is a critical point of* $P(u)$, $\bar{\varsigma}$ *is a critical point of* $P^d(\varsigma)$*, and*

$$P(\bar{u}) = \Xi(\bar{u}, \bar{\varsigma}) = P^d(\bar{\varsigma}). \tag{8.89}$$

This theorem can be easily proved by examining the criticality condition $\delta\Xi(\bar{u}, \bar{\varsigma}) = 0$, which leads to the following canonical Lagrangian equations,

$$\Lambda(\bar{u}) = \delta V^*(\bar{\varsigma}), \qquad \Lambda_t^*(\bar{u})\bar{\varsigma} = \delta U(\bar{u}), \tag{8.90}$$

which are equivalent to the tricanonical forms (8.81) because $V^*(\varsigma)$ is a canonical function on $\mathcal{V}_a^*$. Thus, $\bar{u}$ is a critical point of $P(u)$. By the definition of the canonical dual function, $\bar{\varsigma}$ is also a critical point of $P^d(\varsigma)$. □

Theorem 8.8 shows that there is no duality gap between the primal function and its canonical dual. Actually, in the case where $U(u) = \langle u, f \rangle$ is a linear function, we have

$$U^\Lambda(\varsigma) = G_\varsigma(u) = \langle \Lambda_c(u) \; ; \; \varsigma \rangle = -G_c(u, \varsigma) \;\text{ s.t. }\; \Lambda_t^*(u)\varsigma = f;$$

that is, the duality gap is recovered by the complementary gap function $G_c(u, \varsigma)$. In this case, the function $P^c(u, \varsigma) = -\Xi(u, \varsigma)$ defined by

$$P^c(u, \varsigma) = G_c(u, \varsigma) + V^*(\varsigma) \tag{8.91}$$

is the total complementary energy introduced by Gao and Strang in 1989 (1989a). They proved that if $(\bar{u}, \bar{\varsigma})$ is a critical point of $P^c(u, \varsigma)$, then $\bar{u}$ is a critical point of $P(u)$, and $P(\bar{u}) + P^c(\bar{u}, \bar{\varsigma}) = 0$. The operator $\Lambda(u)$ is usually nonlinear in nonconvex problems, therefore the explicit format of the canonical dual function $P^d(\varsigma)$ will depend on the properties of the function $G_\varsigma(u)$. By the implicit function theory, if $\Lambda(u)$ is twice Gâteaux differentiable and the second Gâteaux differential

$$\delta_u^2 G_\varsigma(u; \delta u^2) \neq 0 \qquad \forall \delta u \neq 0, \tag{8.92}$$

then $U^\Lambda(\varsigma)$ can be formulated explicitly by the Λ-conjugate transformation (8.88). Some simple illustrative examples are given below.

Example 8.1. Recall the nonconvex optimization problem with the double well function (8.47):

$$\min\{P(u) = \frac{1}{2}\alpha(\frac{1}{2}|Bu|^2 - \lambda)^2 - \langle u, f \rangle \;:\; u \in \mathbb{R}^n\},$$

where $W(u)$ is a double-well function and $U(u)$ is a linear function. If we choose $\xi = \Lambda(u) = \frac{1}{2}|Bu|^2$ as a quadratic operator, then we have $\Lambda_t(u) = (Bu)^T B$ and $\Lambda_c(u) = \Lambda(u) - \Lambda_t(u) = -\frac{1}{2}|Bu|^2$. Because for each $\varsigma \neq 0$, $G_\varsigma(u) = \frac{1}{2}|Bu|^2\varsigma - \langle u, f \rangle$ is a quadratic function and $\delta^2 G_\varsigma(\bar{u}) = \varsigma$, the Λ-conjugate U^Λ is well defined by

$$U^\Lambda(\varsigma) = \text{sta}\left\{\langle \frac{1}{2}|Bu|^2; \varsigma\rangle - \langle u, f\rangle \ : \ u \in \mathbb{R}^n\right\} = -\frac{1}{2\varsigma} f^T(B^TB)^{-1}f.$$

The complementary gap function in this case is $G_c(u,\varsigma) = \frac{1}{2}|Bu|^2\varsigma$. Clearly, for any $u \in \mathbb{R}^n$ and $u \neq 0$, $G_c(u,\varsigma) > 0$ if and only if $\varsigma > 0$. Thus, the total complementary function $\Xi(u,\varsigma)$ given by (8.52) is a saddle function for $\varsigma > 0$. This leads to the saddle min-max duality (8.61) in the triality theory.

Example 8.2. In the nonconvex variational problem (8.64), the quadratic differential operator $\xi = \Lambda(u) = \frac{1}{2}u'^2$ has a physical meaning. In finite deformation theory, if u is considered as the displacement of a deformed body, then ξ can be considered as a *Cauchy–Green strain measure* (see the following section). The Gâteaux derivative of the quadratic differential operator $\Lambda(u)$ is $\Lambda_t(u) = u'\mathrm{d}/\mathrm{d}x$. For any given $u \in \mathcal{U}_a$, using integration by parts, we get

$$\langle \Lambda_t(u)u; \varsigma\rangle = \int_0^1 u'^2\varsigma \,\mathrm{d}x = uu'\varsigma|_{x=0}^{x=1} - \int_0^1 u\,[u'\varsigma]' \,\mathrm{d}x = \langle u, \Lambda_t^*(u)\varsigma\rangle,$$

which gives the adjoint operator Λ_t^* via

$$\Lambda_t^*(u)\varsigma = \begin{cases} u'\varsigma & \text{on } x = 1 \\ [u'\varsigma]', & \forall x \in (0,1). \end{cases}$$

For any given $\varsigma \in \mathcal{V}_a$, the Λ-conjugate transformation

$$U^\Lambda(\varsigma) = \text{sta}\{\langle \Lambda(u), \varsigma\rangle - U(u) \ : \ u \in \mathcal{U}_k\} = -\int_0^1 \tau^2\varsigma^{-1}\,\mathrm{d}x.$$

The complementary operator in this problem is $\Lambda_c(u) = \Lambda(u) - \Lambda_t(u)u = -\frac{1}{2}u'^2$, which leads to the complementary gap function

$$G_c(u,\varsigma) = \int_0^1 \frac{1}{2}u'^2\varsigma \,\mathrm{d}x.$$

Clearly, this is positive if $\varsigma \geq 0$.

8.6.2 Extremality Conditions: Triality Theory

In order to study the extremality conditions of the nonconvex problem, we need to clarify the convexity of the canonical function $V(\xi)$. Without loss of generality, we assume that $V : \mathcal{V}_a \to \mathbb{R}$ is convex. Thus, for each $u \in \mathcal{U}_a$, the total complementary function

$$\Xi(u,\varsigma) = \langle \Lambda(u)\ ;\ \varsigma\rangle - V^*(\varsigma) - U(u) : \mathcal{V}_a^* \to \mathbb{R}$$

is concave in $\varsigma \in \mathcal{V}_a^*$. The convexity of $\Xi(\cdot, \varsigma) : \mathcal{U}_a \to \mathbb{R}$ will depend on the geometrical operator $\Lambda(u)$ and the function $U(u)$. We furthermore assume that the function $G_\varsigma(u) = \langle \Lambda(u) \; ; \; \varsigma \rangle - U(u) : \mathcal{U}_a \to \mathbb{R}$ is twice Gâteaux differentiable on $\mathcal{U}_a$ and let

$$\mathcal{G} := \{(u, \varsigma) \in \mathcal{U}_a \times \mathcal{V}_a^* \mid \delta^2 G_\varsigma(u; \delta u^2) \neq 0, \;\; \forall \delta u \neq 0\}, \tag{8.93}$$
$$\mathcal{G}^+ := \{(u, \varsigma) \in \mathcal{U}_a \times \mathcal{V}_a^* \mid \delta^2 G_\varsigma(u; \delta u^2) > 0, \;\; \forall \delta u \neq 0\}, \tag{8.94}$$
$$\mathcal{G}^- := \{(u, \varsigma) \in \mathcal{U}_a \times \mathcal{V}_a^* \mid \delta^2 G_\varsigma(u; \delta u^2) < 0, \;\; \forall \delta u \neq 0\}. \tag{8.95}$$

Theorem 8.9. (Triality Theorem) *Suppose that $(\bar{u}, \bar{\varsigma}) \in \mathcal{G}$ is a critical point of $\Xi(u, \varsigma)$ and $\mathcal{U}_o \times \mathcal{V}_o^* \subset \mathcal{U}_k \times \mathcal{V}_k^*$ is a neighborhood of $(\bar{u}, \bar{\varsigma})$.*

If $(\bar{u}, \bar{\varsigma}) \in \mathcal{G}^+$, then $(\bar{u}, \bar{\varsigma})$ is a saddle point of $\Xi(u, \varsigma)$; that is,

$$\min_{u \in \mathcal{U}_o} \max_{\varsigma \in \mathcal{V}_o^*} \Xi(u, \varsigma) = \Xi(\bar{u}, \bar{\varsigma}) = \max_{\varsigma \in \mathcal{V}_o^*} \min_{u \in \mathcal{U}_o} \Xi(u, \varsigma). \tag{8.96}$$

If $(\bar{u}, \bar{\varsigma}) \in \mathcal{G}^-$, then $(\bar{u}, \bar{\varsigma})$ is a supercritical point of $\Xi(u, \varsigma)$, and we have that either

$$\min_{u \in \mathcal{U}_o} \max_{\varsigma \in \mathcal{V}_o^*} \Xi(u, \varsigma) = \Xi(\bar{u}, \bar{\varsigma}) = \min_{\varsigma \in \mathcal{V}_o^*} \max_{u \in \mathcal{U}_o} \Xi(u, \varsigma) \tag{8.97}$$

holds, or

$$\max_{u \in \mathcal{U}_o} \max_{\varsigma \in \mathcal{V}_o^*} \Xi(u, \varsigma) = \Xi(\bar{u}, \bar{\varsigma}) = \max_{\varsigma \in \mathcal{V}_o^*} \max_{u \in \mathcal{U}_o} \Xi(u, \varsigma). \tag{8.98}$$

Proof. By the assumption on the canonical function $V(\xi)$, we know that $\Xi(u, \varsigma)$ is concave on $\mathcal{V}_a^*$. Because $G_\varsigma(u)$ is twice Gâteaux differentiable on $\mathcal{U}_a$, the theory of implicit functions tells us that if $(\bar{u}, \bar{\varsigma}) \in \mathcal{G}$, then there exists a unique $u \in \mathcal{U}_o \subset \mathcal{U}_k$ such that the dual feasible set $\mathcal{V}_k^*$ is nonempty. If such a point $(\bar{u}, \bar{\varsigma}) \in \mathcal{G}^+$, then $G_\varsigma(u)$ is convex in u and $(\bar{u}, \bar{\varsigma})$ is a saddle point of Ξ on $\mathcal{U}_o \times \mathcal{V}_o^*$. The saddle-Lagrangian duality leads to (8.96). If $(\bar{u}, \bar{\varsigma}) \in \mathcal{G}^-$, then $G_\varsigma(u)$ is locally concave in u and $(\bar{u}, \bar{\varsigma})$ is a supercritical point of $\Xi(u, \varsigma)$ on $\mathcal{U}_o \times \mathcal{V}_o^*$. In this case the biduality theory leads to (8.97) and (8.98). □

If the geometrical operator $\Lambda(u)$ is a quadratic function and $U(u)$ is either quadratic or linear, then the second-order Gâteaux derivative $\delta^2 G_\varsigma(u)$ does not depend on u. In this case, we let

$$\mathcal{V}_+^* := \{\varsigma \in \mathcal{V}_a^* \mid \delta^2 G_\varsigma(u) \text{ is positive definite}\}, \tag{8.99}$$
$$\mathcal{V}_-^* := \{\varsigma \in \mathcal{V}_a^* \mid \delta^2 G_\varsigma(u) \text{ is negative definite}\}. \tag{8.100}$$

The following theorem provides extremality criteria for critical points of $\Xi(u, \varsigma)$.

Theorem 8.10. (Triduality Theorem (Gao, 1998a, 2000a)) *Suppose that $G_\varsigma(u) = \langle \Lambda(u); \varsigma \rangle - U(u)$ is a quadratic function of $u \in \mathcal{U}_a$ and $(\bar{u}, \bar{\varsigma})$ is a critical point of $\Xi(u, \varsigma)$.*

If $\bar{\varsigma} \in \mathcal{V}_+^*$, then $\bar{u}$ is a global minimizer of $P(u)$ on $\mathcal{U}_k$ if and only if $\bar{\varsigma}$ is a global maximizer of $P^d(\varsigma)$ on $\mathcal{V}_+^*$, and

$$P(\bar{u}) = \min_{u \in \mathcal{U}_k} P(u) = \max_{\varsigma \in \mathcal{V}_+^*} P^d(\varsigma) = P^d(\bar{\varsigma}). \tag{8.101}$$

If $\bar{\varsigma} \in \mathcal{V}_-^*$, then on the neighborhood $\mathcal{U}_o \times \mathcal{V}_o^* \subset \mathcal{U}_a \times \mathcal{V}_a^*$ of $(\bar{u}, \bar{\varsigma})$, we have that either

$$P(\bar{u}) = \min_{u \in \mathcal{U}_o} P(u) = \min_{\varsigma \in \mathcal{V}_o^*} P^d(\varsigma) = P^d(\bar{\varsigma}) \tag{8.102}$$

holds, or

$$P(\bar{u}) = \max_{u \in \mathcal{U}_o} P(u) = \max_{\varsigma \in \mathcal{V}_o^*} P^d(\varsigma) = P^d(\bar{\varsigma}). \tag{8.103}$$

This theorem shows that the canonical dual solution $\bar{\varsigma} \in \mathcal{V}_+^*$ provides a global optimality condition for the nonconvex primal problem, whereas the condition $\bar{\varsigma} \in \mathcal{V}_-^*$ provides local extremality conditions.

The triality theory was originally discovered in nonconvex mechanics (Gao, 1997, 1999c). Since then, several modified versions have been proposed in nonconvex parametrical variational problems (for quadratic $\Lambda(u)$ and linear $U(u)$ (Gao, 1998a)), general nonconvex systems (for nonlinear $\Lambda(u)$ and linear $U(u)$ (Gao, 2000a)), global optimization (for general nonconvex functions of type $\Phi(u, \Lambda(u))$ (Gao, 2000c), quadratic $U(u)$ (Gao, 2003a,b)), and dissipative Hamiltonian system (for nonconvex/nonsmooth functions of type $\Phi(u, u_{,t}, \Lambda(u))$ (Gao, 2001c)). In terms of the parametrical function $G_\varsigma(u) = \langle \Lambda(u); \varsigma \rangle - U(u)$, the current version (Theorems 8.9 and 8.10) can be used for solving general nonconvex problem (8.75) with the canonical function $U(u)$.

8.6.3 Complementary Variational Principles in Finite Deformation Theory

In finite deformation theory, the deformation $\mathbf{u}(\mathbf{x})$ is a smooth, vector-valued mapping from an open, simply connected, and bounded domain $\Omega \subset \mathbb{R}^n$ into a deformed domain[2] $\omega \subset \mathbb{R}^m$. Let $\Gamma = \partial\Omega = \Gamma_u \cup \Gamma_t$ be the boundary of Ω such that on Γ_u, the boundary condition $\mathbf{u}(\mathbf{x}) = \bar{\mathbf{u}}$ is prescribed, whereas on the remaining boundary Γ_t, the surface traction (external force) $\bar{\mathbf{t}}(\mathbf{x})$ is applied. Similar to the nonconvex optimization problem (8.48), the primal problem is to minimize the total potential energy functional:

$$\min \left\{ P(\mathbf{u}) = \int_\Omega [W(\nabla \mathbf{u}) - \mathbf{u} \cdot \mathbf{f}] \, \mathrm{d}\Omega - \int_{\Gamma_t} \mathbf{u} \cdot \bar{\mathbf{t}} \, \mathrm{d}\Gamma \ : \ \mathbf{u} = \bar{\mathbf{u}} \ \text{ on } \Gamma_u \right\}, \tag{8.104}$$

where the stored energy $W(\mathbf{F})$ is a Gâteaux differentiable function of $\mathbf{F} = \nabla \mathbf{u}$, and $\mathbf{f}(\mathbf{x})$ is a given force field. Because the deformation gradient $\mathbf{F} = \nabla \mathbf{u} \in$

[2] If $m = n + 1$, then the deformation $\mathbf{u}(\mathbf{x})$ represents a hypersurface in m-dimensional space. Applications of the canonical duality theory in differential geometry were discussed in Gao and Yang (1995).

$\mathbb{R}^{n\times m}$ is a so-called two-point tensor, which is no longer a strain measure in finite deformation theory, the stored energy $W(\mathbf{F})$ is usually nonconvex. Particularly, for St. Venant–Kirchhoff material (see Gao, 2000a), we have

$$W(\epsilon) = \frac{1}{2}\left[\frac{1}{2}(\mathbf{F}^T\mathbf{F} - I)\right] : D : \left[\frac{1}{2}(\mathbf{F}^T\mathbf{F} - I)\right], \tag{8.105}$$

where I is an identity tensor in $\mathbb{R}^{n\times n}$. Due to nonconvexity, the duality relation

$$\boldsymbol{\tau} = \delta W(\mathbf{F})$$

is not one-to-one. Although the two-point tensor $\boldsymbol{\tau} \in \mathbb{R}^{m\times n}$ is called the *first Piola–Kirchhoff stress*, according to Hill's constitutive theory, $(\mathbf{F}, \boldsymbol{\tau})$ is not considered as a work-conjugate (canonical) strain–stress pair (see Gao, 2000a). The Fenchel–Rockafellar type dual variational problem is

$$\max \left\{ P^\sharp(\boldsymbol{\tau}) = \int_{\Gamma_u} \bar{\mathbf{u}} \cdot \boldsymbol{\tau} \cdot \mathbf{n} \,\mathrm{d}\Gamma - \int_\Omega W^\sharp(\boldsymbol{\tau}) \,\mathrm{d}\Omega \right\} \tag{8.106}$$

$$\text{s.t.} \ -\nabla \cdot \boldsymbol{\tau}^T = \mathbf{f} \ \text{ in } \Omega, \ \ \mathbf{n} \cdot \boldsymbol{\tau}^T = \bar{\mathbf{t}} \ \text{ on } \Gamma_t. \tag{8.107}$$

In the case where the stored energy $W(\mathbf{F})$ is convex, then $W^\sharp(\boldsymbol{\tau}) = W^*(\boldsymbol{\tau})$ which is called the *complementary energy* in elasticity. In this case, the functional

$$\Pi^c(\boldsymbol{\tau}) = \int_\Omega W^*(\boldsymbol{\tau}) \,\mathrm{d}\Omega - \int_{\Gamma_u} \bar{\mathbf{u}} \cdot \boldsymbol{\tau} \cdot \mathbf{n} \,\mathrm{d}\Gamma$$

is the well-known *Levinson–Zubov complementary energy*. As discussed before, if the stored energy $W(\mathbf{F})$ is nonconvex, the Legendre conjugate W^* is not uniquely defined. It turns out that the Levinson–Zubov complementary variational principle can be used only for solving convex problems (see Gao, 1992). Although the Fenchel conjugate $W^\sharp(\boldsymbol{\tau})$ can be uniquely defined, the Fenchel–Young inequality $W(\mathbf{F}) + W^\sharp(\boldsymbol{\tau}) \geq \langle \mathbf{F}; \boldsymbol{\tau} \rangle$ leads to a duality gap between the minimal potential variational problem (8.104) and its Fenchel–Rockafellar dual (see Gao, 1992); that is, in general,

$$\min P(\mathbf{u}) \geq \max P^\sharp(\boldsymbol{\tau}). \tag{8.108}$$

By the fact that the criticality condition $\delta P^\sharp(\boldsymbol{\tau}) = 0$ is not equivalent to the primal variational problem and the weak duality is not appreciated in the field of continuum mechanics, the existence of a perfect (i.e., without a duality gap), pure (i.e., involving only stress tensor as variational argument) complementary variational principle in finite elasticity has been argued among well-known scientists for more than three decades (see Hellinger, 1914, Hill, 1978, Koiter, 1973, 1976, Lee and Shield, 1980a,b, Levinson, 1965, Ogden, 1975, 1977, Zubov, 1970). This problem was finally solved by the canonical dual transformation and triality theory in Gao (1992, 1999c).

Similar to the quadratic operator $\Lambda(\mathbf{u}) = \frac{1}{2}|B\mathbf{u}|^2$ (see equation (8.51)) chosen for the nonconvex optimization problem (8.48), we let

$$\mathbf{E} = \Lambda(\mathbf{u}) = \frac{1}{2}[(\nabla \mathbf{u})^T (\nabla \mathbf{u}) - I], \tag{8.109}$$

which is a symmetrical tensor field in $\mathbb{R}^{n\times n}$. In finite deformation theory, $\mathbf{E}$ is the well-known *Green–St. Venant strain tensor*. Thus, in terms of $\mathbf{E}$, the stored energy for St. Venant–Kirchhoff material can be written in the canonical form $W(\nabla \mathbf{u}) = V(\Lambda(\nabla \mathbf{u}))$, and

$$V(\mathbf{E}) = \frac{1}{2}\mathbf{E} : D : \mathbf{E}$$

is a (quadratic) convex function of the symmetrical tensor $\mathbf{E} \in \mathbb{R}^{n\times n}$. The canonical dual variable $\mathbf{E}^* = \delta V(\mathbf{E}) = D \cdot \mathbf{E}$ is called the *second Piola–Kirchhoff stress tensor*, denoted as $\mathbf{T}$. The Legendre conjugate

$$V^*(\mathbf{T}) = \frac{1}{2}\mathbf{T} : D^{-1} : \mathbf{T} \tag{8.110}$$

is also a quadratic function. Let $\mathcal{U}_a = \{\mathbf{u} \in \mathcal{W}^{1,p}(\Omega; \mathbb{R}^3) \mid \mathbf{u} = \bar{\mathbf{u}} \text{ on } \Gamma_u\}$ (where $\mathcal{W}^{1,p}$ is a standard Sobolev space with $p \in (1, \infty)$) and $\mathcal{V}_a^* = \mathcal{C}(\Omega; \mathbb{R}^{n\times n})$. Replacing $W(\nabla \mathbf{u})$ by its canonical dual transformation $V(\Lambda(\mathbf{u})) = \mathbf{E}(\mathbf{u}) : \mathbf{T} - V^*(\mathbf{T})$, the generalized complementary energy $\Xi : \mathcal{U}_a \times \mathcal{V}_a^* \to \mathbb{R}$ has the following format,

$$\Xi(\mathbf{u}, \mathbf{T}) = \int_\Omega [\mathbf{E}(\mathbf{u}) : \mathbf{T} - V^*(\mathbf{T}) - \mathbf{u} \cdot \mathbf{f}] \, \mathrm{d}\Omega - \int_{\Gamma_t} \mathbf{u} \cdot \bar{\mathbf{t}} \, \mathrm{d}\Gamma, \tag{8.111}$$

which is the well-known *Hellinger–Reissner generalized complementary energy* in continuum mechanics.

Furthermore, if we replace $V^*(\mathbf{T})$ by its bi-Legendre transformation $\mathbf{E} : \mathbf{T} - V(\mathbf{E})$, then $\Xi(\mathbf{u}, \mathbf{T})$ can be written as

$$\Xi_{hw}(\mathbf{u}, \mathbf{T}, \mathbf{E}) = \int_\Omega [\Lambda(\nabla \mathbf{u}) - \mathbf{E}) : \mathbf{T} + V(\mathbf{E}) - \mathbf{u} \cdot \mathbf{f}] \, \mathrm{d}\Omega - \int_{\Gamma_t} \mathbf{u} \cdot \bar{\mathbf{t}} \, \mathrm{d}\Gamma. \tag{8.112}$$

This is the well-known *Hu–Washizu* generalized potential energy in nonlinear elasticity. The Hu–Washizu variational principle has important applications in computational analysis of thin-walled structures, where the geometrical equation $\mathbf{E} = \Lambda(u)$ is usually proposed by certain geometrical hypothesis.

Because $\Lambda(\mathbf{u})$ is a quadratic operator, its Gâteaux differential at $\bar{\mathbf{u}}$ in the direction $\mathbf{u}$ is $\delta\Lambda(\bar{\mathbf{u}}; \mathbf{u}) = \Lambda_t(\bar{\mathbf{u}})\mathbf{u} = (\nabla \bar{\mathbf{u}})^T (\nabla \mathbf{u})$ and

$$\Lambda_c(\mathbf{u}) = \Lambda(\mathbf{u}) - \Lambda_t(\mathbf{u})\mathbf{u} = -\frac{1}{2}[(\nabla \mathbf{u})^T (\nabla \mathbf{u}) + I].$$

By using the Gauss–Green theorem, the balance operator $\Lambda_t^*(\mathbf{u})$ can be defined as

$$\Lambda_t^*(\mathbf{u})\mathbf{T} = \begin{cases} -\nabla \cdot [(\nabla \mathbf{u})^T \cdot \mathbf{T}]^T & \text{in } \Omega, \\ \mathbf{n} \cdot [(\nabla \mathbf{u})^T \cdot \mathbf{T}]^T & \text{on } \Gamma. \end{cases}$$

The complementary gap function in this problem is a quadratic functional:

$$G_c(\mathbf{u}, \mathbf{T}) = \langle -\Lambda_c(\mathbf{u}); \, \mathbf{T} \rangle = \int_\Omega \frac{1}{2} \mathrm{tr}[(\nabla \mathbf{u})^T \cdot \mathbf{T} \cdot (\nabla \mathbf{u}) + \mathbf{T}] \, \mathrm{d}\Omega. \tag{8.113}$$

Thus, the complementary variational problem is to find critical (stationary) points $(\bar{\mathbf{u}}, \bar{\mathbf{T}})$ such that

$$P^c(\bar{\mathbf{u}}, \bar{\mathbf{T}}) = \mathrm{sta}\left\{ \int_\Omega \frac{1}{2} \mathrm{tr}[(\nabla \mathbf{u})^T \cdot \mathbf{T} \cdot (\nabla \mathbf{u}) + \mathbf{T}] \, \mathrm{d}\Omega + \int_\Omega V^*(\mathbf{T}) \, \mathrm{d}\Omega \right\} \tag{8.114}$$
$$\text{s.t.} \quad -\nabla \cdot [(\nabla \mathbf{u})^T \cdot \mathbf{T}]^T = \mathbf{f} \ \text{ in } \Omega, \ \ \mathbf{n} \cdot [(\nabla \mathbf{u})^T \cdot \mathbf{T}]^T = \bar{\mathbf{t}} \text{ on } \Gamma_t.$$

The following result is due to Gao and Strang in 1989 (1989a).

Theorem 8.11. (Complementary–Dual Variational Principle (Gao and Strang, 1989a)) *If* $(\bar{\mathbf{u}}, \bar{\mathbf{T}})$ *is a critical point of the complementary variational problem* (8.114), *then* $\bar{\mathbf{u}}$ *is a critical point of the total potential energy* $P(\mathbf{u})$ *defined by* (8.104), *and*

$$P(\bar{\mathbf{u}}) + P^c(\bar{\mathbf{u}}, \bar{\mathbf{T}}) = 0.$$

Moreover, if the complementary gap function

$$G_c(\mathbf{u}, \bar{\mathbf{T}}) \geq 0, \qquad \forall \mathbf{u} \in \mathcal{U}_a, \tag{8.115}$$

then $\bar{\mathbf{u}}$ *is a global minimizer of* $P(\mathbf{u})$ *and*

$$P(\bar{\mathbf{u}}) = \min_{\mathbf{u}} P(\mathbf{u}) = \max_{\mathbf{T}} \min_{\mathbf{u}} \Xi(\mathbf{u}, \mathbf{T}) = -P^c(\bar{\mathbf{u}}, \bar{\mathbf{T}}), \tag{8.116}$$

subject to $\mathbf{T}(\mathbf{x})$ *being positive definite for all* $\mathbf{x} \in \Omega$.

This theorem shows that the positivity of the complementary gap function $G_c(\mathbf{u}, \mathbf{T})$ provides a sufficient condition for a global minimizer, and the equalities (8.11) and (8.116) indicate that there is no duality gap between the total potential $P(\mathbf{u})$ and its complementary energy $P^c(\mathbf{u}, \mathbf{T})$. The physical significance is also clear: a finite deformed material is stable if the second Piola–Kirchhoff stress tensor $\mathbf{T}(\mathbf{x})$ is positive definite everywhere in the domain Ω. The linear operator $B = \nabla$ in this nonconvex variational problem is a partial differential operator, therefore it is difficult to find its inverse. It took more than ten years before the canonical dual problem was finally formulated in Gao (1999a,c). To see this, let us assume that for a given force vector field $\bar{\mathbf{t}}$ on the boundary Γ_t, the first Piola–Kirchhoff stress tensor $\boldsymbol{\tau}(\mathbf{x})$ can be defined by solving the following boundary value problem,

$$-\nabla \cdot \boldsymbol{\tau}^T(\mathbf{x}) = \mathbf{f} \quad \text{in } \Omega, \qquad \mathbf{n} \cdot \boldsymbol{\tau}^T = \bar{\mathbf{t}} \quad \text{on } \Gamma_t. \tag{8.117}$$

Then the canonical dual functional $P^d(\mathbf{T})$ can be formulated as

$$P^d(\mathbf{T}) = -\int_\Omega \frac{1}{2}\mathrm{tr}(\boldsymbol{\tau} \cdot \mathbf{T}^{-1} \cdot \boldsymbol{\tau}^T + \mathbf{T})\, d\Omega - \int_\Omega V^*(\mathbf{T})\, d\Omega. \tag{8.118}$$

The criticality condition $\delta P^d(\mathbf{T}) = 0$ gives the canonical dual equation

$$\mathbf{T} \cdot (2\, \delta V^*(\mathbf{T}) + I) \cdot \mathbf{T} = \boldsymbol{\tau}^T \cdot \boldsymbol{\tau}. \tag{8.119}$$

For St. Venant–Kirchhoff material, $V^*(\mathbf{T}) = \frac{1}{2}\mathbf{T} : D^{-1} : \mathbf{T}$ is a quadratic function and its Gâteaux derivative $\delta V^*(\mathbf{T}) = D^{-1} \cdot \mathbf{T}$ is linear. In this case, the canonical dual equation (8.119) is a cubic equation, which is similar to the dual algebraic equations (8.57) and (8.70).

Theorem 8.12. (Pure Complementary Energy Principle (Gao, 1999a,c)) *Suppose that for a given force field $\bar{\mathbf{t}}(\mathbf{x})$ on Γ_t, the first Piola–Kirchhoff stress field $\boldsymbol{\tau}(\mathbf{x})$ is defined by* (8.117)*. Then each solution $\bar{\mathbf{T}}$ of the canonical dual equation* (8.119) *is a critical point of P^d, the vector defined by the line integral*

$$\bar{\mathbf{u}} = \int \boldsymbol{\tau} \cdot \bar{\mathbf{T}}^{-1} d\mathbf{x} \tag{8.120}$$

is a critical point of $P(\mathbf{u})$, and

$$P(\bar{\mathbf{u}}) = P^d(\bar{\mathbf{T}}).$$

This theorem presents an analytic solution to the nonconvex potential variational problem (8.104). In the finite deformation theory of elasticity, this pure complementary variational principle is also known as the *Gao principle* (Li and Gupta, 2006), which holds also for the general canonical energy function $V(\mathbf{E})$. Similar to Theorem 8.9, the extremality of the critical points can be identified by the complementary gap function. Applications of this pure complementary variational principle for solving nonconvex/nonsmooth boundary value problems are illustrated in Gao (1999c, 2000a) and Gao and Ogden (2008a,b).

8.7 Applications to Semilinear Nonconvex Systems

The canonical dual transformation and the associated triality theory can be used to solve many difficult problems in engineering and science. In this section, we present applications for solving the following nonconvex minimization problem

$$(\mathcal{P}): \quad \min\{P(u) = W(u) + \frac{1}{2}\langle u, Au\rangle - \langle u, f\rangle \ : \ u \in \mathcal{U}_k\}, \tag{8.121}$$

where $W(u) : \mathcal{U}_k \to \mathbb{R}$ is a nonconvex function, and $A : \mathcal{U}_a \subset \mathcal{U} \to \mathcal{U}_a^*$ is a linear operator. If $W(u)$ is Gâteaux differentiable, the criticality condition $\delta P(u) = 0$ leads to a nonlinear Euler equation

$$Au + \delta W(u) = f. \tag{8.122}$$

The abstract form (8.122) of the primal problem $(\mathcal{P})$ covers many situations. In nonconvex mechanics (cf. Gao, Ogden, and Stavroulakis, 2001, Gao, 2003b), where $\mathcal{U}$ is an infinite-dimensional function space, the state variable $u(x)$ is a field function, and $A : \mathcal{U} \to \mathcal{U}^*$ is usually a partial differential operator. In this case, the governing equation (8.122) is a so-called semilinear equation. For example, in the Landau–Ginzburg theory of superconductivity, $A = \Delta$ is the Laplacian over a given space domain $\Omega \subset \mathbb{R}^n$ and

$$W(u) = \int_\Omega \frac{1}{2}\alpha \left(\frac{1}{2}u^2 - \lambda\right)^2 \mathrm{d}\Omega \tag{8.123}$$

is the Landau double-well potential, in which $\alpha, \lambda > 0$ are material constants. Then the governing equation (8.122) leads to the well-known Landau–Ginzburg equation

$$\Delta u + \alpha u(\frac{1}{2}u^2 - \lambda) = f.$$

This semilinear differential equation plays an important role in material science and physics including: ferroelectricity, ferromagnetism, and superconductivity. In a more complicated case where $A = \Delta + \text{curl curl}$, we have

$$\Delta u + \text{curl curl}\ u + \alpha u(\frac{1}{2}u^2 - \lambda) = f,$$

which is the so-called Cahn–Hilliard equation in liquid crystal theory. Due to the nonconvexity of the double-well function $W(u)$, any solution of the semilinear differential equation (8.122) is only a critical point of the total potential $P(u)$. Traditional direct analysis and related numerical methods for finding the global minimizer of the nonconvex variational problem have proven unsuccessful to date.

In dynamical systems, if $A = -\partial_{,tt} + \Delta$ is a wave operator over a given space–time domain $\Omega \subset \mathbb{R}^n \times \mathbb{R}$, then (8.122) is the well-known nonlinear Schrödinger equation

$$-u_{,tt} + \Delta u + \alpha u(\frac{1}{2}u^2 - \lambda) = f.$$

This equation appears in many branches of physics. It provides one of the simplest models of the unified field theory. It can also be found in the theory

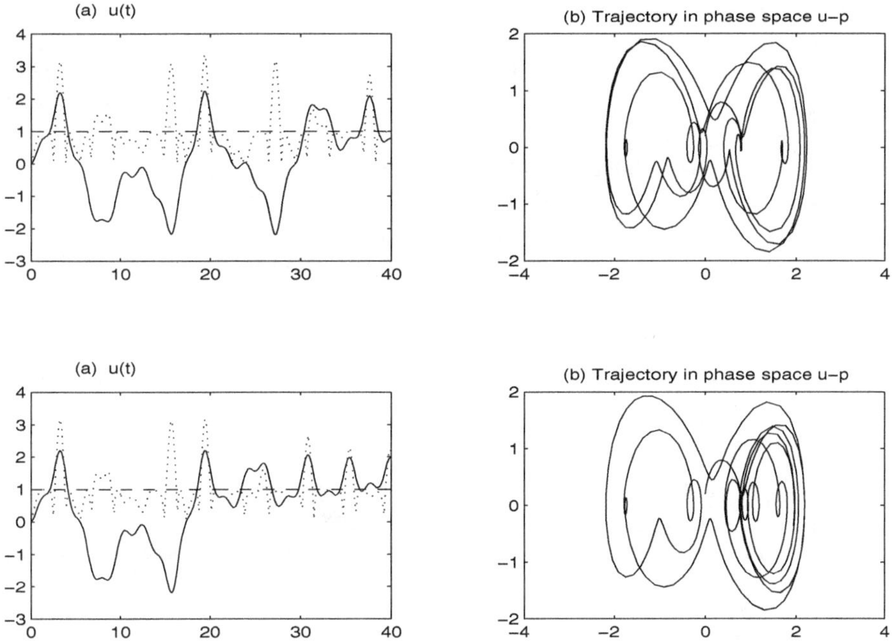

Fig. 8.7 Numerical results by *ode23* (top) and *ode15s* (bottom) solvers in MATLAB.

of dislocations in metals, in the theory of Josephson junctions, as well as in interpreting certain biological processes such as DNA dynamics.

In the most simple case where u depends only on time, the nonlinear Schrödinger equation reduces to the well-known Duffing equation

$$u_{,tt} = \alpha u(\frac{1}{2}u^2 - \lambda) - f.$$

Even for this one-dimensional ordinary differential equation, an analytic solution is still very difficult to obtain. It is known that this equation is extremely sensitive to the initial conditions and the input (driving force) $f(t)$. Figure 8.7 displays clearly that for the same given data, two Runge–Kutta solvers in MATLAB produce very different vibration modes and "trajectories" in the phase space u–p ($p = u_{,t}$). Mathematically speaking, due to the nonconvexity of the function $W(u)$, very small perturbations of the system's initial conditions and parameters may lead the system to different local minima with significantly different performance characteristics, that is, the so-called chaotic phenomena. Numerical results vary with the methods used. This is one of the main reasons why traditional perturbation analysis and direct approaches cannot successfully be applied to nonconvex systems (Gao, 2003b).

Numerical discretization of the nonconvex variational problem $(\mathcal{P})$ in mathematical physics usually leads to a nonconvex optimization problem in finite-dimensional space $\mathcal{U} = \mathbb{R}^n$, where the field variable u is simply a vector $\mathbf{x} \in \mathcal{U}$, the bilinear form $\langle \mathbf{x}, \mathbf{x}^* \rangle = \mathbf{x}^T \mathbf{x}^* = \mathbf{x} \cdot \mathbf{x}^*$ is the dot-product of two vectors, and the operator $A : \mathbb{R}^n \to \mathcal{U}^* = \mathbb{R}^n$ is a symmetrical matrix. In d.c. (difference of convex functions) programming and discrete dynamical systems, the operator $A = A^T \in \mathbb{R}^{n\times n}$ is usually indefinite. The problem (8.121) is then one of global minimization in $\mathbb{R}^n$. In this section, we discuss the canonical dual transformation method for solving this type of problem.

8.7.1 Unconstrained Nonconvex Optimization Problem with Double-Well Energy

First, let us consider an unconstrained global optimization problem in finite-dimensional space $\mathcal{U} = \mathbb{R}^n$, where $A = A^T \in \mathbb{R}^{n\times n}$ is a matrix, and $W(\mathbf{x})$ is a double-well function of the type $W(\mathbf{x}) = \frac{1}{2}(\frac{1}{2}|\mathbf{x}|^2 - \lambda)^2$. Then the primal problem is

$$\min \left\{ P(\mathbf{x}) = \frac{1}{2}\left(\frac{1}{2}|\mathbf{x}|^2 - \lambda\right)^2 + \frac{1}{2}\mathbf{x}^T A \mathbf{x} - \mathbf{x}^T f \ : \ \forall \mathbf{x} \in \mathcal{U}_k = \mathbb{R}^n \right\}. \tag{8.124}$$

The necessary condition $\delta P(\mathbf{x}) = 0$ leads to a coupled nonlinear algebraic system

$$A\mathbf{x} + \left(\frac{1}{2}|\mathbf{x}|^2 - \lambda\right)\mathbf{x} = f. \tag{8.125}$$

Clearly, a direct method for solving this nonlinear equation with n unknown is elusive. By choosing the quadratic operator $\xi = \frac{1}{2}|\mathbf{x}|^2$, the canonical function $V(\xi) = \frac{1}{2}(\xi - \lambda)^2$ is a quadratic function. By the fact that $\frac{1}{2}|\mathbf{x}|^2 = \xi \geq 0, \ \forall \mathbf{x} \in \mathbb{R}^n$, the range of the quadratic mapping $\Lambda(\mathbf{x})$ is

$$\mathcal{V}_a = \{\xi \in \mathbb{R} | \ \xi \geq 0\}.$$

Thus, on $\mathcal{V}_a$, the canonical duality relation $\varsigma = \delta V(\xi) = \xi - \lambda$ is one-to-one and the range of the canonical dual mapping $\delta V : \mathcal{V}_a \to \mathcal{V}^* \subset \mathbb{R}$ is

$$\mathcal{V}_a^* = \{\varsigma \in \mathbb{R} | \ \varsigma \geq -\lambda\}.$$

It turns out that (ξ, ς) is a canonical pair on $\mathcal{V}_a \times \mathcal{V}_a^*$ and the Legendre conjugate V^* is also a quadratic function:

$$V^*(\varsigma) = \text{sta}\{\xi\varsigma - V(\xi) \ : \ \xi \in \mathcal{V}_a\} = \frac{1}{2}\varsigma^2 + \lambda\varsigma.$$

For a given $\varsigma \in \mathcal{V}_a^*$, the Λ-conjugate transformation

$$U^\Lambda(\varsigma) = \text{sta}\left\{\frac{1}{2}\mathbf{x}^2\varsigma - \frac{1}{2}\mathbf{x}^T A\mathbf{x} + \mathbf{x}^T f \ : \ \mathbf{x} \in \mathbb{R}^n\right\}$$
$$= -\frac{1}{2}f^T (A + \varsigma I)^{-1} f$$

is well defined on the canonical dual feasible space $\mathcal{V}_k^*$, given by

$$\mathcal{V}_k^* = \{\varsigma \in \mathbb{R}| \ \det(A + \varsigma I) \neq 0, \ \varsigma \geq -\lambda\}. \tag{8.126}$$

Thus, the canonical dual problem can be proposed as the following (Gao, 2003a):

$$(\mathcal{P}^d): \quad \max\left\{P^d(\varsigma) = -\frac{1}{2}f^T(A+\varsigma I)^{-1}f - \frac{1}{2}\varsigma^2 - \lambda\varsigma \ : \ \varsigma \in \mathcal{V}_k^*\right\}. \tag{8.127}$$

This is a nonlinear programming problem with only one variable! The criticality condition of this dual problem leads to the dual algebraic equation

$$\varsigma + \lambda = \frac{1}{2}f^T(A+\varsigma I)^{-2}f. \tag{8.128}$$

For any given $A \in \mathbb{R}^{n\times n}$ and $f \in \mathbb{R}^n$, this equation can be solved by Mathematica. Extremality conditions of these dual solutions can be identified by the following theorem (see Gao, 2003a).

Theorem 8.13. (Gao, 2003a) *If the matrix A has r distinct nonzero eigenvalues such that*

$$a_1 < a_2 < \cdots < a_r,$$

then the canonical dual algebraic equation (8.128) *has at most* $2r+1$ *roots*

$$\varsigma_1 > \varsigma_2 \geq \varsigma_3 \geq \cdots \geq \varsigma_{2r+1}.$$

For each ς_i, the vector

$$\mathbf{x}_i = (A + \varsigma_i I)^{-1}f, \qquad \forall i = 1, 2, \ldots, 2r+1, \tag{8.129}$$

is a solution to the semilinear algebraic equation (8.125) *and*

$$P(\mathbf{x}_i) = P^d(\varsigma_i), \qquad \forall i = 1, \ldots, 2r+1. \tag{8.130}$$

Particularly, the canonical dual problem has at most one global maximizer $\varsigma_1 > -a_1$ in the open interval $(-a_1, +\infty)$, and $\mathbf{x}_1$ is a global minimizer of $P(\mathbf{x})$ over $\mathcal{U}_k$; that is,

$$P(\mathbf{x}_1) = \min_{\mathbf{x}\in\mathcal{U}_k} P(\mathbf{x}) = \max_{\varsigma > -a_1} P^d(\varsigma) = P^d(\varsigma_1). \tag{8.131}$$

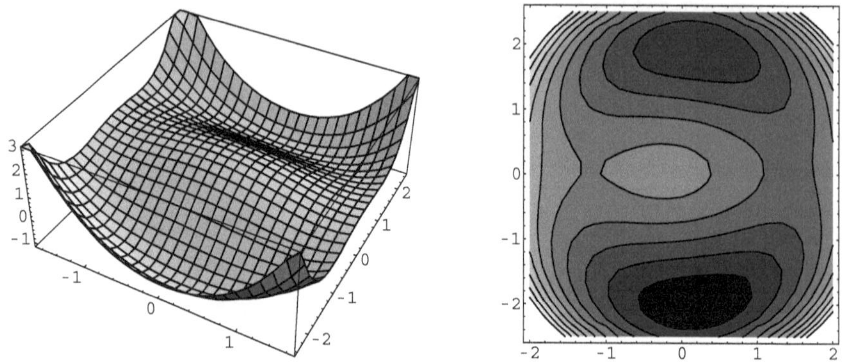

Fig. 8.8 Graph of the primal function $P(x_1, x_2)$ and its contours.

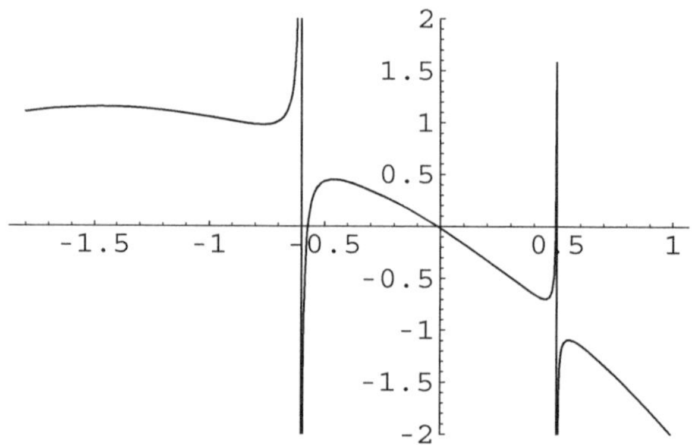

Fig. 8.9 Graph of the dual function $P^d(\varsigma)$.

Moreover, in each open interval $(-a_{i+1}, -a_i)$, *the canonical dual equation* (8.128) *has at most two real roots* $-a_{i+1} < \varsigma_{2i+1} < \varsigma_{2i} < -a_i$, $\forall i = 1, \ldots, 2r+1$, ς_{2i} *is a local minimizer of* P^d, *and* ς_{2i+1} *is a local maximizer of* $P^d(\varsigma)$.

As an example in two-dimensional space, which is illustrated in Figure 8.8, we simply choose $A = \{a_{ij}\}$ with $a_{11} = 0.6$, $a_{22} = -0.5$, $a_{12} = a_{21} = 0$, and $f = \{0.2, -0.1\}$. For a given parameter $\lambda = 1.5$, and $\alpha = 1.0$, the graph of $P(\mathbf{x})$ is a nonconvex surface (see Figure 8.8a) with four potential wells and one local maximizer. The graph of the canonical dual function $P^d(\varsigma)$ is shown in Fig. 8.9. The dual canonical dual algebraic equation (8.128) has a total of five real roots:

$$\bar{\varsigma}_5 = -1.47 < \bar{\varsigma}_4 = -0.77 < \bar{\varsigma}_3 = -0.46 < \bar{\varsigma}_2 = 0.45 < \bar{\varsigma}_1 = 0.55,$$

and we have

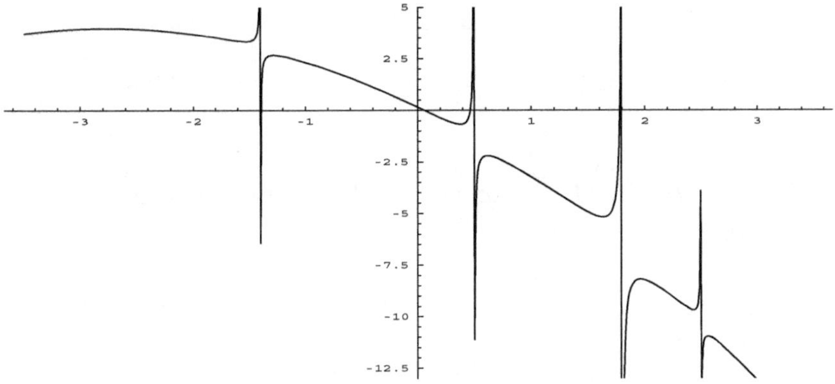

Fig. 8.10 Graph of the dual function $P^d(\varsigma)$ for a four-dimensional problem.

$$P^d(\bar{\varsigma}_5) = 1.15 > P^d(\bar{\varsigma}_4) = 0.98 > P^d(\bar{\varsigma}_3) = 0.44 > P^d(\bar{\varsigma}_2) = -0.70 > P^d(\bar{\varsigma}_1).$$

By the triality theory, we know that $\bar{\mathbf{x}}_1 = (A + \bar{\varsigma}_1 I)^{-1} f = \{0.17, -2.02\}$ is a global minimizer of $P(\bar{\mathbf{x}})$; and accordingly, $P(\bar{\mathbf{x}}_1) = P^d(\bar{\varsigma}_1) = -1.1$; and that $\bar{\mathbf{x}}_5 = \{-0.23, 0.05\}$ and $\bar{\mathbf{x}}_3 = \{1.44, 0.10\}$ are local maximizers, whereas $\bar{\mathbf{x}}_4 = \{-1.21, 0.08\}$ and $\bar{\mathbf{x}}_2 = \{0.19, 1.96\}$ are local minimizers.

The graph of $P^d(\varsigma)$ for a four-dimensional problem is shown in Figure 8.10. It can be easily seen that $P^d(\varsigma)$ is strictly concave for $\varsigma > -a_1$. Within each interval $-a_{i-1} < \varsigma < -a_i$, $\forall i = 1, 2, \ldots, r$, the dual function $P^d(\varsigma)$ has at most one local minimum and one local maximum. These local extrema can be identified by the triality theory (Gao, 2003a).

The nonconvex function $W(\mathbf{x})$ in (8.121) could be in many other forms, for example,

$$W(\mathbf{x}) = \exp\left(\frac{1}{2}|B\mathbf{x}|^2 - \lambda\right),$$

where $B \in \mathbb{R}^{m\times n}$ is a given matrix and $\lambda > 0$ is a constant. In this case, the primal problem $(\mathcal{P})$ is a quadratic-exponential minimization problem

$$\min\left\{P(\mathbf{x}) = \exp\left(\frac{1}{2}|B\mathbf{x}|^2 - \lambda\right) + \frac{1}{2}\mathbf{x}^T A\mathbf{x} - \mathbf{x}^T f \ : \ \mathbf{x} \in \mathbb{R}^n\right\}.$$

By letting $\xi = \Lambda(\mathbf{x}) = \frac{1}{2}|B\mathbf{x}|^2 - \lambda$, the canonical function $V(\xi) = \exp(\xi)$ is convex and its Legendre conjugate is $V^*(\varsigma) = \varsigma(\ln\varsigma - 1)$. The canonical dual problem was formulated in Gao and Ruan (2007):

$$(\mathcal{P}^d): \quad \max\left\{P^d(\varsigma) = -\frac{1}{2}f^T[G(\varsigma)]^{-1}f - (\varsigma\log\varsigma - \varsigma) - \lambda\varsigma \ : \ \varsigma \in \mathcal{V}^*_+\right\},$$

where $G(\varsigma) = A + \varsigma B^T B$ and the dual feasible space is defined by

$$\mathcal{V}_+^* = \{\varsigma \in \mathbb{R} \mid \varsigma > 0, \ \ G(\varsigma) \text{ is positive definite}\}.$$

Detailed study of this case was given in Gao and Ruan (2007).

8.7.2 Constrained Quadratic Minimization over a Sphere

If the function $W(\mathbf{x})$ in problem (8.121) is an indicator of a constraint set $\mathcal{U}_k \subset \mathbb{R}^n$, that is,

$$W(\mathbf{x}) = \begin{cases} 0 & \text{if } \mathbf{x} \in \mathcal{U}_k, \\ +\infty & \text{otherwise,} \end{cases}$$

then the general problem (8.121) becomes a constrained nonconvex quadratic optimization problem, denoted as

$$(\mathcal{P}_q): \quad \min\{P(\mathbf{x}) = \frac{1}{2}\langle \mathbf{x}, A\mathbf{x}\rangle - \langle \mathbf{x}, f\rangle \ : \ \mathbf{x} \in \mathcal{U}_k\}. \tag{8.132}$$

General constrained global optimization problems are discussed in the next section. Here, we consider the following quadratic minimization problem with a nonlinear constraint

$$(\mathcal{P}_q): \qquad \min \ P(\mathbf{x}) = \frac{1}{2}\mathbf{x}^T A\mathbf{x} - f^T\mathbf{x} \tag{8.133}$$
$$\text{s.t.} \ \ |\mathbf{x}| \le r,$$

where $A = A^T \in \mathbb{R}^{n \times n}$ is a symmetric matrix, $f \in \mathbb{R}^n$ is a given vector, and $r > 0$ is a constant. The feasible space $\mathcal{U}_k = \{\mathbf{x} \in \mathbb{R}^n | \ \ |\mathbf{x}| \le r\}$ is a hypersphere in $\mathbb{R}^n$. This problem often arises as a subproblem in general optimization algorithms (cf. Powell, 2002). Often, in the model trust region methods, the objective function in nonlinear programming is approximated locally by a quadratic function. In such cases, the approximation is restricted to a small region around the current iterate. These methods therefore require the solution of quadratic programming problems over spheres.

To solve this constrained nonconvex minimization by using a traditional Lagrange multiplier method, we have

$$L(\mathbf{x}, \lambda) = \frac{1}{2}\mathbf{x}^T A\mathbf{x} - f^T\mathbf{x} + \lambda(|\mathbf{x}| - r). \tag{8.134}$$

For a given $\lambda \ge 0$, the traditional dual function can be defined via the Fenchel–Moreau–Rockafellar duality theory:

$$P^*(\lambda) = \min\{L(\mathbf{x}, \lambda) \ : \ \mathbf{x} \in \mathbb{R}^n\}, \tag{8.135}$$

which is a concave function of λ. However, due to the nonconvexity of $P(\mathbf{x})$, we have only the weak duality relationship

$$\min_{|\mathbf{x}|\leq r} P(\mathbf{x}) \geq \max_{\lambda\geq 0} P^*(\lambda).$$

The duality gap θ given by the slack in the above inequality is typically nonzero indicating that the dual solution does not solve the primal problem. On the other hand, the KKT condition leads to a coupled nonlinear algebraic system

$$A\mathbf{x} + \lambda|\mathbf{x}|^{-1}\mathbf{x} = f,$$
$$\lambda \geq 0, \quad |\mathbf{x}| \leq r, \quad \lambda(|\mathbf{x}| - r) = 0.$$

As indicated by Floudas and Visweswaran (1995), due to the presence of the nonlinear sphere constraint, the solution of $(\mathcal{P}_q)$ is likely to be irrational, which implies that it is not possible to exactly compute the solution. Therefore, many polynomial time algorithms have been suggested to compute the approximate solution to this problem (see Ye, 1992). However, by the canonical dual transformation this problem has been solved completely in Gao (2004b).

First, we need to reformulate the constraint $|\mathbf{x}| \leq r$ in the canonical form

$$\xi = \Lambda(\mathbf{x}) = \frac{1}{2}|\mathbf{x}|^2.$$

Let $\lambda = \frac{1}{2}r^2$, then the canonical function $V(\Lambda(\mathbf{x}))$ can be defined as

$$V(\xi) = \begin{cases} 0 & \text{if } \xi \leq \lambda, \\ +\infty & \text{otherwise,} \end{cases}$$

whose effective domain is $\mathcal{V}_a = \{\xi \in \mathbb{R} \mid \xi \leq \lambda\}$. Letting $U(\mathbf{x}) = \mathbf{x}^T f - \frac{1}{2}\mathbf{x}^T A\mathbf{x}$, the primal problem $(\mathcal{P}_q)$ can be reformulated in the following canonical form,

$$\min\{\Pi(\mathbf{x}) = V(\Lambda(\mathbf{x})) - U(\mathbf{x}) \; : \; \mathbf{x} \in \mathbb{R}^n\}. \tag{8.136}$$

By the Fenchel transformation, the conjugate of $V(\xi)$ is

$$V^\sharp(\varsigma) = \max_{\xi\in\mathcal{V}_a}\{\xi\varsigma - V(\xi)\} = \begin{cases} \lambda\varsigma & \text{if } \varsigma \geq 0, \\ +\infty & \text{otherwise,} \end{cases} \tag{8.137}$$

whose effective domain is $\mathcal{V}_a^* = \{\varsigma \in \mathbb{R} | \; \varsigma \geq 0\}$. The dual feasible space $\mathcal{V}_k^*$ in this problem is

$$\mathcal{V}_k^* = \{\varsigma \in \mathbb{R} \mid \varsigma \geq 0, \; \det(A + \varsigma I) \neq 0\}.$$

Thus, for a given $\varsigma \in \mathcal{V}_a^*$, the Λ-conjugate of U can be formulated as

$$U^\Lambda(\varsigma) = \text{sta}\left\{\frac{1}{2}|\mathbf{x}|^2\varsigma + \frac{1}{2}\mathbf{x}^T A\mathbf{x} - \mathbf{x}^T f \;:\; \mathbf{x} \in \mathbb{R}^n\right\}$$
$$= -\frac{1}{2}f^T(A+\varsigma I)^{-1}f,$$

and the problem $(\mathcal{P}^d)$, which is perfectly dual to $(\mathcal{P}_q)$, is given by

$$(\mathcal{P}_q^d): \quad \max\left\{P^d(\varsigma) = -\frac{1}{2}f^T(A+\varsigma I)^{-1}f - \lambda\varsigma \;:\; \varsigma \in \mathcal{V}_k^*\right\}. \qquad (8.138)$$

The criticality condition $\delta P^d(\bar{\varsigma}) = 0$ leads to a nonlinear algebraic equation

$$\frac{1}{2}f^T(A+\bar{\varsigma}I)^{-2}f = \lambda. \qquad (8.139)$$

Similar to (8.128), this equation can also be solved easily by using Mathematica. Each root $\bar{\varsigma}_i$ is a critical point of $P^d(\varsigma)$. The following theorem presents a complete set of solutions for this dual problem.

Theorem 8.14. (Complete Solution to $(\mathcal{P}_q)$ (Gao, 2004b)) *Suppose that the symmetric matrix A has $p \le n$ distinct eigenvalues, and $i_d \le p$ of them are negative such that*

$$a_1 < a_2 < \cdots < a_{i_d} < 0 \le a_{i_d+1} < \cdots < a_p.$$

Then for a given vector $f \in \mathbb{R}^n$, the canonical dual problem $(\mathcal{P}_q^d)$ has at most $2i_d+1$ critical points $\bar{\varsigma}_i, i = 1, \ldots, 2i_d+1$, satisfying the following distribution law,

$$\bar{\varsigma}_1 > -a_1 > \bar{\varsigma}_2 \ge \bar{\varsigma}_3 > -a_2 > \cdots > -a_{i_d} > \bar{\varsigma}_{2i_d} \ge \bar{\varsigma}_{2i_d+1} > 0. \qquad (8.140)$$

For each $\bar{\varsigma}_i \ge 0, \;\; i = 1, \ldots, 2i_d+1$, the vector defined by

$$\bar{\mathbf{x}}_i = (A + \bar{\varsigma}_i I)^{-1} f \qquad (8.141)$$

is a KKT point of the problem $(\mathcal{P}_q)$ and

$$P(\bar{\mathbf{x}}_i) = P^d(\bar{\varsigma}_i), \qquad i = 1, 2, \ldots, 2i_d+1. \qquad (8.142)$$

Moreover, if $i_d > 0$, then the problem $(\mathcal{P}_q)$ has at most $2i_d+1$ critical points on the boundary of the sphere; that is,

$$\frac{1}{2}|\bar{\mathbf{x}}_i|^2 = \lambda, \qquad i = 1, \ldots, 2i_d+1. \qquad (8.143)$$

Because $A = A^T$, there exists an orthogonal matrix $R^T = R^{-1}$ such that $A = R^T D R$, where $D = (a_i\delta_{ij})$ is a diagonal matrix. For the given vector $f \in \mathbb{R}^n$, let $\mathbf{g} = Rf = (g_i)$, and define

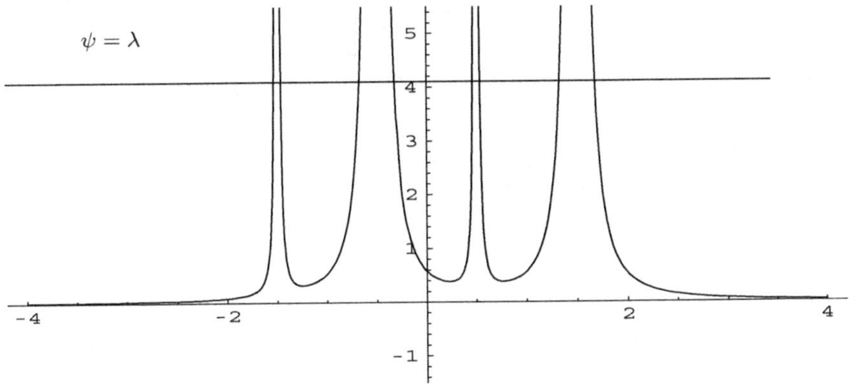

Fig. 8.11 Graph of $\psi(\varsigma)$.

$$\psi(\varsigma) = \frac{1}{2} f^T (A + \varsigma I)^{-2} f$$
$$= \frac{1}{2} \sum_{i=1}^{p} g_i^2 (a_i + \varsigma)^{-2}. \tag{8.144}$$

Clearly, this real-valued function $\psi(\varsigma)$ is strictly convex within each interval $-a_{i+1} < \varsigma < -a_i$, as well as over the intervals $-\infty < \varsigma < -a_p$ and $-a_1 < \varsigma < \infty$ (see Figure 8.11). Thus, for a given parameter $\lambda > 0$, the algebraic equation

$$\psi(\varsigma) = \frac{1}{2} \sum_{i=1}^{p} g_i^2 (a_i + \varsigma)^{-2} = \lambda \tag{8.145}$$

has at most $2p$ solutions $\{\bar{\varsigma}_i\}$ satisfying $-a_{j+1} < \bar{\varsigma}_{2j+1} \leq \bar{\varsigma}_{2j} < -a_j$ for $j = 1, \ldots, p-1$, and $\bar{\varsigma}_1 > -a_1$, $\bar{\varsigma}_{2p} < -a_p$. Because A has only i_d negative eigenvalues, the equality $\psi(\varsigma) = \lambda$ has at most $2i_d + 1$ strictly positive roots $\{\bar{\varsigma}_i\} > 0, \quad i = 1, \ldots, 2i_d + 1$. By the complementarity condition $\bar{\varsigma}_i(\frac{1}{2}|\bar{\mathbf{x}}_i|^2 - \lambda) = 0$, we know that the primal problem $(\mathcal{P}_q)$ has at most $2i_d + 1$ KKT points $\bar{\mathbf{x}}_i$ on the sphere $\frac{1}{2}|\bar{\mathbf{x}}_i|^2 = \lambda$. If $a_{i_d+1} > 0$, the equality $\psi(\varsigma) = \lambda$ may have at most $2i_d$ strictly positive roots.

By using the triality theory, the extremality conditions of the critical points of the problem $(\mathcal{P}_q)$ can be identified by the following result.

Theorem 8.15. (Global and Local Extrema (Gao, 2004b)) *Suppose that a_1 is the smallest eigenvalue of A. Then the dual problem $(\mathcal{P}^d{}_q)$ given in* (8.138) *has a unique solution $\bar{\varsigma}_1$ over the domain $\varsigma > -a_1 \geq 0$, and $\bar{\mathbf{x}}_1$ is a global minimizer of the problem $(\mathcal{P}_q)$; that is,*

$$P(\bar{\mathbf{x}}_1) = \min_{\mathbf{x} \in \mathcal{U}_k} P(\mathbf{x}) = \max_{\varsigma > -a_1} P^d(\varsigma) = P^d(\bar{\varsigma}_1). \tag{8.146}$$

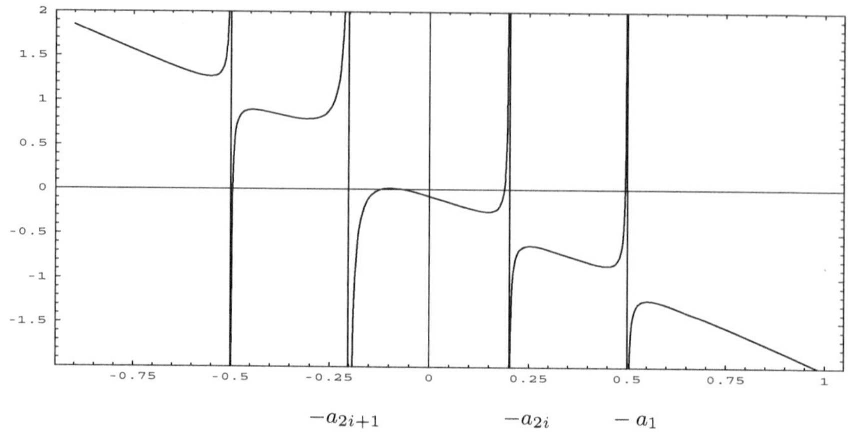

Fig. 8.12 Graph of $P^d(\varsigma)$.

If in each interval $(-a_{i+1}, -a_i)$, $i = 1, \ldots, i_d$, *the dual algebraic equation* (8.139) *has two roots* $-a_{i+1} < \bar{\varsigma}_{2i+1} < \bar{\varsigma}_{2i} < -a_i$, *then* $\bar{\varsigma}_{2i}$ *is a local minimizer of* $P^d(\varsigma)$, *and* $\bar{\varsigma}_{2i+1}$ *is a local maximizer of* $P^d(\varsigma)$ *over the interval* $(-a_{i+1}, -a_i)$.

Proof. Because for any given $\varsigma > -a_1$, the matrix $A + \varsigma I$ is positive definite, that is, the total complementary function $\Xi(\mathbf{x}, \varsigma)$ is a saddle function, the saddle minmax theorem leads to (8.146).

The remaining statements in Theorem 8.15 can be proved by the graph of $P^d(\varsigma)$ (see Figure 8.12). □

It is interesting to note that on the effective domain $\mathcal{V}_a^*$, the Fenchel–Young equality $V(\xi) = \langle \xi; \varsigma \rangle - V^*(\varsigma) = (\xi - \lambda)\varsigma$ holds true. Thus, on $\mathcal{U}_a \times \mathcal{V}_a^*$, the total complementary function

$$\Xi(\mathbf{x}, \varsigma) = \langle \Lambda(\mathbf{x}); \varsigma \rangle - V^*(\varsigma) - U(\mathbf{x})$$

$$= \varsigma \left(\frac{1}{2}|\mathbf{x}|^2 - \lambda \right) + \frac{1}{2}\mathbf{x}^T A \mathbf{x} - \mathbf{x}^T f \tag{8.147}$$

can be viewed as the traditional Lagrangian of the quadratic minimization problem with the reformulated (canonical) quadratic constraint $\frac{1}{2}|\mathbf{x}|^2 \leq \lambda$, which is also called *extended Lagrangian* (see Gao, 2000a). This example exhibits a connection between the nonlinear Lagrange multiplier method and the canonical dual transformation. Based on this observation, the traditional Lagrange multiplier method can be generalized to solve constrained global optimization problems.

8.8 General Constrained Global Optimization Problems

In this section, we present an important application of the canonical duality theory to the following general constrained nonlinear programming problem

$$\min \{-U(\mathbf{x}) \ : \ \mathbf{x} \in \mathcal{U}_k\}, \tag{8.148}$$

where $U(\mathbf{x})$ is a Gâteaux differentiable function, either a linear or canonical function, defined on an open convex set $\mathcal{U}_a \subset \mathbb{R}^n$, and the feasible space $\mathcal{U}_k$ is a convex subset of $\mathcal{U}_a$ defined by

$$\mathcal{U}_k = \{\mathbf{x} \in \mathcal{U}_a \subset \mathbb{R}^n \mid g_i(\mathbf{x}) \le 0, \ i = 1, \dots, p\},$$

in which $g_i(\mathbf{x}) : \mathcal{U}_a \to \mathbb{R}$ are convex functions. We show the connection between the canonical dual transformation and nonlinear Lagrange multiplier methods and how to use the triality theory to identify global and local optima.

8.8.1 Canonical Form and Total Complementary Function

First, we need to put this problem in the framework of the canonical systems. Let the geometrical operator $\boldsymbol{\xi} = \Lambda(\mathbf{x}) = \{g_i(\mathbf{x})\} : \mathcal{U}_a \to \mathcal{V}_a \subset \mathbb{R}^p$ be a vector-valued function. The generalized canonical function

$$V(\boldsymbol{\xi}) = \begin{cases} 0 & \text{if } \boldsymbol{\xi} \le 0 \\ \infty & \text{otherwise} \end{cases}$$

is an indicator of the convex cone $\mathcal{V}_a = \{\boldsymbol{\xi} \in \mathbb{R}^p \mid \boldsymbol{\xi} \le 0\}$. Thus, the canonical form of the constrained problem (8.148) is

$$\min\{\Pi(\mathbf{x}) = V(\Lambda(\mathbf{x})) - U(\mathbf{x}) \ : \ \mathbf{x} \in \mathcal{U}_a\}.$$

By the Fenchel transformation, the conjugate of $V(\xi)$ is an indicator of the dual cone $\mathcal{V}_a^* = \{\varsigma \in \mathbb{R}^p | \ \varsigma \ge 0\}$; that is,

$$V^\sharp(\varsigma) = \max\{\langle \boldsymbol{\xi}; \varsigma\rangle - V(\boldsymbol{\xi}) \ : \ \boldsymbol{\xi} \in \mathbb{R}^p\} = \begin{cases} 0 & \text{if } \varsigma \ge 0 \\ \infty & \text{otherwise.} \end{cases}$$

By the theory of convex analysis we have

$$\varsigma \in \partial^- V(\boldsymbol{\xi}) \ \Leftrightarrow \ \boldsymbol{\xi} \in \partial^- V^\sharp(\varsigma) \ \Leftrightarrow \ \langle \boldsymbol{\xi} \, ; \, \varsigma\rangle = V(\boldsymbol{\xi}) + V^\sharp(\varsigma); \tag{8.149}$$

that is, $(\boldsymbol{\xi}, \varsigma)$ is a generalized canonical pair on $\mathcal{U}_a \times \mathcal{V}_a^*$ (Gao, 2000c). Thus, the extended Lagrangian $\Xi(\mathbf{x}, \varsigma) = \langle \Lambda(\mathbf{x}); \varsigma\rangle - V^\sharp(\varsigma) - U(\mathbf{x})$ in this problem has a very simple form:

$$\Xi(\mathbf{x}, \varsigma) = -U(\mathbf{x}) + \sum_{i=1}^{p} \varsigma_i g_i(\mathbf{x}). \tag{8.150}$$

We can see here that the canonical dual variable $\varsigma \geq 0 \in \mathbb{R}^p$ is nothing but a Lagrange multiplier for the constraints $\Lambda(\mathbf{x}) = \{g_i(\mathbf{x})\} \leq 0$. Let

$$\mathcal{I}(\bar{\mathbf{x}}) := \{i \in \{1, \ldots, p\} | \ g_i(\bar{\mathbf{x}}) = 0\}$$

be the index set of the active constraints at $\bar{\mathbf{x}}$. By the theory of global optimization (cf. Horst et al., 2000) we know that if $\bar{\mathbf{x}}$ is a local minimizer such that $\nabla g_i(\bar{\mathbf{x}}), \ \ i \in \mathcal{I}(\bar{\mathbf{x}})$, are linearly independent, then the KKT conditions hold:

$$g_i(\bar{\mathbf{x}}) \leq 0, \qquad \bar{\varsigma}_i \geq 0, \qquad \bar{\varsigma}_i g_i(\bar{\mathbf{x}}) = 0, \qquad i = 1, \ldots, p, \tag{8.151}$$

$$\nabla U(\bar{\mathbf{x}}) = \sum_{i=1}^{p} \bar{\varsigma}_i \nabla g_i(\bar{\mathbf{x}}). \tag{8.152}$$

Any point $(\bar{\mathbf{x}}, \bar{\varsigma})$ that satisfies (8.151)–(8.152) is called a KKT stationary point of the problem (8.148). However, the KKT conditions (8.151)–(8.152) are only necessary for the minimization problem (8.148). They are sufficient for a constrained global minimum at $\bar{\mathbf{x}}$ provided that, for example, the functions $P(\mathbf{x}) = -U(\mathbf{x})$ and $g_i(\mathbf{x})$, $i = 1, \ldots, p$, are convex. In constrained global optimization problems, the primal problems may possess many local minimizers due to the nonconvexity of the objective function and constraints. Therefore, sufficient optimality conditions play a key role in developing global algorithms. Here we show that the triality theory can provide such sufficient conditions.

The complementary function $V^\sharp(\varsigma) = 0$, $\forall \varsigma \in \mathcal{V}_a^*$, therefore in this constrained optimization problem we have

$$G_\varsigma(\mathbf{x}) = \Xi(\mathbf{x}, \varsigma) = -U(\mathbf{x}) + \varsigma^T \Lambda(\mathbf{x}). \tag{8.153}$$

For a fixed $\varsigma \in \mathcal{V}_a^*$, if the parametric function $G_\varsigma : \mathcal{U}_a \to \mathbb{R}$ is twice Gâteaux differentiable, the space $\mathcal{G}$ can be written as

$$\mathcal{G} = \left\{ (\mathbf{x}, \varsigma) \in \mathcal{U}_a \times \mathcal{V}_a^* \mid \ \det\left(\frac{\partial^2 G_\varsigma(\mathbf{x})}{\partial x_i \partial x_j} \right) \neq 0 \right\}.$$

Clearly for any given $(\mathbf{x}, \varsigma) \in \mathcal{G}$, the dual feasible space $\mathcal{V}_k^*$,

$$\mathcal{V}_k^* = \left\{ \varsigma \in \mathcal{V}_a^* \mid \Lambda_t^*(\mathbf{x})\varsigma = \sum_{i=1}^{p} \varsigma_i \nabla g_i(\mathbf{x}) = \nabla U(\mathbf{x}), \ \ \forall \mathbf{x} \in \mathcal{U}_a \right\} \tag{8.154}$$

is nonempty and the Λ-conjugate transformation

$$U^\Lambda(\varsigma) = \text{sta}\,\{\langle \Lambda(\mathbf{x}); \varsigma\rangle - U(\mathbf{x}) \;:\; \forall \mathbf{x} \in \mathcal{U}_a\}$$

can be well formulated on $\mathcal{V}_k^*$. Thus, the canonical dual problem can be proposed as the following,

$$\max\{P^d(\varsigma) = -U^\Lambda(\varsigma) \;:\; \varsigma \in \mathcal{V}_k^*\}. \tag{8.155}$$

In the following, we illustrate the foregoing results using some examples.

8.8.2 Quadratic Minimization with Quadratic Constraints

Let $U(\mathbf{x}) = \mathbf{x}^T f - \frac{1}{2}\mathbf{x}^T A\mathbf{x}$ and $g(\mathbf{x}) = \frac{1}{2}\mathbf{x}^T C\mathbf{x} - \lambda$ be quadratic functions, where A and C are two symmetrical matrices in $\mathbb{R}^{n\times n}$, $f \in \mathbb{R}^n$ is a given vector, and $\lambda \in \mathbb{R}$ is a given constant. Thus the primal problem is:

$$\min\left\{P(\mathbf{x}) = \frac{1}{2}\mathbf{x}^T A\mathbf{x} - f^T\mathbf{x} \;:\; \frac{1}{2}\mathbf{x}^T C\mathbf{x} \le \lambda, \;\; \mathbf{x} \in \mathbb{R}^n\right\}. \tag{8.156}$$

Because we have only one constraint $g(\mathbf{x}) = \frac{1}{2}\mathbf{x}^T C\mathbf{x} - \lambda$, the extended Lagrangian is simply

$$\Xi(\mathbf{x}, \varsigma) = \frac{1}{2}\mathbf{x}^T(A + \varsigma C)\mathbf{x} - f^T\mathbf{x} - \varsigma\lambda. \tag{8.157}$$

On the dual feasible space

$$\mathcal{V}_k^* = \{\varsigma \in \mathbb{R} \mid \varsigma \ge 0, \;\; \det(A + \varsigma C) \ne 0\},$$

and the canonical dual problem (8.155) can be formulated as (see Gao, 2005a):

$$\max\left\{P^d(\varsigma) = -\frac{1}{2}f^T(A + \varsigma C)^{-1}f - \lambda\varsigma \;:\; \varsigma \in \mathcal{V}_k^*\right\}. \tag{8.158}$$

Because in this problem both $\Lambda(\mathbf{x}) = (\frac{1}{2}\mathbf{x}^T C\mathbf{x} - \lambda)$ and $U(\mathbf{x}) = -\frac{1}{2}\mathbf{x}^T A\mathbf{x} + f^T\mathbf{x}$ are quadratic functions, $\delta^2 G_\varsigma = (A + \varsigma C)$. The following result was obtained recently.

Theorem 8.16. (Gao, 2005a) *Suppose that the matrix C is positive definite, and $\bar{\varsigma} \in \mathcal{V}_a^*$ is a critical point of $P^d(\varsigma)$. If $A + \bar{\varsigma}C$ is positive definite, the vector*

$$\bar{\mathbf{x}} = (A + \bar{\varsigma}C)^{-1}f$$

is a global minimizer of the primal problem (8.156). *However, if $A + \bar{\varsigma}C$ is negative definite, the vector $\bar{\mathbf{x}} = (A + \bar{\varsigma}C)^{-1}f$ is a local minimizer of the primal problem* (8.156).

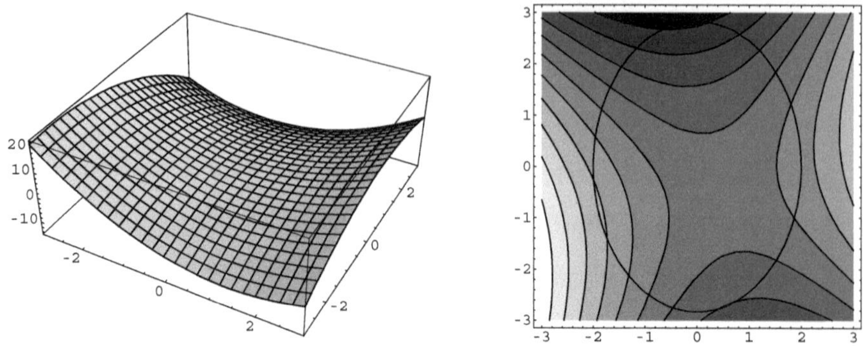

Fig. 8.13 Graph of $P(\mathbf{x})$ (left); contours of $P(\mathbf{x})$ and boundary of $\mathcal{U}_k$ (right).

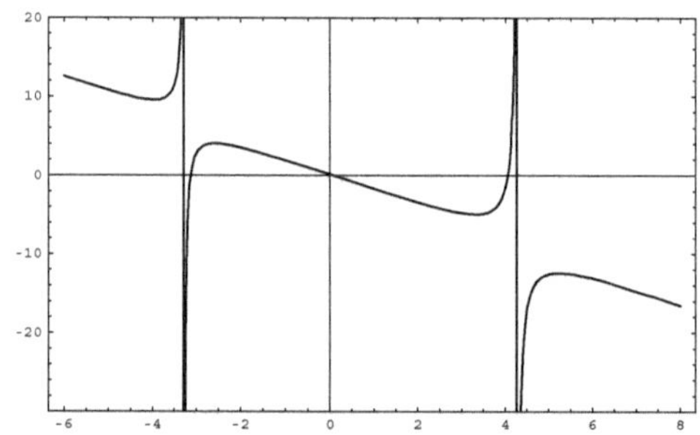

Fig. 8.14 Graphs of $P^d(\varsigma)$.

In two-dimensional space, if we let $a_{11} = 3$, $a_{12} = a_{21} = .5$, $a_{22} = -2.0$, and $c_{11} = 1$, $c_{12} = c_{21} = 0$, $c_{22} = 0.5$, the matrix $A = \{a_{ij}\}$ is indefinite, and $C = \{c_{ij}\}$ is positive definite. Setting $f = \{1, 1.5\}$ and $\lambda = 2$, the graph of the canonical function $P(\mathbf{x}) = \frac{1}{2}\mathbf{x}^T A\mathbf{x} - \mathbf{x}^T f$ is a saddle surface (see Figure 8.13), and the boundary of the feasible set $\mathcal{U}_k = \{\mathbf{x} \in \mathbb{R}^2 \mid \frac{1}{2}\mathbf{x}^T C\mathbf{x} \leq \lambda\}$ is an ellipse (see Figure 8.13). In this case, the dual problem has four critical points (see Figure 8.14):

$$\bar{\varsigma}_1 = 5.22 > \bar{\varsigma}_2 = 3.32 > \bar{\varsigma}_3 = -2.58 > \bar{\varsigma}_4 = -3.97.$$

Because $\bar{\varsigma}_1 \in \mathcal{V}^*_+$ and $\bar{\varsigma}_4 \in \mathcal{V}^*_-$, the triality theory tells us that $\mathbf{x}_1 = \{-0.22, 2.81\}$ is a global minimizer, and $\mathbf{x}_4 = \{-1.90, -0.85\}$ is a local minimizer. From the graph of $P^d(\varsigma)$ we can see that $\mathbf{x}_2 = \{0.59, -2.70\}$ is a local minimizer, and $\mathbf{x}_3 = \{2.0, 0.15\}$ is a local maximizer. We have

$$P(\mathbf{x}_1) = -12.44 < P(\mathbf{x}_2) = -4.91 < P(\mathbf{x}_3) = 4.03 < P(\mathbf{x}_4) = 9.53.$$

8.8.3 Quadratic Minimization with Box Constraints

The primal problem solved in this section is finding a global minimizer of a nonconvex quadratic function over a box constraint:

$$(\mathcal{P}_b): \quad \min\left\{ P(\mathbf{x}) = \frac{1}{2}\mathbf{x}^T A\mathbf{x} - f^T\mathbf{x} \quad : \quad \ell^l \le \mathbf{x} \le \ell^u \right\}, \tag{8.159}$$

where $\mathbf{x} \in \mathbb{R}^n$, and ℓ^l, ℓ^u are two given vectors in $\mathbb{R}^n$. Problems of the form (8.159) appear frequently in partial differential equations, discretized optimal control problems, linear least squares problems, and certain successive quadratic programming methods (cf. Floudas and Visweswaran, 1995). Particularly, if $\ell^l = \mathbf{0}$ and $\ell^u = \mathbf{1}$, the problem $(\mathcal{P}_b)$ is directly related to one of the fundamental problems of combinatorial optimization, namely, a continuous relaxation to the problem of minimizing a quadratic function in 0–1 variables.

In order to solve this problem, we need to reformulate the constraints in canonical form. Without loss of generality, we assume that $\ell^l = -\mathbf{1}$ and $\ell^u = \mathbf{1}$ (if necessary, a simple linear transformation can be used to convert the problem to this form).

$$\min\left\{ P(\mathbf{x}) = \frac{1}{2}\mathbf{x}^T A\mathbf{x} - f^T\mathbf{x} \quad : \quad x_i^2 \le 1, \; i = 1, \dots, n \right\}. \tag{8.160}$$

The constraint in this problem is a vector-valued quadratic function $\Lambda(\mathbf{x}) = \{g_i(\mathbf{x})\} = \{x_i^2 - 1\} \le 0 \in \mathbb{R}^n$. Thus, the canonical dual variable $\varsigma = \{\varsigma_i\}$ should also be a vector in $\mathbb{R}^n$. It has been shown recently that on the dual feasible space,

$$\mathcal{V}_k^* = \{\varsigma \in \mathbb{R}^n | \; \varsigma \ge 0, \;\; \det\,(A + 2\,\mathrm{Diag}\,(\varsigma)) \ne 0\},$$

where $\mathrm{Diag}\,(\varsigma) \in \mathbb{R}^{n\times n}$ represents a diagonal matrix with ς_i, $i = 1, \dots, n$ as its diagonal entries; the canonical dual problem is given by (see Gao, 2007a,b)

$$\max\left\{ P^d(\varsigma) = -\frac{1}{2} f^T (A + 2\,\mathrm{Diag}\,(\varsigma))^{-1} f - \sum_{i=1}^n \varsigma_i \quad : \quad \varsigma \in \mathcal{V}_k^* \right\}. \tag{8.161}$$

This dual problem can be solved to obtain all the critical points $\bar{\varsigma}$. It is shown in Gao (2007a,b) that if

$$\bar{\varsigma} \in \mathcal{V}_+^* = \{\varsigma \in \mathbb{R}^n \mid \varsigma \ge 0, \;\; A + 2\,\mathrm{Diag}\,(\varsigma) \text{ is positive definite}\},$$

then the vector $\bar{\mathbf{x}}(\bar{\varsigma}) = (A + 2\,\mathrm{Diag}\,(\bar{\varsigma}))^{-1} f$ is a global minimizer of the primal problem.

8.8.4 Concave Minimization

The primal problem in this case is given by

$$(\mathcal{P}_c): \quad \min\{P(\mathbf{x}) = -U(\mathbf{x}) \quad : \quad B\mathbf{x} \le \mathbf{b}, \ \mathbf{x} \in \mathbb{R}^n\}, \tag{8.162}$$

where $U(\mathbf{x})$ is a convex, or even nonsmooth function, and where $B \in \mathbb{R}^{m\times n}$ and $\mathbf{b} \in \mathbb{R}^m$ are given. It is well known that this problem is NP-hard. Concave minimization problems constitute one of the most fundamental and intensely studied classes of problems in global minimization. A comprehensive review/survey of the mathematical properties, common applications, and solution methods is given by Benson (1995). By the use of the canonical dual transformation, a perfect dual problem has been formulated in Gao (2005a). In order to provide insights into the connection between the canonical dual transformation and the traditional Lagrange multiplier method, we demonstrate here how this perfect dual formulation can also be reproduced by the classical Lagrangian duality approach when executed in a particular fashion inspired by the canonical duality.

First, let us introduce a parameter μ such that

$$\min\{P(\mathbf{x}) \ : \ B\mathbf{x} \le \mathbf{b}\} \le \mu \le \max\{P(\mathbf{x}) \ : B\mathbf{x} \le \mathbf{b}\}.$$

Then the parameterized canonical form of this problem can be formulated as (see Gao, 2005a)

$$(\mathcal{P}_\mu): \quad \min\{P(\mathbf{x}) = -U(\mathbf{x}) : \ \{U + \mu, B\mathbf{x} - \mathbf{b}\} \le 0 \in \mathbb{R}^{1+m}, \ \mathbf{x} \in \mathbb{R}^n\}. \tag{8.163}$$

In this case, the constraint $g_1(\mathbf{x}) = U(\mathbf{x}) + \mu$ is convex and $\{g_i(\mathbf{x}),\ i = 2, \ldots, m+1\} = B\mathbf{x} - \mathbf{b}$ are linear. By introducing Lagrange multipliers $(\varsigma,\ \mathbf{y}) \in \mathbb{R}^{1+m}$, and letting

$$\mathcal{V}_a^* = \{(\varsigma,\ \mathbf{y}) \in \mathbb{R}^{1+m} \mid \varsigma \ge 0, \ \ \mathbf{y} \ge \mathbf{0} \in \mathbb{R}^m\},$$

the Lagrangian dual to the parameterized canonical problem (8.163) is given by

$$\Xi(\mathbf{x}, \varsigma, \mathbf{y}) = (\varsigma - 1)U(\mathbf{x}) + \mu\varsigma + \mathbf{y}^T(B\mathbf{x} - \mathbf{b}).$$

Thus, by the classical Lagrangian duality, the dual problem to $(\mathcal{P}_\mu)$ is

$$(LD): \quad \max_{(\varsigma,\mathbf{y})\in\mathcal{V}_a^*} \{\mu\varsigma - \mathbf{y}^T\mathbf{b} + \min_{\mathbf{x}}\{(\varsigma - 1)U(\mathbf{x}) + \mathbf{y}^T B\mathbf{x}\}\}. \tag{8.164}$$

Because $U(\mathbf{x})$ is convex, the inner minimization problem in this dual form has a unique solution $\bar{\mathbf{x}}$ if $\varsigma > 1$.

Remark 8.1. Assume that

(1) $U(\mathbf{x})$ is a convex function such that $\mathbf{x}^* = \delta U(\mathbf{x})$ is invertible for each $\mathbf{x} \in \mathbb{R}^n$, and the Legendre conjugate function $U^*(\mathbf{x}^*) = \text{sta}\{\mathbf{x}^T\mathbf{x}^* - U(\mathbf{x}) \ : \ \delta U(\mathbf{x}) = \mathbf{x}^*\}$ is uniquely defined in $\mathbb{R}^n$.
(2) An optimum solution $\bar{\mathbf{x}}$ to the problem $(\mathcal{P}_\mu)$ is a KKT solution with Lagrange multipliers $\bar{\varsigma} > 1$, $\bar{\mathbf{y}} \geq \mathbf{0} \in \mathbb{R}^m$.

Let

$$\mathcal{V}_+^* = \{(\varsigma, \ \mathbf{y}) \in \mathbb{R}^{1+m} \mid \varsigma > 1, \ \ \mathbf{y} \geq \mathbf{0} \in \mathbb{R}^m\}.$$

Under Remark 8.1, thus, we can write (LD) in (8.164) as

$$(LD): \quad \max_{(\varsigma,\mathbf{y})\in\mathcal{V}_+^*} \left\{\mu\varsigma - \mathbf{y}^T\mathbf{b} + (\varsigma - 1)\min_{\mathbf{x}}\left\{\frac{\mathbf{y}^T B\mathbf{x}}{\varsigma - 1} + U(\mathbf{x})\right\}\right\}. \tag{8.165}$$

Observe that the effect of having introduced $U(\mathbf{x}) + \mu \leq 0$ is to convexity the inner minimization problem in (8.165), which, by the assumption of Remark 8.1, reduces (LD) to the following equivalent dual problem.

$$(\mathcal{P}_\mu^d): \quad \max_{(\varsigma,\mathbf{y})\in\mathcal{V}_+^*} \left\{P^d(\varsigma, \mathbf{y}) = \mu\varsigma - \mathbf{y}^T\mathbf{b} + (1-\varsigma)U^*\left(\frac{B^T\mathbf{y}}{1-\varsigma}\right)\right\}. \tag{8.166}$$

This is the dual problem proposed by the canonical dual transformation in Gao (2005a). By the fact that the Legendre conjugate $U^*(\mathbf{x}^*)$ of the convex function $U(\mathbf{x})$ is also convex, this canonical dual is a concave maximization problem over the dual feasible space $\mathcal{V}_+^*$, which can be solved uniquely for a given parameter $\mu \in \mathbb{R}$ if $\mathcal{V}_+^*$ is nonempty.

Under Remark 8.1, note that $\bar{\mathbf{x}}$ solves the primal problem $(\mathcal{P}_\mu)$ because $P(\bar{\mathbf{x}}) = \mu$, and satisfies the KKT conditions

$$(\bar{\varsigma} - 1)\delta U(\bar{\mathbf{x}}) + B^T\bar{\mathbf{y}} = 0, \tag{8.167}$$

$$B\bar{\mathbf{x}} \leq \mathbf{b}, \quad U(\bar{\mathbf{x}}) + \mu = 0, \quad \bar{\mathbf{y}}^T(B\bar{\mathbf{x}} - \mathbf{b}) = 0, \quad \bar{\mathbf{y}} \geq 0, \quad \bar{\varsigma} > 1. \tag{8.168}$$

Writing the (LD) in (8.164) as

$$\max_{(\varsigma,\mathbf{y})\in\mathcal{V}_a^*} P_\theta^d(\varsigma, \mathbf{y}),$$

where

$$P_\theta^d(\varsigma, \mathbf{y}) = \mu\varsigma - \mathbf{y}^T\mathbf{b} + \min_{\mathbf{x}}\{(\varsigma - 1)U(\mathbf{x}) + \mathbf{y}^T B\mathbf{x}\},$$

we get

$$P_\theta^d(\bar{\varsigma}, \bar{\mathbf{y}}) = \bar{\varsigma}\mu - \mathbf{b}^T\bar{\mathbf{y}} + (\bar{\varsigma} - 1)U(\hat{\mathbf{x}}) + \bar{\mathbf{y}}^T B\hat{\mathbf{x}}, \tag{8.169}$$

where $\hat{\mathbf{x}}$ satisfies $\delta U(\hat{\mathbf{x}}) = B^T\bar{\mathbf{y}}/(1-\bar{\varsigma})$. By (8.167) and the assumed invertibility of the canonical dual relation $\mathbf{x}^* = \delta U(\mathbf{x})$, we get $\hat{\mathbf{x}} = \bar{\mathbf{x}}$. Substituting

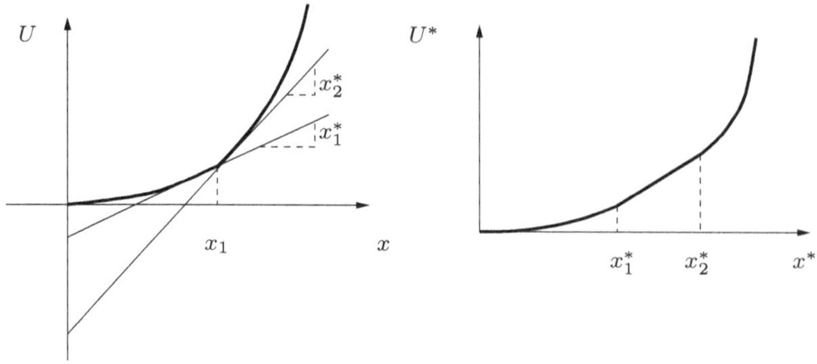

(a) Graph of $U(x)$. (b) Graph of the Legendre conjugate $U^*(x^*)$.

Fig. 8.15 Nonsmooth function and its smooth Legendre conjugate.

this into (8.169) and using (8.168) yields ${P^d}_\theta(\bar{\varsigma}, \bar{\mathbf{y}}) = P(\bar{\mathbf{x}})$; that is, there is zero duality gap. Furthermore, letting

$$\mathcal{U}_\mu = \{\mathbf{x} \in \mathbb{R}^n | \ B\mathbf{x} \leq \mathbf{b}, \ -U(\mathbf{x}) = \mu\},$$

we have the following result.

Theorem 8.17. (KKT Condition and Global Optimality) *Under Remark* 8.1, *for a given parameter* μ, *if* $(\bar{\varsigma}, \bar{\mathbf{y}}) \in \mathcal{V}_a^*$ *is a KKT point of* $(\mathcal{P}_\mu^d)$ *such that*

$$\bar{\mathbf{x}}^* = \frac{B^T \bar{\mathbf{y}}}{1 - \bar{\varsigma}},$$

then the vector $\bar{\mathbf{x}} = \delta U^*(\bar{\mathbf{x}}^*)$ *is a KKT point of* $(\mathcal{P}_\mu)$, *and* $P(\bar{\mathbf{x}}) = P^d(\bar{\varsigma}, \bar{\mathbf{y}})$.

Moreover, if $\bar{\varsigma} > 1$, *then* $(\bar{\varsigma}, \bar{\mathbf{y}})$ *is a global maximizer of* $P^d(\varsigma, \mathbf{y})$ *on* $\mathcal{V}_+^*$, $\bar{\mathbf{x}}$ *is a global minimizer of* $P(\mathbf{x})$ *on the feasible space* $\mathcal{U}_\mu$, *and*

$$\min_{\mathbf{x} \in \mathcal{U}_\mu} P(\mathbf{x}) = \max_{(\varsigma, \mathbf{y}) \in \mathcal{V}_+^*} P^d(\varsigma, \mathbf{y}).$$

This example shows again that when a nonconvex constrained optimization problem can be written in a canonical form, the classical Lagrange multiplier method can be used to formulate a perfect dual problem. A detailed study on the canonical duality theory for solving general constrained nonconvex minimization problems and its connections with Lagrangian duality appears in Gao, Ruan, and Sherali (2008).

One advantage of the canonical duality approach is that if the convex $U(\mathbf{x})$ is nonsmooth on $\mathcal{U}_a$, its Fenchel–Legendre conjugate U^* is a smooth function on $\mathcal{U}_a^*$ (see Figure 8.15). Such an idea has also been used in the study of geometrical dual analysis for solving nonsmooth "shape-preserving" design problems (see Cheng, Fang, and Lavery, 2005, Lavery, 2004, Zhao, Fang, and Lavery, 2006).

8.9 Sequential Canonical Dual Transformation and Solutions to Polynomial Minimization Problems

The canonical dual transformation method can be generalized in different ways to solve the global optimization problem:

$$\min\{P(\mathbf{x}) = W(\mathbf{x}) - U(\mathbf{x}) \;:\; \mathbf{x} \in \mathcal{U}_a\} \tag{8.170}$$

with different types of nonconvex functions $W(\mathbf{x}) = V(\Lambda(\mathbf{x}))$ and geometrical operators Λ. If the geometrical operator $\Lambda : \mathcal{U} \to \mathcal{V}$ is a general nonlinear, nonconvex mapping, we can continue to use the canonical dual transformation such that the general nonconvex function $W(\mathbf{x})$ can be written in the canonical form (see Gao, 2000a):

$$W(\mathbf{x}) = V(\Lambda(\mathbf{x})) = V_n(\xi_n(\xi_{n-1}(\dots(\xi_1(u))\dots))), \tag{8.171}$$

where $\xi_k(\xi_{k-1})$ is either a convex or a concave function of ξ_{k-1}, and we write

$$V_k(\xi_k) = \xi_{k+1}(\xi_k), \qquad k = 1, \dots, n-1.$$

Thus, the geometrical operator $\Lambda : \mathcal{U} \to \mathcal{V}$ in this problem is a sequential composition of nonlinear mappings $\Lambda^{(k)} : \mathcal{V}_{k-1} \to \mathcal{V}_k,\;\; k = 1, \cdots, n,\; \mathcal{V}_0 = \mathcal{U}$, and $\mathcal{V}_n = \mathcal{V}$; that is,

$$\xi_n(\mathbf{x}) = \Lambda(\mathbf{x}) = \left[\Lambda^{(n)} \circ \Lambda^{(n-1)} \circ \dots \circ \Lambda^{(1)}\right](\mathbf{x}).$$

Because each $V_k(\xi_k)$ is a canonical function of ξ_k, the canonical duality relation $\varsigma_k = \delta V_k(\xi_k) : \mathcal{V}_k \to \mathcal{V}_k^*$ is one-to-one. It turns out that the Legendre conjugate

$$V_k^*(\varsigma_k) = \langle \xi_k ; \varsigma_k \rangle - V_k(\xi_k)$$

can be uniquely defined. Letting $\varsigma = \{\varsigma_i\} \in \mathbb{R}^n$, the sequential canonical Lagrangian associated with the general nonconvex problem (8.170) can be written as (see Gao, 2000a)

$$\Xi(\mathbf{x}, \varsigma) = \langle \Lambda^{(1)}(\mathbf{x}); \varsigma_n! \rangle - V_w^*(\varsigma) - U(\mathbf{x}), \tag{8.172}$$

where $\varsigma_p! := \varsigma_p \varsigma_{p-1} \cdots \varsigma_2 \varsigma_1$ and

$$V_w^*(\varsigma) = V_n^*(\varsigma_n) + \varsigma_n V_{n-1}^*(\varsigma_{n-1}) + \dots + \frac{\varsigma_n!}{\varsigma_1} V_1^*(\varsigma_1). \tag{8.173}$$

Thus, the canonical dual problem can be formulated as:

$$\max\{P^d(\varsigma) = U^{\Lambda^{(1)}}(\varsigma) - V_w^*(\varsigma) \;:\; \varsigma \in \mathcal{V}_k^*\}. \tag{8.174}$$

For certain given canonical functions V, and U, and the geometrical operator $\Lambda^{(1)}$, the Λ-conjugate transformation

$$U^{\Lambda^{(1)}}(\varsigma) = \text{sta}\{\langle \Lambda^{(1)}(\mathbf{x}); \varsigma_n ! \rangle - U(\mathbf{x}) \quad : \quad \delta\Lambda^{(1)}(\mathbf{x})\varsigma_n ! = \delta U(\mathbf{x})\}$$

can be well defined on certain dual feasible spaces $\mathcal{V}_k^*$, and the canonical dual variables ς_k linearly depend on ς_1. This canonical dual problem can be solved very easily. Two sequential canonical dual transformation methods have been proposed in Chapter 4 of Gao (2000a). Applications to general nonconvex differential equations and chaotic dynamical systems have been given in Gao (1998a, 2000b).

As an application, let us consider the following polynomial minimization problem

$$\min\{P(\mathbf{x}) = W(\mathbf{x}) - \mathbf{x}^T f : \quad \mathbf{x} \in \mathbb{R}^n\}, \tag{8.175}$$

where $\mathbf{x} = (x_1, x_2, \dots, x_n)^T \in \mathbb{R}^n$ is a real vector, $f \in \mathbb{R}^n$ is a given vector, and $W(\mathbf{x})$ is a so-called *canonical polynomial* of degree $d = 2^{p+1}$ (see Gao, 2000a), defined by

$$W(\mathbf{x}) = \frac{1}{2}\alpha_p \left(\frac{1}{2}\alpha_{p-1} \left(\dots \left(\frac{1}{2}\alpha_1 \left(\frac{1}{2}|\mathbf{x}|^2 - \lambda_1 \right)^2 \dots \right)^2 - \lambda_{p-1} \right)^2 - \lambda_p \right)^2, \tag{8.176}$$

where α_i, λ_i are given parameters. It is known that the general polynomial minimization problem is NP-hard even when $d = 4$ (see Nesterov, 2000). Many numerical methods and algorithms have been suggested recently for finding tight lower bounds of general polynomial optimization problems (see Lasserre, 2001, Parrilo and Sturmfels, 2003).

For the current canonical polynomial minimization problem, the dual problem has been formulated in Gao (2006); that is,

$$(\mathcal{P}^d) : \quad \max_{\varsigma} \left\{ P^d(\varsigma) = -\frac{|f|^2}{2\varsigma_p !} - \sum_{k=1}^{p} \frac{\varsigma_p !}{\varsigma_k !} V_k^*(\varsigma_k) \right\}, \tag{8.177}$$

where

$$\varsigma_1 = \varsigma, \quad \varsigma_k = \alpha_k \left(\frac{1}{2\alpha_{k-1}} \varsigma_{k-1}^2 - \lambda_k \right), \quad k = 2, \dots, p. \tag{8.178}$$

In this case, $V^*{}_k(\varsigma_k)$ is a quadratic function of ς_k defined by

$$V^*{}_k(\varsigma_k) = \frac{1}{2\alpha_k}\varsigma_k^2 + \lambda_k \varsigma_k.$$

The dual problem is a nonlinear program having only one variable $\varsigma \in \mathbb{R}$, which is much easier to solve than the primal problem. Clearly, for any $\varsigma \neq 0$ and $\varsigma_k^2 \neq 2\alpha_k\lambda_{k+1}$, the dual function P^d is well defined and the criticality condition $\delta P^d(\varsigma) = 0$ leads to a dual algebraic equation

$$2(\varsigma_p!)^2(\alpha_1^{-1}\varsigma + \lambda_1) = |f|^2. \tag{8.179}$$

Theorem 8.18. (Complete Solution Set to Canonical Polynomial (Gao, 2006)) *For any parameters α_k, and $\lambda_k, k = 1, \ldots, p$, and input f, the dual algebraic equation* (8.179) *has at most $s = 2^{p+1} - 1$ real solutions: $\bar{\varsigma}^{(i)}$, $i = 1, \ldots, s$. For each dual solution $\bar{\varsigma} \in \mathbb{R}$, the vector $\bar{\mathbf{x}}$ defined by*

$$\bar{\mathbf{x}}(\bar{\varsigma}) = (\bar{\varsigma}_p!)^{-1} f \tag{8.180}$$

is a critical point of the primal problem $(\mathcal{P})$ *and*

$$P(\bar{\mathbf{x}}) = P^d(\bar{\varsigma}).$$

Conversely, every critical point $\bar{\mathbf{x}}$ of the polynomial $P(\mathbf{x})$ can be written in the form (8.180) *for some dual solution $\bar{\varsigma} \in \mathbb{R}$.*

In the case that $p = 1$, the nonconvex function $W(\mathbf{x}) = \frac{1}{2}\alpha_1(\frac{1}{2}|\mathbf{x}|^2 - \lambda_1)^2$ is a double-well function. The global and local extrema can be identified by the triality theory given in Theorem 8.6. For the general case of $p > 1$, the sufficient condition for global minimizer was obtained recently in Gao (2006).

Theorem 8.19. (Sufficient Condition for Global Minimizer) *Suppose that for any arbitrarily given positive parameters α_k, $\lambda_k \geq 0$, $\forall k \in \{1, \ldots, p\}$, $\bar{\varsigma}$ is a solution of the dual algebraic equation* (8.179). *If*

$$\bar{\varsigma} > \varsigma_+ = \sqrt{2\alpha_1\left(\lambda_2 + \sqrt{\frac{2}{\alpha_2}\left(\lambda_3 + \cdots + \sqrt{\frac{2}{\alpha_{p-2}}\left(\lambda_{p-1} + \sqrt{\frac{2}{\alpha_{p-1}}\lambda_p}\right)}\right)}\right)},$$

then $\bar{\varsigma}$ is a global maximizer of P^d on the open domain $(\varsigma_+, +\infty)$, the vector $\bar{\mathbf{x}} = (\bar{\varsigma}_p!)^{-1} f$ is a global minimizer of the polynomial minimization problem (8.175), *and*

$$P(\bar{\mathbf{x}}) = \min_{\mathbf{x} \in \mathbb{R}^n} P(\mathbf{x}) = \max_{\varsigma > \varsigma_+} P^d(\varsigma) = P^d(\bar{\varsigma}). \tag{8.181}$$

In the case of $p = 2$, the nonconvex function $W(\mathbf{x})$ is a canonical polynomial of degree eight. The dual function $P^d(\varsigma)$ has the form of

$$\Pi^d(\varsigma) = -\frac{|f|^2}{2\varsigma\varsigma_2} - \left(\frac{1}{\alpha_2}\varsigma_2^2 + \lambda_2\varsigma_2 + \varsigma_2(\frac{1}{2\alpha_1}\varsigma^2 + \lambda_1\varsigma)\right), \tag{8.182}$$

where $\varsigma_2 = \alpha_2\varsigma^2/(2\alpha_1) - \lambda_2\alpha_2$. In this case, the dual algebraic equation (8.179)

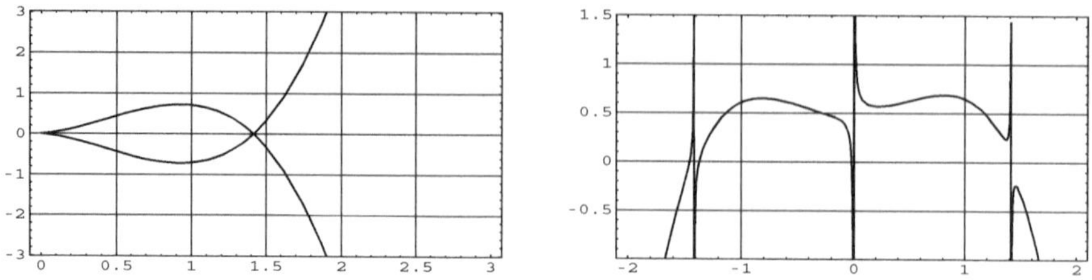

(a) $\lambda_1 = 0$: Three solutions $\varsigma_3 = 0.22 < \varsigma_2 = 1.37 < \varsigma_1 = 1.45$

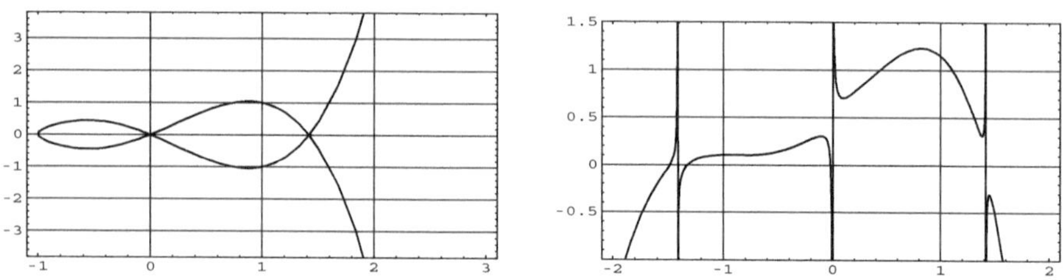

(b) $\lambda_1 = 1$: Five solutions $\{-0.96, -0.11, 0.096, 1.38, 1.45\}$

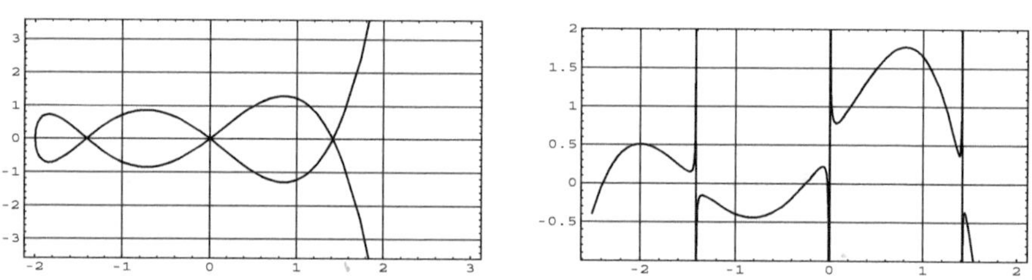

(c) $\lambda_1 = 2$: Seven solutions $\{-2.0, -1.45, -1.35, -0.072, 0.07, 1.39, 1.44\}$

Fig. 8.16 Graphs of the algebraic curve $\phi_2(\varsigma)$ (left) and dual function $P^d(\varsigma)$ (right).

$$2\varsigma^2 \left(\frac{\alpha_2}{2\alpha_1}\varsigma^2 - \lambda_2\alpha_2 \right)^2 \left(\frac{1}{\alpha_1}\varsigma + \lambda_1 \right) = |f|^2 \tag{8.183}$$

has at most seven real roots $\bar{\varsigma}_i$, $i = 1, \ldots, 7$. Let

$$\phi_2(\varsigma) = \pm\varsigma \left(\frac{\alpha_2}{2\alpha_1}\varsigma^2 - \lambda_2\alpha_2 \right) \sqrt{2(\frac{1}{\alpha_1}\varsigma + \lambda_1)},$$

and $f = \{0.1, -0.1\}$, $\alpha_1 = 1$, $\alpha_2 = 1$, and $\lambda_2 = 1$. Then, for different values of λ_1, the graphs of $\phi_2(\varsigma)$ and $P^d(\varsigma)$ are shown in Figure 8.16. The graphs of $P(\mathbf{x})$ are shown in Figure 8.17 (for $\lambda_1 = 0$ and $\lambda_1 = 1$) and Figure 8.18 (for $\lambda_1 = 2$). Because $\varsigma_+ = \sqrt{2\alpha_1\lambda_2} = \sqrt{2}$, we can see that the dual function $P^d(\varsigma)$ is strictly concave for $\varsigma > \varsigma_+ = \sqrt{2}$. The dual algebraic equation

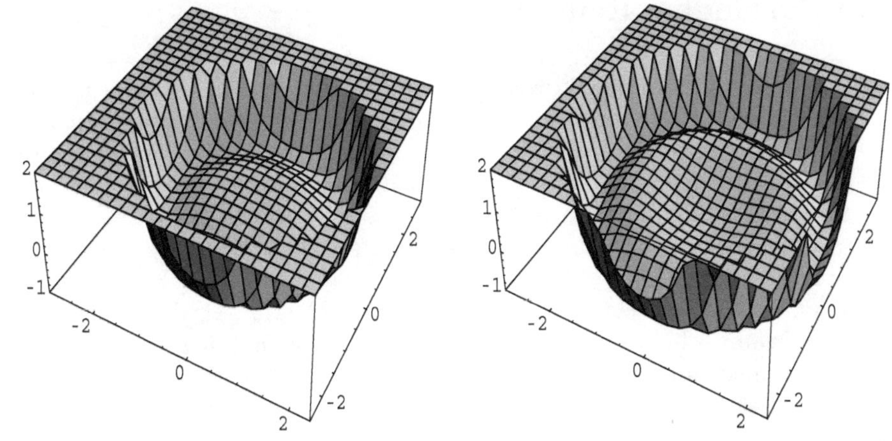

(a) $\lambda_1 = 0$. (b) $\lambda_1 = 1$.

Fig. 8.17 Graphs of $P(\mathbf{x})$.

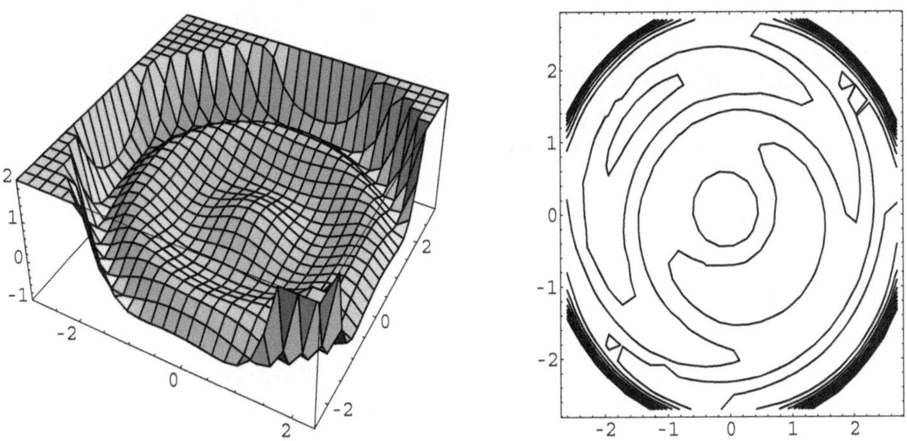

Fig. 8.18 Graph of $P(\mathbf{x})$ with $\lambda_1 = 2$.

(8.183) has a total of seven real solutions when $\lambda_1 = 2$, and the largest $\varsigma_1 = 2.10 > \varsigma_+ = 2$ gives the global minimizer $\mathbf{x}_1 = f/\varsigma_1 = \{2.29, -0.92\}$, and $P(\mathbf{x}_1) = -1.32 = P^d(\varsigma_1)$. The smallest $\varsigma_7 = -4.0$ gives a local maximizer $\mathbf{x}_7 = \{-0.04, 0.02\}$ and $P(\mathbf{x}_7) = 4.51 = P^d(\varsigma_7)$ (see Figure 8.18).

Detailed studies on solving general polynomial minimization problems are given in Gao (2000a, 2006), Lasserre (2001), and Sherali and Tuncbilek (1992, 1997).

8.10 Concluding Remarks

We have presented a detailed review on the canonical dual transformation and its associated triality theory, with specific applications to nonconvex analysis and global optimization problems. Duality plays a key role in modern mathematics and science. The inner beauty of duality theory owes much to the fact that many different natural phenomena can be cast in the unified mathematical framework of Figure 8.1. According to the traditional philosophical principle of ying–yang duality, *The Complementarity of One Ying and One Yang is the Dao* (see Gao, 1996b, Lao Zhi, 400 BC); that is, the constitutive relations in any physical system should be one-to-one. Niels Bohr realized its value in quantum mechanics. His complementarity theory and philosophy laid a foundation on which the field of modern physics was developed (Pais, 1991). In nonconvex analysis and optimization, this one-to-one canonical duality relation serves as the foundation for the canonical dual transformation method. For any given nonconvex problem, as long as the geometrical operator Λ is chosen properly and the tricanonical forms can be characterized correctly, the canonical dual transformation can be used to establish elegant theoretical results and to develop efficient algorithms for robust computations. The extended Lagrangian duality and triality theories show promise of having significance in many diverse fields.

As indicated in Gao (2000a), duality in natural systems is a very broad and rich field. To theoretical scientists and philosophical thinkers as well as great artists, duality has always played a central role in their respective fields. It is really "a splendid feeling to realize the unity of a complex of phenomena that by physical perception appear to be completely separated" (Albert Einstein). It is pleasing to see that more and more knowledgeable researchers and scientists are working in this wonderland and exploring the intrinsic beauty of nature, often revealed via duality theory.

Acknowledgments This work is supported by the National Science Foundation by Grant Numbers DMII-0455807, CCF-0514768, and DMII-0552676.

References

Arthurs, A.M. (1980). *Complementary Variational Principles*, Clarendon Press, Oxford.

Atai, A.A. and Steigmann, D. (1998). Coupled deformations of elastic curves and surfaces, *Int. J. Solids Struct.* **35**, 1915–1952.

Aubin, J.P. and Ekeland, I. (1976). Estimates of the duality gap in nonconvex optimization, *Math. Oper. Res.* **1** (3), 225–245.

Auchmuty, G. (1983). Duality for non-convex variational principles, *J. Diff. Equations* **50**, 80–145.

Auchmuty, G. (1986). Dual variational principles for eigenvalue problems, *Proceedings of Symposia in Pure Math.*, 45, Part 1, 55–71.

Auchmuty, G. (2001). Variational principles for self-adjoint elliptic eigenproblems, in *Nonconvex/Nonsmooth Mechanics: Modelling, Methods and Algorithms*, D.Y. Gao, R.W. Ogden, and G. Stavroulakis, eds., Kluwer Academic.

Benson, H. (1995). Concave minimization: Theory, applications and algorithms, in *Handbook of Global Optimization*, R. Horst and P. Pardalos, eds., Kluwer Academic, pp. 43–148.

Casciaro, R. and Cascini, A. (1982). A mixed formulation and mixed finite elements for limit analysis, *Int. J. Solids Struct.* **19**, 169–184.

Cheng, H., Fang, S.C., and Lavery, J. (2005). Shape-preserving properties of univariate cubic L_1 splines, *J. Comput. Appl. Math.* **174**, 361–382.

Chien, Wei-zang (1980). *Variational Methods and Finite Elements* (in Chinese), Science Press.

Clarke, F.H. (1983). *Optimization and Nonsmooth Analysis*, John Wiley, New York.

Clarke, F.H. (1985). The dual action, optimal control, and generalized gradients, *Mathematical Control Theory*, Banach Center Publ., 14, PWN, Warsaw, pp. 109–119.

Crouzeix, J.P. (1981). Duality framework in quasiconvex programming, in *Generalized Convexity in Optimization and Economics*, S. Schaible and W.T. Ziemba, eds., Academic Press, pp. 207–226.

Dacorogna, D. (1989). *Direct Methods in the Calculus of Variations*, Springer-Verlag, New York.

Ekeland, I. (1977). Legendre duality in nonconvex optimization and calculus of variations, *SIAM J. Control Optim.*, **15**, 905–934.

Ekeland, I. (1990). *Convexity Methods in Hamiltonian Mechanics*, Springer-Verlag, New York.

Ekeland, I. (2003). Nonconvex duality, in *Proceedings of IUTAM Symposium on Duality, Complementarity and Symmetry in Nonlinear Mechanics*, D.Y. Gao, ed., Kluwer Academic, Dordrecht/Boston/London, pp. 13–19.

Ekeland, I. and Temam, R. (1976). *Convex Analysis and Variational Problems*, North-Holland.

Floudas, C.A. and Visweswaran, V. (1995). Quadratic optimization, in *Handbook of Optimization*, R. Horst and P.M. Pardalos, eds., Kluwer Academic, Dordrecht, pp. 217–270.

Gao, D.Y. (1986). *Complementarity Principles in Nonsmooth Elastoplastic Systems and Pan-penalty Finite Element Methods*, Ph.D. Thesis, Tsinghua University, Beijing, China.

Gao, D.Y. (1988a). On the complementary bounding theorems for limit analysis, *Int. J. Solids Struct.* **24**, 545–556.

Gao, D.Y. (1988b). Panpenalty finite element programming for limit analysis, *Computers & Structures* **28**, 749–755.

Gao, D.Y. (1990a). Dynamically loaded rigid-plastic analysis under large deformation, *Quart. Appl. Math.* **48**, 731–739.

Gao, D.Y. (1990b). On the extremum potential variational principles for geometrical nonlinear thin elastic shell, *Science in China (Scientia Sinica) (A)* **33** (1), 324–331.

Gao, D.Y. (1990c). On the extremum variational principles for nonlinear elastic plates, *Quart. Appl. Math.* **48**, 361–370.

Gao, D.Y.(1990d). Complementary principles in nonlinear elasticity, *Science in China (Scientia Sinica) (A) (Chinese Ed.)* **33** (4), 386–394.

Gao, D.Y. (1990e). Bounding theorem on finite dynamic deformations of plasticity, *Mech. Research Commun.* **17**, 33–39.

Gao, D.Y. (1991). Extended bounding theorems for nonlinear limit analysis, *Int. J. Solids Struct.* **27**, 523–531.

Gao, D.Y. (1992). Global extremum criteria for nonlinear elasticity, *Zeit. Angew. Math. Phys.* **43**, 924–937.

Gao, D.Y. (1996a). Nonlinear elastic beam theory with applications in contact problem and variational approaches, *Mech. Research Commun.* **23** (1), 11–17.

Gao, D.Y. (1996b). Complementarity and duality in natural sciences, in *Philosophical Study in Modern Science and Technology* (in Chinese), Tsinghua University Press, Beijing, China, pp. 12–25.

Gao, D.Y. (1997). Dual extremum principles in finite deformation theory with applications to post-buckling analysis of extended nonlinear beam theory, *Appl. Mech. Rev.* **50** (11), November 1997, S64–S71.

Gao, D.Y. (1998a). Duality, triality and complementary extremum principles in nonconvex parametric variational problems with applications, *IMA J. Appl. Math.* **61**, 199–235.

Gao, D.Y. (1998b). Bi-complementarity and duality: A framework in nonlinear equilibria with applications to the contact problems of elastoplastic beam theory, *J. Appl. Math. Anal.* **221**, 672–697.

Gao, D.Y. (1999a). Pure complementary energy principle and triality theory in finite elasticity, *Mech. Res. Comm.* **26** (1), 31–37.

Gao, D.Y. (1999b). Duality-mathematics, *Wiley Encyclopedia of Electrical and Electronics Engineering*, vol. 6, John Wiley, New York, pp. 68–77.

Gao, D.Y. (1999c). General analytic solutions and complementary variational principles for large deformation nonsmooth mechanics, *Meccanica* **34**, 169–198.

Gao, D.Y. (2000a). *Duality Principles in Nonconvex Systems: Theory, Methods and Applications*, Kluwer Academic, Dordrecht.

Gao, D.Y. (2000b). Analytic solution and triality theory for nonconvex and nonsmooth variational problems with applications, *Nonlinear Anal.* **42**, 7, 1161–1193.

Gao, D.Y. (2000c). Canonical dual transformation method and generalized triality theory in nonsmooth global optimization, *J. Global Optim.* **17** (1/4), 127–160.

Gao, D.Y.(2000d). Finite deformation beam models and triality theory in dynamical post-buckling analysis, *Int. J. Non-Linear Mechanics* **5**, 103–131.

Gao, D.Y. (2001a). Bi-Duality in Nonconvex Optimization, in *Encyclopedia of Optimization*, C.A. Floudas and P.D. Pardalos, eds., Kluwer Academic, Dordrecht, vol. 1, pp. 477–482.

Gao, D.Y. (2001b). Gao, D.Y., Tri-duality in Global Optimization, in *Encyclopedia of Optimization*, C.A. Floudas and P.D. Pardalos, eds., Kluwer Academic, Dordrecht, vol. 1, pp. 485–491.

Gao, D.Y. (2001c). Complementarity, polarity and triality in non-smooth, non-convex and non-conservative Hamilton systems, *Phil. Trans. Roy. Soc. Lond. Ser. A Math. Phys. Eng. Sci.* **359**, 2347–2367.

Gao, D.Y. (2002). Duality and triality in non-smooth, nonconvex and nonconservative systems: A survey, new phenomena and new results, in *Nonsmooth/Nonconvex Mechanics with Applications in Engineering*, edited by C. Baniotopoulos, Thessaloniki, Greece, pp. 1–14.

Gao, D.Y. (2003a). Perfect duality theory and complete solutions to a class of global optimization problems, *Optimisation* **52** (4–5), 467–493.

Gao, D.Y. (2003b). Nonconvex semi-linear problems and canonical duality solutions, in *Advances in Mechanics and Mathematics*, vol. II, Kluwer Academic, Dordrecht, pp. 261–312.

Gao, D.Y. (2004a). Complementary variational principle, algorithm, and complete solutions to phase transitions in solids governed by Landau-Ginzburg equation, *Math. Mech. Solids* **9**, 285–305.

Gao, D.Y. (2004b). Canonical duality theory and solutions to constrained nonconvex quadratic programming, *J. Global Optim.* **29**, 377–399.

Gao, D.Y.(2005a). Sufficient conditions and perfect duality in nonconvex minimization with inequality constraints, *J. Indust. Manage. Optim.* **1**, 59–69.

Gao, D.Y. (2005b). Canonical duality in nonsmooth, concave minimization with inequality constraints, in *Advances in Nonsmooth Mechanics, a Special Volume in Honor of Professor J.J. Moreau's 80th Birthday*, P. Alart and O. Maisonneuve, eds., Springer, New York, pp. 305–314.

Gao, D.Y. (2006). Complete solutions to a class of polynomial minimization problems, *J. Global Optim.* **35**, 131–143.

Gao, D.Y. (2007a). Duality-mathematics, *Wiley Encyclopedia of Electrical and Electronics Engineering*, vol. 6 (second edition), John G. Webster, ed., John Wiley, New York.

Gao, D.Y. (2007b). Solutions and optimality to box constrained nonconvex minimization problems, *J. Indust. Manage. Optim.* **3** (2), 293–304.

Gao, D.Y. and Cheung, Y.K. (1989). On the extremum complementary energy principles for nonlinear elastic shells, *Int. J. Solids Struct.* **26**, 683–693.

Gao, D.Y. and Hwang, K.C. (1988). On the complementary variational principles for elastoplasticity, *Scientia Sinica (A)* **31**, 1469–1476.

Gao, D.Y. and Ogden, R.W. (2008a). Closed-form solutions, extremality and nonsmoothness criteria in a large deformation elasticity problem, *Zeit. Angew. Math. Phys.* **59** (3), 498–517.

Gao, D.Y. and Ogden, R.W. (2008b). Multiple solutions to non-convex variational problems with implications for phase transitions and numerical computation, to appear in *Quarterly J. Mech. Appl. Math.*

Gao, D.Y., Ogden, R.W., and Stavroulakis, G. (2001). *Nonsmooth and Nonconvex Mechanics: Modelling, Analysis and Numerical Methods*, Kluwer Academic, Boston.

Gao, D.Y. and Onate, E.T. (1990). Rate variational extremum principles for finite elastoplasticity, *Appl. Math. Mech.* **11** (7), 659–667.

Gao, D.Y. and Ruan, N. (2007). Complete solutions and optimality criteria for nonconvex quadratic-exponential minimization problem, *Math. Meth. Oper. Res.* **67** (3), 479–491.

Gao, D.Y., Ruan, N., and Sherali, H.D. (2008). Canonical duality theory for solving nonconvex constrained optimization problems, to appear in *J. Global Optim.*

Gao, D.Y. and Strang, G. (1989a). Geometric nonlinearity: Potential energy, complementary energy, and the gap function, *Quart. Appl. Math.* **47** (3), 487–504.

Gao, D.Y. and Strang, G. (1989b). Dual extremum principles in finite deformation elastoplastic analysis, *Acta Appl. Math.* **17**, 257–267.

Gao, D.Y. and Wierzbicki, T. (1989). Bounding theorem in finite plasticity with hardening effect, *Quart. Appl. Math.* **47**, 395–403.

Gao, D.Y. and Yang, W.-H. (1995). Multi-duality in minimal surface type problems, *Studies in Appl. Math.* **95**, 127–146.

Gasimov, R.N. (2002). Augmented Lagrangian duality and nondifferentiable optimization methods in nonconvex programming, *J. Global Optim.* **24**, 187–203.

Goh, C.J. and Yang, X.Q. (2002). *Duality in Optimization and Variational Inequalities*, Taylor and Francis.

Greenberg, H.J. (1949). On the variational principles of plasticity, Brown University, ONR, NR-041-032, March.

Guo, Z.H. (1980). The unified theory of variational principles in nonlinear elasticity, *Archive of Mechanics* **32**, 577–596.

Haar, A. and von Kármán, Th. (1909). Zur theorie der spannungszustände in plastischen und sandartigen medien, *Nachr. Ges. Wiss. Göttingen*, 204–218.

Han, Weimin (2005). *A Posteriori Error Analysis via Duality Theory: With Applications in Modeling and Numerical Approximations*, Advances in Mechanics and Mathematics, vol. 8, Springer, New York.

Hellinger, E. (1914). Die allgemeine Ansätze der Mechanik der Kontinua, *Enzyklopädie der Mathematischen Wissenschaften* IV, 4, 602–694.

Hill, R. (1978), Aspects of invariance in solids mechanics, *Adv. in Appl. Mech.* **18**, 1–75.

Hiriart-Urruty, J.-B. (1985). Generalized differentialiability, duality and optimization for problems dealing with difference of convex functions, *Appl. Math. Optim.* **6**, 257–269.

Horst, R., Pardalos, P.M., and Thoai, N.V. (2000). *Introduction to Global Optimization*, Kluwer Academic, Boston.

Hu, H.-C. (1955). On some variational principles in the theory of elasticity and the theory of plasticity, *Scientia Sinica* **4**, 33–54.

Huang, X.X. and Yang, X.Q. (2003). A unified augmented Lagrangian approach to duality and exact penalization, *Math. Oper. Res.* **28**, 524–532.

Koiter, W.T. (1973). On the principle of stationary complementary energy in the nonlinear theory of elasticity, *SIAM J. Appl. Math.* **25**, 424–434.

Koiter, W.T. (1976). On the complementary energy theorem in nonlinear elasticity theory, Trends in Appl. of Pure Math. to Mech., G. Fichera, ed., Pitman.

Lao Zhi (400 BC). *Dao De Jing* (or *Tao Te Ching*), English edition by D.C. Lau, Penguin Classics, 1963.

Lasserre, J. (2001). Global optimization with polynomials and the problem of moments, *SIAM J. Optim.* **11** (3), 796–817.

Lavery, J. (2004). Shape-preserving approximation of multiscale univariate data by cubic L_1 spline fits, *Comput. Aided Geom. Design* **21**, 43–64.

Lee, S.J. and Shield, R.T. (1980a). Variational principles in finite elastostatics, *Zeit. Angew. Math. Phys.* **31**, 437–453.

Lee, S.J. and Shield, R.T. (1980b). Applications of variational principles in finite elasticity, *Zeit. Angew. Math. Phys.* **31**, 454–472.

Levinson, M. (1965). The complementary energy theorem in finite elasticity, *Trans. ASME Ser. E J. Appl. Mech.* **87**, 826–828.

Li, S.F. and Gupta, A. (2006). On dual configuration forces, *J. of Elasticity* **84**, 13–31.

Maier, G. (1969). Complementarity plastic work theorems in piecewise-linear elastoplasticity, *Int. J. Solids Struct.* **5**, 261–270.

Maier, G. (1970). A matrix structural theory of piecewise-linear plasticity with interacting yield planes, *Meccanica* **5**, 55–66.

Maier, G., Carvelli, V., and Cocchetti, G. (2000). On direct methods for shakedown and limit analysis, Plenary lecture at the 4th EUROMECH Solid Mechanics Conference, Metz, France, June 26–30, *European J. Mech. A Solids* **19**, Special Issue, S79–S100.

Marsden, J. and Ratiu, T. (1995). *Introduction to Mechanics and Symmetry*, Springer, New York.

Moreau, J.J. (1966). *Fonctionnelles Convexes*, Séminaire sur les Équations aux Dérivées Partielles II, Collège de France.

Moreau, J.J. (1968). La notion de sur-potentiel et les liaisons unilatérales en élastostatique, *C. R. Acad. Sci. Paris Sér. A* **267**, 954–957.

Moreau, J.J., Panagiotopoulos, P.D., and Strang, G. (1988). *Topics in Nonsmooth Mechanics*, Birkhäuser Verlag, Boston.

Murty, K.G. and Kabadi, S.N. (1987). Some NP-complete problems in quadratic and nonlinear programming, *Math. Program.* **39**, 117–129.

Nesterov, Y. (2000). Squared functional systems and optimization problems, in *High Performance Optimization*, H. Frenk et al., eds., Kluwer Academic, Boston, pp. 405–440.

Noble, B. and Sewell, M.J. (1972). On dual extremum principles in applied mathematics, *IMA J. Appl. Math.* **9**, 123–193.

Oden, J.T. and Lee, J.K. (1977). Dual-mixed hybrid finite element method for second-order elliptic problems, in *Mathematical Aspects of Finite Element Methods* (Proc. Conf., Consiglio Naz. delle Ricerche (C.N.R.), Rome, 1975), Lecture Notes in Math., vol. 606, Springer, Berlin, pp. 275–291.

Oden, J.T. and Reddy, J.N. (1983). *Variational Methods in Theoretical Mechanics*, Springer-Verlag, New York.

Ogden, R.W. (1975). A note on variational theorems in non-linear elastostatics, *Math. Proc. Camb. Phil. Soc.* **77**, 609–615.

Ogden, R.W. (1977). Inequalities associated with the inversion of elastic stress-deformation relations and their implications, *Math. Proc. Camb. Phil. Soc.* **81**, 313–324.

Pais, A. (1991). *Niels Bohr's Times: In Physics, Philosophy, and Polity*, Clarendon Press, Oxford.

Pardalos, P.M. (1991). Global optimization algorithms for linearly constrained indefinite quadratic problems, *Comput. Math. Appl.* **21**, 87–97.

Pardalos, P.M. and Vavasis, S.A. (1991). Quadratic programming with one negative eigenvalue is NP-hard, *J. Global Optim.* **1**, 15–22.

Parrilo, P. and Sturmfels, B. (2003). Minimizing polynomial functions, in *Proceedings of DIMACS Workshop on Algorithmic and Quantitative Aspects of Real Algebraic Geometry in Mathematics and Computer Science*, S. Basu and L. González-Vega, eds., American Mathematical Society, pp. 83–100.

Penot, J.-P. and Volle, M. (1990). On quasiconvex duality, *Math. Oper. Res.* **14**, 597–625.

Pian, T.H.H. and Tong, P. (1980). Reissner's principle in finite element formulations, in *Mechanics Today*, vol. 5, S. Nemat-Nasser, ed., Pergamon Press, Tarrytown, NY, pp. 377–395.

Pian, T.H.H. and Wu, C.C. (2006). *Hybrid and Incompatible Finite Element Methods*, Chapman & Hall/CRC, Boca Raton, FL.

Powell, M.J.D. (2002). UOBYQA: Unconstrained optimization by quadratic approximation, *Math. Program.* **92** (3), 555–582.

Rall, L.B. (1969). *Computational Solution of Nonlinear Operator Equations*, Wiley, New York.

Reissner, E. (1996). *Selected Works in Applied Mechanics and Mathematics*, Jones and Bartlett, Boston.

Rockafellar, R.T. (1967). Duality and stability in extremum problems involving convex functions, *Pacific J. Math.* **21**, 167–187.

Rockafellar, R.T. (1970). *Convex Analysis*, Princeton University Press, Princeton, NJ.

Rockafellar, R.T. (1974). *Conjugate Duality and Optimization*, SIAM, Philadelphia.

Rockafellar, R.T. and Wets, R.J.B. (1998). *Variational Analysis*, Springer, Berlin.

Rowlinson, J.S. (1979). Translation of J. D. van der Waals' "The thermodynamic theory of capillarity under the hypothesis of a continuous variation of density," *J. Statist. Phys.* **20** (2), 197–244.

Rubinov, A.M. and Yang, X.Q. (2003). *Lagrange-Type Functions in Constrained Non-Convex Optimization*, Kluwer Academic, Boston.

Rubinov, A.M., Yang X.Q., and Glover, B.M. (2001). Extended Lagrange and penalty functions in optimization, *J. Optim. Theory Appl.* **111** (2), 381–405.

Sahni, S. (1974). Computationally related problems, *SIAM J. Comput.* **3**, 262–279.

Sewell, M.J. (1987). *Maximum and Minimum Principles*, Cambridge Univ. Press.

Sherali, H.D. and Tuncbilek, C. (1992). A global optimization for polynomial programming problem using a reformulation-linearization technique, *J. Global Optim.* **2**, 101–112.

Sherali, H.D. and Tuncbilek, C. (1997). New reformulation-linearization technique based relaxation for univariate and multivariate polynominal programming problems, *Oper. Res. Lett.* **21** (1), 1–10.

Silverman, H.H. and Tate, J. (1992). *Rational Points on Elliptic Curves*, Springer-Verlag, New York.

Singer, I. (1998). Duality for optimization and best approximation over finite intersections, *Numer. Funct. Anal. Optim.* **19** (7–8), 903–915.

Strang, G. (1979). A minimax problem in plasticity theory, in *Functional Analysis Methods in Numerical Analysis*, M.Z. Nashed, ed., Lecture Notes in Math., 701, Springer, New York, pp. 319–333.

Strang, G. (1982). L^1 and L^∞ and approximation of vector fields in the plane, in *Nonlinear Partial Differential Equations in Applied Science*, H. Fujita, P. Lax, and G. Strang, eds., Lecture Notes in Num. Appl. Anal., 5, Springer, New York, pp. 273–288.

Strang, G. (1983). Maximal flow through a domain, *Math. Program.* **26**, 123–143.

Strang, G. (1984). Duality in the classroom, *Amer. Math. Monthly* **91**, 250–254.

Strang, G. (1986). *Introduction to Applied Mathematics*, Wellesley-Cambridge Press.

Strang, G. and Fix, G. (1973). *An Analysis of the Finite Element Method*, Prentice-Hall, Englewood Cliffs, N.J. Second edition, Wellesley-Cambridge Press (2008).

Tabarrok, B. and Rimrott, F.P.J. (1994). *Variational Methods and Complementary Formulations in Dynamics*, Kluwer Academic, Dordrecht.

Temam, R. and Strang, G. (1980). Duality and relaxation in the variational problems of plasticity, *J. de Mécanique* **19**, 1–35.
Thach, P.T. (1993). Global optimality criterion and a duality with a zero gap in nonconvex optimization, *SIAM J. Math. Anal.* **24** (6), 1537–1556.
Thach, P.T. (1995). Diewert-Crouzeix conjugation for general quasiconvex duality and applications, *J. Optim. Theory Appl.* **86** (3), 719–743.
Thach, P.T., Konno, H., and Yokota, D. (1996). Dual approach to minimization on the set of Pareto-optimal solutions, *J. Optim. Theory Appl.* **88** (3), 689–707.
Toland, J.F. (1978). Duality in nonconvex optimization, *J. Math. Anal. Appl.* **66**, 399–415.
Toland, J.F. (1979). A duality principle for non-convex optimization and the calculus of variations, *Arch. Rat. Mech. Anal.* **71**, 41–61.
Tonti, E. (1972a). A mathematical model for physical theories, *Accad. Naz. dei Lincei*, Serie VIII, **LII**, I, 175–181; II, 350–356.
Tonti, E. (1972b). On the mathematical structure of a large class of physical theories, *Accad. Naz. dei Lincei*, Serie VIII, **LII**, 49–56.
Tuy, H. (1995). D.C. optimization: Theory, methods and algorithms, in *Handbook of Global Optimization*, R. Horst and P. Pardalos, eds., Kluwer Academic, Boston, pp. 149–216.
Vavasis, S. (1990). Quadratic programming is in NP, *Info. Proc. Lett.* **36**, 73–77.
Vavasis, S. (1991). *Nonlinear Optimization: Complexity Issues*, Oxford University Press, New York.
Veubeke, B.F. (1972). A new variational principle for finite elastic displacements, *Int. J. Eng. Sci.* **10**, 745–763.
von Neumann, J. (1932). *Mathematische Grundlagen der Quantenmechanik*, Springer Verlag, Heidelberg.
Walk, M. (1989). *Theory of Duality in Mathematical Programming*, Springer-Verlag, Wien.
Washizu, K. (1955). On the variational principles of elasticity and plasticity, Aeroelastic and Structures Research Laboratory, Technical Report 25-18, MIT, Cambridge.
Wright, M.H. (1998). The interior-point revolution in constrained optimization, in *High-Performance Algorithms and Software in Nonlinear Optimization*, R. DeLeone, A. Murli, P.M. Pardalos, and G. Toraldo, eds., Kluwer Academic, Dordrecht, pp. 359–381.
Ye, Y. (1992). A new complexity result on minimization of a quadratic function with a sphere constraint, in *Recent Advances in Global Optimization*, C. Floudas and P. Pardalos, eds., Princeton University Press, Princeton, NJ, pp. 19–31.
Zhao, Y.B., Fang, S.C., and Lavery, J. (2006). Geometric dual formulation of the first derivative based C^1-smooth univariate cubic L_1 spline functions, to appear in *Complementarity, Duality, and Global Optimization*, a special issue of *J. Global Optim.*, D.Y. Gao and H.D. Sherali, eds.
Zhou, Y.Y. and Yang, X.Q. (2004). Some results about duality and exact penalization, *J. Global Optim.* **29**, 497–509.
Zubov, L.M. (1970). The stationary principle of complementary work in nonlinear theory of elasticity, *Prikl. Mat. Mech.* **34**, 228–232.

Chapter 9
Quantum Computation and Quantum Operations

Stan Gudder

Summary. Quantum operations play an important role in quantum measurement, quantum computation, and quantum information theories. We classify quantum operations according to certain special properties such as unital, tracial, subtracial, self-adjoint, and idempotent. We also consider a type of quantum operation called a Lüders map. Examples of quantum operations that describe noisy quantum channels are discussed. Results concerning iterations and fixed points of quantum operations are presented. The relationship between quantum operations and completely positive maps is discussed and the sequential product of quantum effects is considered.

Key words: Quantum computation, quantum operation, quantum channel, quantum information theory

9.1 Introduction and Basic Definitions

The main arena for studies in quantum computation and quantum information is a finite-dimensional complex Hilbert space which we denote by $\mathcal{H}$. We denote the set of bounded linear operators on $\mathcal{H}$ by $\mathcal{B}(\mathcal{H})$ and we use the notation

$$\begin{aligned}\mathcal{B}(\mathcal{H})^+ &= \{A \in \mathcal{B}(\mathcal{H})\colon A \geq 0\}\\ \mathcal{E}(\mathcal{H}) &= \{A \in \mathcal{B}(\mathcal{H})\colon 0 \leq A \leq I\}\\ \mathcal{D}(\mathcal{H}) &= \left\{\rho \in \mathcal{B}(\mathcal{H})^+\colon \operatorname{tr}(\rho) = 1\right\}.\end{aligned}$$

Stan Gudder
Department of Mathematics, University of Denver, Denver, Colorado 80208
e-mail: sgudder@math.du.edu

D.Y. Gao, H.D. Sherali, (eds.), *Advances in Applied Mathematics and Global Optimization*
Advances in Mechanics and Mathematics 17, DOI 10.1007/978-0-387-75714-8_9,

The elements of $\mathcal{E}(\mathcal{H})$ are called *effects* and the elements of $\mathcal{D}(\mathcal{H})$ are called *states* (or *density operators*). It is clear that $\mathcal{D}(\mathcal{H}) \subseteq \mathcal{E}(\mathcal{H}) \subseteq \mathcal{B}(\mathcal{H})^+$. Effects correspond to quantum yes–no measurements that may be unsharp. If a quantum system is in the state ρ, then the probability that the effect A occurs (has answer yes) is given by $P_\rho(A) = \mathrm{tr}(\rho A)$. As we show, quantum measurements with more than two possible values (not just yes–no) can be described by quantum operations. It is easy to check that $\mathcal{D}(\mathcal{H})$ forms a convex subset of $\mathcal{B}(\mathcal{H})$ and the extreme points of $\mathcal{D}(\mathcal{H})$ are called *pure* states. The pure states have the form P_ψ where P_ψ denotes a one-dimensional projection onto a unit vector $\psi \in \mathcal{H}$. If $\rho = P_\psi$ is a pure state, then

$$P_\rho(A) = \mathrm{tr}(P_\psi A) = \langle A\psi, \psi\rangle.$$

Let $A_i \in \mathcal{B}(\mathcal{H})$, $i = 1, \ldots, n$, and let $\mathcal{A} = \{A_i, A_i^*: i = 1, \ldots, n\}$. We call the map $\phi_\mathcal{A}: \mathcal{B}(\mathcal{H}) \to \mathcal{B}(\mathcal{H})$ given by $\phi_\mathcal{A}(B) = \sum A_i B A_i^*$ a *quantum operation* and we call the operators A_i, $i = 1, \ldots, n$, the *operation elements* of $\phi_\mathcal{A}$. Notice that $\phi_\mathcal{A}: \mathcal{B}(\mathcal{H})^+ \to \mathcal{B}(\mathcal{H})^+$; that is, $\phi_\mathcal{A}$ preserves positivity. Also, $\phi_\mathcal{A}$ is linear and $A \leq B$ implies that $\phi_\mathcal{A}(A) \leq \phi_\mathcal{A}(B)$. We say that $\phi_\mathcal{A}$ is *unital*, *tracial*, or *subtracial*, respectively, in the case $\sum A_i A_i^* = I$, $\sum A_i^* A_i = I$, or $\sum A_i^* A_i \leq I$, respectively. Notice that $\phi_\mathcal{A}$ is a unital if and only if $\phi_\mathcal{A}(I) = I$, $\phi_\mathcal{A}$ is tracial if and only if $\mathrm{tr}\,(\phi_\mathcal{A}(B)) = \mathrm{tr}(B)$ for every $B \in \mathcal{B}(\mathcal{H})$, and $\phi_\mathcal{A}$ is subtracial if and only if $\mathrm{tr}\,(\phi_\mathcal{A}(B)) \leq \mathrm{tr}(B)$ for every $B \in \mathcal{B}(\mathcal{H})^+$. We say that $\phi_\mathcal{A}$ is *self-adjoint* if $A_i = A_i^*$, $i = 1, \ldots, n$. An important type of self-adjoint quantum operation in quantum measurement theory [4, 7, 9] is a *Lüders map* of the form $L(B) = \sum A_i^{1/2} B A_i^{1/2}$ where $A_i \in \mathcal{E}(\mathcal{H})$ with $\sum A_i = I$, $i = 1, \ldots, n$. In this case, L is unital and tracial and $\{A_i: i = 1, \ldots, n\}$ is called a finite POV (*positive operator-valued*) *measure.* We interpret the POV measure $\{A_i: i = 1, \ldots, n\}$ as a quantum measurement with n possible values (which can be taken to be $1, \ldots, n$). Restricting L to $\mathcal{E}(\mathcal{H})$ we have $L: \mathcal{E}(\mathcal{H}) \to \mathcal{E}(\mathcal{H})$ and $L(B)$ is interpreted as the effect resulting from first making the measurement described by $\{A_i: i = 1, \ldots, n\}$ and then measuring B. If we restrict L to $\mathcal{D}(\mathcal{H})$ then $L: \mathcal{D}(\mathcal{H}) \to \mathcal{D}(\mathcal{H})$ is called the *square root dynamics* [2].

Quantum operations have various interpretations in quantum measurement, computation, and information theories [1, 4, 7, 8, 9, 10]. If $\phi_\mathcal{A}$ is tracial, then $\phi_\mathcal{A}: \mathcal{D}(\mathcal{H}) \to \mathcal{D}(\mathcal{H})$ can be thought of as a quantum measurement with possible outcomes $1, 2, \ldots, n$. If the measurement is performed on a quantum system in the state $\rho \in \mathcal{D}(\mathcal{H})$, then the probability of obtaining outcome i is $\mathrm{tr}(A_i \rho A_i^*)$ and the postmeasurement state given that i occurs is $A_i \rho A_i^* / \mathrm{tr}(A_i \rho A_i^*)$. Moreover, the resulting state after the measurement is executed but no observation is made is given by $\phi_\mathcal{A}(\rho)$. Quantum operations can also be interpreted as an interaction of a quantum system with an environment followed by a unitary evolution, a noisy quantum channel, or a quantum error correction map [10]. Depending on the application, at least one of our previous properties is assumed to hold. For illustrative purposes, we mainly consider the noisy quantum channel interpretation.

Notice that if $\phi_{\mathcal{A}}$ and $\phi_{\mathcal{B}}$ are quantum operations on $\mathcal{B}(\mathcal{H})$ with $\mathcal{A} = \{A_i, A_i^*: i = 1, \dots, n\}$, $\mathcal{B} = \{B_j, B_j^*: j = 1, \dots, m\}$, then their composition $\phi_{\mathcal{A}} \circ \phi_{\mathcal{B}}$ is a quantum operation on $\mathcal{B}(\mathcal{H})$ with operation elements A_iB_j, $i = 1, \dots, n$, $j = 1, \dots, m$. If $\mathcal{A} = \mathcal{B}$ we write $\phi_{\mathcal{A}}^2 = \phi_{\mathcal{A}} \circ \phi_{\mathcal{A}}, \dots,$

$$\phi_{\mathcal{A}}^n = \phi_{\mathcal{A}} \circ \cdots \phi_{\mathcal{A}} \quad (n \text{ factors}).$$

A quantum operation $\phi_{\mathcal{A}}$ is *idempotent* if $\phi_{\mathcal{A}}^2 = \phi_{\mathcal{A}}$. We now give some simple basic results.

Lemma 9.1.1. *If $\phi_{\mathcal{A}}$ and $\phi_{\mathcal{B}}$ are both unital, tracial, or subtracial, respectively, then $\phi_{\mathcal{A}} \circ \phi_{\mathcal{B}}$ is unital, tracial, or subtracial, respectively.*

Proof. If $\phi_{\mathcal{A}}$ and $\phi_{\mathcal{B}}$ are both unital, then $\sum A_iA_i^* = \sum B_jB_j^* = I$. Hence,

$$\begin{aligned}\sum_{i,j} A_iB_j(A_iB_j)^* &= \sum_{i,j} A_iB_jB_j^*A_i^* = \sum_i A_i \sum_j B_jB_j^*A_i^* \\ &= \sum_i A_iA_i^* = I.\end{aligned}$$

Therefore, $\phi_{\mathcal{A}} \circ \phi_{\mathcal{B}}$ is unital. In a similar way, if $\phi_{\mathcal{A}}$ and $\phi_{\mathcal{B}}$ are both tracial then $\phi_{\mathcal{A}} \circ \phi_{\mathcal{B}}$ is tracial. Now suppose that $\phi_{\mathcal{A}}$ and $\phi_{\mathcal{B}}$ are both subtracial. Then there exists a $C \in \mathcal{E}(\mathcal{H})$ such that $\sum A_i^*A_i + C = I$. Hence,

$$\begin{aligned}\sum_{i,j} (A_iB_j)^*A_iB_j &= \sum_{i,j} B_j^*A_i^*A_iB_j = \sum_i B_j^* \sum_j A_i^*A_iB_j \\ &= \sum B_j^*B_j - \sum B_j^*CB_j \le \sum B_j^*B_j \le I.\end{aligned}$$

Therefore, $\phi_{\mathcal{A}} \circ \phi_{\mathcal{B}}$ is subtracial.

Lemma 9.1.2. *If $\phi_{\mathcal{A}}$ is subtracial and its operation elements are self-adjoint projection operators, then $\phi_{\mathcal{A}}$ is idempotent.*

Proof. We have that $\phi_{\mathcal{A}}(B) = \sum A_iBA_i$ where $A_i = A_i^* = A_i^2$ and $\sum A_i \le I$, $i = 1, \dots, n$. For $i, j \in \{1, \dots, n\}$, $i \ne j$, we have

$$A_i + A_j \le \sum A_k \le I.$$

It follows that $A_iA_j = A_jA_i = 0$ for $i \ne j$. Hence,

$$\phi_{\mathcal{A}} \circ \phi_{\mathcal{A}}(B) = \sum_{i,j} A_jA_iBA_iA_j = \sum A_iBA_i = \phi_{\mathcal{A}}(B)$$

so that $\phi_{\mathcal{A}}$ is idempotent.

9.2 Completely Positive Maps

In Section 9.1 we defined a quantum operation as a map $\phi\colon \mathcal{B}(\mathcal{H}) \to \mathcal{B}(\mathcal{H})$ of the form

$$\phi(B) = \sum A_i B A_i^* \tag{9.1}$$

and in Section 9.3 we give some simple practical examples of quantum operations. But why do quantum operations have the operator-sum form (9.1)? The present section tries to answer this question in terms of completely positive maps.

We can consider $\mathcal{M}_k = \mathcal{B}(\mathbb{C}^k)$ as the set of all $k \times k$ complex matrices, $k = 1, 2, \ldots$. The set of operators in the tensor product $\mathcal{B}(\mathcal{H}) \otimes \mathcal{M}_k = \mathcal{B}(\mathcal{H} \otimes \mathbb{C}^k)$ can be considered to be the set of $k \times k$ matrices with entries in $\mathcal{B}(\mathcal{H})$. For example if $A, B, C, D \in \mathcal{B}(\mathcal{H})$, then the matrix

$$M = \begin{bmatrix} A & B \\ C & D \end{bmatrix}$$

is an element of $\mathcal{B}(\mathcal{H}) \otimes \mathcal{M}_2$. Of course, $M \in \mathcal{B}(\mathcal{H} \otimes \mathbb{C}^2)$ in the sense that

$$M \begin{bmatrix} x \\ y \end{bmatrix} = \begin{bmatrix} Ax + By \\ Cx + Dy \end{bmatrix}$$

for all $x, y \in \mathcal{H}$. For a linear map $\phi\colon \mathcal{B}(\mathcal{H}) \to \mathcal{B}(\mathcal{H})$ we define the linear maps $\phi_k\colon \mathcal{B}(\mathcal{H}) \otimes \mathcal{M}_k \to \mathcal{B}(\mathcal{H}) \otimes \mathcal{M}_k$ given by

$$\phi_k(M) = [\phi(M_{ij})],$$

where $M = [M_{ij}] \in \mathcal{B}(\mathcal{H}) \otimes \mathcal{M}_k$, $i, j = 1, \ldots, k$. If ϕ_k sends positive operators into positive operators for $k = 1, 2, \ldots$, then ϕ is called *completely positive.* It is easy to check that $\phi\colon \mathcal{B}(\mathcal{H}) \to \mathcal{B}(\mathcal{H})$ is completely positive if and only if $\phi \otimes I_k\colon \mathcal{B}(\mathcal{H}) \otimes \mathcal{M}_k \to \mathcal{B}(\mathcal{H}) \otimes \mathcal{M}_k$ preserves positivity for $k = 1, 2, \ldots$, where I_k is the identity map on $\mathcal{M}_k$.

We have seen that a quantum operation $\phi\colon \mathcal{B}(\mathcal{H}) \to \mathcal{B}(\mathcal{H})$ describes various ways that states are transformed into other states for a quantum system. Because states are positive operators, ϕ must preserve positivity. Now suppose our quantum system interacts (or couples) with an environment such as a noisy quantum channel. If this environment is described by the Hilbert space $\mathbb{C}^k$, then the combined system is described by the tensor product $\mathcal{H} \otimes \mathbb{C}^k$. The natural extension of ϕ to the combined system is given by $\phi \otimes I_k\colon \mathcal{B}(\mathcal{H} \otimes \mathbb{C}^k) \to \mathcal{B}(\mathcal{H} \otimes \mathbb{C}^k)$. The map $\phi \otimes I_k$ just acts on $\mathcal{B}(\mathcal{H})$ like ϕ and leaves the environment unaltered. We would expect $\phi \otimes I_k$ to map states into states so $\phi \otimes I_k$ should also preserve positivity, $k = 1, 2, \ldots$. We conclude that quantum operations should be completely positive maps.

If $x, y \in \mathcal{H}$ we define the linear operator $|x\rangle\langle y| \in \mathcal{B}(\mathcal{H})$ by $|x\rangle\langle y|v = \langle y, v\rangle x$ for every $v \in \mathcal{H}$. If $\{x_1, \ldots, x_n\}$ is an orthonormal basis for $\mathcal{H}$, then any

$A \in \mathcal{B}(\mathcal{H})$ has the form

$$A = \sum a_{ij} |x_i\rangle\langle x_j|, \tag{9.2}$$

where $a_{ij} \in \mathbb{C}$, $i, j = 1, \ldots, n$. Now let $\{y_1, \ldots, y_k\}$ be an orthonormal basis for $\mathbb{C}^k$. Then an orthonormal basis for $\mathcal{H} \otimes \mathbb{C}^k$ is given by

$$\{x_i \otimes y_j \colon i = 1, \ldots, n; j = 1, \ldots, k\}.$$

For an operator $M \in \mathcal{B}(\mathcal{H} \otimes \mathbb{C}^k)$ as in (9.2) we have

$$\begin{aligned} M &= \sum_{r,s,i,j} a_{r,s,i,j} |x_r \otimes y_i\rangle\langle x_s \otimes y_j| \\ &= \sum_{r,s,i,j} a_{r,s,i,j} |x_r\rangle\langle x_s| \otimes |y_i\rangle\langle y_j| \\ &= \sum_{i,j} \left(\sum_{r,s} a_{r,s,i,j} |x_r\rangle\langle x_s| \right) \otimes |y_i\rangle\langle y_j| \\ &= \sum_{i,j} A_{i,j} \otimes |y_i\rangle\langle y_j|, \end{aligned} \tag{9.3}$$

where

$$A_{ij} = \sum_{r,s} a_{rsij} |x_r\rangle\langle x_s| \in \mathcal{B}(\mathcal{H}).$$

If $\phi\colon \mathcal{B}(\mathcal{H}) \to \mathcal{B}(\mathcal{H})$ is a linear map and $M \in \mathcal{B}(\mathcal{H} \otimes \mathbb{C}^k)$ has the representation (9.3), then $\phi \otimes I_k \colon \mathcal{B}(\mathcal{H} \otimes \mathbb{C}^k) \to \mathcal{B}(\mathcal{H} \otimes \mathbb{C}^k)$ satisfies

$$(\phi \otimes I_k)(M) = \sum_{i,j} \phi(A_{ij}) \otimes |y_i\rangle\langle y_j|. \tag{9.4}$$

The following structure theorem is due to Choi [6].

Theorem 9.2.1. *A linear map $\phi\colon \mathcal{B}(\mathcal{H}) \to \mathcal{B}(\mathcal{H})$ is completely positive if and only if there exist a finite number of operators $A_i \in \mathcal{B}(\mathcal{H})$ such that* (9.1) *holds for every $B \in \mathcal{B}(\mathcal{H})$.*

Proof. Suppose ϕ has the representation (9.1). Applying (9.4) we have

$$\begin{aligned} (\phi \otimes I_k)(M) &= \sum_{i,j} \phi(A_{ij}) \otimes |y_i\rangle\langle y_j| \\ &= \sum_{i,j} \sum_r A_r A_{ij} A_r^* \otimes |y_i\rangle\langle y_j|. \end{aligned}$$

Now any $z \in \mathcal{H} \otimes \mathbb{C}^k$ can be represented in the form

$$z = \sum u_s \otimes v_s,$$

where $u_s \in \mathcal{H}$, $v_s \in \mathbb{C}^k$. Writing $z_r = \sum_s A_r^* u_s \otimes v_s$ it is easy to check that

$$\langle (\phi \otimes I_k)(M)z, z\rangle = \sum_r \langle Mz_r, z_r\rangle \geq 0$$

because M is positive.

Conversely, let $\phi\colon \mathcal{B}(\mathcal{H}) \to \mathcal{B}(\mathcal{H})$ be a completely positive map. Let $\{x_1, \dots, x_n\}$ and $\{y_1, \dots, y_n\}$ be two orthonormal bases for $\mathcal{H}$. Now $\phi \otimes I_n$ is positivity preserving. The operator $M \in \mathcal{B}(\mathcal{H} \otimes \mathcal{H})$ defined by

$$\begin{aligned} M &= \sum_{r,s} |x_r\rangle\langle x_s| \otimes |y_r\rangle\langle y_s| \\ &= \sum_{r,s} |x_r \otimes y_r\rangle\langle x_s \otimes y_s| \\ &= \left|\sum_r x_r \otimes y_r\right\rangle\left\langle \sum_s x_s \otimes y_s\right| \end{aligned}$$

is positive because M is a multiple of a one-dimensional projection. Hence,

$$(\phi \otimes I_n)(M) = \sum_{r,s} \phi\left(|x_r\rangle\langle x_s|\right) \otimes |y_r\rangle\langle y_s| \tag{9.5}$$

is a positive operator. By the spectral theorem there exists an orthonormal basis $\{v_1, \dots, v_m\}$ of $\mathcal{H} \otimes \mathcal{H}$ where $m = n^2$ and positive numbers $\lambda_1, \dots, \lambda_m$ such that

$$(\phi \otimes I_n)(M) = \sum_{r,s} \lambda_i |v_i\rangle\langle v_i| = \sum \left|\sqrt{\lambda_i}\, v_i\right\rangle\left\langle \sqrt{\lambda_i}\, v_i\right|. \tag{9.6}$$

If $v = \sum v_{ij} x_i \otimes y_j$ is a vector in $\mathcal{H} \otimes \mathcal{H}$ we associate with v an operator $A_v \in \mathcal{B}(\mathcal{H})$ by

$$A_v = \sum_{i,j} |x_i\rangle\langle x_j|. \tag{9.7}$$

Then a straightforward computation gives

$$|v\rangle\langle v| = \sum_{r,s} A_v |x_r\rangle\langle x_s| A_v^* \otimes |y_r\rangle\langle y_s|. \tag{9.8}$$

Associating with each $\sqrt{\lambda_i}\, v_i$ in (9.6) the operator A_i in (9.7) and using (9.8) we have

$$(\phi \otimes I_n)(M) = \sum_{i,r,s} A_i |x_r\rangle\langle x_s| A_i^* \otimes |y_r\rangle\langle y_s|. \tag{9.9}$$

Applying (9.5) and (9.9) gives

$$\phi(|x_r\rangle\langle x_s|) = \sum_i A_i |x_r\rangle\langle x_s| A_i^*.$$

Because the operators $|x_r\rangle\langle x_s|$ span the whole space $\mathcal{B}(\mathcal{H})$, we conclude that (9.1) holds for every $B \in \mathcal{B}(\mathcal{H})$.

We now show that the operator-sum representation (9.1) is not unique. In other words, the operation elements for a quantum operation are not unique. Let ϕ and ψ be quantum operations acting on $\mathcal{B}(\mathbb{C}^2)$ with operation-sum representations

$$\phi(B) = E_1 B E_1^* + E_2 B E_2^*$$
$$\psi(B) = F_1 B F_1^* + F_2 B F_2^*,$$

where

$$E_1 = \frac{1}{\sqrt{2}}\begin{bmatrix} 1 & 0 \\ 0 & 1 \end{bmatrix} \qquad E_2 = \frac{1}{\sqrt{2}}\begin{bmatrix} 1 & 0 \\ 0 & -1 \end{bmatrix}$$

$$F_1 = \begin{bmatrix} 1 & 0 \\ 0 & 0 \end{bmatrix} \qquad F_1 = \begin{bmatrix} 0 & 0 \\ 0 & 1 \end{bmatrix}.$$

Although ϕ and ψ appear to be quite different, they are actually the same quantum operation. To see this, note that $F_1 = \frac{1}{\sqrt{2}}(E_1 + E_2)$ and $F_2 = \frac{1}{\sqrt{2}}(E_1 - E_2)$. Thus,

$$\psi(B) = \frac{(E_1 + E_2)B(E_1 + E_2) + (E_1 - E_2)B(E_1 - E_2)}{2}$$
$$= E_1 B E_1 + E_2 B E_2 = \phi(B).$$

Notice that in the previous example we could write $F_i = \sum u_{ij} E_j$ where $[u_{ij}]$ is the unitary matrix

$$\frac{1}{\sqrt{2}}\begin{bmatrix} 1 & 1 \\ 1 & -1 \end{bmatrix}.$$

In this sense, the operation elements of ψ are related to the operation elements of ϕ by a unitary matrix. The next theorem, whose proof may be found in [10], shows that this holds in general.

Theorem 9.2.2. *Suppose* $\{E_1, \ldots, E_n\}$ *and* $\{F_1, \ldots, F_m\}$ *are operation elements giving rise to subtracial quantum operations* ϕ *and* ψ, *respectively. By appending zero operators to the shorter list of operation elements we may assume that* $m = n$. *Then* $\phi = \psi$ *if and only if there exist complex numbers* u_{ij} *such that* $F_i = \sum_j u_{ij} E_j$ *where* $[u_{ij}]$ *is an* $m \times m$ *unitary matrix.*

This theorem is important in the development of quantum error-correcting codes [10]. Suppose we have two representations

$$\phi(B) = \sum E_i B E_i^* = \sum F_j B F_j^*$$

for the quantum operation ϕ.

Lemma 9.2.3. *The quantum operation ϕ is unital, tracial, or subtracial, respectively, with respect to the operation elements $\{E_1, \ldots, E_n\}$ if and only if ϕ is unital, tracial, or subtracial, respectively, with respect to the operation elements $\{F_1, \ldots, F_m\}$.*

Proof. If ϕ is unital with respect to $\{E_1, \ldots, E_n\}$, then

$$\sum F_j F_j^* = \phi(I) = \sum E_i E_i^* = I$$

so ϕ is unital with respect to $\{F_1, \ldots, F_m\}$. If ϕ is tracial with respect to $\{F_1, \ldots, F_m\}$, then for any $B \in \mathcal{B}(\mathcal{H})$ we have

$$\operatorname{tr}(B) = \operatorname{tr}(\phi(B)) = \operatorname{tr}\left(\sum F_j B F_j^*\right) = \operatorname{tr}\left(\sum F_j^* F_j B\right).$$

It follows that $\sum F_j^* F_j = I$ so ϕ is tracial with respect to $\{F_1, \ldots, F_m\}$. The subtracial proof is similar.

This last lemma does not apply to self-adjoint quantum operations. For example, if $\phi(B) = \sum A_j B A_j^*$ where the A_j are self-adjoint we can also write $\phi(B) = \sum (iA_j)B(iA_j)^*$ where iA_j are not self-adjoint.

We now give an example which shows that a positivity preserving map need not be completely positive. Define $\phi\colon \mathcal{B}(\mathbb{C}^2) \to \mathcal{B}(\mathbb{C}^2)$ by $\phi(A) = A^T$ where A^T is the transpose of A. Now a matrix

$$A = \begin{bmatrix} a & b \\ c & d \end{bmatrix} \in \mathcal{B}(\mathbb{C}^2)$$

is positive if and only if $a \geq 0$, $d \geq 0$, and $ad - bc \geq 0$. Hence, if $A \geq 0$ then $A^T \geq 0$ so ϕ is positivity preserving. To show that ϕ is not completely positive consider $\phi \otimes I_2$ on $\mathcal{B}(\mathbb{C}^2 \otimes \mathbb{C}^2)$. Let $e_i = (1,0)$, $e_2 = (0,1)$ be the standard basis for $\mathbb{C}^2$ and define the positive operator $A \in \mathcal{B}(\mathbb{C}^2 \otimes \mathbb{C}^2)$ by

$$\begin{aligned}
A &= |e_1 \otimes e_1 + e_2 \otimes e_2\rangle\langle e_1 \otimes e_2 + e_2 \otimes e_2| \\
&= |e_1 \otimes e_1\rangle\langle e_1 \otimes e_1| + |e_1 \otimes e_1\rangle\langle e_2 \otimes e_2| + |e_2 \otimes e_2\rangle\langle e_1 \otimes e_1| \\
&\quad + |e_2 \otimes e_2\rangle\langle e_2 \otimes e_2| \\
&= |e_1\rangle\langle e_1| \otimes |e_1\rangle\langle e_1| + |e_1\rangle\langle e_2| \otimes |e_1\rangle\langle e_2| \\
&\quad + |e_2\rangle\langle e_1| \otimes |e_2\rangle\langle e_1| + |e_2\rangle\langle e_2| \otimes |e_2\rangle\langle e_2| .
\end{aligned}$$

We then have

$$\begin{aligned}(\phi\otimes I_2)(A) &= |e_1\rangle\langle e_1|\otimes|e_1\rangle\langle e_1| + |e_2\rangle\langle e_1|\otimes|e_1\rangle\langle e_2| \\ &\quad + |e_1\rangle\langle e_2|\otimes|e_2\rangle\langle e_1| + |e_2\rangle\langle e_2|\otimes|e_2\rangle\langle e_2| \\ &= |e_1\otimes e_1\rangle\langle e_1\otimes e_1| + |e_1\otimes e_1\rangle\langle e_1\otimes e_2| \\ &\quad + |e_1\otimes e_2\rangle\langle e_2\otimes e_1| + |e_2\otimes e_2\rangle\langle e_2\otimes e_2| \\ &= \begin{bmatrix} 1&0&0&0\\ 0&0&1&0\\ 0&1&0&0\\ 0&0&0&1 \end{bmatrix}.\end{aligned}$$

But letting $v = (0,1,-1,0) \in \mathbb{C}^2 \otimes \mathbb{C}^2$ we have

$$\langle(\phi\otimes I_r)(v), v\rangle = \langle(0,-1,1,0),(0,1,-1,0)\rangle = -2.$$

Hence, $\phi\otimes I_2$ is not positivity preserving so ϕ is not completely positive.

9.3 Noisy Quantum Channels

This section discusses the quantum operation descriptions of some simple noisy quantum channels [10]. A two-dimensional quantum system is called a *qubit.* This is the most basic quantum system studied in quantum computation and quantum information theory. A qubit has a two-dimensional state space $\mathbb{C}^2$ with (computational) basis elements $|0\rangle = (1,0)$ and $|1\rangle = (0,1)$. The *bit flip channel* flips the state of a qubit from $|0\rangle$ to $|1\rangle$ (and vice versa) with probability $1-p$, $0<p<1$. Letting X be the Pauli matrix

$$X = \begin{bmatrix} 0&1\\ 1&0 \end{bmatrix},$$

we can represent the bit flip channel by the quantum operation

$$\phi_{bf}(\rho) = p\rho + (1-p)X\rho X.$$

Notice that ϕ_{bf} has operation elements $\{p^{1/2}I, (1-p)^{1/2}X\}$ and that ϕ_{bf} is self-adjoint and tracial. It is also unital because for any self-adjoint quantum operation tracial and unital are equivalent. Of course, ϕ_{bf} gives a bit flip because $X|0\rangle = |1\rangle$ and $X|1\rangle = |0\rangle$. Hence,

$$\phi_{bf}\left(|0\rangle\langle 0|\right) = p|0\rangle\langle 0| + (1-p)|1\rangle\langle 1|$$

so the pure state $|0\rangle\langle 0|$ is left undisturbed with probability p and is flipped with probability $1-p$. Similarly,

$$\phi_{bj}\left(|1\rangle\langle 1|\right) = p|1\rangle\langle 1| + (1-p)|0\rangle\langle 0|\,.$$

The *phase flip channel* is represented by the quantum operation

$$\phi_{pf}(\rho) = p\rho + (1-p)Z\rho Z,$$

where $0 < p < 1$ and Z is the Pauli matrix

$$Z = \begin{bmatrix} 1 & 0 \\ 0 & -1 \end{bmatrix}.$$

The operation elements for ϕ_{pf} are $\left\{p^{1/2}I, (1-p)^{1/2}Z\right\}$ so again ϕ_{pf} is self-adjoint and tracial. Because $Z|0\rangle = |0\rangle$ and $Z|1\rangle = -|1\rangle$ we see that ϕ_{pf} changes the relative phase of the qubit states with probability $1-p$.

The *bit-phase flip* channel is represented by the quantum operation

$$\phi_{bpf}(\rho) = p\rho + (1-p)Y\rho Y,$$

where $0 < p < 1$ and Y is the Pauli matrix

$$Y = \begin{bmatrix} 0 & -i \\ i & 0 \end{bmatrix}.$$

This gives a combination of a bit flip and a phase flip because $Y = iXZ$. The operation elements for ϕ_{bpf} are $\left\{p^{1/2}I, (1-p)^{1/2}Y\right\}$ so ϕ_{bpf} is self-adjoint and tracial. We obtain an interesting quantum operation by forming the composition $\phi_{bf} \circ \phi_{pf}$. Because $XZ = -iY$ we have

$$\phi_{bf} \circ \phi_{pf}(\rho) = p^2\rho + p(1-p)Z\rho Z + p(1-p)X\rho X + (1-p)^2 Y\rho Y.$$

The operation elements become

$$\left\{pI, \sqrt{p(1-p)}\,Z, \sqrt{p(1-p)}\,X, (1-p)Y\right\}$$

so again, $\phi_{bf} \circ \phi_{pf}$ is self-adjoint and tracial. It is also easy to check that $\phi_{pf} \circ \phi_{bf} = \phi_{bf} \circ \phi_{pf}$.

Another important type of quantum noise is the depolarizing channel given by the quantum operation

$$\phi_{dp}(\rho) = \frac{pI}{2} + (1-p)\rho,$$

where $0 < p < 1$. This channel depolarizes a qubit state with probability p. That is, the state ρ is replaced by the completely mixed state $I/2$ with probability p. By applying the identity

$$\frac{I}{2} = \frac{\rho + X\rho X + Y\rho Y + Z\rho Z}{4}$$

that holds for every $\rho \in \mathcal{D}(\mathbb{C}^2)$ we can write

$$\phi_{dp}(\rho) = \left(1 - \frac{3}{4}p\right)\rho + \frac{p}{4}(X\rho X + Y\rho Y + Z\rho Z).$$

Thus, the operation elements for ϕ_{dp} become

$$\left\{\sqrt{1-3p/4}\,I, \sqrt{p}X/2, \sqrt{p}Y/2, \sqrt{p}Z/2\right\}.$$

As before, ϕ_{dp} is self-adjoint and tracial.

There are practical quantum operations that are not self-adjoint or unital. For example, consider the amplitude damping channel given by the quantum operation

$$\phi_{ad}(\rho) = A_1\rho A_1^* + A_2\rho A_2^*,$$

where

$$A_1 = \begin{bmatrix} 1 & 0 \\ 0 & \sqrt{1-\gamma} \end{bmatrix}, \qquad A_2 = \begin{bmatrix} 0 & \sqrt{\gamma} \\ 0 & 0 \end{bmatrix},$$

and $0 < \gamma < 1$. It is easy to check that ϕ_{ad} is tracial but not self-adjoint nor unital. Although the quantum channels (quantum operations) that we have considered appear to be quite specialized, general quantum channels and quantum operations can be constructed in terms of these simple ones and this is important for the theory of quantum error correction.

9.4 Iterations

It is sometimes important to consider iterations of quantum operations. For example, a measurement may be repeated many times for greater accuracy or quantum data may enter a noisy channel several times. For a quantum operation $\phi_{\mathcal{A}}$, does the sequence of iterations $\phi_{\mathcal{A}}^n(\rho)$, $n = 1, 2, \ldots$, converge for every state $\rho \in \mathcal{D}(\mathcal{H})$? (Because $\mathcal{H}$ is finite-dimensional, all the usual forms of convergence such as norm convergence or matrix entry convergence coincide so we do not need to specify a particular type of convergence.) In general, the answer is no. For example, $\phi(\rho) = X\rho X$ is a self-adjoint, tracial, and unital quantum operation. Because $X^2 = I$ we have $\phi^{2n}(\rho) = \rho$, $n = 1, 2, \ldots$, but $\phi^{2n+1}(\rho) = X\rho X$, $n = 1, 2, \ldots$. Unless $\rho X = X\rho$, the sequence of iterates does not converge.

A state ρ_0 is a fixed point of a quantum operation $\phi_{\mathcal{A}}$ if $\phi_{\mathcal{A}}(\rho_0) = \rho_0$. It is frequently useful to know the fixed points of a quantum operation because these are the states that are not disturbed by a quantum measurement or a noisy quantum channel.

Lemma 9.4.1. *A state ρ_0 is a fixed point of $\phi_{\mathcal{A}}$ if and only if there exists a state ρ such that* $\lim \phi_{\mathcal{A}}^n(\rho) = \rho_0$.

Proof. If $\lim \phi_{\mathcal{A}}^n(\rho) = \rho_0$, by the continuity of $\phi_{\mathcal{A}}$ we have that

$$\rho_0 = \lim \phi_{\mathcal{A}}^{n+1}(\rho) = \lim \phi_{\mathcal{A}} \circ \phi_{\mathcal{A}}^n(\rho) = \phi_{\mathcal{A}}\left(\lim \phi_{\mathcal{A}}^n(\rho)\right) = \phi_{\mathcal{A}}(\rho_0).$$

Hence, ρ_0 is a fixed point of $\phi_{\mathcal{A}}$. Conversely, if ρ_0 is a fixed point of $\phi_{\mathcal{A}}$ we have that

$$\phi_{\mathcal{A}}^n(\rho_0) = \phi_{\mathcal{A}}^{n-1}\left(\phi_{\mathcal{A}}(\rho_0)\right) = \phi_{\mathcal{A}}^{n-1}(\rho_0) = \cdots = \phi_{\mathcal{A}}(\rho_0) = \rho_0.$$

Hence, $\lim \phi_{\mathcal{A}}^n(\rho_0) = \rho_0$.

The next result shows that the iterates of some of the quantum operations considered in Section 9.3 always converge.

Theorem 9.4.2. *For any $\rho \in \mathcal{D}(\mathbb{C}^2)$ we have that*
(a) $\lim \phi_{bf}^n(\rho) = \frac{1}{2}\rho + \frac{1}{2}X\rho X$
(b) $\lim \phi_{pf}^n(\rho) = \frac{1}{2}\rho + \frac{1}{2}Z\rho Z$
(c) $\lim \phi_{bpf}^n(\rho) = \frac{1}{2}\rho + \frac{1}{2}Y\rho Y$
(d) $\lim \phi_{dp}^n(\rho) = \frac{I}{2}$.

Proof. (a) Any $\rho \in \mathcal{D}(\mathbb{C}^2)$ has the Bloch form

$$\rho = \frac{1}{2}\begin{bmatrix} 1 + r_3 & r_1 - ir_2 \\ r_1 + ir_2 & 1 - r_3 \end{bmatrix},$$

where $r_i \geq 0$, $i = 1, 2, 3$, and $r_1^2 + r_2^2 + r_3^2 \leq 1$. Because

$$X\rho X = \frac{1}{2}\begin{bmatrix} 1 - r_3 & r_1 + ir_2 \\ r_1 - ir_2 & 1 + r_3 \end{bmatrix}$$

we have that

$$\phi_{bf}(\rho) = \frac{1}{2}\begin{bmatrix} 1 + (2p-1)r_3 & r_1 - i(2p-1)r_2 \\ r_1 + i(2p-1)r_2 & 1 - (2p-1)r_3 \end{bmatrix}.$$

We can now prove by induction that

$$\phi_{bf}^n(\rho) = \frac{1}{2}\begin{bmatrix} 1 + (2p-1)^n r_3 & r_1 - i(2p-1)^n r_2 \\ r_1 + i(2p-1)^n r_2 & 1 - (2p-1)^n r_3 \end{bmatrix}.$$

Because $0 < p < 1$, we have $-1 < 2p-1 < 1$ so that $\lim (2p-1)^n = 0$. Hence,

$$\lim \phi_{bf}^n(\rho) = \frac{1}{2}\begin{bmatrix} 1 & r_1 \\ r_1 & 1 \end{bmatrix} = \frac{1}{2}\rho + \frac{1}{2}X\rho X.$$

The proofs of (b) and (c) are similar. To prove (d), a simple induction argument shows that for every $\rho \in \mathcal{D}(\mathbb{C}^2)$

$$\phi_{dp}^n(\rho) = \frac{1 - q^n}{2} I + (1-p)^n \rho,$$

where $q = 1 - p$. Because $0 < q, p < 1$, we have that

$$\lim \phi_{bf}^n(\rho) = \frac{I}{2}.$$

We see from Theorem 9.4.2(a) that $\lim \phi_{bf}^n = \phi_{bf}^{1/2}$ where

$$\phi_{bf}^{1/2}(\rho) = \frac{1}{2}\rho + \frac{1}{2}X\rho X$$

and similar results hold for ϕ_{pf} and ϕ_{bpf}. Notice that $\phi_{bf}^{1/2}$ is an idempotent quantum operation. Indeed,

$$\begin{aligned}\phi_{bf}^{1/2} \circ \phi_{bf}^{1/2}(\rho) &= \frac{1}{4}\rho + \frac{1}{4}X\rho X + \frac{1}{4}X\rho X + \frac{1}{4}X^2\rho X^2 \\ &= \frac{1}{2}\rho + \frac{1}{2}X\rho X = \phi_{bf}^{1/2}(\rho)\,.\end{aligned}$$

The next result shows that this always happens.

Theorem 9.4.3. *If there exists a quantum operation ϕ such that $\lim \phi_{\mathcal{A}}^n(\rho) = \phi(\rho)$ for every $\rho \in \mathcal{D}(\mathcal{H})$, then ϕ is idempotent. Moreover, the set of fixed points of $\phi_{\mathcal{A}}$ coincides with the range* $\mathrm{ran}(\phi)$.

Proof. By the continuity of $\phi_{\mathcal{A}}^n$ we have

$$\phi_{\mathcal{A}}^n(\phi(\rho)) = \phi_{\mathcal{A}}^n\left(\lim_{m\to\infty} \phi_{\mathcal{A}}^m(\rho)\right) = \lim_{m\to\infty} \phi_{\mathcal{A}}^{m+n}(\rho) = \phi(\rho)\,.$$

Hence,

$$\phi \circ \phi(\rho) = \lim \phi_{\mathcal{A}}^n(\phi(\rho)) = \phi(\rho)$$

and we conclude that ϕ is idempotent. The last statement follows from Lemma 9.4.1.

9.5 Fixed Points

Let $\phi_{\mathcal{A}}$ be a quantum operation with $\mathcal{A} = \{A_i, A_i^*\colon i = 1, \ldots, n\}$. The commutant $\mathcal{A}'$ of $\mathcal{A}$ is the set

$$\mathcal{A}' = \{B \in \mathcal{B}(\mathcal{H})\colon BA_i = A_iB,\ BA_i^* = A_i^*B,\ i = 1, \ldots, n\}\,.$$

We denote the set of fixed states of $\phi_{\mathcal{A}}$ by $\mathcal{I}(\phi_{\mathcal{A}})$. That is,

$$\mathcal{I}(\phi_{\mathcal{A}}) = \{\rho \in \mathcal{D}(\mathcal{H})\colon \phi_{\mathcal{A}}(\rho) = \rho\}\,.$$

As an example, it is easy to find $\mathcal{I}(\phi_{pf})$. In this case $\rho \in \mathcal{I}(\phi_{pf})$ if and only if $\rho = p\rho + (1-p)Z\rho Z$. This is equivalent to $\rho = Z\rho Z$. Because $Z^2 = I$ we have that $Z\rho = \rho Z$. We conclude that $\rho \in \mathcal{I}(\phi_{pf})$ if and only if $\rho \in \mathcal{A}'$ where $\mathcal{A} = \{I, Z\}$. A similar result holds for ϕ_{bf} and ϕ_{bpf}. In general we have the following result which is a special case of a theorem in [1, 5].

Theorem 9.5.1. *If $\phi_\mathcal{A}$ is a self-adjoint, subtracial quantum operation, then $\mathcal{I}(\phi_\mathcal{A}) \subseteq \mathcal{A}' \cap \mathcal{D}(\mathcal{H})$.*

Proof. Let $\rho \in \mathcal{I}(\phi_\mathcal{A})$ and let h be a unit eigenvector of ρ corresponding to the largest eigenvalue $\lambda_1 = \|\rho\|$. Then $\phi_\mathcal{A}(\rho) = \rho$ implies that

$$\lambda_1 = \sum \langle \rho A_i h, A_i h\rangle \leq \|\rho\| \sum \|A_i h\|^2 = \lambda_1 \sum \langle A_i^2 h, h\rangle \leq \lambda_1 .$$

Because $\langle \rho A_i h, A_i h\rangle \leq \lambda_1 \langle A_i^2 h, h\rangle$, it follows that

$$\langle (\lambda_1 I - \rho) A_i h, A_i h\rangle = 0 .$$

Hence, $(\lambda_1 I - \rho) A_i h = 0$ for every eigenvector h corresponding to λ_1. Thus, A_i leaves the λ_1-eigenspace invariant. Letting P_1 be the corresponding spectral projection of ρ we have $P_1 A_i P_1 = A_i P_1$ which implies that $A_i P_1 = P_1 A_i$, $i = 1, \ldots, n$. Now $\rho = \lambda_1 P_1 + \rho_1$ where ρ_1 is a positive operator with largest eigenvalue. Because

$$\lambda_1 P_1 + \rho_1 = \rho = \phi_\mathcal{A}(\rho) = \lambda_1 \phi_\mathcal{A}(P_1) + \phi_\mathcal{A}(\rho_1) = \lambda_1 P_1 + \phi_\mathcal{A}(\rho_1)$$

we have $\phi_\mathcal{A}(\rho_1) = \rho_1$. Proceeding by induction, $\rho \in \mathcal{A}'$.

Corollary 9.5.2. *If $\phi_\mathcal{A}$ is a self-adjoint, tracial quantum operation, then*

$$\mathcal{I}(\phi_\mathcal{A}) = \mathcal{A}' \cap \mathcal{D}(\mathcal{H}) .$$

As an application of Corollary 9.5.2 we see that $\mathcal{I}(\phi_{dp}) = \{I/2\}$. Indeed, if $\rho = \mathcal{I}(\phi_\mathcal{A})$ then ρ must commute with X,Y, and Z. But any 2×2 matrix is a linear combination of I, X, Y, and Z. It follows that $\rho = I/2$. The next example which is a special case of an example in [3] shows that self-adjointness cannot be deleted from Theorem 9.5.1 or Corollary 9.5.2.

Let $\phi_\mathcal{A}(B) = \sum_{i=1}^4 A_i B A_i^*$ be the quantum operation with

$$A_1 = \begin{bmatrix} 1 & 0 & 0 \\ 0 & 0 & 0 \\ 0 & 0 & 0 \end{bmatrix}, \qquad A_2 = \begin{bmatrix} 0 & 0 & 0 \\ 0 & 1 & 0 \\ 0 & 0 & 0 \end{bmatrix},$$

$$A_3 = \frac{1}{\sqrt{2}} \begin{bmatrix} 0 & 0 & 0 \\ 0 & 0 & 0 \\ 1 & 0 & 0 \end{bmatrix}, \qquad A_4 = \frac{1}{\sqrt{2}} \begin{bmatrix} 0 & 0 & 0 \\ 0 & 0 & 0 \\ 0 & 1 & 0 \end{bmatrix}.$$

It is easy to check that $\phi_\mathcal{A}$ is unital. However, $\phi_\mathcal{A}$ is not self-adjoint and because

$$\sum A_i^* A_i = \frac{3}{2}\begin{bmatrix} 1 & 0 & 0 \\ 0 & 1 & 0 \\ 0 & 0 & 0 \end{bmatrix}$$

we see that $\phi_{\mathcal{A}}$ is not subtracial. Let $\rho \in \mathcal{D}(\mathbb{C}^3)$ be the state

$$\rho = \frac{1}{3}\begin{bmatrix} 2 & 0 & 0 \\ 0 & 0 & 0 \\ 0 & 0 & 1 \end{bmatrix}.$$

It is easy to check that $\rho \in \mathcal{I}(\phi_{\mathcal{A}})$ but $\rho A_3 \neq A_3\rho$ so that $\rho \notin \mathcal{A}'$. If we multiply the A_i, $i = 1, 2, 3, 4$, by $\sqrt{2/3}$ then $\phi_{\mathcal{A}}$ would be subtracial but again $\mathcal{I}(\phi_{\mathcal{A}}) \not\subseteq \mathcal{A}' \cap \mathcal{D}(\mathcal{H})$.

9.6 Idempotents

We showed in Lemma 9.1.2 that if $\phi_{\mathcal{A}}$ is subtracial and its operation elements are self-adjoint projection operators, then $\phi_{\mathcal{A}}$ is idempotent. We conjecture that a weak converse of this result holds. If L is a Lüders map that is idempotent, we conjecture that L can be written in a form so that its operation elements are self-adjoint projections. As a start, our next result shows that this conjecture holds in $\mathbb{C}^2$ if L has two operation elements.

Theorem 9.6.1. *Suppose* $L(B) = A_1^{1/2}BA_1^{1/2} + A_2^{1/2}BA_2^{1/2}$, $A_1, A_2 \geq 0$, $A_1 + A_2 = I$, *is a Lüders map on* $\mathbb{C}^2$ *and* $L^2 = L$. *Then* A_1 *and* A_2 *are self-adjoint projection operators or* L *is the identity map.*

Proof. Because $A_1 + A_2 = I$, A_1 and A_2 commute and because $L^2 = L$ we have

$$A_1^{1/2}BA_1^{1/2} + A_2^{1/2}BA_2^{1/2} = A_1BA_1 + A_2BA_2 + 2A_1^{1/2}A_2^{1/2}BA_2^{1/2}A_1^{1/2} \quad (9.10)$$

for every $B \in \mathcal{B}(\mathbb{C}^2)$. Without loss of generality, we can assume that A_1 is diagonal so that

$$A_1 = \begin{bmatrix} a & 0 \\ 0 & b \end{bmatrix}, \qquad 0 \leq a, b \leq 1.$$

Letting

$$B = \begin{bmatrix} 1 & 1 \\ 1 & 1 \end{bmatrix}$$

in equation (9.10) and equating entries we obtain

$$\sqrt{ab} + \sqrt{(1-a)(1-b)} = (1-a)(1-b) + ab + 2\sqrt{ab(1-a)(1-b)}\,. \quad (9.11)$$

Equation (9.11) can be written as

$$\left(1 - \sqrt{ab} - \sqrt{(1-a)(1-b)}\right)\left(\sqrt{ab} + \sqrt{(1-a)(1-b)}\right) = 0.$$

We conclude that $\sqrt{ab} + \sqrt{(1-a)(1-b)} = 0$ or 1. In the first case $a = 0$, $b = 1$ or $a = 1$, $b = 0$ and we are finished. In the second case, we can square the expression to obtain

$$2\sqrt{ab(1-a)(1-b)} = a + b - 2ab. \tag{9.12}$$

Squaring (9.12) gives

$$(a-b)^2 = a^2 + b^2 - 2ab = 0$$

so that $a = b$. Hence, $A_1 = aI$, $A_2 = (1-a)I$, and $L(B) = B$ for all $B \in \mathcal{B}(\mathbb{C}^2)$.

9.7 Sequential Measurements

This section discusses a topic that is important in quantum measurement theory, namely sequential products of effects. In this section we allow $\mathcal{H}$ to be infinite-dimensional and again denote the set of effects on $\mathcal{H}$ by $\mathcal{E}(\mathcal{H})$. Recall that effects represent yes–no quantum measurements that may be unsharp (imprecise). We may think of effects as fuzzy quantum events. Sharp quantum events are represented by self-adjoint projection operators. Denoting this set by $\mathcal{P}(\mathcal{H})$ we have that $\mathcal{P}(\mathcal{H}) \subseteq \mathcal{E}(\mathcal{H})$.

We mentioned in Section 9.1 that for a quantum system initially in the state $\rho \in \mathcal{D}(\mathcal{H})$, the postmeasurement state given that $A \in \mathcal{E}(\mathcal{H})$ occurs is $A^{1/2}\rho A^{1/2}/\mathrm{tr}(\rho A)$. Assuming that $\mathrm{tr}(\rho A) \neq 0$, it is reasonable to define the conditional probability of $B \in \mathcal{E}(\mathcal{H})$ given $A \in \mathcal{E}(\mathcal{H})$ to be

$$P_\rho(B \mid A) = \frac{\mathrm{tr}(A^{1/2}\rho A^{1/2}B)}{\mathrm{tr}(\rho A)} = \frac{\mathrm{tr}(\rho A^{1/2}BA^{1/2})}{\mathrm{tr}(\rho A)}. \tag{9.13}$$

Now two measurements $A, B \in \mathcal{E}(\mathcal{H})$ cannot be performed simultaneously in general, so they are frequently executed sequentially. We denote by $A \circ B$ a sequential measurement in which A is performed first and B second. It is natural to assume the probabilistic equation

$$P_\rho(A \circ B) = P_\rho(A)P_\rho(B \mid A). \tag{9.14}$$

Combining (9.13) and (9.14) gives

$$\mathrm{tr}(\rho A \circ B) = \mathrm{tr}(\rho A^{1/2}BA^{1/2}). \tag{9.15}$$

Equation (9.15) motivates our definition $A \circ B = A^{1/2}BA^{1/2}$ and we call $A \circ B$ the sequential product of A and B. If $\{A_1, \ldots, A_n\}$ is a finite POV measure, then the Lüders map with operation elements A_i can now be written as $L(B) = \sum A_i \circ B$. Notice that $A \circ B \in \mathcal{E}(\mathcal{H})$ so $\circ$ gives a binary operation on $\mathcal{E}(\mathcal{H})$. Indeed,

$$\begin{aligned} 0 &\le \left\langle A^{1/2}BA^{1/2}x, x\right\rangle = \left\langle BA^{1/2}x, A^{1/2}x\right\rangle \le \left\langle A^{1/2}x, A^{1/2}, x\right\rangle \\ &= \langle Ax, x\rangle \le \langle x, x\rangle \end{aligned} \tag{9.16}$$

so that $0 \le A^{1/2}BA^{1/2} \le I$. It also follows from (9.16) that $A \circ B \le A$.

We say that $A, B \in \mathcal{E}(\mathcal{H})$ are compatible if $AB = BA$. It is clear that the sequential product satisfies

$$\begin{aligned} 0 \circ A &= A \circ 0 = 0 \\ I \circ A &= A \circ I = A \\ A \circ (B + C) &= A \circ B + A \circ C \quad \text{whenever } B + C \le I \\ (\lambda A) \circ B &= A \circ (\lambda B) = \lambda(A \circ B) \quad \text{for } 0 \le \lambda \le 1. \end{aligned}$$

However, $A \circ B$ has practically no other algebraic properties unless compatibility conditions are imposed. To illustrate the fact that $A \circ B$ does not have properties that one might expect, we now show that $A \circ B = A \circ C$ does not imply that $B \circ A = C \circ A$ even for $A, B, C \in \mathcal{P}(\mathcal{H})$. In $\mathcal{H} = \mathbb{C}^2$ consider $A, B, C \in \mathcal{P}(\mathcal{H})$ given by the following matrices,

$$A = \frac{1}{2}\begin{bmatrix} 1 & 1 \\ 1 & 1 \end{bmatrix}, \qquad B = \begin{bmatrix} 1 & 0 \\ 0 & 0 \end{bmatrix}, \qquad C = \begin{bmatrix} 0 & 0 \\ 0 & 1 \end{bmatrix}.$$

We then have

$$A \circ B = ABA = \frac{1}{2}A = ACA = A \circ C.$$

However,

$$B \circ A = BAB = \frac{1}{2}B \ne \frac{1}{2}C = CAC = C \circ A.$$

This example also shows that $A \circ B \not\le B$ in general, even though we always have $A \circ B \le A$.

We say that A, B are sequentially independent if $A \circ B = B \circ A$. It is clear that if A and B are compatible, then they are sequentially independent. To prove the converse, we need the following result due to Fuglede–Putnam–Rosenblum [11].

Theorem 9.7.1. *If $M, N, T \in \mathcal{B}(\mathcal{H})$ with M and N normal, then $MT = TN$ implies that $M^*T = TN^*$.*

Corollary 9.7.2. [8] *For $A, B \in \mathcal{E}(\mathcal{H})$, $A \circ B = B \circ A$ implies $AB = BA$.*

Proof. Because $A \circ B = B \circ A$ we have

$$A^{1/2}B^{1/2}B^{1/2}A^{1/2} = B^{1/2}A^{1/2}A^{1/2}B^{1/2}.$$

Hence, $M = A^{1/2}B^{1/2}$ and $N = B^{1/2}A^{1/2}$ are normal. Letting $T = A^{1/2}$ we have $MT = TN$. Applying Theorem 9.7.1, we conclude that $B^{1/2}A = AB^{1/2}$. Hence,

$$BA = B^{1/2}AB^{1/2} = AB.$$

Sequential independence for three or more effects was considered in [8] and a more general result was proved. Our next result shows that if $A \circ B$ is sharp, then A and B are compatible (and hence, sequentially independent).

Theorem 9.7.3. [8] *For $A, B \in \mathcal{E}(\mathcal{H})$, if $A \circ B \in \mathcal{P}(\mathcal{H})$, then $AB = BA$.*

Proof. Assume that $A \circ B \in \mathcal{P}(\mathcal{H})$. Suppose that $A \circ Bx = x$ where $\|x\| = 1$. We then have $\langle BA^{1/2}x, A^{1/2}x \rangle = 1$. By Schwarz's inequality we have $BA^{1/2}x = A^{1/2}x$ and hence, $Ax = A \circ Bx = x$. Because x is an eigenvector of A with eigenvalue 1, the same holds for $A^{1/2}$. Thus, $A^{1/2}x = x$ so that $BA^{1/2}x = A \circ Bx$. We conclude that $BA^{1/2}x = A \circ Bx$ for all x in the range of $A \circ B$. Now suppose that $A \circ Bx = 0$. We then have

$$\|B^{1/2}A^{1/2}x\|^2 = \left\langle B^{1/2}A^{1/2}x, B^{1/2}A^{1/2}x \right\rangle = \langle A \circ Bx, x \rangle = 0$$

so that $B^{1/2}A^{1/2}x = 0$. Hence, $BA^{1/2}x = 0$ and it follows that $BA^{1/2}x = A \circ Bx$ for all x in the null space of $A \circ B$. We conclude that $BA^{1/2} = A \circ B$. Hence,

$$BA^{1/2} = A \circ B = (A \circ B)^* = A^{1/2}B$$

so that $AB = BA$.

The last theorem shows why it is important to consider unsharp effects. Even if A and B are sharp, then $A \circ B \notin \mathcal{P}(\mathcal{H})$ unless A and B are compatible. Simple examples show that the converse of Theorem 9.7.3 does not hold. However, the converse does hold for sharp effects.

Corollary 9.7.4. *If $A, B \in \mathcal{P}(\mathcal{H})$ then $A \circ B \in \mathcal{P}(\mathcal{H})$ if and only if $AB = BA$.*

It follows from Corollary 9.7.4 that for $A, B \in \mathcal{P}(\mathcal{H})$ we have $A \circ B = B$ if and only if $AB = BA = B$. We now generalize this result to arbitrary effects.

Theorem 9.7.5. [8] *For $A, B \in \mathcal{E}(\mathcal{H})$ the following statements are equivalent.* (a) $A \circ B = B$. (b) $B \circ A = B$. (c) $AB = BA = B$.

Proof. It is clear that (c) implies both (a) and (b). It then suffices to show that (a) and (b) each imply (c). If $A \circ B = B$ we have

$$B^2A = A^{1/2}BA^{1/2}BA = A^{1/2}B(A^{1/2}BA^{1/2})A^{1/2} = A^{1/2}B^2A^{1/2}.$$

Taking adjoints gives $B^2A = AB^2$. It follows that $AB = BA = B$. If $B \circ A = B$ then for every $x \in \mathcal{H}$ we have

$$\left\langle AB^{1/2}x, B^{1/2}x \right\rangle = \langle B \circ Ax, x \rangle = \langle Bx, x \rangle = \|B^{1/2}x\|^2.$$

If $B^{1/2}x \neq 0$ then

$$\left\langle A \frac{B^{1/2}x}{\|B^{1/2}x\|}, \frac{B^{1/2}x}{\|B^{1/2}x\|} \right\rangle = 1.$$

It follows from Schwarz's inequality that $AB^{1/2}x = B^{1/2}x$. Hence, $AB^{1/2} = B^{1/2}$ so $AB^{1/2} = B^{1/2}A = B^{1/2}$. We again conclude that $AB = BA = B$.

Theorem 9.7.5 cannot be strengthened to the case $A \circ B \leq B$. That is $A \circ B \leq B$ does not imply $AB = BA$. For example, in $\mathbb{C}^2$ let

$$A = \frac{1}{4}\begin{bmatrix} 1 & 1 \\ 1 & 1 \end{bmatrix}, \qquad B = \frac{1}{4}\begin{bmatrix} 3 & 0 \\ 0 & 1 \end{bmatrix};$$

then $A \circ B \leq B$ but $AB \neq BA$.

The simplest version of the law of total probability would say that

$$P_\rho(B) = P_\rho(A)P_\rho(B \mid A) + P_\rho(I - A)P_\rho(B \mid I - A), \tag{9.17}$$

where we interpret $I - A$ as the complement (or negation) of $A \in \mathcal{E}(\mathcal{H})$. In terms of the sequential product (9.17) can be written as

$$P_\rho(B) = P_\rho(A \circ B) + P_\rho((I - A) \circ B) = P_\rho[(A \circ B + (I - A) \circ B)]. \tag{9.18}$$

When does (9.18) hold for every $\rho \in \mathcal{D}(\mathcal{H})$? Equivalently, when does the following equation hold?

$$B = A \circ B + (I - A) \circ B. \tag{9.19}$$

This question is also equivalent to finding the fixed points of the Lüders map $L(B) = A \circ B + (I - A) \circ B$ for $B \in \mathcal{E}(\mathcal{H})$.

Theorem 9.7.6. [5, 8] *For $A, B \in \mathcal{E}(\mathcal{H})$, (9.19) holds if and only if $AB = BA$.*

Proof. It is clear that (9.19) holds if $AB = BA$. Conversely, assume that (9.19) holds and write it as

$$B = A^{1/2}BA^{1/2} + (I - A)^{1/2}B(I - A)^{1/2}.$$

Multiplying by $A^{1/2}$ on the left and right, we obtain

$$\begin{aligned}
A^{1/2}BA^{1/2} &= ABA + (I-A)^{1/2}A^{1/2}BA^{1/2}(I-A)^{1/2} \\
&= ABA + (I-A)^{1/2}\left[B-(I-A)^{1/2}B(I-A)^{1/2}\right](I-A)^{1/2} \\
&= ABA - (I-A)B(I-A) + (I-A)^{1/2}B(I-A)^{1/2} \\
&= ABA - (I-A)B(I-A) + B - A^{1/2}BA^{1/2}.
\end{aligned}$$

Hence,

$$2A^{1/2}BA^{1/2} = ABA - (I-A)B(I-A) + B = AB + BA. \tag{9.20}$$

Using the commutator notation $[X,Y] = XY - YX$, (9.20) gives

$$\begin{aligned}
\left[A^{1/2},[A^{1/2},B]\right] &= A^{1/2}(A^{1/2}B - BA^{1/2}) - (A^{1/2}B - BA^{1/2})A^{1/2} \\
&= AB - 2A^{1/2}BA^{1/2} + BA = 0.
\end{aligned}$$

It follows that for every spectral projection E of A we have

$$\left[E,[A^{1/2},B]\right] = 0.$$

By the Jacobi identity

$$\left[E,[A^{1/2},B]\right] + \left[B[E,A^{1/2}]\right] + \left[A^{1/2},[B,E]\right] = 0.$$

We have that $\left[A^{1/2},[E,B]\right] = 0$. As before we obtain $[E,[E,B]] = 0$. Hence,

$$0 = E(EB-BE) - (EB-BE)E = EB + BE - 2BE$$

which we can write as

$$EB = 2EBE - BE.$$

Multiplying on the left by E gives $EB = EBE$. Hence,

$$EB = (EBE)^* = BE.$$

It follows that $AB = BA$.

Although the sequential product is always distributive on the right, Theorem 9.7.6 shows that it is not always distributive on the left. That is, $(A+B)\circ C \neq A\circ C + B\circ C$ in general, when $A+B \leq I$. Indeed, if $AC \neq CA$, then by Theorem 9.7.6 we have

$$A\circ C + (I-A)\circ C \neq C = [A + (I-A)]\circ C.$$

One might conjecture that the following generalization of Theorem 9.7.6 holds. If $A+B \leq I$ and $(A+B)\circ C = A\circ C + B\circ C$, then $CA = AC$ or $CB = BC$. However, this conjecture is false. Indeed, suppose that $CB \neq BC$.

Nevertheless, we have

$$\left(\tfrac{1}{2}B+\tfrac{1}{2}B\right)\circ C = B\circ C = \tfrac{1}{2}B\circ C+\tfrac{1}{2}B\circ C = \left(\tfrac{1}{2}B\right)\circ C+\left(\tfrac{1}{2}B\right)\circ C.$$

We close by considering another generalization of Theorem 9.7.6. Suppose $A_i \in \mathcal{E}(\mathcal{H})$, $i = 1,\dots,n$ with $\sum A_i = I$ and that $B = \sum A_i \circ B$. Does this imply that $BA_i = A_iB$, $i = 1,\dots,n$? Notice that the answer is affirmative if $A_i \in \mathcal{P}(\mathcal{H})$, $i = 1,\dots,n$. In fact, we only need $A_i \in \mathcal{P}(\mathcal{H})$, $i = 1,\dots,n$ and $\sum A_i \le I$. In this case, we have $A_iA_j = A_jA_i = 0$ for $i \ne j$. Hence, if $B = \sum A_i \circ B$, then $A_iB = BA_i = A_i \circ B$, $i = 1,\dots,n$. A proof very similar to that in Theorem 9.5.1 gives an affirmative answer when $\dim \mathcal{H} < \infty$ or when B has discrete spectrum with a strictly decreasing sequence of eigenvalues. However, when $\dim \mathcal{H} = \infty$ the answer is negative in general [1].

References

1. A. Arias, A. Gheondea, and S. Gudder, "Fixed points of quantum operations," *J. Math. Phys.* **43** (2002) 5872–5881.
2. H. Barnum, "Information-disturbance tradeoff in quantum measurement on the uniform ensemble," *Proc. IEEE Intern. Sym. Info. Theor.*, Washington, D.C., 2001.
3. O Bratteli, P. Jorgensen, A. Kishimoto, and R. Werner, "Pure states on $\mathcal{O}_d$," *J. Operator Theory* **43** (2000) 97–143.
4. P. Busch, P. Lahti, and P. Mittelstaedt, *The Quantum Theory of Measurements* (Springer, Berlin, 1996).
5. P. Busch and J. Singh, "Lüders theorem for unsharp quantum effects," *Phys. Lett. A* **249** (1998) 10–24.
6. M.-D. Choi, "Completely positive linear maps on complex matrices," *Linear Alg. Appl.* **10** (1975) 285–290.
7. E. B. Davies, *Quantum Theory of Open Systems* (Academic Press, London, 1976).
8. S. Gudder and G. Nagy, "Sequential quantum measurements," *J. Math. Phys.* **42** (2001) 5212–5222.
9. K. Kraus, *States, Effects, and Operations* (Springer-Verlag, Berlin, 1983).
10. M. Nielsen and I. Chuang, *Quantum Computation and Quantum Information* (Cambridge University Press, Cambridge, 2000).
11. W. Rudin, *Functional Analysis* (McGraw-Hill, New York, 1991).

Chapter 10
Ekeland Duality as a Paradigm

Jean-Paul Penot

Summary. The Ekeland duality scheme is a simple device. We examine its relationships with several classical dualities, such as the Fenchel–Rockafellar duality, the Toland duality, the Wolfe duality, and the quadratic duality. In particular, we show that the Clarke duality is a special case of the Ekeland duality scheme.

Key words: Clarke duality, duality, Ekeland duality, Fenchel transform, Legendre function, Legendre transform, nonsmooth analysis

10.1 Introduction

Duality is a general tool in mathematics. It consists in transforming a difficult problem into a related one which is more tractable; then, when returning to the initial, or "primal", problem, some precious information becomes available. Although such a process is of common use in optimization theory and algorithms (see [23, 41, 45] and their references), it pertains to a much larger field. Cramer, Fourier, Laplace, and Radon transforms give testimonies of the power of such a scheme.

Even in optimization theory, there is a large spectrum of duality processes: linear programming, convex programming, fractional programming [21], geometric programming, generalized convex programming, quadratic programming [13], semidefinite programming, and so on. It is the purpose of the present chapter to show that several classical duality theories can be cast into a simple general framework.

Jean-Paul Penot
Laboratoire de mathématiques appliquées, UMR CNRS 5142, University of Pau, Faculté des Sciences, B.P. 1155, 64013 PAU cédex, France
e-mail: jean-paul.penot@univ-pau.fr

D.Y. Gao, H.D. Sherali, (eds.), *Advances in Applied Mathematics and Global Optimization*
Advances in Mechanics and Mathematics 17, DOI 10.1007/978-0-387-75714-8_10,

A number of physical phenomena can be described by using the minimizers of a suitable potential function; however, it may be sensible to consider that a notion of stationarity is more adapted than minimization or maximization.

In a famous paper [14] I. Ekeland introduced a duality scheme that deals with critical points instead of minimizers and takes advantage of the power of the tools of differential topology. In order to extend the reach of his theory we drop the smoothness properties required in [14], following a track indicated in [15]. For such an aim, we make use of elementary notions of nonsmooth analysis recalled in Section 10.4 below.

We particularly focus our attention on the convex case for which a close link between the classical Fenchel duality and the Ekeland duality can be obtained thanks to a slight extension of the Brønsted–Rockafellar theorem. But we also consider the concave case, the quadratic case, the Toland duality, and the Clarke duality. The Clarke duality deals with the study of the set of critical points of a function f of the form

$$f(x) := \frac{1}{2}\langle Ax, x\rangle + g(x) \qquad x \in X,$$

where X is a Banach space, A is an self-adjoint operator from X into X^* (i.e., $\langle Ax, x'\rangle = \langle x, Ax'\rangle$ for any $x, x' \in X$) and $g\colon X \to \mathbb{R}_\infty := \mathbb{R} \cup \{+\infty\}$ is a closed proper convex function. It has been applied to the study of solutions to the Hamilton equation in [5, 7–10, 16–20].

It is the main purpose of the present chapter to endeavor to cast the Clarke duality in the general framework of the Ekeland duality. Such an aim may enhance the interest for this general approach. We also obtain a slight complement to the Clarke duality. On the other hand, we assume that the operator A is continuous (instead of densely defined). This assumption guarantees that the notion of critical point we adopt corresponds to a general and natural concept and is not just an ad hoc specific notion. This new feature is valid for all usual subdifferentials of nonsmooth analysis. This assumption suffices for the application to Hamiltonian systems.

In Sections 10.2 and 10.3 we recall the Ekeland duality in the framework of normed vector spaces (n.v.s.). In Section 10.4 we present tools from nonsmooth analysis which enable one to give a rigorous treatment without assuming regularity assumptions. In particular, we introduce a concept of extended Legendre function using methods reminiscent of the notion of limiting subdifferentials (Section 10.5). Such a concept encompasses the case of the Fenchel conjugate of a convex function. Therefore we can apply it to convex duality and show in Section 10.6 that the Fenchel–Rockafellar duality is part of the duality scheme we study. The same is shown for the Toland duality in Section 10.7 and for the Wolfe duality in Section 10.8. The last section is devoted to showing that Clarke duality is a special instance of Ekeland duality.

We do not look for completeness but we endeavor to put some light on some significant instances. Duality of integral functionals is considered elsewhere.

Duality in the calculus of variations using the Ekeland's scheme is performed in [14] and [15].

Because, as mentioned above, many phenomena in physics and mechanics can be modeled by using critical point theory rather than minimization, we believe that the extensive approach by D. Gao and his co-authors (see [22–29] and their references) deserves some more attention and should be combined with the present contribution.

In the sequel $\mathbb{P}$ stands for the set of positive real numbers, $B(0, r)$ is the open ball with center 0 and radius r, and $S_X := \{u \in X : \|u\| = 1\}$ is the unit sphere in a normed vector space.

10.2 Preliminaries: The Ekeland–Legendre Transform

The Ekeland duality deals with the search of critical points and critical values of functions or multifunctions. It can be cast in a general framework in which there is no linear structure (see [44]), but here we remain in the framework of normed vector spaces (n.v.s.) in duality.

Definition 10.1. Given two n.v.s. X, X' and a subset J of $X \times X' \times \mathbb{R}$, a pair (x, r) is called a *critical pair* of J if $(x, 0_{X'}, r) \in J$. A point x of X is called a *critical point* of J if there exists some $r \in \mathbb{R}$ such that (x, r) is a critical pair of J. A real number r is called a *critical value* of J if there exists some $x \in X$ such that (x, r) is a critical pair of J.

The extremization of J consists in the determination of the set $\operatorname{ext} J$ of critical pairs of J. When J is a *generalized 1-jet* in the sense that the projection G of J on $X \times \mathbb{R}$ is the graph of a function $j\colon X_0 \to \mathbb{R}$ defined on some subset X_0 of X, the extremization of j is reduced to the search of critical points of J. Note that J is a generalized 1-jet if and only if one has the implication

$$(x_1, x_1', r_1) \in J, \quad (x_2, x_2', r_2) \in J, \quad x_1 = x_2 \implies r_1 = r_2.$$

Example 10.1. In the classical case X' is the topological dual space X^* of X and J is the 1-jet $J^1 j$ of a differentiable function $j\colon X_0 \to \mathbb{R}$, where X_0 is an open subset of X, defined by

$$J^1 j := \{(x, Dj(x), j(x)) : x \in X_0\},$$

where $Dj(x)$ is the derivative of j at x. Then we recover the usual notion. One may also suppose as in [14] that X_0 is a differentiable submanifold in X and replace $Dj(x)$ by dj_x, the restriction to the tangent space to X_0 at x of the 1-form dj.

The fact that J may be different from a 1-jet gives a great versatility to the duality which is exposed.

Example 10.2. Given a convex function $j\colon X \to \mathbb{R}_\infty := \mathbb{R} \cup \{+\infty\}$, let X' be the topological dual space X^* of X and let J be the subjet of j, defined by

$$J := \{(x, x^*, j(x)) : x \in \operatorname{dom} j,\ x^* \in \partial j(x)\},$$

where $\operatorname{dom} j := j^{-1}(\mathbb{R})$ and $\partial j(x) \subset X^*$ is the Fenchel–Moreau subdifferential of j at x given by

$$x^* \in \partial j(x) \Leftrightarrow j(\cdot) \geq x^*(\cdot) + j(x) - x^*(x).$$

Then the extremization of J coincides with the minimization of j.

In view of its importance for the sequel, let us anticipate Section 10.4 by presenting the next example.

Example 10.3. Let J be the subjet $J^\partial j$ of a function $j\colon X \to \mathbb{R}_\infty := \mathbb{R} \cup \{\infty\}$ associated with some subdifferential ∂:

$$J^\partial j := \{(x, x', r) \in X \times X' \times \mathbb{R} : x' \in \partial j(x),\ r = j(x)\}.$$

In such a case, $\operatorname{ext} J$ is the set of pairs (x, r) such that $0_{X'} \in \partial j(x)$, $r = j(x)$. We make clear what we mean by "subdifferential" in Section 10.4. For the moment we may take for ∂j either the proximal subdifferential $\partial^P j$ of j, given by $x^* \in \partial^P j(x)$ iff

$$\exists c, r \in \mathbb{P} : \forall u \in B(0, r) \qquad j(x+u) \geq x^*(u) + j(x) - c\,\|u\|^2,$$

or the Fréchet (or firm) subdifferential $\partial^F j$ of j given by $x^* \in \partial^F f(x)$ iff

$$\exists \alpha \in A : \forall u \in X \qquad j(x+u) \geq x^*(u) + j(x) - \alpha(\|u\|)\,\|u\|,$$

where A is the set of functions $\alpha\colon \mathbb{R}_+ \to \mathbb{R}_+ \cup \{+\infty\}$ satisfying $\lim_{r \to 0} \alpha(r) = 0$, or the Dini–Hadamard (or directional) subdifferential $\partial^D j$ of j given by $x^* \in \partial^D f(x)$ iff

$$\forall u \in S_X,\ \exists \alpha \in A : \forall (v, t) \in X \times \mathbb{R}_+\ j(x+tv) \geq x^*(tv) + j(x) - \alpha(\|u - v\| + t)t,$$

or the Clarke–Rockafellar subdifferential given by $x^* \in \partial^C j(x)$ iff

$$\exists \alpha \in A : \forall (x', v, t) \in X^2 \times \mathbb{R}_+ \quad j(x' + tv) \geq x^*(tv) + j(x') - \alpha(s)t,$$

with $s := \|u - v\| + \|x' - x\| + t$ (in the case where f is continuous). Of course, in the preceding definitions we assume j is finite at x and we take the empty set otherwise.

We can generalize the preceding cases by considering other subdifferentials appropriate for nonconvex functions (here we have chosen the most usual subdifferentials among classical ones).

Example 10.4. Let $j\colon X \to \mathbb{R}$ be a concave function and let J be the subjet $J^{\partial}j$ of j for one of the first three preceding subdifferentials. Then the extremization of J leads to the maximization of j. In fact, if $x^* \in \partial j(x)$, then for all $u \in X$ one has

$$j'(x,u) := \lim_{(t,v)\to(0_+,u)} \frac{1}{t}\left(j(x+tv) - j(x)\right) \geq x^*(u),$$

so that j is Hadamard differentiable at x, with derivative x^*. Thus $-x^* \in \partial(-j)(x)$ and if $x^* = 0$ we get that x is a maximizer of j. If $x^* \in \partial^C j(x)$ and j is continuous, we also have $-x^* \in \partial^C(-j)(x) = \partial(-j)(x)$ because j is locally Lipschitzian.

Example 10.5. Given a subdifferential ∂ and a function $j\colon X \to \mathbb{R}_\infty$, let

$$J := \{(x,x',r) \in X \times X' \times \mathbb{R} : x' \in \Upsilon j(x) := \partial j(x) \cup (-\partial(-j)(x)),\ r = j(x)\}.$$

This choice is justified by the case where j is concave. In such a case, a pair (x,r) is critical if and only if x is a maximizer of j and $r = \max j(X)$: the condition is sufficient because for any maximizer x of j one has $0 \in \partial(-j)(x)$; we have seen that it is necessary when $0 \in \partial j(x)$ and it is obviously necessary when $0 \in -\partial(-j)(x)$ because $-j$ is convex.

Example 10.6. Let j be a d.c. function, that is, a function of the form $j := g - h$, where g and h are convex functions on some convex subset of X. Let

$$J := \{(x,x',r) \in X \times X' \times \mathbb{R} : x' \in \partial g(x) \boxminus \partial h(x),\ r = j(x)\},$$

where, for two subsets C, D of X', $C \boxminus D$ denotes the set of $x' \in X'$ such that $D + x' \subset C$. Some sufficient conditions ensuring that $\partial g(x) \boxminus \partial h(x)$ coincides with the Fréchet subdifferential of j are known [1]; but in general J is different from $J^F j$.

Example 10.7. Let $(S, \mathcal{S}, \sigma)$ be a measured space, let E be a Banach space, and let $\ell\colon S \times E \to \overline{\mathbb{R}}$ be a measurable integrand, with which is associated the integral functional j given by

$$j(x) := \int_S \ell(s, x(s))d\sigma(s) \qquad x \in X,$$

where X is some normed vector space of (classes of) measurable functions from S to E; for instance $X := L_p(S,E)$ for some $p \in [1, +\infty[$. Then, if X' is a space of measurable functions from S to the dual E^* of E (for instance $X' := L_q(S, E^*)$, with $q := (1-p^{-1})^{-1}$) one can take

$$J := \{(x,x',r) \in X \times X' \times \mathbb{R} :\ x'(s) \in \partial\ell_s(x(s)) \text{ a.e. } s \in S, r = j(x)\},$$

where $\ell_s := \ell(s,\cdot)$. One can give conditions ensuring that J is exactly the subjet of j; but in general that is not the case.

Let us present another example of a different kind bearing on mathematical programming.

Example 10.8. Let X and Z be n.v.s. with dual spaces X^* and Z^*, respectively. Given a closed convex cone C in Z and differentiable maps $f\colon X \to \mathbb{R}$, $g\colon X \to Z$, let

$$J := \{(x, f'(x) + z^* \circ g'(x), f(x)) : z^* \in C^0,\ \langle z^*, g(x)\rangle = 0\},$$

where $C^0 := \{z^* \in Z^* : \langle z^*, z\rangle \leq 0\ \forall z \in C\}$ is the polar cone of C. This choice is clearly dictated by the Karush–Kuhn–Tucker optimality conditions. But, as is well known, a solution of the mathematical programming problem

$$(\mathcal{M}) \qquad \text{minimize } f(x) \quad \text{subject to } g(x) \in C$$

is a critical point for J only when some qualification condition is satisfied.

The approach of Ekeland to duality [14, 15] can be extended to the case of an arbitrary coupling (see [44]). Here we limit our study to bilinear couplings. The normed vector space X appearing in the following definition is usually a space of parameters and X' is usually its topological dual space, but other cases may be considered.

Definition 10.2. Given two normed vector spaces X, X' paired by a bilinear coupling function $c\colon X \times X' \to \mathbb{R}$, the *Ekeland* (or *Legendre*) *map* is the mapping $E\colon X \times X' \times \mathbb{R} \to X' \times X \times \mathbb{R}$ given by

$$E(x, x', r) := (x', x, c(x, x') - r).$$

Clearly, E is a kind of involution: denoting by E' the mapping $E'\colon X' \times X \times \mathbb{R} \to X \times X' \times \mathbb{R}$ given by $E'(x', x, r) := (x, x', c(x, x') - r)$, one has $E \circ E' = I$, $E' \circ E = I$, so that $E^{-1} = E'$ and E' has a similar form. In particular, when $X' = X$, one has $E' = E$, and E is a true involution. We show that under appropriate assumptions, the transform E induces a kind of conjugacy between functions on X and on X'. It can also be applied to multifunctions.

Definition 10.3. Given paired n.v.s. X and X', the *Ekeland transform* J^E of a subset J of $X \times X' \times \mathbb{R}$ is the image of J by E: $J^E := E(J)$.

10.3 The Ekeland Duality Scheme

In the present chapter the decision space X and the parameter space W play a symmetric role; it is not the case in [44] where X is supposed to be an arbitrary set. We assume X and W are n.v.s. paired with n.v.s. W' and X', respectively, by couplings denoted by c_W, c_X, or just $\langle \cdot, \cdot \rangle$ if there is no risk of confusion. Then we put $Z := W \times X$ in duality with $X' \times W'$ by the means of the coupling c given by

$$c((w,x),(x',w')) = c_W(w,w') + c_X(x,x'). \tag{10.1}$$

Such an unorthodox coupling is convenient in the sequel.

The following definition is reminiscent of the notion of perturbation which is one of the two main approaches to duality in convex analysis. However, it is taken in a more restrictive sense when J is the subjet of some convex function, unless the convex function is continuous.

Definition 10.4. Given two pairs (W,W'), (X,X') of n.v.s. in duality, and a subset $J \subset X \times X' \times \mathbb{R}$, a subset P of $W \times X \times X' \times W' \times \mathbb{R}$ is said to be a *hyperperturbation* of J if

$$J = \{(x,x',r) \in X \times X' \times \mathbb{R} : \exists w' \in W',\ (0_W,x,x',w',r) \in P\}.$$

A subset P of $W \times X \times X' \times W' \times \mathbb{R}$ is said to be a *critical perturbation* of J if

$$(x,0_{X'},r) \in J \Leftrightarrow \exists w' \in W',\ (0_W,x,0_{X'},w',r) \in P.$$

In other terms, P is a hyperperturbation of J if J coincides with the domain of the slice $P_0\colon X \times X' \times \mathbb{R} \rightrightarrows W'$ of P given by

$$P_0(x,x',r) := \{w' \in W' : (0_W,x,x',w',r) \in P\}.$$

In order to study the extremization problem

$$(\mathcal{P}) \qquad \text{find } (x,r) \in X \times \mathbb{R} \text{ such that } (x,0_{X'},r) \in J,$$

given a critical perturbation P of J and a coupling $c\colon W \times W' \to \mathbb{R}$, following Ekeland [14, 15] one can introduce the transform $P' := E(P) \subset X' \times W' \times W \times X \times \mathbb{R}$ of P given by

$$P' := \{(x',w',w,x,\langle w',w\rangle + \langle x',x\rangle - r) : (w,x,x',w',r) \in P\}.$$

The domain

$$J' = \{(w',w,r') \in W' \times W \times \mathbb{R} : \exists x \in X,\ (0_{X'},w',w,x,r') \in P'\}$$

of the slice $P_0'\colon W' \times W \times \mathbb{R} \rightrightarrows X$ of P' given by

$$P_0'(w',w,r') := \{x \in X : (0_{X'},w',w,x,r') \in P'\}$$

yields the extremization problem

$$(\mathcal{P}') \qquad \text{find } (w',r') \in W' \times \mathbb{R} \text{ such that } (w',0_W,r') \in J'$$

called the adjoint problem. Denoting by $\operatorname{ext} J$ the solution set of $(\mathcal{P})$ (i.e., the set of $(x,r) \in X \times \mathbb{R}$ such that $(x,0_{X'},r) \in J$) and by $\operatorname{ext} J'$ the solution set of $(\mathcal{P}')$, one has the following result.

Theorem 10.1. *Let J be a subset of $X \times X' \times \mathbb{R}$. For any critical perturbation P of J, the set $P' := E(P)$ defined as above is a hyperperturbation of J', hence is a critical perturbation of J'. Moreover, the problems $(\mathcal{P})$ and $(\mathcal{P}')$ are in duality in the following sense.*

(a) If $(w', r') \in \operatorname{ext} J'$, then $P'_0(w', 0_W, r')$ is nonempty and for any $x \in P'_0(w', 0_W, r')$ one has $(x, -r') \in \operatorname{ext} J$.

(b) If $(x, r) \in \operatorname{ext} J$, then $P_0(x, 0_{X'}, r)$ is nonempty and for any $w' \in P_0(x, 0_{X'}, r)$ one has $(w', -r) \in \operatorname{ext} J'$.

(c) The set of critical values of $(\mathcal{P})$ is the opposite of the set of critical values of $(\mathcal{P}')$.

Proof. The first assertion is an immediate consequence of the definition of P' and J': a pair $(w', r') \in W' \times \mathbb{R}$ is in $\operatorname{ext} J'$ if and only if there exists some $x \in X$ such that $(0_{X'}, w', 0_W, x, r') \in P'$; that is, $x \in P'_0(w', 0_W, r')$. For any such x one has $(0_W, x, 0_{X'}, w', -r') \in P$, hence $(x, 0_{X'}, -r') \in J$ or $(x, -r') \in \operatorname{ext} J$. Assertion (b) similarly results from the implications

$$\begin{aligned}(x, r) \in \operatorname{ext} J &\Leftrightarrow (x, 0_{X'}, r) \in J \\ &\Leftrightarrow \exists w' \in W' : (0_W, x, 0_{X'}, w', r) \in P \\ &\Leftrightarrow \exists w' \in W' : (0_{X'}, w', 0_W, x, -r) \in P'\end{aligned}$$

so that for any $w' \in P_0(x, 0_{X'}, r)$ one has $x \in P'_0(w', 0_W, -r)$; that is, $(w', -r) \in \operatorname{ext} J'$. Assertion (c) is part of the preceding analysis. □

The problem

$$(\mathcal{P}^*) \qquad \text{find } (w', r) \in W' \times \mathbb{R} \text{ such that } (w', 0_W, -r) \in J'$$

can be called the *dual problem* of $(\mathcal{P})$.

The preceding result is akin to [15, Proposition 3] which deals with the enlarged problem

$$(\mathcal{E}') \qquad \text{find } (w', x, r') \in W' \times X \times \mathbb{R} \text{ such that } (0_{X'}, w', 0_W, x, r') \in P'.$$

It clearly corresponds to the problem

$$(\mathcal{E}) \qquad \text{find } (x, w', r) \in X \times W' \times \mathbb{R} \text{ such that } (0_W, x, 0_{X'}, w', r) \in P$$

via the relation $r' = -r$. [15, Proposition 3] is subsumed by the following statement. Each of its assertions implies that (x, r) is a solution to $(\mathcal{P})$ and (w', r') is a solution to $(\mathcal{P}')$ for $r = -r'$.

Proposition 10.1. *For an element (w', x, r') of $W' \times X \times \mathbb{R}$ the following assertions are equivalent.*

(a) (w', x, r') is a solution to $(\mathcal{E}')$.

(b) (x, r) with $r := -r'$ is a solution to $(\mathcal{P})$ and $w' \in P_0(x, 0_{X'}, -r')$.

(c) (w', r') is a solution to $(\mathcal{P}')$ and $x \in P'_0(w', 0_W, r')$.

Proof. Each assertion is equivalent to $(0_W, x, 0_{X'}, w', -r) \in P$. □

We notice that applying to P' the same process, we get an enlarged problem $(\mathcal{E}'')$ which coincides with $(\mathcal{E})$. Thus, as for $(\mathcal{P})$ and $(\mathcal{P}')$ we have an appealing symmetry.

10.4 Tools from Nonsmooth Analysis

A case of special interest arises when the perturbation set P is the subjet of some function $p\colon W \times X \to \overline{\mathbb{R}}$. Although its Ekeland transform is not necessarily a subjet, in some cases one can associate a function with it. In such a case, the dual problem becomes close to the classical dual problem, as we show in the following sections. In order to deal with such a nice situation we need to give precise definitions.

Let us first make clear what we mean by "subdifferential." Here, given a n.v.s. X with dual $X' = X^*$, a set $\mathcal{F}(X) \subset \mathbb{R}_\infty^X$ of functions on X with values in $\mathbb{R}_\infty$, a *subdifferential* is a map $\partial\colon \mathcal{F}(X) \times X \to \mathcal{P}(X')$ with values in the space of subsets of X' which associates with a pair $(f, x) \in \mathbb{R}_\infty^X \times X$ a subset $\partial f(x)$ of X' which is empty if x is not in the domain $\operatorname{dom} f := \{x \in X : f(x) \in \mathbb{R}\}$ of f and such that

(M) If x is a minimizer of f, then $0_{X'} \in \partial f(x)$.

Thus, minimizers are critical points. We do not look for a list of axioms, although such lists exist ([4, 30–32, 39] and others). However, we may require some other conditions such as the following ones in which X, Y, Z are n.v.s. and $L(X,Y)$ denotes the space of linear continuous maps from X into Y:

(F) If f is convex, ∂f coincides with the Fenchel–Moreau subdifferential:

$$\partial f(x) := \{x^* \in X^* : f(\cdot) \geq x^*(\cdot) - x^*(x) + f(x)\}.$$

(T) If $f := g + h$, where h is continuously differentiable at x, then $\partial f(x) = \partial g(x) + Dh(x)$.

(T_0) If f is continuously differentiable at x, then $\partial f(x) = \{Df(x)\}$.

(C) If $f := g \circ \ell$, where $\ell \in L(X,Y)$ and $g \in \mathbb{R}_\infty^Y$, then $\partial g(\ell(x)) \circ \ell \subset \partial f(x)$.

(C′) If $f := g \circ \ell$, with $\ell \in L(X,Y)$ open, $g \in \mathbb{R}_\infty^Y$, then $\partial g(\ell(x)) \circ \ell \subset \partial f(x)$.

(O) If $f := g \circ \ell$, where $\ell \in L(X,Y)$ is open and $g\colon Y \to \mathbb{R}$ is locally Lipschitzian, then $\partial f(x) \subset \partial g(\ell(x)) \circ \ell$.

(P) If $f := g \circ p_Y$, where $p_Y : Y \times Z \to Y$ is the canonical projection and $g \in \mathbb{R}_\infty^Y$, then $\partial f(y, z) = \partial g(y) \circ p_Y$.

(D) If $f := g \circ \ell$, where $\ell \in L(X,Y)$ is an isomorphism and $g \in \mathbb{R}_\infty^W$, then $\partial f(x) = \partial g(\ell(x)) \circ \ell$.

Clearly (T_0) is a special case of the translation property (T) and (P) is a special case of the conjunction of the composition properties (C) and (O).

Condition (D) which can be considered as a very special case of (P) is satisfied by all usual subdifferentials. Other relationships are described in the following statement.

Proposition 10.2. *(a) If ∂ is either the Fréchet subdifferential or the Hadamard subdifferential then conditions* (F), (T), (C), *and* (O) *are satisfied.*

(b) If ∂ is either the Clarke subdifferential [6] *or the moderate subdifferential* [33] *then conditions* (F), (T), (C′), *and* (O) *are satisfied.*

Proof. (a) The coincidence with the Fenchel subdifferential (F), the translation property (T), the composition properties (C) and (O) are easy to check. Let us prove the two latest ones. Given $x \in X$, $\ell \in L(X, Y)$, and $y^* \in \partial^D g(y)$, with $y := \ell(x)$, we observe that for every $u \in X$ we have

$$f'(x, u) \geq g'(y, \ell(u)) \geq \langle y^*, \ell(u) \rangle.$$

Thus $y^* \circ \ell \in \partial^D f(x)$. If $y^* \in \partial^F g(y)$, one can find some function $\beta\colon Y \to \mathbb{R}$ such that $\lim_{v \to 0} \beta(v) = 0$ and

$$g(y + v) - g(y) - \langle y^*, v \rangle \geq -\beta(v) \|v\|$$

for v in a neighborhood V of 0 in Y. Then, for $u \in U := \ell^{-1}(V)$ one has

$$f(x + u) - f(x) - \langle y^* \circ \ell, u \rangle \geq -\beta(\ell(u)) \|\ell\| \|u\|,$$

so that $y^* \circ \ell \in \partial^F f(x)$.

Now suppose ℓ is open. Because $B_Y \subset \ell(cB_X)$, for some $c > 0$, where B_X, B_Y are the closed unit balls of X and Y, respectively, for every unit vector $v \in Y$ we can pick some $u \in cB_X$ such that $\ell(u) = v$. By homogeneity, we obtain a map $h\colon Y \to X$ such that $\ell(h(v)) = v$ and $\|h(v)\| \leq c\|v\|$ for all $v \in Y$. Let $x^* \in \partial^D f(x)$. Because g is locally Lipschitzian, for all $u \in X$ we have $\langle x^*, u \rangle \leq f'(x, u) = g'(y, \ell(u))$ and $g'(y, 0) = 0$. Thus, $\langle x^*, u \rangle = 0$ for all u in the kernel N of ℓ. Because ℓ is open, it follows that there exists some $y^* \in Y^*$ such that $x^* = y^* \circ \ell$. From the surjectivity of ℓ and the relation $\langle y^* \circ \ell, u \rangle \leq g'(y, \ell(u))$ for all $u \in X$ we conclude that $y^* \in \partial^D g(y)$. Now let us suppose $x^* \in \partial^F f(x)$. By what precedes we obtain that there exists some $y^* \in \partial^D g(y)$ such that $x^* = y^* \circ \ell$. Let $\alpha\colon X \to \mathbb{R}$ and $r > 0$ be such that $\lim_{u \to 0} \alpha(u) = 0$ and

$$f(x + u) - f(x) - \langle x^*, u \rangle \geq -\alpha(u) \|u\|$$

for $u \in rB_X$. Let $h\colon Y \to X$ be the map constructed above, and let $s := c^{-1}r$. Because for all $v \in sB_Y$ we have $h(v) \in rB_X$, we get, as $\langle y^*, v \rangle = \langle y^*, \ell(h(v)) \rangle = \langle x^*, h(v) \rangle$, $g(y) = f(x)$, and $f(x + h(v)) = g(\ell(x) + \ell(h(v))) = g(y + v)$,

$$g(y + v) - g(y) - \langle y^*, v \rangle \geq -\beta(v) \|v\|$$

with $\beta(v) := c\alpha(h(v)) \to 0$ as $v \to 0$. Thus $y^* \in \partial^F g(y)$.

(b) Again, the assertions concerning (F) and (T) are classical and elementary. For the Clarke subdifferential, the assertions concerning (C′) and (O) are particular cases of [6, Theorem 2.3.10]. The case of the moderate subdifferential is similar. □

Let us insist on the fact that extremization problems are not limited to the examples mentioned in the previous sections. In particular, one may take for J some subset of the closure of a subjet with respect to some topology (or convergence) on $X \times X' \times \mathbb{R}$. Another case of interest appears when X is a n.v.s. and J is the hypergraph of a multifunction $M\colon X \rightrightarrows \mathbb{R}$ associated with a notion of normal cone:

$$H(M) := \{(x, x^*, r) \in X \times X^* \times \mathbb{R} : (x^*, -1) \in N(G(M), (x, r)), r \in M(x)\},$$

where $G(M)$ is the graph of M and $N(G(M), (x, r))$ denotes the normal cone to $G(M)$ at (x, r). The normal cone $N(S, s)$ at s to a subset S of a n.v.s. X can be defined in different ways. Some axiomatic approach can be adopted as in [40]. When one disposes of a subdifferential ∂ on the set of Lipschitzian functions on X one may set $N(S, s) := \mathbb{R}_+ \partial d_S(s)$, where d_S is the distance function to S: $d_S(x) := \inf\{d(x, y) : y \in S\}$. When the subdifferential ∂ is defined over the set $\mathcal{S}(X)$ of lower semicontinuous functions on X, one can also define $N(S, s)$ by $N(S, s) := \partial \iota_S(s)$, where ι_S is the indicator function of S given by $\iota_S(x) = 0$ for $x \in S$, $+\infty$ else.

Introducing the coderivative $D^*M(x, r)$ of M at $(x, r) \in G(M)$ by

$$D^*M(x, r) := \{x^* \in X^* : (x^*, -1) \in N(G(M), (x, r))\},$$

we see that $H(M)$ is the set of $(x, x^*, r) \in X \times X^* \times \mathbb{R}$ such that $x^* \in D^*M(x, r)$. In particular, if M is the epigraph multifunction of a function f, $H(M)$ coincides with $J^\partial f$ whenever $x^* \in \partial f(x)$ if and only if $(x^*, -1) \in N(\operatorname{epi}(f), (x, f(x)))$.

When M is a hypergraph, $E(M)$ is not necessarily a hypergraph. When M is the subjet $J^\partial f$ associated with a function f on X and a subdifferential ∂, the set $E(M)$ is not necessarily the subjet of some function on X'. It is of interest to introduce a notion that implies part of such a requirement. This is the aim of the next section.

10.5 Ekeland and Legendre Functions

We first delineate a class of functions for which a conjugate function can be defined.

Definition 10.5. [42] Given a pairing c between the n.v.s. X and X' and a subdifferential $\partial\colon \mathcal{F}(X) \times X \to \mathcal{P}(X')$, a function $f \in \mathcal{F}(X)$ is an *Ekeland function* with respect to ∂, in short an ∂-Ekeland function, or just an Ekeland function if there is no risk of confusion, if for any $x_1, x_2 \in X$, $x' \in X'$ satisfying $x' \in \partial f(x_1) \cap \partial f(x_2)$ one has $c(x_1, x') - f(x_1) = c(x_2, x') - f(x_2)$.

Then, the *Ekeland transform* of f is the function $f^E\colon X' \to \mathbb{R}_\infty$ given by $f^E(x') := c(x, x') - f(x)$ for $x \in (\partial f)^{-1}(x')$ for $x' \in \partial f(X)$, $f^E(x') = +\infty$ for $x' \in X' \backslash \partial f(X)$.

Thus, the graph of f^E is the projection on $X' \times \mathbb{R}$ of $E(J^\partial f)$.

Example 10.9. Any convex function (on some n.v.s.) is an Ekeland function for any subdifferential satisfying condition (F). In fact, for any given $x' \in X'$, every $x \in (\partial f)^{-1}(x')$ is a maximizer of the function $c(\cdot, x') - f(\cdot)$ so that the value of this function at x is independent of the choice of x.

Example 10.10. Any concave function on some n.v.s. X is an Ekeland function for the Fréchet and the Dini–Hadamard subdifferentials. In fact, for any $x_1, x_2 \in X$, $x^* \in X^*$ satisfying $x^* \in \partial f(x_1) \cap \partial f(x_2)$ one has $\langle x^*, x_1\rangle - f(x_1) = \langle x^*, x_2\rangle - f(x_2)$ because in such a case x^* is the Hadamard derivative of f at x_i $(i = 1, 2)$, hence $\langle x^*, x_i\rangle - f(x_i) = \min\{\langle x^*, x\rangle - f(x) : x \in X\}$. Then f^E is the restriction to $f'(X)$ of the concave conjugate f_* of f. Similar assertions hold when f is continuous.

Example 10.11. Let f be a linear-quadratic function on X; that is, $f(x) := \frac{1}{2}\langle Ax, x\rangle - \langle b, x\rangle + c$ for some continuous symmetric linear map $A\colon X \to X' := X^*$, $b \in X'$, $c \in \mathbb{R}$. Let ∂ be a subdifferential satisfying condition (T_0), such as the Clarke, the Fréchet, the Hadamard or the moderate subdifferential. Then f is an Ekeland function. In fact, given $x' \in X'$, $x_1, x_2 \in X$ such that $f'(x_i) = x'$ one has

$$\langle x', x_i\rangle - f(x_i) = \langle Ax_i - b, x_i\rangle - \frac{1}{2}\langle Ax_i, x_i\rangle + \langle b, x_i\rangle - c = \frac{1}{2}\langle Ax_i, x_i\rangle - c$$

and

$$\langle Ax_1, x_1\rangle - \langle Ax_2, x_2\rangle = \langle A(x_1 - x_2), x_1\rangle + \langle Ax_2, x_1 - x_2\rangle = 0$$

because A is symmetric and $Ax_1 = x' + b = Ax_2$, so that $A(x_1 - x_2) = 0$. Thus, for $x' \in A(X) - b$, we can write $f^E(x') = \frac{1}{2}\langle x' + b, A^{-1}(x' + b)\rangle - c$, even if A is noninvertible.

Example 10.12. Let $f\colon X \to \overline{\mathbb{R}}$ be a partially quadratic function in the sense that there exist a decomposition $X = X_1 \oplus X_2$ as a topological direct sum, an isomorphism $A\colon X_1 \to X_1'$, where $X_1' := X_2^\perp := \{x' \in X' := X^* : x' \mid X_2 = 0\}$, $b \in X_1'$, $c \in \mathbb{R}$ such that $f(x) := \frac{1}{2}\langle Ax, x\rangle - \langle b, x\rangle + c$ for $x \in X_1$, $f(x) = +\infty$ for $x \in X \backslash X_1$. Let ∂ be the Clarke, the Fréchet, the Dini–Hadamard, or the moderate subdifferential. Then, for $x \in X_1$ one has $\partial f(x) = Ax + X_2'$, where $X_2' := X_1^\perp$. Then, as in the preceding example, one sees that for any $x' \in X'$, $x \in (\partial f)^{-1}(x')$ the value of $\langle x', x\rangle - f(x)$ does not depend on the choice of x in $(\partial f)^{-1}(x')$. Thus f is an Ekeland function. □

The following definition stems from the wish to get a concept which is more symmetric than the notion of Ekeland function. It is also motivated by the convex (and the concave) case in which the domain of f^E is the image of ∂f (respectively, f') which is not necessarily convex, whereas a natural extension of f^E is the Fenchel conjugate whose domain is convex and which enjoys nice properties (lower semicontinuity, local Lipschitz property on the interior of its domain, etc.).

Definition 10.6. Let X and X' be n.v.s. paired by a coupling function $c\colon X \times X' \to \mathbb{R}$. A l.s.c. function $f\colon X \to \mathbb{R}_\infty$ is said to be a (generalized) Legendre function for a subdifferential ∂ if there exists a l.s.c. function $f^L\colon X' \to \mathbb{R}_\infty$ such that

(a) f and f^L are Ekeland functions and $f^L \mid \partial f(X) = f^E \mid \partial f(X)$.
(b) For any $x \in \operatorname{dom} f$ there is a sequence $(x_n, x'_n, r_n)_n$ in $J^\partial f$ such that $(x_n, \langle x_n - x, x'_n\rangle, r_n) \to (x, 0, f(x))$.
(b′) For any $x' \in \operatorname{dom} f^L$ there is a sequence $(x'_n, x_n, s_n)_n$ in $J^\partial f^L$ such that $(x'_n, \langle x_n, x'_n - x'\rangle, s_n) \to (x', 0, f^L(x'))$.
(c) The relations $x \in X$, $x' \in \partial f(x)$ are equivalent to $x' \in X'$, $x \in \partial f^L(x')$.

Condition (b) (resp., (b′)) ensures that f (resp., f^L) is determined by its restriction to $\operatorname{dom} \partial f$ (resp., $\operatorname{dom} \partial f^L$). In fact, for any $x \in \operatorname{dom} f$ one has

$$f(x) = \liminf_{u(\in \operatorname{dom} \partial f) \to x} f(u)$$

because $f(x) \le \liminf_{u \to x} f(u)$ and (b) implies $f(x) = \lim_n f(x_n)$ for some sequence $(x_n) \to x$ in $\operatorname{dom} \partial f$. Moreover, conditions (a) and (b′) imply that f^L is determined by f.

Condition (b) can be simplified when ∂f is locally bounded on the domain of f. In that case, condition (b) is equivalent to the simpler condition

(b_0) For any $x \in \operatorname{dom} f$ there exists a sequence $(x_n)_n$ in $\operatorname{dom} \partial f$ such that $(x_n, f(x_n)) \to (x, f(x))$.

Example 10.13. Any classical Legendre function is a (generalized) Legendre function. We say that a function $f\colon U \to \mathbb{R}$ on an open subset U of a n.v.s. X is a *classical Legendre function* if it is of class C^2 on U and if its derivative Df is a diffeomorphism from U onto an open subset U' of X^*. In fact, one can show that it suffices that f be of class C^1 and that its derivative Df be a locally Lipschitzian homeomorphism from U onto an open subset U' of X^* whose inverse is also locally Lipschitzian. See [42, 43] for such refinements.

In particular, let f be the linear-quadratic function on X given by $f(x) := (1/2)\langle Ax, x\rangle - \langle b, x\rangle + c$ for some symmetric isomorphism $A\colon X \to X' := X^*$, $b \in X'$, $c \in \mathbb{R}$. Then f is a classical Legendre function because $Df\colon x \mapsto Ax - b$ is a diffeomorphism.

Example 10.14. A variant is the notion of *Legendre–Hadamard function.* A function $f\colon U \to \mathbb{R}$ on an open subset U of a normed vector space X is a

Legendre–Hadamard function if it is Hadamard differentiable, if its derivative $Df\colon U \to X' := X^*$ is a bijection onto an open subset U' of X' which is continuous when X is endowed with its strong topology and X' is endowed with the topology of uniform convergence on compact subsets, its inverse h satisfying a similar continuity property and the Ekeland transform f^E of f given by

$$f^E(x') := \langle h(x'), x' \rangle - f(h(x')) \qquad x' \in U'$$

being Hadamard differentiable with derivative h. Then f and f^E are of class T^1 in the sense that they are Hadamard differentiable and the functions $df\colon U \times X \to \mathbb{R}$ and $df^E\colon U' \times X' \to \mathbb{R}$ given by $df(u,x) := Df(u)(x)$, $df^E(u',x') := Df^E(u')(x')$ are continuous (see [37]). Then f is a generalized Legendre function for the Dini–Hadamard subdifferential. In fact, if $x' \in \partial f(x)$ for some $x \in U$, one has $x' = Df(x)$, hence $x = h(x')$, $f^E(x') = \langle x, x' \rangle - f(x)$ and $\partial f^E(x') = \{h(x')\} = \{x\}$, so that conditions (a) and (c) of the preceding definition are satisfied. Conditions (b) and (b′) are immediate and in fact, for any $x \in U$ and any sequence $(x_n) \to x$ one has $\langle x_n - x, f'(x_n) \rangle \to 0$ and a similar property for f^E by the assumed continuity property.

Let us give a criterion which has some analogy with the one we gave in the preceding example. Now, the differentiability assumption on f is weaker, but the local Lipschitz condition on the inverse h of Df is changed into the assumption that for any $x' \in U'$ the map h is *directionally compact at* x' in the following sense: for any $v' \in X'$ and any sequences $(v'_n) \to v'$, $(t_n) \to 0_+$ the sequence $(t_n^{-1}(h(x' + t_n v'_n) - h(x')))$ is contained in a compact set. Such an assumption is satisfied when h is Hadamard differentiable at any x' or when X is finite-dimensional and h is locally Lipschitzian.

Proposition 10.3. *Suppose f is of class T^1 and its derivative Df is a bijection from U onto an open subset U' whose inverse h is directionally compact at every point of U'. Suppose the mappings $df\colon (x,v) \mapsto Df(x)(v)$ and $(x',v') \mapsto h(x')(v')$ are continuous from $U \times X$ into $\mathbb{R}$ and from $U' \times X'$ into $\mathbb{R}$, respectively. Then f is a Legendre–Hadamard function.*

Proof. It suffices to prove that f^E is Hadamard differentiable at any $x' \in U'$, with derivative $h(x')$. Let $v' \in X'$ and let $(v'_n) \to v'$, $(t_n) \to 0_+$. Let us set $v_n := t_n^{-1}(h(x' + t_n v'_n) - h(x'))$, $x := h(x')$. By our assumption of directional compactness, (v_n) is contained in a compact subset of X, so that $\alpha_n := t_n^{-1}(f(x + t_n v_n) - f(x) - t_n Df(x)(v_n))$ has limit 0, $x' = Df(x)$ and

$$\begin{aligned}
&f^E(x' + t_n v'_n) - f^E(x') \\
&= \langle x' + t_n v'_n, h(x' + t_n v'_n) \rangle - f(h(x' + t_n v'_n)) - \langle x', h(x') \rangle + f(h(x')) \\
&= \langle x' + t_n v'_n, x + t_n v_n \rangle - f(x + t_n v_n) - \langle x', x \rangle + f(x) \\
&= t_n \langle v'_n, x \rangle + t_n \langle x', v_n \rangle + t_n^2 \langle v'_n, v_n \rangle - t_n Df(x)(v_n) - t_n \alpha_n \\
&= t_n \langle v'_n, x \rangle + t_n \beta_n
\end{aligned}$$

with $\beta_n := t_n\langle v_n', v_n\rangle - \alpha_n \to 0$. This shows that f^E is Hadamard differentiable at x', with derivative $x := h(x')$. $\square$

Example 10.15. Any l.s.c. proper convex function f is a (generalized) Legendre function. In fact, a slight strengthening [38, Proposition 1.1] of the Brønsted–Rockafellar theorem ensures that for any $x \in \operatorname{dom} f$ there exists a sequence (x_n, x_n^*) in the graph of ∂f such that $(\langle x_n - x, x_n^*\rangle) \to 0$ and $(f(x_n)) \to f(x)$. The same is valid for the Fenchel conjugate function $f^E = f^*$. Moreover, as is well known, condition (c) holds in such a case.

Example 10.16. Let $f\colon X \to \mathbb{R} \cup \{-\infty\}$ be a concave function such that $U := \operatorname{dom}(-f)$ and $U' := \operatorname{dom}((-f)^*)$ are open and f and its concave conjugate f_* are differentiable on U and U', respectively; here f_* is given by $f_*(x') = \inf_{x\in X}(\langle x', x\rangle - f(x)) = -(-f)^*(-x')$ and differentiability is taken in the sense of Fréchet (resp., Hadamard) when one takes the Fréchet (resp., Hadamard) subdifferential. Then f is a generalized Legendre function for this subdifferential ∂. In fact, $x' \in \partial f(x)$ if and only if f is Fréchet (resp., Hadamard) differentiable at x and $x' = f'(x)$. Then, for $g := -f$, one has $-x' = g'(x)$, hence $x \in \partial g^*(-x')$. Because f_* is supposed to be differentiable, g^* is also differentiable and $x = (g^*)'(-x') = (f_*)'(x') \in \partial f_*(x')$. Moreover, one has $f^E(x') = \langle x', x\rangle - f(x) = f_*(x')$. Condition (b) is satisfied because for any $x \in U$ one can take $(x_n, x_n', r_n) = (x, f'(x), f(x))$. Because the roles of f and f_* are symmetric, we see that f is a generalized Legendre function.

Remark. Let U be an open convex subset of an Asplund space X with the Radon–Nikodym property. Let f be a continuous concave function on U such that its concave conjugate f_* is finite and continuous on an open convex subset U' of X' and $-\infty$ on $X'\backslash U'$. Let ∂ be either the Fréchet or the Hadamard subdifferential and let $f^L := f_*$. As in the preceding example we see that for any $x' \in \partial f(X)$ one has $f^L(x') = f^E(x')$. By definition of an Asplund space, f is Fréchet differentiable on a dense subset D of U. Because it is also locally Lipschitzian, its derivative is locally bounded on D. Thus, if $x \in U$ and if (x_n) is a sequence of D with limit x, then $(\langle f'(x_n), x_n - x\rangle) \to 0$. Now, because f_* is defined on an open convex subset and is continuous and upper semicontinuous for the weak* topology, it is also Fréchet differentiable on a dense subset of its domain by a result of Collier [11] and by a similar argument, we see that condition (b′) is satisfied. However, condition (c) is not necessarily satisfied. For example, let X be a Hilbert space, and let f be given by $f(x) := -\max(\|x\|, \|x\|^2)$. Then $f_*(x') = -1 - \|x'\|$ for $x' \in 2B'$, where B' is the closed unit ball of X' and $f_*(x') = -\frac{1}{4}\|x'\|^2$ for $x' \in X'\backslash 2B'$. Let u be a unit vector in X and let $u' \in X'$ be such that $\langle u', u\rangle = -2$, $\|u'\| = 2$; then we have $u \in \partial^F f_*(u')$ but $u' \notin \partial^F f(u)$ because f is not Fréchet differentiable at u.

Example 10.17. Let ∂ be a subdifferential such that $\partial(-f)(x) = -\partial f(x)$ when f is locally Lipschitzian around x. For instance ∂ may be the Clarke subdifferential [6], the moderate subdifferential [33], or be given as $\Upsilon f(x) := \partial^F f(x) \cup (-\partial^F(-f)(x))$ or $\partial^D f(x) \cup (-\partial^D(-f)(x))$. Let f be a concave function such that $-f$ and $-f_*$ have open domains and are continuous on their domains. Then f is a generalized Legendre function. In fact, using the notation $g := -f$ and arguments as in the preceding example, we see that if $x' \in \partial f(x)$ we also have $-x' \in \partial g(x)$, hence $x \in \partial g^*(-x') = \partial(-f_* \circ (-I_X))(-x') = \partial f_*(x')$.

10.6 The Fenchel–Rockafellar Duality

A particular case requires some developments. It concerns the case when W and X are n.v.s. with dual spaces W' and X', respectively, and when a subset K of $W \times X \times X' \times W' \times \mathbb{R}$ and a densely defined linear mapping $A\colon X \to W$ with closed graph and transpose $A^\intercal$ are given such that

$$J := \{(x, x', r) : \exists u' \in X', w' \in W',\ (0_W, x, u', w', r) \in K,\ x' = u' + A^\intercal w'\}. \tag{10.2}$$

Again, we consider $W \times X$ and $X' \times W'$ are paired with the coupling c of (10.1) which defines an isomorphism $\gamma\colon (W \times X)^* \to X' \times W'$. Thus the primal problem is

$$(\mathcal{P}) \qquad \text{find } (x, r) \in X \times \mathbb{R} \text{ such that } \exists w' \in W',\ (0_W, x, -A^\intercal w', w', r) \in K.$$

The special case when K is the image by $\widehat{\gamma} := I_{W\times X} \times \gamma \times I_{\mathbb{R}}$ of the subjet of a function $k\colon W \times X \to \overline{\mathbb{R}}$, A is continuous and

$$j(x) := k(Ax, x), \qquad \partial j(x) = \partial k(Ax, x) \circ (A, I_X)$$

deserves some interest and illustrates what follows. More explicitly, in such a case one has

$$K := \{(w, x, x', w', r) :\ (x', w') \in \partial k(w, x),\ r = k(w, x)\}.$$

This case is considered later on. Let us note that when K is the subjet of such a function k and when ∂ satisfies condition (C) the set J contains the subjet of j. But one may have $J \neq J^\partial j$ when $j = k \circ (A, I_X)$. For j of this form, a natural perturbation of j is given by $p(w, x) := k(w + Ax, x)$ for $(w, x) \in W \times X$. Such a perturbation may inspire a hyperperturbation P in the general case to which we return.

Given A, K, and J as in (10.2), we can introduce P by setting

$$\begin{aligned} P &:= \{(w, x, u' + A^\intercal w', w', r) : u' \in X',\ w' \in W',\ (Ax + w, x, u', w', r) \in K\} \\ &= \{(w, x, x', w', r) :\ (Ax + w, x, x' - A^\intercal w', w', r) \in K\}. \end{aligned}$$

Then J is the domain of the slice $P_0 \colon X \times X' \times \mathbb{R} \rightrightarrows W'$ of P given by

$$P_0(x, x', r) := \{w' \in W' : (0_W, x, x', w', r) \in P\},$$

so that P is a hyperperturbation of J. The Ekeland transform $P' := E(P) \subset X' \times W' \times W \times X \times \mathbb{R}$ of P is given by

$$\begin{aligned} P' &:= \{(u' + A^\intercal w', w', w, x, \langle w', w\rangle + \langle u' + A^\intercal w', x\rangle - r) : \\ &\qquad\qquad u' \in X',\ w' \in W',\ (Ax + w, x, u', w', r) \in K\} \\ &= \{(x', w', w, x, r') : (Ax + w, x, x' - A^\intercal w', w', \langle w', w\rangle + \langle x', x\rangle - r') \in K\}, \end{aligned}$$

and the domain J' of the slice $P'_0 \colon W' \times W \times \mathbb{R} \rightrightarrows X$ of P' defined by

$$P'_0(w', w, r') := \{x \in X : (0_{X'}, w', w, x, r') \in P'\}$$

is

$$J' = \{(w', w, r') : \exists x \in X,\ (Ax + w, x, -A^\intercal w', w', \langle w', w\rangle - r') \in K\}$$

and the adjoint problem is

$(\mathcal{P}')$ find $(w', r') \in W' \times \mathbb{R}$ such that $\exists x \in X,\ (Ax, x, -A^\intercal w', w', -r') \in K$.

Equivalently, because $\langle w', Ax\rangle + \langle -A^\intercal w', x\rangle = 0$, we have

$(\mathcal{P}')$ find $(w', r') \in W' \times \mathbb{R}$ such that $\exists x \in X,\ (-A^\intercal w', w', Ax, x, r') \in E(K)$.

Thus, $(\mathcal{P}')$ is obtained from $E(K)$ in a way similar to the one $(\mathcal{P})$ is deduced from K, with $-A^\intercal$, X', W', X, W substituted to A, W, X, W', X', respectively. When A is continuous, k is a generalized Legendre function, and $K := \widehat{\gamma}(J^\partial k)$ for some subdifferential ∂, one has

$$(w', u') \in \partial k(Ax + w, x) \Leftrightarrow (Ax + w, x) \in \partial k^L(u', w')$$

so that P' is obtained from $K' := \widehat{\gamma}'(J^\partial k^L)$ as P is obtained from $K := J^\partial k$ where $\widehat{\gamma}'$ is a transposition similar to $\widehat{\gamma}$. Then $(\mathcal{P}')$ is a substitute for the extremization of the function $j' \colon w' \mapsto k^L(-A^\intercal w', w')$. Under appropriate assumptions, the preceding guideline becomes a precise result.

Lemma 10.1. *Given a function $k \colon W \times X \to \mathbb{R}_\infty$ finite at $(\overline{w}, \overline{x}) \in W \times X$ and a continuous linear map $A \colon X \to W$, let $f \colon W \times X \to \mathbb{R}_\infty$ be given by $f(w, x) := k(Ax + w, x)$. Then, for any subdifferential satisfying condition* (D), *one has*

$$(w, x, w', u' + A^\intercal w', r) \in J^\partial f \Leftrightarrow (Ax + w, x, w', u', r) \in J^\partial k,$$

so that P is the subjet of f up to a transposition.

Proof. The result amounts to

$$(w', u' + A^\intercal w') \in \partial f(w, x) \Leftrightarrow (w', u') \in \partial k(Ax + w, x).$$

It stems from condition (D), because the map $B\colon (w, x) \mapsto (Ax + w, x)$ is an isomorphism with inverse $(z, x) \mapsto (z - Ax, x)$, as a simple computation of the transpose $B^\intercal$ of B shows. □

Proposition 10.4. *Let W and X be reflexive Banach spaces with dual spaces W' and X', respectively, and let $A\colon X \to W$ be linear and continuous. Let $k\colon W \times X \to \mathbb{R}_\infty$ be a generalized Legendre function and let $K := J^\partial k$ be its subjet. Then, for any subdifferential satisfying condition* (D)*, the extremization problem* $(\mathcal{P}')$ *is the extremization problem associated with the hyperperturbation $P' = J^\partial p'$ of J', where*

$$p'(x', w') = k^L(w', x' - A^\intercal w') \qquad (x', w') \in X' \times W'.$$

Proof. Using the preceding lemma with a change of notation, we have

$$\begin{aligned}(x, z - Ax) \in \partial p'(x', w') &\Leftrightarrow (x, z) \in \partial k^L(x' - A^\intercal w', w')\\ &\Leftrightarrow (x' - A^\intercal w', w') \in \partial k(x, z).\end{aligned}$$

Then, the definition of P' given above gives the result. □

Let us observe that when k is convex one gets the generalized Fenchel–Rockafellar duality (see for instance [47, Corollary 2.8.2]):

$$\inf_{x \in X} k(Ax, x) = \max_{w' \in W'} \left(-k^*(w', -A^\intercal w')\right).$$

Proposition 10.5. *Let W and X be reflexive Banach spaces with dual spaces W' and X', respectively, and let $A\colon X \to W$ be linear and continuous. Let $k\colon W \times X \to \mathbb{R}_\infty$ be a l.s.c. proper convex function such that*

$$\mathbb{R}_+(\operatorname{dom} k^* - (I_{W'}, -A^\intercal)(W')) \tag{10.3}$$

is a closed vector subspace of $W' \times X'$. Then the extremization problem $(\mathcal{P}')$ *coincides with the minimization problem*

$$\text{minimize } k^*(w', -A^\intercal w') \qquad w' \in W'.$$

Proof. When k is a l.s.c. proper convex function, it is a generalized Legendre function and $k^L = k^*$, the Fenchel transform of k. Moreover, under the qualification condition (10.3), the Attouch–Brézis theorem ensures that for the convex function $j'\colon w' \mapsto k^L(w', -A^\intercal w')$ one has

$$\begin{aligned}\partial j'(w') &= \{w - Ax : (w, x) \in \partial k^*(w', -A^\intercal w')\}\\ &= \{w - Ax : (w', -A^\intercal w') \in \partial k(w, x)\}.\end{aligned}$$

□

The next result deals with the particular case in which $k(w,x) = g(w-b) + f(x)$ for $(w,x) \in W \times X$, where $f\colon X \to \mathbb{R}_\infty$, $g\colon W \to \mathbb{R}_\infty$ are l.s.c. proper convex functions and $b \in W$ is fixed. It follows from an easy computation of k^*: $k^*(w',x') = g^*(w') + \langle w', b\rangle + f^*(x')$. Then one obtains that condition (10.3) is satisfied if and only if $\mathbb{R}_+(\operatorname{dom} f^* + A^\intercal(W'))$ is a closed vector subspace of X'.

Corollary 10.1. *Let W and X be reflexive Banach spaces with dual spaces W' and X', respectively, let $A\colon X \to W$ be linear and continuous, and let $f\colon X \to \mathbb{R}_\infty$, $g\colon W \to \mathbb{R}_\infty$ be l.s.c. proper convex functions such that*

$$\mathbb{R}_+(\operatorname{dom} f^* + A^\intercal(W')) \tag{10.4}$$

is a closed vector subspace of W'. Then the extremization problem $(\mathcal{P}')$ coincides with the minimization problem

$$\textit{minimize } f^*(-A^\intercal w') + g^*(w') + \langle w', b\rangle \qquad w' \in W'.$$

Let us note that when $\mathbb{R}_+(\operatorname{dom} k - k \circ (A, I_X)(X))$ is a closed vector subspace of $W \times X$, the set J is the subjet of the function j, so that the situation is entirely symmetric. However, such a condition is not required to apply the duality relationships described in the preceding results.

10.7 The Toland Duality

In [15] Ekeland applies his duality scheme to the case of the Toland duality. The primal problem is

$$(\mathcal{T}) \qquad \operatorname{ext} f(x) - g(Ax) \quad x \in X,$$

where $g\colon W \to \mathbb{R}$ and $f\colon X \to \mathbb{R}_\infty$ are l.s.c. proper convex functions and $A\colon X \to W$ is a continuous linear map. We interpret it as the extremization of the set

$$J := \{(x, x', r) \in X \times X' \times \mathbb{R} :\ \exists w' \in \partial g(Ax),\ \exists u' \in \partial f(x),\ x' = u' - A^\intercal w'\}.$$

However, we do not claim that J is the subjet of $j\colon x \mapsto f(x) - g(Ax)$. Thus, instead of using the subjet of $k\colon (w,x) \mapsto f(x) - g(w)$, we introduce the sets

$$K := \{(w, x, x', w', r) :\ -w' \in \partial g(w),\ x' \in \partial f(x),\ r = f(x) - g(w)\}.$$

Now we set

$$P := \{(w, x, x', w', r) : w' \in \partial g(Ax - w),\ \ \exists u' \in \partial f(x),\ x' = u' - A^\intercal w'\}$$

which can be thought of as a similar interpretation of the subjet of $p\colon (w,x) \mapsto f(x) - g(Ax - w)$. Moreover,

$$\begin{aligned} P &:= \{(w, x, u' + A^{\intercal}w', w', r) : u' \in X',\ w' \in W',\ (Ax - w, x, u', w', r) \in K\} \\ &= \{(w, x, x', w', r) :\ (Ax - w, x, x' - A^{\intercal}w', w', r) \in K\}. \end{aligned}$$

Then J is the domain of the slice $P_0\colon X \times X' \times \mathbb{R} \rightrightarrows W'$ of P given by

$$P_0(x, x', r) := \{w' \in W' : (0_W, x, x', w', r) \in P\},$$

so that P is a hyperperturbation of J. The Ekeland transformed set $P' := E(P) \subset X' \times W' \times W \times X \times \mathbb{R}$ of P is given by

$$\begin{aligned} P' := \{(x', w', w, x, r') :& \\ &(Ax - w, x, x' - A^{\intercal}w', w', \langle w', w\rangle + \langle x', x\rangle - r') \in K\} \\ = \{(u' + A^{\intercal}w', w', w, x, &\langle w', w\rangle + \langle u' + A^{\intercal}w', x\rangle - r) : \\ &(Ax - w, x, u', w', r) \in K\} \end{aligned}$$

and the domain J' of the slice $P'_0\colon W' \times W \times \mathbb{R} \rightrightarrows X$ of P' defined by

$$P'_0(w', w, r') := \{x \in X : (0_{X'}, w', w, x, r') \in P'\}$$

is

$$J' = \{(w', w, r') : \exists x \in X,\ (Ax - w, x, -A^{\intercal}w', w', \langle w', w\rangle - r') \in K\}.$$

Thus the adjoint problem is

$(\mathcal{P}')$ find $(w', r') \in W' \times \mathbb{R}$ such that $\exists x \in X,\ (Ax, x, -A^{\intercal}w', w', -r') \in K.$

We observe that, because $\langle x' - A^{\intercal}w', x\rangle + \langle w', Ax - w\rangle = \langle w', -w\rangle + \langle x', x\rangle$, by the Fenchel equality

$$\begin{aligned} &\langle w', w\rangle + \langle x', x\rangle - f(x) + g(Ax + w) \\ &\quad = \langle x' - A^{\intercal}w', x\rangle - \langle -w', Ax + w\rangle - f(x) + g(Ax + w) \\ &\quad = g^*(w') - f^*(A^{\intercal}w' - x'). \end{aligned}$$

Introducing the set $K' := E(K)$,

$$\begin{aligned} K' &= \{(x', w', w, x, r') :\ (-w', x') \in \partial g(w) \times \partial f(x), \\ &\qquad\qquad r' = \langle (w', x'), (w, x)\rangle - f(x) + g(w)\} \\ &= \{(x', w', w, x, r') :\ w \in \partial g^*(-w'),\ x \in \partial f^*(x'), \\ &\qquad\qquad r' = f^*(x') - g^*(-w')\} \end{aligned}$$

which corresponds to the subjet of $(x', w') \mapsto f^*(x') - g^*(-w')$, we have

$$
\begin{aligned}
P' &= \{(x', w', w, x, r') : (x' - A^{\mathsf{T}} w', w', x, w, r') \in K'\} \\
&= \{(x', w', w, x, f^*(x' - A^{\mathsf{T}} w') - g^*(-w')) : \\
&\qquad\qquad Ax - w \in \partial g^*(-w'),\, x \in \partial f^*(x' - A^{\mathsf{T}} w')\}, \\
J' &:= \{(w', w, r') : \exists x \in X,\, Ax - w \in \partial g^*(-w'),\, x \in \partial f^*(-A^{\mathsf{T}} w'), \\
&\qquad\qquad r' = f^*(-A^{\mathsf{T}} w') - g^*(-w')\} \\
&= \{(w', w, r') : \exists u \in \partial \widetilde{g}(A^{\mathsf{T}} w'),\, w + Au \in \partial \widetilde{f}(w'),\, r' = \widetilde{f}(w') - \widetilde{g}(A^{\mathsf{T}} w')\},
\end{aligned}
$$

where $\widetilde{f}(w') := g^*(-w')$, $\widetilde{g}(x') := f^*(-x')$.

Therefore, replacing $\widetilde{f}$, $\widetilde{g}$, A by g^*, f^*, A^{T}, and using a construction similar to the one we have used to pass from j to J, the adjoint problem can be interpreted as

$$
(\mathcal{T}') \qquad \operatorname{ext} \widetilde{f}(w') - \widetilde{g}(A^{\mathsf{T}} w') \quad w' \in W'.
$$

This is the Toland duality. Note that if we use a subdifferential ∂ such that $\partial(-g)(x) = -\partial g(x)$ for a convex function g, and if we dispose of regularity assumptions ensuring a sum rule, the preceding constructions are no more formal.

10.8 The Wolfe Duality

Let us give a version of the Wolfe duality [46, 12, 34, 35, 36] that involves a family of minimization problems rather than a single one; we show that it can be interpreted as an instance of the Ekeland duality.

Given a set U, n.v.s. W, X, a closed convex cone C in W, and mappings $f\colon U \times X \to \mathbb{R}$ and $g\colon U \times X \to W$ which are differentiable in their second variable, let us consider the constrained optimization problem

$$
(\mathcal{M}) \text{ minimize } f(u, x) \text{ under the constraint } g(u, x) \in C.
$$

We consider $(\mathcal{M})$ as a minimization problem with respect to a primary variable x and a second variable u or as a family of partial minimization problems

$$
(\mathcal{M}_u) \text{ minimize } f_u(x) \text{ under the constraint } g_u(x) \in C, \qquad u \in U.
$$

The variant of the Wolfe dual we deal with is the family of partial maximization problems indexed by $u \in U$,

$$
(\mathcal{W}_u) \text{ maximize } \ell_u(x, y) \text{ over } (x, y) \in X \times Y \text{ subject to } \frac{\partial}{\partial x} \ell_u(x, y) = 0,\; y \in C^0,
$$

where $\ell_u(x, y) := f_u(x) + \langle y, g_u(x) \rangle$ is the classical Lagrangian, Y is the dual of W, and $C^0 := \{y \in Y : \forall w \in C\ \langle y, w \rangle \leq 0\}$. We observe that in $(\mathcal{W}_u)$ the implicit constraint $g(u,x) \in C$ which is difficult to deal with has disappeared, and an easier equality constraint appears.

Then one has the following result, whose proof is similar to the one in [12, Theorem 4.7.1].

Theorem 10.2. *Suppose that for all $u \in U$ and all $y \in -C^0$ the functions f_u and $y \circ g_u$ are convex. Then, for all $u \in U$ one has the weak duality relation*

$$\sup(\mathcal{W}_u) \leq \inf(\mathcal{M}_u).$$

If $(\mathcal{M})$ has a solution, then there exists some $u \in U$ such that strong duality holds; that is, the preceding inequality is an equality.

In order to relate this result to the Ekeland scheme, for $u \in U$ we introduce the subset

$$J_u := \{(x,y,x',y',r) : r = f_u(x),\ g_u(x) \in C,\ x' = f_u'(x) + y \circ g_u'(x),\\ y' = g_u(x),\ \langle y, g_u(x)\rangle = 0\}$$

of $X \times C^0 \times X' \times W \times \mathbb{R}$, so that J_u is the intersection of $\{(x,y,x',y',r) \in X \times C^0 \times X' \times W \times \mathbb{R} :\ \langle y, g_u(x)\rangle = 0\}$ with the one-jet

$$J^1\ell_u := \{(x,y,x',y',r) : (x',y') = D\ell_u(x,y),\ r = \ell_u(x,y)\}$$

of the function ℓ_u. The extremization of J_u consists in searching pairs $(x,y) \in X \times C^0$ which are critical points of ℓ_u with respect to $X \times C^0$, that is, which satisfy

$$\frac{\partial}{\partial x}\ell_u(x,y) = 0, \qquad \frac{\partial}{\partial y}\ell_u(x,y) \in C^{00} = C, \qquad \langle y, \frac{\partial}{\partial y}\ell_u(x,y)\rangle = 0.$$

This is exactly the set of solutions of the Kuhn–Tucker system.

It is natural to associate with $(\mathcal{M}_u)$ the perturbed problem by $w \in W$

$$(\mathcal{M}_{u,w}) \text{ minimize } f_u(x) \text{ under the constraint } g_u(x) + w \in C.$$

We associate with this problem the subset P of the set $W \times g_u^{-1}(C) \times C^0 \times X' \times Y' \times W' \times \mathbb{R}$ given by

$$(w,x,y,x',y',w',r) \in P \Leftrightarrow$$
$$x' = Df_u(x) + y \circ Dg_u(x),\ y' = g_u(x) + w,\ w' = y,\ r = f_u(x) + \langle y, g_u(x) + w\rangle.$$

It is clearly a hyperperturbation of J_u. A short computation shows that its Ekeland transform P' is characterized by $(w',x',y',w,x,y,r') \in P'$ if and only if $(w,x,y,x',y',w',r') \in W \times g^{-1}(C) \times C^0 \times X' \times Y' \times W' \times \mathbb{R}$ and

$$r' = \langle w', w\rangle + \langle x', x\rangle - f_u(x),\ w' = y,\ x' = Df_u(x) + y \circ Dg_u(x),\ y' = g_u(x) + w.$$

Thus, considering w as a parameter and (x, y) as the decision variable, we can set

$$J'_u = \{(w', w, r') : \exists (x, y) \in g_u^{-1}(C) \times C^0,\ (w', 0_{X'}, 0_{Y'}, w, x, y, r') \in P'\}.$$

We obtain that $(w', w, r') \in J'_u$ if and only if there exists $(x, y) \in g_u^{-1}(C) \times C^0$ such that $y := w'$,

$$\begin{aligned} Df_u(x) + y \circ Dg_u(x) &= 0, & g_u(x) + w &\in C, \\ \langle y, g_u(x) + w \rangle &= 0, & r' &= \langle w', w \rangle - f_u(x). \end{aligned}$$

Then $r' = \langle y, -g_u(x) \rangle - f_u(x) = -\ell_u(x, y)$.

We see that $\mathrm{ext}(\mathcal{M}'_u)$ corresponds to the search of $(w', r', x) \in Y \times \mathbb{R} \times X$ such that

$$\begin{aligned} Df_u(x) + y \circ Dg_u(x) &= 0, & g_u(x) &\in C, & y &\in C^0, \\ \langle w', g_u(x) \rangle &= 0, & r' &= -\ell_u(x, y), \end{aligned}$$

or, in other terms, to the search of $(x, y, r') \in g_u^{-1}(C) \times C^0 \times \mathbb{R}$ such that $\partial \ell_u(x, y)/\partial x = 0$, $\partial \ell_u(x, y)/\partial y \in C$, $r' = -\ell_u(x, y)$:

$$\begin{aligned} &(y, r') \in \mathrm{ext}(\mathcal{M}'_u) \Leftrightarrow \exists x \in X : g_u(x) \in C, \\ &\quad \langle y, g_u(x) \rangle = 0, \qquad Df_u(x) + y \circ Dg_u(x) = 0, \qquad r' = -\ell_u(x, y). \end{aligned}$$

Now (x, y) is a critical point for the problem

$$\begin{aligned} &(\mathcal{M}'_u) \text{ maximize } \ell_u(x, y) \text{ over } (x, y) \in X \times Y \\ &\text{under the constraints } g_u(x) \in C, \ \frac{\partial}{\partial x} \ell_u(x, y) = 0 \end{aligned}$$

if and only if there exist multipliers $\overline{y} \in C^0$, $\overline{x}^{**} \in X^{**}$ such that for all $(\widehat{x}, \widehat{y}) \in X \times Y$,

$$\begin{aligned} &\langle \overline{y}, g_u(x) \rangle = 0, \\ &-D\ell_u(x, y)(\widehat{x}, \widehat{y}) + \langle \overline{y}, Dg_u(x)(\widehat{x}) \rangle + \langle \overline{x}^{**}, D\frac{\partial}{\partial x} \ell_u(x, y)(\widehat{x}, \widehat{y}) \rangle = 0. \end{aligned}$$

Taking $\overline{y} = 0$, $\overline{x}^{**} = 0$, we see that for any solution (y, r') of $\mathrm{ext}(\mathcal{M}'_u)$ and for any $x \in X$ satisfying the requirements of $\mathrm{ext}(\mathcal{M}'_u)$, one gets a critical point (x, y) of the problem $(\mathcal{M}'_u)$. In turn, considering (u, x) as an auxiliary variable and y as the primary variable, one is led to the maximization problem $(\mathcal{W}_u)$. However, a solution (x, y) of $(\mathcal{W}_u)$ should satisfy the extra conditions $g_u(x) \in C$, $\langle y, g_u(x) \rangle = 0$ in order to yield a solution to $\mathrm{ext}(\mathcal{M}'_u)$.

Note that in the case of the quadratic problem

$$(\mathcal{Q}) \qquad \text{minimize } \frac{1}{2}\langle Qx, x \rangle + \langle q, x \rangle \qquad \text{subject to } Ax - b \in C,$$

where $Q\colon X \to X'$ is linear, continuous, and symmetric (but not necessarily semidefinite positive), $A\colon X \to W$, $q \in X'$, $b \in W$, C being a closed convex cone of W, the Wolfe dual

$$(\mathcal{W})\text{ maximize } \frac{1}{2}\langle Qx, x\rangle + \langle q, x\rangle + \langle y, Ax - b\rangle \text{ over } (x, y) \in X \times Y$$
$$\text{subject to } Qx + q + y \circ A = 0$$

is a simple quadratic problem with linear constraints. It can be given necessary and sufficient optimality conditions provided the map $(x, y) \mapsto Qx + y \circ A$ has a closed range in X'.

10.9 The Clarke Duality

Let X be a reflexive Banach space, let $A\colon X \to X^*$ be a densely defined self-adjoint operator (i.e., such that $\langle Ax_1, x_2\rangle = \langle x_1, Ax_2\rangle$ for any $x_1, x_2 \in \operatorname{dom} A$) and let $g\colon X \to \mathbb{R} \cup \{+\infty\}$ be a l.s.c. proper convex function. Let $X' := X^*$ and let J be given by

$$J := \{(x, x', r) \in X \times X' \times \mathbb{R} : x' + Ax \in \partial g(x),\ r = j(x)\}$$

where

$$j(x) := g(x) - \frac{1}{2}\langle Ax, x\rangle \text{ for } x \in \operatorname{dom} A \cap \operatorname{dom} g,\ j(x) = +\infty \text{ else.}$$

Let us consider the extremization problem of J:

$$(\mathcal{P}) \qquad \text{find } (x, r) \in X \times \mathbb{R} \text{ such that } Ax \in \partial g(x),\ r = j(x).$$

Here we have taken $-A$ instead of A as in [7, 16] and elsewhere in order to get a more symmetric form of the result; of course, this choice is inessential as we make no positiveness assumption on A. When A is continuous, and when the subdifferential ∂ satisfies condition (T) (in particular for the Fréchet, the Hadamard, the moderate, and the Clarke subdifferentials) J is the subjet of j because in that case one has

$$x' \in \partial j(x) \Leftrightarrow x' + Ax \in \partial g(x).$$

In particular, x is a critical point of j in the sense $0 \in \partial j(x)$ iff $Ax \in \partial g(x)$. Then $(\mathcal{P})$ corresponds to the extremization of j.

Let us introduce a hyperperturbation of J by setting $W := X^*$, $W' := X$, $X' := X^*$, and

$$P := \{(w, x, x', x, j(x)) \in W \times \operatorname{dom} j \times X' \times W' \times \mathbb{R} :\ x' + Ax - w \in \partial g(x)\}.$$

In fact, we have

$$\begin{aligned} P_0(x,x',r) &:= \{w' \in W' : (0_W, x, x', w', r) \in P\} \\ &= \{w' \in W' : w' = x,\ x' + Ax \in \partial g(x),\ r = j(x)\}, \end{aligned}$$

hence

$$(x,x',r) \in \operatorname{dom} P_0 \Leftrightarrow x' + Ax \in \partial g(x),\ r = j(x) \Leftrightarrow (x,x',r) \in J,$$

so that P is indeed a hyperperturbation of J in the sense given above. Although we do not need the following result to proceed, it may serve as a guide line.

Lemma 10.2. *When A is continuous and ∂ satisfies conditions* (F), (P), (T), *the set P is the subjet of the function $f\colon W \times X \to \mathbb{R}_\infty$ given by*

$$f(w,x) = g(x) - \frac{1}{2}\langle Ax, x\rangle + \langle w, x\rangle$$

and f is an Ekeland function.

Proof. When A is continuous f is the sum of the continuously differentiable function $(w,x) \mapsto -\frac{1}{2}\langle Ax, x\rangle + \langle w, x\rangle$ with the convex function $(w,x) \mapsto g(x)$, and conditions (T), (P), and (F) ensure that

$$(w',x') \in \partial f(w,x) \Leftrightarrow w' = x,\ x' + Ax - w \in \partial g(x). \tag{10.5}$$

Then, for $(w',x') \in W' \times X'$ and for $(w,x) \in (\partial f)^{-1}(w',x')$ one has

$$\begin{aligned} f^E(w',x') &= \langle w, w'\rangle + \langle x, x'\rangle - \left(g(x) - \frac{1}{2}\langle Ax, x\rangle + \langle w, x\rangle\right) \\ &= \langle w', x'\rangle + \frac{1}{2}\langle Aw', w'\rangle - g(w') \end{aligned}$$

and we see that this value does not depend on the choice of $(w,x) \in (\partial f)^{-1}(w',x')$: f is an Ekeland function. □

Let us return to the general case. In order to describe the dual problem $(\mathcal{P}')$, we observe that

$$\begin{aligned} J' &= \{(w',w,r') \in W' \times W \times \mathbb{R} : \exists x \in X,\ (0_{X'}, w', w, x, r') \in P'\} \\ &= \{(w',w,r') \in W' \times W \times \mathbb{R} : \exists x \in X,\ (w, x, 0_{X'}, w', \langle w, w'\rangle - r') \in P\} \end{aligned}$$

and

$$x \in P_0'(w', 0_W, r') \Leftrightarrow (0_W, x, 0_{X'}, w', -r') \in P$$

so that $(w', 0_W, r') \in J' = \operatorname{dom} P_0'$ iff there exists some $x \in \operatorname{dom} j \subset X$ such that $Ax \in \partial g(x)$, $x = w'$, $r' = -f(0,x)$. Thus, because g is convex and A is symmetric,

$$\begin{aligned}(w',r') \in \operatorname{ext} J' &\Leftrightarrow w' \in \operatorname{dom} j,\ Aw' \in \partial g(w'),\ r' = -f(0,w')\\ &\Leftrightarrow w' \in \operatorname{dom} j,\ w' \in \partial g^*(Aw'),\ r' = -j(w')\\ &\Rightarrow w' \in \operatorname{dom} j,\ Aw' \in A\left(\partial g^*(Aw')\right) \subset \partial\left(g^* \circ A\right)(w'), r' = -j(w').\end{aligned}$$

In particular, when ∂ satisfies conditions (F) and (T) and A is continuous, for any $(w',r') \in \operatorname{ext} J'$, the pair $(w',-r')$ is a critical pair of the function $j' \colon X \to \mathbb{R} \cup \{+\infty\}$ given by

$$j'(x) := g^*(Ax) - \frac{1}{2}\langle Ax, x\rangle.$$

This function is invariant by addition of an element of $\operatorname{Ker} A$, thus we have obtained under these conditions the first part of the following statement which subsumes Clarke duality. In order to prove the second part we introduce the function j'' given by

$$j''(x) := \left(g^* \circ A\right)^*(Ax) - \frac{1}{2}\langle Ax, x\rangle.$$

Theorem 10.3. *Suppose g is l.s.c. proper convex, ∂ satisfies* (F), (T), *and A is continuous. Then,*

(a) For any critical pair (x,r) of J and for any $u \in \operatorname{Ker} A$, the pair $(x+u,-r)$ is a critical pair of J'.

(b) For any critical pair (x',r') of J' and for any $u \in \operatorname{Ker} A$, the pair $(x'+u,-r')$ is a critical pair of j''. If moreover g is convex and

$$\mathbb{R}_+(\operatorname{dom} g^* - A(X)) = X',$$

then there exists $u' \in \operatorname{Ker} A$ such that $(x'+u,-r')$ is a critical pair of j.

Proof. Because J' has the same form as J, with g replaced by $g^* \circ A$, we obtain from part (a) that for any critical pair (x',r') of j' and for any $u \in \operatorname{Ker} A$, the pair $(x'+u,-r')$ is a critical pair of

$$x \mapsto \left(g^* \circ A\right)^*(Ax) - \frac{1}{2}\langle Ax, x\rangle = j''(x).$$

On the other hand, x' is a critical point of j' means that

$$Ax' \in \partial(g^* \circ A)(x').$$

Now, under condition (C), the Attouch–Brézis theorem ensures the equalities $\partial\left(g^* \circ A\right)(x') = A^{\intercal}\left(\partial g^*(Ax')\right) = A(\partial g^*(Ax'))$, so that there exists some $y' \in \partial g^*(Ax')$ such that

$$Ax' = Ay'.$$

Thus, one has $u' := y' - x' \in \operatorname{Ker} A$ and because $y' \in \partial g^*(Ax')$, by the reciprocity formula, we get $Ax' \in \partial g(y')$ or $Ay' \in \partial g(y')$. Therefore, $(x'+u,-r')$ is a critical pair of j. □

References

1. Amahroq, T., Penot, J.-P., and Syam, A., Subdifferentiation and minimization of the difference of two functions, *Set-Valued Anal.* (to appear) DOI: 10.1007/s11228-008-0085-9.
2. Aubin, J.-P., and Ekeland, I., *Applied Nonlinear Analysis*, Wiley, New York (1984).
3. Aubin, J.-P., and Frankowska, H., *Set-Valued Analysis*, Birkhäuser, Boston (1990).
4. Aussel, D., Corvellec, J.-N., and Lassonde, M., Mean value property and subdifferential criteria for lower semicontinuous functions, *Trans. Amer. Math. Soc.* 347, No. 10, 4147–4161 (1995).
5. Blot, J., and Azé, D., *Systèmes Hamiltoniens: Leurs Solutions Périodiques*, Textes Mathématiques Recherche 5, Cedic/Nathan, Paris (1982).
6. Clarke, F., *Optimization and Nonsmooth Analysis*, Wiley (1983), SIAM, Philadelphia (1990).
7. Clarke, F., A classical variational principle for periodic Hamiltonian trajectories, *Proc. Amer. Math. Soc.* 76, 186–188 (1979).
8. Clarke, F.H., Periodic solutions to Hamiltonian inclusions, *J. Diff. Equations* 40, 1–6 (1981).
9. Clarke, F.H., On Hamiltonian flows and symplectic transformations, *SIAM J. Control Optim.* 20, 355–359 (1982).
10. Clarke, F., and Ekeland, I., Hamiltonian trajectories having prescribed minimal period, *Commun. Pure Appl. Math.* 33, 103–116 (1980).
11. Collier, J.B., The dual of a space with the Radon-Nikodym property, *Pacific J. Math.* 64, 103–106 (1976).
12. Craven, B.D., *Mathematical Programming and Control Theory*, Chapman & Hall, London (1978).
13. Dorn, W.S., Duality in quadratic programming, *Quart. Appl. Math.* 18, 155–162 (1960).
14. Ekeland, I., Legendre duality in nonconvex optimization and calculus of variations, *SIAM J. Control Optim.* 15, No. 6, 905–934 (1977).
15. Ekeland, I., Nonconvex duality, *Bull. Soc. Math. France* Mémoire No. 60, Analyse Non Convexe, Pau, 1977, 45–55 (1979).
16. Ekeland, I., *Convexity Methods in Hamiltonian Mechanics*, Ergebnisse der Math. 19, Springer-Verlag, Berlin (1990).
17. Ekeland, I., and Hofer, H., Periodic solutions with prescribed minimal period for convex autonomous Hamiltonian systems, *Invent. Math.* 81, 155–188 (1985).
18. Ekeland, I., and Lasry, J.-M., Principes variationnels en dualité, *C.R. Acad. Sci. Paris* 291, 493–497 (1980).
19. Ekeland, I., and Lasry, J.-M., On the number of periodic trajectories for a Hamiltonian flow on a convex energy surface, *Ann. of Math. (2)* 112, 283–319 (1980).
20. Ekeland, I., and Lasry, J.-M., Duality in nonconvex variational problems, in *Advances in Hamiltonian Systems*, Aubin, Bensoussan, and Ekeland, eds., Birkhäuser, Basel (1983).
21. Frenk, J.B.G., and Schaible, S., Fractional programming, in *Handbook of Generalized Convexity and Generalized Monotonicity*, Hadjisavvas, N., Komlósi, S., and Schaible, S., eds., Nonconvex Optimization and Its Applications 76, Springer, New York, 335–386 (2005).
22. Gao, D.Y., Canonical dual transformation method and generalized triality theory in nonsmooth global optimization, *J. Global Optim.* 17, No. 1–4, 127–160 (2000).
23. Gao, D.Y., *Duality Principles in Nonconvex Systems: Theory, Methods and Applications*, Nonconvex Optimization and Its Applications 39, Kluwer, Dordrecht (2000).
24. Gao, D.Y., Complementarity, polarity and triality in non-smooth, non-convex and non-conservative Hamilton systems, *Phil. Trans. Roy. Soc. Lond. Ser. A Math. Phys. Eng. Sci.* 359, No. 1789, 2347–2367 (2001).

25. Gao, D.Y., Perfect duality theory and complete solutions to a class of global optimization problems, *Optimization* 52, No. 4–5, 467–493 (2003).
26. Gao, D.Y., *Complementary, Duality and Symmetry in Nonlinear Mechanics: Proceedings of the IUTAM Symposium, Shanghai, China, August 13–16, 2002*, Advances in Mechanics and Mathematics 6, Kluwer, Boston, MA (2004).
27. Gao, D.Y., Canonical duality theory and solutions to constrained nonconvex quadratic programming, *J. Global Optim.* 29, No. 3, 377–399 (2004).
28. Gao, D.Y., Ogden, R.W., and Stavroulakis, G., eds., *Nonsmooth/Nonconvex Mechanics: Modeling, Analysis and Numerical Methods*, Nonconvex Optimization and Its Applications 50, Kluwer, Boston (2001).
29. Gao, D.Y., and Teo, K.L., eds., Special issue: On duality theory, methods and applications, *J. Global Optim.* 29, No. 4, 335–516 (2004).
30. Ioffe, A.D., On the local surjection property, *Nonlinear Anal. Theory Meth. Appl.* 11, 565–592 (1987).
31. Ioffe, A.D., Approximate subdifferentials and applications, III: The metric theory, *Mathematika* 36, No. 1, 1–38 (1989).
32. Ioffe, A.D., Metric regularity and subdifferential calculus, *Russ. Math. Surv.* 55, No. 3, 501–558 (2000); translation from *Usp. Mat. Nauk* 55, No. 3, 103–162 (2000).
33. Michel, P., and Penot, J.-P., A generalized derivative for calm and stable functions, *Differential Integral Equat.* 5, No. 2, 433–454 (1992).
34. Mititelu, S., The Wolfe duality without convexity, *Stud. Cercet. Mat.* 38, 302–307 (1986).
35. Mititelu, S., Hanson's duality theorem in nonsmooth programming, *Optimization* 28, No. 3–4, 275–281 (1994).
36. Mittelu, S., Conditions de Kuhn-Tucker et dualité de Wolfe dans la programmation non lipschitzienne, Bull. Math. Soc. Sci. Math. Roum. Nouv. Sér. 37, No. 1–2, 65–74 (1993).
37. Penot, J.-P., Favorable classes of mappings and multimappings in nonlinear analysis and optimization, *J. Convex Anal.* 3, No. 1, 97–116 (1996).
38. Penot, J.-P., Subdifferential calculus without qualification assumptions, *J. Convex Anal.* 3, No. 2, 1–13 (1996).
39. Penot, J.-P., Mean-value theorem with small subdifferentials, *J. Optim. Theory Appl.* 94, No. 1, 209–221 (1997).
40. Penot, J.-P., Mean value theorems for mappings and correspondences, *Acta Math. Vietnamica* 26, No. 3, 365–376 (2002).
41. Penot, J.-P., Unilateral analysis and duality, in *Essays and Surveys in Global Optimization*, Savard, G., et al., eds., Springer, New York, 1–37 (2005).
42. Penot, J.-P., Legendre functions and the theory of characteristics, preprint, University of Pau (2004).
43. Penot, J.-P., The Legendre transform of correspondences, *Pacific J. Optim.* 1, No. 1, 161–177 (2005).
44. Penot, J.-P., Critical duality, *J. Global Optim.* 40, No. 1–3, 319–338 (2008).
45. Penot, J.-P., and Rubinov, A., Multipliers and general Lagrangians, *Optimization* 54, No. 4–5, 443–467 (2005).
46. Wolfe, P., A duality theorem for non-linear programming, *Quart. Appl. Math.* 19, 239–244 (1961).
47. C. Zălinescu, *Convex Analysis in General Vector Spaces*, World Scientific, Singapore (2002).

Chapter 11
Global Optimization in Practice: State of the Art and Perspectives

János D. Pintér

Summary. Global optimization—the theory and methods of finding the best possible solution in multiextremal models—has become a subject of interest in recent decades. Key theoretical results and basic algorithmic approaches have been followed by software implementations that are now used to handle a growing range of applications. This work discusses some practical aspects of global optimization. Within this framework, we highlight viable solution approaches, modeling environments, software implementations, numerical examples, and real-world applications.

Key words: Nonlinear systems analysis and management, global optimization strategies, modeling environments and global solver implementations, numerical examples, current applications and future perspectives

11.1 Introduction

Nonlinearity plays a fundamental role in the development of natural and man-made objects, formations, and processes. Consequently, nonlinear descriptive models are of key relevance across the range of quantitative scientific studies. For related discussions that illustrate this point consult, for instance, Bracken and McCormick (1968), Rich (1973), Mandelbrot (1983), Murray (1983), Casti (1990), Hansen and Jørgensen (1991), Schroeder (1991), Bazaraa et al. (1993), Stewart (1995), Grossmann (1996), Pardalos et al. (1996), Pintér (1996a, 2006a, 2009), Aris (1999), Bertsekas (1999), Corliss and Kearfott (1999), Floudas et al. (1999), Gershenfeld (1999), Papalambros and Wilde (2000), Chong and Zak (2001), Edgar et al. (2001), Gao et

János D. Pintér
Pintér Consulting Services, Inc., Canada, and Bilkent University, Turkey
E-mail: janos.d.pinter@gmail.com; Web site: www.pinterconsulting.com

D.Y. Gao, H.D. Sherali, (eds.), *Advances in Applied Mathematics and Global Optimization*
Advances in Mechanics and Mathematics 17, DOI 10.1007/978-0-387-75714-8_11,

al. (2001), Jacob (2001), Pardalos and Resende (2002), Schittkowski (2002), Tawarmalani and Sahinidis (2002), Wolfram (2002), Diwekar (2003), Stojanovic (2003), Zabinsky (2003), Bornemann et al. (2004), Fritzson (2004), Neumaier (2004), Bartholomew-Biggs (2005), Hillier and Lieberman (2005), Lopez (2005), Nowak (2005), Kampas and Pintér (2009), as well as many other topical works.

Decision support (control, management, or optimization) models that incorporate an underlying nonlinear system description frequently have multiple—local and global—optima. The objective of global optimization (GO) is to find the "absolutely best" solution of nonlinear optimization models under such circumstances.

We consider the general continuous global optimization (CGO) model defined by the following ingredients.

- x decision vector, an element of the real Euclidean n-space $\boldsymbol{R}^n$
- l, u explicit, finite n-vector bounds of x that define a "box" in $\boldsymbol{R}^n$
- $f(x)$ continuous objective function, $f\colon \boldsymbol{R}^n \to \boldsymbol{R}$
- $g(x)$ m-vector of continuous constraint functions, $g\colon \boldsymbol{R}^n \to \boldsymbol{R}^m$

Applying this notation, the CGO model is stated as

$$\min f(x) \tag{11.1}$$

$$x \in D := \{x : l \leq x \leq u \quad g(x) \leq 0\}. \tag{11.2}$$

In (11.2) all vector inequalities are interpreted componentwise (l, x, u, are n-component vectors and the zero denotes an m-component vector). The set of the additional constraints g could be empty, thereby leading to box-constrained GO models. Let us also note that formally more general optimization models that also include $=$ and $\geq$ constraint relations and/or explicit lower bounds on the constraint function values can be simply reduced to the model form (11.1) and (11.2).

The CGO model is very general: in fact, it evidently subsumes linear programming and convex nonlinear programming models, under corresponding additional specifications. Furthermore, CGO also subsumes (formally) the entire class of pure and mixed integer programming problems. To see this, notice that all bounded integer variables can be represented by a corresponding set of binary variables, and then every binary variable $y \in \{0, 1\}$ can be equivalently represented by its continuous extension $y \in [0, 1]$ and the non-convex constraint $y(1 - y) \leq 0$. Of course, this reformulation approach may not be best—or even suitable—for "all" mixed integer optimization models: however, it certainly shows the generality of the CGO model framework. Without going into details, note finally that models with multiple (partially conflicting) objectives are also often deduced to suitably parameterized collections of CGO (or simpler optimization) models: this remark also hints at the interchangeability of the objective f and one of the (active) constraints from g.

Let us observe next that if D is nonempty, then the above-stated basic analytical assumptions guarantee that the optimal solution set X^* in the CGO model is nonempty. This result directly follows by the classical theorem of Weierstrass that states the existence of the global minimizer point—or, in general, a set of such points—of a continuous function over a nonempty, bounded, and closed (compact) set.

For reasons of numerical tractability, the following additional requirements are also often postulated.

- D is a full-dimensional subset ("body") in $\boldsymbol{R}^n$.
- The set of globally optimal solutions to (11.1) and (11.2) is at most countable.
- f and g (componentwise) are Lipschitz-continuous functions on $[l, u]$.

Without going into technical details, notice that the first of these assumptions (the set D is the closure of its nonempty interior) makes algorithmic search easier (or at all possible) within D. The second assumption supports theoretical convergence results: note that in most well-posed practical GO problems the set of global optimizers consists either of a single point x^* or at most of several points. The third assumption is a sufficient condition for estimating $f^* = f(x^*)$ on the basis of a finite set of generated feasible search points. (Recall that the real-valued function h is Lipschitz-continuous on its domain of definition $D \subset \boldsymbol{R}^n$, if $|h(x_1) - h(x_2)| \leq L\|x_1 - x_2\|$ holds for all pairs $x_1 \in D$, $x_2 \in D$; here $L = L(D, h)$ is a suitable Lipschitz-constant of h on the set D.) We emphasize that the exact knowledge of the smallest suitable Lipschitz-constant for each model function is not required, and in practice such information is typically unavailable. At the same time, all models defined by continuously differentiable functions f and g belong to the CGO or even to the Lipschitz model-class.

The notes presented above imply that the CGO model-class covers a very broad range of optimization problems. As a consequence of this generality, it includes also many model instances that are difficult to solve numerically. For illustration, a merely one-dimensional, box-constrained GO model based on the formulation (11.3) is shown in Figure 11.1.

$$\min \cos(x) \sin(x^2 - x) \qquad 0 \leq x \leq 10. \tag{11.3}$$

Model complexity often increases dramatically (in fact, it can grow exponentially) as the model size expressed by the number of variables and constraints grows. To illustrate this point, Figure 11.2 shows the objective function in the model (11.4) that is simply generalized from (11.3) as

$$\min \cos(x) \sin(y^2 - x) + \cos(y) \sin(x^2 - y) \qquad 0 \leq x \leq 10,\ 0 \leq y \leq 10. \tag{11.4}$$

The presented two (low-dimensional, and only box-constrained) models already indicate that GO models—for instance, further extensions of model (11.3), perhaps with added complicated nonlinear constraints—could become

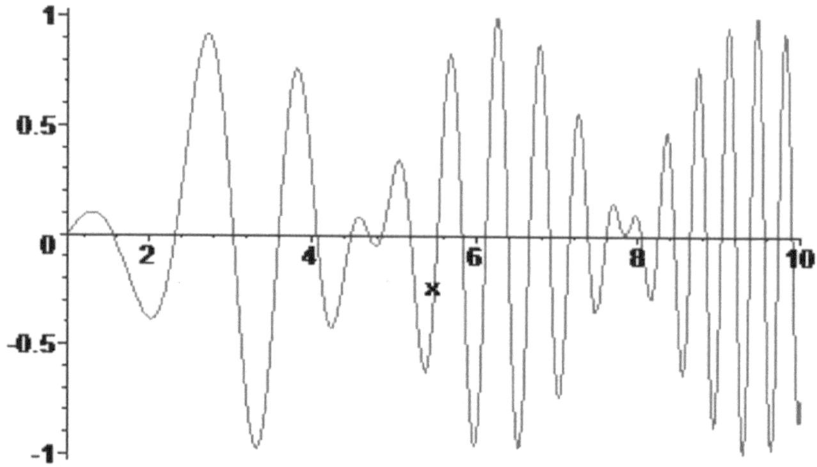

Fig. 11.1 The objective function in model (11.3).

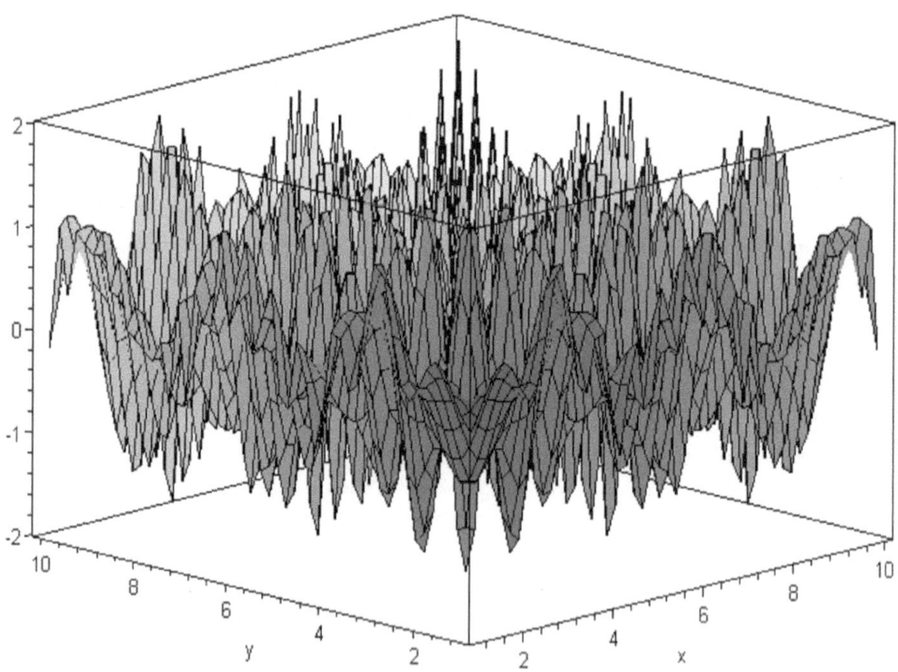

Fig. 11.2 The objective function in model (11.4).

truly difficult to handle numerically. One should also point out here that a direct analytical solution approach is viable only in very special cases, because in general (under further structural assumptions) one should investigate all Kuhn–Tucker points (minimizers, maximizers, and saddle points) of the CGO model. (Think of carrying out this analysis for the model depicted in Figure 11.2, or for its 100-dimensional extension.)

Arguably, not all GO models are as difficult as indicated by Figures 11.1 and 11.2. At the same time, we typically do not have the possibility to directly inspect, visualize, or estimate the overall numerical difficulty of a complicated nonlinear (global) optimization model. A practically important case is when one needs to optimize the parameters of a model that has been developed by someone else. The model may be confidential, or just visibly complex; it could even be presented to the optimization engine as a compiled (object, library, or similar) software module. In such situations, direct model inspection and structure verification are not possible. In other practically relevant cases, the evaluation of the optimization model functions may require the numerical solution of a system of embedded differential and/or algebraic equations, the evaluation of special functions, integrals, the execution of other deterministic computational procedures or stochastic simulation modules, and so on.

Traditional nonlinear optimization methods (discussed in most topical textbooks such as Bazaraa et al., 1993, Bertsekas, 1999, Chong and Zak, 2001, and Hillier and Lieberman, 2005) search only for local optima. This generally followed approach is based on the tacit assumption that a "sufficiently good" initial solution (that is located in the region of attraction of the "true" global solution) is available. Figures 11.1 and 11.2 and the practical situations mentioned above suggest that this may not always be a realistic assumption. Nonlinear models with less "dramatic" difficulty, but in (perhaps much) higher dimensions may also require global optimization. For instance, in advanced engineering design, optimization models with hundreds, thousands, or more variables and constraints are analyzed and need to be solved. In similar cases, even an approximately completed, but genuinely global scope search strategy may (and typically will) yield better results than the most sophisticated local search approach "started from the wrong valley". This fact has motivated research to develop practical GO strategies.

11.2 Global Optimization Strategies

As of today, well over a hundred textbooks and an increasing number of Web sites are devoted (partly or completely) to global optimization. Added to this massive amount of information is a very substantial body of literature on combinatorial optimization (CO), the latter being, at least in theory, a "subset of GO." The most important global optimization model types and (mostly exact, but also several prominent heuristic) solution approaches are discussed in detail by the *Handbook of Global Optimization* volumes, edited by Horst and Pardalos (1995), and by Pardalos and Romeijn (2002). We also refer to the topical Web site of Neumaier (2006), with numerous links to other useful information sources. The concise review of GO strategies presented here draws on these sources, as well as on the more detailed expositions in Pintér (2001a, 2002b). Let us point out that some of the methods listed below are more often used in solving CGO models, whereas others have been mostly applied so far to handle CO models. Because CGO formally includes CO, it should not be surprising that approaches suitable for certain specific CO model-classes can (or could) be put to good use to solve CGO models.

Instead of a more detailed (but still not unambiguous) classification, here we simply classify GO methods into two primary categories: exact and heuristic. Exact methods possess theoretically established (deterministic or stochastic) global convergence properties. That is, if such a method could be carried out completely as an infinite iterative process, then the generated limit point(s) would belong to the set of global solutions X^*. (For a single global solution x^*, this would be the only limit point.) In the case of stochastic GO methods, the above statement is valid "only" with probability one. In practice—after a finite number of algorithmic search steps—one can only expect a numerically validated or estimated (deterministic or stochastic) lower bound for the global optimum value $z^* = f(x^*)$, as well as a best feasible or near-feasible global solution estimate. We emphasize that to produce such estimates is not a trivial task, even for implementations of theoretically well-established algorithms. As a cautionary note, one can conjecture that there is no GO method, and never will be one, that can solve "all" CGO models with a certain number of variables to an arbitrarily given precision (in terms of the argument x^*), within a given time frame, or within a preset model function evaluation count. To support this statement, please recall Figures 11.1 and 11.2: both of the objective functions displayed could be made arbitrarily more difficult, simply by changing the frequencies and amplitudes of the embedded trigonometric terms. We do not attempt to display such "monster" functions, because even the best visualization software will soon become inadequate: think for instance of a function such as $1000\cos(1000x)\sin(1000(x^2 - x))$. For a more practically motivated example, one can also think of solving a difficult system of nonlinear equations: here, after a prefixed finite number of model function evaluations, we may not have an "acceptable" approximate numerical solution.

Heuristic methods do not possess similar convergence guarantees to those of exact methods. At the same time, they may provide good quality solutions in many difficult GO problems, assuming that the method in question suits the specific model type (structure) solved. Here a different cautionary note is in order. Because such methods are often based on some generic metaheuristics, overly optimistic claims regarding the "universal" efficiency of their implementations are often not supported by results in solving truly difficult, especially nonlinearly constrained, GO models. In addition, heuristic metastrategies are often more difficult to adjust to new model types than some of the solver implementations based on exact algorithms. Exact stochastic methods based on direct sampling are a good example for the latter category, because these can be applied to "all" GO models directly, without a need for essential code adjustments and tuning. This is in contrast, for example, to most population-based search methods in which the actual steps of generating new trial solutions may depend significantly on the structure of the model-instance solved.

11.2.1 Exact Methods

- "Naïve" approaches (grid search, pure random search): these are obviously convergent, but in general "hopeless" as the problem size grows.
- Branch-and-bound methods: these include interval-arithmetic-based strategies, as well as customized approaches for Lipschitz global optimization and for certain classes of difference of convex functions (D.C.) models. Such methods can also be applied to constraint satisfaction problems and to (general) pure and mixed integer programming.
- Homotopy (path following, deformation, continuation, trajectory, and related other) methods: these are aimed at finding the set of global solutions in smooth GO models.
- Implicit enumeration techniques: examples are vertex enumeration in concave minimization models, and generic dynamic programming in the context of combinatorial optimization.
- Stochastically convergent sequential sampling methods: these include adaptive random searches, single- and multistart methods, Bayesian search strategies, and their combinations.

For detailed expositions related to deterministic GO techniques in addition to the *Handbooks* mentioned earlier, consult, for example, Horst and Tuy (1996), Kearfott (1996), Pintér (1996a), Tawarmalani and Sahinidis (2002), Neumaier (2004), and Nowak (2005). On stochastic GO strategies, consult, for example, Zhigljavsky (1991), Boender and Romeijn (1995), Pintér (1996a), and Zabinsky (2003).

11.2.2 Heuristic Methods

- Ant colony optimization is based on individual search steps and "ant-like" interaction (communication) between search agents.
- Basin-hopping strategies are based on a sequence of perturbed local searches, in an effort to find improving optima.

- Convex underestimation attempts are based on a limited sampling effort that is used to estimate a postulated (approximate) convex objective function model.
- Evolutionary search methods model the behavioral linkage among the adaptively changing set of candidate solutions ("parents" and their "children," in a sequence of "generations").
- Genetic algorithms emulate specific genetic operations (selection, crossover, mutation) as these are observed in nature, similarly to evolutionary methods.
- Greedy adaptive search strategies (a metaheuristics often used in combinatorial optimization) construct "quick and promising" initial solutions which are then refined by a suitable local optimization procedure.
- Memetic algorithms are inspired by analogies to cultural (as opposed to natural) evolution.
- Neural networks are based on a model of the parallel architecture of the brain.
- Response surface methods (directed sampling techniques) are often used in handling expensive "black box" optimization models by postulating and then gradually adapting a surrogate function model.
- Scatter search is similar in its algorithmic structure to ant colony, genetic, and evolutionary searches, but without their "biological inspiration."
- Simulated annealing methods are based on the analogy of cooling crystal structures that will attain a (low-energy level, stable) physical equilibrium state.
- Tabu search forbids or penalizes search moves which take the solution in the next few iterations to points in the solution space that have been previously visited. (Tabu search as outlined here has been typically applied in the context of combinatorial optimization.)
- Tunneling strategies, filled function methods, and other similar methods attempt to sequentially find an improving sequence of local optima, by gradually modifying the objective function to escape from the solutions found.

In addition to the earlier mentioned topical GO books, we refer here to several works that discuss mostly combinatorial (but also some continuous) global optimization models and heuristic strategies. For detailed discussions of theory and applications, consult, for example, Michalewicz (1996), Osman and Kelly (1996), Glover and Laguna (1997), Voss et al. (1999), Jacob (2001), Ferreira (2002), Rothlauf (2002), and Jones and Pevzner (2004). It is worth pointing out that Rudolph (1997) discusses the typically missing theoretical foundations for evolutionary algorithms, including stochastic convergence studies. (The underlying key convergence results for adaptive stochastic search methods are discussed also in Pintér (1996a).) The topical chapters in Pardalos and Resende (2002) also offer expositions related to both exact and heuristic GO approaches.

To conclude this very concise review, let us emphasize again that numerical GO can be tremendously difficult. Therefore it can be good practice to try several—perhaps even radically different—search approaches to tackle GO models, whenever this is possible. To do this, one needs ready-to-use model development and optimization software tools.

11.3 Nonlinear Optimization in Modeling Environments

Advances in modeling techniques, solver engine implementations and computer technology have led to a rapidly growing interest in modeling environments. For detailed discussions consult, for example, the topical *Annals of Operations Research* volumes edited by Maros and Mitra (1995), Maros et al. (1997), Vladimirou et al. (2000), Coullard et al. (2001), as well as the volumes edited by Voss and Woodruff (2002) and by Kallrath (2004). Additional useful information is provided by the Web sites of Fourer (2006), Mittelmann (2006), and Neumaier (2006), with numerous further links. Prominent examples of widely used modeling systems that are focused on optimization include AIMMS (Paragon Decision Technology , 2006), AMPL (Fourer et al., 1993), the Excel Premium Solver Platform (Frontline Systems , 2006), GAMS (Brooke et al., 1988), ILOG (2004), the LINDO Solver Suite (LINDO Systems, 2006), MPL (Maximal Software, 2006), and TOMLAB (2006). (Please note that the literature references cited may not always reflect the current status of the modeling systems discussed here: for the latest information, contact the developers and/or visit their Web sites.)

There also exist a large variety of core compiler platform-based solver systems with more or less built-in model development functionality: in principle, such solvers can be linked to the modeling languages listed above.

At the other end of the spectrum, there is also notable development in relation to integrated scientific and technical computing (ISTC) systems such as Maple (Maplesoft, 2006), *Mathematica* (Wolfram Research, 2006), Mathcad (Mathsoft, 2006), and MATLAB (The MathWorks, 2006). From among the many hundreds of books discussing ISTC systems, we mention here as examples the works of Birkeland (1997), Bhatti (2000), Parlar (2000), Wright (2002), Wilson et al. (2003), Moler (2004), Wolfram (2003), Trott (2004), and Lopez (2005). The ISTC systems offer a growing range of optimization-related features, either as built-in functions or as add-on products.

The modeling environments listed above are aimed at meeting the needs of different types of users. User categories include educational (instructors and students); research scientists, engineers, consultants, and other practitioners (possibly, but not necessarily equipped with an in-depth optimization-related background); optimization experts, software application developers, and other "power users." (Observe that the user categories listed are not necessarily disjoint.) The pros and cons of the individual software products—in terms of their hardware and software demands, ease of usage, model prototyping options, detailed code development and maintenance features, optimization model checking and processing tools, availability of solver options and other auxiliary tools, program execution speed, overall level of system integration, quality of related documentation and support, customization options, and communication with end users—make the corresponding modeling and solver approaches more or less attractive for the various user groups.

Given the almost overwhelming amount of topical information, in short, which are the currently available platform and solver engine choices for the GO researcher or practitioner? The more than a decade-old software review (Pintér, 1996b; also available at the Web site of Mittelmann, 2006) listed a few dozen individual software products, including several Web sites with further software collections. Neumaier's (2006) Web page currently lists more than 100 software development projects. Both of these Web sites include general-purpose solvers, as well as application-specific products. (It is noted that quite a few of the links in these software listings are now obsolete, or have been changed.)

The user's preference obviously depends on many factors. A key question is whether one prefers to use "free" (noncommercial, research, or even open source) code, or looks for a "ready-to-use" professionally supported commercial product. There is a significant body of freely available solvers, although the quality of solvers and their documentation arguably varies. (Of course, this remark could well apply also to commercial products.)

Instead of trying to impose personal judgment on any of the products mentioned in this work, the reader is encouraged to do some Web browsing and experimentation, as his or her time and resources allow. Both Mittelmann (2006) and Neumaier (2006) provide more extensive information on noncommercial, as opposed to commercial, systems. Here we mention several software products that are part of commercial systems, typically as an add-on option, but in some cases as a built-in option. Needless to say, although this author (being also a professional software developer) may have opinions, the alphabetical listing presented below is strictly matter-of-fact. We list only currently available products that are explicitly targeted towards global optimization, as advertised by the Web sites of the listed companies. For this reason, nonlinear (local) solvers are, as a rule, not listed here; furthermore, we do not list modeling environments that currently have no global solver options.

AIMMS, by Paragon Decision Technology (www.aimms.com). The BARON and LGO global solver engines are offered with this modeling system as add-on options.

Excel Premium Solver Platform (PSP), by Frontline Systems (www.solver .com): The developers of the PSP offer a global presolver option to be used with several of their local optimization engines: these currently include LSGRG, LSSQP, and KNITRO. Frontline Systems also offers (as genuine global solvers) an Interval Global Solver, an Evolutionary Solver, and OptQuest.

GAMS, by the GAMS Development Corporation (www.gams.com). Currently, BARON, DICOPT, LGO, MSNLP, OQNLP, and SBB are offered as solver options for global optimization.

LINDO, by LINDO Systems (www.lindo.com). Both the LINGO modeling environment and What'sBest! (the company's spreadsheet solver) have built-in global solver functionality.

Maple, by Maplesoft (www.maplesoft.com) offers the Global Optimization Toolbox as an add-on product.

Mathematica, by Wolfram Research (www.wolfram.com) has a built-in function (called NMinimize) for numerical global optimization. In addition, there are several third-party GO packages that can be directly linked to Mathematica: these are Global Optimization, MathOptimizer, and MathOptimizer Professional.

MPL, by Maximal Software (www.maximal-usa.com). The LGO solver engine is offered as an add-on.

TOMLAB, by TOMLAB Optimization AB (www.tomopt.com) is an optimization platform for solving MATLAB models. The TOMLAB global solvers include CGO, LGO, MINLP, and OQNLP. Note that MATLAB's own Genetic Algorithm and Direct Search Toolboxes also have heuristic global solver capabilities.

To illustrate the functionality and usage of global optimization software, next we review the key features of the LGO solver engine, and then apply its Maple platform-specific implementation in several numerical examples.

11.4 The LGO Solver Suite and Its Implementations

11.4.1 LGO: Key Features

The Lipschitz Global Optimizer (LGO) solver suite has been developed and used for more than a decade. The top-level design of LGO is based on the seamless combination of theoretically convergent global and efficient local search strategies. Currently, LGO offers the following solver options.

- Adaptive partition and search (branch-and-bound) based global search (BB)
- Adaptive global random search (single-start) (GARS)
- Adaptive global random search (multistart) (MS)
- Constrained local search by the generalized reduced gradient (GRG) method (LS).

In a typical LGO optimization run, the user selects one of the global (BB, GARS, MS) solver options; this search phase is then automatically followed by the LS option. It is also possible to apply only the LS solver option, making use of an automatically set (default) or a user-supplied initial solution.

The global search methodology implemented in LGO is based on the detailed exposition in Pintér (1996a), with many added numerical features. The well-known GRG method is discussed in numerous articles and textbooks; consult for instance Edgar et al. (2001). Therefore only a very brief overview of the LGO component algorithms is provided here.

BB, GARS, and MS are all based on globally convergent search methods. Specifically, in Lipschitz-continuous models with suitable Lipschitz-constant (over)estimates for all model functions BB theoretically generates a sequence of search points that will converge to the global solution point. If there is a countable set of such optimal points, then a convergent search point sequence will be generated in association with each of these.

In a GO model with a continuous structure (but without postulating access to Lipschitz information), both GARS and MS are globally convergent, with probability one (w.p. 1). In other words, the sequence of points that is associated with the generated sequence of global optimum estimates will converge to a point which belongs to X^*, with probability one. (Again, if several such convergent point sequences are generated by the stochastic search procedure, then each of these sequences has a corresponding limit point in X^*, w.p. 1.)

The LS method (GRG) is aimed at finding a locally optimal solution that satisfies the Karush–Kuhn–Tucker system of necessary local optimality conditions, assuming standard model smoothness and regularity conditions.

In all three global search modes the model functions are aggregated by an exact penalty (merit) function. By contrast, in the local search phase all model functions are considered and handled individually. The global search phases incorporate both deterministic and stochastic sampling procedures: the latter support the usage of statistical bound estimation methods, under basic continuity assumptions. All LGO component algorithms are derivative-free. In the global search phase, BB, GARS, and MS use only direct sampling information based on generated points and corresponding model function values. In the LS phase central differences are used to approximate function gradients (under a postulated locally smooth model structure). This direct search approach reflects our objective to handle also models defined by merely computable, continuous functions, including completely "black box" systems.

In numerical practice—with finite runs, and user-defined or default option settings—the LGO global solver options generate a global solution estimate that is subsequently refined by the local search mode. If the LS mode is used without a preceding global search phase, then LGO serves as a general-purpose local solver engine. The expected practical outcome of using LGO to solve a model (barring numerical problems which could impede any numerical method) is a global-search-based feasible solution that meets at least the local optimality conditions. Extensive numerical tests and a range of practical applications demonstrate that LGO can locate the global solution not only in the usual academic test problems, but also in more complicated, sizeable GO models: this point is illustrated later on in Sections 11.5 and 11.6. (At the same time, keep in mind the caveats mentioned earlier regarding the performance of any global solver: nothing will "always" work satisfactorily, under resource limitations.)

11.4.2 LGO Implementations

The current platform-specific implementations include the following.

- LGO with a text input/output interface, for C and FORTRAN compiler platforms
- LGO integrated development environment with a Microsoft Windows style menu interface, for C and FORTRAN compiler platforms
- AIMMS /LGO solver engine
- AMPL /LGO solver engine
- GAMS /LGO solver engine
- Global Optimization Toolbox for Maple (the LGO solver linked to Maple as a callable add-on package)
- MathOptimizer Professional, with an LGO solver engine link to Mathematica
- MPL /LGO solver engine
- TOMLAB /LGO, for MATLAB users

Technical descriptions of these software implementations, including detailed numerical tests and a range of applications, have appeared elsewhere. For implementation details and illustrative results, consult Pintér (1996a, 1997, 2001a,b, 2002a,b, 2003b, 2005), as well as Pintér and Kampas (2003) and Pintér et al. (2004, 2006).

The compiler-based LGO solver suite can be used in standalone mode, and also as a solver option in various modeling environments. In its core (text input/output based) implementation version, LGO reads an input text file that contains application-specific (model descriptor) information, as well as a few key solver options (global solver type, precision settings, resource and time limits). During the program run, LGO makes calls to an application-specific model function file that returns function values for the algorithmically chosen sequence of arguments. Upon completing the LGO run, automatically generated summary and detailed report files are available. As can be expected, this LGO version has the lowest demands for hardware; it also runs fastest, and it can be directly embedded into various decision support systems, including proprietary user applications. The same core LGO system is also available in directly callable form, without reading and writing text file: this version is frequently used as a built-in solver module in other (general-purpose or customized modeling) systems.

LGO can also be equipped, as a readily available (implemented) option, with a Microsoft Windows style menu interface. This enhanced version is referred to as the LGO Integrated Development Environment (IDE). The LGO IDE supports model development, compilation, linking, execution, and the inspection of results, together with built-in basic help facilities.

In the two LGO implementations mentioned above, models can be connected to LGO using one of several programming languages that are available on personal computers and workstations. Currently supported platforms include, in principle, "all" professional FORTRAN 77/90/95 and C/C++ compilers. Examples of supported compilers include Compag, Intel, Lahey, and Salford FORTRAN, as well as g77 and g95, and Borland and Microsoft C/C++. Other customized versions (to use with other compilers or software applications) can also be made available upon request.

In the optimization modeling language (AIMMS, AMPL, GAMS, and MPL) or ISTC (Maple, Mathematica, and TOMLAB) environments the core LGO solver engine is seamlessly linked to the corresponding modeling platform, as a dynamically callable or shared library, or as an executable program. The key advantage of using LGO within a modeling or ISTC environment is the combination of modeling-system-specific features, such as model prototyping and detailed development, model consistency checking, integrated documentation, visualization, and other platform-specific features, with a numerical performance comparable to that of the standalone LGO solver suite.

For peer reviews of several of the listed implementations, the reader is referred to Benson and Sun (2000) on the core LGO solver suite, Cogan (2003) on MathOptimizer Professional, and Castillo (2005), Henrion (2006), and Wass (2006) on the Global Optimization Toolbox for Maple. Let us also mention here that LGO serves to illustrate global optimization software (in connection with a demo version of the MPL modeling system) in the prominent O.R. textbook by Hillier and Lieberman (2005).

11.5 Illustrative Examples

In order to present some small-scale, yet nontrivial numerical examples, in this section we illustrate the functionality of the LGO software as it is implemented in the Global Optimization Toolbox (GOT) for Maple.

Maple (Maplesoft, 2006) enables the development of interactive documents called worksheets. Maple worksheets can incorporate technical model description, combined with computing, programming, and visualization features. Maple includes several thousands of built-in (directly callable) functions to support the modeling and computational needs of scientists and engineers. Maple also offers a detailed online help and documentation system with ready-to-use examples, topical tutorials, manuals, and Web links, as well as a built-in mathematical dictionary. Application development is assisted by debugging tools, and automated (ANSI C, FORTRAN 77, Java, Visual Basic, and MATLAB) code generation. Document production features include HTML, MathML, TeX, and RTF converters. These capabilities accelerate and expand the scope of the optimization model development and solution process. Maple, similarly to other modeling environments, is portable across all major hardware platforms and operating systems (including Windows, Macintosh, Linux, and UNIX versions).

Without going into further details on Maple itself, we refer to the Web site www.maplesoft.com that offers in-depth topical information, including product demos and downloadable technical materials.

The core of the Global Optimization Toolbox for Maple is a customized implementation of the LGO solver suite (Maplesoft, 2004) that, as an add-on product, upon installation, can be fully integrated with Maple. The advantage of this approach is that, in principle, the GOT can readily handle "all" continuous model functions that can be defined in Maple, including also new (user-defined) functions.

We do not wish to go into programming details here, and assume that the key ideas shown by the illustrative Maple code snippets are easily understandable to all readers with some programming experience. Maple commands are typeset in **`Courier bold`** font, following the so-called classic Maple input format. The input commands are typically followed by Maple output lines, unless the latter are suppressed by using the symbol ":" instead of ";" at the end of an input line.

In the numerical experiments described below, an AMD Athlon 64 (3200+, 2GHz) processor-based desktop computer has been used that runs under Windows XP Professional (Version 2002, Service Pack 2).

11.5.1 Getting Started with the Global Optimization Toolbox

To illustrate the basic usage of the Toolbox, let us revisit model (11.3). The Maple command

```
> with(GlobalOptimization);
```

makes possible the direct invocation of the subsequently issued, GOT related, commands. Then the next Maple command numerically solves model (11.3): the response line below the command displays the approximate optimum value, and the corresponding solution argument.

```
> GlobalSolve(cos(x)*sin(x^2-x), x=1..10);
```

[-.990613849411236758, [x = 9.28788130421885682]]

The detailed runtime information not shown here indicates that the total number of function evaluations is 1262; the associated runtime is a small fraction of a second.

Recall here Figure 11.1 which, after careful inspection, indicates that this is indeed the (approximate) global solution. (One can also see that the default visualization—similarly to other modeling environments—has some difficulties to depict this rapidly changing function.) There are several local solutions that are fairly close to the global one: two of these numerical solutions are

[-.979663995439954860, [x = 3.34051270473064265]],

and

[-.969554320487729716, [x = 6.52971402762202757]].

Similarly, the next statement returns an approximate global solution in the visibly nontrivial model (11.4):

```
> GlobalSolve(cos(x)*sin(y^2-x)+cos(y)*sin(x^2-y),
x=1..10, y=1..10);
```

[-1.95734692335253380,
[x = 3.27384194476651214, y = 6.02334184076140478]].

The result shown above has been obtained using GOT default settings: the total number of function evaluations in this case is 2587, and the runtime is still practically zero. Recall now also Figure 11.2 and the discussion related to the possibly numerical difficulty of GO models. The solution found by the GOT is global-search-based, but without a rigorous deterministic guarantee of its quality. Let us emphasize that to obtain such guarantees (e.g., by using interval-arithmetic-based solution techniques) can be a very resource-demanding exercise, especially in more complex and/or higher-dimensional models, and that it may not be possible, for example, in "black box" situations. A straightforward way to attempt finding a better quality solution is to increase the allocated global search effort. Theoretically, using an "infinite" global search effort will lead to an arbitrarily close numerical estimate of the global optimum value. In the next statement we set the global search effort to 1000000 steps (this limit is applied only approximately, due to the possible activation of other stopping criteria):

```
> GlobalSolve(cos(x)*sin(y^2-x)+cos(y)*sin(x^2-y),
x=1..10, y=1..10, evaluationlimit=1000000,
noimprovementlimit=1000000);
```

[–1.98122769882222882,
[x = 9.28788128193757068, y = 9.28788127177065270]].

Evidently, we have found an improved solution, at the expense of a significantly increased global search effort. (Now the total number of function evaluations is 942439, and the runtime is approximately 5 seconds.) In general, more search effort can always be added, in order to verify or perhaps improve the incumbent numerical solution.

Comparing now the solution obtained to that of model (11.3), and observing the obvious formal connection between the two models, one can deduce that now we have found a close numerical approximation of the true global solution. Simple modeling insight also tells us that the global solution in model (11.4) is bounded from below by –2. Hence, even without Figures 11.1 and 11.2 we would know that the solution estimates produced above must be fairly close to the best possible solution.

The presented examples illustrate several important points.

- Global optimization models can be truly difficult to solve numerically, even in (very) low dimensions.
- It is not always possible to "guess" the level of difficulty. One cannot always (or at all) generate model visualizations similar to Figures 11.1 and 11.2, even in chosen variable subspaces, because it could be too expensive numerically, even if we have access to suitable graphics facilities. Insight and model-specific expertise can help significantly, and these should be used whenever possible.
- There is no solver that will handle all possible instances from the general CGO model class within an arbitrary prefixed amount of search effort. In practice, one needs to select and recommend default solver parameters and options that "work well in most cases, based on an acceptable amount of effort." Considering the fact that practically motivated modeling studies are often supported only by noisy and/or scarce data, this pragmatic approach is justifiable in many practical situations.
- The default solver settings should return a global-search-based high-quality feasible solution (arguably, the models (11.3) and (11.4) can be considered as difficult instances for their low dimensionality). Furthermore, it should be easy to modify the default solver settings and to repeat runs, if this is deemed necessary.

The GOT software implementation automatically sets default parameter values for its operations, partly based on the model to solve. These settings are suitable in most cases, but the user can always assign (i.e., override) them. Specifically, one can select the following options and parameter values.

- Minimization or maximization model
- Search method (BB+LS, GARS+LS, MS+LS, or standalone LS)

- Initial solution vector setting (used by the LS operational mode), if available
- Constraint penalty multiplier: this is used by BB, GARS, and MS, in an aggregated merit function (recall that the LS method handles all model functions individually)
- Maximal number of merit function evaluations in the selected global search mode
- Maximal number of merit function evaluations in the global search mode, without merit function value improvement
- Acceptable target value for the merit function, to trigger an "operational switch" from global to local search mode
- Feasibility tolerance used in LS mode
- Karush–Kuhn–Tucker local optimality tolerance in LS mode
- Solution (computation) time limit

For further information regarding the GOT, consult the product Web page (Maplesoft, 2004), the article (Pintér et al., 2006), and the related Maple help system entries. The product page also includes links to detailed interactive demos, as well as to downloadable application examples.

11.5.2 Handling (General) Constrained Global Optimization Models

Systems of nonlinear equations play a fundamental role in quantitative studies, because equations are often used to characterize the equilibrium states and optimality conditions of physical, chemical, biological, or other systems. In the next example we formulate and solve a system of equations. At the same time, we also illustrate the use of a general model development style that is easy to follow in Maple, and—mutatis mutandis—also in other modeling systems. Consider the equations

```
> eq1 := exp(x-y)+sin(2*x)-cos(y+z)=0:                    (11.5)
  eq2 := 4*x-exp(z-y)+5*sin(6*x-y)+3*cos(3*x*y)=0:
  eq3 := x*y*z-10=0:
```

To solve this system of equations, let us define the optimization model components as shown below (notice the dummy objective function).

```
> constraints := eq1,eq2,eq3:
> bounds := x=-2..2, y=-1..3, z=2..4:
> objective:=0:
```

Then the next Maple command is aimed at generating a numerical solution to (11.5), if such solution exists.

```
> solution:=
GlobalSolve(objective, constraints, bounds);
```

solution:=[0.,
[x=1.32345978290539557,y=2.78220763578413344,z=2.71581206431678090]].

This solution satisfies all three equations with less than 10^{-9} error, as verified by the next statement:

```
> eval(constraints, solution[2]);
```

$$\{-0.1 \cdot 10^{-9} = 0,\ -0.6 \cdot 10^{-9} = 0,\ 0 = 0\}$$

Without going into details, let us note that multiple solutions to (11.5) can be found (if such solutions exist), for example, by iteratively adding constraints that will exclude the solution(s) found previously. Furthermore, if a system of equations has no solutions, then using the GOT we can obtain an approximate solution that has globally minimal error over the box search region, in a given norm: consult Pintér (1996a) for details.

Next, we illustrate the usage of the GOT in interactive mode. The statement shown below directly leads to the Global Optimization Assistant dialog, see Figure 11.3.

```
> solution:=
Interactive(objective, constraints, bounds);
```

Using the dialog, one can also directly edit (modify) the model formulation if necessary. The figure shows that the default (MS+LS) GOT solver mode returns the solution presented above. Let us point out here that none of the local solver options indicated in the Global Optimization Assistant (see the radio buttons under Solver) is able to find a feasible solution to this model. This finding is not unexpected: rather, it shows the need for a global scope search approach to handle this model and many other similar problems.

Following the numerical solution step, one can press the Plot button (shown in the lower right corner in Figure 11.3). This will invoke the Global Optimization Plotter dialog shown in Figure 11.4. In the given subspace (x, y) that can be selected by the GOT user, the surface plot shows the identically zero objective function. Furthermore, on its surface level one can see the constraint curves and the location of the global solution found: in the original color figure this is a light green dot close to the boundary as indicated by the numerical values found above. Notice also the option to select alternative subspaces (defined by variable pairs) for visualization.

The figures can be rotated, thereby offering the possibility of detailed model function inspection. Such inspection can help users to increase their understanding of the model.

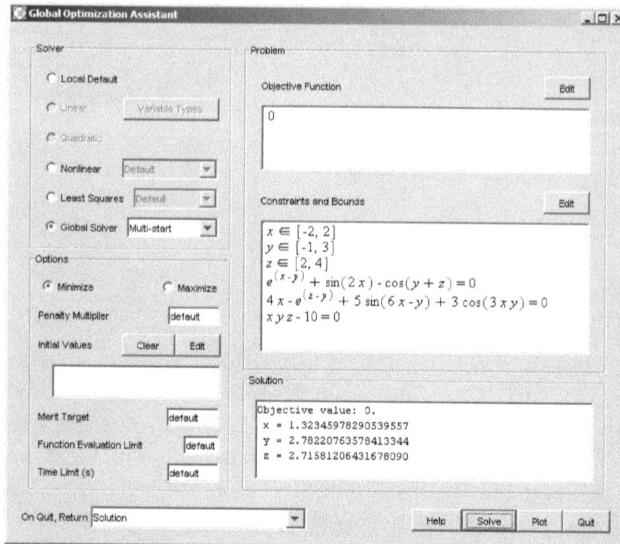

Fig. 11.3 Global Optimization Assistant dialog for model (11.5).

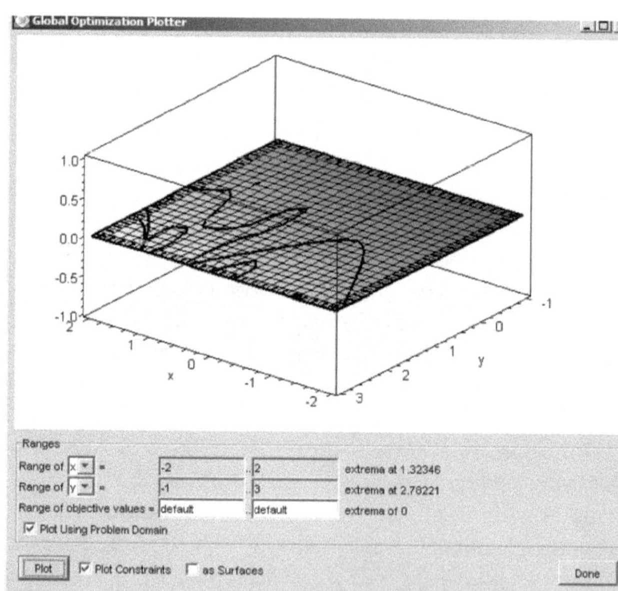

Fig. 11.4 Global Optimization Plotter dialog for model (11.5).

11.5.3 Optimization Models with Embedded Computable Functions

It was pointed out earlier (in Section 11.1) that in advanced decision models some model functions may require the execution of various computational procedures. One of the advantages of using an ISTC system such as Maple is that the needed functionality to perform these operations is often readily available, or directly programmable. To illustrate this point, in the next example we show the globally optimized argument value of an objective function defined by Bessel functions. As it is known, the function BesselJ(ν, x) satisfies Bessel's differential equation

$$x^2y'' + xy' + (x^2 - \nu^2)y = 0. \tag{11.6}$$

In (11.6) x is the function argument, and the real value ν is the order (or index parameter) of the function. The evaluation of BesselJ requires the solution function of the differential equation (11.6), for the given value of ν, and then the calculation of the corresponding function value for argument x. For example, BesselJ(0, 2)~0.2238907791; consult Maple's help system for further details. Consider now the optimization model defined and solved below:

```
> objective:=BesselJ(2,x)*BesselJ(3,y)-                        (11.7)
BesselJ(5,y)*BesselJ(7,x):
> bounds := x=-10..20, y=-15..10:
> solution:=GlobalSolve(objective, bounds);
```

solution := [–.211783151218360000,
[x = –3.06210564091438720, y = –4.20467390983796196]].

The corresponding external solver runtime is about 4 seconds. The next figure visualizes the box-constrained optimization model (11.7). Here a simple inspection and rotation of Figure 11.5 helps to verify that the global solution is found indeed. Of course, this would not be directly possible in general (higher-dimensional or more complicated) models: recall the related earlier discussion and recommendations from Section 11.5.1.

11.6 Global Optimization: Applications and Perspectives

In recent decades, global optimization gradually has become an established discipline that is now taught worldwide at leading academic institutions. GO methods and software are also increasingly applied in various research contexts, including industrial and consulting practice. The currently available

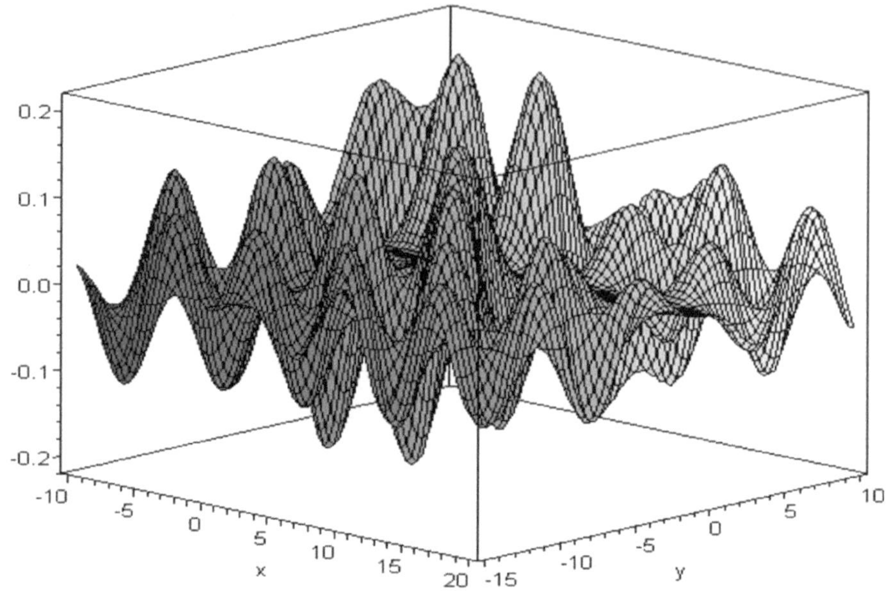

Fig. 11.5 Optimization model objective defined by Bessel functions.

professional software implementations are routinely used to solve models with tens, hundreds, and sometimes even thousands of variables and constraints. Recall again the caveats mentioned earlier regarding the potential numerical difficulty of model instances: if one is interested in a guaranteed high-quality solution, then the necessary runtimes could become hours (or days, or more), even on today's high-performance computers. One can expect further speed-up due to both algorithmic improvements and progress in hardware/software technology, but the theoretically exponential "curse of dimensionality" associated with the subject of GO will always be there.

In the most general terms, global optimization technology is well suited to analyze and solve models in advanced (acoustic, aerospace, chemical, control, electrical, environmental, and other) engineering, biotechnology, econometrics and financial modeling, medical and pharmaceutical studies, process industries, telecommunications, and other areas.

For detailed discussions of examples and case studies consult, for example, Grossmann (1996), Pardalos et al. (1996), Pintér (1996a), Corliss and Kearfott (1999), Papalambros and Wilde (2000), Edgar et al. (2001), Gao et al. (2001), Schittkowski (2002), Tawarmalani and Sahinidis (2002), Zabinsky (2003), Neumaier (2006), Nowak (2005), and Pintér (2006a), as well as other topical works.

For example, recent numerical studies and applications in which LGO implementations have been used are described in the following works:

- Cancer therapy planning (Tervo et al., 2003)

- Combined finite element modeling and optimization in sonar equipment design (Pintér and Purcell, 2003)
- Laser equipment design (Isenor et al., 2003)
- Model calibration (Pintér, 2003a, 2006b)
- Numerical performance analysis on a collection of test and "real-world" models (Pintér, 2003b, 2006b)
- Physical object configuration analysis and design (Kampas and Pintér, 2006)
- Potential energy models in computational chemistry (Pintér, 2000, 2001b, Stortelder et al., 2001)
- Circle packing models and their industrial applications (Kampas and Pintér, 2004, Pintér and Kampas, 2005a,b, Castillo et al., 2008)

The forthcoming volumes by Kampas and Pintér (2009) and Pintér (2009) also discuss a large variety of GO applications, with extensive references.

11.7 Conclusions

Global optimization is a subject of growing practical interest as indicated by recent software implementations and by an increasing range of applications. In this work we have discussed some of these developments, with an emphasis on practical aspects.

In spite of remarkable progress, global optimization remains a field of extreme numerical challenges, not only when considering "all possible" GO models, but also in practical attempts to handle complex and sizeable problems within an acceptable timeframe. The present discussion advocates a practical solution approach that combines theoretically rigorous global search strategies with efficient local search methodology, in integrated, flexible solver suites. The illustrative examples presented here, as well as the applications referred to above, indicate the practical viability of such an approach.

The practice of global optimization is expected to grow dynamically. We welcome feedback regarding current and future development directions, new test challenges, and new application areas.

Acknowledgments First of all, I wish to thank David Gao and Hanif Sherali for their kind invitation to the CDGO 2005 conference (Blacksburg, VA, August 2005), as well as for the invitation to contribute to the present volume dedicated to Gilbert Strang on the occasion of his 70th birthday. Thanks are due to an anonymous reviewer for his/her careful reading of the manuscript, and for the suggested corrections and modifications.

I also wish to thank my past and present developer partners and colleagues—including AMPL LLC, Frontline Systems, the GAMS Development Corporation, Frank Kampas, Lahey Computer Systems, LINDO Systems, Maplesoft, Mathsoft, Maximal Software, Paragon Decision Technology, The Mathworks, TOMLAB AB, and Wolfram Research—for cooperation, quality software and related documentation, and technical support.

In addition to professional contributions and in-kind support offered by developer partners, the research work summarized and reviewed in this chapter has received partial financial support in recent years from the following organizations: DRDC Atlantic Region, Canada (Contract W7707-01-0746), the Dutch Technology Foundation (STW Grant CWI55.3638), the Hungarian Scientific Research Fund (OTKA Grant T 034350), Maplesoft, the National Research Council of Canada (NRC IRAP Project 362093), the University of Kuopio, and Wolfram Research.

Special thanks are due to our growing clientele, and to all reviewers and testers of our various software implementations, for valuable feedback, comments, and suggestions.

References

Aris, R. (1999) *Mathematical Modeling: A Chemical Engineer's Perspective.* Academic Press, San Diego, CA.

Bartholomew-Biggs, M. (2005) *Nonlinear Optimization with Financial Applications.* Kluwer Academic, Dordrecht.

Bazaraa, M.S., Sherali, H.D., and Shetty, C.M. (1993) *Nonlinear Programming: Theory and Algorithms.* Wiley, New York.

Benson, H.P., and Sun, E. (2000) LGO — Versatile tool for global optimization. *OR/MS Today* 27 (5), 52–55. See www.lionhrtpub.com/orms/orms-10-00/swr.html.

Bertsekas, D.P. (1999) *Nonlinear Programming.* (2nd Edition) Athena Scientific, Cambridge, MA.

Bhatti, M. A. (2000) *Practical Optimization Methods with Mathematica Applications.* Springer-Verlag, New York.

Birkeland, B. (1997) *Mathematics with Mathcad.* Studentlitteratur / Chartwell Bratt, Lund.

Boender, C.G.E., and Romeijn, H.E. (1995) Stochastic methods. In: Horst and Pardalos, Eds. *Handbook of Global Optimization.* Volume 1, pp. 829–869. Kluwer Academic, Dordrecht.

Bornemann, F., Laurie, D., Wagon, S., and Waldvogel, J. (2004) *The SIAM 100-Digit Challenge. A Study in High-Accuracy Numerical Computing.* SIAM, Philadelphia.

Bracken, J., and McCormick, G.P. (1968) *Selected Applications of Nonlinear Programming.* Wiley, New York.

Brooke, A., Kendrick, D., and Meeraus, A. (1988) *GAMS: A User's Guide.* The Scientific Press, Redwood City, CA. (Revised versions are available from the GAMS Corporation.) See also www.gams.com.

Casti, J.L. (1990) *Searching for Certainty.* Morrow, New York.

Castillo, I. (2005) Maple and the Global Optimization Toolbox. *ORMS Today,* 32 (6) 56–60. See also www.lionhrtpub. com/orms/orms-12-05/frswr.html.

Castillo, I., Kampas, F.J., and Pintér, J.D. (2008) Solving circle packing problems by global optimization: Numerical results and industrial applications. *European Journal of Operational Research* 191, 786–802.

Chong, E.K.P., and Zak, S.H. (2001) *An Introduction to Optimization.* (2nd Edition) Wiley, New York.

Cogan, B. (2003) How to get the best out of optimization software. *Scientific Computing World* 71 (2003) 67–68. See also www.scientific-computing.com/scwjulaug03review_optimisation.html.

Corliss, G.F., and Kearfott, R.B. (1999) Rigorous global search: Industrial applications. In: Csendes, T., ed. *Developments in Reliable Computing,* pp. 1–16. Kluwer Academic, Dordrecht.

Coullard, C., Fourer, R., and Owen, J. H., Eds. (2001) *Annals of Operations Research Volume 104: Special Issue on Modeling Languages and Systems.* Kluwer Academic, Dordrecht.

Diwekar, U. (2003) *Introduction to Applied Optimization.* Kluwer Academic, Dordrecht.

Edgar, T.F., Himmelblau, D.M., and Lasdon, L.S. (2001) *Optimization of Chemical Processes.* (2nd Edition) McGraw-Hill, New York.

Ferreira, C. (2002) *Gene Expression Programming.* Angra do Heroísmo, Portugal.

Floudas, C.A., Pardalos, P.M., Adjiman, C., Esposito, W.R., Gümüş, Z.H., Harding, S.T., Klepeis, J.L., Meyer, C.A., and Schweiger, C.A. (1999) *Handbook of Test Problems in Local and Global Optimization.* Kluwer Academic, Dordrecht.

Fourer, R. (2006) *Nonlinear Programming Frequently Asked Questions.* Optimization Technology Center of Northwestern University and Argonne National Laboratory. See www-unix.mcs.anl.gov/otc/Guide/faq/nonlinear-programming-faq.html.

Fourer, R., Gay, D.M., and Kernighan, B.W. (1993) *AMPL — A Modeling Language for Mathematical Programming.* The Scientific Press, Redwood City, CA. (Reprinted by Boyd and Fraser, Danvers, MA, 1996.) See also www.ampl.com.

Fritzson, P. (2004) *Principles of Object-Oriented Modeling and Simulation with Modelica 2.1.* IEEE Press, Wiley-Interscience, Piscataway, NJ.

Frontline Systems (2006) *Premium Solver Platform — Solver Engines. User Guide.* Frontline Systems, Inc. Incline Village, NV. See www.solver.com.

Gao, D.Y., Ogden, R.W., and Stavroulakis, G.E., Eds. (2001) *Nonsmooth/Nonconvex Mechanics: Modeling, Analysis and Numerical Methods.* Kluwer Academic, Dordrecht.

Gershenfeld, N. (1999) *The Nature of Mathematical Modeling.* Cambridge University Press, Cambridge.

Glover, F., and Laguna, M. (1997) *Tabu Search.* Kluwer Academic, Dordrecht.

Grossmann, I.E., Ed. (1996) *Global Optimization in Engineering Design.* Kluwer Academic, Dordrecht.

Hansen, P.E., and Jørgensen, S.E., Eds. (1991) *Introduction to Environmental Management.* Elsevier, Amsterdam.

Henrion, D. (2006) A review of the Global Optimization Toolbox for Maple. *IEEE Control Syst. Mag.* 26 (October 2006 issue), 106–110.

Hillier, F.J., and Lieberman, G.J. (2005) *Introduction to Operations Research.* (8th Edition) McGraw-Hill, New York.

Horst, R., and Pardalos, P.M., Eds. (1995) *Handbook of Global Optimization.* Volume 1. Kluwer Academic, Dordrecht.

Horst, R., and Tuy, H. (1996) *Global Optimization - Deterministic Approaches.* (3rd Edition) Springer, Berlin.

ILOG (2004) ILOG OPL Studio and Solver Suite. www.ilog.com.

Isenor, G., Pintér, J.D., and Cada, M. (2003) A global optimization approach to laser design. *Optim. Eng.* 4, 177–196.

Jacob, C. (2001) *Illustrating Evolutionary Computation with Mathematica.* Morgan Kaufmann, San Francisco.

Jones, N.C., and Pevzner, P.A. (2004) *An Introduction to Bioinformatics Algorithms.* MIT Press, Cambridge, MA.

Kallrath, J., Ed. (2004) *Modeling Languages in Mathematical Optimization.* Kluwer Academic, Dordrecht.

Kampas, F.J., and Pintér, J.D. (2004) Generalized circle packings: Model formulations and numerical results. *Proceedings of the International Mathematica Symposium* (Banff, AB, Canada, August 2004).

Kampas, F.J., and Pintér, J.D. (2006) Configuration analysis and design by using optimization tools in Mathematica. *The Mathematica Journal* 10 (1), 128–154.

Kampas, F.J., and Pintér, J.D. (2009) *Advanced Optimization: Scientific, Engineering, and Economic Applications with Mathematica Examples.* Elsevier, Amsterdam. (To appear)

Kearfott, R.B. (1996) *Rigorous Global Search: Continuous Problems.* Kluwer Academic, Dordrecht.

Lahey Computer Systems (2006) *Fortran 95 User's Guide.* Lahey Computer Systems, Inc., Incline Village, NV. www.lahey.com.

LINDO Systems (1996) *Solver Suite.* LINDO Systems, Inc., Chicago, IL. See also www.lindo.com.

Lopez, R.J. (2005) *Advanced Engineering Mathematics with Maple.* (Electronic book edition.) Maplesoft, Inc., Waterloo, ON. See www.maplesoft.com/products/ebooks/AEM/.

Mandelbrot, B.B. (1983) *The Fractal Geometry of Nature.* Freeman, New York.

Maplesoft (2004) *Global Optimization Toolbox for Maple.* Maplesoft, Inc. Waterloo, ON. See www.maplesoft.com/products/toolboxes/globaloptimization/.

Maplesoft (2006) *Maple.* Maplesoft, Inc., Waterloo, ON. www.maplesoft.com.

Maros, I., and Mitra, G., Eds. (1995) *Annals of Operations Research Volume 58: Applied Mathematical Programming and Modeling II (APMOD 93).* J.C. Baltzer AG, Science, Basel.

Maros, I., Mitra, G., and Sciomachen, A., Eds. (1997) *Annals of Operations Research Volume 81: Applied Mathematical Programming and Modeling III (APMOD 95).* J.C. Baltzer AG, Science, Basel.

Mathsoft (2006) *Mathcad.* Mathsoft Engineering & Education, Inc., Cambridge, MA.

Maximal Software (2006) *MPL Modeling System.* Maximal Software, Inc. Arlington, VA. www.maximal-usa.com.

Michalewicz, Z. (1996) *Genetic Algorithms + Data Structures = Evolution Programs.* (3rd Edition) Springer, New York.

Mittelmann, H.D. (2006) *Decision Tree for Optimization Software.* See plato.la.asu.edu/guide.html. (This Web site was started and maintained jointly for several years with Peter Spellucci.)

Moler, C.B. (2004) *Numerical Computing with Matlab.* SIAM, Philadelphia, 2004.

Murray, J.D. (1983) *Mathematical Biology.* Springer-Verlag, Berlin.

Neumaier, A. (2004) Complete search in continuous global optimization and constraint satisfaction. In: Iserles, A., Ed. *Acta Numerica 2004,* pp. 271–369. Cambridge University Press, Cambridge.

Neumaier, A. (2006) *Global Optimization.* www.mat.univie.ac.at/~neum/glopt.html.

Nowak, I. (2005) *Relaxation and Decomposition Methods for Mixed Integer Nonlinear Programming.* Birkhäuser, Basel.

Osman, I.H., and Kelly, J.P., Eds. (1996) *Meta-Heuristics: Theory and Applications.* Kluwer Academic, Dordrecht.

Papalambros, P.Y., and Wilde, D.J. (2000) *Principles of Optimal Design.* Cambridge University Press, Cambridge.

Paragon Decision Technology (2006) AIMMS. Paragon Decision Technology BV, Haarlem, The Netherlands. See www.aimms.com.

Pardalos, P.M., and Resende, M.G.C., Eds. (2002) *Handbook of Applied Optimization.* Oxford University Press, Oxford.

Pardalos, P.M., and Romeijn, H.E., Eds. (2002) *Handbook of Global Optimization.* Volume 2. Kluwer Academic, Dordrecht.

Pardalos, P.M., Shalloway, D., and Xue, G., Eds. (1996) *Global Minimization of Nonconvex Energy Functions: Molecular Conformation and Protein Folding.* DIMACS Series, Vol. 23, American Mathematical Society, Providence, RI.

Parlar, M. (2000) *Interactive Operations Research with Maple.* Birkhäuser, Boston.

Pintér, J.D. (1996a) *Global Optimization in Action.* Kluwer Academic, Dordrecht.

Pintér, J.D. (1996b) Continuous global optimization software: A brief review. *Optima* 52, 1–8. (Web version is available at plato.la.asu.edu/gom.html.)

Pintér, J.D. (1997) LGO — A program system for continuous and Lipschitz optimization. In: Bomze, I.M., Csendes, T., Horst, R., and Pardalos, P.M., Eds. *Developments in Global Optimization,* pp. 183–197. Kluwer Academic, Dordrecht.

Pintér, J.D. (2000) Extremal energy models and global optimization. In: Laguna, M., and González-Velarde, J.-L., Eds. *Computing Tools for Modeling, Optimization and Simulation,* pp. 145–160. Kluwer Academic, Dordrecht.

Pintér, J.D. (2001a) *Computational Global Optimization in Nonlinear Systems.* Lionheart, Atlanta, GA.

Pintér, J.D. (2001b) Globally optimized spherical point arrangements: Model variants and illustrative results. *Annals of Operations Research* 104, 213–230.

Pintér, J.D. (2002a) *MathOptimizer — An Advanced* Modeling *and Optimization System for Mathematica Users. User Guide.* Pintér Consulting Services, Inc., Halifax, NS. For a summary, see also www.wolfram.com/products/ applications/mathoptimizer/.

Pintér, J.D. (2002b) Global optimization: Software, test problems, and applications. In: Pardalos and Romeijn, Eds. *Handbook of Global Optimization.* Volume 2, pp. 515–569. Kluwer Academic, Dordrecht.

Pintér, J.D. (2003a) Globally optimized calibration of nonlinear models: Techniques, software, and applications. *Optim. Meth. Softw.* 18, 335–355.

Pintér, J.D. (2003b) GAMS /LGO nonlinear solver suite: Key features, usage, and numerical performance. Available at www.gams.com/solvers/lgo.

Pintér, J.D. (2005) *LGO — A Model Development System for Continuous Global Optimization. User's Guide.* (Current revision) Pintér Consulting Services, Inc., Halifax, NS. For summary information, see www.pinterconsulting.com.

Pintér, J.D., Ed. (2006a) *Global Optimization — Scientific and Engineering Case Studies.* Springer Science + Business Media, New York.

Pintér, J.D. (2006b) *Global Optimization with Maple: An Introduction with Illustrative Examples.* An electronic book published and distributed by Pintér Consulting Services Inc., Halifax, NS, Canada and Maplesoft, a division of Waterloo Maple Inc., Waterloo, ON, Canada.

Pintér, J.D. (2009) *Applied Nonlinear Optimization in Modeling Environments.* CRC Press, Boca Raton, FL. (To appear)

Pintér, J.D., and Kampas, F.J. (2003) *MathOptimizer Professional — An Advanced Modeling and Optimization System for Mathematica Users with an External Solver Link. User Guide.* Pintér Consulting Services, Inc., Halifax, NS, Canada. For a summary, see also www.wolfram.com/products/applications/mathoptpro/.

Pintér, J.D., and Kampas, F.J. (2005a) Model development and optimization with Mathematica. In: Golden, B., Raghavan, S., and Wasil, E., Eds. *Proceedings of the 2005 INFORMS Computing Society Conference* (Annapolis, MD, January 2005), pp. 285–302. Springer Science + Business Media, New York.

Pintér, J.D., and Kampas, F.J. (2005b) Nonlinear optimization in *Mathematica* with *MathOptimizer Professional. Mathematica Educ. Res.* 10, 1–18.

Pintér, J.D., and Purcell, C.J. (2003) Optimization of finite element models with *MathOptimizer* and *ModelMaker.* Presented at the *2003 Mathematica Developer Conference,* Champaign, IL. Available at library.wolfram.com/infocenter/Articles/5347/.

Pintér, J.D., Holmström, K., Göran, A.O., and Edvall, M.M. (2004) User's Guide for TOMLAB /LGO. TOMLAB Optimization AB, Västerås, Sweden. See www.tomopt.com/docs/TOMLAB_LGO.pdf.

Pintér, J.D., Linder, D., and Chin, P. (2006) Global Optimization Toolbox for Maple: An introduction with illustrative applications. *Optim. Meth. Softw.* 21 (4) 565–582.

Rich, L.G. (1973) *Environmental Systems Engineering.* McGraw-Hill, Tokyo.

Rothlauf, F. (2002) *Representations for Genetic and Evolutionary Algorithms.* Physica-Verlag, Heidelberg.

Rudolph, G. (1997) *Convergence Properties of Evolutionary Algorithms.* Verlag Dr. Kovac, Hamburg.

Schittkowski, K. (2002) *Numerical Data Fitting in Dynamical Systems.* Kluwer Academic, Dordrecht.

Schroeder, M. (1991) *Fractals, Chaos, Power Laws.* Freeman, New York.

Stewart, I. (1995) *Nature's Numbers.* Basic Books / Harper and Collins, New York.

Stojanovic, S. (2003) *Computational Financial Mathematics Using Mathematica.* Birkhäuser, Boston.

Stortelder, W.J.H., de Swart, J.J.B., and Pintér, J.D. (2001) Finding elliptic Fekete point sets: Two numerical solution approaches. *J. Comput. Appl. Math.* 130, 205–216.

Tawarmalani, M., and Sahinidis, N.V. (2002) *Convexification and Global Optimization in Continuous and Mixed-integer Nonlinear Programming.* Kluwer Academic, Dordrecht.

Tervo, J., Kolmonen, P., Lyyra-Laitinen, T., Pintér, J.D., and Lahtinen, T. (2003) An optimization-based approach to the multiple static delivery technique in radiation therapy. *Ann. Oper. Res.* 119, 205–227.

The MathWorks (2006) *MATLAB.* The MathWorks, Inc., Natick, MA. See www.mathworks.com.

TOMLAB Optimization (2006) TOMLAB. TOMLAB Optimization AB, Västerås, Sweden. See www.tomopt.com.

Trott, M. (2004) *The Mathematica GuideBooks,* Volumes 1–4. Springer Science + Business Media, New York.

Vladimirou, H., Maros, I., and Mitra, G., Eds. (2000) *Annals of Operations Research Volume 99: Applied Mathematical Programming and Modeling IV (APMOD 98).* J.C. Baltzer AG, Science, Basel.

Voss, S., and Woodruff, D.L., Eds. (2002) *Optimization Software Class Libraries.* Kluwer Academic, Dordrecht.

Voss, S., Martello, S., Osman, I.H., and Roucairol, C., Eds. (1999) *Meta-Heuristics: Advances and Trends in Local Search Paradigms for Optimization.* Kluwer Academic, Dordrecht.

Wass, J. (2006) Global Optimization with Maple — An add-on toolkit for the experienced scientist. *Sci. Comput.*, June 2006 issue.

Wilson, H.B., Turcotte, L.H., and Halpern, D. (2003) *Advanced Mathematics and Mechanics Applications Using MATLAB.* (3rd Edition) Chapman and Hall/CRC Press, Boca Raton, FL.

Wolfram, S. (2003) *The Mathematica Book.* (4th Edition) Wolfram Media, Champaign, IL, and Cambridge University Press, Cambridge.

Wolfram Research (2006) *Mathematica.* Wolfram Research, Inc., Champaign, IL. www.wolfram.com.

Wright, F. (2002) *Computing with Maple.* Chapman and Hall/CRC Press, Boca Raton, FL.

Zabinsky, Z.B. (2003) *Stochastic Adaptive Search for Global Optimization.* Kluwer Academic, Dordrecht.

Zhigljavsky, A.A. (1991) *Theory of Global Random Search.* Kluwer Academic, Dordrecht.

Chapter 12
Two-Stage Stochastic Mixed-Integer Programs: Algorithms and Insights

Hanif D. Sherali and Xiaomei Zhu

Summary. Stochastic (mixed-) integer programs pose a great algorithmic and computational challenge in that they combine two generally difficult classes of problems: stochastic programs and discrete optimization problems. Exploring its dual angular structure, various decomposition methods have been widely studied, including Benders' decomposition, Lagrangian relaxation, and test-set decomposition. These decomposition methods are often combined with search procedures such as branch-and-bound or branch-and-cut. Within the confines of these broad frameworks, fine-tuned algorithms have been proposed to overcome obstacles such as nonconvexity of the second-stage value functions under integer recourse, and to take advantage of the many similar structured scenario subproblems using variable transformations. In this chapter, we survey some recent algorithms developed to solve two-stage stochastic (mixed-) integer programs, as well as provide insights into and results concerning their interconnections, particularly for alternative convexification techniques.

Key words: Two-stage stochastic mixed-integer programs, L-shaped method, Benders' decomposition, branch-and-cut, convexification, reformulation-linearization technique (RLT), disjunctive programming

12.1 Introduction

In this chapter, we discuss two-stage stochastic mixed-integer programs (SMIP) of the following form.

Hanif D. Sherali · Xiaomei Zhu
Grado Department of Industrial and Systems Engineering, Virginia Tech, Blacksburg, VA 24061, U.S.A., e-mail: hanifs@vt.edu, xzhu@vt.edu

D.Y. Gao, H.D. Sherali, (eds.), *Advances in Applied Mathematics and Global Optimization*
Advances in Mechanics and Mathematics 17, DOI 10.1007/978-0-387-75714-8_12,

$$\textbf{SMIP:}\quad \text{Minimize}\ \ cx + E[f(x,\tilde{\omega})] \tag{12.1a}$$

$$\text{subject to}\ \ Ax \geq b \tag{12.1b}$$

$$x \geq 0 \tag{12.1c}$$

$$x_i \text{ binary},\ \forall i \in I_b \subseteq I \equiv \{1, ..., n_1\} \tag{12.1d}$$

$$x_i \text{ integer},\ \forall i \in I_{int} \subseteq I \setminus I_b, \tag{12.1e}$$

where $\tilde{\omega}$ is a random variable defined on a probability space $(\tilde{\Omega}, \tilde{\mathcal{A}}, \tilde{\mathcal{P}})$ (with $\tilde{\Omega}$, $\tilde{\mathcal{A}}$, and $\tilde{\mathcal{P}}$, respectively, denoting the set of all outcomes, a collection of random variables, and their associated probability distributions), and where for any given realization ω of $\tilde{\omega}$, we have

$$f(x,\omega) = \text{minimum}\ \ g(\omega)y \tag{12.1f}$$

$$\text{subject to}\ \ W(\omega)y \geq r(\omega) - T(\omega)x \tag{12.1g}$$

$$y \geq 0 \tag{12.1h}$$

$$y_j \text{ binary},\ \forall j \in J_b \subseteq J \equiv \{1, ..., n_2\} \tag{12.1i}$$

$$y_j \text{ integer},\ \forall j \in J_{int} \subseteq J \setminus J_b. \tag{12.1j}$$

In the above, A is an $m_1 \times n_1$ matrix, and for each ω, $W(\omega)$ is an $m_2 \times n_2$ *recourse matrix*, $T(\omega)$ is an $m_2 \times n_1$ *technology matrix*, and the other defined vectors are of corresponding conformable sizes. We assume that the elements of A, $T(\omega)$, and $W(\omega)$ are rational (so that they are scalable to integers, if necessary). For computational viability, a finite number of scenarios, denoted as a set S and indexed by s, is often considered based on some discretization of the possible realizations of $\tilde{\omega}$, each with an associated probability of occurrence p_s, $s \in S$. (See [29] for a justification on approximating continuously distributed scenario parameters by a discrete distribution having a finite support.) Accordingly, the realizations of $g(\omega)$, $W(\omega)$, $T(\omega)$, and $r(\omega)$ are correspondingly denoted as g_s, W_s, T_s, and r_s, respectively, for $s \in S$. In this chapter, we assume such a discrete probability distribution.

Arguably, SMIP is among the most challenging of optimization problems because it combines two generally difficult classes of problems: stochastic programs and discrete optimization problems. Extended from theories particularly in large scale optimization and integer programming ([25, 24]), researchers have actively studied the properties and solution approaches for such problems. We refer the reader to the extensive survey papers by Schultz et al. [31], Klein Haneveld and van der Vlerk [21], Schultz [30], and Sen [34], and an annotated bibliography by Stougie and van der Vlerk [42] for a discussion on the principal properties of SMIPs and some earlier algorithmic developments in this area. A group of important stochastic integer programs are problems having simple integer recourse. Research in this area, such as by Klein Haneveld et al. [19, 20], have been included in detail in various aforementioned surveys. In this chapter, we focus on certain recent algorithmic advances in solving two-stage stochastic (mixed-) integer programs. We also provide insights into their interconnections, and present some results relating the convexification strategies afforded by the reformulation-linearization technique (RLT) and disjunctive programming methods for solving mixed-integer 0–1 SMIPs.

The remainder of this chapter is organized as follows. In Section 12.2, we survey different decomposition frameworks that have been used to solve SMIPs, including decomposition by stage (primal decomposition), decomposition by scenario (dual decomposition), and test-set decomposition. In Section 12.3, we exhibit certain insightful relationships between some convexification methods that have been used for solving mixed-integer 0–1 stochastic problems, particularly, employing disjunctive programming and RLT-based cutting plane methods. In Section 12.4, we discuss three enumerative methods using tender variables when the technology matrix is fixed. We conclude this chapter in Section 12.5.

12.2 Stage, Scenario, and Test-Set Decomposition

Collecting the problems for the two stages together, and denoting the constraints (12.1b)–(12.1e) as $x \in X$ and the constraints (12.1h)–(12.1j) written for scenario s as $y^s \in Y^s$, the deterministic equivalent form for (12.1) is given as follows.

$$\text{Minimize} \quad cx + \sum_{s \in S} p_s g_s y^s \tag{12.2a}$$

$$\text{subject to} \quad T_s x + W_s y^s = r_s, \quad \forall s \in S \tag{12.2b}$$

$$x \in X, \; y^s \in Y^s, \quad \forall s \in S. \tag{12.2c}$$

This representation reveals a dual angular structure that lends itself well to decomposition schemes. We discuss two major groups of decomposition methods, by stages and by scenarios, as well as a novel approach called test-set decomposition. Decomposition by stages, also known as primal decomposition, essentially adopts the framework of Benders' decomposition (Benders [9]), and is more popularly known as the L-shaped method (Van Slyke and Wets [43]) in the context of stochastic programming. In this approach, for each first-stage solution $\bar{x}$ produced by a *master program*, one solves the corresponding second-stage problem (also called a *scenario subproblem*, or simply *subproblem*) based on Problem (12.2) with x fixed at $\bar{x}$. We discuss these methods in Section 12.2.1. Scenario decomposition methods, also known as dual decomposition, work with relaxing and restoring the *nonanticipativ-*

ity condition, which, in a two-stage problem setting, simply means that all scenario outcomes should be based on some identical first-stage solution. When this condition is relaxed, smaller-scaled problems corresponding to scenarios are obtained that are easier to solve, but they also yield different first-stage solutions, which need to be reconciled. These methods are covered in Section 12.2.2. In Section 12.2.3, we describe the test-set decomposition approach by Hemmecke and Schultz [17], which decomposes the problem's Graver test-sets, instead of the problem itself.

12.2.1 Stage Decomposition – L-Shaped Methods

The simplest form of stochastic mixed-integer programs contains purely binary first-stage variables and purely continuous second-stage variables. In this case, Benders' decomposition, or the L-shaped method, can be easily applied using some form of enumeration on the binary first-stage variables, as in Wollmer [44]. Alternatively, and also more generally, if the second-stage problems are easy to solve (not necessarily purely continuous), then these problems can be solved using the integer L-shaped method of Laporte and Louveaux [23]. This is a branch-and-cut (B&C) procedure that is implemented in the projected space of the first-stage variables. At each node, the two-stage problem is solved using Benders' decomposition, which generates feasibility and/or optimality cuts for the first-stage solution, as necessary. The Benders master problem (or the "current problem" of the L-shaped method) at some node q then takes the form:

$$\text{Minimize} \quad cx + \eta \tag{12.3a}$$

$$\text{subject to} \quad D_k x \geq d_k, \quad \forall k = 1, \ldots, F_q \tag{12.3b}$$

$$L_k x \geq l_k, \quad \forall k = 1, \ldots, O_q \tag{12.3c}$$

$$x \in X, \ \eta \text{ unrestricted}, \tag{12.3d}$$

where $k = 1, \ldots, F_q$ and $k = 1, \ldots, O_q$, respectively, index the feasibility cuts (12.3b) and the optimality cuts (12.3c) at node q, and η is a variable representing an approximation of the expected second-stage objective value (or the second-stage *value function*). If feasibility of the subproblems is guaranteed for any fixed x, or $x \in X$, respectively known as the case of *complete recourse* or *relatively complete recourse*, then feasibility cuts will not be generated from the second-stage problems, and (12.3b) will only include certain relevant constraints that fix particular components of x at 0 or 1 as per the branching restrictions. A nodal problem at node q in the B&C tree is resolved whenever a feasibility or optimality cut is added. It is fathomed by the infeasibility of (12.3) or by the bounding processes, and is otherwise partitioned via branching if the solution to the continuous relaxation does not satisfy the binary restrictions. Unlike in a typical branch-and-bound (B&B) process, whenever integrality is satisfied and reveals a potentially better incumbent solution to the current (relaxed) master program, the node is not necessarily fathomed. Instead, the subproblem is solved to possibly generate an optimality cut.

Theoretically, this framework of embedding the L-shaped method in a B&C process applies to two-stage SMIPs in which the first stage contains binary or integer variables and valid optimality cuts are obtainable from the second-stage problems. In practice, however, valid optimality cuts are not easy to derive unless the second stage contains purely continuous variables, or the first stage contains purely binary variables. In the former case, the second-stage value functions $f_s(\cdot)$ are piecewise linear and convex, and so, valid optimality cuts can be derived using the dual variables directly. In the latter case, for any binary feasible solution x^r, defining $I_r \equiv \{i \in I : \bar{x}_i^r = 1\}$, the following is a set of valid optimality cuts.

$$\eta \geq (Q(x^r) - L) \left[\sum_{i \in I_r} x_i - \sum_{i \notin I_r} x_i - |I_r| + 1\right] + L, \forall r = 1, \ldots, R, \qquad (12.4)$$

where $Q(x) \equiv \sum_s p_s f_s(x)$, L is a lower bound on Q, and $1, \ldots, R$ are indices of all the binary feasible solutions that have been encountered thus far.

When the second-stage problems contain integer variables, their value functions $f_s(\cdot)$ are lower semicontinuous (Blair and Jeroslow [11]) and in general nonconvex and discontinuous. Thus optimality cuts are not readily available from dual variables as in the continuous case. In such cases, when the two-stage program has a deterministic cost vector g and recourse matrix W, Carøe and Tind [14] propose to use a subset $\mathcal{F}$ of the *dual price functions* of the second-stage integer programs (see Nemhauser and Wolsey [25] for dual price functions for integer programs) to derive feasibility cuts and optimality cuts. These functions are, however, nonlinear in general, resulting in a nonlinear form of Problem (12.3). Moreover, the dual price function class $\mathcal{F}$ has to be sufficiently large so that the duality gap between the second-stage problem and its ($\mathcal{F}$-) dual is closed. The authors show that this is achieved when the second-stage problems are solved using cutting plane techniques or branch-and-bound. In these cases, finite termination is also established, given that (12.3) can be solved finitely. In particular, when Gomory's cuts are applied, the dual price functions can be transformed into linear functions having mixed-integer variables, and (12.3) is then a mixed-integer linear problem in lieu of a nonlinear problem.

The above decomposition methods work with the primal second-stage problems, given any fixed first-stage solution. Within this framework, alternative methods have been developed to obtain optimality cuts in the case of integer recourse. We discuss some of these methods in Section 12.3. In Section 12.2.2 below, we turn to another type of decomposition approach that relaxes the nonanticipativity condition and works with copies of the first-stage solution to restore this condition.

12.2.2 Scenario Decomposition – Relaxation of the Nonanticipativity Condition

The second-stage problems are related to each other through the first-stage solution only. If we make a copy of the first-stage variable x for each scenario, say x^s for $s \in S$, and enforce the nonanticipativity condition, $x^1 = x^s$, $\forall s \in S$, conveniently written as $\sum_{s\in S} H^s x^s = 0$ for some suitable $H^s_{l\times n_1}$, then the deterministic equivalent problem shown in (12.2) can be rewritten as follows,

$$\min\{\sum_{s\in S} p_s(cx^s + g_s y^s) : (12.2\text{b}), (12.2\text{c}), \sum_{s\in S} H^s x^s = 0\}. \quad (12.5)$$

The Lagrangian dual of (12.5) is $\max\limits_{\lambda\in\mathbb{R}^l}\{\sum_{s\in S} D_s(\lambda)\}$, where

$$D_s(\lambda) \equiv \min\{p_s(cx^s + g_s y^s) + \lambda(H^s x^s) : (12.2\text{b}), (12.2\text{c})\}, \forall s \in S.$$

The Lagrangian dual value yields a tighter lower bound for Problem (12.2) than its LP-relaxation. Carøe and Schultz [12] accordingly propose a dual decomposition algorithm that uses the values of the Lagrangian dual as lower bounds in a B&B process. The tradeoff here is having to solve a nonsmooth Lagrangian dual problem compared to a linear program. At each nodal problem, after solving the associated Lagrangian dual problem, the solutions x^s, $\forall s \in S$, are averaged and rounded to obtain some $\bar{x}$ satisfying the integrality restrictions. The objective value of (12.2) obtained after fixing x at $\bar{x}$ is used to update the upper bound. At the branching step, for a selected branching variable x_i (component of x), constraints $x_i \leq \lfloor \bar{x}_i \rfloor$ and $x_i \geq \lceil \bar{x}_i \rceil$, or $x_i \leq \bar{x}_i - \epsilon$ and $x_i \geq \bar{x}_i + \epsilon$ for some tolerance $\epsilon > 0$, are applied at the two child nodes, respectively, depending on whether x_i is an integer or continuous variable. This algorithm is finitely convergent if X is bounded and the x variables are purely integer-restricted. Schultz and Tiedemann [33] extend this approach to solve stochastic programs that include an additional objective term based on the probability of a risk function exceeding a prespecified threshold value. A survey for SMIPs having risk measures can be found in Schultz [30].

Alonso-Ayuso et al. [3] also relax the nonanticipativity condition in their branch-and-fix coordination algorithm (BFC) for pure 0–1 multistage stochastic programs, or mixed 0–1 two-stage stochastic programs having purely binary variables in the first stage. They assume deterministic technology and recourse matrices. In this algorithm, both the nonanticipativity condition and the binary restrictions on x are relaxed, resulting in a linear problem for each scenario. A B&B tree is developed for each scenario at a terminal node of the scenario tree, and the branch and fix operations are performed on the so-called "twin node families." In a two-stage setting, a twin node family includes the nodal problems in the B&B trees for all scenarios that have the already fixed x-variables equal to a same value (0 or 1). To enforce the nonanticipativity condition, the same branching variable is selected for the active nodes in a twin node family. Note that these nodes belong to different scenarios. Two new twin node families are then formed by fixing the selected branching variable at 0 and 1, respectively. Lower bounds on the objective value are updated by calculating the deterioration of the objective value due to the fixing. Alonso-Ayuso et al. [4] demonstrated this approach in a set of supply chain planning problems, having up to 3933 constraints and 3768 variables (including 114 binary variables), using an 800 MHz Pentium III Processor with 512 Mb of RAM. However, no comparison with other algorithms or commercial software packages was made.

12.2.3 Decomposition of Test-Sets

When the cost vector g, the technology matrix T, and the recourse matrix W are all deterministic, and randomness only appears in the right-hand-side values r_s, Hemmecke and Schultz [17] take advantage of the dual angular structure in (12.2) to develop a test-set decomposition-based approach for two-stage stochastic integer programs. If the problem contains continuous variables, certain inherited computational difficulties from such an approach demand extra care.

A finite *universal test-set* for a family of integer problems

$$(IP)_{c,b}: \quad \min\{cz : Az = b, z \in \mathbb{Z}_+^d\}, \tag{12.6}$$

where $A \in \mathbb{Q}^{l \times d}$ is fixed, and where $c \in \mathbb{R}^d$ and $b \in \mathbb{R}^l$ vary, contains a set of vectors that can be used to solve any problem in this family $(IP)_{c,b}$ for a given c and b (Hemmecke [16]). In this process, an initial feasible solution is first found using the test-set vectors, then an optimal solution is obtained by searching along improving directions, again using the vectors (called improving vectors) in the test-set. The IP Graver test-set (Graver [15]) is one such finite universal test-set, and can be computed using the kernel of A, $ker(A)$, as shown in Hemmecke [16].

A major drawback of using the test-set approach to solve integer programs is that a large number of vectors need to be stored, even for small-sized problems. Hence, for the typically large-sized stochastic integer programs, directly applying this approach is impractical. Exploring the dual angular structure of (12.2), Hemmecke and Schultz [17] show that Graver test-set vectors for this type of problem can be decomposed into, and then constructed from, a small number of building-block vectors, and the latter can then be used to solve large-scaled stochastic integer programs.

Let

$$A_{|S|} \equiv \begin{pmatrix} A & 0 & 0 & \dots & 0 \\ T & W & 0 & \dots & 0 \\ T & 0 & W & \dots & 0 \\ \vdots & \vdots & \vdots & \ddots & \vdots \\ T & 0 & 0 & \dots & W \end{pmatrix} \quad \text{and } A_1 \equiv \begin{pmatrix} A & 0 \\ T & W \end{pmatrix}. \tag{12.7}$$

Observing that $(u, v_1, \dots, v_{|S|}) \in ker(A_{|S|}) \Leftrightarrow (u, v_1), \dots, (u, v_{|S|}) \in ker(A_1)$, Hemmecke and Schultz [17] propose to use the individual vectors $u, v_1, \dots,$ and $v_{|S|}$ as building blocks of the vectors $(u, v_1, \dots, v_{|S|})$ in the Graver test-set of $A_{|S|}$. These building blocks, in lieu of the test-set vectors, are collected into a set $\mathcal{H}_\infty$ and arranged in pairs (u, V_u), where V_u is the set of vectors v such that $(u, v) \in ker(A)$. The set of building blocks $\mathcal{H}_\infty$ is shown to be finite, and can be computed using a finite algorithm. Although the computation of $\mathcal{H}_\infty$ is expensive, because it depends only on A, W, and T, and is independent of $|S|$, cost coefficients, and right-hand-side values, the proposed approach is insensitive to the number of scenarios once $\mathcal{H}_\infty$ is obtained. After computing and storing the building blocks, the test-set vectors are constructed during the two steps of obtaining an initial feasible solution and finding improving directions to solve (12.2). The finiteness of $\mathcal{H}_\infty$ guarantees the finiteness of the proposed algorithm.

12.3 Disjunctive Programming and RLT Cutting Plane Methods for Mixed 0–1 Stochastic Programs

In this section, we assume the following.

A1. No general integer variable exists in either stage; that is, $I_{int} = J_{int} = \varnothing$.
A2. The continuous variables in both stages are bounded. Moreover, for the purpose of exposition, the continuous variables in the second stage are scaled onto [0,1], with the corresponding bounding restrictions $y \leq 1$ being absorbed within (or implied by) (12.1g).
A3. For any feasible first-stage variable $x \in X$, the second-stage problems are feasible (relative complete recourse).

We study two groups of cutting plane approaches for solving two-stage SMIPs having 0–1 mixed-integer variables, and focus on the idea of sharing cut coefficients among scenarios. The first group is based on disjunctive programming, represented by the research of Carøe and Tind [13], Sen and Higle [35], Sen and Sherali [36], and Ntaimo and Sen [27]. The second group applies the reformulation-linearization technique (RLT) to derive cutting planes, and includes the papers by Sherali and Fraticelli [39] and Sherali and Zhu [41]. Both disjunctive cuts and RLT cuts are convexification cutting planes applied to deal with the binary variables in the second-stage problems. In solving stochastic programs, particularly when the number of scenarios is large, memory space is crucial. If some cut coefficient can be shared between scenarios or iterations, we can gain much in memory savings and algorithmic efficacy. In Section 12.3.1, we therefore focus on the convexification process and cut-sharing properties for solving mixed-integer 0–1 second-stage problems. In Section 12.3.2, we then discuss solution approaches for problems containing continuous first-stage variables, as these problems pose an extra challenge on assuring convergence. In Section 12.3.3, we relate the cuts generated using disjunctive programming and the RLT technique to show their interconnections.

12.3.1 Solving Mixed-Integer 0–1 Second-Stage Problems

Given an index set H and some polyhedral sets $\mathcal{P}_h$, for $h \in H$, a disjunctive set $\mathcal{P} = \bigcup_{h \in H} \mathcal{P}_h$ is said to be in *disjunctive normal form*. Using disjunctive programming, we can characterize the closure of the convex hull of $\mathcal{P}$ and generate valid inequalities and even facets for this representation (see, for example, Blair and Jeroslow [10], Balas [5], and Sherali and Shetty [40]). The class of 0–1 mixed-integer programs (MIP) belongs to a special type of disjunctive programs called *facial disjunctive programs*, written in *conjunctive normal form*, that is, a conjunction over $i \in I_b$ of the disjunction $\{x_i = 0 \vee x_i = 1\}$. For this type of programs, we can obtain the convex hull of the feasible solution set using a sequential convexification process.

The lift-and-project algorithm proposed by Balas et al. [7, 8, 6] solves 0–1 mixed-integer programs using disjunctive programming. Carøe and Tind [13] modified this approach and applied it to the deterministic equivalent form of SMIP (Problem (12.2)) in which the first-stage variables are purely continuous (i.e., $I_b = \varnothing$) and the second stage contains both continuous and binary variables. Let $\mathcal{P}^s$ be the set of solutions in the space of $(x, y^h, \forall h \in S)$ when only the constraints on x and y^s are considered. A direct application of the lift-and-project method would generate cuts in the $(x, y^h, \forall h \in S)$ space, sequentially treating a single variable y_j^s for $j \in J_b$ and $s \in S$ as binary, and the remaining y-variables as continuous. Exploring the dual angular structure of the deterministic equivalent form, Carøe and Tind [13] proposed the generation of cuts for $\mathcal{P}^s$ in the (x, y^s) space for each scenario $s \in S$. This is valid because all the y^s, $s \in S$, are independent, and by the assumption of relatively complete recourse, the projection of $\mathcal{P}^s$ and the projection of the original solution set on the (x, y^s) space are the same.

To generate such cuts for a current solution $(\bar{x}, \bar{y}^s)$ for which the value for $\bar{y}_q^s$, $q \in J_b$, is fractional, consider the disjunction $\mathcal{P}^s = \mathcal{P}_0^s \cup \mathcal{P}_1^s$, where

$$\mathcal{P}_0^s = \{Ax \geq b \qquad \leftarrow \tau_0 \tag{12.8a}$$

$$T_s x + W_s y^s \geq r_s \qquad \leftarrow \lambda_0 \tag{12.8b}$$

$$-y_q^s \geq 0\} \qquad \leftarrow \varsigma_0 \tag{12.8c}$$

and

$$\mathcal{P}_1^s = \{Ax \geq b \qquad \leftarrow \tau_1 \tag{12.8d}$$

$$T_s x + W_s y^s \geq r_s \qquad \leftarrow \lambda_1 \tag{12.8e}$$

$$y_q^s \geq 1\} \qquad \leftarrow \varsigma_1 \tag{12.8f}$$

with the associated multipliers as shown above, and where we assume that the nonnegativity and the bounding constraints for x and y are absorbed in $Ax \geq b$ and $T_s x + W_s y^s \geq r_s$, respectively. Define e_q as the qth unit row vector. A convexification cutting plane for the disjunction (12.8) is generated by solving the following linear program.

$$\text{Minimize} \quad \alpha\bar{x} + \beta\bar{y}^s - \gamma \tag{12.9a}$$

$$\text{subject to:} \quad \alpha - \tau_h A - \lambda_h T_s \geq 0, \quad \forall h = 0, 1 \tag{12.9b}$$

$$\beta + \varsigma_0 e_q - \lambda_0 W_s \geq 0 \tag{12.9c}$$

$$\beta - \varsigma_1 e_q - \lambda_1 W_s \geq 0 \tag{12.9d}$$

$$\tau_0 b + \lambda_0 r_s - \gamma \geq 0 \tag{12.9e}$$

$$\varsigma_1 + \tau_1 b + \lambda_1 r_s - \gamma \geq 0 \tag{12.9f}$$

$$\alpha\bar{x} + \beta\bar{y}^s - \gamma \geq -1 \tag{12.9g}$$

$$\tau, \lambda, \varsigma \geq 0, \;\; \alpha, \beta, \gamma \text{ unrestricted.} \tag{12.9h}$$

Let $(\bar{\alpha}, \bar{\beta}, \bar{\gamma}, \bar{\lambda}, \bar{\tau}, \bar{\varsigma})$ be an optimal solution of (12.9). If the objective value obtained for (12.9a) is negative (i.e., equals -1 due to (12.9g)), the inequality

$$\bar{\alpha}x + \bar{\beta}y^s \geq \bar{\gamma} \tag{12.10}$$

eliminates the current fractional solution.

Carøe and Tind observe that if the recourse matrix and technology matrix are fixed, (12.9b)–(12.9d) will be the same for all scenarios. Hence, $\bar{\alpha}$ and $\bar{\beta}$ obtained for one scenario will satisfy these constraints for all scenarios, and the cuts generated for one scenario are valid for all scenarios by updating only the right-hand sides. In particular, $\bar{\alpha}x + \bar{\beta}y^{s'} \geq \gamma_{s'}$ is valid for scenario s', $\forall s' \in S$, where $\gamma_{s'} = \bar{\gamma} + \min\{\bar{\lambda}_0(r_{s'} - r_s), \bar{\lambda}_1(r_{s'} - r_s)\}$. When only the recourse matrix is fixed but the technology matrix is stochastic, constraints (12.9c) and (12.9d) are satisfied by $\bar{\beta}$ for all $s' \in S$, and the following linear program can be solved to obtain coefficients $\tilde{\alpha}$ and $\tilde{\gamma}$ for deriving a valid inequality for scenario s'.

$$\text{Minimize} \quad \alpha\bar{x} - \gamma \tag{12.11a}$$

$$\text{subject to:} \quad \alpha - \tau_h A \geq \bar{\lambda}_h T_{s'}, \quad \forall h = 0, 1 \tag{12.11b}$$

$$\tau_0 b - \gamma \geq -\bar{\lambda}_0 r_{s'} \tag{12.11c}$$

$$\tau_1 b - \gamma \geq -\bar{\varsigma}_1 - \bar{\lambda}_1 r_{s'} \tag{12.11d}$$

$$\alpha\bar{x} - \gamma \geq -1 \tag{12.11e}$$

$$\tau \geq 0, \ \alpha, \gamma \text{ unrestricted.} \tag{12.11f}$$

Comparing this with (12.9b)–(12.9h), because the $\bar{\beta}$-value is already at hand, the terms and constraints related to β and W_s are dropped, and this reduces the size of the problem. Letting $(\tilde{\alpha}, \tilde{\gamma}, \tilde{\tau})$ be an optimal solution to (12.11), the cut $\tilde{\alpha}x + \bar{\beta}y^{s'} \geq \tilde{\gamma}$ is valid for scenario s', $s' \in S$.

In a similar fashion, if T and r are fixed and W is stochastic, or if W and r are fixed and T is stochastic, then a valid inequality for scenario s' can be obtained by revising the valid inequality (12.10) for scenario s as above.

Proposition 12.1. *Let* (12.10) *be a valid inequality obtained for scenario s by solving* (12.9). *We have:*

(a) If the technology matrix T and the right-hand side r are fixed, and the recourse matrix W is stochastic, then

$$\bar{\alpha}x + \beta_{s'}y^{s'} \geq \bar{\gamma} \tag{12.12a}$$

is a valid inequality for scenario s', where

$$\beta_{s'} = \bar{\beta} + \max\{\bar{\lambda}_0(W_{s'} - W_s), \bar{\lambda}_1(W_{s'} - W_s)\}, \tag{12.12b}$$

and where the max-operation for vectors is performed componentwise.
(b) Similarly, if the recourse matrix W and the right-hand side r are fixed, and the technology matrix T is stochastic, then

$$\alpha_{s'}x + \bar{\beta}y^{s'} \geq \bar{\gamma} \tag{12.12c}$$

is a valid inequality for scenario s', where

$$\alpha_{s'} = \bar{\alpha} + \max\{\bar{\lambda}_0(T_{s'} - T_s), \bar{\lambda}_1(T_{s'} - T_s)\}. \tag{12.12d}$$

Proof. The proof follows the same line as that of the fixed T and W case in Carøe and Tind [13]. In the case of fixed T and r, we have that the coefficients

$\bar{\alpha}$ and $\bar{\gamma}$ satisfy constraints (12.9b), (12.9e), and (12.9f) for s, where T_s and r_s are indistinguishable for all scenarios. Furthermore, from (12.12b),

$$\beta_{s'} \geq -\bar{\varsigma}_0 e_q + \bar{\lambda}_0 W_s + \bar{\lambda}_0(W_{s'} - W_s) = -\bar{\varsigma}_0 e_q + \lambda_0 W_{s'}$$

$$\text{and } \beta_{s'} \geq \bar{\varsigma}_1 e_q + \bar{\lambda}_1 W_s + \bar{\lambda}_1(W_{s'} - W_s) = \bar{\varsigma}_1 e_q + \lambda_1 W_{s'},$$

which satisfies (12.9c) and (12.9d) for scenario s'. This proves Part (a), and Part (b) can be proved similarly. □

Note that the cuts (12.12a) and (12.12c) preserve the fixed technology matrix and recourse matrix structure, respectively, in addition to having the same right-hand side. Linear programs similar to (12.11) can be constructed to generate valid inequalities when only one of T, W, and r is deterministic and the other two are stochastic.

The above approach is based on solving the deterministic equivalent form of the two-stage SMIP. If a subproblem having the set of variables (x, y^s) is solved for each scenario s sequentially, then the validity of the above cuts still holds true.

When using the L-shaped method (or Benders' decomposition) to solve two-stage stochastic programs, we need the value functions of the Benders subproblems (i.e., the second-stage problems) to be convex. However, if the second-stage problems contain binary or integer variables, their value functions are in general nonconvex and discontinuous. This then requires a relaxation of the binary restrictions in the second-stage problems and an accompanying convexification process.

When solving the second-stage problems, a given solution $\bar{x}$ is available from the first-stage problem and becomes part of the right-hand sides of the second-stage problems. Valid inequalities obtained in this context will be in the form of $\bar{\beta}_s y^s \geq \bar{\gamma}_s$. It is helpful to lift these (possibly facetial) cuts to be valid in the (x, y^s) space, in the form of $\alpha_s x + \bar{\beta}_s y^s \geq \bar{\gamma}_s + \alpha_s \bar{x}$. By treating x as a variable, these cuts can be reused in subsequent Benders iterations by updating the first-stage solutions from the Benders master problem, as detailed in Sherali and Fraticelli [39]. This cut lifting is not well posed when x is continuous. If x is restricted to be binary, then the feasible x-solutions are always facial to the convex hull of the two-stage constraint set, and we can lift a valid inequality using the disjunction

$$\left\{\begin{array}{l} Ax \geq b \\ T_s x + W_s y^s \geq r_s \\ -x_i \geq 0 \end{array}\right\} \bigvee \left\{\begin{array}{l} Ax \geq b \\ T_s x + W_s y^s \geq r_s \\ x_i \geq 1 \end{array}\right\}. \tag{12.13}$$

The lifted cut can be obtained via the solution to the following linear program, where $\bar{\beta}_s$ and $\bar{\gamma}_s$ are cut coefficients already obtained in the y^s space.

$$\text{Minimize} \quad \alpha_s \bar{x} \tag{12.14a}$$

$$\text{subject to:} \quad \alpha_s + \varsigma_0 e_i - \tau_0 A - \lambda_0 T_s \geq 0 \tag{12.14b}$$

$$\alpha_s - \varsigma_1 e_i - \tau_1 A - \lambda_1 T_s \geq 0 \tag{12.14c}$$

$$-\lambda_h W_s \geq -\bar{\beta}_s, \quad \forall h = 0, 1 \tag{12.14d}$$

$$-\alpha_s \bar{x} + \tau_0 b + \lambda_0 r_s \geq \bar{\gamma}_s \tag{12.14e}$$

$$-\alpha_s \bar{x} + \varsigma_1 + \tau_1 b + \lambda_1 r_s \geq \bar{\gamma}_s \tag{12.14f}$$

$$\alpha_s \bar{x} \geq -1 \tag{12.14g}$$

$$\tau, \lambda, \varsigma \geq 0, \quad \alpha_s \text{ unrestricted.} \tag{12.14h}$$

Also, in the purely binary first-stage problem setting, if additionally the recourse matrix is fixed, Sen and Higle [35] propose to lift cuts for second-stage problems in the (x, y^s)-space, such that the cuts share the same coefficients $\bar{\beta}$ for the y-variables. These cuts are then in the form of $\bar{\beta} y^s \geq \gamma_s - \alpha_s x$. This property is named common cut coefficients (C^3), and is established as follows.

Given a first-stage solution $\bar{x}$ and a fixed recourse matrix W, consider the disjunction

$$\left\{ \begin{array}{l} W y^s \geq r_s - T_s \bar{x} \\ -y_q^s \geq 0 \end{array} \right\} \bigvee \left\{ \begin{array}{l} W y^s \geq r_s - T_s \bar{x} \\ y_q^s \geq 1 \end{array} \right\}. \tag{12.15}$$

The common cut coefficients $\bar{\beta}$ are obtained via the optimal value for β given by the following linear program, where $\bar{y}^s$, $\forall s \in S$, are the current second-stage solutions.

$$\text{Minimize} \quad \sum_s p_s (\beta \bar{y}^s - \pi_s) \tag{12.16a}$$

$$\text{subject to:} \quad \beta + \varsigma_0 e_q - \lambda_0 W \geq 0 \tag{12.16b}$$

$$\beta - \varsigma_1 e_q - \lambda_1 W \geq 0 \tag{12.16c}$$

$$\lambda_0 (r_s - T_s \bar{x}) - \pi_s \geq 0, \quad \forall s \in S \tag{12.16d}$$

$$\varsigma_1 + \lambda_1 (r_s - T_s \bar{x}) - \pi_s \geq 0, \quad \forall s \in S \tag{12.16e}$$

$$\sum_s p_s (\beta \bar{y}^s - \pi_s) \geq -1, \tag{12.16f}$$

$$\pi, \lambda, \varsigma \geq 0, \quad \beta \text{ unrestricted.} \tag{12.16g}$$

Letting $\bar{\lambda}$ be an optimal value for λ in Problem (12.16), the cut

$$\bar{\beta}y \geq \min\{\bar{\lambda}_0 r_s - \bar{\lambda}_0 T_s x, \bar{\lambda}_1 r_s + \varsigma_1 - \bar{\lambda}_1 T_s x\} \tag{12.17}$$

is then valid for scenario s, and its piecewise linear concave right-hand side can be convexified using its epigraph over the bounded region $\bar{X} \equiv \{x \in \mathbb{R}_+^{n_1} | Ax \geq b\}$. Denote this epigraph as

$$\Pi_{\bar{X}}^s \equiv \{(\varpi, x) | x \in \bar{X}, \varpi \geq \mathcal{R}^s(x)\}, \tag{12.18}$$

where $\mathcal{R}^s(x) \equiv \min\{\bar{\lambda}_0 r_s - \bar{\lambda}_0 T_s x, \bar{\lambda}_1 r_s + \varsigma_1 - \bar{\lambda}_1 T_s x\}$.

In this process, for $x \in \bar{X}$, we can find a lower bound for $\mathcal{R}^s(x)$, say, $\mathcal{L}$. Each affine function in $\mathcal{R}^s(x)$ is then translated by $\mathcal{L}$, if necessary, to ensure that $\varpi \geq 0$. This will translate the convex hull of the epigraph as well. After convexification cuts are obtained, they can be translated back by $-\mathcal{L}$ to recover the original values. So without loss of generality, assume that $\mathcal{L} = 0$.

We can then represent $\Pi_{\bar{X}}^s$ as the following disjunction

$$\left\{ \begin{array}{l} \varpi \geq \bar{\lambda}_0 r_s - \bar{\lambda}_0 T_s x \\ Ax \geq b \end{array} \right\} \bigvee \left\{ \begin{array}{l} \varpi \geq \bar{\lambda}_1 r_s + \bar{\varsigma}_1 - \bar{\lambda}_1 T_s x \\ Ax \geq b \end{array} \right\}. \tag{12.19}$$

The right-hand side of (12.17) can then be convexified based on this disjunction by solving the following linear program for each scenario s.

$$\text{Minimize} \quad \sigma_s \bar{x} + \upsilon_s - \delta_s \tag{12.20a}$$

$$\text{subject to:} \quad \sigma_s - \tau_h A - \theta_h(\bar{\lambda}_h T_s) \geq 0, \quad \forall h = 0, 1 \tag{12.20b}$$

$$\upsilon_s - \theta_h \geq 0, \quad \forall h = 0, 1 \tag{12.20c}$$

$$\tau_0 b + \theta_0(\bar{\lambda}_0 r_s) - \delta_s \geq 0 \tag{12.20d}$$

$$\tau_1 b + \theta_1(\bar{\lambda}_1 r_s) + \theta_1 \bar{\varsigma}_1 - \delta_s \geq 0 \tag{12.20e}$$

$$\theta_0 + \theta_1 = 1 \tag{12.20f}$$

$$\theta, \tau \geq 0 \quad \sigma, \delta, \upsilon \text{ unrestricted.} \tag{12.20g}$$

Note that from (12.20c) and (12.20f), we have that $\upsilon_s > 0$. For an optimal extreme point solution $(\bar{\sigma}_s, \bar{\upsilon}_s, \bar{\delta}_s)$, we then have $clconv(\Pi_{\bar{X}}^s) = \{(\varpi, x) | x \in \bar{X}, \varpi \geq (\bar{\delta}_s / \bar{\upsilon}_s) - (\bar{\sigma}_s / \bar{\upsilon}_s) x\}$. This completes the convexification of the piecewise linear concave $\mathcal{R}^s(x)$. Letting $\bar{\alpha}_s = \bar{\sigma}_s / \bar{\upsilon}_s$ and $\bar{\gamma}_s = \bar{\delta}_s / \bar{\upsilon}_s$, the inequality

$$\bar{\beta}y \geq \bar{\gamma}_s - \bar{\alpha}_s x \tag{12.21}$$

is then valid for scenario s, $\forall s \in S$, and all scenarios share the same coefficient $\bar{\beta}$ for y in the new cuts, which need to be stored only once.

The disjunctive decomposition (D^2) method by Sen and Higle [35] thus applies Benders' decomposition to solve two-stage stochastic programs, using the above C^3-scheme embodied by (12.16) and (12.20) to solve the second-stage subproblems.

The two approaches by Carøe and Tind [13] and Sen and Higle [35] directly stem from disjunctive programming. Using another approach, the reformulation-linearization technique (RLT) (see Sherali and Adams [37, 38]), we can sequentially construct (partial) convex hulls for the second-stage subproblems. Sherali and Fraticelli [39] proposed a method for solving two-stage problems having purely binary first-stage variables using a modified Benders'-decomposition scheme, where the subproblems are solved by adding RLT cuts generated in the (x, y^s) space. The idea is to store the cuts as functions of x, so that at each Benders iteration, when new x-values are obtained from the master (first-stage) problem, the cutting planes obtained previously from the second-stage subproblem solutions can be reused in the subsequent subproblems of the same scenario simply by updating the x-values.

In a Benders'-decomposition setting, each second-stage problem is solved independently; hence, we omit the superscript s for y-variables when no confusion arises. Furthermore, denote the qth column of matrix W_s as W_s^q, and the matrix formed by the remaining columns of W_s as $\tilde{W}_s$. Correspondingly, denote variable y without the qth element as $\tilde{y}$ and let $\tilde{\beta}$ be its associated coefficient; that is, $\beta y \equiv \tilde{\beta}\tilde{y} + \beta_q y_q$. To derive cuts in the (x, y) space, that is, to introduce x as a variable into the cut generation process, we include the bounding constraint $0 \leq x \leq e$ into the second stage. By the RLT process, given a solution $(\bar{x}, \bar{y})$, which has $\bar{y}_q$ fractional for some $q \in J_b$, we multiply $0 \leq x \leq e$ and the constraints in the second stage by the factors y_q and $(1 - y_q)$ to obtain a system in a higher-dimensional space (x, y, z^x, z^y), including the new RLT variables z^x and z^y to represent the resulting nonlinear products; that is,

$$z^x \equiv x y_q \text{ and } z^y \equiv \tilde{y} y_q. \tag{12.22}$$

Denote the resulting system as

$$T_s z^x + \tilde{W}_s z^y \geq (r_s - W_s^q) y_q \qquad \leftarrow \phi_0 \tag{12.23a}$$

$$-T_s z^x - \tilde{W}_s z^y \geq r_s - T_s x - \tilde{W}_s \tilde{y} - r_s y_q \qquad \leftarrow \phi_1 \tag{12.23b}$$

$$\Gamma_z z^x \geq h_b - \Gamma_x x - \Gamma_y y_q, \qquad \leftarrow \phi_b. \tag{12.23c}$$

The constraints (12.23a) and (12.23b) are obtained by multiplying (12.1g) (for scenario s) by y_q and $(1 - y_q)$, respectively. The constraints (12.23c) are obtained by multiplying the bounding constraints of $0 \leq x \leq e$ by y_q and $(1 - y_q)$, where Γ_z, Γ_x, and Γ_y are used to denote the resulting coefficient matrices for z^x, x, and y_q, respectively. Associating the dual multipliers ϕ as indicated in (12.23), the solution $(\bar{\alpha}_s, \bar{\beta}_s, \bar{\gamma}_s)$ of the following linear program yields the valid inequality $\bar{\beta}_s y \geq \bar{\gamma}_s - \bar{\alpha}_s x$ for scenario s in the projected space of the original variables (x, y).

$$\begin{aligned}
\text{Minimize } & \tilde{\beta}_s \tilde{\bar{y}} + \beta_{qs} \bar{y}_q + \alpha_s \bar{x} - \gamma_s && (12.24a)\\
\text{subject to: } & (\phi_0 - \phi_1) T_s + \phi_b \Gamma_z = 0 && (12.24b)\\
& (\phi_0 - \phi_1) \tilde{W}_s = 0 && (12.24c)\\
& \alpha_s = \phi_1 T_s + \phi_b \Gamma_x && (12.24d)\\
& \tilde{\beta}_s = \phi_1 \tilde{W}_s && (12.24e)\\
& \beta_{qs} = \phi_0 (W_s^q - r_s) + \phi_1 r_s + \phi_b \Gamma_y && (12.24f)\\
& \gamma_s = \phi_1 r_s + \phi_b h_b && (12.24g)\\
& \tilde{\beta}_s \tilde{\bar{y}} + \beta_{qs} \bar{y}_q + \alpha_s \bar{x} - \gamma_s \geq -1 && (12.24h)\\
& \phi \geq 0,\ \alpha_s, \beta_s, \gamma_s \text{ unrestricted.} && (12.24i)
\end{aligned}$$

Proposition 12.2. *If the technology matrix T and recourse matrix W are fixed, and $\bar{\tilde{\beta}}_s \tilde{y} + \bar{\beta}_{qs} y_q \geq \bar{\gamma}_s - \bar{\alpha}_s x$ is a valid inequality for scenario s obtained from* (12.24)*, then*

$$\bar{\tilde{\beta}}_s \tilde{y} + \beta_{qs'} y_q \geq \gamma_{s'} - \bar{\alpha}_s x \qquad (12.25)$$

is valid for scenario s', where $\beta_{qs'} = \bar{\beta}_{qs} + \bar{\phi}_0 (r_s - r_{s'})$ and $\gamma_{s'} = \bar{\gamma}_s + \bar{\phi}_1 (r_{s'} - r_s)$.

Proof. Follows from the feasibility of (12.24b)–(12.24i). □

Note that the cut (12.25) is valid for scenario s', but it disturbs the fixed recourse structure. Hence, if applied, it should not be included in the cut generation process in later iterations, which should continue to use the fixed W. However, ultimately, to obtain a convexification of the second-stage problem, such cuts would need to be included in the fashion discussed by Sherali and Fraticelli [39]. The following proposition provides a way to derive valid inequalities and retain a fixed recourse, in a more general setting of a stochastic technology matrix.

Proposition 12.3. *Let the recourse matrix W be fixed, and the technology matrix T_s and right-hand side r_s be stochastic. Denote $(\bar{\alpha}_s, \bar{\beta}_s, \bar{\gamma}_s, \bar{\phi})$ as an optimal solution obtained for Problem* (12.24)*. Solve the following linear program corresponding to another scenario $s' \neq s$.*

$$\begin{aligned}
\textit{Minimize } & 0 && (12.26a)\\
\textit{subject to: } & \phi T_{s'} = \bar{\phi}_1 T_{s'} - \bar{\phi}_b \Gamma_z && (12.26b)\\
& \phi \tilde{W} = \bar{\phi}_1 \tilde{W} && (12.26c)\\
& \phi (W^q - r_{s'}) = \bar{\phi}_0 (W^q - r_{s'}) && (12.26d)\\
& \phi \geq 0. && (12.26e)
\end{aligned}$$

If (12.26) *is feasible, then*

$$\bar{\beta}_s y \geq \gamma_{s'} - \alpha_{s'} x \qquad (12.27a)$$

is a valid inequality for scenario s'*, where*

$$\alpha_{s'} = \bar{\phi}_1 T_{s'} + \bar{\phi}_b \Gamma_x \quad \text{and} \quad \gamma_{s'} = \bar{\phi}_1 r_{s'} + \bar{\phi}_b h_b. \tag{12.27b}$$

Proof. The proof again follows from the feasibility to (12.24b)–(12.24i) for scenario s' by revising ϕ_0 to ϕ and retaining $\phi_1 = \bar{\phi}_1$ and $\phi_b = \bar{\phi}_b$, where (12.24e) remains the same, and (12.24d) and (12.24g) are satisfied by (12.27b). Furthermore, (12.24b), (12.24c), and (12.24f), are satisfied by (12.26b), (12.26c), and (12.26d), respectively, and the normalization constraint (12.24h) is inconsequential. □

The purpose of constraint (12.26d) is to retain the same β_q value for all new cuts generated, so as to keep the fixed recourse structure if the cuts are to be inherited. It can be replaced by other normalization constraints, and a valid cut can be generated using the correspondingly computed $(\alpha_{s'}, \beta_{s'}, \gamma_{s'})$ values. However, we may no longer have $\beta_{s'} = \beta$ in the resulting cut, thereby losing the fixed recourse structure.

In a Benders'-decomposition context where subproblems are solved using a cutting plane method, and where these cutting planes are valid in the (x, y)-space for $conv\{(x, y) \mid T_s x + Wy \geq r_s, 0 \leq x \leq e, y \geq 0, y_j \in \{0, 1\}, \forall j \in J_b\}$, we can append these cuts along with any possibly added bounding constraints for the y-variables to $T_s x + Wy \geq r_s$, and Sherali and Fraticelli [39] have shown that Benders cuts generated using the dual solution of the augmented LP relaxation system are valid optimality cuts in terms of the first-stage variables.

Using this idea and disjunctive programming, Sen and Sherali [36] develop a disjunctive decomposition-based branch-and-cut (D^2-BAC) approach, where subproblems are partially solved using branch-and-cut, and where optimality cuts applied in the first-stage master problem are generated using the disjunctive programming concept. In the B&B tree for any subproblem in this process, each node is an LP relaxation of the subproblem that includes some node-dependent bounding constraints for y. Assuming that all the nodes are associated with feasible LP relaxations and are fathomed only by the bound computations, then at least one terminal node corresponds to an optimal solution. Let Q_s denote the set of terminal nodes of the tree that have been explored for the subproblem for scenario s, and let z_{qls} and z_{qus} denote the lower and upper bounds for y in the nodal subproblem q, for $q \in Q_s$. Let $\bar{\lambda}_{qs}$, $\bar{\psi}_{qls}$, and $\bar{\psi}_{qus}$ be dual solutions associated with the constraint set $Wy \geq r_s - T_s x$ and the lower and upper bounding constraints for y, respectively, in the nodal subproblem for node q. We then obtain the following disjunction:

$$\eta \geq \bar{\lambda}_{qs}[r_s - T_s x] + \bar{\psi}_{qls} z_{qls} - \bar{\psi}_{qus} z_{qus} \quad \text{for at least one } q \in Q_s. \tag{12.28}$$

Similar to the convexification process for (12.17), we can use the following disjunction to generate a disjunctive cut for (12.28),

$$\bigvee_{q\in Q_s} \left\{ \begin{array}{l} \eta + \bar{\lambda}_{qs}T_s x \geq \bar{\lambda}_{qs}r_s + \bar{\psi}_{qls}z_{qls} - \bar{\psi}_{qus}z_{qus} \\ Ax \geq b \end{array} \right\}. \tag{12.29}$$

Denote $\bar{\eta}$ as a current estimate lower bound for the second-stage value function, possibly obtained from previous iterations, and let $(\bar{\sigma}_s, \bar{\upsilon}_s, \bar{\delta}_s)$ be an optimal extreme point solution of the following linear program.

$$\text{Minimize } \sigma_s\bar{x} + \upsilon_s\bar{\eta} - \delta_s \tag{12.30a}$$

$$\text{subject to: } \sigma_s - \tau_q A - \theta_q\bar{\lambda}_{qs}T_s \geq 0, \quad \forall q \in Q_s \tag{12.30b}$$

$$\upsilon_s - \theta_q \geq 0, \quad \forall q \in Q_s \tag{12.30c}$$

$$\tau_q b + \theta_q(\bar{\lambda}_{qs}r_s + \bar{\psi}_{qls}z_{qls} - \bar{\psi}_{qus}z_{qus}) - \delta_s \geq 0, \quad \forall q \in Q_s \tag{12.30d}$$

$$\sum_{q\in Q_s} \theta_q = 1 \tag{12.30e}$$

$$\theta, \tau \geq 0 \quad \sigma, \delta, \upsilon \text{ unrestricted.} \tag{12.30f}$$

Again, due to (12.30c) and (12.30e), we have that $\bar{\upsilon}_s > 0$. A disjunctive cut that provides a lower bound on the second-stage value function can then be obtained in the form of $\eta_s \geq \bar{\gamma}_s - \bar{\alpha}_s x$, where $\bar{\alpha}_s = \bar{\sigma}_s/\bar{\upsilon}_s$ and $\bar{\gamma}_s = \bar{\delta}_s/\bar{\upsilon}_s$.

12.3.2 Pure Continuous and Mixed 0–1 First-Stage Problems

Following the cutting plane game concept of Jeroslow [18], the disjunctive and RLT cut generation processes finitely solve the second-stage subproblems. Therefore, finite convergence of the D^2 algorithm of Sen and Higle [35], and of the modified Benders'-decomposition algorithm of Sherali and Fraticelli [39], is afforded by the finite number of feasible first-stage solutions and by the finite cutting plane generation processes for solving subproblems. If the first stage contains continuous variables, however, a feasible first-stage solution $\bar{x}$ may not be facial with respect to its bounding constraints, and the algorithms of Section 12.3.1 would no longer assure convergence.

To retain the facial property of the first-stage solutions, Ntaimo and Sen [27] and Sherali and Zhu [41] propose to build a B&B tree via a partitioning process in the projected space of the bounded first-stage variables so as to ultimately induce the aforementioned facial property.

Using the D^2 algorithm, when $\bar{x}$ is at a vertex of its bounding region, for the right-hand side of the cut (12.21) generated by (12.16) and (12.20), we will have

$$\bar{\gamma}_s - \bar{\alpha}_s x = \min\{\bar{\lambda}_0 r_s - \bar{\lambda}_0 T_s x, \bar{\lambda}_1 r_s + \varsigma_1 - \bar{\lambda}_1 T_s x\} (\equiv \mathcal{R}^s(x)); \tag{12.31}$$

otherwise, this equality may be violated. Ntaimo and Sen [27] then propose a finitely convergent D^2-based branch-and-cut (D^2-CBAC) algorithm for problems containing purely continuous first-stage variables. This algorithm constructs a B&B tree defined on the first-stage feasible region, applies the D^2 algorithm at each node, selects some scenario s and iteration k such that (12.31) is mostly violated, and partitions based on the disjunction prompted by (12.31), enforcing $\bar{\lambda}_0 r_s - \bar{\lambda}_0 T_s x \geq \bar{\lambda}_1 r_s + \varsigma_1 - \bar{\lambda}_1 T_s x$ on one child node and $\bar{\lambda}_0 r_s - \bar{\lambda}_0 T_s x \leq \bar{\lambda}_1 r_s + \varsigma_1 - \bar{\lambda}_1 T_s x$ on the other. The convexification cut (12.21) at the parent node is accordingly updated to $\beta y \geq \bar{\lambda}_0^k r_s^k - \bar{\lambda}_0^k T_s^k x$ and $\beta y \geq \bar{\lambda}_1^k r_s + \varsigma_1^k - \bar{\lambda}_1^k T_s^k x$, at the two child nodes, respectively.

Because there are finitely many disjunctive variables in the second stage, for some iteration of the embedded D^2 algorithm, there are a finite number of right-hand sides of (12.17) that can be constructed; hence, there are finitely many partitions of the first-stage feasible region to consider, which leads to finite convergence of the D^2-CBAC algorithm.

If the first stage further contains both continuous and binary variables, Sherali and Zhu [41] propose a decomposition-based B&B (DBAB) algorithm that is guaranteed to converge to an optimal solution. They assume relative complete recourse with respect to some bounding hyperrectangle of x. The branch-and-bound tree is again defined on the projected space of the bounded first-stage variables, where lower bounds for the nodal problems are computed by applying a modified Benders' method extended from Sherali and Fraticelli [39], but defined on a subdivision of the original bounding hyperrectangle for x, and the Benders subproblems are derived based on partial convex hull representations in the (x, y^s)-spaces using the second-stage constraints and the current bounding constraints for x. For some given feasible first-stage solution $\bar{x}$, because $x = \bar{x}$ may not be facial with respect to its bounding constraints, the Benders subproblems are shown to define lower bounds for second-stage value functions. Therefore, any resulting Benders master problem provides a lower bound for the original stochastic program defined over the same hyperrectangle, and yields the same objective value if $\bar{x}$ is a vertex of the defining region.

In the branch-and-bound process of the DBAB algorithm, a node yielding the least lower bound is selected for branching at each partitioning step. Hence, the nodal objective value provides a lower bound for the original two-stage problem. In the partitioning step, the first-stage continuous and binary variables are dealt with differently. A variable x_p whose current value is most in-between its bounds is selected as the partitioning variable. If $p \in I_b$, then x_p is fixed at 0 and 1 in the two child nodes, respectively; otherwise, x_p is a continuous variable, and its current value $\bar{x}_p$ (or the midpoint of the current bounding interval for x_p) is used as the lower and upper bounds for the two child nodes, respectively. Therefore, barring finite convergence, along any infinite branch of the branch-and-bound tree, there will exist a subsequence of the selected nodes such that the bounding interval for some continuous variable x_p is partitioned infinitely many times and such that the limiting value $\bar{x}_p$ coincides with one of its bounds. As x_p was selected whenever its value was most in-between its bounds, in the limit, all $\bar{x}_i$, $\forall i = 1, \ldots, n_1$ would coincide with one of their bounds, and hence, $\bar{x}$ would be a vertex of the limiting bounding hyperrectangle, thereby providing an upper bounding solution for the original two-stage stochastic program. Together with the node selection rule, the partitioning process therefore guarantees convergence of the algorithm to an optimal solution.

A difficulty in directly implementing the modified Benders' method by Sherali and Fraticelli [39] at the nodal problems arises due to the fact that when $\bar{x}$ is not extremal with respect to its bounds, the solution $\bar{y}$ obtained for a Benders subproblem defined on a partial convex hull representation in the (x, y)-space may not satisfy its binary restrictions. We thus need to be able to detect whether a Benders subproblem is already solved by such a fractional $\bar{y}$. This can be achieved as follows. If $\bar{y}_j \in \{0, 1\}$, $\forall j \in J_b$, or if $(\bar{x}, \bar{y})$ can be represented as a convex combination of some extreme points of the current partial convex hull defining the Benders subproblem such that these extreme points have binary y_j-variables for all $j \in J_b$, then $\bar{y}$ solves the Benders subproblem. Sherali and Zhu [41] describe a procedure to check this situation in their overall algorithmic scheme.

The RLT cuts generated for any given subhyperrectangle are reusable by updating the x-values in subsequent Benders iterations at the same node, and are also inheritable by the subproblems of the child nodes. Likewise, the Benders cuts derived for a given subhyperrectangle can also be inherited by the lower bounding master programs solved for its child nodes.

12.3.3 Connections Between Disjunctive Cuts and RLT Cuts

In this section, we demonstrate the connections between the two types of convexification cuts, namely, the disjunctive cuts generated by Problems (12.16) and (12.20) (or the counterpart of problem (12.9) in a stochastic setting), and the RLT cuts generated by Problem (12.24) under fixed recourse. We discuss these cuts in the context of fixed recourse stochastic programs having continuous first-stage variables, and the analysis is similar for the case of discrete first-stage variables.

Let x be continuous and bounded by $l \leq x \leq u$ at some node of the B&B process performed on the projected x-space as described in Section 12.3.2. Upon multiplying the constraints

$$T_s x + \tilde{W}\tilde{y} + W^q y_q \geq r_s \quad \text{and} \quad l \leq x \leq u \tag{12.32}$$

by y_q and $(1 - y_q)$, linearizing upon substituting the resulting nonlinear terms using (12.22), the higher-dimensional system (12.23) obtained is as follows.

$$
\begin{aligned}
& T_s z^x + \tilde{W} z^y \geq (r_s - W^q) y_q && \leftarrow \phi_0 && \text{(12.33a)} \\
& -T_s z^x - \tilde{W} z^y \geq r_s - T_s x - \tilde{W}\tilde{y} - r_s y_q && \leftarrow \phi_1 && \text{(12.33b)} \\
& z^x \geq l y_q && \leftarrow \phi_{xl0} && \text{(12.33c)} \\
& -z^x \geq -u y_q && \leftarrow \phi_{xu0} && \text{(12.33d)} \\
& -z^x \geq -l y_q + l - x && \leftarrow \phi_{xl1} && \text{(12.33e)} \\
& z^x \geq u y_q - u + x && \leftarrow \phi_{xu1}. && \text{(12.33f)}
\end{aligned}
$$

The cut generation problem as a counterpart to (12.24) then takes the following form.

$$
\begin{aligned}
\text{Minimize } \; & \tilde{\beta}_s \bar{\tilde{y}} + \beta_{qs} \bar{y}_q + \alpha_s \bar{x} - \gamma_s && \text{(12.34a)} \\
\text{subject to: } \; & (\phi_0 - \phi_1) T_s + \phi_{xl0} - \phi_{xu0} - \phi_{xl1} + \phi_{xu1} = 0 && \text{(12.34b)} \\
& (\phi_0 - \phi_1) \tilde{W} = 0 && \text{(12.34c)} \\
& \alpha_s = \phi_1 T_s + \phi_{xl1} - \phi_{xu1} && \text{(12.34d)} \\
& \tilde{\beta}_s = \phi_1 \tilde{W} && \text{(12.34e)} \\
& \beta_{qs} = \gamma_s + \phi_0 (W^q - r_s) - \phi_{xl0} l + \phi_{xu0} u && \text{(12.34f)} \\
& \gamma_s = \phi_1 r_s + \phi_{xl1} l - \phi_{xu1} u && \text{(12.34g)} \\
& \tilde{\beta}_s \bar{\tilde{y}} + \beta_{qs} \bar{y}_q + \alpha_s \bar{x} - \gamma_s \geq -1 && \text{(12.34h)} \\
& \phi \geq 0, \; \alpha_s, \tilde{\beta}_s, \beta_{qs}, \gamma_s \text{ unrestricted,} && \text{(12.34i)}
\end{aligned}
$$

where $(\bar{x}, \bar{y})$ is the current solution having $\bar{y}_q$, $q \in J_b$, fractional. Note that the resulting cut is of the form

$$\bar{\alpha}_s x + \bar{\tilde{\beta}}_s \tilde{y} + \bar{\beta}_{qs} y_q \geq \bar{\gamma}_s, \tag{12.35}$$

where $(\bar{\alpha}_s, \bar{\tilde{\beta}}_s, \bar{\beta}_{qs}, \bar{\gamma}_s, \bar{\phi})$ solves Problem (12.34).

Proposition 12.4. *Problem* (12.34) *is equivalent to*

$$
\begin{aligned}
\textit{Minimize } \; & \tilde{\beta}_s \bar{\tilde{y}} + \beta_{qs} \bar{y}_q + \alpha_s \bar{x} - \gamma_s && \text{(12.36a)} \\
\textit{subject to: } \; & \alpha_s \geq \phi_0 T_s + \phi_{xl0} - \phi_{xu0} && \text{(12.36b)} \\
& \alpha_s \geq \phi_1 T_s + \phi_{xl1} - \phi_{xu1} && \text{(12.36c)} \\
& \tilde{\beta}_s \geq \phi_0 \tilde{W} && \text{(12.36d)} \\
& \tilde{\beta}_s \geq \phi_1 \tilde{W} && \text{(12.36e)} \\
& \beta_{qs} \geq \gamma_s + \phi_0 (W^q - r_s) - \phi_{xl0} l + \phi_{xu0} u && \text{(12.36f)} \\
& \gamma_s \leq \phi_1 r_s + \phi_{xl1} l - \phi_{xu1} u && \text{(12.36g)} \\
& \tilde{\beta}_s \bar{\tilde{y}} + \beta_{qs} \bar{y}_q + \alpha_s \bar{x} - \gamma_s \geq -1 && \text{(12.36h)} \\
& \phi \geq 0, \; \alpha_s, \tilde{\beta}_s, \beta_{qs}, \gamma_s \; \textit{unrestricted.} && \text{(12.36i)}
\end{aligned}
$$

Proof. Noting the objective coefficients of β_{qs} and γ_s in Problem (12.34), we can equivalently write the equalities of (12.34f) and (12.34g) as the inequalities (12.36f) and (12.36g), respectively, as these constraints will always hold as equalities in any optimal solution. Substituting for the terms involving ϕ_1 in (12.34d) and (12.34e) into (12.34b) and (12.34c), respectively, yields $\alpha_s = \phi_0 T_s + \phi_{xl0} - \phi_{xu0}$ and $\tilde{\beta}_s = \phi_0 \tilde{W}$, that is, (12.36b) and (12.36d) in equality forms.

To complete the proof, we now show that at an optimal solution to (12.36), constraints (12.36b)–(12.36e) will hold as equalities as in (12.34b)–(12.34e). Suppose that, on the contrary, we have an optimal solution

$$(\hat{\alpha}, \hat{\tilde{\beta}}, \hat{\beta}_{qs}, \hat{\gamma}, \hat{\phi}_{xl1}, \hat{\phi}_{xl0}, \hat{\phi}_{-}),$$

where ϕ_{-} represent the vector ϕ except for the elements ϕ_{xl1} and $\hat{\phi}_{xl0}$, such that

$$\hat{\alpha}_{si} = \hat{\phi}_0 T_{si} + \hat{\phi}_{xl0i} - \hat{\phi}_{xu0i} > \hat{\phi}_1 T_{si} + \hat{\phi}_{xl1i} - \hat{\phi}_{xu1i}, \quad \text{or} \tag{12.37a}$$

$$\hat{\alpha}_{si} = \hat{\phi}_1 T_{si} + \hat{\phi}_{xl1i} - \hat{\phi}_{xu1i} > \hat{\phi}_0 T_{si} + \hat{\phi}_{xl0i} - \hat{\phi}_{xu0i} \tag{12.37b}$$

for some $i \in I$, where T_{si} is the ith column of T_s.

If (12.37a) occurs, then let $\phi_{\epsilon i} = [\hat{\phi}_0 T_{si} + \hat{\phi}_{xl0i} - \hat{\phi}_{xu0i}] - [\hat{\phi}_1 T_{si} + \hat{\phi}_{xl1i} - \hat{\phi}_{xu1i}] > 0$. Obtain a new solution $(\hat{\alpha}, \hat{\tilde{\beta}}, \beta'_{qs}, \gamma'_s, \phi'_{xl1}, \hat{\phi}_{xl0}, \hat{\phi}_{-})$ such that

$$\phi'_{xl1i} = \hat{\phi}_{xl1i} + \phi_{\epsilon i}, \qquad \phi'_{xl1j} = \hat{\phi}_{xl1j}, \qquad \forall j \neq i,$$
$$\gamma'_s = \hat{\gamma}_s + l_i \phi_{\epsilon i}, \qquad \text{and} \qquad \beta'_{qs} = \hat{\beta}_{qs} + l_i \phi_{\epsilon i}.$$

This new solution is feasible, satisfies $\hat{\alpha}_{si} = \hat{\phi}_0 T_{si} + \hat{\phi}_{xl0i} - \hat{\phi}_{xu0i} = \hat{\phi}_1 T_{si} + \phi'_{xl1i} - \hat{\phi}_{xu1i}$, and reduces the objective value by $\phi_{\epsilon i} l_i (1 - \bar{y}_q) > 0$. Hence, if (12.37a) holds, $(\hat{\alpha}, \hat{\tilde{\beta}}, \hat{\beta}_{qs}, \hat{\gamma}, \hat{\phi}_{xl1}, \hat{\phi}_{xl0}, \hat{\phi}_{-})$ cannot be an optimal solution. On the other hand, suppose that (12.37b) occurs. Then let $\phi_{\epsilon i} = [\hat{\phi}_1 T_{si} + \hat{\phi}_{xl1i} - \hat{\phi}_{xu1i}] - [\hat{\phi}_0 T_{si} + \hat{\phi}_{xl0i} - \hat{\phi}_{xu0i}] > 0$. Similarly, we can obtain a new solution $(\hat{\alpha}, \hat{\tilde{\beta}}, \beta'_{qs}, \hat{\gamma}_s, \hat{\phi}_{xl1}, \phi'_{xl0}, \hat{\phi}_{-})$ such that

$$\phi'_{xl0i} = \hat{\phi}_{xl0i} + \phi_{\epsilon i}, \ \phi'_{xl0j} = \hat{\phi}_{xl0j}, \ \forall j \neq i, \text{ and } \beta'_{qs} = \hat{\beta}_{qs} - l_i \phi_{\epsilon i}.$$

Again, this new solution is feasible, satisfies $\hat{\alpha}_{si} = \hat{\phi}_1 T_{si} + \hat{\phi}_{xl1i} - \hat{\phi}_{xu1i} = \hat{\phi}_0 T_{si} + \phi'_{xl0i} - \hat{\phi}_{xu0i}$, and reduces the objective value by $\phi_{\epsilon i} l_i \bar{y}_q > 0$. Hence, if (12.37b) holds, then $(\hat{\alpha}, \hat{\tilde{\beta}}, \hat{\beta}_{qs}, \hat{\gamma}, \hat{\phi}_{xl1}, \hat{\phi}_{-})$ cannot be an optimal solution, either. Therefore, we always have (12.36b) and (12.36c) tight at an optimal solution. Similarly, (12.36d) and (12.36e) always hold as equalities in an optimal solution. □

Now, consider the disjunctive cut generation where $Ax \geq b$ contains only the bounding constraints $l \leq x \leq u$ for x. Applying $y_q^s \in \{0,1\}$ in the system (12.32), we obtain the following disjunction for scenario s.

$$\begin{matrix} \phi_1 \to \\ \phi_{xl1} \to \\ \phi_{xu1} \to \\ \lambda_1 \to \end{matrix} \left\{ \begin{matrix} T_s x + \tilde{W}\tilde{y}^s \geq r_s \\ x \geq l \\ -x \geq -u \\ -y_q^s \geq 0 \end{matrix} \right\} \bigvee \left\{ \begin{matrix} T_s x + \tilde{W}\tilde{y}^s \geq r_s - W^q \\ x \geq l \\ -x \geq -u \\ y_q^s \geq 1 \end{matrix} \right\}. \begin{matrix} \leftarrow \phi_0 \\ \leftarrow \phi_{xl0} \\ \leftarrow \phi_{xu0} \\ \leftarrow \lambda_0 \end{matrix} \tag{12.38}$$

Using the associated multipliers, a disjunctive cut can be generated by solving the following problem.

$$\text{Minimize } \tilde{\beta}_s \bar{\tilde{y}} + \beta_{qs}\bar{y}_q + \alpha_s \bar{x} - \gamma_s \tag{12.39a}$$
$$\text{subject to: constraints (12.36b)–(12.36e), (12.36h), and} \tag{12.39b}$$
$$\beta_{qs} \geq -\lambda_1 \tag{12.39c}$$
$$\beta_{qs} \geq \lambda_0 \tag{12.39d}$$
$$\gamma_s \leq \phi_0(r_s - W^q) + \phi_{xl0}l - \phi_{xu0}u + \lambda_0 \tag{12.39e}$$
$$\gamma_s \leq \phi_1 r_s + \phi_{xl1}l - \phi_{xu1}u \tag{12.39f}$$
$$\phi, \lambda \geq 0,\ \alpha_s, \tilde{\beta}_s, \beta_{qs}, \gamma_s \text{ unrestricted.} \tag{12.39g}$$

Because β_{qs} appears only in constraints (12.39c) and (12.39d), we will have $\beta_{qs} \equiv \lambda_0 \geq -\lambda_1$ in an optimal solution. Constraints (12.39c)–(12.39e) then directly reduce to (12.36f), and Problem (12.39) is exactly the same as Problem (12.36). Therefore, the cut (12.35) generated using a RLT process is indeed also a disjunctive cut.

Applying the C^3 result from Sen and Higle [35], we can then generate disjunctive cuts that are valid for the disjunctions (12.38) for all scenarios, so that the cuts contain a common coefficient β for y. Similar to (12.16), the cut coefficients can be obtained by collecting the coefficient matrices of all scenarios and solving the following problem.

$$\text{Minimize } \sum_s p_s(\tilde{\beta}\bar{\tilde{y}}^s + \beta_q \bar{y}_q^s + \alpha_s \bar{x} - \gamma_s) \tag{12.40a}$$
$$\text{subject to: } \alpha_s \geq \phi_h T_s + \phi_{xlhs} - \phi_{xuhs}, \quad \forall h = 0,1,\ \forall s \in S \tag{12.40b}$$
$$\tilde{\beta} \geq \phi_h \tilde{W}, \quad \forall h = 0,1 \tag{12.40c}$$
$$\beta_q \geq \gamma_s + \phi_0(W^q - r_s) - \phi_{xl0s}l + \phi_{xu0s}u, \quad \forall s \in S \tag{12.40d}$$
$$\gamma_s \leq \phi_1 r_s + \phi_{xl1s}l - \phi_{xu1s}u, \quad \forall s \in S \tag{12.40e}$$
$$\sum_s p_s(\tilde{\beta}\bar{\tilde{y}}^s + \beta_q \bar{y}_q^s + \alpha_s \bar{x} - \gamma_s) \geq -1 \tag{12.40f}$$
$$\phi \geq 0,\ \alpha_s, \tilde{\beta}, \beta_q, \gamma_s \text{ unrestricted.} \tag{12.40g}$$

For an optimal solution $(\bar{\bar{\beta}}, \bar{\beta}_q, \bar{\alpha}_s, \bar{\gamma}_s, \forall s)$, $\bar{\bar{\beta}}\tilde{y}^s + \bar{\beta}_q y_q^s \geq \bar{\gamma}_s - \bar{\alpha}_s x$, $\forall s \in S$ are then valid inequalities for disjunctions (12.38) for all scenarios $s \in S$, and they share the same coefficient $\bar{\beta} \equiv (\bar{\bar{\beta}}, \bar{\beta}_q)$ for y.

We close this section by emphasizing that the efficacy of cutting plane methods depends on problem structures. They are most efficient when tight representations are available using relatively few cuts. Ntaimo and Sen [26] have successfully applied the disjunctive decomposition (D^2) method ([35]) on stochastic server location problems having as many as 1,000,010 binary variables and 120,010 constraints, using a Sun 280R with UltraSPARC-III+ cpus running at 900 MHz. For most of the nontrivial problems, the benchmarking commercial MIP software package CPLEX 7.0 failed to solve the problems. Another reason for this method to be so effective is due to the cut coefficient sharing property. As we have mentioned, in practice, cut coefficient sharing and reuse are important in saving computation, and greatly enhance the efficacy of cutting plane methods.

12.4 Structural Enumeration Using a Fixed Technology Matrix

In this section, we assume the following.

A1. The first-stage feasible region X is bounded, and thus compact.

A2. For any $x \in X$, the second-stage problems are feasible (relative complete recourse).

A3. For any $x \in X$, the second-stage value functions are bounded.

A4. The second-stage variables are purely integer; that is, $y \in \mathbb{Z}_+^{n_2}$.

A5. The technology matrix T is fixed (i.e., deterministic).

A6. All elements in the recourse matrices W_s are integral. (Rational elements can be scaled to obtain integral elements.)

Assumption A1 assures finite termination for the enumeration schemes used in the algorithms in this section. Assumptions A2 and A3 imply that $f_s(x)$ and $u(r_s - Tx)$, $\forall u \in \mathbb{R}_+^{m_2}$, are finite. Using Assumption A5, we can transform the second-stage value functions as defined on the space of the so-called "tender variables," $-Tx$, which we denote as χ. That is,

$$F_s(\chi) = \min\{g_s y \mid W_s y \geq r_s + \chi, y \in \mathbb{Z}_+^{n_2}\} = f_s(x). \tag{12.41}$$

From Assumptions A4 and A6, we have that $W_s y \geq t$ implies $W_s y \geq \lceil t \rceil$, where $\lceil \cdot \rceil$ and $\lfloor \cdot \rfloor$, respectively, denote the componentwise rounded-up and rounded-down forms of a vector. Therefore, we have that $f_s(x)$ is constant on the sets

$$\{x \in \mathbb{R}_+^{n_1} : \lceil r_s - Tx \rceil = \kappa\} = \{x : \kappa - r_s - e < -Tx \leq \kappa - r_s\}, \quad \forall \kappa \in \mathbb{Z}^{m_2}, \tag{12.42a}$$

and $F_s(\chi)$ is constant on the hyperrectangles

$$\prod_{j=1}^{m_2} (\kappa_j - r_{sj} - 1, \kappa_j - r_{sj}], \qquad \forall \kappa \in \mathbb{Z}^{m_2}. \tag{12.42b}$$

Therefore, for some integral vector $k \equiv (k_{11}, \ldots, k_{sj}, \ldots, k_{|S|m_2})^T \in \mathbb{Z}^{|S|m_2}$, the expectation $Q(x) \equiv \sum_s p_s f_s(x)$ is constant over the intersection of the sets (12.42a):

$$C(k) \equiv \bigcap_{s \in S} \bigcap_{j=1}^{m_2} \{x : k_{sj} - r_{sj} - 1 < -T_j x \leq k_{sj} - r_{sj}\}, \qquad \forall k \in \mathbb{Z}^{|S|m_2} \tag{12.43a}$$

and $\mathcal{Q}(\chi) \equiv \sum_s p_s F_s(\chi)$ is constant over the intersection of the hyperrectangles (12.42b):

$$\mathcal{C}(k) \equiv \bigcap_{s \in S} \prod_{j=1}^{m_2} (k_{sj} - r_{sj} - 1, k_{sj} - r_{sj}], \qquad \forall k \in \mathbb{Z}^{|S|m_2}. \tag{12.43b}$$

Based on the above observation that function values are constant over regions, Schultz et al. [32] and Ahmed et al. [1] developed algorithms that construct partitions of the first-stage feasible region using $C(\cdot)$ and $\mathcal{C}(\cdot)$, respectively, and evaluate expected values, $Q(\cdot)$ and $\mathcal{Q}(\cdot)$, on these partitions. The algorithm in the former paper operates over the x-space, whereas that in the latter works in the χ-space.

Under Assumption A1 that the first-stage feasible region X is compact, Schultz et al. [32] show that there are a finite number of vertices of the sets $C(\cdot) \cap X$, and these vertices contain an optimal solution for x. (If X is not compact, this vertex set is only countable, and a level set is used to bound the feasible region to guarantee finite termination.) Aside from A1–A6, Schultz et al. [32] further assume the following.

A7. The recourse matrix W is fixed.

A8. The first-stage variables are continuous; that is, $I_b = I_{int} = \varnothing$.

The purpose of these assumptions is to use the Gröbner-basis method from computational algebra to evaluate the second-stage value functions. Although the computation of Gröbner bases is expensive, it only uses the T and W matrices, and does not rely on the right-hand-side values. Hence, for fixed T and W, Gröbner bases need to be computed only once, and then for each different right-hand-side value, the second-stage function evaluation, which reduces to a single generalized division, is very easy. Using this idea, the objective value $cx + Q(x)$ is evaluated at each candidate point from the finite vertex set using Gröbner bases, and after enumerating all these candidate points, the one that yields the lowest function value is an optimal solution to the problem. Improved versions of this algorithm are proposed to reduce the number of candidate points that need to be evaluated.

The explicit enumeration approach is still quite expensive for nontrivial problems. Using a systematic enumeration scheme such as B&B to evaluate the partitions is more practical. However, in the x-space, the discontinuities between these partitions are not orthogonal to the (x-) variable axes, thus branching on x-variables would result in discontinuities in the interior of the hyperrectangles defining the nodal problems. To circumvent this difficulty, Ahmed et al. [1] propose to transform the problem into the space of the tender variable vector χ:

$$\text{TP: } \min\{cx + \sum_s p_s F_s(\chi) \mid x \in X,\ Tx = \chi, \text{ and } (12.41)\}, \tag{12.44}$$

so that the partitioned hyperrectangles are of the form (12.43b), having constant second-stage function values, and the discontinuities between these hyperrectangles are orthogonal to the (χ-) variable axes. They show that if χ^* is an optimal solution to (12.44), then $x^* \in \operatorname{argmin}\{cx \mid x \in X,\ Tx = \chi^*\}$ is an optimal solution of (12.1).

Their B&B algorithm can handle general first-stage variables and scenario-dependent recourse. Hence, Assumptions A7 and A8 are not needed. For each hyperrectangular partition $P^k \equiv \prod_{j=1}^{m_2}(l_j^k, u_j^k]$ in the form of (12.42b), because $F_s(\chi)$ is lower semicontinuous and nondecreasing, we can obtain a lower bound on the two-stage problem defined on P^k by solving the following problem.

$$\text{Minimize } cx + \eta \tag{12.45a}$$
$$\text{subject to } x \in X,\ Tx = \chi \tag{12.45b}$$
$$l^k \le \chi \le u^k \tag{12.45c}$$
$$\eta \ge \sum_{s \in S} p_s F_s(l^k + \epsilon), \tag{12.45d}$$

where $F_s(\cdot)$ is as defined in (12.41), and ϵ is calculated a priori as a sufficiently small number such that $F_s(\cdot)$ is constant over $(l^k, l^k + \epsilon]$. The value for ϵ is decided as follows. Along each axis j, $\forall j = 1, \ldots, m_2$, within the unit interval $(k_{1j} - r_{1j} - 1, k_{1j} - r_{1j}]$ for some $k_{1j} \in \mathbb{Z}$, the candidate point of discontinuity for each scenario s is identified as $\lfloor k_{1j} - r_{1j} + r_{sj} \rfloor - r_{sj}$, $\forall s \in S$. These points of discontinuity repeat the same pattern to the right of $k_{1j} - r_{1j}$ with a unit period. It is then sufficient to sort these points, and obtain the smallest interval as ϵ_j along axis j. The final ϵ-value is chosen as strictly smaller than each ϵ_j.

At the node selection step, the partition that yields the least lower bound is selected for further branching. At the branching step, the branching variable $\chi_{j'}$ is selected such that $\chi_{j'} + r_{sj'}$ is an integer greater than $l_{j'}$ and $\mathcal{Q}(\chi)$ is discontinuous at $\chi_{j'}$ for some scenario s. Let $\bar{y}^s$ be the solution obtained from (12.45) for scenario s. For each $j = 1, \ldots, m_2$, let the $p_j \equiv \min_{s \in S}\{(W_s\bar{y}^s)_j - r_{sj}\}$, and let the variable $\chi_{j'}$ be chosen such that $p_{j'}$ lies most in-between its bounds $l_{j'}^k$ and $u_{j'}^k$. Partition P^k is then branched along axis j' at the value $p_{j'}$. Using such a B&B scheme, the algorithm therefore avoids an exhaustive enumeration of the partitions, yet guarantees finite convergence.

This algorithm is applied by Ahmed and Garcia [2] to a dynamic acquisition and assignment problem. Problems up to the size of 24,027 variables (including 24,009 binary variables), 10,518 constraints, and 46,545 nonzeros were randomly generated to test the efficacy of the algorithm. On a Sun Sparc Ultra60 workstation, 85% of the problems were solved within the specified tolerance. The remaining prematurely terminated problems reached an average gap of 2.05% for those due to time limitation and 1.10% for those due to memory or node limitation. Using CPLEX 7.0, however, only 27% of the problems were solved.

Similar to Schultz et al. [32], Kong et al. [22] examine integer programs that differ only in their right-hand sides. In lieu of Gröbner bases, their algorithms use stored value functions of parameterized integer programs. In addition to Assumptions A1–A7, they further assume the following.

A9. The second-stage objective coefficient vector g is fixed.

A10. The first-stage variables are purely integer-restricted; that is, $x \in \mathbb{Z}_+^{n_1}$.

A11. All elements in A, T, b, and r_s, $\forall s \in S$, are integral.

The first- and second-stage problems now both belong to the type of pure integer programs that have integral coefficient matrices and right-hand-side vectors. In general, given a coefficient matrix $G \in \mathbb{Z}_+^{m \times n}$, this type of integer program can be expressed as

$$(VF) : z(\zeta) = \min\{dx \mid Gx \geq \zeta, x \in \mathbb{Z}_+^n\}, \quad \text{for } \zeta \in \mathbb{Z}^m, \tag{12.46}$$

where $z(\cdot) : \mathbb{Z}^m \mapsto \mathbb{Z}$ is its value function. $z(\cdot)$ is nondecreasing over $\mathbb{Z}^m$ and subadditive over $D = \{\zeta \in \mathbb{Z}^m \mid \{x \in \mathbb{Z}_+^n \mid Gx \geq \zeta\} \neq \varnothing\}$. That is, for ζ_1, $\zeta_2 \in D$, $\zeta_1 + \zeta_2 \in D$ implies that $z(\zeta_1) + z(\zeta_2) \geq z(\zeta_1 + \zeta_2)$. (The authors use the superadditivity property for maximization problems. We chose to use subadditivity for minimization problems to maintain consistency in exposition.)

Using these properties, Kong et al. [22] develop an integer programming-based algorithm and a dynamic programming-based algorithm that compute the value of $z(\zeta)$ for each ζ in a finite set Υ. In the integer programming-based algorithm, at each iteration k, some $\zeta^k \in \Upsilon$ is selected, and the corresponding integer program $\min\{dx \mid Gx \geq \zeta^k, x \in \mathbb{Z}_+^n\}$ is solved to obtain a solution $\hat{x}^k$. Based on the value of $\hat{x}^k$, other $\zeta \in \Upsilon$ are then selected to have their lower bounds, $l(\zeta)$, and upper bounds, $u(\zeta)$, updated using the nondecreasing and subadditivity properties of $z(\zeta)$, until $l(\zeta) = u(\zeta)$, for all $\zeta \in \Upsilon$, at which point, $z(\zeta) = l(\zeta) = u(\zeta)$ are available for all $\zeta \in \Upsilon$. The dynamic programming-based approach does not solve any integer program, but is only applicable when G is nonpositive. Denote Υ_j to be the set of $\zeta \in \Upsilon$ such that $\zeta \leq G_j$. This algorithm uses the initial condition $z(\zeta) = 0$, $\forall \zeta \in \Upsilon \setminus \bigcup_{j=1}^n \Upsilon_j$, and the recursive function $z(\zeta) = \min\{d_j + z(\zeta + G_j) : G_j \in \Upsilon, j = 1, \ldots, n\}$, $\forall \zeta \in \bigcup_{j=1}^n \Upsilon_j$. Naturally, both algorithms work better when the size of Υ is small.

To use these value function evaluation algorithms to solve two-stage SMIPs, denote the first-stage value function as

$$z_1(\zeta_1) = \min\{cx \mid Ax \geq b,\, Tx \geq \zeta_1,\, x \in \mathbb{Z}_+^{n_1}\},$$

$$\forall \zeta_1 \in \Upsilon^1 \equiv \{\zeta_1 \in \mathbb{Z}^{m_2} \mid \zeta_1 = Tx \text{ for some } x \in X\},$$

and denote the second-stage value function as

$$z_2(\zeta_2) = \min\{gy \mid Wy \geq \zeta_2,\, y \in \mathbb{Z}_+^{n2}\},$$

$$\forall \zeta_2 \in \Upsilon^2 \equiv \bigcup_{\zeta_1 \in \Upsilon^1} \bigcup_{s \in S} \{r_s - \zeta_1\}.$$

These function values for the first- and second-stage problems are stored for all possible right-hand-side values in Υ^1 and Υ^2, respectively. (Due to the storage requirement, this approach is more suitable for problems having a relatively small number of rows.) After storing these value function responses, the feasible region for χ is systematically searched to obtain an optimal value χ^*. A finitely convergent B&B search scheme is proposed. At node k defined on a hyperrectangle $[l^k, u^k]$, a lower bound LB^k and an upper bound UB^k are computed as

$$LB^k = z_1(l^k) + \sum_s p_s z_2(r_s - u^k),$$

and

$$UB^k = z_1(u^k) + \sum_s p_s z_2(r_s - l^k).$$

Branching is performed by partitioning the current hyperrectangle by bisecting some selected axis, which ensures convergence.

12.5 Conclusion

In this chapter, we have reviewed some recent advances in solving two-stage stochastic (mixed-) integer programs, and have provided some insights and results that exhibit certain interconnections between the methods. Due to the dual angular structure, these solution approaches apply various versions of decomposition, branch-and-bound, and convexification techniques. We have studied these methods from the viewpoint of the adopted decomposition

Table 12.1 Literature on solving stochastic mixed-integer programs

First Stage (x)	Second Stage (y)	Literature	Approach	Section
Binary	Continuous	[23]	L-shaped	12.2.1
		[3]	Branch-and-Fix	12.2.2
Continuous	Integer	[32]	Gröbner basis	12.4
Integer	Integer	[17]	Test-set decomposition	12.2.3
		[22]	Value function	12.4
Mixed-integer	Integer	[1]	χ-transformation	12.4
		[14]	Dual pricing	12.2.1
Continuous	0–1 mixed-integer	[13]	Disj. prog.	12.3.1
		[27]	Disj. prog.	12.3.2
Binary	0–1 mixed-integer	[39]	Benders & RLT	12.3.1
		[36]	Disj. prog.	12.3.1
		[35]	Disj. prog.	12.3.1
0–1 mixed-integer	0–1 mixed-integer	[41]	Convexification	12.3.2
Mixed-integer	Mixed-integer	[12]	Lagrangian dual	12.2.2
		[33]	Lagrangian dual	12.2.2

framework, the convexification approach used in solving problems having integer recourse, and the enumeration scheme employed when the technology matrix is deterministic.

Table 12.1 lists the literature we have covered, grouped by the type of variables appearing in the two stages, which is most relevant to the algorithmic developments. Many of these methods are theoretically extendable to multistage SMIPs. However, scalability becomes a major issue here, which requires further research. For an introduction on multistage stochastic integer programs, we refer the reader to Römisch and Schultz [28].

Acknowledgment This research is supported by the *National Science Foundation* under Grant Numbers DMI-0552676, DMI-0245643, and DMI-0455807.

References

1. Ahmed, S., Tawarmalani, M., and Sahinidis, N. V. (2004). A finite branch and bound algorithm for two-stage stochastic integer programs. *Mathematical Programming* **100**(2), 355–377.
2. Ahmed, S. and Garcia, R. (2004). Dynamic capacity acquisition and assignment under uncertainty. *Annals of Operations Research* **124**, 267–283.
3. Alonso-Ayuso, A., Escudero, L. F., and Ortuño, M. T. (2003). BFC, a branch-and-fix coordination algorithmic framework for solving some types of stochastic pure and mixed 0–1 programs. *European Journal of Operational Research* **151**, 503–519.

4. Alonso-Ayuso, A., Escudero, L. F., Garín, A., Ortuño, M. T., and Pérez, G. (2003). An approach for strategic supply chain planning under uncertainty based on stochastic 0–1 programming. *Journal of Global Optimization* **26**, 97–124.
5. Balas, E. (1979). Disjunctive programming. *Annals of Discrete Mathematics* **5**, 3–51.
6. Balas, E. (1997). A modified lift-and-project procedure. *Mathematical Programming* **79**(1–3), 19–31.
7. Balas, E., Ceria, S., and Cornuéjols, G. (1993). A lift-and-project cutting plane algorithm for mixed 0–1 programs. *Mathematical Programming* **58**, 295–324.
8. Balas, E., Ceria, S., and Cornuéjols, G. (1996). Mixed 0–1 programming by lift-and-project in a branch-and-cut framework. *Management Science* **42**(9), 1229–1246.
9. Benders, J. F. (1962). Partitioning procedures for solving mixed-variables programming problems. *Numerische Mathematik* **4**, 238–252.
10. Blair, C. and Jeroslow, R. (1978). A converse for disjunctive constraints. *Journal of Optimization Theory and Applications* **25**, 195–206.
11. Blair, C. and Jeroslow, R. (1982). The value function of an integer program. *Mathematical Programming* **23**, 237–273.
12. Carøe, C. C. and Schultz, R. (1999). Dual decomposition in stochastic integer programming. *Operations Research Letters* **24**(1–2), 37–45.
13. Carøe, C. C. and Tind, J. (1997). A cutting-plane approach to mixed 0–1 stochastic integer programs. *European Journal of Operational Research* **101**(2), 306–316.
14. Carøe, C. C. and Tind, J. (1998). L-shaped decomposition of two-stage stochastic programs with integer recourse. *Mathematical Programming* **83a**(3), 451–464.
15. Graver, J. E. (1975). On the foundation of linear and integer programming I. *Mathematical Programming* **9**, 207–226.
16. Hemmecke, R. (2000). On the positive sum property of Graver test sets. Preprint SM-DU-468, University of Duisburg, http://www.uni-duisburg.de/FB11/disma/ramon/articles/preprint2.ps.
17. Hemmecke, R. and Schultz, R. (2003). Decomposition of test sets in stochastic integer programming. *Mathematical Programming, Ser. B* **94**, 323–341.
18. Jeroslow, R. G. (1980). A cutting plane game for facial disjunctive programs. *SIAM Journal of Control and Optimization* **18**, 264–280.
19. Klein Haneveld, W. K., Stougie, L., and van der Vlerk, M. H. (1995) On the convex-hull of the simple integer recourse objective function. *Annals of Operations Research* **56**, 209 224.
20. Klein Haneveld, W. K., Stougie, L., and van der Vlerk, M. H. (1996) An algorithm for the construction of convex hulls in simple integer recourse programming. *Annals of Operations Research* **64**, 67–81.
21. Klein Haneveld, W. K. and van der Vlerk, M. H. (1999). Stochastic integer programming: general models and algorithms. *Annals of Operations Research* **85**, 39–57.
22. Kong, N., Schaefer, A. J., and Hunsaker, B. (2006). Two-stage integer programs with stochastic right-hand sides: a superadditive dual approach. *Mathematical Programming*, **108** (2–3), 275–296.
23. Laporte, G. and Louveaux, F. V. (1993). The integer L-shaped method for stochastic integer programs with complete recourse. *Operations Research Letters* **13**(3), 133–142.
24. Martin, R. K. (1998). *Large Scale Linear and Integer Optimization: A Unified Approach*, Kluwer Academic, Norwell, MA.
25. Nemhauser, G. L. and Wolsey, L. A. (1999). *Integer and Combinatorial Optimization*, Wiley-Interscience, New York, 2nd edition.
26. Ntaimo, L. and Sen, S. (2005). The million-variable "march" for stochastic combinatorial optimization. *Journal of Global Optimization* **32**(3), 385–400.
27. Ntaimo, L. and Sen, S. (2008). A branch-and-cut algorithm for two-stage stochastic mixed-binary programs with continuous first-stage variables. *International Journal of Computational Science and Engineering* **3**(6), 232–241.

28. Römisch, W. and Schultz, R. (2001). Multi-stage stochastic integer programs: An introduction. In: M. Grötschel, S. O. Krumke, and J. Rambau (ed), *Online Optimization of Large Scale Systems*, 581–600. Springer, Berlin.
29. Schultz, R. (1995). On structure and stability in stochastic programs with random technology matrix and complete integer recourse. *Mathematical Programming* **70**(1), 73–89.
30. Schultz, R. (2003). Stochastic programming with integer variables. *Mathematical Programming* **97**(1–2), 285–309.
31. Schultz, R., Stougie, L., and van der Vlerk, M. H. (1996). Two-stage stochastic integer programming: a survey. *Statistica Neerlandica* **50**(3), 404–416.
32. Schultz, R., Stougie, L., and van der Vlerk, M. H. (1998). Solving stochastic programs with integer recourse by enumeration: A framework using Gröbner basis reductions. *Mathematical Programming* **83**(2), 229–252.
33. Schultz, R. and Tiedemann, S. (2003). Risk aversion via excess probabilities in stochastic programs with mixed-integer recourse. *SIAM Journal on Optimization* **14**(1), 115–138.
34. Sen, S. (2005). Algorithms for stochastic mixed-integer programming models. In: K. Aardal, G. Nemhauser, and R. Weismantel (ed), *Handbooks in Operations Research and Management Science: Discrete Optimization*, Vol. 12, Chapter 9, 511–558. Elsevier, Dordrecht.
35. Sen, S. and Higle, J. L. (2005). The C^3 theorem and a D^2 algorithm for large scale stochastic mixed-integer programming: Set convexification. *Mathematical Programming* **104**(1), 1–20.
36. Sen, S. and Sherali, H. D. (2006). Decomposition with branch-and-cut approaches for two-stage stochastic mixed-integer programming. *Mathematical Programming* **106**(2), 203–223.
37. Sherali, H. D. and Adams, W. P. (1990). A hierarchy of relaxations between the continuous and convex hull representations for zero–one programming problems. *SIAM Journal on Discrete Mathematics* **3**(3), 411–430.
38. Sherali, H. D. and Adams, W. P. (1994). A hierarchy of relaxations and convex hull characterizations for mixed-integer zero–one programming problems. *Discrete Applied Mathematics* **52**(1), 83–106.
39. Sherali, H. D. and Fraticelli, B. M. P. (2002). A modification of Benders' decomposition algorithm for discrete subproblems: An approach for stochastic programs with integer recourse. *Journal of Global Optimization* **22**, 319–342.
40. Sherali, H. D. and Shetty, C. M. (1980). *Optimization with Disjunctive Constraints. Lecture Notes in Economics and Mathematical Systems*, Vol. 181. Springer-Verlag, Berlin.
41. Sherali, H. D. and Zhu, X. (2006). On solving discrete two-stage stochastic programs having mixed-integer first- and second-stage variables. *Mathematical Programming*, **108** (2–3), 597–616.
42. Stougie, L. and van der Vlerk, M. H. (1997). Stochastic integer programming. In: M. Dell'Amico, F. Maffioli, and S. Martello (ed), *Annotated Bibliographies in Combinatorial Optimization*, 127–141. Wiley, Chichester.
43. Van Slyke, R. and Wets, R. (1969). L-shaped linear programs with applications to control and stochastic programming. *SIAM Journal on Applied Mathematics* **17**, 638–663.
44. Wollmer, R. M. (1980). Two-stage linear programming under uncertainty with 0–1 first-stage variables. *Mathematical Programming* **19**, 279–288.

Chapter 13
Dualistic Riemannian Manifold Structure Induced from Convex Functions

Jun Zhang and Hiroshi Matsuzoe

Key words: Legendre–Fenchel duality, biorthogonal coordinates, Riemannian metric, conjugate connections, equiaffine geometry, parallel volume form, affine immersion, Hessian geometry

13.1 Introduction

Convex analysis has wide applications in science and engineering, such as mechanics, optimization and control, theoretical statistics, mathematical economics and game theory, and so on. It offers an analytic framework to treat systems and phenomena that depart from linearity, based on an elegant mathematical characterization of the notion of "duality" (Rockafellar, 1970, 1974, Ekeland and Temam, 1976). Recent work of David Gao (2000) further provided a comprehensive and unified treatment of duality principles in convex and nonconvex systems, greatly enriching the theoretical foundation and scope of applications.

Central to convex analysis is the Legendre–Fenchel transform, and duality between two sets of variables defined on a pair of vector spaces that are dual with respect to each other. When the convex functions involved are smooth, these variables are in one-to-one correspondence; they can actually be viewed as two coordinate systems on a certain Riemannian manifold. This is the viewpoint from the so-called information geometry (Amari, 1985, Amari and Nagaoka, 2000), and it is investigated at great length in this chapter.

Jun Zhang · Department of Psychology, University of Michigan, Ann Arbor, MI 48109, U.S.A., e-mail: junz@umich.edu

Hiroshi Matsuzoe · Department of Computer Science and Engineering, Nagoya Institute of Technology, Gokiso-cho, Showa-ku, Nagoya 466-8555, Japan
e-mail: matsuzoe@nitech.ac.jp

D.Y. Gao, H.D. Sherali, (eds.), *Advances in Applied Mathematics and Global Optimization*
Advances in Mechanics and Mathematics 17, DOI 10.1007/978-0-387-75714-8_13,

The link between convex functions and Riemannian geometry is shown to be severalfold. First, the pair of convex functions conjugate to one another are the potential functions that induce the Riemannian metric. Second, the two sets of variables are special coordinate systems of the manifold in that they are "biorthogonal;" that is, the Jacobian of coordinate transformation between them is precisely the Riemannian metric. It turns out that biorthogonal coordinates are global coordinates for a pair of dually flat connections on the Riemannian manifold. Third, the Fenchel inequality provides a natural way to construct directed ("pseudo-") distance over the convex point set; this is the Bregman divergence (a.k.a. canonical divergence), which gives rise to the dually flat connections. Finally, the geometric structure (Riemannian metric, conjugate/dual connections) can be induced from graph immersions of a convex function into a higher-dimensional affine space.

Our goal in this chapter is to review such a geometric view of convex functions and the associated conjugacy/duality, as well as provide some new results. We review the background of convex analysis and Riemannian geometry (and affine hypersurface theory) in Section 13.2, with attention to the well-established relation between biorthogonal coordinates and dually flat (also called "Hessian") manifolds. In Section 13.3, we develop the full-fledged α-Hessian geometry, which extends the dually flat Hessian manifold ($\alpha = \pm 1$), and give an example from theoretical statistics when such geometry arises; this parallels the generalization of the convex-induced divergence function with arbitrary α (Zhang, 2004) from Bregman divergence ($\alpha = \pm 1$). To close, we give a summary and discuss some open problems in Section 13.4.

13.2 Convex Functions and Riemannian Geometry

13.2.1 Convex Functions and the Associated Divergence Functions

A strictly convex (or simply "convex") function $\Phi\colon V \subseteq \mathbb{R}^n \to \mathbb{R}, x \mapsto \Phi(x)$ is defined by

$$\frac{1-\alpha}{2}\Phi(x) + \frac{1+\alpha}{2}\Phi(y) - \Phi\left(\frac{1-\alpha}{2}x + \frac{1+\alpha}{2}y\right) > 0 \qquad (13.1)$$

for all $x \neq y$ for any $|\alpha| < 1$ (the inequality sign is reversed when $|\alpha| > 1$). In this chapter, V (and $\widetilde{V}$ below) identifies a subset of $\mathbb{R}^n$ both as a point set and as a vector space. We assume Φ to be sufficiently smooth (differentiable up to fourth order). Define

$$\mathcal{B}_\Phi(x, y) = \Phi(x) - \Phi(y) - \langle x - y, \partial\Phi(y)\rangle, \qquad (13.2)$$

where $\partial\Phi = [\partial_1\Phi, \ldots, \partial_n\Phi]$ with $\partial_i \equiv \partial/\partial x^i$ denotes the gradient valued in the co-vector space $\widetilde{V} \subseteq \mathbb{R}^n$, and $\langle \cdot, \cdot \rangle_n$ denotes the canonical pairing of a point/vector $x = [x^1, \ldots, x^n] \in V$ and a point/co-vector $u = [u_1, \ldots, u_n] \in \widetilde{V}$ (dual to V):

$$\langle x, u \rangle_n = \sum_{i=1}^{n} x^i u_i. \tag{13.3}$$

(Where there is no danger of confusion, the subscript n in $\langle \cdot, \cdot \rangle_n$ is often omitted.) A basic fact in convex analysis is that the necessary and sufficient condition for a smooth function Φ to be convex is

$$\mathcal{B}_\Phi(x, y) > 0 \tag{13.4}$$

for $x \neq y$. We remark that $\mathcal{B}_\Phi$ is sometimes called "Bregman divergence" (Bregman, 1967), widely used in convex optimization literature (Della Pietra et al., 2002, Bauschke, 2003, Bauschke and Combettes, 2003, Bauschke et al., 2003).

Zhang (2004) introduced the following family of functions on $V \times V$ as indexed by $\alpha \in \mathbb{R}$,

$$\mathcal{D}_\Phi^{(\alpha)}(x, y) = \frac{4}{1-\alpha^2}\left(\frac{1-\alpha}{2}\Phi(x) + \frac{1+\alpha}{2}\Phi(y) - \Phi\left(\frac{1-\alpha}{2}x + \frac{1+\alpha}{2}y\right)\right). \tag{13.5}$$

Here $\mathcal{D}_\Phi^{(\pm 1)}(x, y)$ is defined by taking $\lim_{\alpha \to \pm 1}$:

$$\begin{aligned} \mathcal{D}_\Phi^{(1)}(x, y) &= \mathcal{D}_\Phi^{(-1)}(y, x) = \mathcal{B}_\Phi(x, y), \\ \mathcal{D}_\Phi^{(-1)}(x, y) &= \mathcal{D}_\Phi^{(1)}(y, x) = \mathcal{B}_\Phi(y, x). \end{aligned}$$

Note that $\mathcal{D}_\Phi^{(\alpha)}(x, y)$ satisfies the relation (called "referential duality" in Zhang, 2006a)

$$\mathcal{D}_\Phi^{(\alpha)}(x, y) = \mathcal{D}_\Phi^{(-\alpha)}(y, x);$$

that is, exchanging the asymmetric status of the two points (in the directed distance) amounts to $\alpha \leftrightarrow -\alpha$.

From its construction, $\mathcal{D}_\Phi^{(\alpha)}(x, y)$ is nonnegative for $|\alpha| < 1$ due to equation (13.1), and for $|\alpha| = 1$ due to equation (13.4). For $|\alpha| > 1$, assuming $(((1-\alpha)/2)\, x + ((1+\alpha)/2)\, y) \in V$, the nonnegativity of $\mathcal{D}_\Phi^{(\alpha)}(x, y)$ can also be proven due to the inequality (13.1) reversing its sign. Therefore, we have

Lemma 13.1. *For a smooth function Φ: $V \subseteq \mathbb{R}^n \to \mathbb{R}$, the following conditions are equivalent (for $x, y \in V$).*

(i) *Φ is strictly convex.*
(ii) $\mathcal{D}_\Phi^{(1)}(x, y) \geq 0$.
(iii) $\mathcal{D}_\Phi^{(-1)}(x, y) \geq 0$.

(iv) $\mathcal{D}_{\Phi}^{(\alpha)}(x,y) \geq 0$ *for all* $|\alpha| < 1$.
(v) $\mathcal{D}_{\Phi}^{(\alpha)}(x,y) \geq 0$ *for all* $|\alpha| > 1$.

Recall that, when Φ is convex, its convex conjugate $\widetilde{\Phi}\colon \widetilde{V} \subseteq \mathbb{R}^n \to \mathbb{R}$ is defined through the Legendre–Fenchel transform:

$$\widetilde{\Phi}(u) = \langle (\partial\Phi)^{-1}(u), u\rangle - \Phi((\partial\Phi)^{-1}(u)), \tag{13.6}$$

with $\widetilde{\widetilde{\Phi}} = \Phi$ and $(\partial\Phi) = (\partial\widetilde{\Phi})^{-1}$. The function $\widetilde{\Phi}$ is also convex, and through which (13.4) precisely expresses the Fenchel inequality

$$\Phi(x) + \widetilde{\Phi}(u) - \langle x, u\rangle \geq 0$$

for any $x \in V$, $u \in \widetilde{V}$, with equality holding if and only if

$$u = (\partial\Phi)(x) = (\partial\widetilde{\Phi})^{-1}(x) \longleftrightarrow x = (\partial\widetilde{\Phi})(u) = (\partial\Phi)^{-1}(u), \tag{13.7}$$

or, in component form,

$$u_i = \frac{\partial\Phi}{\partial x^i} \longleftrightarrow x^i = \frac{\partial\widetilde{\Phi}}{\partial u_i}. \tag{13.8}$$

With the aid of conjugate variables, we can introduce the "canonical divergence" $\mathcal{A}_{\Phi}\colon V \times \widetilde{V} \to \mathbb{R}_+$ (and $\mathcal{A}_{\widetilde{\Phi}}\colon \widetilde{V} \times V \to \mathbb{R}_+$) where $\mathbb{R}_+ = \mathbb{R}^+ \cup \{0\}$

$$\mathcal{A}_{\Phi}(x,v) = \Phi(x) + \widetilde{\Phi}(v) - \langle x, v\rangle = \mathcal{A}_{\widetilde{\Phi}}(v,x).$$

They are related to the Bregman divergence (13.2) via

$$\mathcal{B}_{\Phi}(x, (\partial\Phi)^{-1}(v)) = \mathcal{A}_{\Phi}(x,v) = \mathcal{B}_{\widetilde{\Phi}}((\partial\widetilde{\Phi})(x), v).$$

Bregman (or canonical) divergence[1] provides a measure of directed distance between two points; that is, it is nonnegative for all values of $x, y \in V$, and vanishes only when $x = y$. More formally, a divergence function $\mathcal{D}\colon V \times V \to \mathbb{R}_+$ is a smooth function (differentiable up to third order) that satisfies

(i) $\mathcal{D}(x,y) \geq 0 \ \forall x, y \in V$ with equality holding if and only if $x = y$,
(ii) $\partial_{x^i}\mathcal{D}(x,y)\big|_{x=y} = \partial_{y^i}\mathcal{D}(x,y)\big|_{x=y} = 0$,
(iii) $\partial_{x^i}\partial_{y^j}\mathcal{D}(x,y)\big|_{x=y}$ is negative definite.

Here ∂_{x^i} denotes partial derivative with respect to the ith component of the x-variable only.[2]

---

[1] The divergence function, also called the "contrast function," is a terminology arising out of the theoretical statistics literature. It has nothing to do with the divergence operation in vector calculus.

[2] The reader should not confuse the shorthand notations ∂_i with ∂_{x^i} (or ∂_{y^i}): the former operates on a function defined on V such as $\Phi\colon x \mapsto \Phi(x) \in \mathbb{R}$, whereas the latter operates on a function defined on $V \times V$ such as $\mathcal{D}\colon (x,y) \mapsto \mathcal{D}(x,y) \in \mathbb{R}_+$.

13.2.2 Differentiable Manifold: Metric and Connection Structures

A differentiable manifold $\mathfrak{M}$ is a space that locally "looks like" a Euclidean space $\mathbb{R}^n$. By "looks like," we mean that for any base (reference) point $p \in \mathfrak{M}$, there exists a bijective mapping ("coordinate functions") between the neighborhood of p (i.e., a patch of the manifold) and a subset V of $\mathbb{R}^n$. By locally, we mean that various such mappings must be smoothly related to one another (if they are centered at the same reference point) or consistently glued together (if they are centered at different reference points) and globally cover the entire manifold. Below, we assume that a coordinate system is chosen such that each point is indexed by $x \in V$, with the origin as the reference point.

A manifold is specified with certain structures. First, there is an inner-product structure associated with tangent spaces of the manifold. This is given by the metric tensor field g which is, when evaluated at each location x (omitted in our notation), a symmetric bilinear form $g(\cdot,\cdot)$ of tangent vectors $X, Y \in T_p(\mathfrak{M}) \simeq \mathbb{R}^n$ such that $g(X,X)$ is always positive for all nonzero vectors X. In local coordinates with bases $\partial_i \equiv \partial/\partial x^i$, $i = 1,\ldots,n$ (i.e., X, Y are expressed as $X = \sum_i X^i\partial_i$, $Y = \sum_i Y^i\partial_i$), the components of g are denoted as

$$g_{ij}(x) = g(\partial_i, \partial_j). \tag{13.9}$$

The metric tensor allows us to define distance on a manifold as the shortest curve (called "geodesic") connecting two points. It also allows the measurement of angles and hence defines orthogonality of a vector to a submanifold. Projections of vectors to a lower-dimensional submanifold become possible once a metric is given.

Second, there is a structure associated with the notion of "parallelism" of vector fields on a manifold. This is given by the affine (linear) connection (or simply "connection") ∇, mapping two vector fields X and Y to a third one denoted by $\nabla_Y X$: $(X,Y) \mapsto \nabla_Y X$. Intuitively, it represents the "intrinsic" difference of the vector field X from its value at point x and its value at a nearby point connected to x (in the direction given by Y). Here "intrinsic" means that vector comparison at two neighboring locations of the manifold is through a process called "parallel transport," whereby a vector's components are adjusted as it moves across points on the base manifold. Under the local coordinate system with bases $\partial_i \equiv \partial/\partial x^i$, components of ∇ can be written out in "contravariant" form denoted Γ_{ij}^l (which is a collection of n^3 functions of x),

$$\nabla_{\partial_i}\partial_j = \sum_l \Gamma_{ij}^l\,\partial_l. \tag{13.10}$$

Under coordinate transform $x \mapsto \tilde{x}$, the new set of functions $\widetilde{\Gamma}$ are related to old ones Γ via

$$\tilde{\Gamma}^l_{mn}(\tilde{x}) = \sum_k \left(\sum_{i,j} \frac{\partial x^i}{\partial \tilde{x}^m} \frac{\partial x^j}{\partial \tilde{x}^n} \Gamma^k_{ij}(x) + \frac{\partial^2 x^k}{\partial \tilde{x}^m \partial \tilde{x}^n} \right) \frac{\partial \tilde{x}^l}{\partial x^k}. \tag{13.11}$$

A curve whose tangent vectors are intrinsically parallel along it is called an "auto-parallel curve."

As a primitive on a manifold, affine connections can be characterized in terms of their torsion and curvature. The torsion T of a connection Γ, which is a tensor itself, is given by the asymmetric part of the connection $T(\partial_i, \partial_j) = \nabla_{\partial_i}\partial_j - \nabla_{\partial_j}\partial_i = \sum_k T^k_{ij}\partial_k$, where T^k_{ij} is its local representation[3] given as

$$T^k_{ij}(x) = \Gamma^k_{ij}(x) - \Gamma^k_{ji}(x).$$

The curviness/flatness of a connection Γ is described by the curvature tensor R, defined as

$$R(\partial_i, \partial_j)\partial_k = (\nabla_{\partial_i}\nabla_{\partial_j} - \nabla_{\partial_j}\nabla_{\partial_i})\partial_k.$$

Writing $R(\partial_i, \partial_j)\partial_k = \sum_l R^l_{kij}\partial_l$ and substituting (13.10), the components of the curvature tensor are[4]

$$R^l_{kij}(x) = \frac{\partial \Gamma^l_{jk}(x)}{\partial x^i} - \frac{\partial \Gamma^l_{ik}(x)}{\partial x^j} + \sum_m \Gamma^l_{im}(x)\Gamma^m_{jk}(x) - \sum_m \Gamma^l_{jm}(x)\Gamma^m_{ik}(x).$$

By definition, R^l_{kij} is antisymmetric when $i \leftrightarrow j$. A connection is said to be flat when $R^l_{kij}(x) \equiv 0$. Note that this is a tensorial condition, so that the flatness of a connection ∇ is a coordinate-independent property even though the local expression of the connection (in terms of Γ) is highly coordinate-dependent. For any flat connection, there exists a local coordinate system under which $\Gamma^k_{ij}(x) \equiv 0$ in a neighborhood; this is the affine coordinate for a flat connection.

In the above discussions, metric and connections are treated as inducing separate structures on a manifold. On a manifold where both are defined, then it is convenient to express a connection Γ in its "covariant" form

$$\Gamma_{ij,k} = g(\nabla_{\partial_i}\partial_j, \partial_k) = \sum_l g_{lk}\Gamma^l_{ij}. \tag{13.12}$$

Although Γ^k_{ij} is the more primitive quantity that does not involve the metric, $\Gamma_{ij,k}$ represents the projection of an intrinsically differentiated vector field onto the manifold spanned by the bases ∂_k. The covariant form of the curva-

---

[3] Here and below, we restrict to holonomic coordinate systems in $\mathbb{R}^n$ only, where all coordinate bases commute $[\partial_i, \partial_j] = 0$ for $i \neq j$.

[4] This componentwise notation of curvature tensor here follows standard differential geometry textbooks, such as Nomizu and Sasaki (1994). On the other hand, information geometers, such as Amari and Nagaoka (2000), adopt the notation $R(\partial_i, \partial_j)\partial_k = \sum_l R^l_{ijk}\partial_l$, with $R_{ijkl} = \sum_l R^m_{ijk} g_{ml}$.

ture tensor is (cf. footnote 4)

$$R_{lkij} = \sum_m g_{lm} R^m_{kij}.$$

When the connection is torsion free, R_{lkij} is antisymmetric when $i \leftrightarrow j$ or when $k \leftrightarrow l$, and symmetric when $(i,j) \leftrightarrow (l,k)$. It is related to the Ricci tensor Ric (to be defined in (13.27) below) via $\text{Ric}_{kj} = \sum_{i,l} R_{lkij} g^{il}$.

13.2.3 Dualistic Structure on a Manifold: Compatibility Between Metric and Connection

A fundamental theorem of Riemannian geometry states that given a metric, there is a unique connection (among the class of torsion-free connections) that "preserves" the metric; that is, the following condition is satisfied:

$$\partial_k g(\partial_i, \partial_j) = g(\widehat{\nabla}_{\partial_k} \partial_i, \partial_j) + g(\partial_i, \widehat{\nabla}_{\partial_k} \partial_j). \tag{13.13}$$

Such a connection, denoted as $\widehat{\nabla}$, is known as the Levi-Civita connection. Its component forms, called Christoffel symbols, are determined by the components of the metric tensor as ("Christoffel symbols of the second kind")

$$\widehat{\Gamma}^k_{ij} = \sum_l \frac{g^{kl}}{2} \left(\frac{\partial g_{il}}{\partial x^j} + \frac{\partial g_{jl}}{\partial x^i} - \frac{\partial g_{ij}}{\partial x^l} \right)$$

and ("Christoffel symbols of the first kind")

$$\widehat{\Gamma}_{ij,k} = \frac{1}{2} \left(\frac{\partial g_{ik}}{\partial x^j} + \frac{\partial g_{jk}}{\partial x^i} - \frac{\partial g_{ij}}{\partial x^k} \right).$$

The Levi-Civita connection is compatible with the metric, in the sense that it treats tangent vectors of the shortest curves on a manifold as being parallel (or equivalently, auto-parallel curves are also geodesics).

It turns out that one can define a kind of "compatibility" relation more general than expressed by (13.13), by introducing the notion of "conjugacy" (denoted by $*$) between two connections. A connection ∇^* is said to be "conjugate" to ∇ with respect to g if

$$\partial_k g(\partial_i, \partial_j) = g(\nabla_{\partial_k} \partial_i, \partial_j) + g(\partial_i, \nabla^*_{\partial_k} \partial j). \tag{13.14}$$

Clearly, $(\nabla^*)^* = \nabla$. Moreover, $\widehat{\nabla}$, which satisfies (13.13), is special in the sense that it is self-conjugate $(\widehat{\nabla})^* = \widehat{\nabla}$.

Because metric tensor g provides a one-to-one mapping between points in the tangent space (i.e., vectors) and points in the cotangent space (i.e., co-vectors), (13.14) can also be seen as characterizing how co-vector fields are to be parallel-transported in order to preserve their dual pairing $\langle \cdot, \cdot \rangle$ with vector fields.

Writing out (13.14) explicitly,

$$\frac{\partial g_{ij}}{\partial x^k} = \Gamma_{ki,j} + \Gamma^*_{kj,i}, \tag{13.15}$$

where analogous to (13.10) and (13.12),

$$\nabla^*_{\partial_i} \partial_j = \sum_l \Gamma^{*l}_{ij} \, \partial_l$$

so that

$$\Gamma^*_{kj,i} = g(\nabla^*_{\partial_j} \partial_k, \partial_i) = \sum_l g_{il} \Gamma^{*l}_{kj}.$$

In the following, a manifold $\mathfrak{M}$ with a metric g and a pair of conjugate connections Γ, Γ^* with respect to g is called a "Riemannian manifold with dualistic structure," and denoted by $\{\mathfrak{M}, g, \Gamma, \Gamma^*\}$. Obviously, Γ and Γ^* satisfy the relation (in either covariant or contravariant forms)

$$\widehat{\Gamma} = \frac{1}{2}(\Gamma + \Gamma^*).$$

More generally, in information geometry, a one-parameter family of affine connections $\Gamma^{(\alpha)}$, called "α-connections" ($\alpha \in \mathbb{R}$), is introduced (Amari, 1985, Amari and Nagaoka, 2000)

$$\Gamma^{(\alpha)} = \frac{1+\alpha}{2}\Gamma + \frac{1-\alpha}{2}\Gamma^*. \tag{13.16}$$

Obviously, $\Gamma^{(0)} = \widehat{\Gamma}$.

It can be shown that the curvatures R_{lkij}, R^*_{lkij} for the pair of conjugate connections Γ, Γ^* satisfy

$$R_{lkij} = R^*_{lkij}.$$

So, Γ is flat if and only if Γ^* is flat. In this case, the manifold is said to be "dually flat." When Γ, Γ^* are dually flat, then $\Gamma^{(\alpha)}$ is called "α-transitively flat" (Uohashi, 2002). In such case, $\{\mathfrak{M}, g, \Gamma^{(\alpha)}, \Gamma^{(-\alpha)}\}$ is called an "α-Hessian manifold," or a manifold with α-Hessian structure.

13.2.4 Biorthogonal Coordinate Transformation

Consider coordinate transform $x \mapsto u$,

$$\partial^i \equiv \frac{\partial}{\partial u_i} = \sum_l \frac{\partial x^l}{\partial u_i}\frac{\partial}{\partial x^l} = \sum_l J^{li}\partial_l,$$

where the Jacobian matrix J is given by

$$J_{ij}(x) = \frac{\partial u_i}{\partial x^j}, \qquad J^{ij}(u) = \frac{\partial x^i}{\partial u_j}, \qquad \sum_l J_{il}J^{lj} = \delta_i^j, \tag{13.17}$$

where δ_i^j is the Kronecker delta (taking the value of 1 when $i = j$ and 0 otherwise). If the new coordinate system $u = [u_1, \dots, u_n]$ (with components expressed by subscripts) is such that

$$J_{ij}(x) = g_{ij}(x), \tag{13.18}$$

then the x-coordinate system and the u-coordinate system are said to be "biorthogonal" to each other because, from the definition of a metric tensor (13.9),

$$g(\partial_i, \partial^j) = g(\partial_i, \sum_l J^{lj}\partial_l) = \sum_l J^{lj} g(\partial_i, \partial_l) = \sum_l J^{lj} g_{il} = \delta_i^j.$$

In such a case, denote

$$g^{ij}(u) = g(\partial^i, \partial^j), \tag{13.19}$$

which equals $J^{ij}(u)$, the Jacobian of the inverse coordinate transform $u \mapsto x$. Also introduce the (contravariant version) of the affine connection Γ under the u-coordinate system and denote it by an unconventional notation Γ_t^{rs} defined by

$$\nabla_{\partial^r}\partial^s = \sum_t \Gamma_t^{rs}\partial^t;$$

similarly Γ_t^{*rs} is defined via

$$\nabla^*_{\partial^r}\partial^s = \sum_t \Gamma_t^{*rs}\partial^t.$$

The covariant version of the affine connections is denoted by superscripted Γ and Γ^*:

$$\Gamma^{ij,k}(u) = g(\nabla_{\partial^i}\partial^j, \partial^k), \qquad \Gamma^{*ij,k}(u) = g(\nabla^*_{\partial^i}\partial^j, \partial^k). \tag{13.20}$$

As in (13.11), the affine connections in u-coordinates (expressed in superscript) and in x-coordinates (expressed in subscript) are related via

$$\Gamma_t^{rs}(u) = \sum_k \left(\sum_{i,j} \frac{\partial x^r}{\partial u_i}\frac{\partial x^s}{\partial u_j}\Gamma_{ij}^k(x) + \frac{\partial^2 x^k}{\partial u_r \partial u_s} \right) \frac{\partial u_k}{\partial x^t} \tag{13.21}$$

and

$$\Gamma^{rs,t}(u) = \sum_{i,j,k} \frac{\partial x^r}{\partial u_i} \frac{\partial x^s}{\partial u_j} \frac{\partial x^t}{\partial u_k} \Gamma_{ij,k}(x) + \frac{\partial^2 x^t}{\partial u_r \partial u_s}. \tag{13.22}$$

Similar relations hold between $\Gamma_t^{*rs}(u)$ and $\Gamma_{ij}^{*k}(x)$, and between $\Gamma^{*rs,t}(u)$ and $\Gamma^*_{ij,k}(x)$.

Analogous to (13.15), we have the following identity,

$$\frac{\partial^2 x^t}{\partial u_s \partial u_r} = \frac{\partial g^{rt}(u)}{\partial u_s} = \Gamma^{rs,t}(u) + \Gamma^{*ts,r}(u),$$

which leads to

Proposition 13.1. *Under biorthogonal coordinates, the component forms of the metric tensor satisfy*

$$\sum_k g^{ik}(u) g_{kj}(x) = \delta^i_j$$

while the pair of conjugate connections Γ, Γ^* satisfies

$$\Gamma^{*ts,r}(u) = -\sum_{i,j,k} g^{ir}(u) g^{js}(u) g^{kt}(u) \Gamma_{ij,k}(x) \tag{13.23}$$

and

$$\Gamma_r^{*\,ts}(u) = -\sum_j g^{js}(u) \Gamma^t_{jr}(x). \tag{13.24}$$

Next, we discuss the conditions under which biorthogonal coordinates exist on an arbitrary Riemannian manifold. From its definition (13.18), we can easily show that

Lemma 13.2. *A Riemannian manifold $\mathfrak{M}$ with metric g_{ij} admits biorthogonal coordinates if and only if $\partial g_{ij}/\partial x^k$ is totally symmetric,*[5]

$$\frac{\partial g_{ij}(x)}{\partial x^k} = \frac{\partial g_{ik}(x)}{\partial x^j}. \tag{13.25}$$

That (13.25) is satisfied for biorthogonal coordinates is evident by virtue of (13.17) and (13.18). Conversely, given (13.25), there must be n functions $u_i(x)$, $i = 1, 2, \ldots, n$ such that

$$\frac{\partial u_i(x)}{\partial x^j} = g_{ij}(x) = g_{ji}(x) = \frac{\partial u_j(x)}{\partial x^i}.$$

---

[5] Note that $(\partial g_{ij}/\partial x^k) \equiv \partial_k (g(\partial_i, \partial_j)) \neq (\partial_k g)(\partial_i, \partial_j)$, the latter is necessarily totally symmetric whenever there exist a pair of torsion-free connections Γ, Γ^* that are conjugate with respect to g.

The above identity, in turn, implies that there exists a function Φ such that $u_i = \partial_i \Phi$ and, by positive definiteness of g_{ij}, Φ would have to be a strictly convex function! In this case, the x- and u-variables satisfy (13.7), and the pair of convex functions, Φ and its conjugate $\widetilde{\Phi}$, is related to g_{ij} and g^{ij} by

$$g_{ij}(x) = \frac{\partial^2 \Phi(x)}{\partial x^i \partial x^j} \longleftrightarrow g^{ij}(u) = \frac{\partial^2 \widetilde{\Phi}(u)}{\partial u_i \partial u_j}.$$

It follows from Lemma 13.2 that a necessary and sufficient condition for a Riemannian manifold to admit biorthogonal coordinates is that its Levi-Civita connection is given by

$$\widehat{\Gamma}_{ij,k}(x) \equiv \frac{1}{2}\left(\frac{\partial g_{ik}}{\partial x^j} + \frac{\partial g_{jk}}{\partial x^i} - \frac{\partial g_{ij}}{\partial x^k}\right) = \frac{1}{2}\frac{\partial g_{ij}}{\partial x^k}.$$

From this, the following can be shown.

Proposition 13.2. *A Riemannian manifold $\{\mathfrak{M}, g\}$ admits a pair of biorthogonal coordinates x and u if and only if there exists a pair of conjugate connections γ and γ^* such that $\gamma_{ij,k}(x) = 0$, $\gamma^{*rs,t}(u) = 0$. In other words, biorthogonal coordinates are affine coordinates for dually flat conjugate connections.*

In fact, we can now define a pair of torsion-free connections by

$$\gamma_{ij,k}(x) = 0, \qquad \gamma^{*}_{ij,k}(x) = \frac{\partial g_{ij}}{\partial x^k}$$

and show that they are conjugate with respect to g; that is, they satisfy (13.14). This is to say that we select an affine connection γ such that x is its affine coordinate. From (13.22), when γ^* is expressed in u-coordinates,

$$\begin{aligned}
\gamma^{*rs,t}(u) &= \sum_{i,j,k} g^{ir}(u) g^{js}(u) \frac{\partial x^k}{\partial u_t}\frac{\partial g_{ij}(x)}{\partial x^k} + \frac{\partial g^{ts}(u)}{\partial u_r} \\
&= \sum_{i,j} g^{ir}(u)\left(-\frac{\partial g^{js}(u)}{\partial u_t} g_{ij}(x)\right) + \frac{\partial g^{ts}(u)}{\partial u_r} \\
&= -\sum_{j} \delta_j^r \frac{\partial g^{js}(u)}{\partial u_t} + \frac{\partial g^{ts}(u)}{\partial u_r} = 0.
\end{aligned}$$

This implies that u is an affine coordinate system with respect to γ^*. Therefore, biorthogonal coordinates are affine coordinates for a pair of dually flat connections. Such a manifold $\{\mathfrak{M}, g, \gamma, \gamma^*\}$ is called a "Hessian manifold" (Shima, 2007, Shima and Yagi, 1997). It is a special case of the α-Hessian manifold (introduced in Section 13.3.2).

13.2.5 Equiaffine Structure and Parallel Volume Form on a Manifold

For a restrictive class of connections, called "equiaffine" connections, the manifold $\mathfrak{M}$ may admit uniquely a parallel volume form $\omega(x)$. Here, a volume form is a skew-symmetric multilinear map from n linearly independent vectors to a nonzero scalar, and "parallel" is in the sense that $(\partial_i \omega)(\partial_1, \ldots, \partial_n) = 0$ where

$$(\partial_i \omega)(\partial_1, \ldots, \partial_n) \equiv (\nabla_{\partial_i} \omega)(\partial_1, \ldots, \partial_n)$$
$$= \partial_i(\omega(\partial_1, \ldots, \partial_n)) - \sum_{l=1}^{n} \omega(\ldots, \nabla_{\partial_i} \partial_l, \ldots).$$

Applying (13.10), the equiaffine condition becomes

$$\partial_i(\omega(\partial_1, \ldots, \partial_n)) = \sum_{l=1}^{n} \omega\left(\ldots, \sum_{k=1}^{n} \Gamma_{il}^{k} \partial_k, \ldots\right)$$
$$= \sum_{l=1}^{n} \sum_{k=1}^{n} \Gamma_{il}^{k}\, \delta_k^l\, \omega(\partial_1, \ldots, \partial_n) = \omega(\partial_1, \ldots, \partial_n) \sum_{l=1}^{n} \Gamma_{il}^{l}$$

or

$$\sum_{l} \Gamma_{il}^{l}(x) = \frac{\partial \log \omega(x)}{\partial x^i}. \tag{13.26}$$

Whether a connection is equiaffine is related to the so-called Ricci tensor Ric, defined as the contraction of the curvature tensor R,

$$\mathrm{Ric}_{ij}(x) = \sum_{k} R_{ikj}^{k}(x). \tag{13.27}$$

For a torsion-free connection $\Gamma_{ij}^{k} = \Gamma_{ji}^{k}$, applying the definition of the curvature tensor R to the above yields

$$\mathrm{Ric}_{ij} - \mathrm{Ric}_{ji} = \frac{\partial}{\partial x^i}\left(\sum_{l} \Gamma_{jl}^{l}(x)\right) - \frac{\partial}{\partial x^j}\left(\sum_{l} \Gamma_{il}^{l}(x)\right) \tag{13.28}$$
$$= \sum_{k} R_{kij}^{k}.$$

One immediately sees that the existence of a function ω satisfying (13.26) is equivalent to the right side of (13.28) being identically zero. In other words, the necessary and sufficient condition for a torsion-free connection to be equiaffine is that its Ricci tensor is symmetric, $\mathrm{Ric}_{ij} = \mathrm{Ric}_{ji}$, or equivalently, $\sum_k R^k_{kij} = 0$.

Making use of (13.26), it is easy to show that the parallel volume form of a Levi-Civita connection $\widehat{\Gamma}$ is given by

$$\widehat{\omega}(x) = \sqrt{\det[g_{ij}(x)]} \longleftrightarrow \widehat{\omega}(u) = \sqrt{\det[g^{ij}(u)]}.$$

The parallel volume forms ω, ω^* associated with Γ and Γ^* satisfy (apart from a positive, multiplicative constant)

$$\omega(x)\,\omega^*(x) = (\widehat{\omega}(x))^2 = \det[g_{ij}(x)], \tag{13.29}$$
$$\omega(u)\,\omega^*(u) = (\widehat{\omega}(u))^2 = \det[g^{ij}(u)]. \tag{13.30}$$

Let us now consider the parallel volume forms under biorthogonal coordinates. Contracting the indices t with r in (13.24), and invoking (13.26), we obtain

$$\frac{\partial \log \omega^*(u)}{\partial u_s} + \sum_j \frac{\partial x^j}{\partial u_s}\frac{\partial \log \omega(x)}{\partial x^j} = \frac{\partial \log \omega^*(u)}{\partial u_s} + \frac{\partial \log \omega(x(u))}{\partial u_s} = 0.$$

After integration,

$$\omega^*(u)\,\omega(x) = const. \tag{13.31}$$

From (13.29)–(13.31),

$$\omega(u)\,\omega^*(x) = const. \tag{13.32}$$

The relations (13.31) and (13.32) indicate that the volume forms of the pair of conjugate connections, when expressed in biorthogonal coordinates respectively, are inversely proportional to each other. Note that $\omega(x) = \omega(\partial_1, \ldots, \partial_n)$ and $\omega^*(x) = \omega^*(\partial_1, \ldots, \partial_n)$, as skew-symmetric multilinear maps, transform to $\omega(u) = w(\partial^1, \ldots, \partial^n)$ and $\omega^*(u) = \omega^*(\partial^1, \ldots, \partial^n)$ via

$$\omega(x) = \det[J_{ij}(x)]\omega(u) \longleftrightarrow \omega^*(x) = \det[J^{ij}(u)]\omega^*(u),$$

where $\det[J_{ij}(x)] = \det[g_{ij}(x)] = (\det[J^{ij}(u)])^{-1} = (\det[g^{ij}(u)])^{-1}$.

When the pair of equiaffine connections Γ, Γ^* are further assumed to be dually flat, then the entire family of α-connections $\Gamma^{(\alpha)}$ given by (13.16) are equiaffine (Takeuchi and Amari, 2005, Matsuzoe et al., 2006, Zhang, 2007). The $\Gamma^{(\alpha)}$-parallel volume element $\omega^{(\alpha)}$ can be shown to be given by

$$\omega^{(\alpha)} = \omega^{(1+\alpha)/2}(\omega^*)^{(1-\alpha)/2}.$$

Clearly,

$$\omega^{(\alpha)}(x)\omega^{(-\alpha)}(x) = \det[g_{ij}(x)] \longleftrightarrow \omega^{(\alpha)}(u)\omega^{(-\alpha)}(u) = \det[g^{ij}(u)].$$

13.2.6 Affine Hypersurface Immersion (of Co-Dimension One)

We next discuss dualistic geometry from convex functions as related to hypersurfaces in affine space, which is the subject of study in affine differential geometry (Simon et al., 1991, Nomizu and Sasaki, 1994).

Let $\mathbb{A}^{n+1}$ be the standard affine space of dimension $n+1$, and $\mathfrak{M}$ an n-dimensional manifold immersed into $\mathbb{A}^{n+1}$ as a hypersurface with affine coordinates $f = [f^1, \ldots, f^{n+1}]$; that is, $f\colon \mathfrak{M} \to \mathbb{A}^{n+1}$. Assume that the local coordinate system on $\mathfrak{M}$ is $x = [x^1, \ldots, x^n]$. Let $\xi = [\xi^1, \ldots, \xi^{n+1}]$ be a vector field defined on $\mathfrak{M}$ that is "transversal," that is, nowhere tangential to $\mathfrak{M}$. Denote the vector space associated with $\mathbb{A}^{n+1}$ as V, with $\dim(V) = n+1$, and the canonical pairing of V with its dual vector space $\widetilde{V}$ (with $\dim(\widetilde{V}) = n+1$) as $\langle\ ,\ \rangle_{n+1}$; see (13.3). The duplet $\{f, \xi\}$ is called an "affine immersion." In local coordinates, they can be explicitly written as functions of x: $\{f(x), \xi(x)\}$, where f is valued in $\mathbb{A}$ and ξ is valued in V.

Because the tangent space $T_p(\mathfrak{M})$ is spanned by

$$\left\{ \left[\frac{\partial f^a}{\partial x^1}, \ldots, \frac{\partial f^a}{\partial x^n} \right],\ a = 1, \ldots, n+1 \right\},$$

we may decompose the second derivatives of f as

$$\frac{\partial^2 f^a}{\partial x^i \partial x^j} = \sum_{k=1}^{n} \Gamma_{ij}^k \frac{\partial f^a}{\partial x^k} + h_{ij} \xi^a \qquad (i, j = 1, \ldots, n), \tag{13.33}$$

where $h_{ij} = h_{ji}$ (called "induced bilinear form" or "affine fundamental form"); if f is convex, then h_{ij} is positive definite. The set of coefficients Γ_{ij}^k is called the "induced connection" on $\mathfrak{M}$, because it is induced by a flat connection on $\mathbb{A}^{n+1}$. Under coordinate transform, these coefficients can be shown to transform according to (13.21). Similarly, decompose the derivative of ξ^a as

$$\frac{\partial \xi^a}{\partial x^i} = -\sum_{k=1}^{n} S_i^k \frac{\partial f^a}{\partial x^k} + \tau_i \xi^a, \tag{13.34}$$

where S_i^k is known as the "affine shape operator," and τ_i is a 1-form on $\mathfrak{M}$ called the "transversal connection form;" when $\tau = 0$ everywhere on $\mathfrak{M}$, the affine immersion $\{f, \xi\}$ is called "equiaffine."

We define a volume form ω on $\mathfrak{M}$ arising out of the immersion of $\{f, \xi\}$,

$$\omega(\partial_1, \ldots, \partial_n) = \mathrm{Det}(\partial_1 f, \ldots, \partial_n f, \xi),$$

where Det is the determinant form on $\mathbb{A}^{n+1}$, and $\partial_i f$ is the vector field $\partial_i f = [\partial_i f^1, \ldots, \partial_i f^{n+1}]$. The covariant derivative of ω is given as follows (see Nomizu and Sasaki, 1994):

$$(\nabla_{\partial_i}\omega)(\partial_1,\ldots,\partial_n)=\tau_i\omega(\partial_1,\ldots,\partial_n).$$

This implies that the induced volume form ω is parallel with respect to the induced connection ∇ if and only if $\{f,\xi\}$ is equiaffine: $\tau=0$.

In order to consider the geometry induced from convex functions and biorthogonal coordinates, we consider a special kind of affine immersion called "graph immersion:"

$$f=[x^1,\ldots,x^n,\Phi(x)],\qquad \xi=[0,\ldots,0,1], \tag{13.35}$$

where Φ is some nondegenerate (in particular, convex) function. Applying (13.33), we obtain the induced connection $\Gamma_{ij}^k(x)=0$ and the affine fundamental form $h_{ij}(x)$ as the Hessian of Φ,

$$h_{ij}(x)=\frac{\partial^2\Phi(x)}{\partial x^i\partial x^j}.$$

Thus the geometry of a graph affine immersion coincides with the Hessian geometry induced from a convex function. Because the transversal vector field ξ is parallel along f, from (13.34), obviously $\mathfrak{M}$ has an equiaffine structure.

We can define the "dual" of graph immersion, $\{\widetilde{f},\widetilde{\xi}\}$, mapping $\mathfrak{M}$ to $\mathbb{A}^{n+1}$ as another graph. Here $\widetilde{f}=[u_1,\ldots,u_n,\widetilde{\Phi}(u)]$, with $\widetilde{\Phi}$ and u given by (13.6) and (13.7), respectively. The transversal vector field $\widetilde{\xi}=(0,\ldots,0,1)$ is valued in $\widetilde{V}$, the dual vector space. The affine fundamental form $\widetilde{h}$ is

$$\widetilde{h}^{ij}(u)=\frac{\partial^2\widetilde{\Phi}(u)}{\partial u_i\partial u_j}.$$

Because of the identity

$$\frac{\partial^2\widetilde{\Phi}(u)}{\partial u_i\partial u_j}=\sum_{k,l}\frac{\partial x^k}{\partial u_i}\frac{\partial x^l}{\partial u_j}\frac{\partial^2\Phi(x)}{\partial x^k\partial x^l},$$

such affine fundamental form transforms as a 0–2 tensor

$$\widetilde{h}^{ij}(u)=\sum_{k,l}\frac{\partial x^k}{\partial u_i}\frac{\partial x^l}{\partial u_j}h_{kl}(x)$$

(even though second derivatives in general do not transform in a tensorlike fashion). This means that for dual graph immersions $\{f,\xi\}$ and $\{\widetilde{f},\widetilde{\xi}\}$, the induced affine fundamental form is one and the same $\widetilde{h}=h$. The induced objects $\{\mathfrak{M},h,\Gamma,\Gamma^*\}$ form a Hessian structure (i.e., induced connections are dually flat).

More generally, for an arbitrary affine immersion, we can introduce the notion of "co-normal mapping," defined as $\zeta\colon\mathfrak{M}\to\widetilde{V}$ as

$$\langle \xi(x), \zeta(x) \rangle_{n+1} = 1, \tag{13.36}$$
$$\langle \partial_i f(x), \zeta(x) \rangle_{n+1} = 0 \quad (i = 1, \dots, n); \tag{13.37}$$

that is,

$$\sum_{a=1}^{n+1} \xi^a(x)\, \zeta_a(x) = 1, \qquad \sum_{a=1}^{n+1} \frac{\partial f^a}{\partial x^i}(x)\, \zeta_a(x) = 0 \qquad (i = 1, \dots, n).$$

Intuitively, the co-normal map is a uniquely defined "normal" vector of the tangent hyperplane at $f(x)$. (This property comes from (13.37).) The co-normal map is not a unit vector; the "length" of the map is normalized by (13.36). Note that the word "length" and "normal" are in quotation marks because no metric has ever been introduced on V or $\widetilde{V}$; normalization is through the pairing operation $\langle \cdot, \cdot \rangle$.

When $\{f, \xi\}$ is equiaffine, then the co-normal map ζ can be viewed as an immersion from $\mathfrak{M}$ to $\mathbb{A}^{n+1}$ (Nomizu and Sasaki, 1994, p. 57). Specifically, $\zeta(\mathfrak{M})$ is taken to be (the negative of) the positional vector field (with respect to a center point) in addition to being the transversal vector field. In this case $\{\widetilde{f}, \zeta\} = \{-\zeta, \zeta\}$ is an affine immersion, called the "co-normal immersion" of $\{f, \xi\}$. We also call $\{-\zeta, \zeta\}$ a "centroaffine immersion" because the immersion has a center, with the position vector $-\zeta$ (the first element in the duplet) transversal to its image $\mathfrak{M}$. We denote by $\widetilde{\Gamma}$, $\widetilde{h}$, $\widetilde{\tau}$, $\widetilde{S}$, ... the induced objects of $\{-\zeta, \zeta\}$. Then we have the following formulae (see Simon et al., 1991);

$$\widetilde{\Gamma}_{kj,i} = -\Gamma_{ki,j} + \partial_k h_{ij}, \tag{13.38}$$
$$\widetilde{h}_{ij} = \sum_{k=1}^{n} S_i^k h_{kj}, \tag{13.39}$$
$$\widetilde{\tau}_i = 0,$$
$$\widetilde{S}_j^i = \delta_j^i.$$

Equation (13.38) implies that ∇ and $\widetilde{\nabla}$ are mutually conjugate with respect to h. Note that Γ and $\widetilde{\Gamma}$ are, respectively, the induced connections when $\mathfrak{M}$ is immersed into $\mathbb{A}^{n+1}$ in two distinct ways, $\{f, \xi\}$ and $\{-\zeta, \zeta\}$.

Suppose that $\{f, \xi\}$ is a graph affine immersion with respect to some convex function, and $\{-\zeta, \zeta\}$ is the co-normal immersion of $\{f, \xi\}$. From (13.34), the affine shape operator S of $\{f, \xi\}$ vanishes. This implies that $\widetilde{h} = 0$ from (13.39). Thus, although the co-normal map of an equiaffine immersion is a centroaffine hypersurface in $\mathbb{A}^{n+1}$, the co-normal map of graph immersion has its image lie on an affine hyperplane in $\mathbb{A}^{n+1}$.

For an affine immersion $\{f, \xi\}$ and the co-normal immersion $\{-\zeta, \zeta\}$, we define the "geometric divergence" $\mathcal{G}$ on any two points on $\mathfrak{M}$ by

$$\mathcal{G}(x, y) = \langle f(x) - f(y), \zeta(y) \rangle_{n+1} = \sum_{a=1}^{n+1} (f^a(x) - f^a(y)) \zeta_a(y).$$

For a graph immersion given by (13.35), we can explicitly solve for ζ from (13.36) and (13.37):

$$\zeta = [-\partial_1 \Phi, \ldots, -\partial_n \Phi, 1].$$

Therefore, the expression for geometric divergence becomes

$$\mathcal{G}(x, y) = -\langle x - y, (\partial \Phi)(y) \rangle_n + \Phi(x) - \Phi(y) \equiv \mathcal{B}_\Phi(x, y);$$

geometric divergence is nothing but Bregman divergence (13.2), see Kurose (1994) and Matsuzoe (1998).

13.2.7 Centroaffine Immersion of Co-Dimension Two

Now we consider affine immersion of $\mathfrak{M}$ (with $\dim(\mathfrak{M}) = n$) into a co-dimension two affine space $\mathbb{A}^{n+2}$ (rather than the co-dimension one affine space $\mathbb{A}^{n+1}$ as discussed in the last section). In this case, in addition to specifying the immersion, denoted by $f\colon \mathfrak{M} \to \mathbb{A}^{n+2}$, we need to specify two noncollinear vector fields, both "transversal" on $\mathfrak{M}$. The vector space is denoted as V with $\dim(V) = n+2$; the dual vector space is denoted as $\widetilde{V}$ with $\dim(\widetilde{V}) = n+2$. To simply the situation, we consider centroaffine immersion such that one of the transversal vector fields is the (negative of the) positional vector $-f$ and the other is, as before, denoted ξ, that is, the affine immersion is denoted as $\{f, -f, \xi\}$; the elements are valued in $\mathbb{A}^{n+2}$, V, V, respectively. The second derivatives of f and ξ are decomposed as follows (for $i, j = 1, \ldots, n$; $a = 1, \ldots, n+2$);

$$\frac{\partial^2 f^a}{\partial x^i \partial x^j} = \sum_{k=1}^{n} \Gamma_{ij}^k \frac{\partial f^a}{\partial x^k} + h_{ij} \xi^a - t_{ij} f^a,$$

$$\frac{\partial \xi^a}{\partial x^i} = -\sum_{k=1}^{n} S_i^k \frac{\partial f^a}{\partial x^k} + \tau_i \xi^a - \kappa_i f^a.$$

As in affine immersion of co-dimension one, we call Γ_{ij}^k the "induced connection," h_{ij} the "affine fundamental form," τ_i the "transversal connection form," and S_i^k the "affine shape operator." Below, we assume that h is positive definite (i.e., f is strictly convex) and $\tau = 0$ (the centroaffine immersion is equiaffine).

We denote the "dual map" of $\{f, -f, \xi\}$ as another centroaffine map taking the form of $\{\tilde{f}, -\tilde{f}, \zeta\}$, where the elements are valued in $\mathbb{A}^{n+2}$, $\widetilde{V}$, $\widetilde{V}$, respectively; $\tilde{f}$ and ζ are specified by

$$\langle \tilde{f}(x), \xi(x) \rangle_{n+2} = 1, \quad \langle \zeta(x), \xi(x) \rangle_{n+2} = 0,$$
$$\langle \tilde{f}(x), f(x) \rangle_{n+2} = 0, \quad \langle \zeta(x), f(x) \rangle_{n+2} = 1,$$
$$\langle \tilde{f}(x), \partial_i f(x) \rangle_{n+2} = 0, \quad \langle \zeta(x), \partial_i f(x) \rangle_{n+2} = 0 \quad (i = 1, \ldots, n),$$

or explicitly

$$\sum_{a=1}^{n+2} \tilde{f}_a(x) \xi^a(x) = 1, \qquad \sum_{a=1}^{n+2} \zeta_a(x) \xi^a(x) = 0,$$
$$\sum_{a=1}^{n+2} \tilde{f}_a(x) f^a(x) = 0, \qquad \sum_{a=1}^{n+2} \zeta_a(x) f^a(x) = 1, \tag{13.40}$$
$$\sum_{a=1}^{n+2} \tilde{f}_a(x) \frac{\partial f^a}{\partial x^i}(x) = 0, \qquad \sum_{a=1}^{n+2} \zeta_a(x) \frac{\partial f^a}{\partial x^i}(x) = 0 \qquad (i = 1, \ldots, n). \tag{13.41}$$

Denote the induced objects as $\widetilde{\Gamma}$, $\widetilde{h}$, $\widetilde{\tau}$, ...; we have the following formulae (see Nomizu and Sasaki, 1994, Matsuzoe, 1998);

$$\begin{aligned} \partial_k h_{ij} &= \Gamma_{ki,j} + \widetilde{\Gamma}_{kj,i}, \\ \widetilde{h}_{ij} &= h_{ij}, \\ \widetilde{\tau}_i &= 0. \end{aligned} \tag{13.42}$$

We remark that (13.42) is different from (13.39) of the co-dimension one case. If a centroaffine immersion $\{f, -f, \xi\}$ induces $\{g, \Gamma\}$ on $\mathfrak{M}$, then the dual map $\{\tilde{f}, -\tilde{f}, \zeta\}$ induces $\{g, \widetilde{\Gamma}\}$ on $\mathfrak{M}$. This implies that the theory of centroaffine immersions of co-dimension two is more useful than that of affine immersions of co-dimension one when we discuss the duality of statistical manifold.

Consider the special case of graph immersion (of co-dimension two) $\{f, -f, \xi\}$; that is,

$$f = [x^1, \ldots, x^n, \Phi(x), 1], \qquad \xi = [0, \ldots, 0, 1, 0], \tag{13.43}$$

where $\Phi(x)$ is some convex function. If $\{f, -f, \xi\}$ has other representations, they are centroaffinely congruent (linearly congruent) to (13.43); hence it suffices to consider (13.43).

From straightforward calculations, the dual map $\{\tilde{f}, -\tilde{f}, \zeta\}$ of $\{f, -f, \xi\}$ takes the form

$$\tilde{f} = [-u_1, \ldots, -u_n, 1, \widetilde{\Phi}(u)], \qquad \zeta = [0, \ldots, 0, 0, 1]. \tag{13.44}$$

The left side equation in (13.40) then gives

$$-\sum_{i=1}^{n} x^i u_i + \Phi(x) + \widetilde{\Phi}(u) = 0,$$

and the left side equation in (13.41) is

$$-u_i + \frac{\partial \Phi}{\partial x^i}(x) = 0.$$

Thus, $\widetilde{\Phi}$ is the convex conjugate of Φ as in (13.6), and $u = [u_1, \ldots, u_n]$ is the conjugate variable as in (13.7). For graph immersion, it is easy to check that $\Gamma_{ki,j} = 0$, $S_i^k = 0$, $t_{ij} = 0$, $\tau_i = 0$, $\kappa_i = 0$ for all indices and

$$h_{ij}(x) = \frac{\partial^2 \Phi(x)}{\partial x^i \partial x^j}.$$

The same is true for induced objects in dual immersion.

Just as in the case of equiaffine immersion $\{f, \xi\}$ of co-dimension one and the associated co-normal map $\{-\zeta, \zeta\}$, we can construct the geometric divergence $\mathcal{G}$ on $\mathfrak{M}$ for centroaffine immersion $\{f, -f, \xi\}$ of co-dimension two and the associated dual map $\{\tilde{f}, -\tilde{f}, \zeta\}$:

$$\begin{aligned}\mathcal{G}(x,y) &= \langle \tilde{f}(y), f(x) - f(y) \rangle_{n+2} \\ &= \langle \tilde{f}(y), f(x) \rangle_{n+2} \\ &= \sum_{a=1}^{n+2} \tilde{f}_a(y) f^a(x).\end{aligned}$$

For graph immersion, we substitute f and $\tilde{f}$ in (13.43) and (13.44) to yield

$$\begin{aligned}\mathcal{G}(x,y) &= -\langle x, (\partial\Phi)(y) \rangle_n + \Phi(x) + \widetilde{\Phi}((\partial\Phi)(y)) \\ &\equiv \mathcal{B}_\Phi(x,y).\end{aligned}$$

In both the equiaffine immersion of co-dimension one (discussed in Section 13.2.6) and centroaffine immersion of co-dimension two (discussed here), the notion of geometric divergence is a generalization of the Bregman (canonical) divergence on a dually flat space.

Proposition 13.3. (Kurose, 1994, Matsuzoe, 1998) *Let Φ be a strictly convex function on $\mathbb{R}^n$. Then geometric divergence $\mathcal{G}(x,y)\colon V \times V \to \mathbb{R}$ induced by the affine immersion of Φ as a graph in $\mathbb{A}^{n+1}$ or by the centroaffine immersion of Φ as a graph in $\mathbb{A}^{n+2}$ equals the Bregman divergence $\mathcal{B}_\Phi(x,y)$.*

13.3 The α-Hessian Structure Associated with Convex-Induced Divergence

The discussion at the end of the last section anticipate a close relation between convex functions and the Riemannian structure on a differentiable manifold whose coordinates are the variables of the convex functions. On such a manifold, divergence functions take the role of pseudo-distance functions that are nonnegative but need not be symmetric. That dualistic Riemannian manifold structures can be induced from a divergence function was first demonstrated by S. Eguchi.

Lemma 13.3. (Eguchi, 1983, 1992) *A divergence function induces a Riemannian metric g and a pair of conjugate connections Γ, Γ^* given as*

$$g_{ij}(x) = -\partial_{x^i}\partial_{y^j}\mathcal{D}(x,y)\big|_{y=x}; \tag{13.45}$$

$$\Gamma_{ij,k}(x) = -\partial_{x^i}\partial_{x^j}\partial_{y^k}\mathcal{D}(x,y)\big|_{y=x}; \tag{13.46}$$

$$\Gamma^*_{ij,k}(x) = -\partial_{y^i}\partial_{y^j}\partial_{x^k}\mathcal{D}(x,y)\big|_{y=x}. \tag{13.47}$$

It is easily verifiable that $g_{ij}, \Gamma_{ij,k}, \Gamma^*_{ij,k}$ as given above satisfy (13.15). Furthermore, under arbitrary coordinate transform, these quantities behave properly as desired. Equations (13.45)–(13.47) link a divergence function $\mathcal{D}$ to the dualistic Riemannian structure $\{\mathfrak{M}, g, \Gamma, \Gamma^*\}$.

Applying Lemma 13.3 to Bregman divergence $\mathcal{B}_\Phi(x,y)$ given by (13.2) yields

$$g_{ij}(x) = \frac{\partial^2\Phi(x)}{\partial x^i \partial x^j}$$

and

$$\Gamma_{ij,k}(x) = 0, \qquad \Gamma^*_{ij,k}(x) = \frac{\partial^3\Phi(x)}{\partial x^i \partial x^j \partial x^k}.$$

Calculating their curvature tensors shows the pair of connections are dually flat. It is commonly referred to, in affine geometry literature, as the "Hessian manifold" (see Section 13.2.4), although in the study by Shima (2007), the potential function Φ need not be convex but only semidefinite. In u-coordinates, these geometric quantities can be expressed as

$$g^{ij}(u) = \frac{\partial^2\widetilde{\Phi}(u)}{\partial u_i \partial u_j}, \qquad \Gamma^{*ij,k}(u) = 0, \qquad \Gamma^{ij,k}(u) = \frac{\partial^3\widetilde{\Phi}(u)}{\partial u_i \partial u_j \partial u_k},$$

where $\widetilde{\Phi}$ is the convex conjugate of Φ. Below, this link from convex functions to Riemannian manifold is explored in greater detail.

13.3.1 The α-Hessian Geometry

We start by reviewing a main result from Zhang (2004) linking the divergence function $\mathcal{D}^{(\alpha)}_\Phi(x,y)$ defined in (13.5) and the α-Hessian structure.

Proposition 13.4. (Zhang, 2004) *The manifold* $\{\mathfrak{M}, g_x, \Gamma_x^{(\alpha)}, \Gamma_x^{(-\alpha)}\}$[6] *associated with* $\mathcal{D}_\Phi^{(\alpha)}(x,y)$ *is given by*

$$g_{ij}(x) = \Phi_{ij} \tag{13.48}$$

and

$$\Gamma_{ij,k}^{(\alpha)}(x) = \frac{1-\alpha}{2}\,\Phi_{ijk}, \qquad \Gamma_{ij,k}^{*(\alpha)}(x) = \frac{1+\alpha}{2}\,\Phi_{ijk}. \tag{13.49}$$

Here, Φ_{ij}, Φ_{ijk} *denote, respectively, second and third partial derivatives of* $\Phi(x)$

$$\Phi_{ij} = \frac{\partial^2 \Phi(x)}{\partial x^i \partial x^j}, \qquad \Phi_{ijk} = \frac{\partial^3 \Phi(x)}{\partial x^i \partial x^j \partial x^k}.$$

Recall that an α-Hessian manifold is equipped with an α-independent metric and a family of α-transitively flat connections $\Gamma^{(\alpha)}$ (i.e., $\Gamma^{(\alpha)}$ satisfying (13.16) and $\Gamma^{(\pm 1)}$ are dually flat). From (13.49),

$$\Gamma_{ij,k}^{*(\alpha)} = \Gamma_{ij,k}^{(-\alpha)},$$

with the Levi-Civita connection given as:

$$\widehat{\Gamma}_{ij,k}(x) = \frac{1}{2}\Phi_{ijk}.$$

Straightforward calculation shows that:

Corollary 13.1. *For* α*-Hessian manifold* $\{\mathfrak{M}, g_x, \Gamma_x^{(\alpha)}, \Gamma_x^{(-\alpha)}\}$,

(i) *The curvature tensor of the* α*-connection is given by*

$$R_{\mu\nu ij}^{(\alpha)}(x) = \frac{1-\alpha^2}{4}\sum_{l,k}(\Phi_{il\nu}\Phi_{jk\mu} - \Phi_{il\mu}\Phi_{jk\nu})\Psi^{lk} \;=\; R_{ij\mu\nu}^{*(\alpha)}(x),$$

with Ψ^{ij} *being the matrix inverse of* Φ_{ij},

(ii) *All* α*-connections are equiaffine, with the* α*-parallel volume forms (i.e., the volume forms that are parallel under* α*-connections) given by*

$$\omega^{(\alpha)}(x) = \det[\Phi_{ij}(x)]^{(1-\alpha)/2}.$$

The reader is reminded that the metric and conjugated connections in the forms (13.48) and (13.49) are induced from (13.5). Using the convex conjugate $\widetilde{\Phi}\colon \widetilde{V} \to \mathbb{R}$ given by (13.6), we introduce the following family of divergence functions $\widetilde{\mathcal{D}}_{\widetilde{\Phi}}^{(\alpha)}(x,y)\colon V \times V \to \mathbb{R}_+$ defined by

---

[6] The subscript in x (or u below) indicates that the x-coordinate system (or u-coordinate system, resp.) is being used. Recall from Section 13.2.4 that under x (u, resp.) local coordinates g and Γ, in component forms, are expressed by lower (upper, resp.) indices.

$$\widetilde{\mathcal{D}}_{\widetilde{\Phi}}^{(\alpha)}(x,y) \equiv \mathcal{D}_{\widetilde{\Phi}}^{(\alpha)}((\partial\Phi)(x),(\partial\Phi)(y)).$$

Explicitly written, this new family of divergence functions is

$$\widetilde{\mathcal{D}}_{\widetilde{\Phi}}^{(\alpha)}(x,y) = \frac{4}{1-\alpha^2}\left(\frac{1-\alpha}{2}\,\widetilde{\Phi}(\partial\Phi(x)) + \frac{1+\alpha}{2}\,\widetilde{\Phi}(\partial\Phi(y)) - \widetilde{\Phi}\left(\frac{1-\alpha}{2}\,\partial\Phi(x) + \frac{1+\alpha}{2}\,\partial\Phi(y)\right)\right).$$

Straightforward calculation shows that $\widetilde{\mathcal{D}}_{\widetilde{\Phi}}^{(\alpha)}(x,y)$ induces the α-Hessian structure $\{\mathfrak{M}, g_x, \Gamma_x^{(-\alpha)}, \Gamma_x^{(\alpha)}\}$ where $\Gamma^{(\mp\alpha)}$ are given by (13.49); that is, the pair of α-connections are themselves "conjugate" (in the sense of $\alpha \leftrightarrow -\alpha$) to those induced by $\mathcal{D}_{\Phi}^{(\alpha)}(x,y)$.

13.3.2 Biorthogonal Coordinates on α-Hessian Manifold

If, instead of choosing $x = [x^1, \ldots, x^n]$ as the local coordinates for the manifold $\mathfrak{M}$, we use its biorthogonal counterpart $u = [u_1, \ldots, u_n]$ to index points on $\mathfrak{M}$. Under this u-coordinate system, the divergence function $\mathcal{D}_{\Phi}^{(\alpha)}$ between the same two points on $\mathfrak{M}$ becomes

$$\widetilde{\mathcal{D}}_{\Phi}^{(\alpha)}(u,v) \equiv \mathcal{D}_{\Phi}^{(\alpha)}((\partial\widetilde{\Phi})(u),(\partial\widetilde{\Phi})(v)).$$

Explicitly written,

$$\widetilde{\mathcal{D}}_{\Phi}^{(\alpha)}(u,v) = \frac{4}{1-\alpha^2}\left(\frac{1-\alpha}{2}\,\Phi((\partial\Phi)^{-1}(u)) + \frac{1+\alpha}{2}\,\Phi((\partial\Phi)^{-1}(v)) - \Phi\left(\frac{1-\alpha}{2}\,(\partial\Phi)^{-1}(u) + \frac{1+\alpha}{2}(\partial\Phi)^{-1}(v)\right)\right).$$

Recalling our notation (13.19) and (13.20), we have

Corollary 13.2. *The α-Hessian manifold $\{\mathfrak{M}, g_u, \Gamma_u^{(\alpha)}, \Gamma_u^{(-\alpha)}\}$ associated with $\widetilde{\mathcal{D}}_{\Phi}^{(\alpha)}(u,v)$ is given by*

$$g^{ij}(u) = \widetilde{\Phi}^{ij}(u), \tag{13.50}$$

$$\Gamma^{(\alpha)ij,k}(u) = \frac{1+\alpha}{2}\,\widetilde{\Phi}^{ijk}, \qquad \Gamma^{*(\alpha)ij,k}(u) = \frac{1-\alpha}{2}\,\widetilde{\Phi}_{ijk}. \tag{13.51}$$

Here, $\widetilde{\Phi}^{ij}$, $\widetilde{\Phi}^{ijk}$ denote, respectively, second and third partial derivatives of $\widetilde{\Phi}(u)$,

Table 13.1 Divergence functions and induced geometry

Divergence Function	Defined as	Induced Geometry
$\mathcal{D}^{(\alpha)}_{\Phi}(x,y)$	$V \times V \to \mathbb{R}_+$	$\left\{\Phi_{ij}, \Gamma^{(\alpha)}_x, \Gamma^{(-\alpha)}_x\right\}$
$\mathcal{D}^{(\alpha)}_{\widetilde{\Phi}}((\partial\Phi)(x),(\partial\Phi)(y))$	$V \times V \to \mathbb{R}_+$	$\left\{\Phi_{ij}, \Gamma^{(-\alpha)}_x, \Gamma^{(\alpha)}_x\right\}$
$\mathcal{D}^{(\alpha)}_{\widetilde{\Phi}}(u,v)$	$\widetilde{V} \times \widetilde{V} \to \mathbb{R}_+$	$\left\{\widetilde{\Phi}^{ij}, \Gamma^{(-\alpha)}_u, \Gamma^{(\alpha)}_u\right\}$
$\mathcal{D}^{(\alpha)}_{\Phi}((\partial\widetilde{\Phi})(u),(\partial\widetilde{\Phi})(v))$	$\widetilde{V} \times \widetilde{V} \to \mathbb{R}_+$	$\left\{\widetilde{\Phi}^{ij}, \Gamma^{(\alpha)}_u, \Gamma^{(-\alpha)}_u\right\}$

$$\widetilde{\Phi}^{ij}(u) = \frac{\partial^2 \widetilde{\Phi}(u)}{\partial u_i \partial u_j}, \qquad \widetilde{\Phi}^{ijk}(u) = \frac{\partial^3 \widetilde{\Phi}(u)}{\partial u_i \partial u_j \partial u_k}.$$

We remark that the same metric (13.50) and the same α-connections (13.51) are induced by $\mathcal{D}^{(-\alpha)}_{\widetilde{\Phi}}(u,v) \equiv \mathcal{D}^{(\alpha)}_{\widetilde{\Phi}}(v,u)$; this follows as a simple application of Lemma 13.3.

An application of (13.23) gives rise to the following relations.

$$\Gamma^{(\alpha)mn,l}(u) = -\sum_{i,j,k} g^{im}(u) g^{jn}(u) g^{kl}(u) \Gamma^{(-\alpha)}_{ij,k}(x),$$

$$\Gamma^{*(\alpha)mn,l}(u) = -\sum_{i,j,k} g^{im}(u) g^{jn}(u) g^{kl}(u) \Gamma^{(\alpha)}_{ij,k}(x),$$

$$R^{(\alpha)klmn}(u) = \sum_{i,j,\mu,\nu} g^{ik}(u) g^{jl}(u) g^{\mu m}(u) g^{\nu n}(u) R^{(\alpha)}_{ij\mu\nu}(x).$$

The volume form associated with $\Gamma^{(\alpha)}$ is

$$\omega^{(\alpha)}(u) = \det[\widetilde{\Phi}^{ij}(u)]^{(1+\alpha)/2}.$$

When $\alpha = \pm 1$, $\widetilde{\mathcal{D}}^{(\alpha)}_{\Phi}(u,v)$, as well as $\widetilde{\mathcal{D}}^{(\alpha)}_{\widetilde{\Phi}}(x,y)$ introduced earlier, take the form of Bregman divergence (13.2). In this case, the manifold is dually flat, with curvature tensor $R^{(\pm 1)}_{ij\mu\nu}(x) = R^{(\pm 1)klmn}(u) = 0$.

We summarize the relations between the convex-induced divergence functions and the geometry they generate in Table 13.1.

13.3.3 Applications of α-Hessian Geometry

Finally, we give an application of the α-Hessian geometry in mathematical statistics. A statistical model is a set of (what we call) ζ-functions $\zeta \mapsto p(\zeta)$, where a ζ-function is an element of some function space $\mathbb{B} = \{p(\cdot) \colon \mathcal{X} \to$

$\mathbb{R}, p(\zeta) > 0\}$ over a σ-finite set $\mathcal{X}$ with dominant measure μ. A parametric model $\mathfrak{M}_\theta$ is defined as

$$\mathfrak{M}_\theta = \{p(\cdot|\theta) \in \mathbb{B}, \theta \in V \subseteq \mathbb{R}^n\}.$$

That is, $\mathfrak{M}_\theta$ forms a smooth manifold with θ as coordinates.

One can define divergence functionals to measure the directed distance between two ζ-functions p and q. The most familiar is the Kullback–Leibler divergence. With the aid of a smooth and strictly convex function $f: \mathbb{R} \to \mathbb{R}$ and a strictly increasing function $\rho: \mathbb{R} \to \mathbb{R}$, one can show that the following is a general form of convex-induced divergence functional.

$$\frac{4}{1-\alpha^2} \int_{\mathcal{X}} \frac{1-\alpha}{2} f(\rho(p)) + \frac{1+\alpha}{2} f(\rho(q)) - f\left(\frac{1-\alpha}{2}\rho(p) + \frac{1+\alpha}{2}\rho(q)\right) d\mu, \tag{13.52}$$

since it is nonnegative and equals zero if and only if $p(\zeta) = q(\zeta)$ almost surely. A parametric model $p(\cdot|\theta) \in \mathfrak{M}_\theta$ is said to be "ρ-affine" if there exists a set of linearly independent functions $\lambda_i(\zeta) \in \mathbb{B}$ such that

$$\rho(p(\zeta|\theta)) = \sum_i \theta^i \lambda_i(\zeta).$$

The parameter $\theta = [\theta^1, \ldots, \theta^n]$ is called the "natural parameter" of a ρ-affine parametric model, and the functions $\lambda_1(\zeta), \ldots, \lambda_n(\zeta)$ are the affine basis functions. Examples of ρ-affine manifold include the so-called "alpha-affine manifolds" (Amari, 1985, Amari and Nagaoka, 2000), where $\rho(\cdot)$ takes on the following form (indexed by $\beta \in [-1, 1]$),

$$l^{(\beta)}(t) = \begin{cases} \log t & \beta = 1, \\ \dfrac{2}{1-\beta} t^{(1-\beta)/2} & \beta \in [-1, 1). \end{cases}$$

When a parametric model is ρ-affine, the function

$$\Phi(\theta) = \int_{\mathcal{X}} f(\rho(p(\zeta|\theta)))\, d\mu = \int_{\mathcal{X}} f\left(\sum_i \theta^i \lambda_i(\zeta)\right) d\mu$$

can be shown to be strictly convex. Therefore, the divergence functional in (13.52) takes the form of the divergence function $\mathcal{D}_\Phi^{(\alpha)}(\theta_p, \theta_q)$ on $V \times V$ given by

$$\mathcal{D}_\Phi^{(\alpha)}(\theta_p, \theta_q) = \frac{4}{1-\alpha^2}\left(\frac{1-\alpha}{2}\Phi(\theta_p) + \frac{1+\alpha}{2}\Phi(\theta_q) - \Phi\left(\frac{1-\alpha}{2}\theta_p + \frac{1+\alpha}{2}\theta_q\right)\right).$$

This is exactly (13.5)! An immediate consequence is that a ρ-affine manifold is the α-Hessian manifold, with metric and affine connections given by Proposition 13.4.

For any ζ-function $\zeta \mapsto p(\zeta)$, we now define

$$\eta_i = \int_{\mathcal{X}} f'(\rho(p(\zeta)))\, \lambda_i(\zeta)\, d\mu$$

such that $\eta = [\eta_1, \ldots, \eta_n] \in \widetilde{V} \subseteq \mathbb{R}^n$. We call η the "expectation parameter" of $p(\zeta)$ with respect to the set of (affine basis) functions $\lambda_1(\zeta), \ldots, \lambda_n(\zeta)$. It can be easily verified that for the ρ-affine parametric models,

$$\eta_i = \frac{\partial \Phi(\theta)}{\partial \theta^i}.$$

Define

$$\Phi^*(\theta) = \int_{\mathcal{X}} \tilde{f}(f'(\rho(p(\zeta|\theta))))\, d\mu,$$

where $\tilde{f} \colon \mathbb{R} \to \mathbb{R}$ is the Fenchel conjugate of f; then $\widetilde{\Phi}(\eta) \equiv \Phi^*((\partial\Phi)^{-1}(\eta))$ is the Fenchel conjugate of $\Phi(\theta)$. The pair of convex functions Φ, Φ^* induces η, θ via:

$$\frac{\partial \Phi(\theta)}{\partial \theta^i} = \eta_i \longleftrightarrow \frac{\partial \widetilde{\Phi}(\eta)}{\partial \eta_i} = \theta^i \,.$$

In theoretical statistics, we can call $\Phi(\theta)$ the generalized cumulant generating function (or partition function), and $\widetilde{\Phi}(\eta)$ the generalized entropy function. Natural parameter θ and expectation parameter η, which form biorthogonal coordinates, play important roles in statistical inference.

13.4 Summary and Open Problems

For two smooth, strictly convex functions $\Phi, \widetilde{\Phi}$ that are mutually conjugate, the variables $u = \partial\Phi(x)$ and $x = \partial\widetilde{\Phi}(u)$ are in one-to-one correspondence. It has been shown in this chapter that such a pair of variables can be viewed as biorthogonal coordinate systems on a Riemannian manifold whose metric is the second derivative of Φ when the x-coordinate system is used (or of $\widetilde{\Phi}$ when the u-coordinate system is used). Furthermore, a family of affine connections (indexed by α) can be defined with nonzero curvatures except for $\alpha = \pm 1$, the dually flat case (the so-called "Hessian manifold"). Each of these α-connections is equiaffine and admits a parallel volume form, and the entire family is induced from the divergence function $\mathcal{D}^{(\alpha)}_{\Phi}$ (or $\widetilde{\mathcal{D}}^{(\alpha)}_{\widetilde{\Phi}}$) associated with any convex function Φ (or $\widetilde{\Phi}$).

Our analysis revealed that the conjugate ($\pm\alpha$)-connections reflect two kinds of duality embodied by the convex-induced divergence function. The first is referential duality related to the choice of the reference and the comparison status for the two points (x versus y) for computing the value of the divergence $\mathcal{D}_\Phi^{(\alpha)}(x,y) = \mathcal{D}_\Phi^{(-\alpha)}(y,x)$. The second is representational duality related to the construction of two families of divergence functions, $\mathcal{D}_\Phi^{(\alpha)}(x,y)$ versus $\mathcal{D}_{\tilde{\Phi}}^{(\alpha)}((\partial\Phi)(x),(\partial\Phi)(y))$, using conjugate convex functions (see Table 13.1 in Section 13.3). The geometric quantities expressed in x-coordinates and expressed in u-coordinates are related to each other via Proposition 13.1. When $\alpha = \pm 1$, the two members of divergence functions coincide (and become Bregman divergence), so that the two kinds of duality reveal themselves as biduality:

$$\mathcal{D}_\Phi^{(-1)}(x,y) = \mathcal{D}_{\tilde{\Phi}}^{(-1)}(\partial\Phi(y),\partial\Phi(x)) = \mathcal{D}_{\tilde{\Phi}}^{(1)}(\partial\Phi(x),\partial\Phi(y)) = \mathcal{D}_\Phi^{(1)}(y,x) \ ,$$

which is compactly written in the form of canonical divergence as

$$\mathcal{A}_\Phi(x,v) = \mathcal{A}_{\tilde{\Phi}}(v,x) \ .$$

The relation between convex-induced divergence functions and α-connections is intriguing; that α as a convex mixture parameter coincides with α as indexing the family of connections is remarkable! We know that, in general, there may be many families of divergence functions that could yield the same α-connections. An explicit construction is as follows. Take the families of divergence functions ($\gamma \in \mathbb{R}$, $\beta \in [-1,1]$)

$$\frac{1+\beta}{2}\mathcal{D}_\Phi^{(\gamma)}(x,y) + \frac{1-\beta}{2}\mathcal{D}_\Phi^{(-\gamma)}(x,y) \,,$$

which induce an α-Hessian structure whose metric and conjugate connections are given in the forms (13.48) and (13.49), with α taking the value of $\beta\gamma$. The nonuniqueness of divergence functions giving rise to the family of α-connections invites the question of how to characterize the convex-induced divergence functions from the perspective of α-Hessian geometry. There is reason to believe that such axiomatization is possible because (i) the form of divergence function for the dually flat manifold ($\alpha = \pm 1$) is unique, namely, the Bregman divergence $\mathcal{B}_\Phi$; (ii) Lemma 13.1 gives that $\mathcal{D}^{(\alpha)} \geq 0$ if and only if $\mathcal{B}_\Phi \geq 0$ for any smooth function Φ. This hints at a deeper connection yet to be understood between convexity of a function and the α-Hessian geometry.

Another topic that needs further investigation is with respect to affine hypersurface realization of the α-Hessian manifold. We know that in affine immersion, geometric divergence is a generalization of the canonical divergence of dually flat (i.e., Hessian) manifolds. How to model the nonflat manifold with a general α value remains an open question. In particular, is there a generalization of geometric divergence that mirrors the way a convex-induced divergence function $\mathcal{D}_\Phi^{(\alpha)}$ generalizes Bregman divergence $\mathcal{B}_\Phi$ (or equivalently, the canonical divergence $\mathcal{A}_\Phi$)?

Finally, how do we extend the above analysis to an infinite-dimensional setting? The use of convex analysis (in particular, Young function and Orlicz space) to model the infinite-dimensional probability manifold yields fruitful insights for understanding difficult topological issues (Pistone and Sempi, 1995). It would thus be a worthwhile effort to extend the notion of biorthogonal coordinates to the infinite-dimensional manifold to study nonparametric information geometry. To this end, it would also be useful to extend the affine hypersurface theory to the infinite-dimensional setting and provide the formulation for co-dimension one affine immersion and co-dimension two centroaffine immersion. Here, affine hypersurfaces are submanifolds (resulting from normalization and positivity constraints on probability density functions; see, e.g., Zhang and Hasto, 2006) of an ambient manifold of unrestricted Banach space functions. Preliminary analyses (Zhang, 2006b) show that such an ambient manifold is flat for all α-connections, $\alpha \in \mathbb{R}$. So it provides a natural setting (i.e., affine space) in which probability densities can be embedded as an affine hypersurface. The value of such a viewpoint for statistical inference remains a topic for future exploration.

References

Amari, S. (1985). *Differential Geometric Methods in Statistics.* Lecture Notes in Statistics 28, Springer-Verlag, New York. Reprinted in 1990.

Amari, S. and Nagaoka, H. (2000). *Method of Information Geometry.* AMS Monograph, Oxford University Press.

Bauschke, H.H. (2003). Duality for Bregman projections onto translated cones and affine subspaces. *J. Approx. Theory* **121**, 1–12.

Bauschke, H.H. and Combettes, P.L. (2003). Iterating Bregman retractions. *SIAM J. Optim.* **13**, 1159–1173.

Bauschke, H.H., Borwein, J.M., and Combettes, P.L. (2003). Bregman monotone optimization algorithms. *SIAM J. Control Optim.* **42**, 596–636.

Bregman, L.M. (1967). The relaxation method of finding the common point of convex sets and its application to the solution of problems in convex programming. *USSR Comput. Math. Physics* **7**, 200–217.

Della Pietra, S., Della Pietra, V., and Lafferty, J. (2002). Duality and auxiliary functions for Bregman distances. *Technical Report CMU-CS-01-109*, School of Computer Science, Carnegie Mellon University.

Eguchi, S. (1983). Second order efficiency of minimum contrast estimators in a curved exponential family. *Ann. Statistics* **11**, 793–803.

Eguchi, S. (1992). Geometry of minimum contrast. *Hiroshima Math. J.* **22**, 631–647.

Ekeland, I. and Temam, R. (1976). *Convex Analysis and Variational Problems.* SIAM, Amsterdam.

Gao, D.Y. (2000). *Duality Principles in Nonconvex Systems: Theory, Methods and Applications.* Kluwer Academic, Dordrecht, xviii+454 pp.

Kurose, T. (1994). On the divergences of 1-conformally flat statistical manifolds. *Tohoku Math. J.* **46**, 427–433.

Matsuzoe, H. (1998). On realization of conformally-projectively flat statistical manifolds and the divergences. *Hokkaido Math. J.* **27**, 409–421.

Matsuzoe, H., Takeuchi, J., and Amari. S (2006). Equiaffine structures on statistical manifolds and Bayesian statistics. *Differential Geom. Appl.* **24**, 567–578.

Nomizu, K. and Sasaki, T. (1994). *Affine Differential Geometry – Geometry of Affine Immersions.* Cambridge University Press.

Pistone, G. and Sempi, C. (1995). An infinite dimensional geometric structure on the space of all the probability measures equivalent to a given one. *Ann. Statistics* **33**, 1543–1561.

Rockafellar, R.T. (1970). *Convex Analysis.* Princeton University Press.

Rockafellar, R.T. (1974). *Conjugate Duality and Optimization.* SIAM, Philadelphia.

Shima, H. (2007). *The Geometry of Hessian Structures.* World Scientific, Singapore.

Shima, H. and Yagi, K. (1997). Geometry of Hessian manifolds. *Differential Geom. Appl.* **7**, 277–290.

Simon, U., Schwenk-Schellschmidt, A., and Viesel, H. (1991). *Introduction to the Affine Differential Geometry of Hypersurfaces.* Lecture Notes, Science University of Tokyo.

Takeuchi, J. and Amari, S. (2005). α-Parallel prior and its properties. *IEEE Trans. Inf. Theory* **51**, 1011–1023.

Uohashi, K. (2002). On α-conformal equivalence of statistical manifolds. *J. Geom.* **75**, 179–184.

Zhang, J. (2004). Divergence function, duality, and convex analysis. *Neural Comput.* **16**, 159–195.

Zhang, J. (2006a). Referential duality and representational duality in the scaling of multidimensional and infinite-dimensional stimulus space. In: Dzhafarov, E. and Colonius, H. (Eds.) *Measurement and Representation of Sensations: Recent Progress in Psychological Theory.* Lawrence Erlbaum, Mahwah, NJ.

Zhang, J. (2006b). Referential duality and representational duality on statistical manifolds. *Proceedings of the Second International Symposium on Information Geometry and Its Applications*, Tokyo (pp. 58–67).

Zhang, J. (2007). A note on curvature of α-connections on a statistical manifold. *Ann. Inst. Statist. Math.* **59**, 161–170.

Zhang, J. and Hasto, P. (2006). Statistical manifold as an affine space: A functional equation approach. *J. Math. Psychol.* **50**, 60–65.

Chapter 14
NMR Quantum Computing

Zhigang Zhang, Goong Chen, Zijian Diao, and Philip R. Hemmer

Summary. Quantum computing is at the forefront of scientific and technological research and development of the 21st century. NMR quantum computing is one the most mature technologies for implementing quantum computation. It utilizes the motion of spins of nuclei in custom-designed molecules manipulated by RF pulses. The motion is on a nano- or microscopic scale governed by the Schrödinger equation in quantum mechanics. In this chapter, we explain the basic ideas and principles of NMR quantum computing, including basic atomic physics, NMR quantum gates, and operations. New progress in optically addressed solid-state NMR is expounded. Examples of Shor's algorithm for factorization of composite integers and the quantum lattice-gas algorithm for the diffusion partial differential equation are also illustrated.

14.1 Nuclear Magnetic Resonance

Many chapters in this book are concerned with mathematical problems in mechanics, elasticity, fluid mechanics, materials, and so on, which are on the macroscale. At the other extreme is the study of problems in atoms and molecules, photonics, nanotechnology, and the like, which are of the micro- or nanoscale governed chiefly by the Schrödinger equation. This area has un-

Zhigang Zhang · Goong Chen
Department of Mathematics, Texas A&M University, College Station, TX 77843
e-mail: zgzhang@math.tamu.edu, gchen@math.tamu.edu

Zijian Diao
Department of Mathematics, Ohio University - Eastern, St. Clairsville, OH 43950
e-mail: diao@ohio.edu

P.R. Hemmer
Department of Electrical and Computer Engineering, Texas A&M University, College Station, TX 77843, e-mail: prhemmer@ece.tamu.edu

D.Y. Gao, H.D. Sherali, (eds.), *Advances in Applied Mathematics and Global Optimization*
Advances in Mechanics and Mathematics 17, DOI 10.1007/978-0-387-75714-8_14,

dergone rapid advancement during the past ten years, in large part due to the stimuli from laser applications, quantum computing and quantum technology, and nanoelectronics. Most of the practitioners in this area are physicists and it appears that this area has not drawn enough attention from mathematicians. Here, we wish to describe one such development, namely, nuclear magnetic resonance (NMR) quantum computing. There already exist many papers on this topic (see, e.g., [18, 19, 16, 56, 58, 109]) written by physicists and computer scientists. Our chapter here describes the same interest, but perhaps from a more mathematical point of view.

14.1.1 Introduction

As of today, NMR is the most mature technology for the implementation of quantum computing. Naturally, this area is rife with papers. A good Internet resource for looking up NMR quantum computing references, both old and new, is the U.S. Los Alamos National Laboratory's Web site http://xxx.lanl.gov/quant-ph.

At present, several types of elementary quantum computing devices have been developed, based on AMO (atomic, molecular, and optical) or semiconductor physics and technologies. We may roughly classify them into the following.

Atomic — ion and atom traps, cavity QED [13].
Molecular — NMR.
Semiconductor — coupled quantum dots [12], silicon (Kane) [59].
Crystal structure — nitrogen-vacancy (NV) diamond.
Superconductivity — SQUID.

The above classification is not totally rigorous as new types of devices, such as quantum dots, or ion traps embedded in cavity-QED, have emerged which are of a hybrid nature. Also, laser pulse control, which is of an optical nature, seems to be omnipresent. In [3], a total of 12 types of quantum computing proposals have been listed.[1] Nevertheless, it is clear that NMR quantum computing belongs to the class of molecular computing where we use molecules as a small computer. The logic bits are the nuclear spins of atoms in custom-designed molecules. Spin flips are achieved through the application of radio-frequency (RF) fields on resonance at the nuclear spin frequencies. The system can be initialized by cooling the system down to the ground state or known low-entropy state, or using a special technology called averaging, especially for liquid NMR working in room temperature. Measurement or readout is carried out by measuring the magnetic induction signal generated by the

[1] The additional proposals not listed above but given in [3] are quantum Hall qubits, electrons in liquid helium, and spin spectroscopies.

precessing spin on the receiver coil. Numerous experiments have been successfully tried for different algorithms, mostly using liquid NMR technology. The algorithms tested include Grover's search algorithm [108, 58, 46, 122], other generalized search algorithms [76], quantum Fourier transforms [26, 114], Shor's algorithm [111], Deutsch–Jozsa algorithm [15, 74, 27, 84, 24], order finding [107, 100], error correcting code [65], and dense coding [33]. There are also other implementations reported such as the cat-code benchmark [64], information teleportation [87], and quantum system simulation [98].

NMR is an important tool in chemistry which has been in use for the determination of molecular structure and composition of solids, liquid, and gases since the mid-1940s, by research groups in Stanford and MIT independently, led by F. Bloch and E.M. Purcell, both of whom shared the Nobel prize in physics in 1952 for the discovery.

There are many excellent monographs on NMR [31, 91, 82]. There are also many other nice Internet Web-site resources offering concise but highly useful information about NMR (cf., e.g., [28, 51, 115]). Let us briefly explain the physics of NMR by following Edwards [28]. The NMR phenomenon is based on the fact that the spin of nuclei of atoms have magnetic properties that can be utilized to yield chemical, physical, and biological information. Through the famous Stern–Gerlach experiment in the earlier development of quantum mechanics, it is known that subatomic particles (protons, neutrons, and electrons) have spins. Nuclei with spins behave as a bar magnet in a magnetic field. In some atoms, for example, ^{12}C (carbon-12), ^{16}O (oxygen-16), and ^{32}S (sulphur-32), these spins are paired and cancel each other out so that the nucleus of the atom has no overall spin. However, in many atoms (^{1}H, ^{13}C, ^{31}P, ^{15}N, ^{19}F, etc.) the nucleus does possess an overall spin. To determine the spin of a given nucleus one can use the following rules.

1. If the number of neutrons and the number of protons are both even, the nucleus has no spin.
2. If the number of neutrons plus the number of protons is odd, then the nucleus has a half-integer spin (i.e., 1/2, 3/2, 5/2).
3. If the number of neutrons and the number of protons are both odd, then the nucleus has an integer spin (i.e., 1, 2, 3).

In quantum mechanical terms, the nuclear magnetic moment of a nucleus can align with an externally applied magnetic field of strength B_0 in only $2I+1$ ways, either with or against the applied field B_0, where I is the nuclear spin given in (1), (2), and (3) above. For example, for a single nucleus with $I = 1/2$, only one transition is possible between the two energy levels. The energetically preferred orientation has the magnetic moment aligned parallel with the applied field (spin $m = +1/2$) and is often denoted as α, whereas the higher-energy anti-parallel orientation (spin $m = -1/2$) is denoted as β. See Figure 14.1. In NMR quantum computing, these spin-up and spin-down quantum states resemble the two binary states 0 and 1 in a classical computer. Such a nuclear spin can serve as a quantum bit, or *qubit*. The

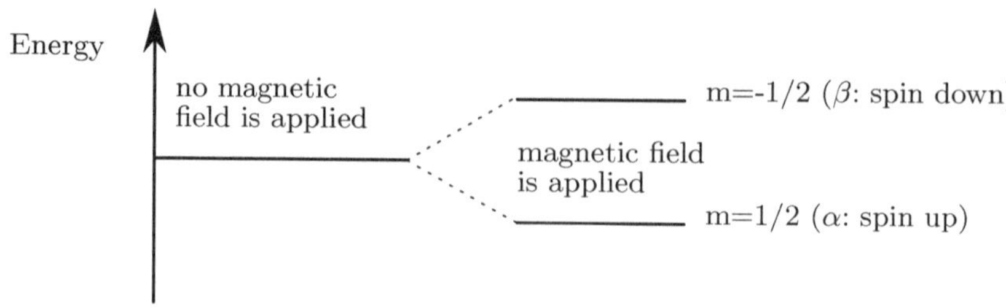

Fig. 14.1 Splitting of energy levels of a nucleus with spin quantum number 1/2.

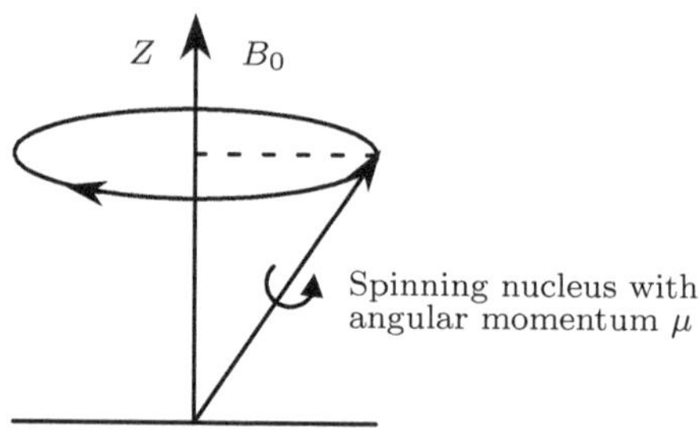

Fig. 14.2 A magnetic field B_0 is applied along the z-axis, causing the spinning nucleus to precess around the applied magnetic field.

rotational axis of the spinning nucleus cannot be oriented exactly parallel (or anti-parallel) with the direction of the applied field B_0 (aligned along the z-axis) but must *precess* (motion similar to a gyroscope) about this field at an angle, with an angular velocity ω_0, given by the expression $\omega_0 = \gamma B_0$. The precession rate ω_0 is called the *Larmor frequency* (cf. Figure 14.2). See more discussion of ω_0 below. The constant γ is called the *magnetogyric ratio*. This precession process generates a magnetic field with frequency ω_0. If we irradiate the sample with radio waves (MHz), then the proton can absorb the energy and be promoted to the higher-energy state. This absorption is called *resonance* because the frequencies of the applied radiation and the precession coincide at that frequency, leading to resonance.

There is another technique related to NMR, called *electron spin resonance* (ESR), that deals with the spins of electrons instead of those of the nuclei. The principles for ESR are nevertheless similar.

Quantum entanglement is accomplished through spin–spin coupling from the electronic bonds between the nuclei within the molecule and special RF pulse manipulations.

We now examine some fundamentals of atomic physics that are essential in any quantitative study of the manipulation of the quantum behavior of atoms. A complete description of the Hamiltonian (i.e., energy) of an atom contains nine terms as follows [31];

$$H = H_{el} + H_{CF} + H_{LS} + H_{SS} + H_{Ze} + H_{HF} + H_{Zn} + H_{II} + H_Q. \quad (14.1)$$

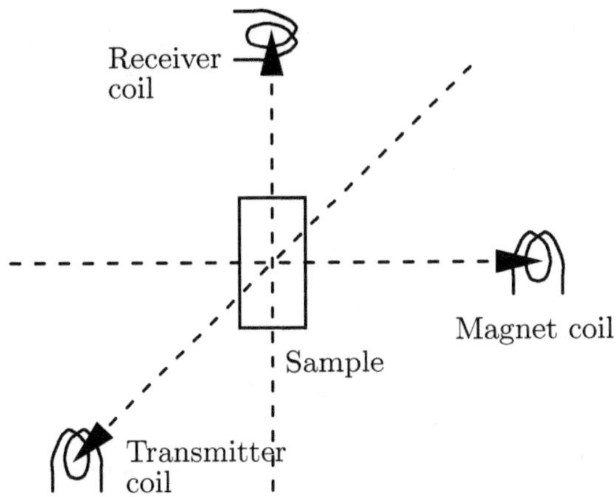

Fig. 14.3 Schematic diagram of an NMR apparatus. A sample which has nonzero-spin nuclei is put in a static magnetic field regulated by the current through the magnet coil. A transmitter coil provides the perpendicular field and a receiver coil picks up the signal. We can change the current through the magnet coil or change the frequency of the current in the transmitter coil to reach resonance.

The first three terms have the highest order, called the atomic Hamiltonian. They are the *electronic Hamiltonian term*, *crystal field term*, and the *spin–orbit interaction term*, respectively. The electronic Hamiltonian consists of kinetic energy of all electrons, $mv_i^2/2 = p_i^2/2m$, and two Coulomb terms: the potential energy of electrons relative to the nuclei, $-z_n e^2/r_{ni}$, and the interelectronic repulsion energy, e^2/r_{ij}:

$$H_{el} = \sum_i \frac{p_i^2}{2m} - \sum_{i,n} \frac{z_n e^2}{r_{ni}} + \sum_{i>j} \frac{e^2}{r_{ij}},$$

where r_{ni} denote the distance between the ith electron with the nth nucleus, and r_{ij} denote the interelectronic distance between the ith and the jth electrons.

The term H_{CF} is called the crystal field term. It comes from the interaction between the electron and the electronically charged ions forming the crystal, and is essentially a type of electrical potential energy:

$$V = -\sum_{i,j} \frac{Q_j e}{r_{ij}},$$

where Q_j is the ionic charge and r_{ij} is the distance from the electron to the ion. Normally, only those ions nearest to the electron are considered.

The third in the atomic Hamiltonian is the interaction between spin and orbit:

$$H_{LS} = \lambda \mathbf{L} \cdot \mathbf{S},$$

where $\mathbf{L}$ and $\mathbf{S}$ are the angular momenta of the orbit and spin, respectively, and λ is the coupling constant. In this section, we use $\mathbf{S}$ for the electron spin and $\mathbf{I}$ for the nuclear spin.

The remaining six terms are called *spin Hamiltonians*. Terms H_{Ze} and H_{Zn} are two that result from the application of an external magnetic field:

$$H_{Ze} = \beta \mathbf{B} \cdot (\mathbf{L} + \mathbf{S}),$$
$$H_{Zn} = -\sum_i g_{ni} \beta_n \mathbf{B} \cdot \mathbf{I}_i,$$

where $\mathbf{B}$ is the magnetic field strength. These two terms are called *Zeeman terms*, and they play major roles in NMR and ESR.

The nuclear spin–spin interaction term H_{II} is also important in quantum computation:

$$H_{II} = \sum_{i>j} \mathbf{I}_i \cdot J_{ij} \cdot \mathbf{I}_j,$$

because it provides a mechanism for the interaction between qubits. Hyperfine interaction arises from the interaction between the nuclear magnetic moments and the electron:

$$H_{HF} = \mathbf{S} \cdot \sum_i A_i \cdot \mathbf{I}_i.$$

In (14.1), by letting the z-axis be the privileged direction of spin measurement, the spin–spin interaction term H_{SS} is expressed as

$$H_{SS} = D[S_z^2 - \frac{1}{3}S(S+1)] + E(S_x^2 - S_y^2).$$

The very last term in (14.1) is called the quadrupolar energy:

$$H_Q = \frac{e^2 Q}{4I(2I-1)} \left(\frac{\partial^2 V}{\partial Z^2} \right) (3I_z^2 - I(I+1) + \eta(I_x^2 - I_y^2)).$$

For a specific system, only the Hamiltonian playing major roles is needed in the final model. For example, in the study of ESR, only three terms are retained and the Hamiltonian is written as

$$H = H_{Ze} + H_{HF} + H_{SS},$$

whereas in the NMR case,

$$H = H_{Zn} + H_{II}.$$

14.1.2 More about the Hamiltonian of NMR

A classical way to explain NMR is to regard it as a rotating charged particle that acts as a current circulating in a loop ([31, 10]), which creates a magnet with magnetic moment μ, $\mu = qvr/2$, where q is the electronic charge. The particle is rotating at $v/2\pi r$ revolutions per second.

Converting μ to electromagnetic units by dividing it by the velocity of light, and using angular momentum of the particle rather than the velocity of the particle, we obtain

$$\boldsymbol{\mu} = (q/2Mc)\mathbf{p},$$

where $\mathbf{p}$ is the angular momentum oriented along the rotating axis. The ratio μ/p is called the magnetogyric ratio, denoted by γ. A static magnetic field with strength B will apply a torque, which is equal to $\boldsymbol{\mu} \times \mathbf{B}$, on this particle. Newton's law states that the angular momentum will change according to a differential equation

$$\frac{d\mathbf{p}}{dt} = \boldsymbol{\mu} \times \mathbf{B} = \frac{q}{2Mc}\mathbf{p} \times \mathbf{B}.$$

Computation shows that $\mathbf{p}$ will rotate around the direction of $\mathbf{B}$ with frequency ω_0 defined by

$$\omega_0 = \frac{q}{2Mc}B.$$

The above is called the *Larmor equation*, and the frequency ω_0 is called the Larmor frequency, the precession frequency, or the resonance frequency as mentioned previously in Figure 14.2.

The above classical considerations are now modified by quantization to incorporate the quantum-mechanical behaviors of the nuclear spin. The vector variable $\mathbf{p}$ is quantized with quantum number $(I(I+1))^{1/2}$, and its projection to the z-axis (the direction of the magnetic field) is $m\hbar$. In total, there are $2I+1$ valid values of m evenly distributed from $-I$ to I; that is, $m = -I$, $-I+1, \ldots, I-1, I$. A factor g is introduced to include both the spin and orbital motion in the total angular momentum, called the Landé or spectroscopic splitting factor. For a free electron and proton, the magnetic momenta can be given as

$$\mu_e = \frac{g_e}{2}\left(\frac{he}{4\pi M_e c}\right) = \frac{g_e\beta}{2},$$

$$\mu_n = g_n I\left(\frac{he}{4\pi M_N c}\right) = g_n\, I\, \beta_N,$$

where $g_e = 2.0023$, $g_n = 5.58490$. Numbers β and β_N are called, respectively, the Bohr and the nucleus magneton where $\beta = 9.27 \times 10^{-21}$ erg gauss^{-1} and $\beta_N = 5.09 \times 10^{-24}$ erg gauss^{-1}. These values vary for different particles. In NMR, it is convenient to use the resonance frequency ω_0:

$$\hbar\omega_0 = g_e \beta B_0,$$
$$\hbar\omega_0 = g_N I \, \beta_N B_0.$$

Now we can write the Hamiltonian of a free nucleus as

$$H = -\boldsymbol{\mu} \cdot \mathbf{B} = -\hbar\gamma \mathbf{I} \cdot \mathbf{B}, \tag{14.2}$$

where γ is the magnetogyric ratio defined by $\gamma = \mu / I\hbar$ just as in the classical case. It is a characteristic constant for every type of nuclei; different nuclei have different magnetogyric ratios. Vector $\mathbf{I}$ after quantization, becomes the operator of angular momentum. The eigenvalues of this system, or the energy levels are

$$E = \gamma\hbar m B, \qquad m = -I, -I+1, \ldots, I-1, I. \tag{14.3}$$

The difference between two neighboring energy levels is $\gamma\hbar B$, which defines the resonance frequency depending on the magnetic field B and the particle.

There are other factors to be considered. The resonance frequency changes with the chemical environment of the nucleus. An example is the fluorine resonance spectrum of perfluorioisopropyl iodide. Two resonance lines of fluorine are observed in the spectrum, and the intensities ratio 6:1 agrees with the population ratio of the two groups of fluorine atoms. This phenomenon, called the *chemical shift*, is proportional to the strength of the magnetic field applied. This effect comes up because electrons close to the nucleus change the magnetic field around it; in other words, they create a diamagnetic shielding surrounding the nucleus. If the static field applied is B_0, then the electrons precessing around the magnetic field direction produce an induced magnetic field opposing B_0. The total effective magnetic field around the nucleus is then

$$\mathbf{B} = \mathbf{B}_0 - \mathbf{B}' = (1 - \sigma)\mathbf{B}_0,$$

where the parameter σ is called the *shielding coefficient.* In some cases σ is dependent on the temperature.

High-resolution NMR spectroscopy has found that the chemical shifted peaks are also composed of several lines, a result of the spin–spin coupling, which is the second term in the NMR Hamiltonian:

$$H_{II} = \sum_{i>j} \mathbf{I_i} \cdot J_{ij} \cdot \mathbf{I_j}.$$

14.1.3 Organization of the Chapter

Section 14.1 thus far has introduced some basic facts of nuclear spins and atomic physics.

In Section 14.2, we give background on what quantum computing is about, and introduce universal quantum gates based on liquid NMR.

Section 14.3 describes the most recent progress in solid-state NMR quantum gate controls and designs.

Sections 14.4 and 14.5 explain applications of the NMR quantum computer to Shor's algorithm and a lattice-gas algorithm.

14.2 Basic Technology Used in Quantum Computation with NMR

14.2.1 Introduction to Quantum Computation

Quantum mechanics is one of the revolutionary scientific discoveries of the 20th century. The field of quantum computation, our emphasis in this chapter, was born when the principles of quantum mechanics were introduced to modern computer science. Quantum computation mainly studies the analysis and construction of quantum algorithms with an eye toward surpassing the classical counterparts. Another tightly connected field is quantum information, which deals more with the storage, compression, encryption, and communication of information by quantum mechanical means [40, 7]. Quantum teleportation [6, 11] and quantum cryptography [5, 29] are two of the most known subjects of this field.

Modern computer science emerged when the eminent British mathematician Alan Turing invented the concept of the Turing machine (TM) in 1936 [103]. Although very simple and primitive, the TM captures the essence of computation. It serves as the universal model for all known physical computation devices. For many years, quantum effects had never been considered in the theory of computation, until the early 1980s. Benioff [4] first coined the term of quantum Turing machine (QTM). Motivated by the problem that classical computers cannot simulate quantum systems efficiently, Feynman [35] posed the quantum computer as a solution. Now we know that, in terms of computability, quantum computers and classical computers possess exactly the same computational power. But in terms of computational complexity, which measures the efficiency of computation, there are many exciting examples confirming that quantum computers do solve certain problems faster. The two most significant ones are Shor's factorization algorithm [96] and Grover's search algorithm [46], among other examples such as the Deutsch–Jozsa problem [24], the Bernstein–Vazirani problem [9], and Simon's problem [97].

Current physical realization of quantum computers follows the quantum circuit model [23], instead of the QTM model. The quantum circuit model is another fundamental model of computation, which is equivalent to the QTM model [118], but easier to implement. This model shares many common features of classical computers. In a classical computer, information is encoded

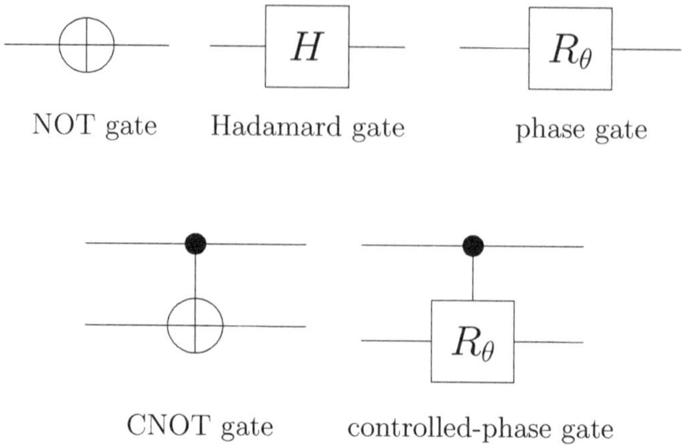

Fig. 14.4 Circuit diagrams of the NOT/Hadamard/phase/CNOT/controlled-phase gate.

in multibit binary states (0 or 1), transferred from one register to another, and processed by logic gates in concatenation. In a quantum computer, information is represented by the quantum states of the qubits, and manipulated by various quantum control mechanisms. Those control mechanisms trigger quantum operations to process information in a way resembling the gates in a classical computer. Such quantum operations are called *quantum gates* and a series of quantum gates in concatenation constitutes a *quantum circuit* [112]. However, because of the special effects of quantum mechanics, major distinctions exist.

In contrast to a classical system, a quantum system can exist in different states at the same time, an interesting phenomenon called *superposition*. Superposition enables quantum computers to process data in parallel. That is why a quantum computer can solve certain problems faster than a classical computer. From now on, we use the Dirac bra–ket notation. In this notation a pure one-qubit quantum state can be written as $|\phi\rangle = a|0\rangle + b|1\rangle$. Here $|0\rangle$ and $|1\rangle$ are the two basis states of the qubit, for example, in NMR, the spin-up and spin-down states, and $a, b \in \mathbf{C}$ with $|a|^2 + |b|^2 = 1$. When we make a measurement of a qubit, the result might be either $|0\rangle$ or $|1\rangle$, with probabilities $|a|^2$ and $|b|^2$, respectively. More generally, a string of n qubits can exist in any state of the form $|\psi\rangle = \sum_{x=00\ldots0}^{11\ldots1} \psi_x|x\rangle$, where $\psi_x \in \mathbf{C}$ and $\sum |\psi_x|^2 = 1$. When we make a measurement on $|\psi\rangle$, it collapses to $|x\rangle$, one of the 2^n basis states, with probability $|\psi_x|^2$. This indeterministic nature makes the design of efficient quantum algorithms highly nontrivial.

Another distinctive feature of the quantum circuit is that the operations performed by quantum gates must be unitary ($U^\dagger U = I$). It is the natural consequence of the unobserved quantum systems evolving according to the Schrödinger equation. A quantum gate may operate on any number of qubits. Here are some examples (cf. Figure 14.4 for the circuit diagrams).

1. NOT gate Λ_0: $\Lambda_0|0\rangle = |1\rangle$, $\Lambda_0|1\rangle = |0\rangle$, or $\Lambda_0 = \begin{bmatrix} 0 & 1 \\ 1 & 0 \end{bmatrix}$.
2. The Hadamard gate H: $H|0\rangle = \frac{1}{\sqrt{2}}(|0\rangle + |1\rangle)$, $H|1\rangle = \frac{1}{\sqrt{2}}(|0\rangle - |1\rangle)$, or
$$H = \frac{1}{\sqrt{2}} \begin{bmatrix} 1 & 1 \\ 1 & -1 \end{bmatrix}.$$
3. One-qubit phase gate R_θ: $R_\theta|0\rangle = |0\rangle$, $R_\theta|1\rangle = e^{i\theta}|1\rangle$, or
$$R_\theta = \begin{bmatrix} 1 & 0 \\ 0 & e^{i\theta} \end{bmatrix}.$$
4. Two-qubit controlled-NOT (CNOT) gate Λ_1: $\Lambda_1|00\rangle = |00\rangle$, $\Lambda_1|01\rangle = |01\rangle$, $\Lambda_1|10\rangle = |11\rangle$, $\Lambda_1|11\rangle = |10\rangle$, or
$$\Lambda_1 = \begin{bmatrix} 1 & 0 & 0 & 0 \\ 0 & 1 & 0 & 0 \\ 0 & 0 & 0 & 1 \\ 0 & 0 & 1 & 0 \end{bmatrix}.$$
5. Two-bit controlled-phase gate $\Lambda_1(R_\theta)$, where R_θ is the one-bit phase gate: $\Lambda_1(R_\theta)|00\rangle = |00\rangle$, $\Lambda_1(R_\theta)|01\rangle = |01\rangle$, $\Lambda_1(R_\theta)|10\rangle = |10\rangle$, $\Lambda_1(R_\theta)|11\rangle = e^{i\theta}|11\rangle$, or
$$\Lambda_1(R_\theta) = \begin{bmatrix} 1 & 0 & 0 & 1 \\ 0 & 1 & 0 & 0 \\ 0 & 0 & 1 & 0 \\ 0 & 0 & 0 & e^{i\theta} \end{bmatrix}.$$

The one-qubit and two-qubit quantum gates are of particular importance to the construction of a quantum computer, because of the following universality result.

Theorem 14.2.1. *(DiVincenzo [2, 25]) The collection of all the one-qubit gates and the two-qubit CNOT gate suffice to generate any unitary operations on any number of qubits.*

Figure 14.5 illustrates, as an example, how to generate the two-qubit controlled-phase gate using 2 CNOT gates and 3 one-qubit phase gates. The controlled-phase gate is an important building block for the quantum Fourier transform (cf. Figures 14.14 and 14.15).

The standard procedure of executing a quantum algorithm on a quantum circuit usually follows these steps.

1. Initialize the qubits.
2. Apply a proper sequence of quantum gates on the qubits.
3. Measure the qubits.

We address the details of these steps in the scenario of NMR technology.

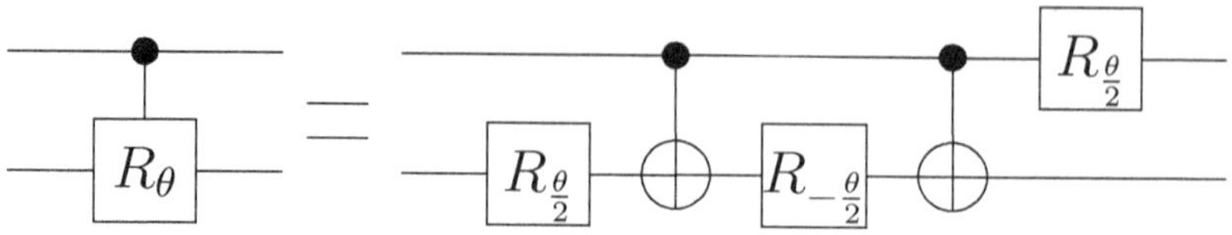

Fig. 14.5 Construction of the controlled-phase gate with CNOT gates and phase gates.

14.2.2 Realization of a Qubit

As mentioned in Section 14.1, NMR quantum computing is accomplished by using the spin-up and spin-down states of a spin-$\frac{1}{2}$ nucleus. A molecule with several nuclear spins may work as a quantum computer where each spin constitutes a qubit. In fact, NMR has a long history in information science. Back in the 1950s, nuclear spins were already used for information storage in computers.

Liquid NMR receives more interest due to its mature technology and readiness for application. For now, spin-$\frac{1}{2}$ nuclei such as protons and ^{13}C are preferred because they naturally represent a qubit, but multilevel qubits formed by spin-n nuclei, $n = 1, 2, \ldots$, may provide more freedom in the future. Through careful design, the potential qubits or nuclei are configured with different resonance frequencies and can be distinguished from each other. In a low-viscosity liquid, dipolar coupling between nuclei is averaged away by the random motion of the molecules. The J-coupling (scalar coupling) dominates the spin–spin interaction, which is an indirect through-bond electronic interaction. Previously, a very difficult part of the system operation was to set the quantum system to a special state (or to initialize it). Now a very complicated technology has been developed to solve this problem.

Figure 14.6 shows the structure of a trichloroethylene (TCE) molecule and a chloroform molecule used in NMR quantum computers. The hydrogen nucleus (proton) and two ^{13}C nuclei in a TCE molecule form three qubits which can be manipulated, and the chloroform molecule provides two qubits. The sample used by an NMR quantum computer has a large number ($\sim 10^{23}$) of such molecules. This is also called a *bulk* quantum computer. Although most molecules are in a totally random state at room temperature, there are still a small amount of spins standing out and serving our purpose. Theoretically, we use a statistical spin state called a pseudo-pure state, which has the same transformation property as that of a pure quantum state.

Let $|\phi\rangle = a|0\rangle + b|1\rangle$ be the state of a single qubit, $|0\rangle$ for spin-up and $|1\rangle$ for spin-down. We also assume that a is real because only the relative phase is important. Thus this state can be represented using two angles θ and ψ:

$$|\phi\rangle = \cos\frac{\theta}{2}|0\rangle + e^{i\psi}\sin\frac{\theta}{2}|1\rangle, \tag{14.4}$$

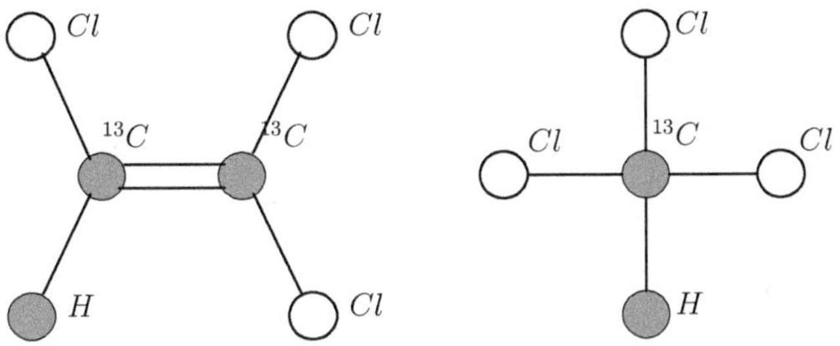

Fig. 14.6 The molecule structure of a candidate 3-qubit quantum system, trichloroethylene (left), and a candidate 2-qubit quantum system, chloroform. The trichloroethylene molecule has two labelled ^{13}C and a proton, all having one-half-spin nuclei. By considering the static magnetic field and spin–spin interaction, its Hamiltonian can be written as $H = -\sum_{i=1}^{3} g_{ni}\beta_{ni}\mathbf{I_i}\cdot\mathbf{B} + \sum_{i=1}^{2}\sum_{j=i+1}^{3}\mathbf{I_i}\cdot J_{i,j}\cdot\mathbf{I_j}$. The chloroform has one labelled ^{13}C and one proton.

where $\theta \in [0, \pi]$ and $\psi \in [0, 2\pi)$. If we think of $|0\rangle$ and $|1\rangle$ as the standard basis in $\mathbf{C}^2$, the quantum state corresponds to a unit vector in $\mathbf{C}^2$.

For the study of NMR spectroscopy with many nuclei, density matrices are preferred and are often written as the linear combination of product operators [83]:

$$\begin{aligned}\rho &= |\phi\rangle\langle\phi| \\ &= \begin{bmatrix} \cos^2\frac{\theta}{2} & e^{-i\psi}\frac{\sin\theta}{2} \\ e^{i\psi}\frac{\sin\theta}{2} & \sin^2\frac{\theta}{2}\end{bmatrix} \\ &= I_0 + \sin\theta\cos\psi I_x + \sin\theta\sin\psi I_y + \cos\theta I_z,\end{aligned} \tag{14.5}$$

where the product operators are defined as

$$I_0 = \frac{1}{2}\begin{bmatrix}1\,0\\0\,1\end{bmatrix},\quad I_x = \frac{1}{2}\begin{bmatrix}0\,1\\1\,0\end{bmatrix},\quad I_y = \frac{1}{2}\begin{bmatrix}0 & -i\\ i & 0\end{bmatrix},\quad I_z = \frac{1}{2}\begin{bmatrix}1 & 0\\0 & -1\end{bmatrix}. \tag{14.6}$$

They are different from the Pauli matrices only by a constant factor and share the similar commutative law. Upon collecting all the coefficients of I_x, I_y, and I_z together, we obtain a vector

$$\mathbf{v} = [\sin\theta\cos\psi \quad \sin\theta\sin\psi \quad \cos\theta]^T,$$

which is called a *Bloch vector*. In essence, we have defined a mapping from the set of unit vectors $|\phi\rangle \in \mathbf{C}^2$ to the set of unit vectors $\mathbf{v} \in \mathbf{R}^3$. We have good reasons to ignore the coefficient of I_0, because it has no effect on the spectroscopy and remains unchanged under any unitary transformation. Each Bloch vector determines a point on the unit sphere, called the *Bloch sphere*, which is displayed in Figure 14.7 [69, 83]. Bloch vectors have proven to be a very good tool for NMR quantum operations.

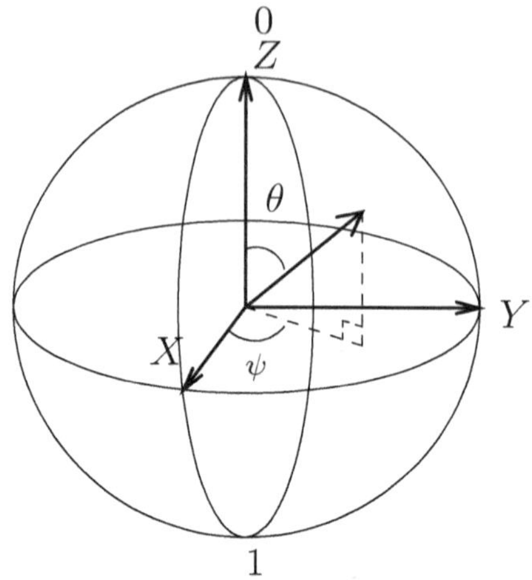

Fig. 14.7 The Bloch sphere representation of a quantum state.

The mapping defined above is surjective, because every point on the Bloch sphere gives rise to a unit vector $\mathbf{v} = [\sin\theta\cos\psi \;\; \sin\theta\sin\psi \;\; \cos\theta]^T$ for some pair of (θ, ψ). Conversely, if $\mathbf{v}(\theta', \psi') = \mathbf{v}(\theta, \psi)$, we get

$$\begin{cases} \cos\theta &= \cos\theta', \\ \sin\theta\cos\psi &= \sin\theta'\cos\psi', \\ \sin\theta\sin\psi &= \sin\theta'\cos\psi', \end{cases} \tag{14.7}$$

which can be used to show that the mapping is also injective if we identify all pairs of $(0, \psi)$ with one point and all pairs of (π, ψ) with another point. In fact, these two sets correspond to two states $|0\rangle$ and $|1\rangle$, respectively.

14.2.3 Transformation of Quantum States: SU(2) *and* SO(3)

When a quantum operation is applied to a quantum system, it may change the quantum state of the system from one to another. The representation of the operation depends on how the quantum state is represented. For example, (14.4) leads to an operator or matrix U which connects the new and old states of a single-spin quantum system:

$$|\phi'\rangle = U|\phi\rangle,$$

where $|\phi'\rangle$ and $|\phi\rangle$ are the quantum states after and before the operation, respectively. The fact that both states are unit vectors implies that U is a 2×2 unimodular complex matrix. Moreover, U is also unitary; that is, $U \in \mathbf{SU}(2)$,[2] a Lie group endowed with a certain topology.

If the quantum state is represented by a three-dimensional Bloch vector, the effect of a unitary operation can be viewed as that of a rotation which rotates the Bloch sphere, and the operator is represented by a 3×3 real matrix S. If the quantum system has states $\mathbf{v}$ and $\mathbf{v}'$ in Bloch vector form before and after the operation, respectively, then

$$\mathbf{v}' = S\mathbf{v}.$$

The matrix S is a proper rotation matrix; that is, $S \in \mathbf{SO}(3)$.[3] It is isometric and preserves the threefold product.

If both S and U represent the same physical operation, such as a transformation induced by a series of pulses in NMR, there must be a connection between them. One can show that there is a mapping R from $\mathbf{SU}(2)$ to $\mathbf{SO}(3)$ such that $S = R(U)$, for any $U \in \mathbf{SU}(2)$ and its corresponding Bloch-sphere representation S [88]. Simple computation shows that the entry of matrix $S = R(U)$ at the kth row and ith column is given as

$$S_{ki} = \mathrm{Tr}(\sigma_k \, U \, I_i \, U^{\dagger}), \tag{14.8}$$

where σ_k are the Pauli matrices,[4] and Tr is the trace operator. It can also be shown that R is a two-to-one homomorphism between $\mathbf{SU}(2)$ and $\mathbf{SO}(3)$ with kernel $\ker(R) = \{I, -I\}$. It coincides with the fact that U and $-U$ in $\mathbf{SU}(2)$ represent the same operation because only the relative phase matters. This mapping is also surjective, so it defines an isomorphism from the quotient group $\mathbf{SU}(2)/\ker(R)$ to $\mathbf{SO}(3)$. We provide a more detailed discussion about this isomorphism in the appendix.

It is known that any $U \in \mathbf{SU}(2)$ can be written into an exponential form parameterized by an angle $\theta \in [0, 2\pi)$ and a unit vector $\mathbf{n}$ such that

$$\begin{aligned} U(\theta, \mathbf{n}) &= e^{-i(\theta/2)\mathbf{n}\cdot\boldsymbol{\sigma}} \\ &= \begin{bmatrix} \cos\frac{\theta}{2} - in_3 \sin\frac{\theta}{2} & -\sin\frac{\theta}{2}(n_2 + in_1) \\ \sin\frac{\theta}{2}(n_2 - in_1) & \cos\frac{\theta}{2} + in_3 \sin\frac{\theta}{2} \end{bmatrix} \\ &= \cos\tfrac{\theta}{2} I - i \sin\tfrac{\theta}{2} \mathbf{n}\cdot\boldsymbol{\sigma}, \end{aligned} \tag{14.9}$$

where $\boldsymbol{\sigma} = [\sigma_x, \sigma_y, \sigma_z]$. With this parameterization of $\mathbf{SU}(2)$, entries of $S = R(U)$ can be computed using (14.8) as

---

[2] $\mathbf{SU}(n)$ is the special unitary group of $n \times n$ matrices. An $n \times n$ matrix $A \in \mathbf{SU}(n)$ if and only if A is unitary (i.e., $A \cdot A^{\dagger} = I_n$, where $A^{\dagger}$ is the Hermitian adjoint of A, and $\det A = 1$).

[3] $\mathbf{SO}(3)$ denotes the special orthogonal group of 3×3 matrices. An $n \times n$ matrix $A \in \mathbf{SO}(n)$ if and only if A is real, $AA^T = I_n$, and $\det A = 1$.

[4] The Pauli matrices are $\sigma_x = \left[\begin{smallmatrix} 0 & 1 \\ 1 & 0 \end{smallmatrix}\right]$, $\sigma_y = \left[\begin{smallmatrix} 0 & -i \\ i & 0 \end{smallmatrix}\right]$, and $\sigma_z = \left[\begin{smallmatrix} 1 & 0 \\ 0 & -1 \end{smallmatrix}\right]$.

$$S_{ij} = R(U)_{ij} = \cos\theta\,\delta_{ij} + (1-\cos\theta)n_i n_j + \sum_{k=1}^{3} \sin\theta\,\epsilon_{ikj} n_k. \tag{14.10}$$

It should be noted now that S coincides with a rotation about the axis along $\mathbf{n}$ with an angle θ in the three-dimensional Euclidean space after comparing S_{ij} with the standard formula of a rotation matrix. This interpretation is important in understanding the terminologies used in NMR. For example, the rotations around x-, y-, and z-axes ($x/y/z$-rotations) with an arbitrary angle θ define the following three unitary operators in $\mathbf{SU}(2)$, respectively.

$$X_\theta = e^{-i\theta\sigma_x/2} = \begin{bmatrix} \cos\frac{\theta}{2} & -i\sin\frac{\theta}{2} \\ -i\sin\frac{\theta}{2} & \cos\frac{\theta}{2} \end{bmatrix}, \tag{14.11}$$

$$Y_\theta = e^{-i\theta\sigma_y/2} = \begin{bmatrix} \cos\frac{\theta}{2} & -\sin\frac{\theta}{2} \\ \sin\frac{\theta}{2} & \cos\frac{\theta}{2} \end{bmatrix}, \tag{14.12}$$

$$Z_\theta = e^{-i\theta\sigma_z/2} = \begin{bmatrix} e^{-i\theta/2} & 0 \\ 0 & e^{i\theta/2} \end{bmatrix}. \tag{14.13}$$

14.2.4 Construction of Quantum Gates

From Theorem 14.2.1, we know that the collection of all the one-qubit gates and the two-qubit CNOT gate are universal. In addition, the following fact [85, p. 175] holds for one-qubit quantum gates.

Theorem 14.2.2. *Suppose U is a unitary operation on a single qubit. Then there exist real numbers α, β, γ, and δ such that*

$$U = e^{i\alpha} Z_\beta Y_\gamma Z_\delta.$$

For example, the Hadamard gate H can be decomposed as $H = e^{i\pi/2} Y_{\pi/2} Z_\pi$. Clearly, the $x/y/z$-rotation gates provide building blocks sufficient to construct any one-qubit unitary gate. In this section, we show how to realize these one-qubit rotation gates and the two-qubit CNOT gate using NMR. We also show how to decouple the interaction between two spins, a process called *refocusing* [85].

14.2.4.1 One-Qubit Gates

A single-spin system has Hamiltonian $H = -\boldsymbol{\mu}\cdot\mathbf{B}$, where $\boldsymbol{\mu}$ is the magnetic moment, and

$$\mathbf{B} = B_0\mathbf{e}_z + B_1(\mathbf{e}_x\cos(\omega t) + \mathbf{e}_y\sin(\omega t)) \tag{14.14}$$

is the magnetic field applied. B_0, a large constant, is the amplitude of the static magnetic field, and B_1 is the amplitude of the oscillating magnetic field in the x–y plane. When $B_1 = 0$, the Hamiltonian and Schrödinger equations can be obtained as ([85])

$$H = \frac{\omega_0}{2}\sigma_z \tag{14.15}$$

and

$$i\partial_t|\psi(t)\rangle = H|\psi(t)\rangle, \tag{14.16}$$

respectively, where $\hbar$ has been divided from both sides in the second equation and we take $\hbar$ away from H in the first one just for simplicity. The Larmor frequency $\omega_0 = -B_0\gamma$ is defined by the nuclei and the magnetic field; see (14.3). Assume that the initial state is $|\psi_0\rangle = a_0|0\rangle + b_0|1\rangle$. Then the evolution of the quantum state of the spin and the density matrix can be solved directly and given as

$$\begin{aligned}
|\psi(t)\rangle &= e^{-i\omega_0\sigma_z t/2}|\psi_0\rangle \\
&= \begin{bmatrix} e^{-i\omega_0 t/2} & 0 \\ 0 & e^{i\omega_0 t/2} \end{bmatrix} \begin{bmatrix} a_0 \\ b_0 \end{bmatrix} \\
&= e^{-i\omega_0 t/2} \begin{bmatrix} 1 & 0 \\ 0 & e^{i\omega_0 t} \end{bmatrix} |\psi_0\rangle, \\
\rho(t) &= e^{-itH}\rho(0)e^{itH}.
\end{aligned}$$

This evolution is also called a *chemical shift evolution*, resembling the precessing of a magnet in a static field. Recall the Bloch vector on the Bloch sphere. It is exactly Z_θ, the rotation operator around the z-axis with $\theta = \omega_0 t$.

To achieve an x-rotation operator, we need a small magnetic field transverse to the z-direction to control the evolution of the quantum state. The Hamiltonian is given as in (14.14) by choosing B_1 different from zero:

$$H = -\boldsymbol{\mu}\cdot\mathbf{B} = \frac{\omega_0}{2}\sigma_z + \frac{\omega_1}{2}\left(\sigma_x\cos(\omega t) + \sigma_y\sin(\omega t)\right),$$

where ω_1 depends on the x–y plane component B_1 of the magnetic field, $\omega_1 = -B_1\gamma$. To solve the Schrödinger equation, we put $|\psi(t)\rangle$ in a "frame" rotating with the magnetic field around the z-axis at frequency ω, $|\phi(t)\rangle = e^{i\omega t\sigma_z/2}|\psi(t)\rangle$. With this substitution, the Schrödinger equation (14.16) becomes

$$i\partial_t|\phi(t)\rangle = (e^{i\omega\sigma_z t/2}He^{-i\omega\sigma_z t/2} - \frac{\omega}{2}\sigma_z)|\phi(t)\rangle. \tag{14.17}$$

Using properties

$$\begin{aligned}
e^{i\omega\sigma_z t/2}\sigma_z e^{-i\omega\sigma_z t/2} &= \sigma_z, \\
e^{i\omega\sigma_z t/2}\sigma_x e^{-i\omega\sigma_z t/2} &= \sigma_x\cos(\omega t) - \sigma_y\sin(\omega t), \\
e^{i\omega\sigma_z t/2}\sigma_y e^{-i\omega\sigma_z t/2} &= \sigma_x\sin(\omega t) + \sigma_y\cos(\omega t),
\end{aligned} \tag{14.18}$$

we obtain

$$\begin{aligned} i\partial_t|\phi(t)\rangle &= \left(\frac{\omega_0-\omega}{2}\sigma_z + \frac{\omega_1}{2}\sigma_x\right)|\phi(t)\rangle, \\ |\phi(t)\rangle &= e^{-i((\omega_0-\omega)\sigma_z/2+\omega_1\sigma_x/2)t}|\phi(0)\rangle. \end{aligned} \tag{14.19}$$

We know from (14.9) that this is a rotation around the axis

$$\mathbf{n} = \frac{1}{\sqrt{1+(\frac{\omega_1}{\omega_0-\omega})^2}}\left(\mathbf{z} + \frac{\omega_1}{\omega_0-\omega}\mathbf{x}\right). \tag{14.20}$$

An important case is $\omega_0 = \omega$, also called the *resonance* case where its name comes from the zero denominator in (14.20). By (14.19), we see that a relatively weak transverse magnetic field causes a rotation around the x-axis:

$$|\psi(t)\rangle = e^{-i\omega_0\sigma_z t/2}|\phi(t)\rangle = e^{-i\omega_0 t\sigma_z/2}e^{-i\omega_1 t\sigma_x/2}|\phi(0)\rangle = Z_\theta X_\beta|\psi(0)\rangle, \tag{14.21}$$

where $X_\beta = e^{-i\omega_1 t\sigma_x/2}$, $\beta = \omega_1 t$. By applying another $Z_{-\theta}$, we obtain a rotation X_β as desired. Because the frequency of the precession is in the radio frequency band, the field applied is called an *RF pulse*.

When $|\omega_0 - \omega| \gg \omega_1$, the rotation axis direction is almost along $\mathbf{z}$ and the RF pulse has no effect on it:

$$|\psi(t)\rangle = e^{-i\omega\sigma_z t/2}|\phi(t)\rangle \approx e^{-i\omega_0 t\sigma_z/2}|\psi(0)\rangle = Z_{\omega_0 t}|\psi(0)\rangle,$$

thus we can tell one qubit from another because their resonance frequencies are designed to be different. There are still cases where the difference of resonance frequencies between spins is not large enough. The RF pulse may cause similar rotations on all those spins. To avoid or at least minimize it, a soft pulse is applied instead of the so-called hard pulse. It is a pulse with a longer time span and weaker magnetic field, in other words, a smaller ω_1. This strategy makes these "close" qubits fall into the $|\omega_0 - \omega| \gg \omega_1$ case.

If we change the magnetic field to

$$\mathbf{B} = B_0\mathbf{e}_z + B_1(\mathbf{e}_x\cos(\omega_0 t+\alpha) + \mathbf{e}_y\sin(\omega_0 t+\alpha)), \tag{14.22}$$

the Hamiltonian will become

$$H = \frac{\omega_0}{2}\sigma_z + \frac{\omega_1}{2}(\sigma_x\cos(\omega_0 t+\alpha) + \sigma_y\sin(\omega_0+\alpha)), \tag{14.23}$$

where ω_1 is defined as before. The RF field is almost the same as (14.14) in the resonance case except for a phase shift. Using the same rotation frame as before with $\omega = \omega_0$, we obtain

$$i\partial_t|\phi(t)\rangle = \frac{\omega_1}{2}(\sigma_x\cos(\alpha) + \sigma_y\sin(\alpha))|\phi(t)\rangle, \tag{14.24}$$

after simplification. After time duration t, the new system state is given as

$$|\phi(t)\rangle = e^{-i(\omega_1/2)(\sigma_x \cos(\alpha)+\sigma_y \sin(\alpha))t}|\phi(0)\rangle, \tag{14.25}$$

and the evolution operator can be computed using (14.9) as

$$\begin{aligned} U_{\theta/2,\alpha} &= e^{-i(\omega_1/2)(\sigma_x \cos(\alpha)+\sigma_y \sin(\alpha))t} \\ &= \begin{bmatrix} \cos(\frac{\theta}{2}) & -i\sin(\frac{\theta}{2})e^{-i\alpha} \\ -i\sin(\frac{\theta}{2})e^{i\alpha} & \cos(\frac{\theta}{2}) \end{bmatrix}, \end{aligned} \tag{14.26}$$

where $\theta = \omega_1 t$. This is a one-qubit rotation operator, and is sometimes called a Rabi rotation gate. When $\alpha = \pi/2$,

$$\begin{aligned} U_{\theta/2,\pi/2} &= \begin{bmatrix} \cos(\frac{\theta}{2}) & -\sin(\frac{\theta}{2}) \\ \sin(\frac{\theta}{2}) & \cos(\frac{\theta}{2}) \end{bmatrix} \\ &= Y_\theta. \end{aligned} \tag{14.27}$$

We have achieved a y-rotation operator just by adding a phase shift to the RF field.

14.2.4.2 Two-Qubit Gates

The construction of a two-qubit gate requires the coupling of two spins. In a liquid sample of NMR, *J-coupling* is the dominating coupling between spins. Under the assumption that the resonance frequency difference between the coupled spins is much larger than the strength of the coupling (a so-called weak coupling regime), the total Hamiltonian of a two-spin system without transverse field may be given as

$$H = \frac{1}{2}\omega_1\sigma_z^1 + \frac{1}{2}\omega_2\sigma_z^2 + \frac{1}{2}J\sigma_z^1\sigma_z^2, \tag{14.28}$$

where ω_i is the frequency corresponding to spin i, σ_z^i is the z-projection operator of spin i, for $i = 1, 2$, and J is the coupling coefficient. Take the chloroform in Figure 14.6, for example [16, 69]. In an 11.7 T magnetic field, the precession frequency of ^{13}C is about $2\pi \times 500$ MHz and the precession frequency of the proton is about $2\pi \times 125$ MHz. The coupling constant J is about $2\pi \times 100$ Hz. Here we set $B_1 = 0$, which means no transverse magnetic field is applied and those terms such as σ_x, σ_y do not appear. The remaining terms in the Hamiltonian only contain operators σ_z^1 or σ_z^2, which are commutative. Thus, we can obtain the eigenstates and eigenvalues of this two-spin system and we map the set of eigenstates to the standard basis of $\mathbf{C}^4$, as follows.

$$|00\rangle = \begin{bmatrix} 1\\0\\0\\0 \end{bmatrix}, \quad |01\rangle = \begin{bmatrix} 0\\1\\0\\0 \end{bmatrix}, \quad |10\rangle = \begin{bmatrix} 0\\0\\1\\0 \end{bmatrix}, \quad |00\rangle = \begin{bmatrix} 0\\0\\0\\1 \end{bmatrix}; \tag{14.29}$$

$$\begin{aligned} H|00\rangle &= k_{00}|00\rangle, \quad k_{00} = \tfrac{1}{2}\omega_1 + \tfrac{1}{2}\omega_2 + \tfrac{1}{2}J; \\ H|01\rangle &= k_{01}|01\rangle, \quad k_{01} = \tfrac{1}{2}\omega_1 - \tfrac{1}{2}\omega_2 - \tfrac{1}{2}J; \\ H|10\rangle &= k_{10}|10\rangle, \quad k_{10} = -\tfrac{1}{2}\omega_1 + \tfrac{1}{2}\omega_2 - \tfrac{1}{2}J; \\ H|11\rangle &= k_{11}|11\rangle, \quad k_{11} = -\tfrac{1}{2}\omega_1 - \tfrac{1}{2}\omega_2 + \tfrac{1}{2}J. \end{aligned} \tag{14.30}$$

Because the matrix is diagonal, the evolution of this two-spin system can be easily derived as

$$|\psi(t)\rangle = e^{-iHt}|\psi(0)\rangle = \begin{bmatrix} e^{-ik_{00}t} & & & \\ & e^{-ik_{01}t} & & \\ & & e^{-ik_{10}t} & \\ & & & e^{-ik_{11}t} \end{bmatrix} |\psi(0)\rangle. \tag{14.31}$$

We can also rewrite the one-qubit rotation operators for this two-$\frac{1}{2}$-spin system in matrix form with respect to the same basis:

$$Z^1_{\pi/2} = \begin{bmatrix} e^{-i\pi/4} & & & \\ & e^{-i\pi/4} & & \\ & & e^{i\pi/4} & \\ & & & e^{i\pi/4} \end{bmatrix}, \tag{14.32}$$

$$Z^2_{-\pi/2} = \begin{bmatrix} e^{i\pi/4} & & & \\ & e^{-i\pi/4} & & \\ & & e^{i\pi/4} & \\ & & & e^{-i\pi/4} \end{bmatrix}, \tag{14.33}$$

$$Y^2_{\pi/2} = \frac{\sqrt{2}}{2} \begin{bmatrix} 1 & -1 & & \\ 1 & 1 & & \\ & & 1 & -1 \\ & & 1 & 1 \end{bmatrix}, \tag{14.34}$$

$$Y^2_{-\pi/2} = \frac{\sqrt{2}}{2} \begin{bmatrix} 1 & 1 & & \\ -1 & 1 & & \\ & & 1 & 1 \\ & & -1 & 1 \end{bmatrix}, \tag{14.35}$$

where Z^i_θ is the rotation operator for spin i with angle θ around the z-axis while keeping another spin unchanged, and all Y^i_θ are similarly defined operators about the y-axis; see (14.12). A careful reader may raise issues about the one-qubit gate we have obtained in Section 14.2.4.1 because the coupling between two qubits always exists and has not been considered. We need to turn off the coupling when we only want to operate one spin but the coupling is nonnegligible. This is in fact one of the major characteristic difficulties associated with the NMR quantum computing technology. A special technology called *refocusing* is useful. It works as follows. We apply a soft π pulse on the spare spin that we don't want to change at the middle point of the operation time duration while we are working on the target spin. The effect is that the coupling before the pulse cancels the one after the pulse, so the result of no-coupling is achieved. Another π pulse will be needed to turn the spin back. All pulses are soft.

This technology is so important that we now state it here as a theorem.

Theorem 14.2.3. *Let $H = (\omega_1/2)\,\sigma_z^1 + (J/2)\,\sigma_z^1\sigma_z^2 + A$ be a given Hamiltonian, where A is a Hamiltonian that does not act on spin 1 and commutes with σ_z^2. Then the evolution operators of A and H satisfy*

$$e^{-iAt} = -X_\pi^1 e^{-iHt/2} X_\pi^1 e^{-iHt/2}; \tag{14.36}$$

that is, the collective evolution of the quantum system with Hamiltonian H and additional two X_π^1-pulses at the middle and the end of the time duration, equals that of a system with Hamiltonian A (up to a global phase shift π, or a factor -1).

Proof. Assume that the time duration is t and denote U for

$$U = X_\pi^1 e^{-iHt/2} X_\pi^1 e^{-iHt/2}. \tag{14.37}$$

Note that $X_\pi^1 = e^{-i(\pi/2)\sigma_x^1}$ and it commutes with A which contains no operators acting on spin 1, thus

$$U = X_\pi^1 e^{-i((\omega_1/2)\sigma_z^1+(J/2)\sigma_z^1\sigma_z^2)t/2} X_\pi^1 e^{-i((\omega_1/2)\sigma_z^1+(J/2)\sigma_z^1\sigma_z^2)t/2} e^{-iAt}. \tag{14.38}$$

It suffices to prove that the part before e^{-iAt} satisfies

$$B = X_\pi^1 e^{-i((\omega_1/2)\sigma_z^1+(J/2)\sigma_z^1\sigma_z^2)t/2} X_\pi^1 e^{-i((\omega_1/2)\sigma_z^1+(J/2)\sigma_z^1\sigma_z^2)t/2} = -I. \tag{14.39}$$

We first check the effect of B on the four basis vectors. We have

$$\begin{aligned}
B|11\rangle &= X_\pi^1 e^{-i((\omega_1/2)\sigma_z^1+(J/2)\sigma_z^1\sigma_z^2)t/2} X_\pi^1 e^{-i((\omega_1/2)\sigma_z^1+(J/2)\sigma_z^1\sigma_z^2)t/2}|11\rangle \\
&= e^{-i((-\omega_1+J)/4)t}(-i) X_\pi^1 e^{-i((\omega_1/2)\sigma_z^1+(J/2)\sigma_z^1\sigma_z^2)t/2}|01\rangle \\
&= (-i)e^{-i((-\omega_1+J)/4)t} X_\pi^1 e^{-i((\omega_1-J)/4)t}|01\rangle \\
&= (-i)^2|11\rangle \\
&= -|11\rangle,
\end{aligned} \tag{14.40}$$

$$\begin{aligned}
B|01\rangle &= X_\pi^1 e^{-i((\omega_1/2)\sigma_z^1+(J/2)\sigma_z^1\sigma_z^2)t/2} X_\pi^1 e^{-i((\omega_1/2)\sigma_z^1+(J/2)\sigma_z^1\sigma_z^2)t/2}|01\rangle \\
&= e^{-i((\omega_1-J)/4)t}(-i)X_\pi^1 e^{-i((\omega_1/2)\sigma_z^1+(J/2)\sigma_z^1\sigma_z^2)t/2}|11\rangle \\
&= (-i)e^{-i((-\omega_1+J)/4)t} X_\pi^1 e^{-i((-\omega_1+J)/4)t}|11\rangle \\
&= (-i)^2|01\rangle \\
&= -|01\rangle, \qquad (14.41)
\end{aligned}$$

and similarly,

$$\begin{aligned}
B|10\rangle &= -|10\rangle, \\
B|00\rangle &= -|00\rangle.
\end{aligned} \qquad (14.42)$$

In the computation above, we have used the fact that X_π^1 has no effect on the second spin and the four basis vectors $|00\rangle$, $|01\rangle$, $|10\rangle$ and $|11\rangle$ are the eigenstates of the operator $(\omega_1/2)\,\sigma_z^1 + (J/2)\,\sigma_z^2\sigma_z^1$. The result shows that $B = -I$, and we are done. □

When the Hamiltonian is given in the form as (14.28), the above theorem tells us that both the chemical shift evolution (precession) and the J-coupling effect on spin 1 are removed and only the term $(\omega_2/2)\,\sigma_z^2$ remains. We obtain a z-rotation of spin 2 while freezing spin 1. By combining it with several hard pulses, we can also achieve any arbitrary rotation on spin 2 with the motion of spin 1 frozen [73]. Similar computation shows that a hard π pulse applied at the middle point of the time duration cancels the chemical shift evolution of both spins. This can be seen by checking the identity

$$e^{-iHt/2}X_\pi^1 X_\pi^2 e^{-iHt/2} = \begin{bmatrix} & & & e^{-iJt/2} \\ & & e^{iJt/2} & \\ & e^{iJt/2} & & \\ e^{-iJt/2} & & & \end{bmatrix}. \qquad (14.43)$$

Another hard π pulse can rotate two spins back, so we have achieved an evolution which has only the J-coupling effect, denoted by $\mathbb{Z}_\theta$:

$$\mathbb{Z}_\theta = \begin{bmatrix} e^{-i\theta/2} & & & \\ & e^{i\theta/2} & & \\ & & e^{i\theta/2} & \\ & & & e^{-i\theta/2} \end{bmatrix},$$

and when $\theta = \pi/2$,

$$\mathbb{Z}_{\pi/2} = \begin{bmatrix} e^{-i\pi/4} & & & \\ & e^{i\pi/4} & & \\ & & e^{i\pi/4} & \\ & & & e^{-i\pi/4} \end{bmatrix}. \qquad (14.44)$$

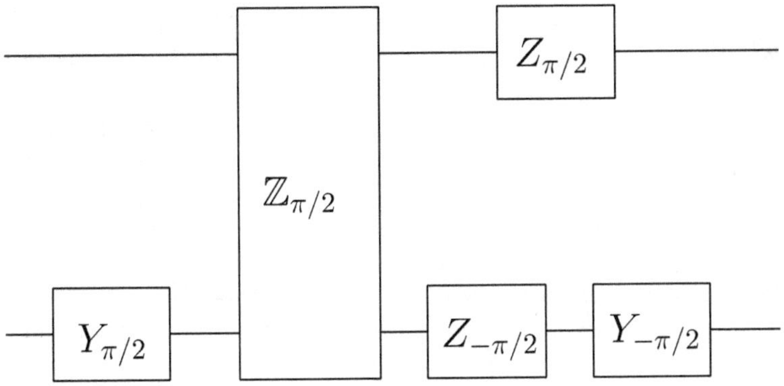

Fig. 14.8 The quantum circuit used to realize a quantum controlled-not gate.

Although we give only an example of the 2-qubit system in the above, the reader should note that a general method is available to reserve only the couplings wanted while keeping all the others cancelled for multiqubit systems [57, 71, 73]. Combining operators in (14.32) through (14.35) and (14.44), we can now construct a CNOT gate as in Figure 14.8 which includes four 1-qubit $\pi/2$ rotations around y- or z-axes and one 2-qubit $\pi/2$ rotation. The total operator, denoted by CN, can be computed as

$$CN = Z^1_{\pi/2} Y^2_{-\pi/2} Z^2_{-\pi/2} \mathbb{Z}_{\pi/2} Y^2_{\pi/2} = e^{-\frac{\pi}{4}i} \begin{bmatrix} 1 & & & \\ & 1 & & \\ & & 0 & 1 \\ & & 1 & 0 \end{bmatrix}, \tag{14.45}$$

which is a CNOT gate up to a phase of $-\pi/4$ [69].

We have shown how to construct one-qubit gates and the two-qubit CNOT gate using the NMR technology. The simple pulse design works fine in ideal situations. In practice, errors arise from various factors. Decoherence causes the lost of quantum information with time. Thus, all operations should be completed within a short time, roughly constrained by the energy relaxation time T_1 and the phase randomization time T_2. Again, take the chloroform for an example. For protons, $T_1 \approx 7$ sec and $T_2 \approx 2$ sec; for carbons, $T_1 \approx 16$ sec and $T_2 \approx 0.2$ sec [16, 69]. The pulses have to be short enough so that all the pulses can be jammed in the time window. Ideally, a pulse can be completed quite fast, but this may incur undesirable rotations in other qubits because the frequency bandwidth is inversely proportional to the time length of the pulse. A shorter and stronger pulse will have a wider frequency band that may cover the resonance frequency of another spin, called *cross-talking*. It should also be noted that both T_1 and T_2 are defined and measured in a simplified situation, and they can only be used as an approximation of the decoherence rate for the quantum computation. Coupling is also a problem that makes the pulse design much more complicated. Finally, any experimental facility is not perfect, which may introduce more errors. Typical error resources include inhomogeneities in the static and RF field, pulse length calibration errors, frequency offsets, and pulse timing/phase imperfections.

If the quantum circuit can be simplified and the number of gates needed is reduced, the requirements on the pulses can be alleviated. Mathematicians are looking for methods to find time-optimal pulse sequences [43, 60, 61, 99], with the goal of finding the shortest path between the identity and a point in the space of $\mathbf{SU}(n)$ allowed by the system and the control Hamiltonians. In addition, NMR spectroscopists have already developed advanced pulse techniques to deal with system errors such as cross-talking and coupling. They turn out to work well and are now widely used in NMR quantum computation. Such techniques include composite pulses [21, 36, 72, 54, 55, 108] and pulse shaping. The latter consists mainly of two methods: phase profiles [89] and amplitude profiles [39, 66].

14.2.5 Initialization

An NMR sample eventually will go into its equilibrium state when no RF pulse is applied for a long time. Then the density matrix is proportional to $e^{-H/kT}$, according to the Boltzmann distribution, where $k = 1.381 \times 10^{-23}$ J/K and T is the absolute temperature. Normally, the environment temperature is far larger than the energy difference between the up and down states of the spin, and H/kT is very small, about 10^{-4}. We also make the assumption that the coupling terms are small enough compared with the resonant frequency, thus we can make a reasonable approximation of the equilibrium state density matrix of a system with n spins:

$$\rho_{eq} = \frac{e^{-H/kT}}{tr(e^{-H/kT})} \approx I - \frac{1}{kT}(\epsilon_1\sigma_z^1 + \epsilon_2\sigma_z^2 + \cdots + \epsilon_n\sigma_z^n). \tag{14.46}$$

In the four operators appearing in the density matrix (14.5), only those with zero traces can be observed in NMR. The operator I_0 is invisible, and moreover, it remains invariant under any unitary similarity transformation. Therefore, we only need to take care of the zero-trace part of the initial density matrix, noting that only that part (called the deviation) is effective. Most algorithms prefer an initial state such as

$$\rho_0 = \frac{1-\epsilon}{2^n} I + \epsilon|00\cdots 0\rangle\langle 0\cdots 00|,$$

which is an example of the so-called pseudo-pure states, corresponding to the pure state $|00\cdots 0\rangle$.

To initialize the system to a pseudo-pure state as above, we may use a scheme called *averaging*. Let us explain this for a 2-spin system. Suppose we have three 2-spin subsystems with density matrices

$$\rho_1 = \begin{bmatrix} a\,0\,0\,0 \\ 0\,b\,0\,0 \\ 0\,0\,c\,0 \\ 0\,0\,0\,d \end{bmatrix}, \qquad \rho_2 = \begin{bmatrix} a\,0\,0\,0 \\ 0\,c\,0\,0 \\ 0\,0\,d\,0 \\ 0\,0\,0\,b \end{bmatrix}, \qquad \rho_3 = \begin{bmatrix} a\,0\,0\,0 \\ 0\,d\,0\,0 \\ 0\,0\,b\,0 \\ 0\,0\,0\,c \end{bmatrix}, \tag{14.47}$$

respectively, where a, b, c, and d are nonnegative, and $a+b+c+d=1$. These are three diagonal matrices with three of their diagonal elements in cyclic permutation.

Now, we mix these three subsystems together (for an n-qubit system, we may have 2^n-1 subsystems) and assume that the three subsystems have the same signal scale. Because the readout is linear with respect to the initial state, we are in fact working on a system with an effective initial density matrix

$$\begin{aligned} \frac{1}{3}\sum_{i=1}^{3}\rho_i &= \frac{1}{3}\begin{bmatrix} 3a & & & \\ & b+c+d & & \\ & & b+c+d & \\ & & & b+c+d \end{bmatrix} \\ &= \frac{b+c+d}{3}I + \frac{1}{3}\begin{bmatrix} 4a-1 & 0 & 0 & 0 \\ 0 & 0 & 0 & 0 \\ 0 & 0 & 0 & 0 \\ 0 & 0 & 0 & 0 \end{bmatrix}, \end{aligned} \tag{14.48}$$

which is a pseudo-pure state corresponding to $|00\cdots0\rangle$.

Various methods have been developed to achieve this effect of averaging. Because ρ_1, ρ_2, and ρ_3 differ only by a permutation of the diagonal elements, a sequence of CNOT pulses can be used to transform one to another. In most cases, we only have one sample; the same algorithm can be repeated on that sample three times but with different initial states ρ_1, ρ_2, and ρ_3, respectively. At last, after all three outputs are obtained and added together (average), we achieve the same result as what we will get when the algorithm is employed on a system with the expected initial state $|00\cdots0\rangle$. This is called "temporal averaging" [63]. Gradient fields can also be used to divide the sample into different slices in space which are prepared in different initial states, and the averaging is realized spatially, called "spatial averaging" [19]. The number of the experiments and pulses needed grows very large when the number of qubits increases. For example, nine experiments are combined in order to prepare one pseudo-pure state for a 5-qubit system and 48 pulses are used to form one pseudo-pure state in a 7-qubit system [41] after modifications such as logical labeling [42, 106] and selective saturation [62].

14.2.6 Measurement

An NMR computer differs from other quantum computers in that it works on an ensemble of spins instead of just a single one. It produces an observable macroscopic signal that can be picked up by a set of coils positioned on the x–y plane, as shown in Figure 14.3. The signal measures the change rate of the magnetic field created by a large number of spins in the sample rotating around the z-axis, called *free induction decay* (FID). Due to relaxation, peaks of the Fourier transform of the signal, or spectra, have width. However, we do not need to worry about that because it does not make any substantial difference in our discussion here. One disadvantage is that the readout from NMR is an average of all the possible states, in contrast to most existing quantum algorithms that ask for the occurrence of only a single state. But it is possible for one to modify ordinary quantum algorithms to make NMR results usable.

The magnetization detected by the coil in Figure 14.3 is proportional to the trace of the product of the density matrix with $\sigma_+ = \sigma_x + i\sigma_y$:

$$M_x + iM_y = nV\langle \mu_x + i\mu_y \rangle = nV\gamma\hbar\,\mathrm{Tr}(\rho(\sigma_x + i\sigma_y)), \tag{14.49}$$

where γ is the magnetogyric ratio as in (14.3) and ρ is the density matrix. When the external RF magnetic field is removed, the density matrix will change according to the system's Hamiltonian as we discussed earlier. If we decompose the density matrix into a sum of product operators as in (14.5), only I_x and I_y contribute to the readout. We cannot "see" the coefficients of I_0 and I_z. Recall (14.18): if a one-spin system begins from density matrix $\rho_0 = I_0 + \sin\theta\cos\psi I_x + \sin\theta\sin\psi I_y + \cos\theta I_z$, the magnetization will rotate with the resonant frequency as

$$\begin{aligned} M_z + iM_y &= C\,\mathrm{Tr}(e^{-iHt}\rho_0 e^{iHt}\sigma_+) \\ &= C\,\mathrm{Tr}(e^{-iHt}(I_0 + \sin(\theta)\cos(\psi)I_x \\ &\quad + \sin(\theta)\sin(\psi)I_y + \cos(\theta)I_z)e^{iHt}\sigma_+) \\ &= C\,\mathrm{Tr}((\sin\theta\cos\psi(\cos(\omega t)I_x + \sin(\omega t)I_y) \\ &\quad + \sin\theta\sin\psi(\cos(\omega t)I_y - \sin(\omega t)I_x))\sigma_+) \\ &= C\,\sin\theta\; e^{i(\omega t+\psi)}, \end{aligned} \tag{14.50}$$

where $C = nV\gamma\hbar$. This rotating magnetization will introduce an oscillating electric potential in the receiver coils, which will be processed by a computer to generate the spectra. Note that the signal is proportional to $\sin\theta$. If an x rotation with angle $\pi/2$ is applied on the spin before the measurement, the magnetization will become

$$M_z + iM_y = \frac{\sqrt{2}}{2}C(\sin\theta \;-\; i\cos\theta)e^{i\omega t}.$$

For simplicity, we have chosen $\psi = 0$. The imaginary part is proportional to the population difference:

$$\cos\theta = \cos^2\frac{\theta}{2} - \sin^2\frac{\theta}{2}.$$

Computation of a two-spin system is complicated, so we only give some partial results here. The purpose is to point out what methodology is used. We still use the basis given by (14.29) and the Hamiltonian in (14.28). The system begins from a density matrix as

$$\rho_0 = \begin{bmatrix} \rho_{11} & \rho_{12} & \rho_{13} & \rho_{14} \\ \rho_{21} & \rho_{22} & \rho_{23} & \rho_{24} \\ \rho_{31} & \rho_{32} & \rho_{33} & \rho_{34} \\ \rho_{41} & \rho_{42} & \rho_{43} & \rho_{44} \end{bmatrix}. \qquad (14.51)$$

The operator σ_+ is a summation of operators from the two subsystems:

$$\begin{aligned} \sigma_+ &= \sigma_+^1 + \sigma_+^2 \\ &= \begin{bmatrix} 0 & 2 & 2 & 0 \\ 0 & 0 & 0 & 2 \\ 0 & 0 & 0 & 2 \\ 0 & 0 & 0 & 0 \end{bmatrix}. \end{aligned} \qquad (14.52)$$

The magnetization in the x–y plane is composed of four frequencies:

$$\begin{aligned} M_x + iM_y &= C\,\mathrm{Tr}(e^{-iHt}\rho_0 e^{iHt}\sigma_+) \\ &= C\left(\rho_{31}e^{i(\omega_1+J)t} + \rho_{42}e^{i(\omega_1-J)t} + \rho_{43}e^{i(\omega_2-J)t} + \rho_{21}e^{i(\omega_2+J)t}\right). \end{aligned} \qquad (14.53)$$

The spectrum has two pairs of peaks, one pair around the precession frequency ω_1 and another pair around ω_2. See Figure 14.9. The splitting is a result of coupling. If the system has more than two spins, the coupling will split a peak into up to 2^{n-1} peaks where n is the number of spins. We also combine all the constants in C to make the formula concise. Only four of the elements out of the density matrix appear in this spectrum, so we need to design certain control pulses to move the expected information to these four positions where numbers can be shown via free induction signal. If multitests are allowed, theoretically, all the elements of the density matrix can be retrieved [15, 14]. It is also possible to transport the desired information (computational results) to the four positions where the observer can see.

A typical pulse used in reading out is a hard $X_{\pi/2}$ pulse which rotates all the spins about the x-axis with angle $\pi/2$. Let us still use two-spin systems as an example. The operation is the tensor product of two x-rotation operators; that is, $X_{\pi/2} = X^1_{\pi/2}X^2_{\pi/2}$. The imaginary parts of the four effective elements of the density matrix ρ' after the operation, utilizing the fact that the density

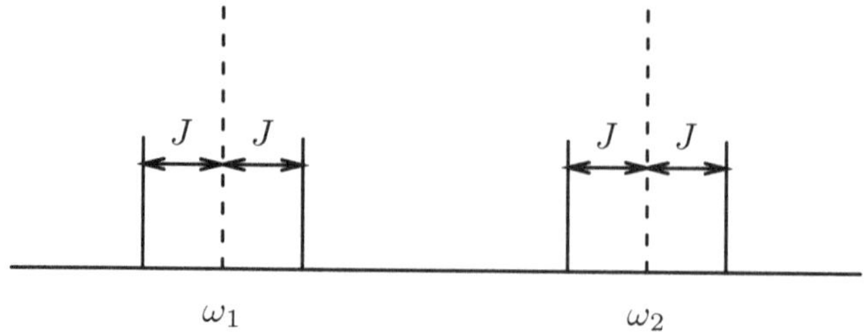

Fig. 14.9 Simplified stick spectra of a two-qubit molecule. The two dotted lines show two peaks at ω_1 and ω_2, respectively, when no coupling is applied ($J = 0$). After coupling, every peak is split into two small peaks with the intensities reduced to half.

matrix is Hermitian, are

$$\begin{aligned}
\mathrm{Im}(\rho'_{31}) &= \tfrac{1}{4}(\rho_{33} + \rho_{44} - \rho_{11} - \rho_{22} - 2\,\mathrm{Im}(\rho_{21}) - 2\,\mathrm{Im}(\rho_{34})),\\
\mathrm{Im}(\rho'_{42}) &= \tfrac{1}{4}(\rho_{33} + \rho_{44} - \rho_{11} - \rho_{22} + 2\,\mathrm{Im}(\rho_{21}) + 2\,\mathrm{Im}(\rho_{34})),\\
\mathrm{Im}(\rho'_{43}) &= \tfrac{1}{4}(\rho_{22} + \rho_{44} - \rho_{11} - \rho_{33} + 2\,\mathrm{Im}(\rho_{31}) + 2\,\mathrm{Im}(\rho_{24})),\\
\mathrm{Im}(\rho'_{21}) &= \tfrac{1}{4}(\rho_{22} + \rho_{44} - \rho_{11} - \rho_{33} - 2\,\mathrm{Im}(\rho_{31}) - 2\,\mathrm{Im}(\rho_{24})).
\end{aligned} \tag{14.54}$$

Find the sum of $\mathrm{Im}(\rho'_{31})$ and $\mathrm{Im}(\rho'_{42})$ and that of $\mathrm{Im}(\rho'_{43})$ and $\mathrm{Im}(\rho'_{21})$:

$$\begin{aligned}
\mathrm{Im}(\rho'_{31} + \rho'_{42}) &= -\tfrac{1}{2}(\rho_{11} + \rho_{22} - \rho_{33} - \rho_{44}),\\
\mathrm{Im}(\rho'_{43} + \rho'_{21}) &= -\tfrac{1}{2}(\rho_{11} - \rho_{22} + \rho_{33} - \rho_{44}).
\end{aligned} \tag{14.55}$$

Because what the coils pick up is the change rate of the magnetic field rather than the magnetic field itself, the imaginary parts we have listed above are reflected in the real part of the spectra. The computation above shows that the sum of the real parts of each pair of peaks in the spectra is proportional to the population difference between the spin-up and the spin-down states of the corresponding spin.

14.3 Solid-State NMR

Liquid NMR, discussed in Section 14.2, has several constraints that make a liquid NMR quantum computer not scalable. At first, as the result of the pseudo-pure state preparation, the signal–noise ratio decreases exponentially when the number of qubits increases, limiting its ability to realize more qubits. Another difficulty arises when we want to control the system as accurately as desired. Because the range of the chemical shift is limited by nature, the number of qubits represented by the same type of nuclei, such as carbon, is constrained as the resonance frequency gaps between any two qubits must be large enough so that we can distinguish the qubits easily and control them with great precision. It is estimated that a quantum computer realized by liquid-state NMR can have at most 10 to 20 qubits.

Solid-state NMR has the potential to overcome many of the problems of its liquid-state counterpart as in the preceding paragraph. These advantages are derived partly from the lack of motion of the molecules and partly from the ability to cool to low temperatures. As with many potential solutions, there are tradeoffs to consider. Here we summarize.

1. At low temperatures, near or below that of liquid helium, it is possible to initialize electron spins using the thermal Boltzman distribution. Nuclear spins do not become significantly oriented until much lower temperatures because of their $\sim$1000 times lower energies, but there are existing pulse RF sequences that can transfer an electron spin orientation to nearby nuclear spins using their mutual spin–spin interaction. In principle, this solves the problem of qubit initialization. In practice, the thermal initialization process can be slow because it depends on the electron spin population lifetime. It is possible to find systems with short electron spin lifetimes, but this will tend to result in faster decoherence of the nuclear spins, because they must be coupled to the electron in order to initialize in the first place.

2. Because the molecules in a solid are usually not tumbling, the dipole coupling between nearby spins does not average out. This has the advantage of making multiqubit gates faster, because the dipole coupling is much larger than the scalar coupling. The orientation-dependent chemical shifts also do not average out, in principle making individual qubits easier to address so that more qubits can be used. Here, it should be noted that custom molecules containing electron spins [111] can be used to enhance this effect. There is a tradeoff to consider, in that the faster interaction with nearby spins provided by dipole coupling can also lead to faster decoherence times.

3. Spin lifetimes in solids can be much longer than in liquids. Lack of molecular motion eliminates the spatial diffusion of spins which is a problem in liquid NMR for times in the range of milliseconds or longer [34]. Phonons can cause decoherence in solids at room temperature, but this can be strongly suppressed at temperatures achievable in liquid helium. It is not unusual to see spin population lifetimes of minutes in solids, especially at low temperatures. Unfortunately, spin coherence times are usually somewhat shorter due to dephasing caused by mutual spin flips through the strong dipole coupling. To eliminate this decoherence mechanism, there are two main approaches. One is to disperse the active molecule, as a dopant in a spin-free host. Actually the host does not need to be completely spin-free provided its spins are far enough off resonance with those of the active molecule. Another technique is to use stoichiometric materials consisting of relatively large unit cells containing many spin-free atoms. The idea for both these approaches is to keep the active nuclei relatively far apart, except for nearest neighbors.

In addition to the above differences, nuclei with nonzero spin in solid state can also be used for quantum computation [67] and manipulated similarly to the liquid-state NMR. Because all the nuclei are fixed in space, a static magnetic field with a strong gradient in one direction separates the nuclei into different layers along the direction. Every layer of nuclei can be regarded as a qubit and the qubits have different resonance frequencies as the magnetic field is different from one layer to another. Readouts also can be made to take advantage of the bulk quantum computer much as with the liquid NMR. The signal is picked up using methods such as magnetic resonance force microscopy.

There are two types of methods to make such a nuclei arrangement. Crystal, such as cerium-monophosphide (CeP), is a natural choice, where the 1/2 spin ^{31}P nuclei form periodical layers in the crystal with interlayer distance about 12 Å [45, 116]. Another method is to grow a chain of ^{29}Si that has 1/2 spin along the static field direction on a base of pure ^{28}Si or ^{30}Si which are both 0 spin nuclei [1, 68]. The last one combines the mature crystal growth and processing technology for silicon from the semiconductor industry. Liquid crystal [117] or solid-state samples [70] are also candidates for realizing an NMR quantum computer.

Recently, there has been considerable progress made in the area of optically addressed spins in solids. As a result some highly scalable designs have recently come forward that have the potential to eliminate all of the limitations of NMR. Aside from potentially solving NMR's problems, optical addressing has the important advantage that it would provide an interface between spin qubits and optical qubits, which is essential to interface with existing quantum communication systems, and for quantum networking in general.

Optically addressed spins are better known in the literature as spectral hole burning (SHB) materials [81]. Most of these are dopant–host systems that exhibit strong zero-phonon optical absorption lines at low temperature. Due to the inherent inhomogeneity of dopant–host systems it is often found that this optical zero-phonon linewidth is much larger than that of the individual atoms. Furthermore, when these transitions are excited with a narrowband laser, the resonant atoms can be optically pumped into a different ground state, making the material more transparent at the laser frequency. This is known as burning a spectral hole, hence the name SHB. In many SHB materials, the optical pumping is into different ground-state spin sublevels, and hence the hole burning process can initialize spin qubits, as illustrated in Figure 14.10. This type of spin qubit initialization can be much faster than Boltzmann initialization, especially in spin systems with long spin population lifetimes, because the tradeoff between spin lifetime and initialization speed is removed.

In addition to optical initialization of spins, the hole burning process can also be used to read out the spin state of the qubits. This happens when the quantum algorithm returns some of the spin qubits to the state that was initially emptied by the laser, resulting in a temporary increase in laser absorption and/or fluorescence that is proportional to the final population of this spin state. Of course, the readout process also reinitializes, so one must

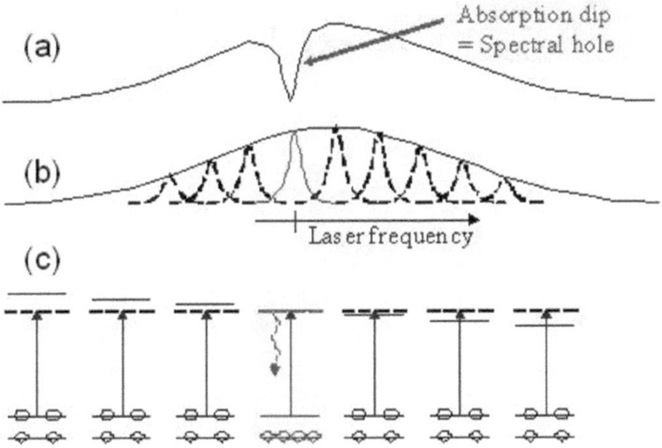

Fig. 14.10 (a) The signature of spectral hole burning is a narrowband dip in the optical absorption spectrum. (b) This dip occurs when an optical laser bleaches out an ensemble of atoms at a particular transition frequency. (c) In the case when bleaching is due to spin sublevel optical pumping, it can be used to initialize qubits.

take care to work with a large enough ensemble to achieve the desired readout fidelity. In general, optical readout is orders of magnitude more sensitive than the typical NMR coil, and so it is possible to work with small ensembles consisting of very dilute dopant–host systems that can have very long spin coherence lifetimes.

Spin qubit coherence lifetimes in SHB materials can be lengthened by a variety of techniques. In dilute dopant–host systems, the choice of a spin-free or low-spin host has the largest benefit. Examples include praseodymium [50] or europium [32] doped in an yttrium-silicate host (Pr:YSO or Eu:YSO) and nitrogen-vacancy [110] (NV) color centers doped in diamond; see Figure 14.11. In Pr:YSO only the yttrium host nuclei have spin but the magnetic moment is very weak. In NV diamond, the only host spins are $\sim$1% abundant ^{13}C which can be virtually eliminated with isotopically pure material. In dopant–host systems dephasing due to host spins is reduced by the so-called *frozen core effect* [101], wherein the magnetic field generated by the active (qubit) spin system tunes nearby host nuclei out of resonance with the rest of the crystal up to a distance that defines the frozen core radius. This suppresses the energy-conserving mutual spin flips that are the main source of spin decoherence.

In Pr:YSO the spin Hamiltonian is given by [77]:

$$H = \mathbf{B}\cdot\left(g_J^2\mu_B^2\overleftrightarrow{\Lambda}\right)\cdot\mathbf{B}+\mathbf{B}\cdot\left(\gamma_N\overleftrightarrow{E}+2A_Jg_J\mu_B\overleftrightarrow{\Lambda}\right)\cdot\mathbf{I}+\mathbf{I}\cdot\left(A_J^2\overleftrightarrow{\Lambda}+\overleftrightarrow{T}_Q\right)\cdot\mathbf{I}, \tag{14.56}$$

where the tensor $\overleftrightarrow{\Lambda}$ is given by

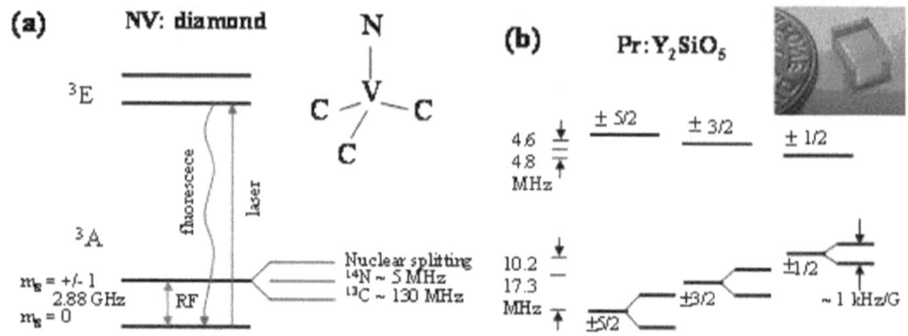

Fig. 14.11 (a) Spin sublevels of nitrogen-vacancy (NV) color center in diamond. (b) Spin sublevels of Pr: YSO.

$$\Lambda_{\alpha\beta} = \sum_{n=1}^{2J+1} \frac{\langle 0|J_\alpha|n\rangle\langle n|J_\beta|0\rangle}{\Delta E_{n,0}} \tag{14.57}$$

In (14.56), $\overleftrightarrow{E}$ is the 3×3 identity matrix, **B** is the magnetic field, **I** is the nuclear spin vector, g_J is the Lande g, γ_N is the nuclear magnetogyric ratio, and A_J is the hyperfine interaction. The term $\mathbf{I} \cdot \overleftrightarrow{T}_Q \cdot \mathbf{I}$ describes the nuclear electric quadrupole interaction and $A_J^2 \mathbf{I} \cdot \overleftrightarrow{\Lambda} \cdot \mathbf{I}$ is the second-order magnetic hyperfine or pseudoquadrupole interaction.

Recently, a spin coherence lifetime of 1/2 minute has been observed in Pr: YSO [37]. This impressive result is made possible by combining two techniques. The first technique involves magnetically tuning the qubit spin to a level anticrossing [38]. This is common in systems with spin 1 or larger. Near such an anticrossing there is no first-order magnetic Zeeman shift. Consequently, spin flips of nearby host and active spins, which ordinarily introduce coherence by perturbing the local magnetic field of the qubit, no longer have a first-order effect. The complication is that the magnetic field is a vector so that the level anticrossing must exist in all three directions. Nonetheless such global level-crossings were found in Pr: YSO and were used to lengthen the qubit spin coherence lifetime by orders of magnitude. More important, the residual spin decoherence was found to decay as a quadratic exponential in time, meaning it decays as $e^{-(t/\tau)^2}$. This is critical because most quantum error correction schemes require the short-time decay to be slower than the usual linear exponential decay. Because this condition was satisfied in Pr: YSO, a version of bang-bang error correction was successfully applied to give the observed half-minute coherence times.

Manipulation of spin qubits in SHB materials is generally done using RF coils similar to those used in liquid NMR. Recently, optical Raman transitions have been explored as an alternative to this, in which case the spin-qubits are manipulated by lasers instead of an RF coil. The advantages of this are twofold. First, the gate time can be made faster because it depends on the optical Rabi frequency, rather than that of the spin transition. One reason for this is that spin transitions are generally magnetic dipole allowed transitions, whereas optical transitions are often electric dipole allowed. Another reason is that it is often easier to insert strong laser fields into a cryostat than strong RF fields. Second, the selectivity of qubit excitation can be improved considerably because only spins with the correct optical transition frequency and spin transition frequency are manipulated. Additional spatial selectivity exists because the optical laser beams can be focused down to microns, and only the illuminated part undergoes qubit manipulations. This is especially important for algorithms like those designed for a Type II quantum computer; see Section 14.5.

The real power of optically addressed NMR lies in multiqubit manipulations. The optical "handle" allows several options to increase scalability. First, the relatively long range of optical interactions frees NMR from the restrictions imposed by near-neighbor interactions. An example of this is an ensemble-based two-qubit gate demonstration in Eu:YSO involving ions separated by ~100 nanometers [79], which is orders of magnitude larger than distances required by conventional NMR. In this demonstration, a series of optical pulses refocuses a "target" optical qubit with a different phase depending on whether the "control" qubit is excited. Because these qubits are defined only by their transition frequency, neither the exact location nor number of spins located in between is unimportant. This demonstration also illustrates the interesting fact that optical transitions in some SHB materials have a coherence lifetime that is similar to that of many room temperature NMR transitions.

In principle, long-range optical interactions such as in the Eu:YSO example are scalable. In practice, however, the Eu:YSO demonstration experiment is not very scalable because well-defined pairs of qubits must be distilled out of a random ensemble [94], and this incurs an exponential penalty with number of qubits. To make this technique more scalable one approach would be to apply it to special solid-state pseudo-molecules. These pseudo-molecules exist in a number of stoichiometric crystals, the most interesting of which are those containing europium, for example $EuVO_4$ or Eu_2O_3 because of their narrow optical transitions at low temperatures [48, 78]. In these pseudo-molecules, localized defects have a large effect on the optical transition frequency of the Eu ions. Up to 50 optical transitions can be easily resolved in these materials. Assuming that all the defects are identical, each optical transition would correspond to a Eu spin system in a well-defined location near the defect center, thereby producing a pseudo-molecule. By using the long-range optical coupling demonstrated in Eu:YSO, one could in principle construct up to a 50-qubit quantum computer without most of the usual scaling limitations of NMR.

To achieve scalability beyond 100 qubits, single-spin manipulation is preferred. The excitation and especially detection of single spins in a solid is a very active area of research. Much of this research is based on a proposal to build a quantum computer using qubits consisting of nuclear spins of a phosphorus atom implanted in a silicon host [59]; see Figure 14.12. ^{31}P nuclei in spin-free ^{28}Si have a spin population lifetime on the order of hours at ultracold temperatures, with coherence lifetimes on the scale of tens of milliseconds so far. Single-qubit manipulations would be done with the usual NMR pulse sequences. However, to avoid driving all the qubits at once, an off-resonant RF field is applied and the active qubit is tuned into resonance when desired by using the interaction with its electron spin. The electron spin in turn is controlled by distorting the P electron cloud with a voltage applied to a nearby gate, called the A-electrode. To achieve two-qubit logic, the electron clouds of two neighboring P atoms are overlapped using a J-electrode. The resulting exchange interaction between the two electrons can then be transferred to the P nuclei using RF and/or gate pulse sequences. Because the P atom has electron spin, Boltzmann initialization can be used, although a number of faster alternatives such as spin injection are being explored. For readout, the nuclear spin state is transferred to the electron spin which in turn is converted into a charge state via spin exchange interactions with a nearby readout atom. The charge state is then detected with a single-electron transistor.

The Hamiltonian for P in Si is given by [59]:

$$H = \mu_B B \sigma_z^e - g_n \mu_n B \sigma_z^n + A \sigma^e \cdot \sigma^n, \tag{14.58}$$

where μ_B is the Bohr magneton, μ_n is the nuclear magneton, g_n is the nuclear g-factor, B is the applied magnetic field (assumed parallel to z), and σ are the Pauli matrices, with superscripts e and n for electron and nuclear. For two coupled qubits the Hamiltonian becomes:

$$H = H(B) + A_1 \sigma^{1n} \cdot \sigma^{1e} + A_2 \sigma^{2n} \cdot \sigma^{2e} + J \sigma^{1n} \cdot \sigma^{2e}, \tag{14.59}$$

where $H(B)$ is the magnetic field part and the superscripts 1 and 2 refer to the two spins,

$$4J \cong 1.6 \frac{e^2}{\varepsilon a_B} \left(\frac{r}{a_B} \right)^{5/2} e^{(-2r/a_B)}, \tag{14.60}$$

where r is the distance between P atoms, ε is the dielectric constant of silicon, and a_B is the effective Bohr radius.

Many of the more challenging operations proposed for the P in Si quantum computer have recently been demonstrated in quantum wells/dots using individual electrons and/or excitons [102]. In most of these experiments electron spins, rather than nuclear spins, are the qubits. Unfortunately, these experiments are usually done in GaAs which is not a spin-free host and so decoherence is a problem. Spin-free semiconductor hosts exist, but fabrication in these systems is not yet as mature (except for silicon).

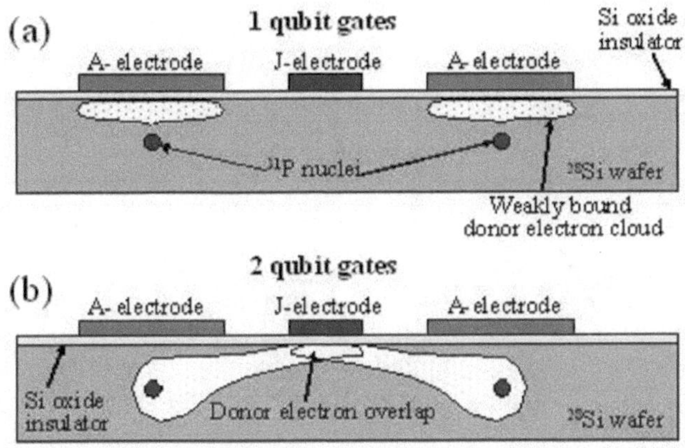

Fig. 14.12 Operation of the P doped Si quantum computer. (a) Single qubits are tuned into resonance with applied microwave field using voltages on A-electrodes to distort weakly bound electron cloud. (b) Two-qubit gates are enabled by using voltages on J-electrodes to overlap neighboring electron clouds, thereby switching on spin exchange coupling.

Although the exciting prospect of single electron and/or nuclear spin detection is currently a technical challenge for most solids, it was done long ago in nitrogen-vacancy (NV) diamond [47]. In this optically active SHB material, single-electron spin qubits are routinely initialized and read out with high fidelity at liquid helium temperature. In fact, the fidelity is so high that it begins to compare to trapped ions [52]. The electron spin coherence has also been transferred to nearby nuclear spins to perform (nonscalable) two-qubit logic [53]. Single-qubit logic is usually done with RF pulses, but optical Raman transitions between spin sublevels have been observed in NV diamond under certain experimental conditions [49]. Two-qubit gates can be performed using the electron spin coupling between adjacent qubits. Initialization of such a two-qubit system consisting of a NV and nearby N atom has recently been demonstrated using optical pumping. Scalability of this system can be achieved using an electron spin resonance (ESR) version of a Raman transition to transfer this spin coupling to the nuclear spins [113]. This ESR Raman has already been demonstrated for single NV spins.

The spin Hamiltonian for NV diamond is given by [92]

$$H = D(S_z^2 - \frac{1}{3}S^2) + \beta_e \mathbf{B} \overleftrightarrow{g}_e \mathbf{S} + \mathbf{S} \overleftrightarrow{A} \mathbf{I} + P(I_z^2 - \frac{1}{3}I^2), \qquad (14.61)$$

where D is the zero-field splitting, B is the applied magnetic field, $\overleftrightarrow{A}$ is the electron–nuclear hyperfine coupling tensor, S and I are the electron and nuclear spin vectors, and P is the nuclear quadrupole contribution.

With optical Raman, it should be possible to use long-range optical dipole–dipole coupling [80], or eventually cavity-based quantum electrodynamic

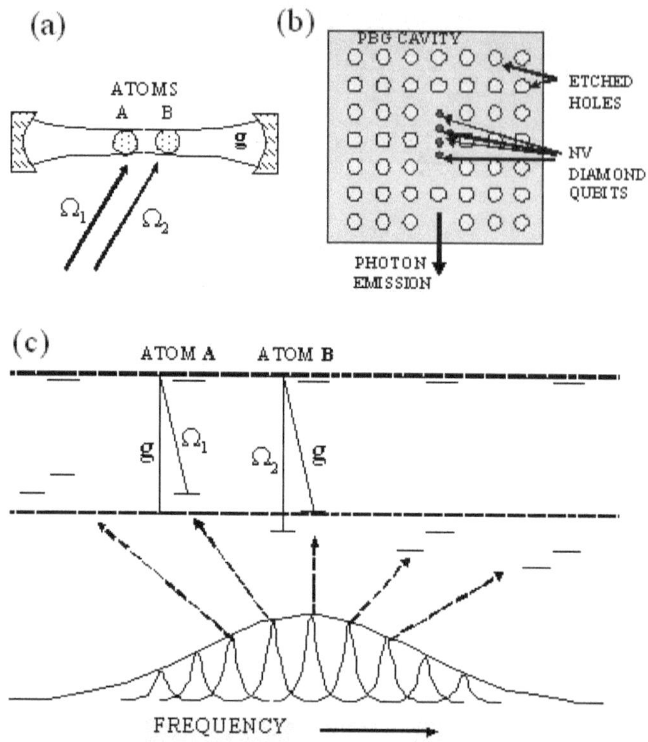

Fig. 14.13 (a) Two distant qubits coupled by vacuum mode of cavity using cavity QED. (b) Possible implementation of a multiqubit bus using a photon bandgap cavity in NV diamond. (c) Illustration of the use of optical Raman with cavity QED to couple spectrally adjacent qubits.

(QED) coupling, to perform two-qubit spin logic; see Figure 14.13. If successful, this will be highly scalable because the optical transition frequency can be tuned where desired using dc electric Stark shifts created by gate electrodes (A-electrodes). By requiring both optical laser frequency and electrode voltages to be correctly tuned, qubit excitation becomes very selective and two-qubit coupling only exists when needed, in contrast to the usual case in NMR.

Electron spin population lifetimes up to minutes have so far been observed at low temperature, with coherence lifetimes up to 0.3 milliseconds at room temperature [52]. Interestingly, the optical initialization and readout still works at room temperature (although with less fidelity), as well as the ESR Raman transitions. This raises the intriguing question of whether a room-temperature solid-state NMR quantum computer can eventually be built using NV diamond.

14.4 Shor's Algorithm and Its Experimental Realization

Through the rest of the chapter, we describe two applications of the NMR quantum computer: Shor's algorithm and a lattice algorithm. Entangled states are extremely important in quantum computation. Entanglement, together with superposition, gives a quantum computer the power to perform massively parallel computation and thus makes it particularly suitable for computing certain complex problems. Shor's algorithm for the factorization of integers aimed at decryption is a special example of a "killer ap" of quantum computing [30, 96]. Recently, a successful experiment has shown the potential capability of the implementation of Shor's algorithm, although it is still very simple and tentative. In [111], Vandersypen et al. factor 15 into 3 times 5. That work has shown the liquid NMR quantum computer to be the most successful quantum computer so far.

14.4.1 Shor's Algorithm

It is not difficult to factor a composite integer (i.e., nonprime) into prime numbers when that integer is small, but the computation burden grows rapidly when the number increases. The currently most efficient algorithm, the number field sieve, requires an exponential running time $e^{c(\log n)^{1/3}(\log\log n)^{2/3}}$, where n is the number to be factored and clearly $\log n$ is proportional to the number of the bits needed to store this number. This makes it practically impossible to factor a large number using a classical computer. This difficulty is used to construct several cryptosystems, such as the RSA public key cryptosystem [95]. Peter W. Shor has shown that this problem can be solved in polynomial running time instead of exponential time by using the quantum computer. A more accessible account of Shor's algorithm is given by Lomonaco [75].

Let n be an odd integer to be factored, and choose another random integer x less than n. We require x to be coprime with n; otherwise, we find a factor of n immediately by the Euclidean method. It is then known that function $f(s) = x^s \bmod n$ is periodic. The period of f (and also of x) is the smallest integer r such that $x^r = 1 \bmod n$. For example, when $n = 15$ and $x = 3$, the moduli of x^s, with s being 1, 2, 3, ..., are 3, 9, 12, 6, 3, 9, 12, 6, ..., and the period is 4.

Now we check the period r. If r is even, $r = 2t$, then $x^{2t} - 1 = (x^t + 1)(x^t - 1) = 0 \bmod n$, so either $x^t - 1$ or $x^t + 1$ has a common factor with n. A classical computer can use the Euclidean algorithm to compute the greatest common divisors, denoted as $\gcd(x^t + 1, n)$ and $\gcd(x^t - 1, n)$, in polynomial time. It is possible that we only obtain the trivial factors 1 or n using the x we choose. This happens only when $x^t = -1 \bmod n$, because $x^t - 1 = 0 \bmod n$ cannot happen with r being already the smallest integer such that $x^r = 1 \bmod n$.

Fortunately it has been proved that the probability to meet such a bad x is at most $1/2^k$, where k is the number of distinct prime factors of n. Because k is at least 2, the probability is still large enough for us to find a good x, which has an even period r and $x^t \neq -1 \bmod n$.

A quantum computer can find the period r because of the speedup afforded by the quantum Fourier transform (QFT). Let us have two b-qubit registers. We select b large enough such that we can observe many periods. At the beginning, we set the two registers to state $|0\rangle$. Then we randomize the first register to a new state

$$|\psi_1\rangle = \frac{1}{\sqrt{S}} \sum_{k=0}^{S-1} |k\rangle|0\rangle, \tag{14.62}$$

where $S = 2^b$, the number of the total b-qubit states of the first register, with b large enough such that $2n^2 > S > n^2$.

We now design a certain series of pulses to compute $f(k) = x^k \bmod n$, and change the quantum state to

$$|\psi_2\rangle = \frac{1}{\sqrt{S}} \sum_{k=0}^{S-1} |k\rangle|f(k)\rangle. \tag{14.63}$$

Now, we apply QFT [90] to the first register in (14.63), which is a unitary transform mapping every $|k\rangle$ to another state:

$$|k\rangle \to \frac{1}{\sqrt{S}} \sum_{k=0}^{S-1} e^{2\pi i uk/S}|u\rangle. \tag{14.64}$$

Then the quantum state of the system changes to

$$|\psi_3\rangle = \frac{1}{S} \sum_{u=0}^{S-1} |u\rangle \sum_{k=0}^{S-1} e^{2\pi i uk/S}|f(k)\rangle. \tag{14.65}$$

Assume that $f(k)$ has period r, and we write $k = d + jr$ such that $0 \le d < r$, where d is the remainder of k after it is divided by r and j ranges from 0 to A, the largest integer such that $Ar < S$. This way, we can write $|\psi_3\rangle$ as

$$|\psi_3\rangle = \frac{1}{S} \sum_{u=0}^{S-1} |u\rangle \sum_{d=0}^{r-1} |f(d)\rangle e^{2\pi i ud/S} \sum_{j=0}^{A} e^{2\pi i urj/S} I_{(d+rj<S)},$$

where $I_{(d+rj<S)} = 1$ when $d + rj < S$, and 0 otherwise. If $S = (A+1)r$, $I_{(d+rj<S)} = 1$ for every d and j. If $S \neq (A+1)r$, it is still reasonable to ignore the difference and let $I_{(d+rj<S)} = 1$ everywhere because we have chosen S large enough. In this case, we let

$$b_u = \frac{1}{S}\sum_{j=0}^{A} e^{2\pi iurj/S} = \frac{1}{S}\left(\frac{1-e^{2\pi iur(A+1)/S}}{1-e^{2\pi iur/S}}\right); \tag{14.66}$$

thus our quantum state is now

$$|\psi_3\rangle = \frac{1}{S}\sum_{u=0}^{S-1}\sum_{d=0}^{r-1} b_u e^{2\pi iud/S}|u\rangle|f(d)\rangle.$$

We can now measure the first register, and we want to find such a u, for which there is an l satisfying

$$\left|\frac{u}{S} - \frac{l}{r}\right| \leq \frac{1}{2S}. \tag{14.67}$$

There are about r such us, and it has been estimated that the probability to find such a u is at least 0.4 [90]. Because $1/2S < 1/2n^2$, and we know that $r < n$, there is at most one fraction k/r satisfying the condition and we can use continued fraction expansions to find the fraction. If k and r are coprime, we obtain r as the denominator of the fraction. If not, we only find a factor of r. If r is odd or $x^{r/2}$ does not give a useful result, we choose another x and try again. It may be necessary to try several (of the order $\mathcal{O}(\log\log n)$) times until r is successfully found, but the overall running time is still reasonable.

14.4.2 Circuit Design for Shor's Algorithm

Before we introduce the experiment by Vandersypen et al. [111], we extend the above discussion a little further to the case when r divides S. Now S/r becomes an integer and (14.66) always holds so that S doesn't have to be very large. Moreover, (14.67) becomes an identity

$$u = \frac{l \cdot S}{r}; \tag{14.68}$$

that is, r is the denominator of the fraction u/S after canceling the common factor between u and S if l and r are coprime. The integer 15 falls into this situation. The possible x can be 2, 4, 7, 8, 11, or 13. When we choose x to be 2, 7, 8, or 13, the period r is 4. In other cases, r is 2. The period r divides $S = 2^b$ in both cases. Only two qubits at most are required to compute one period of f. In the experiment, three qubits are used to obtain more periods.

Vandersypen et al. used liquid NMR to realize Shor's algorithm in factorizing 15. The sample in the experiment is a custom-synthesized material whose molecules have five ^{19}F and two ^{13}C, so it has seven qubits ready for use. Those seven qubits are divided into two registers, three to store the number

k (the first register, represented by $|k_2k_1k_0\rangle$) and four to store the modular exponentiation y (the second register, represented by $|y_3y_2y_1y_0\rangle$); see Figures 14.14 and 14.15. The total Hamiltonian is

$$H = \sum_{i=1}^{7} \frac{1}{2}\omega_i\sigma_z^i + \sum_{i<j} 2\pi J_{ij}\sigma_z^i\sigma_z^j.$$

Each run of the experiment consists of four steps. In the first step, the sample is initialized to a certain pseudo-pure state; in the second step, a series of specially designed pulses is applied to realize the computation of modular exponentiation; in the third step, QFT is applied to the first register; finally, the period was obtained through the reading of the spectrum. The system begins from thermal equilibrium, where the density matrix is given by $\rho_0 = e^{-H/kT} \approx I - (H/kT)$. A suitable initial pseudo-pure state $|\psi_1\rangle = |0000001\rangle$ is obtained by the temporal averaging method.

Although it is difficult to design a general circuit for the modular exponentiation, it is easy to "hard-wire" for this special case in consideration. As the exponent k can be written as $k = k_0 + 2k_1 + 4k_2$, we can change the modular exponential x^k mod 15 into successive operations of modular multiplications by $x^{2^i k_i}$, with $i = 0,\ 1,\ 2$, applied to the second register y beginning from $y = 1$.

When $i = 0$, $y \cdot x = x = 1 + (x-1)$, so the multiplication is actually a controlled-addition with $(x-1)$ in the case $k_0 = 1$. For $x = 7 = (0111)_2$, it is equal to flip the states of y_1 and y_2 ($y = (0001)_2$ before the multiplication). For $x = 11 = (1011)_2$, the same reasoning shows that we only have to flip the states of y_3 and y_1, depending on k_0. Gates A and B in Figures 14.14 and 14.15 accomplish the modular multiplication x^{k_0}.

The situation is a little more complicated for $i = 1$. We only discuss the situation when $k_1 = 1$, because y will not change when $k_1 = 0$. Different strategies are needed for $x = 7$ and $x = 11$. When $x = 11$, because $11^2 = 121 = 15 \times 8 + 1$, $y \times 11^2 = y \pmod{15}$. We need to do nothing and the same result holds for the third qubit k_2. When $x = 7$, we can design the circuit by first investigating the following identity.

$$\begin{aligned}
y \cdot 7^2 &= y \cdot 4 \mod 15 \\
&= (y_0 + 2y_1 + 4y_2 + 8y_3) \cdot 4 \mod 15 \\
&= (4y_0 + 8y_1 + 16y_2 + 32y_3) \mod 15 \\
&= (y_2 + 2y_3 + 4y_0 + 8y_1) \mod 15 \\
&= (y_2 \cdot 2^0 + y_3 \cdot 2^1 + y_0 \cdot 2^2 + y_1 \cdot 2^3) \mod 15.
\end{aligned}$$

It shows that the modular multiplication can be achieved by exchanging the first qubit y_0 with the third qubit y_2, and the second qubit y_1 with the fourth qubit y_3. In Figure 14.14, gates C, D, and E are used to accomplish the former, and gates F, G, and H the latter. Further simplification of the

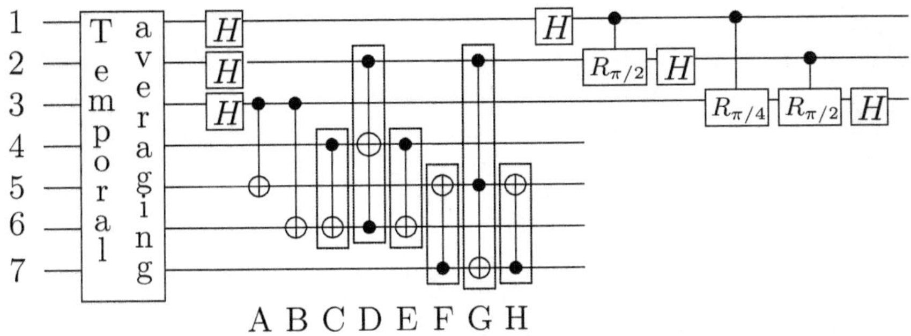

Fig. 14.14 The quantum circuit for the (hard) case for the realization of Shor's algorithm ($x = 7$). From top to bottom, the qubits are k_2, k_1, k_0, y_3, y_2, y_1, and y_0, respectively, in sequential order.

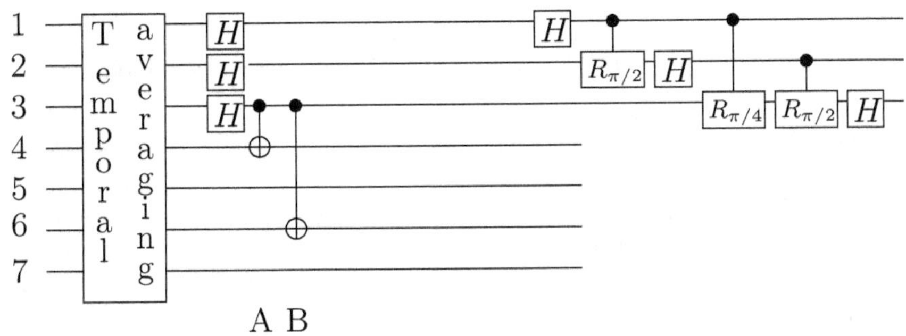

Fig. 14.15 The quantum circuit for the (easy) case for realization of Shor's algorithm ($x = 11$). From top to bottom, the qubits are k_2, k_1, k_0, y_3, y_2, y_1, and y_0, respectively, in sequential order.

circuit can be made. Because the control bit y_3 is $|0\rangle$ before gate C, that gate can just be omitted. Gates H and E have no effect on the period; they can be omitted, too.

The circuit design for the quantum Fourier transform is just a standard design; see, for example, [30, Figure 5]. It has three Hadamard gates and three controlled-phase gates. Figures 14.14 and 14.15 show the circuit designs for $y = 7$ and $y = 11$. About 300 pulses are used in the experiment in total and it takes about 700 ms to accomplish all steps in the case of $x = 7$.

14.4.3 Experimental Result

Readout of the experiment needs a careful interpretation of the data. Because an NMR sample consists of many molecules, the readout is the average value of u from all molecules instead of the reading from a single molecule.

Both qubits k_0 and k_1 are found to be in state $|0\rangle$ after the extraction of the spectra [111] for the easy case of $x = 11$, whereas qubit k_2 is in an equal mixture state of $|0\rangle$ and $|1\rangle$. Thus the possible u can be 0 and 4, that is, 000 and 100 in binary form. From (14.68), r can be obtained as $r = 8/4 = 2$, and the greatest common divisors are computed as $\gcd(11^{2/2} + 1, 15) = 3$ and $\gcd(11^{2/2} - 1, 15) = 5$.

In the case of $x = 7$, the spectra in [111] indicate that only qubit k_0 is in state $|0\rangle$, and both qubits k_1 and k_2 are in an equal mixture of states $|0\rangle$ and $|1\rangle$. Thus u is in a mixture of states $|0\rangle$, $|2\rangle$, $|4\rangle$, and $|6\rangle$. We can see that the period of u is 2, thus the period of the modular exponent r is $8/2 = 4$. The factors of 15 can finally be obtained as $\gcd(7^{4/2} - 1, 15) = 3$ and $\gcd(7^{4/2} + 1, 15) = 5$.

14.5 Quantum Algorithm for Lattice-Gas Systems

In the previous sections, we have explained how to construct a quantum computer using liquid NMR and illustrated a successful experiment. We have taken it for granted that the coherence can be maintained long enough and different qubits can be entangled even if they are separated far apart in space. Unfortunately, these assumptions are not always practical and in fact they constitute great obstacles to overcome. The problem becomes more serious when more qubits are involved. *Type-II quantum computers* are proposed to alleviate this problem. A Type-II quantum computer is composed of a network or array of small quantum computers interconnected by classical communication channels [120]. Instead of the global coherence and entanglement, only local coherence and entanglement within every small quantum computer, called a *node*, are required, and the difficulty faced by the centralized quantum computer is dramatically eased.

The wave function of the whole Type-II quantum computer system is a tensor product over all the nodes:

$$|\psi(t)\rangle = |\psi(x_1, t)\rangle \otimes \quad \cdots \quad \otimes |\psi(x_N, t)\rangle, \tag{14.69}$$

where N is the number of nodes. The lattice-gas algorithm (LGA) is specially suited for this structure. Every computation cycle can be broken up into three steps with two intermediate states $|\psi'\rangle$ and $|\psi''\rangle$:

$$\begin{aligned} |\psi'\rangle &= \hat{C}\,|\psi(t)\rangle\,, \\ |\psi''\rangle &= \Gamma\,|\psi'\rangle\,, \\ |\psi(t+1)\rangle &= T\,|\psi''\rangle\,, \end{aligned} \tag{14.70}$$

where $\hat{C}$ is a unitary operator acting locally on every node, Γ is a projection operator, such as a measurement, and T is the streaming operator that exchanges information among nodes. The Type-II quantum computer takes advantage of parallelism in two ways: one classical, where all the nodes work simultaneously; the other quantum, where quantum entanglement is still kept inside every node. Because measurement is applied and the system is reset at the end of every computation cycle, the coherence only needs to be maintained for a short time.

14.5.1 Quantum Algorithm for a Lattice-Gas Model

Consider a one-dimensional diffusion equation without boundary condition

$$\frac{\partial \rho}{\partial t} = \frac{\partial^2 \rho}{\partial x^2}, \tag{14.71}$$

where ρ is the mass density or temperature function along the x-axis. Using the finite difference method, we can write a finite difference approximation to solve the above partial differential equation numerically:

$$\frac{\rho(x, t+\tau) - \rho(x, t)}{\tau} = \frac{\rho(x+l, t) - 2\rho(x, t) + \rho(x-l, t)}{l^2}, \tag{14.72}$$

where τ is the time step size and l is the space step size. From physics, we know that the above equation may be studied by a lattice-gas algorithm. Without loss of generality, we assume that τ and l are normalized so that the difference equation can be written as

$$\rho(x_i, k+1) - \rho(x_i, k) = \frac{1}{2}(\rho(x_{i+1}, k) - 2\rho(x_i, k) + \rho(x_{i-1}, k)).$$

Points x_i are evenly distributed along the x-axis, also called *nodes*. To study the above equation, two functions, $f_1(x_i, k)$ and $f_2(x_i, k)$, called *channels*, are defined for each node x_i at time k. The set of values of these two functions is called the *state* of node x_i. Any physical observable, such as the density function $\rho(x_i, k)$, is a function of the state at the node. The evolution of the lattice, or the state of all nodes, consists of two operations: *collision* and *propagation*. A collision is a local operator only defined by the state of the node itself. The propagation operator transfers information from one node to another and the state at one node changes according to the state of other nodes. This is completed by defining a velocity vector for every channel which gives the information flow a direction. In our special example here, information in the two channels flows in opposite directions. After propagation, one channel gets its new value from its left neighbor, and the other from its right neighbor. This LGA is completed with a Type-II quantum computer by J. Yepez of the Air Force Research Laboratory and M.A. Pravia et al. of the Department of Nuclear Engineering at MIT [8, 93, 119, 120]. The actual result is not as good as desired, but improvement is still possible.

To store a floating-point number, a classical computer uses a register with 32 or 64 bits, depending on the machine. Quantum computers presently have difficulty in doing it the same way as classical computers because there is not yet the technology for 32 or 64 qubits. In this quantum lattice-gas algorithm, a two-qubit system is proposed for every node. The two qubits are represented by $|q_1(x_i,k)\rangle$ and $|q_2(x_i,k)\rangle$, respectively, and

$$|q_1(x_i,k)\rangle = \sqrt{f_1(x_i,k)}|0\rangle + \sqrt{1-f_1(x_i,k)}|1\rangle,$$

$$|q_2(x_i,k)\rangle = \sqrt{f_2(x_i,k)}|0\rangle + \sqrt{1-f_2(x_i,k)}|1\rangle.$$

The state of the whole system $|\psi(x_i,k)\rangle$ at node x_i and time k is a tensor product:

$$\begin{aligned}|\psi(x_i,k)\rangle &= |q_1(x_i,k)\rangle|q_2(x_i,k)\rangle\\ &= \sqrt{f_1(x_i,k)f_2(x_i,k)}|00\rangle + \sqrt{(1-f_1(x_i,k))f_2(x_i,k)}|10\rangle\\ &\quad+\sqrt{f_1(x_i,k)(1-f_2(x_i,k))}|01\rangle\\ &\quad+\sqrt{(1-f_1(x_i,k))(1-f_2(x_i,k))}|11\rangle.\end{aligned}$$

Quantities $f_1(x_i,k)$ and $f_2(x_i,k)$ are the probabilities of occurrence of the state $|0\rangle$ for qubit 1 and 2, respectively, corresponding to the two channels, and $1-f_{1,2}(x_i,k)$ are the occurrence probabilities of the state $|1\rangle$. Because the states are normalized, $0 \le f_{1,2}(x_i,k) \le 1$, and we let $\rho(x_i,k) = f_1(x_i,k)+f_2(x_i,k)$. It is noted that our Type-II quantum computer assigns ρ a continuous value (a function of the occurrence probabilities) rather than a discrete value as a digital computer does. An array of two-qubit systems is used in the computation, corresponding to a series of nodes.

The quantum LGA here has three steps in every cycle that complete a step of the finite difference algorithm computation: *collision*, *measurement*, and *reinitialization*. The last two composed are equal to one propagation operation in a normal lattice-gas algorithm. Because the propagation needs information exchange among different nodes, measurement and classical communication are needed to accomplish one operation. We map the quantum state to a vector in $\mathbf{C}^4$ as that given in (14.29).

In the collision step, a unitary operator is applied simultaneously to all nodes:

$$|\overline{\psi}(x_i,k)\rangle = U|\psi(x_i,k)\rangle,$$

where

$$U = \begin{bmatrix}1 & 0 & 0 & 0\\ 0 & \frac{1}{2}-\frac{i}{2} & \frac{1}{2}+\frac{i}{2} & 0\\ 0 & \frac{1}{2}+\frac{i}{2} & \frac{1}{2}-\frac{i}{2} & 0\\ 0 & 0 & 0 & 1\end{bmatrix}. \tag{14.73}$$

The new occurrence probabilities of the state $|0\rangle$ of the two qubits after the operation, $\overline{f}_1$ and $\overline{f}_2$, can be computed using

$$\begin{aligned} \overline{f}_1 &= \langle\overline{\psi}|n_1|\overline{\psi}\rangle, \qquad & n_1 &= \begin{bmatrix} 1\,0\,0\,0 \\ 0\,1\,0\,0 \\ 0\,0\,0\,0 \\ 0\,0\,0\,0 \end{bmatrix}, \\ \overline{f}_2 &= \langle\overline{\psi}|n_2|\overline{\psi}\rangle, \qquad & n_2 &= \begin{bmatrix} 1\,0\,0\,0 \\ 0\,0\,0\,0 \\ 0\,0\,1\,0 \\ 0\,0\,0\,0 \end{bmatrix}, \end{aligned} \tag{14.74}$$

leading to

$$\overline{f}_1(x_i, k) = \tfrac{1}{2}(f_1(x_i, k) + f_2(x_i, k)),$$
$$\overline{f}_2(x_i, k) = \tfrac{1}{2}(f_1(x_i, k) + f_2(x_i, k)).$$

The collision operator is actually doing a job of averaging. The state after the collision is also called the local equilibrium.

In the second step, a measurement is applied at every node and $\overline{f}_{1,2}(x_i, k)$ of all the nodes are retrieved for future use.

In the third step, using information from the measurement from the previous step, the states of all the nodes are reinitialized to a separable state

$$\begin{aligned} |q_1(x_i, k+1)\rangle &= \sqrt{\overline{f}_1(x_{i+1}, k)}|0\rangle + \sqrt{1 - \overline{f}_1(x_{i+1}, k)}|1\rangle, \\ |q_2(x_i, k+1)\rangle &= \sqrt{\overline{f}_2(x_{i-1}, k)}|0\rangle + \sqrt{1 - \overline{f}_2(x_{i-1}, k)}|1\rangle. \end{aligned} \tag{14.75}$$

It can be seen that the second and third steps have accomplished the propagation operation. At node x_i, the new state of channel one is acquired from the same channel of its right neighbor node x_{i+1}, and channel two acquires its state from its left neighbor. It is complicated here only because the communication between two quantum systems is difficult.

To see how this LGA works, let us begin from a local equilibrium state, $f_1(x_i, k) = f_2(x_i, k) = \rho(x_i, k)/2$, where the states come from a collision operation (step 2). We list the $f_{1,2}$ around position x_i before the third step in two rows

$$\begin{array}{llllll} f_1: & \cdots & \frac{\rho(x_{i-2},k)}{2} & \frac{\rho(x_{i-1},k)}{2} & \frac{\rho(x_i,k)}{2} & \frac{\rho(x_{i+1},k)}{2} & \cdots \\ f_2: & \cdots & \frac{\rho(x_{i-2},k)}{2} & \frac{\rho(x_{i-1},k)}{2} & \frac{\rho(x_i,k)}{2} & \frac{\rho(x_{i+1},k)}{2} & \cdots \end{array}$$

and after the third step

$$\begin{array}{llllll} f_1: & \cdots & \frac{\rho(x_{i-1},k)}{2} & \frac{\rho(x_i,k)}{2} & \frac{\rho(x_{i+1},k)}{2} & \frac{\rho(x_{i+2},k)}{2} & \cdots \\ f_2: & \cdots & \frac{\rho(x_{i-3},k)}{2} & \frac{\rho(x_{i-2},k)}{2} & \frac{\rho(x_{i-1},k)}{2} & \frac{\rho(x_i,k)}{2} & \cdots. \end{array}$$

We can see that the row of f_1 (channel one) moves left and the row of f_2 (channel two) moves right. According to our definition, the new value of ρ is the sum of $f_1(x_i, k+1)$ and $f_2(x_i, k+1)$; that is, $\rho(x_i, k+1) = \frac{1}{2}(\rho(x_{i+1}, k) + \rho(x_{i-1}, k))$. It is easy to check that

$$\rho(x_i, k+1) - \rho(x_i, k) = \frac{1}{2}(\rho(x_{i+1}, k) - 2\rho(x_i, k) + \rho(x_{i-1}, k)),$$

as desired.

Applications of the Type-II quantum computer with quantum LGA also have been reported in the simulation of the time-dependent evolution of a many-body quantum-mechanical system [121], solution of a one-dimensional magnetohydrodynamic turbulence [105], representation of solitons [104], and other equations.

14.5.2 Physical Realization and Result

The experiment in Section 14.5.1 uses a two-qubit molecule, chloroform, whose structure is shown in Figure 14.6. The hydrogen and carbon nuclei serve as qubits 1 and 2, respectively.

The actual results obtained from the experiment are compared with simulation results [93]. After 12 steps, the error becomes very large. Imperfection in the decoupling sequences is blamed and it is believed that the problem can be mitigated when the technology is improved in the future. Extreme requirement of high accuracy in the control pulse and readout is a disadvantage of this Type-II quantum computer, because it uses a continuous representation (the probability of occurrence) instead of a discrete one. Thus, it is more vulnerable to the inaccuracy in the NMR operation. Small errors in every step accumulate and finally become intolerable. Repeated measurement and reinitialization ease the requirement for coherence time, but place a high requirement on the fidelity at the same time.

14.6 Conclusion

In this chapter, we present the basic technology used to construct a quantum computer with liquid-state NMR. The successful experiments for many algorithms have shown that liquid-state NMR is capable of simulating a quantum computer and forms a testbed for quantum algorithms. It is so far the only technology available to realize a seven-qubit algorithm in the laboratory. One reason for its success is the robustness of the spin system which only interacts with the external magnetic field, and it is possible to maintain the coherence for a long time (from seconds to hours). Besides, over the 60-year history of NMR spectroscopy, analytic tools have been developed for the purpose of chemical and medical applications, and exact description and dedicated coherence control of the dynamics of the quantum spin system are now available to achieve high accuracy in the pulse design and application. In fact, the experimental techniques established in NMR, especially the coherence control technology, can be easily transferred to other quantum systems if they have a similar Hamiltonian, and the research in NMR is therefore helpful for the development of other more complicated and powerful quantum computers.

Liquid NMR has played a pioneering role in quantum computer technology development. But its lack of scalability has constituted a severe obstacle to its future applicability. However, in Section 14.3 we show new technology of solid-state NMR that has the potential to overcome liquid NMR's difficulties. For solid-state NMR, under low temperature, the relaxation times of spins are typically very long, and the coupling between qubits is strong so that the control can be fast and easy. The small ratio of the gate time and the decoherence time makes more gates available, and more complicated algorithms can be tested. The nuclei can be cooled down easily and the spin system is highly polarized. The signal is much stronger so that fewer nuclei are needed. Even without the help of the gradient field and silicon technology, as we have mentioned, a quantum computer with 30 to 40 qubits is envisioned with designed molecules similar to that of the liquid-state NMR computer except that the ensemble is in a solid crystal state. This is already a quantum system that reaches the limit a classical computer can simulate. Although it is still not scalable and not a standard quantum computer, these small- and medium-scale quantum computers will help in the building of a scalable and working quantum computer.

Acknowledgment We wish to thank the reviewer for constructive comments that have improved the quality of this chapter.

Appendix

The Homeomorphism from SU(2) to SO(3)

We use the same notation as in Section 14.2.3. Recall that U is a complex matrix in $\mathbf{SU}(2)$, and we want to find a mapping R from $\mathbf{SU}(2)$ to the space of 3×3 real matrices, so that $R(U)$ and U represent the same physical operation. Let $\mathbf{v}$ be a Bloch vector corresponding to the old state $|\phi\rangle$ and $\mathbf{v}'$ to the new one. Then the following identity must be satisfied,

$$\mathbf{v}' = R(U)\mathbf{v}, \tag{14.76}$$

where $U \in \mathbf{SU}(2)$. Using the definition of Bloch vector and density matrix, we obtain

$$\begin{aligned}\mathbf{v}' \cdot \mathbf{A} &= U|\psi\rangle\langle\psi|U^\dagger - I_0 \\ &= U\mathbf{v} \cdot \mathbf{A}U^\dagger,\end{aligned} \tag{14.77}$$

where $\mathbf{A} = [I_x, I_y, I_z]$ and we use the dot to denote the inner product between vectors.

The product of I_i and σ_j satisfies

$$\mathrm{Tr}(I_i\sigma_j) = \delta_{ij}, \qquad i,j \in \{x,y,z\},$$

where σ_j are the Pauli matrices. By multiplying both sides of (14.77) with σ_k and taking the trace of both sides, we obtain

$$\begin{aligned}v'_k &= \mathrm{Tr}(\mathbf{v}' \cdot \mathbf{A}\sigma_k) \\ &= \mathrm{Tr}(\sigma_k\, U\, I_i\, U^\dagger)v_i,\end{aligned} \tag{14.78}$$

where v'_k and v_i are, respectively, the kth and ith entry of the corresponding vectors. We apply the summation convention in this section, where summing over repeated indices is implied unless otherwise stated. Comparing the above result with equation

$$v'_k = R(U)_{ki}v_i,$$

we obtain the desirable matrix $R(U)$:

$$R(U)_{ki} = \mathrm{Tr}(\sigma_k\, U\, I_i\, U^\dagger). \tag{14.79}$$

Thus we have constructed a mapping from $\mathbf{SU}(2)$ to the set of 3×3 matrices. Let us now tentatively accept the fact that the target matrix is real which we prove later. We first show that $R(U)$ is in fact a proper rotation matrix, that is, $R(U) \in \mathbf{SO}(3)$, by proving that it is isometric and preserves the threefold product (cf. (14.82)).

Let $\mathbf{v} = [a\ \ b\ \ c]^T \in R^3$. First note that we can find its norm by computing the determinant of a special matrix:

$$\begin{aligned}\det(\mathbf{v} \cdot \mathbf{A}) &= \tfrac{1}{4}\begin{vmatrix} c & a - bi \\ a + bi & -c \end{vmatrix} \\ &= -\tfrac{1}{4}(c^2 + a^2 + b^2) \\ &= -\tfrac{1}{4}\|\mathbf{v}\|^2.\end{aligned} \tag{14.80}$$

Together with (14.77), we see that the transformation is isometric:

$$\begin{aligned}\|\mathbf{v}'\|^2 &= -4\det(\mathbf{v}'\cdot\mathbf{A})\\ &= -4\det(U\,\mathbf{v}\cdot\mathbf{A}U^\dagger)\\ &= -4\det(\mathbf{v}\cdot\mathbf{A})\\ &= \|\mathbf{v}\|^2.\end{aligned} \tag{14.81}$$

Direct computation also shows the preservation of the threefold product, as follows. Let $\mathbf{v}^l$, $l = 1, 2, 3$, be three vectors on the Bloch sphere, $\mathbf{v}'^{\,l} = R(u)\mathbf{v}^l$. We have

$$\begin{aligned}\mathrm{Tr}\left((\mathbf{v}^1\cdot\mathbf{A})(\mathbf{v}^2\cdot\mathbf{A})(\mathbf{v}^3\cdot\mathbf{A})\right) &= \mathrm{Tr}(v_i^1 v_j^2 v_k^3 I_i I_j I_k)\\ &= \tfrac{i}{4} v_i^1 v_j^2 v_k^3 \epsilon_{ijk}\\ &= \tfrac{i}{4}\mathbf{v}^1\cdot(\mathbf{v}^2\times\mathbf{v}^3).\end{aligned} \tag{14.82}$$

The above identity can be used to show the preservation of the threefold product:

$$\begin{aligned}\mathbf{v}'^{\,1}\cdot(\mathbf{v}'^{\,2}\times\mathbf{v}'^{\,3}) &= \tfrac{4}{i}\,\mathrm{Tr}\left((\mathbf{v}'^{\,1}\cdot\mathbf{A})(\mathbf{v}'^{\,2}\cdot\mathbf{A})(\mathbf{v}'^{\,3}\cdot\mathbf{A})\right)\\ &= \tfrac{4}{i}\,\mathrm{Tr}\left((U\mathbf{v}^1\cdot\mathbf{A}U^\dagger)(U\mathbf{v}^2\cdot\mathbf{A}U^\dagger)(U\mathbf{v}^3\cdot\mathbf{A}U^\dagger)\right)\\ &= \tfrac{4}{i}\,\mathrm{Tr}\left((\mathbf{v}^1\cdot\mathbf{A})(\mathbf{v}^2\cdot\mathbf{A})(\mathbf{v}^3\cdot\mathbf{A})\right)\\ &= \mathbf{v}^1\cdot(\mathbf{v}^2\times\mathbf{v}^3),\end{aligned} \tag{14.83}$$

and we see that $R(U) \in \mathbf{SO}(3)$. To prove that the mapping is surjective, we introduce a parameterization of $\mathbf{SU}(2)$ in some exponential form. For any $U \in \mathbf{SU}(2)$, we can find a $\theta \in [0, 2\pi)$ and a unit vector $\mathbf{n}$ such that

$$\begin{aligned}U(\theta,\mathbf{n}) &= e^{-i(\theta/2)\mathbf{n}\cdot\boldsymbol{\sigma}}\\ &= \begin{bmatrix}\cos\frac{\theta}{2} - in_3\sin\frac{\theta}{2} & -\sin\frac{\theta}{2}(n_2 + in_1)\\ \sin\frac{\theta}{2}(n_2 - in_1) & \cos\frac{\theta}{2} + in_3\sin\frac{\theta}{2}\end{bmatrix}\\ &= \cos\tfrac{\theta}{2}I - i\sin\tfrac{\theta}{2}\mathbf{n}\cdot\boldsymbol{\sigma},\end{aligned} \tag{14.84}$$

where $\boldsymbol{\sigma} = [\sigma_x, \sigma_y, \sigma_z]$. Every combination of θ and $\mathbf{n}$ also corresponds to a complex matrix in $\mathbf{SU}(2)$. By using (14.8) and the equality $U(\theta,\mathbf{n})^{-1} = U(\theta, -\mathbf{n})$, the element of $R(U)$ can be computed as

$$\begin{aligned}R(U)_{ij} &= \mathrm{Tr}(\sigma_i\, U\, I_j\, U^\dagger)\\ &= \mathrm{Tr}\left(\sigma_i(\cos\tfrac{\theta}{2}I - i\sin\tfrac{\theta}{2}\mathbf{n}\cdot\boldsymbol{\sigma})I_j(\cos\tfrac{\theta}{2}I + i\sin\tfrac{\theta}{2}\mathbf{n}\cdot\boldsymbol{\sigma})\right)\\ &= \mathrm{Tr}(\sigma_i I_j\cos^2\tfrac{\theta}{2} + i\sigma_i I_j\mathbf{n}\cdot\boldsymbol{\sigma}\sin\tfrac{\theta}{2}\cos\tfrac{\theta}{2}\\ &\quad - i\sigma_i\mathbf{n}\cdot\boldsymbol{\sigma}I_j\sin\tfrac{\theta}{2}\cos\tfrac{\theta}{2} + \sigma_i\mathbf{n}\cdot\boldsymbol{\sigma}I_j\mathbf{n}\cdot\boldsymbol{\sigma}\sin^2\tfrac{\theta}{2}).\end{aligned} \tag{14.85}$$

We divide the trace into four parts and compute them separately:

$$\mathrm{Tr}(\sigma_i I_j) = \delta_{ij}, \quad \mathrm{Tr}(\sigma_i I_j \mathbf{n}\cdot\boldsymbol{\sigma}) = i\epsilon_{ijk}n_k, \quad \mathrm{Tr}(\sigma_i \mathbf{n}\cdot\boldsymbol{\sigma} I_j) = i\epsilon_{ikj}n_k, \tag{14.86}$$

$$\begin{aligned}
&\mathrm{Tr}(\sigma_i \mathbf{n}\cdot\boldsymbol{\sigma} I_j \mathbf{n}\cdot\boldsymbol{\sigma}) = \mathrm{Tr}(\sigma_i\sigma_k I_j \sigma_l n_k n_l)\\
&= \tfrac{1}{2}\,\mathrm{Tr}\left((\delta_{ik}I + i\epsilon_{ikm}\sigma_m)(\delta_{jl}I + i\epsilon_{jln}\sigma_n)n_k n_l\right)\\
&= \delta_{ik}\delta_{jl}n_k n_l - \epsilon_{ikm}\epsilon_{jlm}n_k n_l\\
&= \delta_{ij}n_i n_j + (2-\delta_{ij})n_i n_j - \delta_{ij} = 2n_i n_j - \delta_{ij}.
\end{aligned} \tag{14.87}$$

Substituting (14.87) and (14.86) back into (14.85), we obtain

$$\begin{aligned}
R(U)_{ij} &= \cos^2\tfrac{\theta}{2}\delta_{ij} + \sin^2\tfrac{\theta}{2}(2n_i n_j - \delta_{ij})\\
&\quad - \epsilon_{ijk}n_k \sin\tfrac{\theta}{2}\cos\tfrac{\theta}{2} + \epsilon_{ikj}n_k \sin\tfrac{\theta}{2}\cos\tfrac{\theta}{2}\\
&= \cos\theta\,\delta_{ij} + (1-\cos\theta)n_i n_j + \sin\theta\,\epsilon_{ikj}n_k,
\end{aligned} \tag{14.88}$$

showing that $R(U)_{ij}$ is real. We also claim that $R(U)$ is a rotation about the axis $\mathbf{n}$ with angle θ in the three-dimensional Euclidean space by comparing it with the standard formula of a rotation matrix. Because every matrix in $\mathbf{SO}(3)$ can be regarded as a rotation about a certain axis with a certain angle, that matrix is now shown to be an image of some $U \in \mathbf{SU}(2)$. Thus the mapping is surjective.

Finally, we want to show that this mapping from $\mathbf{SU}(2)$ to $\mathbf{SO}(3)$ is a homomorphism and to investigate the multiplicity. Let U and T be two matrices in $\mathbf{SU}(2)$, $\mathbf{v}$ be an arbitrary vector on the Bloch sphere, and $\mathbf{v}' = R(U)\mathbf{v}$, $\mathbf{v}'' = R(T)\mathbf{v}' = R(T)R(U)\mathbf{v}$. Using (14.77) repeatedly, we obtain

$$\mathbf{v}''\cdot\mathbf{A} = T(\mathbf{v}'\cdot\mathbf{A})T^\dagger = TU(\mathbf{v}\cdot\mathbf{A})U^\dagger T^\dagger = R(TU)\mathbf{v}\cdot\mathbf{A},$$

thus $\mathbf{v}'' = R(TU)\mathbf{v}$. Because $\mathbf{v}$ is arbitrary, $R(VU) = R(V)R(U)$, implying that the mapping is a homomorphism.

For the multiplicity of this mapping, we investigate the kernel of R, $\ker(R)$, which is an invariant subgroup of $\mathbf{SU}(2)$, and the quotient group $\mathbf{SU}(2)/\ker(R)$ will then be isomorphic to $\mathbf{SO}(3)$. Suppose that $U \in \mathbf{SU}(2)$ and $R(U) = I_3$, the identity matrix in $\mathbf{SO}(3)$. Then for any $\mathbf{v}$ on the Bloch sphere or R^3, we have equality $\mathbf{v}\cdot\mathbf{A} = U(\mathbf{v}\cdot\mathbf{A})U^\dagger$, because $\mathbf{v} = I\mathbf{v}$. Multiplying both sides with U from the right and using (14.84), we have

$$\mathbf{v}\cdot\mathbf{A}(\cos\frac{\theta}{2}I - i\sin\frac{\theta}{2}\mathbf{n}\cdot\boldsymbol{\sigma}) = (\cos\frac{\theta}{2}I - i\sin\frac{\theta}{2}\mathbf{n}\cdot\boldsymbol{\sigma})\mathbf{v}\cdot\mathbf{A}.$$

Subtract from both sides the term $\mathbf{v}\cdot\mathbf{A}\cos(\theta/2)\,I$,

$$(\mathbf{v}\cdot\mathbf{A})\,(\mathbf{n}\cdot\boldsymbol{\sigma})\sin\frac{\theta}{2} = (\mathbf{n}\cdot\boldsymbol{\sigma})\,(\mathbf{v}\cdot\mathbf{A})\sin\frac{\theta}{2}.$$

Assuming that $\sin(\theta/2) \neq 0$, we can divide this factor from both sides to obtain

$$(\mathbf{v}\cdot\mathbf{A})\,(\mathbf{n}\cdot\boldsymbol{\sigma}) = (\mathbf{n}\cdot\mathbf{A})\,(\mathbf{v}\cdot\boldsymbol{\sigma}).$$

Let $\mathbf{v} = [1, 0, 0]^T$. Using properties of Pauli matrices, we obtain

$$n_2 I_z - n_3 I_y = 0.$$

The above is possible only when $n_2 = n_3 = 0$. Trying other different $\mathbf{v}$ leads to $n_1 = 0$, too. This is a contradiction, because we know that $\mathbf{n}$ is a unit vector. Thus we need $\sin(\theta/2) = 0$. In this case $U = I$ or $U = -I$ and it is easy to verify that these two are really mapped to the identity in $\mathbf{SO}(3)$. Now, we can conclude that the mapping we defined by (14.79) is a *two-to-one homomorphism* from $\mathbf{SU}(2)$ to $\mathbf{SO}(3)$ with kernel $\ker(R) = \{I, -I\}$. The mapping is also surjective, so it defines an isomorphism from the quotient group $\mathbf{SU}(2)/\ker(R)$ to $\mathbf{SO}(3)$. The two elements in the kernel, $\pm I$, are in fact the same transformation for quantum systems because only the relative phase matters for a quantum system. For any $O \in \mathbf{SO}(3)$, the two elements in $R^{-1}(O)$, U and $-U$ for some $U \in \mathbf{SU}(2)$, represent the same transformation, too. Thus, nothing is lost if we employ $\mathbf{SO}(3)$ to represent the transformations of a one-spin quantum system.

References

1. E. Abe, K.M. Itoh, T.D. Ladd, J.R. Goldman, F. Yamaguchi, and Y. Yamamoto, Solid-state silicon NMR quantum computer, *J. Superconduct. Incorporating Novel Magnetism* **16**(2003), 1, 175–178.
2. A. Barenco, C.H. Bennett, R. Cleve, D.P. DiVincenzo, N. Margolus, P. Shor, T. Sleator, J. Smolin, and H. Weinfurter, Elementary gates for quantum computation, *quant-ph*/9503016.
3. S. Bartlett, *http://www.physics.usyd.edu.au/~bartlett/ISIT2005/*.
4. P. Benioff, The computer as a physical system: A microscopic quantum mechanical Hamiltonian model of computers as represented by Turing machines, *J. Stat. Phys.* **22** (1980), 563–591.
5. C.H. Bennett, G. Brassard, Quantum cryptography: Public key distribution and coin tossing, *Proceedings of IEEE International Conference on Computers, Systems and Signal Processing*(1984), IEEE, New York, 175–179.
6. C.H. Bennett, G. Brassard, C. Crepeau, R. Jozsa, A. Peres, and W. Wootters, Teleporting an unknown quantum state via dual classical and EPR channels, *Phys. Rev. Lett.* **70**(1993), 1895–1899.
7. C.H. Bennett, P. Shor, Quantum information theory, *IEEE Trans. Inf. Theory* **44**(1998), 2724–2742.
8. G.P. Berman, A.A. Ezhov, D.I. Kamenev, and J. Yepez, Simulation of the diffusion equation on a type-II quantum computer, *Phys. Rev. A* **66**(2002), 012310.
9. E. Bernstein and U. Vazirani, Quantum complexity theory, *SIAM J. Comput.* **26**(1997), 1411–1473.
10. F. Bloch, Nuclear induction, *Phys. Rev.* **70**(1946), 460–474.
11. D. Bouwmeester, J.W. Pan, K. Mattle, M. Eibl, H. Weinfurter, and A. Zeilinger, Experimental quantum teleportation, *Nature* **390**(1997), 575–579.
12. G. Burkard, D. Loss and D. DiVincenzo, Coupled quantum dots as quantum gates, *Phys. Rev. B* **59** (1999), 2070–2078.
13. G. Chen, D.A. Church, B-G. Englert and M.S. Zubairy, Mathematical models of contemporary elementary quantum computing devices, in *Quantum Control: Math-*

ematical and Numerical Challenges (A.D. Bandrauk, M.C. Delfour and C. Le Bris, ed.), CRM Proc. & Lecture Notes, Vol. 33, Amer. Math. Soc., Providence, R.I., 2003, 79–117.
14. A.M. Childs, I.L. Chuang, and D.W. Leung, Realization of quantum process tomography in NMR, *Phys. Rev. A* **64**(2001), 012314.
15. I.L. Chuang, L.M.K. Vandersypen, X. Zhou, D.W. Leung, and S. Lloyd, Experimental realization of a quantum algorithm, *Nature* **393**(1998), 143–146.
16. I.L. Chuang, Quantum computation with nuclear magnetic resonance, in *Introduction to Quantum Computation and Information*, edited by H. Lo, S. Popescu, and T. Spiller, World Scientific, Singapore, 1998.
17. I.L. Chuang, Y. Yamamoto, Simple quantum computer, *Phys. Rev. A* **52**(1995), 5, 3489–3496.
18. D.G. Cory, M.D. Price, and T.F. Havel, Nuclear magnetic resonance spectroscopy: An experimentally accessible paradigm for quantum computing, *Physica D* **120**(1998), 82–101.
19. D.G. Cory, A.F. Fahmy, and T.F. Havel, Ensemble quantum computing by NMR spectroscopy, *Proc. Natl. Acad. Sci. USA* **94**(1997), 1634–1639.
20. D.G. Cory, R. Laflamme, and E. Knill et al., NMR based quantum information processing: Achievement and prospect, *Fortschr. Phys.* **48**(2000), 875–907.
21. H.K. Cummings, and J.A. Jones, Use of composite rotations to correct systematic errors in NMR quantum computation, *New J. Phys.* **2**(2000) 6.1–6.12.
22. D. Deutsch, Quantum theory, the Church-Turing principle and the universal quantum computer, *Proc. Roy. Soc. Lond. A* **400**(1985), 97.
23. D. Deutsch, Quantum computational networks, *Proc. Roy. Soc. Lond. A* **425**(1989), 73.
24. D. Deutsch and R. Jozsa, Rapid solution of problems by quantum computer, *Proc. Roy. Soc. Lond. A* **439**(1992), 553–558.
25. D.P. DiVincenzo, Two-bit quantum gates are universal for quantum computation, *Phys. Rev. A* **51**(1995), 1015–1022.
26. K. Dorai, and D. Suter, Efficient implementations of the quantum Fourier transform: an experimental perspective, *Int. J. Quant. Inf.* **3**(2005), 2, 413–424.
27. K. Dorai, Arvind, and A. Kumar, Implementing quantum-logic operations, pseudopure states, and the Deutsch-Jozsa algorithm using noncommuting selective pulses in NMR, *Phys. Rev. A* **61**(2000), 042306.
28. J.C. Edwards, *http://www.process-nmr.com/nmr.htm*
29. A. Ekert, Quantum cryptography based on Bell's Theorem, *Phys. Rev. Lett.* **67**(1991), 661–663.
30. A. Ekert and R. Jozsa, Quantum computation and Shor's factoring algorithm, *Rev. Modern Phys.* **68**(1996), 3, 733–753.
31. J.W. Emsley, J. Feeney and L.H. Sutcliffe, *High Resolution Nuclear Magnetic Resonance Spectroscopy*, Pergamon Press, Oxford, 1965.
32. R.W. Equall, Y. Sun, R.M. Macfarlane, Ultraslow optical dephasing In Eu-3+-Y2SiO5, *Phys. Rev. Lett.* **72**(1994), 2179–2181.
33. X. Fang, X. Zhu, M. Feng, X. Mao, and F. Du, Experimental implementation of dense coding using nuclear magnetic resonance, *Phys. Rev. A* **61**(2000), 022307.
34. S. Fernbach, W.g. Proctor, Spin-Echo Memory Device, *J. Appl. Phys.* **26**(1955), 2, 170–181.
35. R.P. Feynman, Simulating physics with computers, *Int. J. Theor. Phys.* **21**(1982), 467.
36. E.M. Fortunato, M.A. Pravia, and N. Boulant et al., Design of strongly modulating pulses to implement precise effective Hamiltonians for quantum information processing, *J. Chem. Phys.* **116**(2002), Issue 17, 7599–7606.
37. E. Fraval, M.J. Sellars, J.J. Longdell, Dynamic decoherence control of a solid-state nuclear-quadrupole qubit, *Phys. Rev. Lett.* **95**(2005), 030506.

38. E. Fraval, M.J. Sellars, J.J. Longdell, Method of extending hyperfine coherence times in Pr3+: Y2SiO5, *Phys. Rev. Lett.* **92**(2004), 077601.
39. R. Freeman, Shaped radiofrequency pulses in high resolution NMR, *J. Prog. Nucl. Magnetic Resonance Spectroscopy* **32**(1998), 59–106.
40. C.A. Fuchs, Distinguishability and accessible information in quantum theory, Ph.D. Dissertation, University of New Mexico; *quant-ph*/9601020.
41. B. Fung, and V.L. Ermakov, A simple method for the preparation of pseudopure states in nuclear magnetic resonance quantum information processing, *J. Chem. Phys.* **121**(2004), 17, 8410–8414.
42. N. Gershenfeld, and I.L. Chuang, Bulk spin-resonance quantum computation, *Science* **275**(1997), 350–356.
43. S.J. Glaser, T. Schulte-Herbruggen, M. Sieveking, O. Schedletzky, N.C. Nielsen, O.W. Sorensen, and C. Griesinger, Unitary control in quantum ensembles: Maximizing signal intensity in coherent spectroscopy, *Science* **280**(1998), Issue 5362, 421–424.
44. D. Gottesman, and I.L. Chuang, Demonstrating the viability of universal quantum computation using teleportation and single-qubit operations, *Nature* **402**(1999), 390–393.
45. J.R. Goldman, T.D. Ladd, F. Yamaguchi, and Y. Yamamoto, Magnet designs for a crystal-lattice quantum computer, *Appl. Phys. A* **71**(2000), 11–17.
46. L.G. Grover, Quantum mechanics helps in searching for a needle in a haystack, *Phys. Rev. Lett.* **79**(1997), 325.
47. A. Gruber, A. Drabenstedt, C. Tietz, L. Fleury, J. Wrachtrup, C. vonBorczyskowski, Scanning confocal optical microscopy and magnetic resonance on single defect centers, *Science* **276**(1997), Issue 5321, 2012–2014.
48. P.C. Hansen, M.J.M. Leask, B.M. Wanklyn, Y. Sun, R.L. Cone, M.M. Abraham, Spectral hole burning and optically detected nuclear quadrupole resonance in flux-grown stoichiometric europium vanadate crystals, *Phys. Rev. B* **56**(1997), 7918–7929.
49. P.R. Hemmer, A.V. Turukhin, M.S. Shahriar, and J.A. Musser, Raman excited spin coherences in N-V diamond, *Opt. Lett.* **26**(2001), 361–363.
50. K. Holliday, M. Croci, E. Vauthey,U.P. Wild, Spectral hole-burning and holography in an Y2SiO5Pr3+ crystal, *Phys. Rev. B* **47**(1993), 14741–14752.
51. J.P. Hornak, *http://www.cis.rit.edu/htbooks/nmr/inside.htm.*
52. F. Jelezko, J. Wrachtrup, Quantum information processing in diamond, *quant-ph*/0510152.
53. F. Jelezko, T. Gaebel, I. Popa, M. Domhan, A. Gruber, J. Wrachtrup, Observation of coherent oscillation of a single nuclear spin and realization of a two-qubit conditional quantum gate, *Phys. Rev. Lett.* **93**(2004), 130501.
54. J.A. Jones, Robust quantum information processing with techniques from liquid-state NMR, *Phil. Trans. Roy. Soc. Lond. Ser. A Math. Phys. Eng. Sci.* **361**(2003), 1429–1440.
55. J.A. Jones, Robust Ising gates for practical quantum computation, *Phys. Rev. A* **67**(2003), 012317.
56. J.A. Jones, NMR quantum computation: A critical evaluation, *Fortschr. Phys.* **48**(2000), 909–924.
57. J.A. Jones, and E. Knill, Efficient refocusing of one-spin and two-spin interaction for NMR quantum computation, *J. Magnetic Resonance* **141**(1999), 322–325.
58. J.A. Jones and M. Mosca, Implementation of a quantum algorithm on a nuclear magnetic resonance quantum computer, *J. Chem. Phys.* **109**(1998), 5, 1648–1653. Another short version can be found at *Nature* **393**(1998), 344–346.
59. B.E. Kane, A silicon-based nuclear spin quantum computer, *Nature* **393**(1998), 133–137.
60. N. Khaneja, R. Brockett, and S.J. Glaser, Time optimal control in spin system, *Phys. Rev. A* **63**(2001), 032308.

61. N. Khaneja, S.J. Glaser, and R. Brockett, Sub-Riemannian geometry and time optimal control of three spin systems: quantum gates and coherent transfer, *Phys. Rev. A* **65**(2002), 032301.
62. A.K. Khitrin, H. Sun, and B.M. Fung, Method of multifrequency excitation for creating pseudopure states for NMR quantum computing, *Phys. Rev. A* **63**(2001), 020301.
63. E. Knill, I. Chuang, and R. Laflamme, Effective pure states for bulk quantum computation, *Phys. Rev. A* **57**(1998), 3348.
64. E. Knill, R. Laflamme, R. Martinez, and C.H. Tseng, An algorithm benchmark for quantum information processing, *Nature* **404**(2000), 368–370.
65. E. Knill, R. Laflamme, R. Martinez, and C. Negrevergne, Benchmarking quantum computers: The five-qubit error correcting code, *Phys. Rev. Lett.* **86**(2001), 5811–5814.
66. E. Kupce, and R. Freeman, Close encounters between soft pulses, *J. Magnetic Resonance Ser. A* **112**(1995), 261–264.
67. T.D. Ladd, D. Maryenko, Y. Yamamoto, E. Abe, and K.M. Itoh, Coherence time of decoupled nuclear spins in silicon, *Phys. Rev. B* **71**(2005), 014401.
68. T.D. Ladd, J.R. Goldman, F. Yamaguchi, and Y. Yamamoto, All-silicon quantum computer, *Phys. Rev. Lett.* **89**(2002), No. 1, 017901.
69. R. Laflamme, E. Knill, C. Negrevergne, R. Martinez, S. Sinha, and D.G. Cory, Introduction to NMR quantum information processing, in *Experimental Quantum Computation and Information, Proceeding of the International School of Physics "Enrico Fermi,"* edited by F. De Martini and C. Monroe, IOS Press, Amsterdam, Netherlands, 2002.
70. G.M. Leskowitz, N. Ghaderi, R.A. Olsen, and L.J. Mueller, Three-qubit nuclear magnetic resonance quantum information processing with a single-crystal solid, *J. Chem. Phys.* **119**(2003), 1643–1649.
71. D.W. Leung, I.L. Chuang, F. Yamaguchi, and Y. Yamamoto, Efficient implementation of coupled logic gates for quantum computation, *Phys. Rev. A* **61**(2000), 042310.
72. M.H. Levitt, Composite pulses, *Prog. Nucl. Magnetic Resonance Spectroscopy* **18**(1986), 61–122.
73. N. Linden, B. Herve, R.J. Carbajo and R. Freeman, Pulse sequences for NMR quantum computers: How to manipulate nuclear spins while freezing the motion of coupled neighbors, *Chem. Phys. Lett.* **305**(1999), 28–34.
74. N. Linden, H. Barjat, and R. Freeman, An implementation of the Deutsch-Jozsa algorithm on a three-qubit NMR quantum computer, *Chem. Phys. Lett.* **296**(1998), 61–67.
75. S.J. Lomonaco, Shor's Quantum Factoring Algorithm, *quant-ph*/0010034.
76. G. Long, H. Yan, Y. Li, C. Lu, J. Tao, and H. Chen et al., Experimental NMR realization of a generalized quantum search algorithm, *Phys. Lett. A* **286**(2001), 121–126.
77. J.J. Longdell, M.J. Sellars, and N.B. Manson, Hyperfine interaction in ground and excited states of praseodymium-doped yttrium orthosilicate, *Phys. Rev. B* **66**(2002), 035101.
78. J.J. Longdell and M.J. Sellars, Selecting ensembles for rare earth quantum computation, quant-ph/0310105, 2003.
79. J.J. Longdell, M.J. Sellars, N.B. Manson, Demonstration of conditional quantum phase shift between ions in a solid, *Phys. Rev. Lett.* **93**(2004), 130503.
80. M.D. Lukin and P.R. Hemmer, Quantum entanglement via optical control of atom–atom interactions, *Phys. Rev. Lett.* **84**(2000), 2818–2821.
81. R.M. Macfarlane, High-resolution laser spectroscopy of rare-earth doped insulators: A personal perspective, *J. Luminescence* **100**(2002), 1–4, 1–20.
82. M.L. Martin, and G.J. Martin, *Practical NMR Spectroscopy*, Heyden, London, UK, 1980.
83. G.D. Mateescu and A. Valeriu, *2D NMR : Density Matrix and Product Operator Treatment*, PTR Prentice-Hall, New Jersey, 1993.

84. R. Marx, A.F. Fahmy, J.M. Myers, W. Bermel, and S.J. Glaser, Approaching five-bit NMR quantum computing, *Phys. Rev. A* **62**(2000), 012310.
85. M.A. Nielsen and I.L. Chuang, *Quantum Computation and Quantum Information*, Cambridge University Press, Cambridge, UK, 2000.
86. M.A. Nielson and I.L. Chuang, Programmable quantum gate arrays, *Phys. Rev. Lett.* **79**(1997), 2, 321–323.
87. M.A. Nielson, E. Knill, and R. Laflamme, Complete quantum teleportation using NMR, *Nature* **396**(1995), 52–55.
88. J. Normand, *A Lie Group: Rotations in Quantum Mechanics*, North-Holland, New York, 1980.
89. S.L. Patt, Single- and multiple-frequency-shifted laminar pulses, *J. Magnetic Resonance* **96**(1992), 1, 94–102.
90. A.O. Pittenger, *An Introduction to Quantum Computing Algorithms*, Birkhäuser, Boston, 2000.
91. C.P. Poole, and H.A. Farach, *Theory of Magnetic Resonance*, Wiley, New York, 1987.
92. I. Popa, T. Gaebel, M. Domhan, C. Wittmann, F. Jelezko, and J. Wrachtrup, Energy levels and decoherence properties of single electron and nuclear spins in a defect center in diamond, *quant-ph*/0409067, 2004
93. M.A. Pravia, Z. Chen, J. Yepez, and D.G. Cory, Towards a NMR implementation of a quantum lattice gas algorithm, *Comput. Phys. Commun.* **146**(2002), Issue 3, 339–344.
94. G.J. Pryde, M.J. Sellars, N.B. Manson, Solid state coherent transient measurements using hard optical pulses, *Phys. Rev. Lett.* **84**(2000), 1152–1155.
95. R.L. Rivest, A. Shamir, and L. Adleman, A method of obtaining digital signature and public-key cryptosystems, *Commun. ACM* **21**(1978), 120–126.
96. P.W. Shor, Polynomial-time algorithms for prime factorization and discrete logarithms on a quantum computer, *SIAM J. Comput.* **26**(1997), 5, 1484–1509.
97. D. Simon, On the power of quantum computation, *Proceedings, 35th Annual Symposium on Foundations of Computer Science* (1994), IEEE Press, Los Alamitos, CA, 116–123.
98. S. Somaroo, C.H. Tseng, T.F. Havel, R. Laflamme, and D.G. Cory, Quantum simulation on a quantum computer, *Phys. Rev. Lett.* **82**(1999), 5381–5384.
99. O.W. Sorenson, Polorization transfer experiments in high-resolution NMR spectroscopy, *Prog. Nucl. Magnetic Resonance Spectroscopy* **21**(1989), Issue 6, 503–569.
100. M. Steffen, L.M.K. Vandersypen, I.L. Chuang, Toward quantum computation: A five-qubit quantum processor, *IEEE Micro*, 24–34, 2001.
101. A. Szabo, Spin dependence of optical dephasing in ruby - The frozen core, *Opt. Lett.* **8**(1983), 9, 486–487.
102. J.M. Taylor, H.A. Engel, W. Dur, A. Yacoby, C.M. Marcus, P. Zoller, M.D. Lukin, Fault-tolerant architecture for quantum computation using electrically controlled semiconductor spins, *Nature Phys.* **1**(2005), 3, 177–183.
103. A.M. Turing, On computable numbers, with an application to the Entscheidungs problem, *Proc. Lond. Math. Soc. 2* **42**(1936), 230.
104. G. Vahala, J. Yepez, and L. Vahala, Quantum lattice gas representation of some classical solitons, *Phys. Lett. A* **310**(2003), 187–196.
105. L. Vahala, G. Vahala, and J. Yepez, Lattice Boltzmann and quantum lattice gas representations of one-dimensional magnetohydrodynamic turbulence, *Phys. Lett. A* **306**(2003), 227–234.
106. L.M.K. Vandersypen, C.S. Yannoni, M.H. Sherwood, and I.L. Chuang, Realization of logically labeled effective pure states for bulk quantum computation, *Phys. Rev. Lett.* **83**(1999), 3085.
107. L.M.K. Vandersypen, M. Steffen, G. Breyta, C. Yannoni, R. Cleve, and I.L. Chuang, Experimental realization of an order-finding algorithm with an NMR quantum computer, *Phys. Rev. Lett.* **85**(2000), 5452.

108. L.M.K. Vandersypen, and M. Steffen, M.H. Sherwood, C. Yannoni, G. Breyta, and I.L. Chuang, Implementation of a three-quantum-bit search algorithm, *Appl. Phys. Lett.* **76**(2000), 5, 646–648.
109. L.M.K. Vandersypen, I.L. Chuang, NMR techniques for quantum control and computation, *Rev. Modern Phys.* **76**(2004), 4, 1037–1069.
110. E. VanOort, N.B. Manson, M. Glasbeek, Optically detected spin coherence of the diamond N-V center in its triplet ground-state, *J. Phys. C Solid State Phys.* **21**(1988), 4385–4391.
111. L.M.K. Vandersypen, M. Steffen, G. Breyta, et al., Experimental realization of Shor's quantum factoring algorithm using nuclear magnetic resonance. *Nature* **414**(2001), 883–887.
112. V. Vedral, A. Barenco, and A. Ekert, Quantum network for elementary arithmetic operations, *Phys. Rev. A* **54**(1996), 147–153.
113. C. Wei, N.B. Manson, Observation of electromagnetically induced transparency within an electron spin resonance transition, *J. Opt. B Quantum Semiclassical Opt.* **1**(1999), 464–468.
114. Y.S. Weinstein, M.A. Pravia, E.M. Fortunato, S. Lloyd, and D.G. Cory, Implementation of the quantum Fourier transform, *Phys. Rev. Lett.* **86**(2001), 1889–1891.
115. Wikipedia, *http://en.wikipedia.org/wiki/Nuclear_magnetic_resonance.*
116. F. Yamaguchi, and Y. Yamamoto, Crystal lattice quantum computer, *Microelectron. Eng.* **47**(1999), 273–275.
117. C. Yannoni, M.H. Sherwood, D.C. Miller, I.L. Chuang, L.M.K. Vandersypen, and M.G. Kubines, Nuclear magnetic resonance quantum computing using liquid crystal solvents, *Appl. Phys. Lett.* **75**(1999), No. 22, 3563–3562.
118. C.C. Yao, Quantum circuit complexity, *Proc. of the 34th Ann. IEEE Symp. on Foundations of Computer Science* (1993), 352–361.
119. J. Yepez, Quantum lattice-gas model for the diffusion equation, *Int. J. Modern Phys. C* **12**(2001), 9, 1285–1303.
120. J. Yepez, Type-II quantum computers, *Int. J. Modern Phys. C* **12**(2001), 9, 1273–1284.
121. J. Yepez, and B. Boghosian, An efficient and accurate quantum lattice-gas model for the many-body Schrödinger wave equation, *Comput. Phys. Commun.* **146**(2002), Issue 3, 280–294.
122. J. Zhang, Z. Lu, L. Shan, and Z. Deng, Synthesizing NMR analogs of Einstein-Podolsky-Rosen state using the generalized Grover's algorithm, *Phys. Rev. A* **66** (2002), 044308.

Kolibri 360 GmbH
Ferdinand-Jühlke-Straße 7
99095 Erfurt, Deutschland
produktsicherheit@kolibri360.de